P9-BVM-859

Protein Interactions

PROTEIN REVIEWS

Recent Volumes in this Series

VIRAL MEMBRANE PROTEINS: STRUCTURE, FUNCTION, AND DRUG DESIGN
Edited by Wolfgang B. Fischer

THE p53 TUMOR SUPPRESSOR PATHWAY AND CANCER
Edited by General P. Zambetti

PROTEOMICS AND PROTEIN-PROTEIN INTERACTIONS: BIOLOGY, CHEMISTRY, BIOINFORMATICS, AND DRUG DESIGN
Edited by Gabriel Waksman

PROTEIN MISFOLDING, AGGREGATION AND CONFORMATIONAL DISEASES PART A: PROTEIN AGGREGATION AND CONFORMATIONAL DISEASES
Edited by Vladimir N. Uversky and Anthony L. Fink

PROTEIN INTERACTIONS: BIOPHYSICAL APPROACHES FOR THE STUDY OF COMPLEX REVERSIBLE SYSTEMS
Edited by Peter Schuck

PROTEIN MISFOLDING, AGGREGATION AND CONFORMATIONAL DISEASES PART B: MOLECULAR MECHANISMS OF CONFORMATIONAL DISEASES
Edited by Vladimir N. Uversky and Anthony L. Fink

A Continuation Order Plan is available for this series. A continuation order will bring delivery of each new volume immediately upon publication. Volumes are billed only upon actual shipment. For further information please contact the publisher.

Protein Interactions

Biophysical Approaches for the Study of Complex Reversible Systems

Edited by

Peter Schuck

National Institutes of Health, Bethesda, Maryland

Peter Schuck
National Institutes of Health
Bethesda, MD 20892
USA

Library of Congress Control Number: 2006928099

ISBN-10: 0-387-35965-6 e-ISBN-10: 0-387-35966-4
ISBN-13: 978-0-387-35965-6 e-ISBN-13: 978-0-387-35966-3

Printed on acid-free paper.

9 8 7 6 5 4 3 2 1

springer.com

Contents

Contributors

Claire A. Adams
Department of Molecular and Cellular Biochemistry
University of Kentucky College of Medicine
Lexington, KY, USA

Peter S. Andersen
Symphogen A/S,
Copenhagen, Denmark

Peter S. Backlund Jr
National Institute of Child Health and Human Development
National Institutes of Health
Bethesda, MD, USA

Hacène Boukari
National Institute of Child Health and Human Development
National Institutes of Health
Bethesda, MD, USA

Christine Ebel
Institut de Biologie Structurale UMR 5075 CEA-CNRS-UJF
Grenoble, France

Dorothy A. Erie
Department of Chemistry
University of North Carolina at Chapel Hill
Chapel Hill, NC, USA

Michael G. Fried
Department of Molecular and Cellular Biochemistry
University of Kentucky College of Medicine
Lexington, KY, USA

Wolfram Gronwald
Institute of Biophysics and Physical Biochemistry
University of Regensburg
Regensburg, Germany

Myun K. Han
Excimus Biotech, Inc.
Baltimore, MD, USA

John J. Harvey
Excimus Biotech, Inc.
Baltimore, MD, USA

Hans Robert Kalbitzer
Institute of Biophysics and Physical Biochemistry
University of Regensburg
Regensburg, Germany

Elizabeth A. Komives
Department of Chemistry and Biochemistry
University of California, San Diego
La Jolla, CA, USA

Jay R. Knutson
National Heart, Lung, and Blood Institute
National Institutes of Health
Bethesda, MD, USA

Werner Kremer
Institute of Biophysics and Physical Biochemistry
University of Regensburg
Regensburg, Germany

John E. Ladbury
Department of Biochemistry & Molecular Biology
University College London
London, UK

Emmanuel Margeat
INSERM U554
CNRS UMR 5048
Montpellier, France

Jacob Piehler
Institute of Biochemistry
Johann Wolfgang Goethe-University
Frankfurt am Main, Germany

Carol V. Robinson
University of Cambridge
Department of Chemistry
Cambridge, UK

Catherine A. Royer
INSERM U554
CNRS UMR 5048
Montpellier, France

Alan M. Sandercock
Department of Chemistry
University of Cambridge
Cambridge, UK

Peter Schuck
National Institute of Biomedical Imaging and Bioengineering
National Institutes of Health
Bethesda, MD, USA

Frank Schumann
Institute of Biophysics and Physical Biochemistry
University of Regensburg
Regensburg, Germany

Michael Spörner
Institute of Biophysics and Physical Biochemistry
University of Regensburg
Regensburg, Germany

Eric J. Sundberg
Boston Biomedical Research Institute
Watertown, MA, USA

Dmitri I. Svergun
European Molecular Biology Laboratory
Hamburg, Germany

Patrice Vachette
Institut de Biochimie et Biophysique Moléculaire et Cellulaire
UMR 8619 CNRS-Université Paris-Sud
Orsay Cedex, France

Hong Wang
National Institute of Environmental Health Sciences
National Institutes of Health
Research Triangle Park, NC, USA

Mark A. Williams
Department of Biochemistry and Molecular Biology
University College London
London, UK

Kristi Wojtuszewski
National Heart, Lung, and Blood Institute
National Institutes of Health
Bethesda, MD, USA

Yong Yang
Department of Chemistry
University of North Carolina at Chapel Hill
Chapel Hill, NC, USA

Alfred L. Yergey
National Institute of Child Health and Human Development
National Institutes of Health
Bethesda, MD, USA

Preface

When I was invited to edit this volume, I wanted to take the opportunity to assemble reviews of different biophysical methodologies for protein interactions at a level sufficiently detailed to understand how complex systems can be studied. There are several excellent introductory texts for biophysical methodologies, many with hands-on descriptions or embedded in general introductions to physical biochemistry. The goal of the present volume was to present state-of-the-art reviews that do not necessarily enable the reader to carry out these techniques, but to gain a deep understanding of the biophysical observables, to stimulate creative thought on how the techniques may be applied to study a particular biological system, and to foster collaboration and multidisciplinary work.

Reversible protein interactions involve noncovalent chemical bonds, producing protein complexes with free energies not far from the order of magnitude of the thermal energy kT. As a consequence, they can be highly dynamic and may be controlled, for example, by protein expression levels and changes in the intracellular or microenvironment. Reversible protein complexes may have sufficient stability to be purified for study, but frequently their short lifetime essentially limits their existence to solutions of mixtures of the binding partners in which they remain populated through dissociation and reassociation processes. To understand the function of such protein complexes, it is important to study their structure and dynamics. Even when these studies take place *in vitro*, they elucidate the principles of the interactions imposed by the protein structures, principles which may be quantitatively modulated but have to be followed *in vivo*.

Maps of protein interaction networks display the interdependence of protein interactions and highlight the importance of interactions of more than two proteins. It is probably safe to assume that we currently know only a small fraction of the protein interactome, in particular, triple or higher-order complexes. Proteins that are able to interact with multiple protein binding partners or other ligands can exhibit higher functionality. This can include, for example, logical switches, with cooperativity steepening the isotherms of binding, resulting in highly sensitive and discriminate response to a cellular stimulus. In many systems, this involves interplay of multiprotein complexes, binding of small ligands, covalent protein modifications, and conformational changes in proteins.

Techniques to elucidate such linked protein interactions and/or multiprotein complexes, with regard to thermodynamics, kinetics, conformation, and flexibility, and the possible role of spatial confinements to surfaces define the scope of the present volume. The side-by-side presentation of different approaches highlights aspects they have in common and orthogonal viewpoints that may provide opportunities for a synergistic combination. For example, an important recurring theme is the role of protein solvation, which is addressed in many chapters from different perspectives. To illustrate the use of the techniques, some applications are described, which, at the same time, are also aimed at providing a kaleidoscopic view of different biological systems and principles of protein interactions.

The list of biophysical methods reviewed in the present volume is far from complete. The selection of topics should not be understood in the sense of merit or importance of the different methods, but rather is a reflection of practical limitations in the scope and assembly of this work.

I want to thank the authors for their contributions. I also want to thank the series editor Dr. Zou Atassi, the publisher, and the National Institutes of Health for making this work possible. Finally, I want to thank my beloved wife Teresa for her patience and support.

Bethesda, 2005

1

The Characterization of Biomolecular Interactions Using Fluorescence Fluctuation Techniques

Emmanuel Margeat, Hacène Boukari, and Catherine A. Royer*

1.1. INTRODUCTION

Fluorescence correlation spectroscopy (FCS) has become one of the most popular methods available for investigating the physical properties of biomolecular complexes ever since it was first proposed by the groups of Elliot Elson and Watt Webb in the 1970s (Aragon and Pecora, 1974; Magde *et al.*, 1974; Webb, 1976; Icenogle and Elson, 1983; Elson, 2001). Over the years, a number of thorough reviews have appeared of the now extensive literature (Madge, 1976; Webb, 1976; Berland *et al.*, 1995; Rigler, 1995; Schwille, 2001; Hess *et al.*, 2002; Thompson *et al.*, 2002; Haustein and Schwille, 2003; Muller *et al.*, 2003; Elson, 2004; Haustein and Schwille, 2004). While the early work in the field was carried out on dyes and purified biological molecules, it became quickly apparent that FCS was well suited for cellular applications as well (Berland *et al.*, 1995; Schwille, 2001; Hess *et al.*, 2002; Krivensky and Bonnet, 2002; Bacia and Schwille, 2003; Vukojevic *et al.*, 2005). FCS has been used to study lipid diffusion and protein associations in model and biological membranes, nucleic acid hybridization and protein–nucleic acid interactions, protein–protein interactions, both homologous and heterologous, in solution and in cells, and protein and DNA associations with

E. MARGEAT • INSERM, U554, Montpellier, 34090 Cedex, France; CNRS-UM-UM2, UMR5048. H. BOUKARI • National Institute of Child Health and Human Development, National Institutes of Health, Bethesda, MD, USA. C. A. ROYER • INSERM, U554, Montpellier, 34090 Cedex, France; CNRS-UM1-UM2, UMR5048 and *Corresponding author: *e*-mail: royer@cbs.cnrs.fr.

small ligands. There are literally hundreds of articles in the PubMed database on the subject of the use of FCS to study biomolecular interactions. Therefore, it is outside the scope of the present work to review this abundant literature in detail. Rather, this review touches on some of the more practical aspects of the use of FCS in binding studies. The advantages vis-à-vis other techniques are discussed, and some examples of the applications of FCS using simple diffusion time measurements, photon statistics, and cross-correlation measurements are presented. Finally, some of the more problematic artifacts are described, along with approaches designed to minimize their contributions.

We restrict the discussion here to experimental setups in which the fluorescence signal from diffusing molecules is detected in a very small, open volume at concentrations that are sufficiently low such that significant fluctuations in the fluorescence intensity are detected as molecules enter and leave the volume, i.e., at or near the single molecule level. In general, an FCS setup (see Figure 1.1) is based on laser excitation into an inverted microscope equipped with an objective featuring a high numerical aperture. Detection is accomplished through dichroic mirrors and appropriate filters by photocounting detectors such as avalanche photodiodes or photomultiplier tubes. The time sequence of the detected intensity is either directly processed by a fast digital autocorrelator (hardware) or stored as a raw photon versus time stream and processed using autocorrelation software.

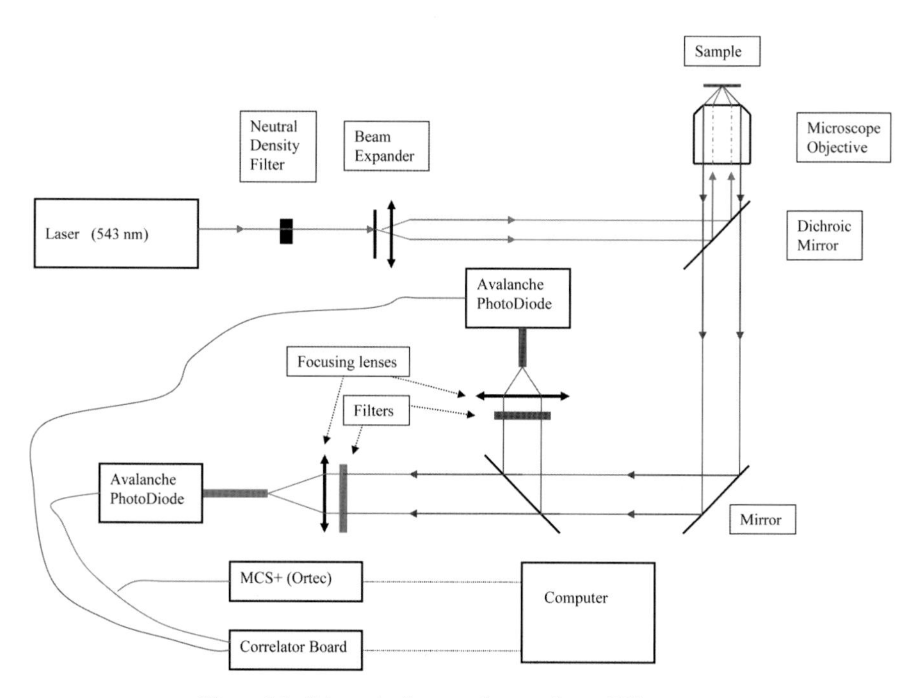

Figure 1.1. Schematic diagram of a one-photon FCS system.

Examples of various experimental setups can be found in the reviews cited earlier. The small open volume (typically <1 fL) can be achieved by either using visible laser excitation coupled with a confocal pinhole (Eigen and Rigler, 1994) or small diameter optical fibers (10–50 μm) to limit the detection volume, or by two-photon excitation with a femtosecond pulsed IR laser beam through a high numerical aperture objective (Denk *et al.*, 1990), which focuses the beam to a diffraction-limited spot with a diameter equal to $\sim \lambda/2$. The point spread function is typically considered to exhibit a 3-D Gaussian profile. In two-photon excitation (TPE), molecules simultaneously absorb two photons whose combined energy allows the transition to an excited electronic state. While in one-photon excitation, the emission is at longer wavelengths than the excitation, in TPE, the emission is of higher energy than the excitation. The probability of TPE depends on the square of the incident photon flux. For a discussion of the two-photon excitation theory see the review by So *et al.* (2000) and references therein.

1.2. ADVANTAGES OF FLUCTUATION SPECTROSCOPY

FCS and related techniques benefit from the fundamental advantage that fluorescence in general presents over a number of other approaches to the quantitative study of biomolecular interactions, namely its high sensitivity. This high sensitivity allows for the fast detection of emission from as little as a single molecule. Fast detection renders accessible the analysis of events on fast timescales. The ultimate lower limit of the timescale for fluorescence detection is set by the rate at which emission occurs, typically on the nanosecond timescale. Another important advantage of fluorescence, in principle, is that it can be measured in solution, in gels, or in live cells. Since the observable is fluorescence, the background physical properties of the medium are much less important than in other techniques, such as dynamic light scattering (DLS), for example. Conceptually, DLS and FCS use a similar approach, namely time-correlating a fluctuating signal, which is further analyzed to extract the size of the macromolecules. Both techniques are relatively noninvasive as they use the interaction of an optical beam with the bio-macromolecules of interest. The underlying physical mechanism for these fluctuations is, however, different for each technique. As noted previously, FCS fluctuations are attributed to changes in intensity induced by the fluorescent bio-macromolecules moving in and out of an excitation volume (we ignore the internal photophysical fluctuations). In contrast, the scattered intensity fluctuations in DLS result from the constructive and destructive interference of the electric fields emanating coherently from the bio-macromolecules present in a small probed volume. Each technique has its own advantages and limitations. As described earlier, FCS requires nanomolar concentration of fluorescent bio-macromolecules, which could be an important factor in bioassays of scarce materials, whereas micromolar or higher concentrations are generally needed in DLS experiments. Because of fluorescence, the bio-macromolecules of interest can be specifically distinguished from a host of other particles, hence allowing the study of their dynamical behavior in the presence of the other particles. This is the main reason that FCS has become attractive to cellular studies, where the cells are notoriously known for their crowded environment. In DLS all the particles present in the sample are potential scatterers and contribute to the scattered and detected signal, making it challenging to analyze the resulting correlation function and to extract the characteristics of the macromolecules of interest. An advantage of DLS is the possibility of studying dynamical processes at various length-scales by performing angle-dependence measurements; the angle, commonly defined as the scattering angle, subtends the direction of the incident beam and that of detection. In FCS, there is practically one length-scale, limited by the diffraction limit of the focused beam. Finally, the macromolecule of interest does not need to be labeled in DLS experiments, whereas FCS requires a genetic, enzymatic, or chemical labeling of the studied macromolecule.

Consequently, fluorescence allows for the specific observation of the fluorescent species in relatively heterogeneous media, which is advantageous even in *in vitro* experiments. It allows one to monitor associations involving the fluorescent species in the presence of unlabeled competitor molecules, for example. Multicolor experiments can give access to the separate observation of the interactions of different proteins, with each other or competitors. Moreover, this specificity of observation allows detection in a background such as gels and cells, in which many other molecules with diverse physical properties are present.

When measuring molecular interactions by the observation of some fluorescence signal (in the present case, intensity fluctuations) there is no need to separate bound from free species because their respective signals can be differentiated in the overall fluorescence parameter. This means that fluorescence fluctuation measurements can be made under conditions of true equilibrium, with target molecules present at concentrations usually well below the dissociation constant for the complex. Measurements in solution also allow for the relatively simple modulation of solution conditions, such as temperature, concentration of interacting species, pH, ionic strength, and ligand concentration. Thus the role of these parameters in modulating the interaction can be assessed in a straightforward manner.

Fluctuation spectroscopy presents another practical advantage over a number of other techniques used in the study of biomolecular complexation, its relatively small requirement for sample. For example, given the fact that the excitation volumes involved are $\sim$1 fL or less, the volume of sample required for FCS measurements is even smaller than that required for classical fluorescence studies, in the 10-μL range for practical reasons. Such a small observation volume can be achieved using a sandwich of two coverslips separated by a silicone isolator (Molecular Probes, Eugene, OR) for example. The single molecule sensitivity means that the fluorescently labeled molecule is present at nanomolar concentrations or lower in these small volumes, which for a protein of 50 kDa corresponds to less than 0.5 ng of material. Such low levels of materials allow access to dissociation constants for high affinity interactions not accessible to analytical ultracentrifugation and calorimetric approaches. They also render possible experiments that otherwise cannot be carried out for lack of enough material.

Fluctuation spectroscopy can also complement fluorescence anisotropy-based studies of protein–protein, protein–ligand, or protein–nucleic acid interactions. The size range of accessible biomolecular complexes in measurements of the rotational correlation time based on fluorescence anisotropy is limited to correlation times about tenfold longer than the fluorescence lifetime ($\tau_c \sim 40$–80 ns or MW sphere $\sim$100–200 kDa). This is because significant rotation must occur during the excited state lifetime, to reliably determine the rotational correlation. However, there is no theoretical upper limit in fluctuation spectroscopy, other than that associated with diffusion being so slow that photobleaching becomes a problem. While the translational diffusion coefficient is less sensitive to molecular size than is the rotational diffusion coefficient, we shall see subsequently that the resolution levels are relatively high. Moreover, while the interpretation of both time-resolved and steady-state anisotropy is complicated by the need to distinguish local probe mobility from global macromolecular tumbling, the local mobility of the probe is not an issue in the measurement of translational diffusion or molecular brightness by fluctuation spectroscopy. Finally, since fluctuation spectroscopy is also a particle-counting technique, even when molecular weight changes are small, one can rely on changes in the molecular brightness of complexes to monitor interactions.

1.3. FLUORESCENCE CORRELATION SPECTROSCOPY, AND MOLECULAR DIFFUSION

The general theory of fluorescence fluctuation spectroscopy has been presented at length in the reviews cited earlier. Therefore, in the present review only a succinct introduction to FCS theory is given and we limit our discussion to number fluctuations, since these are most useful in the study of biomolecular interactions. We note that there exists a wide range of applications of fluctuation spectroscopy based on the measurement of fluctuations due to changes in conformation (through FRET or quenching). These measurements allow for the characterization of the amplitude and timescale of these molecular motions and have been used in the study of protein folding, enzyme kinetics, and other conformational fluctuations. These molecular phenomena are not the subject of the present review, and are not treated here. We refer the reader to the following articles for further information on this subject (Kettling *et al.*, 1998; Grunwell *et al.*, 2001; Chattopadhyay *et al.*, 2002; Cotlet *et al.*, 2004; Joo *et al.*, 2004; Li *et al.*, 2004; Chattopadhyay *et al.*, 2005; Karymov *et al.*, 2005; Lee *et al.*, 2005; Li and Yeung, 2005; Sato *et al.*, 2005; Schuler, 2005; Slaughter *et al.*, 2005; Wilson *et al.*, 2005).

For those interested in a more detailed introduction to correlation spectroscopy, the recent review by Haustein and Schwille (Haustein and Schwille, 2004) provides a clear presentation of FCS theory. Briefly, if we consider typical measurements of fluorescence intensity of a sample of fluorescent molecules excited in a small illuminated volume, one can define the fluorescence fluctuations as $\delta F(t) = F(t) - \langle F \rangle$, where $F(t)$ is the fluorescence intensity measured at time,

t, and $<F>$ denotes the time-averaged fluorescence intensity. These fluctuations are then time-correlated to generate an autocorrelation function $G(\tau)$, defined as

$$G(\tau) = \frac{\langle \delta F(t)\ \delta F(t+\tau) \rangle}{\langle F \rangle^2} \tag{1.1}$$

with τ being the lag time. The intensity fluctuations are assumed to be directly related to fluctuations in the concentration of the fluorescent molecules in the illuminated volume and can be expressed as

$$\delta F(t) = A \int W(\vec{r})\ \delta c(\vec{r}, t) \mathrm{d}\vec{r}, \tag{1.2}$$

where $W(\vec{r})$ denotes the profile of the excitation volume (usually the laser-beam profile), δc the concentration fluctuation around the average concentration, and A, a constant. The concentration fluctuations are induced by a number of mechanisms, the most studied one being the diffusion of fluorescent molecules in and out of the small excitation volume. For an ideal case of monodisperse, uniformly bright, and freely diffusing fluorescent molecules, a closed-form expression for Eq. (1.1) was derived

$$G(\tau) = \frac{1}{N}\left(1 + \frac{\tau}{\tau_{\mathrm{d}}}\right)^{-1}\left(1 + \frac{r_{\mathrm{o}}^2 \tau}{z_{\mathrm{o}}^2 \tau_{\mathrm{d}}}\right)^{-1/2}, \tag{1.3}$$

which is commonly used to analyze measured autocorrelation functions, more precisely extract two parameters: the diffusion time, τ_{d}, and the average number of molecules, N. Here it is assumed that the fluorescent molecules are excited by a 3D Gaussian beam such that,

$$W(r, z) = \mathrm{Be}^{(-2\mathrm{r}^2/\mathrm{r}_{\mathrm{o}}^2)}\mathrm{e}^{(-2\mathrm{z}^2/\mathrm{z}_{\mathrm{o}}^2)} \tag{1.4}$$

characterized by r_{o} and z_{o}, respectively, the $1/e^2$ Gaussian intensity beam waists in the radial and axial dimensions as defined by the direction of the laser beam. The extrapolated values of the autocorrelation function in Eq. (1.1) at $\tau = 0$ can be rewritten as

$$G(0) = \frac{1}{\langle C \rangle V_{\mathrm{eff}}}, \tag{1.5}$$

where $V_{\mathbf{eff}}$ is an effective observation volume (Nagy and Schwabe, 2004b). Generally, this volume is expressed in terms of that derived from the intensity profile (the point-spread function) $V_{\mathbf{psf}}$ such that

$$G(0) = \frac{\gamma}{\langle C \rangle V_{\mathrm{psf}}} \tag{1.6}$$

with

$$V_{\mathrm{psf}} = \frac{\int W(\vec{r})\ \mathrm{d}\vec{r}}{W(0)}$$

being a mathematical normalization factor for the chosen excitation profile. For example, in the 3D Gaussian profile we have

$$V_{\mathrm{psf}} = \left(\frac{\pi}{2}\right)^{3/2} (r_{\mathrm{o}}^{2} z_{\mathrm{o}}). \tag{1.7}$$

In Eq. (1.6) the γ factor provides a measure of the uniformity of the fluorescence intensity observed for molecules at different positions within the volume and the abruptness of the boundaries of the latter (Nagy and Schwabe, 2004a). The γ factor is typically less than 1; that is, the effective observation volume is larger than the volume of the spread function. The γ factor is only equal to 1 for a true physical volume with well-defined boundaries.

In principle, one can determine the effective volume, V_{eff}, from the intercept ($\tau \to 0$) of measured correlation functions from standard fluorophores with known concentration. In practice, however, possible experimental artifacts from both the studied sample and the instrument, which are discussed later, need to be taken into account for an absolute calibration. Thereafter, for unknown samples, in addition to the diffusion time, the concentration, C, of the fluorescent species in solution can also be determined.

The second relevant parameter in Eq. (1.3) is the diffusion time, τ_{d}, which is related to the translational diffusion coefficient as follows for one- and two-photon excitations, respectively,

$$\tau_{\mathrm{d}} = \frac{r_{\mathrm{o}}^{2}}{4D} \quad \text{and} \quad \tau_{\mathrm{d}} = \frac{r_{\mathrm{o}}^{2}}{8D} \tag{1.8}$$

the factor 8 in the denominator of the second expression arising because of the quadratic dependence of fluorescence intensity on excitation intensity in two-photon excitation. Furthermore, in dilute solutions, one can apply the Stokes–Einstein relation of the diffusion coefficient, D,

$$D = \frac{k_{\mathrm{b}} T}{6\pi \eta r_{\mathrm{h}}} \tag{1.9}$$

to determine the hydrodynamic radius, r_{h}. In Eq. (1.9) k_b denotes the Boltzman constant ($1.38 \times 10^{-23}\,\mathrm{kg\,m^2/s^2/K}$), T is the temperature in Kelvin, and η is the the solvent viscosity. The viscosity of water at 20°C is ~1 cp (10^{-3} Pa s or 10^{-3} kg / (m $\times$ s). The radius of spherical molecules is the hydrodynamic radius, r_{h}, which can be related to the molecular weight, M (g/mol), Avogadro's number, N (molecules/mol), and the hydrated volume, V_{h}, of the protein ($\sim 1.03\,\mathrm{cm^3/g}$).

$$r_{\mathrm{h}} = \left(\frac{3MV_{\mathrm{h}}}{4\pi N}\right)^{1/3}. \tag{1.10}$$

For a spherical protein of 10 kDa, the hydrodynamic radius is 1.5 nm, while for a protein of 100 kDa, r_{h} is 3.4 nm. Substituting the value of r_{h} into Eq. (1.9) yields diffusion coefficients of $143 \times 10^{-12}\,m^2/s$ or $143\,\mu m^2/s$ for the 10 kDa spherical protein and $63\,\mu m^2/s$ for the 100 kDa spherical protein. Thus, the smaller the fluorescent molecule, the faster will be its diffusion, and the autocorrelation function will decay to zero more quickly (e.g., Figure 1.2). However, the function is not a strong one owing to the cube root dependence of the radius of the protein on the molecular weight. Hence a factor difference of 10 in size yields only a little more than a factor difference of 2 in the diffusion coefficient.

An intuitive picture of the autocorrelation function can be imagined as follows. If the molecule is large, then the fluorescence signal remains high (similar to that at time, t) for a longer period than if the molecule is small, since the larger molecule takes a longer time to leave the volume once it has entered. Thus, the amplitude of the autocorrelation function remains high for a longer period before decaying to zero.

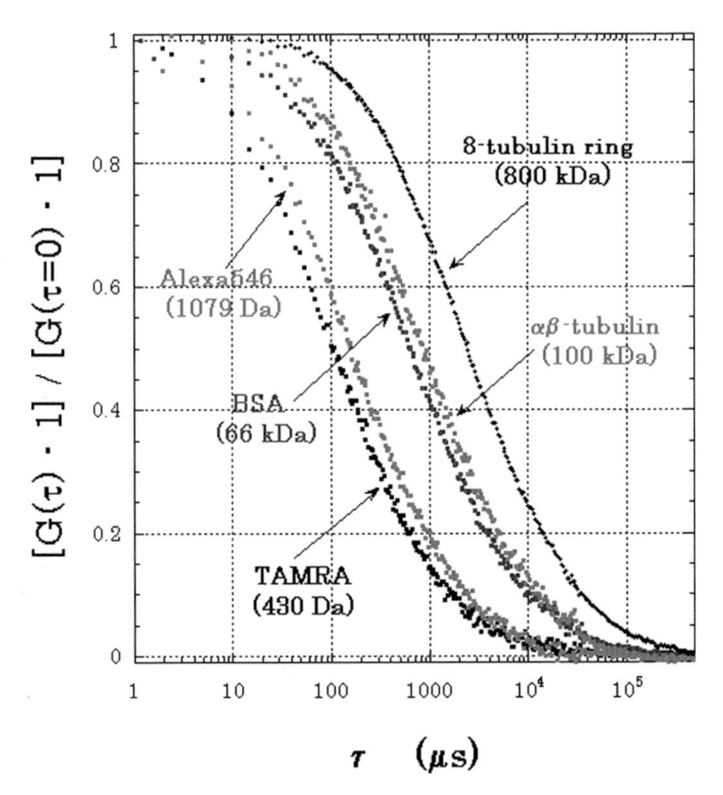

Figure 1.2. One-photon autocorrelation profiles of TAMRA (0.43 kDa), Alexa 546 (1.079 kDa), TAMRA - labeled BSA (66 kDa), the TAMRA - labeled αβ tubulin dimer (100 kDa), and the TAMRA - labeled 8 tubulin ring (800 kDa). It can be seen that a difference of much less than a factor of 2 (66 and 100 kDa) can be resolved if care is taken in the measurements.

Moreover, since the amplitude of the correlation is inversely proportional to the number of molecules, the higher the concentration used in the experiment, the lower the maximum value [or $G(0)$] of the autocorrelation function. This relationship can be intuitively understood by considering that the entrance or exit of a fluorescent particle into or from the detection volume when no other fluorescent molecules are present gives rise to a large relative fluctuation, whereas, if the concentration is high, such that many molecules are present on average in the volume, then the fluctuation due to the entrance or exit of a single molecule is quite small.

Autocorrelation functions for a fluorescent species interacting with itself or another nonfluorescent species can be analyzed in terms of the diffusion times of the bound and free species, their molecular brightness values, and their respective fractional populations.

$$G(\tau) = \frac{1}{N}\left(\left(Y_{\mathrm{f}}\left(1+\frac{\tau}{\tau_{\mathrm{df}}}\right)^{-1}\left(1+\frac{r_{\mathrm{o}}^{2}\tau}{z_{\mathrm{o}}^{2}\tau_{\mathrm{df}}}\right)^{-1/2}\right)+\left(Y_{\mathrm{b}}\left(1+\frac{\tau}{\tau_{\mathrm{db}}}\right)^{-1}\left(1+\frac{r_{\mathrm{o}}^{2}\tau}{z_{\mathrm{o}}^{2}\tau_{\mathrm{db}}}\right)^{-1/2}\right)\right), \tag{1.11}$$

where

$$Y_{\mathrm{f}} = \frac{\varepsilon_{\mathrm{f}}^{2}X_{\mathrm{f}}}{(\varepsilon_{\mathrm{f}}X_{\mathrm{f}}+\varepsilon_{\mathrm{b}}X_{\mathrm{b}})^{2}} \quad \text{and} \quad Y_{\mathrm{b}} = \frac{\varepsilon_{\mathrm{b}}^{2}X_{\mathrm{b}}}{(\varepsilon_{\mathrm{f}}X_{\mathrm{f}}+\varepsilon_{\mathrm{b}}X_{\mathrm{b}})^{2}} \tag{1.12}$$

with ϵ_{f} and ϵ_{b}, τ_{df} and τ_{db}, X_{df} and X_{db} corresponding to the molecular brightness diffusion time and the fractional population of the free and bound species, respectively. N here represents an effective number of total molecules. If the brightness of the free and bound species is the same, then Y_{f} and Y_{b} correspond directly to the fractional populations of the two species. In carrying out such experiments it is important to use a global analysis of the entire family of autocorrelation curves for the multiple concentrations tested in order to recover the diffusion times, molecular brightness values, and populations with a reasonable degree of certainty.

1.4. CROSS-CORRELATION AND HETEROLOGOUS ASSOCIATIONS

While the interpretation of the autocorrelation function can be hindered by anomalous or hindered diffusion, or small changes in molecular weight, two-color cross-correlation experiments provide a much clearer indication of interaction between biomolecules. The previous discussion of the autocorrelation function pertains to a single-channel correlation instrument. However, if one can arrange to label two partners in a biomolecular interaction with two different fluorophores that emit at different wavelengths, to excite simultaneously in time and in space the two fluorophores, and finally to simultaneously detect their emission on two separate channels, i and j, then it is possible to cross-correlate the traces of intensity versus time of from the two channels. These fluorophores should be chosen so as to avoid any energy transfer since FRET is anti-correlated, and also any cross-talk (also referred to as bleed-through) of the emission between the two channels. The amplitude of this cross-correlation signal is directly related to the degree of interaction between the two labeled molecules (Figure 1.3). If the two dyes are in the same complex, then as they diffuse in and out of the observation (or excitation volume) the fluctuations in their intensity will be correlated, whereas if the two biomolecules do not interact then the intensity fluctuations of their two dyes due to their diffusion in and out of the effective volume will have no relationship to each other in time.

We consider here, as earlier, only intensity fluctuations due to number fluctuations. Under ideal conditions in which there is no cross-talk and the intensity measured in channel i only emanates from species 1 and 12, while that in channel j only arises from species 2 and 12, the cross-correlation function then becomes

$$G_{ij}(\tau) = \frac{\langle F_i(t)\, F_j(t+\tau)\rangle}{\langle F_i(t)\rangle\, \langle F_j(t)\rangle}, \tag{1.13}$$

$$G_{ij}(\tau) = \frac{(\langle C_{12}\rangle\, M_{12}(\tau))}{V_{\mathrm{eff}}(\langle C_1\rangle + \langle C_{12}\rangle)(\langle C_2\rangle + \langle C_{12}\rangle)}, \tag{1.14}$$

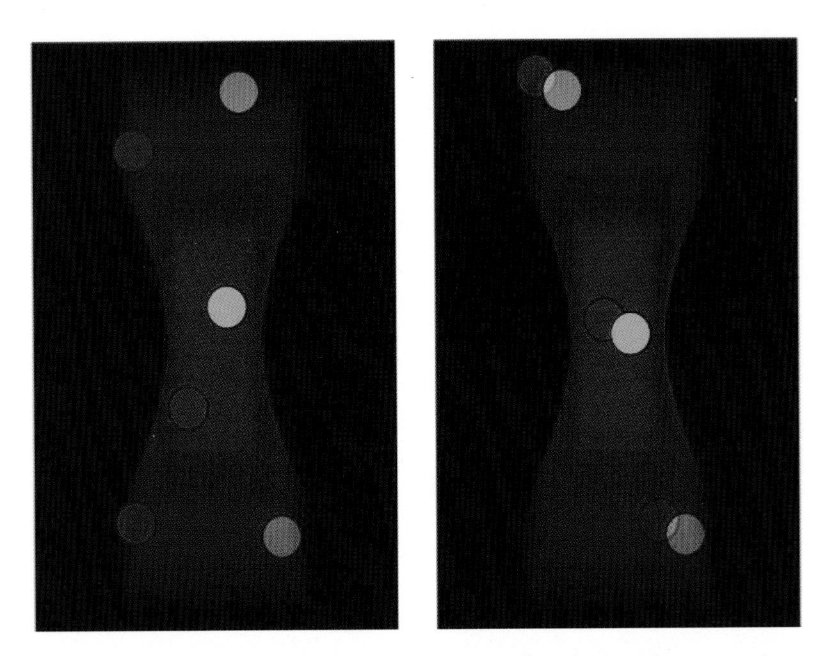

Figure 1.3. Schematic representation of cross-correlation studies of a heterologous protein–protein interaction. One protein here is labeled with a green dye and the other with a red dye. The colors of the dyes are chosen to avoid cross-talk in the detection channels. The fluctuations in fluorescence intensity due to self-diffusion in the detection volume detected on two separate channels are only correlated with each other in time if the two proteins are in interaction (*right*).

where C_1, C_2, and C_{12} are the concentrations of the free and interacting species, respectively, and M_{12} is the term describing the diffusion of the complex.

$$M_{12} = \left(1 + \frac{\tau}{\tau_{d12}}\right)^{-1} \left(1 + \frac{r_o^2 \tau}{z_o^2 \tau_{d12}}\right)^{-1/2}. \quad (1.15)$$

The autocorrelation functions from each of the channels can be expressed as

$$G_{ii}(\tau) = \frac{(\langle C_{11} \rangle M_{11}(\tau) + \langle C_{12} \rangle M_{12}(\tau))}{V_{\text{eff}}(\langle C_{11} \rangle + \langle C_{12} \rangle)},$$

$$G_{jj}(\tau) = \frac{(\langle C_{22} \rangle M_{22}(\tau) + \langle C_{12} \rangle M_{12}(\tau))}{V_{\text{eff}}(\langle C_{22} \rangle + \langle C_{12} \rangle)}. \quad (1.16)$$

If the two fluorescent species are noninteracting and hence do not diffuse together, then the fluctuations in intensity due to their number fluctuations will be entirely uncorrelated. In this case, the M_{12} term is null and the cross-correlation function becomes zero at all times. If on the other hand, the two fluorescent species are in complex with each other, then their fluctuations will be 100% correlated, and the amplitude of the cross-correlation function will reach that of the lower of the two autocorrelation functions.

Simultaneous excitation of two fluorophores can be achieved either using two co-axially aligned laser beams of different wavelengths (Kettling *et al.*, 1998) or alternately through two-photon excitation using a pulsed IR laser at a single wavelength (Heinze *et al.*, 2000). The broad two-photon cross-sections of many fluorophores are due to the different selection rules for the two-photon transition and allow for simultaneous excitation of fluorophores that exhibit large wavelength differences in their one-photon absorption spectra. Two-photon excitation for two-color cross-correlation is easier to align since there is only one excitation source and no emission pinholes. Thus the observation volume is exactly the same for the two detection channels. In one-photon excitation, the two excitation lasers must be exactly aligned so that the excitation volume is equivalent. Moreover, if each of the detectors has its own emission pinhole, then these must be perfectly aligned such that the focal volume defined by the two pinholes is exactly the same. Alternately, only one pinhole can be used before splitting the emission between the two channels.

The equations above pertain to ideal conditions in which the singly labeled species are detected only in their respective channels, i.e, there is no cross-talk, and only the doubly labeled species are detected in both channels. Otherwise, one must take into account the contributions of all of the species in each of the channels. Considering three species 1, 2, and 12 corresponding to the free species labeled with a short wavelength emitting dye, the free species labeled with a long wavelength emitting dye, and the complex between the two, then in channel i, the fluorescence fluctuations will include contributions from species 1 of concentration C_1 with ε_{1i}, its brightness in channel i, from species 2 of concentration C_2 with ε_{2i}, its brightness in channel i, and species 12 of concentration C_{12} with ε_{12i}, its brightness in channel i.

Similar contributions will hold in channel j. Thus, the amplitude in channel i (or j) at time zero can be expressed as follows:

$$G_i(0) = \frac{(\varepsilon_{i1}^2 C_1 + \varepsilon_{i2}^2 C_2 + \varepsilon_{i12}^2 C_{12})}{V_{\text{eff}}(\varepsilon_{i1} C_1 + \varepsilon_{i2} C_2 + \varepsilon_{i12} C_{12})^2} \tag{1.17}$$

while the amplitude of the cross-correlation function at time zero will be

$$G_{ij}(0) = \frac{(\varepsilon_{i1}\varepsilon_{j1} C_1 + \varepsilon_{i1}\varepsilon_{j2} C_2 + \varepsilon_{i12}\varepsilon_{j12} C_{12})}{V_{\text{eff}}(\varepsilon_{i1} C_1 + \varepsilon_{i2} C_2 + \varepsilon_{i12} C_{12})(\varepsilon_{j1} C_1 + \varepsilon_{j2} C_2 + \varepsilon_{j12} C_{12})}. \tag{1.18}$$

Indeed, while the interpretation of the time-dependent part of the cross-correlation poses the same difficulties as that for autocorrelation, the analysis of the amplitude of the cross-correlation function at time zero allows for a quantitative determination of the degree of interaction, even in the presence of cross-talk.

Two-color two-photon fluorescence cross-correlation spectroscopy (TCTPFCCS) was first applied to the study of biomolecular interactions *in vitro* (Kettling *et al.*, 1998; Heinze *et al.*, 2000; Rippe, 2000; Rarbach *et al.*, 2001; Heinze *et al.*, 2002; Kohl *et al.*, 2002; Berland, 2004; Heinze *et al.*, 2004; Jahnz and Schwille, 2005). In addition to having contributed significantly to the use of this technique for *in vitro* applications, Schwille's group has recently pioneered its use in live cells, as well. Since they first measured cholera toxin subunit interactions after endocytosis by TCTPFCCS (Bacia *et al.*, 2002), several applications either based on fluorescent protein fusions or microinjections of labeled proteins have appeared recently in which quantitative measurements of protein–protein interactions have been made in live cells using this approach (Kim *et al.*, 2004; Kim *et al.*, 2005; Kohl *et al.*, 2005; Larson *et al.*, 2005). The two-photon cross-sections allow for simultaneous excitation of any pair of fluorescent proteins (FPs) at wavelengths between 800 and 1,000 nm (Blab *et al.*, 2005), which can be achieved using tunable femtosecond IR lasers. Applications in live cells based on simultaneous two-color excitation with visible lasers have been published recently, as well (Vamosi *et al.*, 2004; Baudendistel *et al.*, 2005; Sato *et al.*, 2005; Thews *et al.*, 2005). As noted earlier, it is best to work under conditions in which there is no cross-talk between channels, and indeed it is important to avoid FRET between the two dyes, since FRET is anti-correlated. The most appropriate fluorescent protein pairs that would fit these requirements are eCFP or monomeric Cerulean with a red fluorescent protein. Recently monomeric forms of this latter have become available (Campbell *et al.*, 2002), allowing for the study of AP1 subunit interactions in live cells in the absence of any cross-talk (Baudendistel *et al.*, 2005). Finally, one should note that cross-correlation with a single visible laser is also possible using fluorophores such as quantum dots, which despite similar excitation profiles exhibit distinct emission maxima (Hwang and Wohland, 2005; Swift *et al.*, 2006).

We note that while two-color cross-correlation measurements are best used for studying heterologous associations, since the partners can be separately labeled with different dyes, it is also possible to use the technique for homologous oligomerization studies. While the dynamic range of the value of $G(0)$ is limited to 50% of the total change that could be observed for a heterologous interaction, mixing equal amounts of two preparations of the same protein labeled with different dyes also allows to discern and characterize interactions between protein monomers.

1.5. PHOTON STATISTICS

One of the thorniest issues in studying biomolecular complexation is that of stoichiometry. While measurement of the rotational or translational diffusion coefficient using fluorescence anisotropy, analytical ultracentrifugation, light scattering, or even correlation spectroscopy can sometimes provide limits for the stoichiometry, or its value under certain limiting concentration conditions, the interpretation of data from any of these approaches is complicated by issues of molecular shape, heterogeneous populations, and hydration considerations. In fluctuation spectroscopy however, one has the added advantage that it is a particle-counting technique, which allows for the measure of molecular brightness. Molecular brightness corresponds to the number of counts per second per molecule (cpspm) observed from the fluorescent species. One can imagine for example (Figure 1.4) that a protein–DNA complex containing a fluorescently labeled protein (1 dye molecule/monomer) emits twice as many photons as it moves through the excitation or

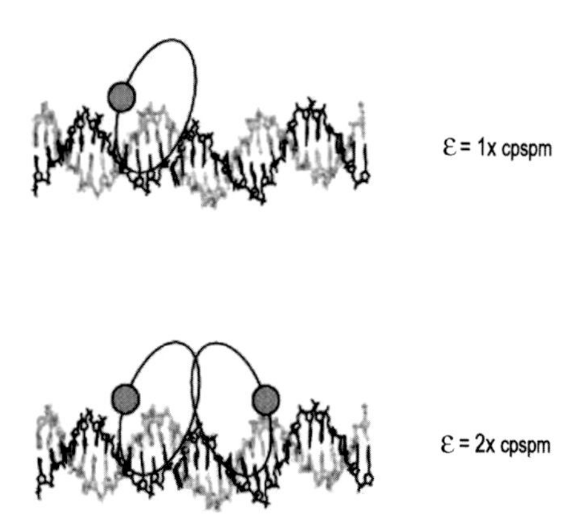

Figure 1.4. Schematic representation of a molecular brightness experiment. A protein, labeled with a green fluorophore binds to a target DNA sequence. In order to determine whether the protein binds as a monomer or a dimer, the fluorescent protein (with a 1:1 fluorophore to monomer labeling ratio) and DNA are observed at various concentration ratios (at least tenfold above the Kd of the interaction). If the complex formed is a 1:1 monomer to DNA complex, then the molecular brightness will be equal to that of the protein monomer, measured separately. If on the other hand, the complex is a 2:1 monomer to DNA complex, then the molecular brightness will be twice that of the protein monomer.

observation volume in an FCS experiment, if the protein binds as a dimer, rather than as a monomer. Thus, in addition to allowing for the determination of affinity, which is often more easily attained by other techniques, at least *in vitro*, fluctuation spectroscopy can offer a rather unambiguous measure of complex stoichiometry. Moreover, particle counting (Berland *et al.*, 1996), molecular brightness (Chen *et al.*, 1999b; Margeat *et al.*, 2001), or distribution analysis (Kask *et al.*, 2000; Palo *et al.*, 2000) (discussed subsequently), all of which are capabilities of fluorescence fluctuation measurements, may be more appropriate for monitoring complexation events involving small relative changes in molecular weight.

While the autocorrelation function describes the temporal fluctuation of the fluorescence signal, several other methods rely on the analysis of the amplitude of these fluctuations. These analysis procedures allow the extraction from the fluorescence fluctuation data of the average number $\bar{N}$ and the molecular (or specific) brightness ε of the molecule in the observation volume, related to each other through the relation

$$\varepsilon = \frac{\langle k \rangle}{\bar{N}}, \tag{1.19}$$

where $\langle k \rangle$ is the average number of photon counts per unit of time, and ε is expressed as the mean count rate per molecule (in cpspm). It is especially useful to determine this latter parameter when, during the interaction process under study, the molecular weight of the species, and thus the diffusion coefficient, does not change dramatically, while the molecular brightness is strongly affected. This is the case for example during the dimerization of a labeled species, where the diffusion coefficient increases on average by only 26%; whereas the molecular brightness increases by 100% (if the two fluorophores do not interact). It is important to remember that in order to compare the values of molecular brightness, one must not vary any of the instrument parameters. That is to say that the focus, the excitation intensity, and the detection efficiency must remain constant.

We review here the different methods that have been developed to extract from the fluorescence fluctuation data either the molecular brightness alone or the molecular brightness and the diffusion coefficient at the same time, thus allowing to resolve complex mixtures of biomolecules based on these different parameters. The theories presented here take into account in various ways the geometry of the observation volume and the effects of shot noise that arises from the randomness of the fluorescence emission and detection processes. We limit ourselves to the analysis of the fluctuation data in the "low ensemble" concentration regime, where several molecules can be present in the observation volume. For the "single molecule" regime, i.e., when there is always less than 1 molecule at the time in the observation volume (i.e., typically when the fluorophore concentration is less than 0.2 nM), the methods presented here are still valid with few adjustments, but it becomes possible to analyze individually the signal from each single molecule, which appears as a "burst" of fluorescence above the background (see Deniz *et al.*, 1999 and references therein).

1.5.1. Moment Analysis

This analysis method, introduced by Qian and Elson (Qian and Elson, 1990a,b), uses the first order moment $\langle k \rangle$ (average) and the second order moment $\langle \Delta k^2 \rangle$ (variance) of the photon counts to calculate the fluctuation amplitude, $G(0)$. This $G(0)$ value, which depends on the average number of molecules in the excitation volume $\bar{N}$, is related to the average and the variance of the fluorescence intensity through the relation

$$G(0) = \frac{\langle \Delta F^2 \rangle}{\langle F \rangle^2} = \frac{\langle F^2 \rangle - \langle F \rangle^2}{\langle F \rangle^2}. \quad (1.20)$$

The first two moments of the fluorescence intensity can be related to the moments of the photon counts

$$\langle F \rangle = \langle k \rangle \quad (1.21)$$

and

$$\langle F^2 \rangle = \langle k^2 \rangle - \langle k \rangle. \quad (1.22)$$

Thus, $G(0)$ can be rewritten in terms of photon counts

$$G(0) = \frac{\langle \Delta k^2 \rangle - \langle k \rangle}{\langle k \rangle^2}. \quad (1.23)$$

With this method, $G(0)$ can be calculated in a fast and model-independent manner. However, in order to recover the parameters, $\bar{N}$ and ε, it is necessary to know the geometric factor γ, see Eq. (1.6) and thus the calculation is not model-independent anymore. The factor γ, which depends on the shape of the PSF, equals $1/2\sqrt{2}$ for a 3D Gaussian profile and $3/4\pi^2$ for a Gaussian–Lorentzian profile. For the case of a single species, $\bar{N}$ is given by

$$\bar{N} = \frac{\gamma}{G(0)} \quad (1.24)$$

and ε is calculated using Eq. (1.19) (Qian and Elson, 1990b).

Thus, for the simple case of a single species in solution, it is straightforward to recover its brightness and concentration in a computationally simple and rapid manner, which provides a convenient means of checking the quality of a data set. However, when more species are present in solution, it is necessary to take into account the higher-order moments of the photon counts. Although suggested as early as 1990 and applied to the detection of large fluorescent beads, this approach has not been fully explored until recently, with the introduction of the factorial cumulants method (see later).

1.5.2. PCH and FIDA: Fitting the Photon Counts Distribution

Instead of calculating the moments of the photon-counts distribution, it is possible to fit this histogram directly and thus to use more information to extract the molecular brightness and occupancy. The various sources of fluctuation that account for the shape of the distribution have to be explicitly taken into account, i.e., the shot noise, the fluctuation in fluorescence intensity caused by the diffusion of the molecules in an inhomogeneous detection profile, and the fluctuation of the number of particles within this observation volume. Two methods have been developed quasi-simultaneously to describe these distributions, the photon-counting histogram (PCH) (Chen *et al.*, 1999b) and the fluorescence intensity distribution analysis (FIDA) (Kask *et al.*, 1999), which differ mainly in their approach to the treatment of the observation volume. In the original description of the PCH, the observation volume generated by the two-photon excitation process was adequately modeled with a 2D Gaussian–Lorentzian function. The parameters of the function were directly recovered from the fit, and thus no calibration is needed. In the case of one-photon excitation, however, a 3D Gaussian function fails to describe the histogram correctly, and thus additional parameters have to be included in the fit to take into account the contribution to the histogram of the photons coming from the out-of-focus regions (Huang *et al.*, 2004). In the case of FIDA, also developed for one-photon excitation, the observation volume is described with a polynomial function with up to three parameters. This approach, although fast and versatile, introduces several other fitting parameters, without any specific physical meaning, and thus it is necessary to calibrate the observation volume using solutions of known dyes before the analysis of unknown samples.

We must note that one condition must be fulfilled for these types of analysis to be valid; that is, that the molecules be quasi-immobile within the observation volume during the counting time interval used to build the histogram. Practically, this means that the counting time interval has to be at least ten times smaller than the diffusion time of the molecule. If this sampling interval is too long relative to the diffusion time, the PCH/FIDA theory breaks down. Thus, for an accurate evaluation of the occupancy and the brightness of the species under study, it is necessary either to determine the diffusion time using FCS analysis or to perform a PCH/FIDA analysis at various counting time intervals, and check that the recovered parameters are constant.

All these methods have the ability to extract from the histograms the molecular brightness and concentration of a mixture of species with different brightness, provided that the signal statistics are sufficient (Muller *et al.*, 2000; Huang *et al.*, 2004). They have been successfully applied to study ligand–protein (Chen *et al.*, 2000; Rudiger *et al.*, 2001; Scheel *et al.*, 2001), protein–protein (Margeat *et al.*, 2001), or polymer–oligonucleotides interactions (Van *et al.*, 2001).

1.5.3. Fluorescence Intensity Multiple Distribution Analysis

Since temporal fluctuation analysis (i.e., FCS) and amplitude fluctuation analysis (PCH, FIDA, and others) are performed on the same original data set, it is possible in principle to combine these two types of analyses to extract simultaneously, and with a better accuracy, the diffusion time, concentration, and brightness parameters of the various species present in solution.

In the FIMDA, photon-counting histograms are built for various lengths of counting intervals (binning times), and analyzed simultaneously (Palo *et al*., 2000). We remind the reader that for the analysis of the photon-count distributions presented earlier, one fundamental assumption was that the binning time is chosen to be short enough, so that the molecules can be safely assumed to be immobile in the observation volume during the integration time. However, for the FIMDA analysis, this hypothesis is not valid anymore, and thus, for each binning time, only an *apparent* brightness is recovered; then, the dependence of the variation of this apparent brightness on the width of the time window allows an estimation of the diffusion time. Although this estimation is indirect, as compared with the direct fitting of the correlation function $G(t)$ performed in an FCS analysis, it has been shown that the accuracy of the recovered diffusion time by both methods is equivalent (Palo *et al*., 2000).

1.5.4. Fluorescence Cumulant Analysis

Recently, a new theory called Fluorescence Cumulant Analysis (FCA) has been developed and tested to extract directly without any fitting, the concentration and brightness parameters of several species, by exploiting the factorial cumulants of the photon counts (Muller, 2004). The cumulants are related to the moments of the photon counts, but are more convenient from a mathematical point of view. Indeed, cumulants of the sum of statistically independent variables are simply given by the sum of the cumulants of the individual variables. Since two cumulants are necessary to determine the brightness and concentration for one species, $2n$ statistically significant cumulants (i.e., statistically significant cumulants up to the $2n$th order) will be necessary to resolve a mixture of n species (Muller, 2004). An error analysis method was developed for the factorial photon-count cumulants, allowing the evaluation of the relative error of each experimental cumulant and thus its statistical significance. Knowing the number of significant cumulants, it is straightforward to determine how many species can be resolved. If necessary, the acquisition time can be increased to obtain the $2n$ statistically significant cumulants necessary to resolve the n species. An extension of this theory, called time-integrated fluorescence cumulant analysis (TIFCA), allows one to recover the diffusion time of the diffusing species in addition to their brightness values and concentrations. This is achieved by rebinning the data taken at short sampling time, calculating the experimental cumulants of the photon counts as a function of binning time, and then comparing them to theoretical models. This analysis procedure allows not only determination of the diffusion time, but also improves considerably the accuracy of the determination of the brightness and concentration, without increasing the acquisition time (Wu and Muller, 2005).

1.5.5. Photon-Count Distribution Analysis in Multiple Channels: Dual-Color PCH and 2D-FIDA

To examine interactions involving several partners, it is often desirable to label them with spectrally distinct fluorophores. Like cross-correlation analysis, dual color fluorescence fluctuation data can be analyzed in terms of photon-count distributions. Two similar approaches have been developed for this purpose, allowing to extract concentration and brightness parameters in each detection channel : two-dimensional fluorescence intensity distribution analysis (2D-FIDA) (Kask *et al.*, 2000) and the dual-color photon counting histogram (dual-color PCH) (Chen *et al.*, 2005). So far, 2D-FIDA has been mainly used in high throughput screening studies of protein–ligand interaction (Kask *et al.*, 2000; Schilb *et al.*, 2004). The introduction of dual-color PCH allows us to resolve CFP/YFP mixtures *in vitro*, based on their relative brightness in each detection channel, an usually difficult task due to the high spectral overlap between these two probes (Chen *et al.*, 2005).

1.5.6. Monitoring Diffusion Time, Brightness, Concentration, and Dual-Color Coincidence with Photon Arrival-Time Interval Distribution

In order to analyze the photon streams in one or two channels in the ''single molecule'' and ''low ensemble'' regime, Laurence *et al.* took a different approach than analyzing the photon-counts distribution binned in evenly spaced intervals (Laurence *et al.*, 2004). Instead, in their method called photon arrival-time interval distribution (PAID), a 2D histogram is built, based on the observation of time interval between photons, which emphasizes ''photon-rich'' time intervals, where molecules are present in the detection volume. The x-axis is the time interval between two photons (the *start* and *stop* photons, not necessarily consecutive), while the y-axis represents the number of photons in the *monitor* channel counted in the time interval between the *start* and *stop* photons. For a one-color experiment, the *start*, *stop*, and *monitor* channels are the same, while in a two-color experiment, each channel can be ''yellow'' or ''red,'' and thus eight different PAID histograms are built with the different combinations. Interestingly, a simple collapse of the histogram on the time axis provides the FCS correlation function. The PAID histogram is fitted using a model that includes a numerically approximated, non-Gaussian detection volume. The various parameters characterizing the diffusing species are recovered with an accuracy comparable to FIMDA. However, PAID is also suitable for dual-color experiments, and thus provides a unique tool for the sorting of a heterogeneous mixture in the ''small ensemble'' regime. As a demonstration, a mixture of various components of transcription complexes has been successfully resolved, showing the presence of : (1) free DNA (labeled with a single red dye, Cy5), (2) species associated to σ^{70}, the initiation factor (labeled with a single yellow dye, Tetramethylrhodamine), (3) DNA–RNA Polymerase-σ^{70} complexes (labeled with a single yellow and a single red dye), and (4) σ^{70}-aggregates (labeled with several yellow dyes) (Laurence *et al.*, 2004).

As shown previously, a large number of methods that allow the analysis of the same sets of fluorescence fluctuation data have been developed, to extract the brightness, concentration, and diffusion time of the species present in solution. Most of them are relatively recent, and have not been applied to solve problems of biological interest, except for PCH and FIDA. These two methods are the simplest to implement and are already included in the software of commercial FCS systems. However, great care has to be taken in the interpretation of the data, and it is important to be able to check that the data set used contains a sufficient amount of information to resolve the various species in solution without ambiguity (Muller *et al.*, 2000). From this point of view, the TIFCA method, which facilitates the calculation of the statistical significance of each of the factorial photon-count cumulants, seems to be the most advanced approach to avoid any ambiguity (Wu and Muller, 2005).

1.6. EXAMPLES OF THE USE OF FLUCTUATION SPECTROSCOPY TO STUDY BIOMOLECULAR INTERACTIONS

1.6.1. Resolution in the Measurement of Biomolecular Diffusion Times

Even with a well-characterized FCS setup and without inclusion of other photophysical processes (i.e., triplet-state emission, blinking, etc.) fitting correlation functions measured from a generic sample with the expression in Eq. (1.3) is not generally straightforward as possible polydispersity in size (D) and brightness (ε) of the fluorescent particles introduces a multiparameter fit that is difficult to handle (see Starchev *et al.*, 1999; Chen *et al.*, 1999a; Krichevsky and Bonnet, 2002, for discussion). Indeed, as in dynamic light scattering and similar scattering techniques, extracting the distribution of sizes from the measured FCS correlation functions requires solving a mathematical inverse problem, a challenging task since the problem is ill-posed (Meseth *et al.*, 1999; Starchev *et al.*, 1999). For example, a group of diffusing particles with three distinctly different sizes might yield a correlation function that is indistinguishable from that deriving from a continuous distribution spanning the same size range. Moreover, for a complete and consistent fitting of the FCS correlations, *a priori* knowledge of the distribution of brightness of the diffusing fluorescent particles is required. This information cannot be derived from the correlation functions, but rather must be obtained separately from other measurements such as photon histograms (Chen *et al.*, 1999a; Krichevsky and Bonnet, 2002). These requirements impose some limitations on the technique, as shown by Meseth *et al.* (1999), who demonstrated that the resolution limit of FCS depends on several factors including difference in size between particles as well as their concentration and brightness.

In Figure 1.2 we include normalized correlation function measured from solutions of bovine serum albumin (BSA) (66 kDa), αβ–tubulin dimers (100 kDa), and 8-tubulin rings (~800 kDa) in PBS or PIPES buffer. All functions in Figure 1.2 were collected at room temperature. Note the uniform time shift of the correlation functions with increasing molecular weight of the particles, demonstrating first the time-resolution of the FCS setup. Though not shown in Figure 1.2, the measurements on TAMRA and Alexa 546 were extended to 0.1 μs at high-count rates ($>$ 500,000 count/s) to determine the diffusion coefficient of each of these molecules more reliably (see Figure 1.1). Note that the sizes of the fluorophores are close to nanometer, a range difficult to probe with standard light scattering techniques. Unlike the individual TAMRA and Alexa 546 molecular fluorophores, the BSA and tubulin samples are labeled with an exogenous fluorophore (TAMRA). The labeling protocol, which can be elaborate and complicated, may yield variations in the brightness of the labeled particles (number of fluorophores/particle). However, for diffusion coefficient measurements, it is much more important to clean the sample of the excess of the labeling fluorophores, which, otherwise, may contribute to the correlation function. The BSA and tubulin samples are characterized by their narrow dispersity in size as indicated by the uniform shift of the correlation functions with respect to those of TAMRA and Alexa 546 (see Figure 1.2); in a generic sample, this is not generally the case. A fit of the data with the expression of Eq. (1.3) for one diffusing component yields the following values for the apparent diffusion coefficient: $D_{\text{BSA}} = 47\,\mu\text{m}^2/\text{s}$; $D_{\text{tubulin}} = 35\,\mu\text{m}^2/\text{s}$ $D_{\text{ring}} = 12\,\mu\text{m}^2/\text{s}$. It can be seen here that the difference in molecular weight between BSA (66 kDa) and tubulin (100 kDa) is easily resolved in these measurements. Thus, protein dimerization can be resolved by FCS diffusion measurements, but only under conditions in which the samples are sufficiently bright and homogeneous.

1.6.2. Hydrodynamic Modeling

Often forgotten in typical analyses of FCS measurements is the relation between the translational diffusion coefficient and the structure of a biomolecule, although new theories and computational methodologies have been developed for the calculation of hydrodynamic properties of arbitrarily shaped biomolecules (Garcia de La Torre and Bloomfield, 1981; Douglas *et al.*, 1994; Garcia de la Torre *et al.*, 1994, 2000; Kang *et al.*, 2004). In particular, these theories appear to yield proper values for the translational diffusion if an adequate structural model is constructed for the biomolecule. The model is made, generally, with N elements, named beads, of arbitrary diameter [the bead is commonly assumed rigid and spherical although other shapes for the beads such as cylinders are now included in the ZENO code (Kang *et al.*, 2004)]. The size and shape of the model should be as close as possible to that of the biomolecule. If available, the X-ray structure can be used to guide the model construction. One can then apply a generic code such as HYDRO or ZENO to calculate the hydrodynamic properties of the model (Garcia de la Torre *et al.*, 1994; Kang *et al.*, 2004). Note that for simple well-defined geometries such as dimers, rings, and cylinders, closed-form expressions for the hydrodynamic properties have been derived (Douglas *et al.*, 1994 and references therein). This approach was successfully applied by Boukari *et al.* (2004), who showed that hydrodynamic modeling was useful in the analysis of FCS and sedimentation measurements on the polymerization of tubulin dimers into closed tubulin rings. In particular, the average number of tubulin dimers per ring was determined. This example suggests that hydrodynamic modeling may turn to be valuable for complete and detailed analysis of FCS measurements of biomolecular diffusion.

1.6.3. Tubulin Self-Assembly

Boukari *et al.* (2004) applied FCS to investigate tubulin oligomers that result from interactions of a novel class of antimitotic natural peptides, particularly cryptophycin 1, dolastatin 10, and hemiasterlin. Cytotoxity assays on various cell lines showed that these peptides exhibit potent cytotoxicity with IC50 in the pico- to nanomolar concentration, a range accessible by FCS. More importantly, the availability of the materials (the peptides) is limited. *In vitro*, the peptides tend to cause depolymerization of microtubules and induce, instead, the formation of rings. Analysis of FCS and sedimentation velocity (see Chapter 16) revealed differences in the interactions of the peptides with tubulin, though it was shown that the peptides bind to the same tubulin domain. The cryptophycin-tubulin rings were found to be rigid, exhibiting circular geometry, were highly monodisperse in size (8 tubulin dimers/ring, Figure 1.2), and appear stable even with tubulin concentration as low as 1 nM. In contrast, the dolastatin-tubulin rings were composed of 14 tubulin dimers and appeared unstable on dilution, with significant dissociation below 10 nM. Moreover, the dolastatin-tubulin rings tended to aggregate at micromolar concentration. Finally, the hemiasterlin-tubulin rings were found to be the most unstable.

1.6.4. Cross-Correlation Study of a Complex Protein–DNA Interaction

Transcriptional regulation, even in bacteria, involves complex interactions between transcription factors, ligands, DNA, and RNA polymerase. Rippe and coworkers (Rippe, 2000) published a nice application of two-color cross-correlation spectroscopy several years ago, which allowed them to propose a model for how the NtrC transcriptional activator from *E. coli* promotes transcription. Using two DNA targets labeled, respectively, with 6-carboxy-fluorescein and 6-carboxy-X-rhodamine and one-photon excitation with an argon krypton ion laser with lines at 488 and 568 nm, they showed (Figure 1.5) that titration of a 1:1 mixture of the two by the NtrC protein leads to a significant increase in the cross-correlation amplitude for the two detection channels. This demonstrated that one NtrC octamer could bind two molecules of DNA and allowed them to propose a looping model for the complex, explaining how the NtrC protein could interact with and activate RNA polymerase bound to a relatively distant promoter.

1.6.5. Diffusion in Live Cells, Concentrated Solution, and Gels

The expression in Eq. (1.3), which is often used to fit measured FCS time-correlation data, was derived under the assumption that the fluorescent molecules are in a steady-state, freely diffusing in an open 3D Gaussian excitation beam,

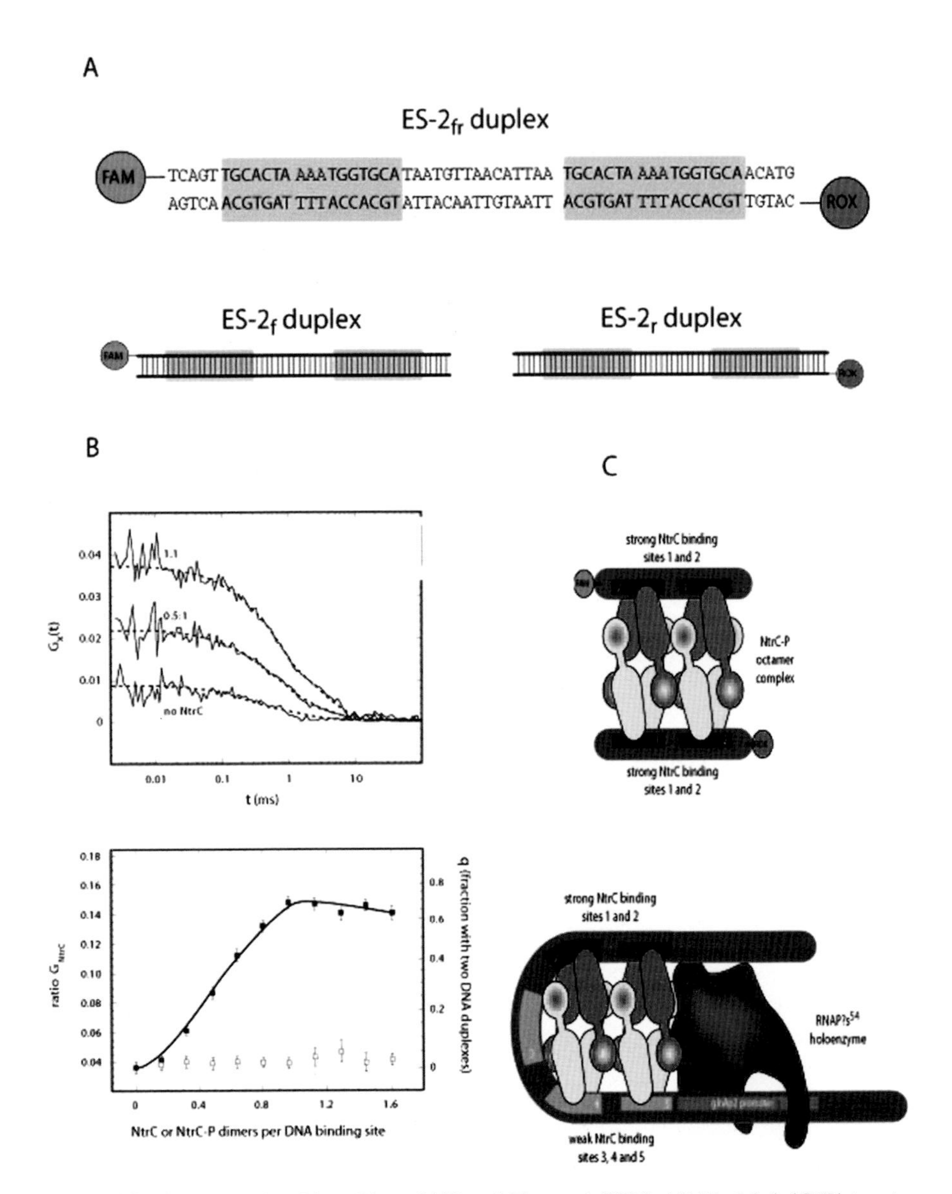

Figure 1.5. Figure reproduced from Rippe (2000) and Kim *et al.* (2005). (A) The labeled DNA targets used in the experiment, ES-2f (labeled with fluorescein) and ES-2r (labeled with rhodamine X). (B) The cross-correlation curves obtained with increasing concentrations of the NtrC protein. The increase in the cross-correlation signal indicates that the two labeled DNA molecules are diffusing in the same complex. (C) The model derived from the experiments showing how DNA looping and binding of the NtrC protein to two different sites could allow it to activate the RNA Polymerase bound at the promoter.

a condition typically satisfied with dilute solutions. Recently, FCS has been applied to the study of biomolecular diffusion in live cells, which are notorious for their crowded intracellular environment and its heterogeneous structure (Webb, 1976; Brock and Jovin, 1998; Schwille *et al.*, 1999a,b; Gennerich and Schild, 2000; Elson, 2001; Pramanik and Rigler, 2001; Schwille, 2001; Hess *et al.*, 2002; Hink *et al.*, 2002; Bacia and Schwille, 2003; Bulseco and Wolf, 2003; Clamme *et al.*, 2003; Maertens *et al.*, 2005; Vukojevic *et al.*, 2005). Moreover, cells are dynamic in nonequilibrium state. Major complications arise then in the analysis and interpretation of the measured correlation functions (Gennerich and Schild, 2000; Wachsmuth *et al.*, 2000; Clamme *et al.*, 2003; Weiss *et al.*, 2003, 2004; Masuda *et al.*, 2005). It is sometimes difficult to ascertain whether interaction between biomolecules is occurring or not. In such cases it is not clear whether the expression in Eq. (1.3) is valid. In Schwille *et al.* (1999b), the idea of anomalous diffusion was invoked and a modified expression to Eq. (1.3) was derived to account for the effects of the intracellular environment on the transport of molecules in cells.

Because of similarities in the mechanical and dynamical properties between cells and concentrated solutions or gels, FCS investigations on biomolecular diffusion in crowded/concentrated solutions or gels may provide insight into the behavior of diffusing particles in such environments. Their results may be helpful in the interpretation of FCS measurements on biomolecules in live cells. Here one should acknowledge that elucidating the factors affecting transport properties of particles in crowded solutions or seemingly random, inhomogeneous gels is also of interest to fundamental research and applied engineering (drug delivery, biomolecule separation). Whatever the purpose, FCS appears well suited to studies of transport properties of fluorescent probes in nonfluorescent—hence invisible—crowded/concentrated solutions or gels (Dauty and Verkman, 2004; Michelman-Ribeiro *et al.*, 2004; Kang *et al.*, 2005). The technique was recently applied to measure the diffusion coefficients of various fluorescent probes (TAMRA, Dextran, albumin, DNA) as a function of concentration of different crowding agents (Ficoll-70, Glycerol, Dextran) or polymers (Poly(Vinyl) Alcohol (PVA)) in solutions (Dauty and Verkman, 2004; Michelman-Ribeiro *et al.*, 2004). It was pointed out that the expression in Eq. (1.3) appears to fit the measured correlation data well, indicating a simple (nonanomalous) diffusion mechanism for the diffusing probes under the studied conditions (unlike results from work on cells). It was shown that the reduction of the diffusion coefficient appears exponential or stretched exponential, consistent with the behavior derived from transport models described in polymer physics (Phillies, 1991). When chemically cross-linked, PVA polymer gels appear to slow the diffusion of the TAMRA molecules further, despite the smallness of TAMRA relative to the correlation length of the polymer gels and the size of the superstructures induced by the cross-links. More interestingly, the study indicates a linear correlation between the diffusion times of TAMRA (dynamics) and the elastic modulus of the gel (static) (Michelman-Ribeiro *et al.*, 2004). More experimental and theoretical work is needed to elucidate these interesting observations.

1.6.6. Cross-Correlation in Live Cells

Schwille's group has recently extended their two-color, two-photon cross-correlation studies (Kim et al. 2004) of the interactions between calmodulin and CaM kinase in live cells to include a quantitative determination of the stoichiometry of the complexes formed (Kim et al. 2005). The calmoduline labeled with Alexa 568 was electroporated into the cells, while the CaM kinase partner was expressed as a GFP fusion. The top panel in Figure 1.6 shows their confocal images of stably transfected eGFP-CaM-kinase II in HEK293 cells electroporated with A568 CaM in

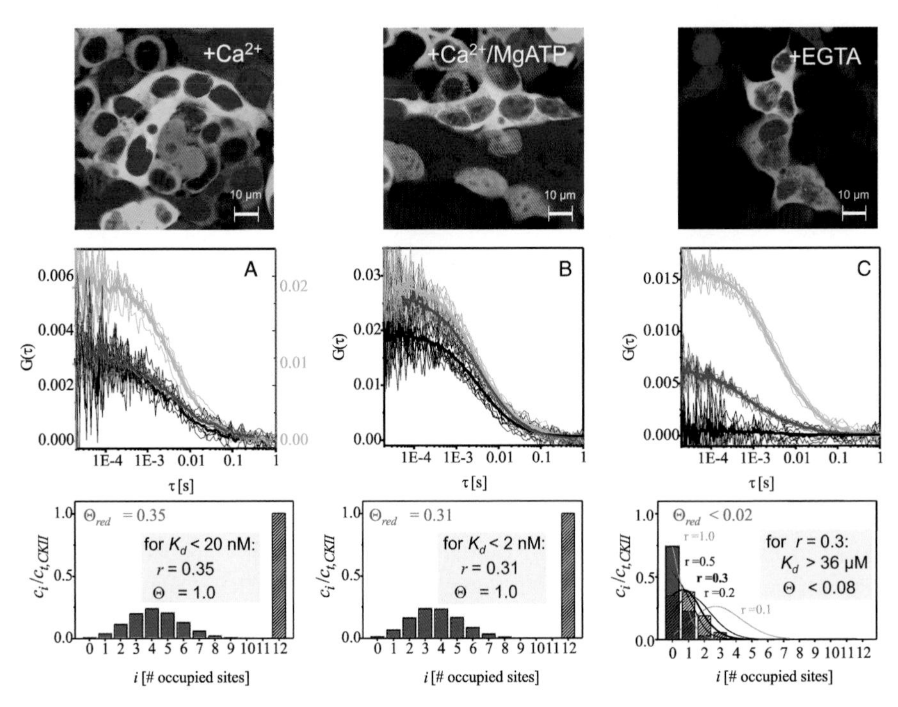

Figure 1.6. Figure reproduced from Kim *et al.* (2005). Cross-correlation studies of Alexa-568-labeled Calmodulin binding to GFP-labeled CaM kinase in life cells (*Top*). Confocal images of stably transfected eGFP-CaM-kinase II HEK293 cells electroporated with A568(C2)CaM are shown under 10 mM Ca^{2+} (*left*) followed by 10 mM Ca^{2+} and 1 mM MgATP (*middle*) and 200 mM EGTA (*right*) in the presence of 15 mg/mL a-hemolysin in the same dish (scale bar = 10 μm) (*Middle*). Intracellular auto- and cross-correlation curves were measured simultaneously under each of the conditions but at lower intracellular protein concentrations than those used for the LSM images here. Plotted are the six subsequent 10-s acquisitions (*thin lines*) for both the auto- and cross-correlation measurements and their respective average curves with data fits (*corresponding thick lines*). The hatched shaded bars represent the distribution of CaM irrespective of the label, and the red bars denote the distribution of Alexa-568-labeled CaM molecules only. In (A) and (B), an assumption of a high binding affinity yields virtually full binding (*hatched shaded bar*) and the indicated values for the labeled fractions r. In (C) the hatched bars exemplify a CaM distribution for $r = 0.3$ based on (A) and (B). Distributions for conditions with other r fractions are also denoted (*shaded lines*).

the presence of 10 mM Ca^{2+}, Ca^{2+}/MgATP, or with 200 mM EGTA. Under slightly lower intracellular concentration conditions, the auto- and cross-correlation curves were measured as well. The authors convincingly demonstrate that the relative amplitude of the cross-correlation curve increases under elevated Ca^{2+} conditions and is nearly null in the absence of Ca^{2+}. The stoichiometry of the complexes (given the known binding affinities) were also determined.

1.6.7. *In vitro* PCH Analysis of the Stoichiometry of a Heterologous Protein–Protein Interaction

In the study of the interaction of the estrogen receptor α dimer (ERα) ($\sim$130 kDa) with a labeled fragment of a coactivator protein (SRC-1) (26 kDa) (Margeat *et al.*, 2001), discriminating between one- or two-bound coactivator molecules corresponding to a 16% difference in molecular weight was not possible by classical autocorrelation of the fluctuation signal, but was successful based on the fact that the twice bound complex would exhibit a molecular brightness in cpspm of a complex twice as large as the once bound species. Figure 1.7 shows the photon-count histograms for the Alexa-488-labeled SRC-1 fragment in the presence of varying concentrations of ERα and saturating agonist ligand in all cases. The affinity constant measured by fluorescence anisotropy indicated that if the 2:1 complex were formed it should have been detected with a 96% degree of confidence in these experiments. However, under all cases, the PCH profiles in the presence of ERα were identical to those of the labeled coactivator alone, demonstrating convincingly that under most concentration conditions, the stoichiometry of the complex is 1 coactivator molecule per ERα dimer.

1.7. LIMITATIONS OF FCS

Despite its multiple advantages, there are, with fluctuation spectroscopy, as with any technique, limitations and problems that should be considered. Technical issues linked to the instrumentation and photophysics of the dyes are described subsequently. Here we mention just a few general limitations of which one should be aware when considering the use of fluctuation spectroscopy to measure interactions. First of all, while the measurement of the translational diffusion time is not fraught with the uncertainty inherent in anisotropy measurements due to limited fluorescence lifetimes and the convolution of local and global rotational motions, translational diffusion is much less sensitive to changes in molecular size on complexation. Indeed since the diffusion time increases with the cube root of the molecular weight, to observe a doubling of the diffusion time one needs an eightfold increase in molecular weight (for a sphere). Hence, dimerization reactions are difficult to study (Meseth *et al.*, 1999), although by using global analysis of curves obtained at multiple concentrations, it is possible to obtain good values for the dimerization constant. In the study of heterologous interactions, it is important

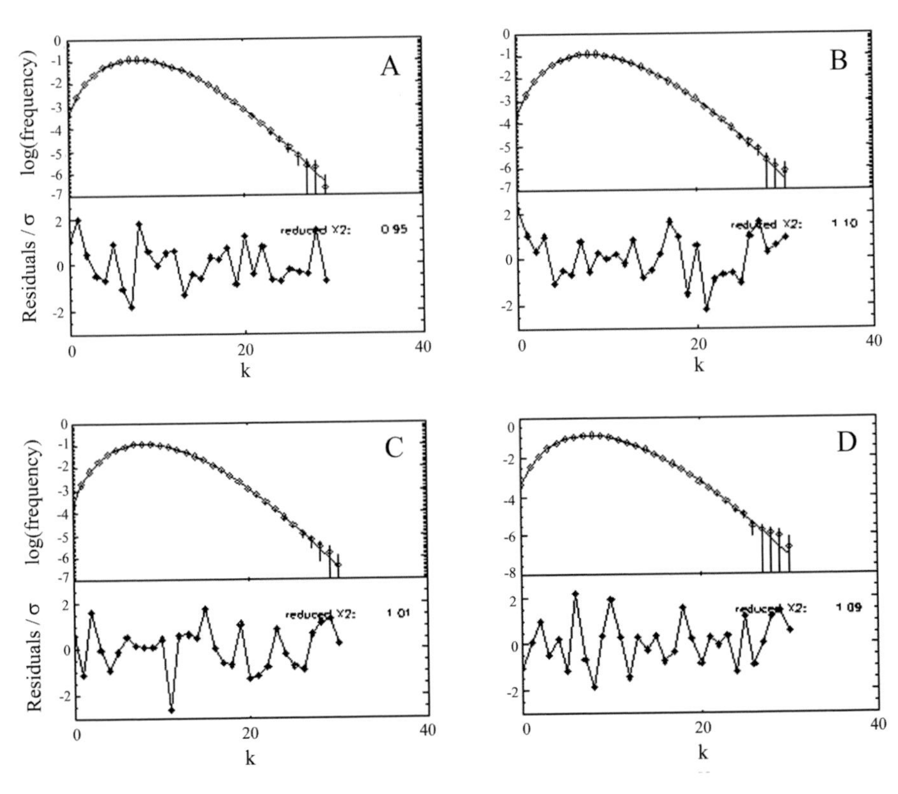

Figure 1.7. Figure reproduced from Margeat *et al.* (2001) with permission from Elsevier. Photon-counting histograms of SRC-1 Alexa 488 (200 nM) in the presence of estradiol, with (A) 0 nM, (B) 10.5 nM, (C) 42 nM, or (D) 105 nM, full-length ERα dimer. Data [including values at 5, 21, and 84 nM (not shown)] were analyzed globally with a single species model. The upper panels shows the PCH raw data and the fit, and the lower panels are a plot of the residuals.

therefore, as in anisotropy studies, to label the smallest of the partners, to ensure that the change in diffusion time upon complex formation be as large as possible.

1.7.1. Surface Adsorption

The small solution volumes and concentrations used in fluctuation spectroscopy, while limiting the amount of material necessary for the experiments, introduce the problem of loss of material because of the adsorption of sample to the slide surface. The use of BSA in excess or the prior treatment of the slide cover used in the measurement by silanization methods can alleviate these problems, but rarely are they eliminated entirely. Note that one has to be careful about possible interactions of BSA with the macromolecules of interest. Nucleic acids show little susceptibility to such surface activity linked sample loss, but many proteins pose such problems. Indeed, if the surface active protein itself is labeled, at least the FCS measurement can identify its real concentration in solution. The deposit of the protein on the surface is often visible upon focusing on the surface and using the ocular or a video camera attachment to the microscope. However, if the surface active partner in the complex is unlabeled, then the experimenter is unaware of its true solution concentration and this can lead to serious artifacts in the calculation of the dissociation constants of the complex. In cases where the labeled protein is surface active, it is also important to make measurements at a distance from the surface at which the fluorescence from the immobile particles will not contribute to the signal, as these particles are not diffusing and thus the fluorescence detected is uncorrelated. Uncorrelated photons detected in the sample channel have the effect of lowering the apparent correlation amplitude and thus cause an overestimation of the concentration of the sample.

1.7.2. Afterpulsing and Dead Time

In fluorescence fluctuation experiments, the detectors, generally avalanche photodiodes (APD), exhibit two effects that can plague the measurements: dead time, which is a fixed period of time after the registration of a photon during which the detector cannot accept another photon, and afterpulsing, which is a spurious pulse that can follow genuine output pulses. These spurious afterpulsing and dead-time effects tend to distort the autocorrelation at short timescales ($\leq 10\,\mu s$). This distortion can be misleading as it can be erroneously construed as a real photo-physical phenomenon such as excitation of triplet states of the probed fluorophores. Further, for small fluorescent molecules such as TAMRA (MW = 430 Da), the effects may limit precise determination of the diffusion time and the amplitude, the latter being related to the apparent numbers of particles in the excitation volume. To overcome this problem one can split the detected beam into two signals, which are directed onto two similar but separate detectors. The two signals are then cross-correlated. Since the afterpulsing parts of the signals of the detectors are independent—hence uncorrelated—the resulting cross-correlation function is decoupled from the afterpulsing effects.

In Figure 1.8 we show correlation functions measured from Alexa 546 molecules (MW = 1,079 Da, Molecular Probes, Oregon) in solution with FCS setup in auto- and cross-correlation modes. In the autocorrelation mode the function, which is measured with a 10-μm pinhole, shows two time regimes: one below $10\,\mu s$ caused mainly by afterpulsing followed by a regime due to translational diffusion of Alexa 546. However, when the emitted beam is split into two beams detected by two separate detectors, the correlation function generated by cross-correlating the two detected intensities is significantly different. The first time regime is significantly reduced, indicating the removal of afterpulsing signals. As a result, the diffusion time and the amplitude are well fit with the expression of Eq. (1.3) for one type of molecule. In systems set up for two-channel two-color detection, the use of a filter combination that allows some bleed-through of the high-energy emitter into the low-energy channel can also be used to check for the afterpulsing tendencies of the system detectors.

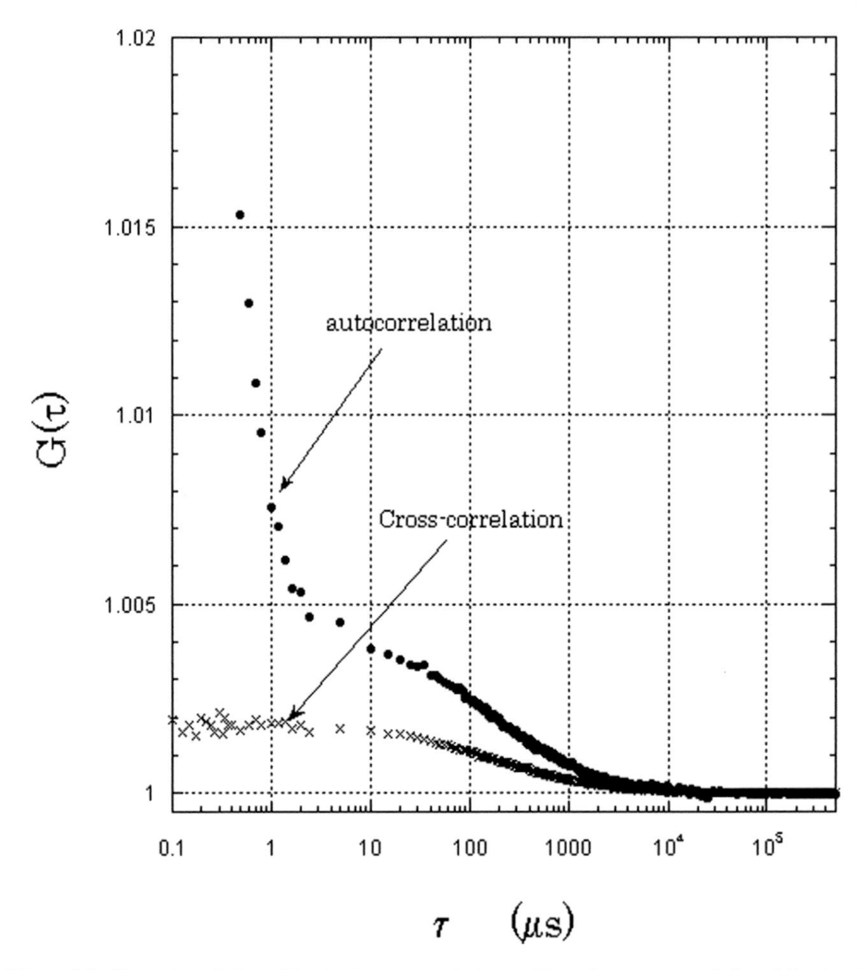

Figure 1.8. Correction of afterpulsing in the autocorrelation profiles using cross-correlation of the signal detected on two avalanche photodiodes. What is often taken to be triplet or other photophysics may simply be afterpulsing that is not correlated. Thus, it can be eliminated by cross-correlation of the signal.

For some specific cases in photon statistical analysis, such as PCH, it is also important to take into account in the model of analysis the nonideality of the photodetectors. Typically, for a one-color experiment, it has been shown that afterpulsing can be safely ignored, but dead-time effect leads to an underestimation of the molecular brightness when the concentration of the fluorescent species increases (Hillesheim and Muller, 2003). This is understandable since at high concentrations, many photons are lost during the dead time, where the detector is ''blind''. This effect has been explicitly taken into account in a new PCH theory (Hillesheim and Muller, 2003), and, together with the afterpulsing contribution, in a dual-color PCH theory (Hillesheim and Muller, 2005). Using this improved PCH theory, fluorescence fluctuation experiments of EGFP have been reported in cells up to concentrations of 10 μM (Chen *et al.*, 2003). The ability of improved PCH to analyze fluctuation data at these high concentrations is of first importance to study low-affinity interactions (in the 0.1–10 μM range) and for cellular work, where the expression level of the protein of interest can vary greatly from cell to cell. The improved PCH theory integrates the dead-time and afterpulsing effects into the analysis. In addition, since this improved PCH has been developed for two-photon-based microscopy, it appears as the method of choice to characterize macromolecular interactions in cells, based on the molecular brightness of the interacting partners.

1.7.3. Observation Volume Artifacts

In one-photon FCS, if a pinhole of very large diameter is used, then the observation volume can deviate considerably from the Gaussian approximation (Hess and Webb, 2002). This leads to residuals in the fitting function that can be misinterpreted as arising from photophysical events or conformational fluctuations. A non-Gaussian volume can be caused by the existence of diffraction fringes from overfilling the optical back apertures of the focusing objectives of the setup. One remedy is to underfill the back aperture of the objective, which is just the opposite of what is required to obtain a small excitation volume and, in the case of two-photon excitation, to increase the photon flux. In the absence of this artifact, the count rate should increase with increasing detector aperture. However, as one approaches a non-Gaussian volume, the count rate begins to decrease with increasing aperture. This is because in the presence of diffraction fringes at a large aperture the integral of the square of the normalized spatial intensity profile grows more slowly than the integral of its value and thus the ratio between aperture and count rate is no longer constant (Hess and Webb, 2002). Thus, the experimenter in one-photon FCS must find the appropriate compromise between underfilling the objective and decreasing the aperture size on the one hand and an acceptable signal to noise ratio on the other.

The maximum excitation rate of a fluorophore has a limit, the saturation limit, defined by the time required for the excited fluorophore to relax back to the ground state. Excitation saturation can distort the observation volume in both one- and two-photon excitation. A full treatment of this limitation can be found in the work of Berland's group (Berland and Shen, 2003; Nagy *et al.*, 2005). However, the phenomenon can be understood intuitively by considering that at excitation intensities that exceed the saturation threshold, the molecules at the central regions of the focal volume become saturated before those in the edges of the profile. Thus, the shape of the profile becomes flatter at the top and broader. This results in decreased amplitudes of the autocorrelation function because the volume is actually larger, and thus the fluctuations are apparently smaller than in the absence of saturation. To avoid this effect, the autocorrelation should be measured as a function of excitation intensity, and the intensity set at a level for which no further changes in the amplitude of the autocorrelation function, $G(0)$, are observed.

1.7.4. Photobleaching

Photobleaching is another artifact that can significantly alter the autocorrelation function (Eggeling *et al.*, 2005). If the fluorophore is irreversibly photobleached before exiting the detection volume, then this has the effect of increasing the apparent diffusion coefficient, since the autocorrelation function decays more quickly than it would in absence of photobleaching. If photobleaching occurs for the fluorophore used in the calibration of the observation volume, then the result is a significant underestimation of the volume. For example, if a volume radius smaller than the diffraction limit is calculated from the autocorrelation function of the standard, then the experimenter is probably in the presence of photobleaching. Whether one is performing a calibration or a measurement, it is good practice to carry out a series of acquisitions at different excitation power values to ensure that data are obtained under conditions where changes in the excitation power cause no changes in either the timescale or the amplitude of the autocorrelation function.

1.7.5. Dark States

Reversible transitions to nonfluorescent states such as the triplet state (Bacia and Schwille, 2003), protonation or deprotonation transitions, and isomerizations (Dickson *et al.*, 1997; Haupts *et al.*, 1998; Moerner *et al.*, 1999; Prendergast, 1999; Weber *et al.*, 1999; Garcia-Parajo *et al.*, 2000) can give rise to intensity fluctuations on multiple timescales that require additional fluctuation terms in the analysis function for the autocorrelation profile (Palmer and Thompson, 1987). Modification of the expression for the autocorrelation function is applicable only if the fluctuations occur on a timescale much faster than the diffusion. In the case of one additional process, the expression for G(τ) becomes

$$G(\tau) = G(\tau_{\mathrm{d}})\left(\frac{1}{1-f}\right)\left(1-f-f\left(e^{\tau/\tau_c}\right)\right), \tag{1.25}$$

where $G(\tau_{\mathrm{d}})$ corresponds to the diffusion part of the autocorrelation function in Eq. (1.3), f corresponds to the fractional population of the dark state, and τ_c is the characteristic residence time in the dark state. Experiments can also be designed to ensure that particular conformational fluctuations of the biomolecules give rise to fluctuations in the fluorescence intensity, and then the timescales of these fluctuations can be characterized using FCS. Depending on the timescale of the fluctuations, it is wise to slow diffusion sufficiently (by encapsulation in vesicles or by changing the viscosity for example) such that the approximation for the use of Eq. (1.25) still holds. Alternately, the biomolecules can be immobilized and then the only intensity fluctuations will be those arising from the conformational fluctuations. We note, however, that in all cases it is important to ascertain that the fast events present in the autocorrelation curves do not correspond to detector afterpulsing. Moreover, it is difficult to distinguish between fluctuations due to probe photophysics and those due to conformational fluctuations, and typically experiments designed to perturb one class of events, and not the other, must be performed. Indeed, for the issues here, namely the study of biomolecular associations, the basis for any observed fast kinetics may not be of interest. In order to facilitate the analysis of the autocorrelation profiles in terms of diffusion times, it is preferable to eliminate these fast processes, to the extent possible. As we mentioned earlier, afterpulsing can be eliminated by cross-correlating the signal detected on two different detectors. Protonation/deprotonation transitions are pH sensitive, such that if one suspects this as the major contributing event to the fast processes observed, then testing at a series of pH values is advisable. Light-induced blinking is dependent on the excitation wavelength and the excitation power, such that varying these two parameters may help to diminish this contribution. Finally it has been noted by many practitioners of two-photon FCS that the population of the triplet state is less prominent than in one-photon excitation, for reasons that have not been elucidated. Moreover, if it appears that the fast processes arise from protein fluctuations, then any number of changes in the solution conditions (addition of ligand, denaturant, salt, mutations, etc.) may alter the intrinsic protein dynamics. Only a complete study will allow one to ascertain the reasons for the decrease in the autocorrelation function at short times (1–10 μs).

1.7.6. Focus Position

In studies of dynamical processes on surfaces, the amplitude and the timescale of FCS correlation functions also depend on the relative position of the focus positioning with respect to the surface. For example, in studying membrane samples containing fluorescent lipids some of the lipid partitions into the aqueous phase above the lipid layer (Bacia and Schwille, 2003) and so depending upon the position of the focus with respect to the supported mono- or bilayer, the fraction of freely diffusing species can vary. In addition, depending on the dichroic and emission filters and the excitation wavelength of the laser, some laser excitation may be detected in the sample channel. Since the fluctuations in the exciting light are uncorrelated, they tend to lower the amplitude of the correlation function. While ideally one must work to eliminate their contribution, this is not always possible. In particular, the scattered exciting light from the surface of the microscope slide will contribute more as one approaches the surface. Thus, it is extremely important to test the focus dependence of the autocorrelation function.

1.7.7. Singular Events

Biological preparations and, to an even greater extent, observation of biomolecules in live cells involve rather heterogeneous mixtures and environments. Often due to small amounts of aggregate material, the autocorrelation function can be dominated by large singular intensity spikes (e.g., see Bacia *et al.*, 2002). These events are all the more probable with longer acquisition times. It is therefore important to be able to eliminate these spurious occurrences from the photon stream. This is most easily accomplished using systems in which the raw data are stored and then autocorrelation (or alternative) analyses are carried out afterward. Note that certain commercially available correlator boards (i.e., BI-9000T, Brookhaven Instruments Corporation) have built-in cut-off options that allow the removal of spurious spikes above a defined threshold.

1.8. CONCLUSION

In this review we have attempted to provide an overview of the possibilities afforded by fluorescence fluctuation techniques, to illustrate these approaches with a few examples (clearly nonexhaustive), and most importantly to draw the readers' attention to some of the pitfalls inherent in such measurements and how to diminish their contributions. If applied with care, these approaches can be powerful tools in deciphering the mechanisms of biomolecular interactions, both *in vitro* and in live cells. It is apparent from this review that the experience of the field in general in correlation techniques is greater than in photon statistics. This is in large part because many fewer studies have been carried out by a relatively small number of investigators, using photon statistics to approach issues of interaction and stoichiometry. It will take more time and the involvement of more research groups to provide the sort of rule of thumb experience level for these statistical approaches. Moreover, the development of user-friendly software for the simultaneous analysis of the brightness, the number of particles, and their time-dependent behavior should help to advance the field.

While this review was in press, a number of interesting studies have appeared describing fluorescence cross-correlation in live cells (Rosales *et al.*, 2006; Muto *et al.*, 2006), new molecules and FCS methods (Kogure *et al.*, 2006; Hwang *et al.*, 2006; Leutenegger *et al.*, 2006; Oyama *et al.*, 2006; Swift *et al.*, 2006; Gosch *et al.*, 2005) and a new simultaneous analysis approaches for evaluating both diffusion and brightness (Wu *et al.*, 2006).

REFERENCES

Aragon, S. R. and Pecora, R. (1974). Fluorescence correlation spectroscopy. *J Chem Phys* 1792–1803.

Bacia, K., Majoul, I. V., and Schwille, P. (2002). Probing the endocytic pathway in live cells using dual-color fluorescence cross-correlation analysis. *Biophys J* 83:1184–1193.

Bacia, K. and Schwille, P. (2003). A dynamic view of cellular processes by in vivo fluorescence auto- and cross-correlation spectroscopy. *Methods* 29:74–85.

Baudendistel, N., Muller, G., Waldeck, W., Angel, P., and Langowski, J. (2005). Two-hybrid fluorescence cross-correlation spectroscopy detects protein–protein interactions in vivo. *Chemphyschem* 6:984–990.

Berland, K. and Shen, G. (2003). Excitation saturation in two-photon fluorescence correlation spectroscopy. *Appl Opt* 42:5566–5576.

Berland, K. M. (2004). Detection of specific DNA sequences using dual-color two-photon fluorescence correlation spectroscopy. *J Biotechnol* 108:127–136.

Berland, K. M., So, P. T., and Gratton, E. (1995). Two-photon fluorescence correlation spectroscopy: method and application to the intracellular environment. *Biophys J* 68:694–701.

Berland, K. M., So, P. T., Chen, Y., Mantulin, W. W., and Gratton, E. (1996). Scanning two-photon fluctuation correlation spectroscopy: particle counting measurements for detection of molecular aggregation. *Biophys J* 71:410–420.

Blab, G. A., Lommerse, P. H. M., Cognet, L., Harms, G. S., and Schmidt, T. (2005). Two-photon excitation action cross-sections of the autofluorescent proteins. *Chem Phys Letters* 350:71–77.

Boukari, H., Nossal, R., Sackett, D. L., and Schuck, P. (2004). Hydrodynamics of nanoscopic tubulin rings in dilute solutions. *Phys Rev Lett* 93:098106.

Brock, R. and Jovin, T. M. (1998). Fluorescence correlation microscopy (FCM)-fluorescence correlation spectroscopy (FCS) taken into the cell. *Cell Mol Biol (Noisy-le-grand)* 44:847–856.

Bulseco, D. A. and Wolf, D. E. (2003). Fluorescence correlation spectroscopy: molecular complexing in solution and in living cells. *Methods Cell Biol* 72:465–498.

Campbell, R. E., Tour, O., Palmer, A. E., Steinbach, P. A., Baird, G. S., Zacharias, D. A., and Tsien, R. Y. (2002). A monomeric red fluorescent protein. *Proc Natl Acad Sci USA* 99:7877–7882.

Chattopadhyay, K., Saffarian, S., Elson, E. L., and Frieden, C. (2002). Measurement of microsecond dynamic motion in the intestinal fatty acid binding protein by using fluorescence correlation spectroscopy. *Proc Natl Acad Sci USA* 99:14171–14176.

Chattopadhyay, K., Saffarian, S. Elson, E. L. and Frieden, C. (2005). Measuring unfolding of proteins in the presence of denaturant using fluorescence correlation spectroscopy. *Biophys J* 88:1413–1422.

Chen, Y., Muller, J. D., Berland, K. M., and Gratton, E. (1999a). Fluorescence fluctuation spectroscopy. *Methods* 19:234–252.

Chen, Y., Muller, J. D., So, P. T., and Gratton, E. (1999b). The photon counting histogram in fluorescence fluctuation spectroscopy. *Biophys J* 77:553–567.

Chen, Y., Muller, J. D., Tetin, S. Y., Tyner, J. D., and Gratton, E. (2000). Probing ligand protein binding equilibria with fluorescence fluctuation spectroscopy. *Biophys J* 79:1074–1084.

Chen, Y., Tekmen, M., Hillesheim, L., Skinner, J., Wu, B., and Muller, J. D. (2005). Dual-color photon-counting histogram. *Biophys J* 88:2177–2192.

Chen, Y., Wei, L. N., and Muller, J. D., (2003). Probing protein oligomerization in living cells with fluorescence fluctuation spectroscopy. *Proc Natl Acad Sci USA* 100:15492–15497.

Clamme, J. P., Krishnamoorthy, G., and Mely, Y. (2003). Intracellular dynamics of the gene delivery vehicle polyethylenimine during transfection: investigation by two-photon fluorescence correlation spectroscopy. *Biochim Biophys Acta* 1617:52–61.

Cotlet, M., Masuo, S., Luo, G., Hofkens, J., van der, A. M., Verhoeven, J., Mullen, K., Xie, X. S., and De, S. F. (2004). Probing conformational dynamics in single donor–acceptor synthetic molecules by means of photoinduced reversible electron transfer. *Proc Natl Acad Sci USA* 101:14343–14348.

Dauty, E. and Verkman, A. S. (2004). Molecular crowding reduces to a similar extent the diffusion of small solutes and macromolecules: measurement by fluorescence correlation spectroscopy. *J Mol Recognit* 17:441–447.

Deniz, A. A., Dahan, M., Grunwell, J. R., Ha, T., Faulhaber, A. E., Chemla, D. S., Weiss, S., and Schultz, P. G. (1999). Single-pair fluorescence resonance energy transfer on freely diffusing molecules: observation of Forster distance dependence and subpopulations. *Proc Natl Acad Sci USA* 96: 3670–3675.

Denk, W., Strickler, J. H., and Webb, W. W. (1990). Two-photon laser scanning fluorescence microscopy. *Science* 248:73–76.

Dickson, R. M., Cubitt, A. B., Tsien, R. Y., and Moerner, W. E. (1997). On/off blinking and switching behaviour of single molecules of green fluorescent protein. *Nature* 388:355–358.

Douglas, J. F., Zhou, H. X., and Hubbard, J. B. (1994). Hydrodynamic friction and the capacitance of arbitrarily shaped objects. *Phys Rev E Stat Phys Plasmas Fluids Relat Interdiscip Topics* 49:5319–5331.

Eggeling, C., Volkmer, A., and Seidel, C. A. (2005). Molecular photobleaching kinetics of Rhodamine 6G by one- and two-photon induced confocal fluorescence microscopy. *Chemphyschem* 6:791–804.
Eigen, M. and Rigler, R. (1994). Sorting single molecules: application to diagnostics and evolutionary biotechnology. *Proc Natl Acad Sci USA* 91:5740–5747.
Elson, E. L. (2001). Fluorescence correlation spectroscopy measures molecular transport in cells. *Traffic* 2:789–796.
Elson, E. L. (2004) Quick tour of fluorescence correlation spectroscopy from its inception. *J Biomed Opt* 9:857–864.
Garcia de La Torre, J. G. and Bloomfield, V. A. (1981). Hydrodynamic properties of complex, rigid biological macromolecules: theory and applications. *Quart Rev Biophys* 14:81–139.
Garcia de la Torre, J., Huertas, M. L., and Carrasco, B. (2000). Calculation of hydrodynamic properties of globular proteins from their atomic-level structure. *Biophys J* 78:719–730.
Garcia de la Torre, J., Navarro, S., Lopez Martinez, M. C., Diaz, F. G., and Lopez Cascales, J. J. (1994). HYDRO: a computer program for the prediction of hydrodynamic properties of macromolecules. *Biophys J* 67:530–531.
Garcia-Parajo, M. F., Segers-Nolten, G. M., Veerman, J. A., Greve, J., and van Hulst, N. F. (2000). Real-time light-driven dynamics of the fluorescence emission in single green fluorescent protein molecules. *Proc Natl Acad Sci USA* 97:7237–7242.
Gennerich, A. and Schild, D. (2000). Fluorescence correlation spectroscopy in small cytosolic compartments depends critically on the diffusion model used. *Biophys J* 79:3294–3306.
Grunwell, J. R., Glass, J. L., Lacoste, T. D., Deniz, A. A., Chemla, D. S., and Schultz, P. G. (2001). Monitoring the conformational fluctuations of DNA hairpins using single-pair fluorescence resonance energy transfer. *J Am Chem Soc* 123:4295–4303.
Gosch M, Blom H, Anderegg S, Korn K, Thyberg P, Wells M, Lasser T, Rigler R, Magnusson A, Hard S. (2005) Parallel dual-color fluorescence cross-correlation spectroscopy using diffractive optical elements. *J Biomed Opt* 10(5):054008.
Haupts, U., Maiti, S., Schwille, P., and Webb, W. W. (1998). Dynamics of fluorescence fluctuations in green fluorescent protein observed by fluorescence correlation spectroscopy. *Proc Natl Acad Sci USA* 95:13573–13578.
Haustein, E. and Schwille, P. (2003). Ultrasensitive investigations of biological systems by fluorescence correlation spectroscopy. *Methods* 29:153–166.
Haustein, E. and Schwille, P. (2004) Single-molecule spectroscopic methods. *Curr Opin Struct Biol* 14:531–540.
Heinze, K. G., Jahnz, M., and Schwille, P. (2004). Triple-color coincidence analysis: one step further in following higher order molecular complex formation. *Biophys J* 86:506–516.
Heinze, K. G., Koltermann, A., and Schwille, P. (2000). Simultaneous two-photon excitation of distinct labels for dual-color fluorescence crosscorrelation analysis. *Proc Natl Acad Sci USA* 97:10377–10382.
Heinze, K. G., Rarbach, M., Jahnz, M., and Schwille, P. (2002). Two-photon fluorescence coincidence analysis: rapid measurements of enzyme kinetics. *Biophys J* 83:1671–1681.
Hess, S. T., Huang, S., Heikal, A. A., and Webb, W. W. (2002). Biological and chemical applications of fluorescence correlation spectroscopy: a review. *Biochemistry* 41:697–705.
Hess, S. T. and Webb, W. W. (2002). Focal volume optics and experimental artifacts in confocal fluorescence correlation spectroscopy. *Biophys J* 83:2300–2317.
Hillesheim, L. and Muller, J. D. (2005). The dual color photon counting histogram with non-ideal detectors. *Biophys J* 89:3491–3507.
Hillesheim, L. N. and Muller, J. D. (2003). The photon counting histogram in fluorescence fluctuation spectroscopy with non-ideal photodetectors. *Biophys J* 85:1948–1958.
Hink, M. A., Bisselin, T., and Visser, A. J. (2002). Imaging protein–protein interactions in living cells. *Plant Mol Biol* 50:871–883.
Huang, B., Perroud, T. D., and Zare, R. N. (2004). Photon counting histogram: one–photon excitation. *Chemphyschem* 5:1523–1531.

Hwang, L. C., Gosch, M., Lasser, T., and Wohland, T. (2006) Simultaneous multicolor fluorescence cross-correlation spectroscopy to detect higher order molecular interactions using single wavelength laser excitation. *Biophys J* 91(2):715–27.

Hwang, L. C. and Wohland, T. (2005). Single wavelength excitation fluorescence cross-correlation spectroscopy with spectrally similar fluorophores: resolution for binding studies. *J Chem Phys* 122:114708.

Icenogle, R. D. and Elson, E. L. (1983). Fluorescence correlation spectroscopy and photobleaching recovery of multiple binding reactions. I. Theory and FCS measurements. *Biopolymers* 22:1919–1948.

Jahnz, M. and Schwille, P. (2005). An ultrasensitive site-specific DNA recombination assay based on dual-color fluorescence cross-correlation spectroscopy. *Nucleic Acids Res* 33:e60.

Joo, C., McKinney, S. A., Lilley, D. M., and Ha, T. (2004). Exploring rare conformational species and ionic effects in DNA Holliday junctions using single-molecule spectroscopy. *J Mol Biol* 341:739–751.

Kang, K., Gapinski, J., Lettinga, M. P., Buitenhuis, J., Meier, G., Ratajczyk, M., Dhont, J. K., and Patkowski, A. (2005). Diffusion of spheres in crowded suspensions of rods. *J Chem Phys* 122:44905.

Kang, E. H., Mansfield, M. L., and Douglas, J. F. (2004). Numerical path integration technique for the calculation of transport properties of proteins. *Phys Rev E Stat Nonlin Soft Matter Phys* 69:031918.

Karymov, M., Daniel, D., Sankey, O. F., and Lyubchenko, Y. L. (2005). Holliday junction dynamics and branch migration: single-molecule analysis. *Proc Natl Acad Sci USA* 102:8186–8191.

Kask, P., Palo, K., Fay, N., Brand, L., Mets, U., Ullmann, D., Jungmann, J., Pschorr, J., and Gall, K. (2000). Two-dimensional fluorescence intensity distribution analysis: theory and applications. *Biophys J* 78:1703–1713.

Kask, P., Palo, K., Ullmann, D., and Gall, K. (1999). Fluorescence-intensity distribution analysis and its application in biomolecular detection technology. *Proc Natl Acad Sci USA* 96:13756–13761.

Kettling, U., Koltermann, A., Schwille, P., and Eigen, M. (1998). Real-time enzyme kinetics monitored by dual-color fluorescence cross-correlation spectroscopy. *Proc Natl Acad Sci USA* 95:1416–1420.

Kim, S. A., Heinze, K. G., Bacia, K., Waxham, M. N., and Schwille, P. (2005). Two-photon cross correlation analysis of intracellular reactions with variable stoichiometry. *Biophys J* 88:4319–4336.

Kim, S. A., Heinze, K. G., Waxham, M. N., and Schwille, P. (2004). Intracellular calmodulin availability accessed with two-photon cross-correlation. *Proc Natl Acad Sci USA* 101:105–110.

Kogure, T., Karasawa, S., Araki, T., Saito, K., Kinjo, M., and Miyawaki, A. (2006) A fluorescent variant of a protein from the stony coral Montipora facilitates dual-color single-laser fluorescence cross-correlation spectroscopy. *Nat Biotechnol* 24(5):577–81.

Kohl, T., Haustein, E., and Schwille, P. (2005). Determining protease activity in vivo by fluorescence cross-correlation analysis. *Biophys J* 89:2770–2782.

Kohl, T., Heinze, K. G., Kuhlemann, R., Koltermann, A., and Schwille, P. (2002). A protease assay for two-photon crosscorrelation and FRET analysis based solely on fluorescent proteins. *Proc Natl Acad Sci USA* 99:12161–12166.

Krichevsky, O. and Bonnet, G. (2002). Fluorescence correlation spectroscopy: the technique and its applications. Rep. Prog. Phys. 251–297.

Larson, D. R., J. A. Gosse, D. A. Holowka, B. A. Baird, and W. W. Webb. (2005). Temporally resolved interactions between antigen-stimulated IgE receptors and Lyn kinase on living cells. *J Cell Biol* 171:527–536.

Laurence, T. A., Kapanidis, A. N., Kong, X., Chemla, D. S., and Weiss, S. (2004). Photon arrival time interval distribution (PAID): a novel tool for analyzing molecular interactions. *J Phys Chem B* 108:3051–3067.

Lee, N. K., Kapanidis, A. N., Wang, Y., Michalet, X., Mukhopadhyay, J., Ebright, R. H., and Weiss, S. (2005). Accurate FRET measurements within single diffusing biomolecules using alternating-laser excitation. *Biophys J* 88:2939–2953.

Leutenegger, M., Blom, H., Widengren, J., Eggeling, C., Gosch, M., Leitgeb, R. A., and Lasser, T. (2006) Dual-color total internal reflection fluorescence cross-correlation spectroscopy. *J Biomed Opt* 11(4):040502.

Li, H., Ren, X., Ying, L., Balasubramanian, S., and Klenerman, D. (2004). Measuring single-molecule nucleic acid dynamics in solution by two-color filtered ratiometric fluorescence correlation spectroscopy. *Proc Natl Acad Sci USA* 101:14425–14430.

Li, H. W. and Yeung, E. S. (2005). Direct observation of anomalous single-molecule enzyme kinetics. *Anal Chem* 77:4374–4377.

Madge, D. (1976). Chemical kinetics and fluorescence correlation spectroscopy. *Quart Rev Biophys* 9:35–47.

Maertens, G., Vercammen, J., Debyser, Z., and Engelborghs, Y. (2005). Measuring protein–protein interactions inside living cells using single color fluorescence correlation spectroscopy. Application to human immunodeficiency virus type 1 integrase and LEDGF/p75. *Faseb J* 19:1039–1041.

Magde, D., Elson, E. L., and Webb, W. W. (1974). Fluorescence correlation spectroscopy. II. An experimental realization. *Biopolymers* 13:29–61.

Margeat, E., Poujol, N., Boulahtouf, A., Chen, Y., Muller, J. D., Gratton, E., Cavailles, V., and Royer, C. A. (2001). The human estrogen receptor alpha dimer binds a single SRC-1 coactivator molecule with an affinity dictated by agonist structure. *J Mol Biol* 306:433–442.

Masuda, A., Ushida, K., and Okamoto, T. (2005). New fluorescence correlation spectroscopy enabling direct observation of spatiotemporal dependence of diffusion constants as an evidence of anomalous transport in extracellular matrices. *Biophys J* 88:3584–3591.

Merkle, D., Block, W. D., Yu, Y., Lees-Miller, S. P., and Cramb, D. T., (2006) Analysis of DNA-dependent protein kinase-mediated DNA end joining by two-photon fluorescence cross-correlation spectroscopy. *Biochemistry* 45(13):4164–72.

Meseth, U., Wohland, T., Rigler, R., and Vogel, H. (1999). Resolution of fluorescence correlation measurements. *Biophys J* 76:1619–1631.

Michelman-Ribeiro, A., Boukari, H., Nossal, R., and Horkay, F. (2004). Structural changes in polymer gels probed by fluorescence correlation spectroscopy. *Macromol* 37:10212–10214.

Moerner, W. E., Peterman, E. J., Brasselet, S., Kummer, S., and Dickson, R. M. (1999). Optical methods for exploring dynamics of single copies of green fluorescent protein. *Cytometry* 36:232–238.

Muto, H., Nagao, I., Demura, T., Fukuda, H., Kinjo, M., Yamamoto, K. T. 2006 Fluorescence cross-correlation analyses of the molecular interaction between an Aux/IAA protein, MSG2/IAA19, and protein-protein interaction domains of auxin response factors of arabidopsis expressed in HeLa cells. *Plant Cell Physiol* 47(8):1095–101.

Muller, J. D. (2004). Cumulant analysis in fluorescence fluctuation spectroscopy. *Biophys J* 86:3981–3992.

Muller, J. D., Chen, Y., and Gratton, E. (2000). Resolving heterogeneity on the single molecular level with the photon-counting histogram. *Biophys J* 78:474–486.

Muller, J. D., Chen, Y., and Gratton, E. (2003). Fluorescence correlation spectroscopy. *Methods Enzymol* 361:69–92.

Nagy, L. and Schwabe, J. W. (2004a). Mechanism of the nuclear receptor molecular switch. *Trends Biochem Sci* 29:317–324.

Nagy, L. and Schwabe, J. W. (2004b). Mechanism of the nuclear receptor molecular switch. *Trends Biochem Sci* 29:317–324.

Nagy, A., Wu, J., and Berland, K. M. (2005). Observation volumes and gamma factors in two-photon fluorescence fluctuation spectroscopy. *Biophys J* 89:2077–2090.

Oyama, R., Takashima, H., Yonezawa, M., Doi, N., Miyamoto-Sato, E., Kinjo, M., and Yanagawa, H. (2006) Protein-protein interaction analysis by C-terminally specific fluorescence labeling and fluorescence cross-correlation spectroscopy. *Nucleic Acids Res* 34(14):e102

Palmer, A. G., III and Thompson, N. L. (1987). Theory of sample translation in fluorescence correlation spectroscopy. *Biophys J* 51:339–343.

Palo, K., Mets, U., Jager, S., Kask, P., and Gall, K. (2000). Fluorescence intensity multiple distributions analysis: concurrent determination of diffusion times and molecular brightness. *Biophys J* 79:2858–2866.

Phillies, G. D. J. (1991). The hydrodynamic scaling model for polymer dynamics. Non-Crystalline Solids 131–132, 612–619.

Pramanik, A. and Rigler, R. (2001). Ligand-receptor interactions in the membrane of cultured cells monitored by fluorescence correlation spectroscopy. *Biol Chem* 382:371–378.

Prendergast, F. G. (1999). Biophysics of the green fluorescent protein. *Methods Cell Biol* 58:1–18.

Qian, H. and Elson, E. L. (1990a). Distribution of molecular aggregation by analysis of fluctuation moments. *Proc Natl Acad Sci USA* 87:5479–5483.

Qian, H. and Elson, E. L. (1990b). On the analysis of high order moments of fluorescence fluctuations. *Biophys J* 57:375–380.

Rarbach, M., Kettling, U., Koltermann, A., and Eigen, M. (2001). Dual-color fluorescence cross-correlation spectroscopy for monitoring the kinetics of enzyme-catalyzed reactions. *Methods* 24:104–116.

Rigler, R. (1995). Fluorescence correlations, single molecule detection and large number screening. Applications in biotechnology. *J Biotechnol* 41:177–186.

Rippe, K. (2000). Simultaneous binding of two DNA duplexes to the NtrC-enhancer complex studied by two-color fluorescence cross-correlation spectroscopy. *Biochemistry* 39:2131–2139.

Rosales, T., Georget, V., Malide, D., Smirnov, A., Xu, J., Combs, C., Knutson, J. R., Nicolas, J. C., and Royer, C. A. (2006) Quantitative detection of the ligand-dependent interaction between the androgen receptor and the co-activator, Tif2, in live cells using two color, two photon fluorescence cross-correlation spectroscopy. *Eur Biophys J* Oct 5; [Epub ahead of print]

Rudiger, M., Haupts, U., Moore, K. J., and Pope, A. J. (2001). Single-molecule detection technologies in miniaturized high throughput screening: binding assays for g protein-coupled receptors using fluorescence intensity distribution analysis and fluorescence anisotropy. *J Biomol Screen* 6:29–37.

Sato, Y. T., Hamada, T., Kubo, K., Yamada, A., Kishida, T., Mazda, O., and Yoshikawa, K. (2005). Folding transition into a loosely collapsed state in plasmid DNA as revealed by single-molecule observation. *FEBS Lett* 579:3095–3099.

Scheel, A. A., Funsch, B., Busch, M., Gradl, G., Pschorr, J., and Lohse, M. J. (2001). Receptor–ligand interactions studied with homogeneous fluorescence-based assays suitable for miniaturized screening. *J Biomol Screen* 6:11–18.

Schilb, A., Riou, V., Schoepfer, J., Ottl, J., Muller, K., Chene, P., Mayr, L. M., and Filipuzzi, I. (2004). Development and implementation of a highly miniaturized confocal 2D-FIDA-based high-throughput screening assay to search for active site modulators of the human heat shock protein 90beta. *J Biomol Screen* 9:569–577.

Schuler, B. (2005). Single-molecule fluorescence spectroscopy of protein folding. *Chemphyschem* 6:1206–1220.

Schwille, P. (2001). Fluorescence correlation spectroscopy and its potential for intracellular applications. *Cell Biochem Biophys* 34:383–408.

Schwille, P., Haupts, U., Maiti, S., and Webb, W. W. (1999a). Molecular dynamics in living cells observed by fluorescence correlation spectroscopy with one- and two-photon excitation. *Biophys J* 77:2251–2265.

Schwille, P., Korlach, J., and Webb, W. W. (1999b). Fluorescence correlation spectroscopy with single-molecule sensitivity on cell and model membranes. *Cytometry* 36:176–182.

Slaughter, B. D., Unruh, J. R., Allen, M. W., Bieber Urbauer, R. J., and Johnson, C. K. (2005). Conformational substates of calmodulin revealed by single-pair fluorescence resonance energy transfer: influence of solution conditions and oxidative modification. *Biochemistry* 44:3694–3707.

So, P. T., Dong, C. Y., Masters, B. R., and Berland, K. M. (2000). Two-photon excitation fluorescence microscopy. *Annu Rev Biomed Eng* 2:399–429.

Starchev, K., Buffle, J., and Perez, E. (1999). Applications of fluorescence correlation spectroscopy: polydispersity measurements. *J Colloid Interface Sci* 213:479–487.

Swift, J. L., Heuff, R., and Cramb, D. T., (2006). A two-photon excitation fluorescence cross-correlation assay for a model ligand-receptor binding system using quantum dots. *Biophys J* 90:1396–1410.

Thews, E., Gerken, M., Eckert, R., Zapfel, J., Tietz, C., and Wrachtrup, J. (2005). Cross Talk Free fluorescence cross-correlation spectroscopy in live cells. *Biophys J* 89:2069–2076.

Thompson, N. L., Lieto, A. M., and Allen, N. W. (2002). Recent advances in fluorescence correlation spectroscopy. *Curr Opin Struct Biol* 12:634–641.

Vamosi, G., Bodnar, A., Vereb, G., Jenei, A., Goldman, C. K., Langowski, J., Toth, K., Matyus, L., Szollosi, J., Waldmann, T. A., and Damjanovich, S. (2004). IL-2 and IL-15 receptor alpha-subunits are coexpressed in a supramolecular receptor cluster in lipid rafts of T cells. *Proc Natl Acad Sci USA* 101:11082–11087.

Van, R. E., Chen, Y., Muller, J. D., Gratton, E., Van, C. E., Engelborghs, Y., De, S. S., and Demeester, J. (2001). Fluorescence fluctuation analysis for the study of interactions between oligonucleotides and polycationic polymers. *Biol Chem* 382:379–386.

Vukojevic, V., Pramanik, A., Yakovleva, T., Rigler, R., Terenius, L., and Bakalkin, G. (2005). Study of molecular events in cells by fluorescence correlation spectroscopy. *Cell Mol Life Sci* 62:535–550.

Wachsmuth, M., Waldeck, W., and Langowski, J. (2000). Anomalous diffusion of fluorescent probes inside living cell nuclei investigated by spatially-resolved fluorescence correlation spectroscopy. *J Mol Biol* 298:677–689.

Webb, W. W. (1976). Applications of fluorescence correlation spectroscopy. *Quart Rev Biophys* 9:49–68.

Weber, W., Helms, V., McCammon, J. A., and Langhoff, P. W. (1999). Shedding light on the dark and weakly fluorescent states of green fluorescent proteins. *Proc Natl Acad Sci USA* 96:6177–6182.

Weiss, M., Hashimoto, H., and Nilsson, T. (2003). Anomalous protein diffusion in living cells as seen by fluorescence correlation spectroscopy. *Biophys J* 84:4043–4052.

Weiss, M., Elsner, M., Kartberg, F., and Nilsson, T. (2004). Anomalous subdiffusion is a measure for cytoplasmic crowding in living cells. *Biophys J* 87:3518–3524.

Wilson, T. J., Nahas, M., Ha, T., and Lilley, D. M. (2005). Folding and catalysis of the hairpin ribozyme. *Biochem Soc Trans* 33:461–465.

Wu, B., Chen, Y., Muller, J. D. (2006). Dual-color time-integrated fluorescence cumulant analysis. *Biophys J* 91(7):2687–98.

Wu, B. and Muller, J. D. (2005). Time-Integrated Fluorescence Cumulant Analysis in Fluorescence Fluctuation Spectroscopy. *Biophys J* 89:2721–2735.

2

Characterization of Protein–Protein Interactions Using Atomic Force Microscopy

Hong Wang, Yong Yang, and Dorothy A. Erie*

2.1. INTRODUCTION

Atomic force microscopy (AFM), invented in 1986, expanded the application of scanning tunneling microscopy to nonconductive, soft, and live biological samples (Binnig *et al.*, 1986; Hansma *et al.*, 1988; Marti *et al.*, 1988; Drake *et al.*, 1989). AFM has several capabilities, including characterizing topographic details of surfaces from the submolecular level to the cellular level (Radmacher *et al.*, 1992), and monitoring the dynamic process of single molecules in physiological relevant solutions (Drake *et al.*, 1989; Engel and Muller, 2000). More excitingly, AFM not only extends our ''vision,'' but also extends our ability to ''touch and manipulate'' during our exploration of the biological system at the molecular level. For example, AFM can be used to manipulate macromolecules (Zlatanova and Leuba, 2003; Bockelmann, 2004; Gutsmann *et al.*, 2004), monitor the unfolding of proteins, RNA, and protein fibers (Carrion-Vazquez *et al.*, 1999; Fisher *et al.*, 2000; Zhuang and Rief, 2003; Rounsevell *et al.*, 2004), and measure the forces between interacting molecules (Chilkoti *et al.*, 1995; Dammer *et al.*, 1995; Ros *et al.*, 2004). Over the past two decades, the application of AFM has advanced our knowledge in many areas of the biological sciences including DNA (Fritzsche *et al.*, 1997; Hansma, 2001; Hansma

H. WANG • Laboratory of Molecular Genetics, National Institute of Environmental Health Sciences, National Institutes of Health, Research Triangle Park, NC, 27709, USA. Y. YANG • Department of Chemistry, University of North Carolina at Chapel Hill, Chapel Hill, NC, 27599, USA. D. A. ERIE • Department of Chemistry, University of North Carolina at Chapel Hill, Chapel Hill, NC 27599, USA and *Corresponding author: Tel: 919 962-6370; fax: 919 966-3675; e-mail: derie@email.unc.edu

et al., 2004), RNA (Lyubchenko *et al.*, 1992; Liphardt *et al.*, 2001; Abels *et al.*, 2005), chromatin (Bustamante *et al.*, 1997; Tamayo, 2003a,b; Zlatanova and Leuba, 2003; Leuba *et al.*, 2004), proteins (Ratcliff and Erie, 2001; Stahlberg *et al.*, 2001), lipids (Ikai and Afrin, 2003; Henderson *et al.*, 2004), carbohydrates (Bucior and Burger, 2004), polysaccharides (Abu-Lail and Camesano, 2003), various biomolecular complexes (Lyubchenko *et al.*, 1995; Bustamante and Rivetti, 1996; Bonin *et al.*, 2000; Willemsen *et al.*, 2000; Henn *et al.*, 2001; Safinya, 2001; Janicijevic *et al.*, 2003a), and cellular (Ohnesorge *et al.*, 1997) and subcellular (Henderson *et al.*, 2004; Jena, 2004) structures.

The main focus of this review is on the application of AFM imaging in air, which is the most widely used imaging mode. However, imaging in liquids, force spectroscopy imaging, and lateral force manipulation using AFM are briefly discussed. AFM imaging is a single molecule technique that can resolve individual protein–protein and DNA–protein complexes. For example, for studying DNA–protein interactions, an ensemble of DNA–protein complexes visualized by AFM can provide snapshots of the whole dynamic process. Furthermore, using AFM, the distribution of conformations within a complex population of molecules can be characterized (Bustamante and Rivetti, 1996). Meanwhile, multiple information, such as oligomeric state of proteins, protein-induced conformational changes in DNA, DNA-binding specificities, and DNA–protein binding constants (Yang *et al.*, 2005) can be deducted simultaneously from AFM images.

This chapter focuses only on the application of AFM for investigation of protein–protein interactions free in solution and on substrates. Biological pathway events are normally implemented by protein oligomers or multiprotein assemblies rather than single proteins. If we imagine proteins as a team of workers who have jobs to do, AFM can help us understand how the players come together to bring about functions. Specifically, AFM imaging can be used to study (1) stoichiometry and protein–protein association constant (the partnership between proteins); (2) the architecture of a protein and a multiprotein complex; (3) recognition specificity of a protein complex on nucleic acids or matrix protein (the job site for a particular protein); (4) the mechanism of action of a protein, such as DNA bending or wrapping (How is the job done?); and (5) complex actions of the same or different proteins on multiple sites on DNA that result in protein filament formation, DNA looping, DNA condensation, DNA supercoiling, nucleosome remodeling, or joining of two distinct DNA molecules (How is the job done collectively?).

2.2. USE OF AFM

2.2.1. Principles of AFM

The principle of AFM varies with the different modes of AFM operation, such as contact mode, oscillating mode, and force spectroscopy mode. In the contact mode, the AFM cantilever is deflected by the sample surface. A fixed deflection is maintained during an *X–Y* dimensional scan by adjusting the *Z* position of piezo (Figure 2.1A). The AFM image is generated by plotting the *Z* movement of the piezo as a function of the *X–Y* position. In the oscillating mode, the cantilever is oscillated by a vibration piezo. The sample surface is brought into contact with the oscillating cantilever such that it clips the amplitude of oscillation. The amplitude of this clipped oscillation, which is monitored by the laser projected on the photodiode, is maintained constant during the scan by adjusting the *Z* position of the piezo using a feedback loop. The AFM image captured in oscillating mode is generated similarly to that in contact mode; that is, by plotting the *Z* movement of the piezo as a function of the *X–Y* position. Although oscillating mode is similar to contact mode, in that the tip–surface interaction is maintained constant during an AFM scan, oscillating mode generates smaller lateral forces on the sample, which improves the lateral resolution of the AFM image on nondensely packed samples.

It is well known that due to the finite size of AFM tip, AFM imaging is a result of a tip dilation (or convolution) of the sample by the imaging tip (Figure 2.1B). Tip dilation, commonly called tip convolution, mainly refers to the contribution of size and shape of tip to the AFM image. It should be kept in mind that the spatial resolution of AFM depends on the properties of the instrument, imaging conditions, and characteristics of the samples (Bustamante and Rivetti, 1996). Consequently, there is no general definition of resolution in AFM (Bustamante and Rivetti, 1996). This argument can be supported by the fact that two objects, that can be resolved when they have nearly equal height, may not be resolved when their heights are not equal (Figure 3 in Bustamante *et al.*, 1996; Yin *et al.*, 1995).

2.2.2. Substrates for Sample Preparation

Flatness and biocompatibility are two basic requirements for substrates used to prepare samples for AFM imaging. Glass, mica, gold, and silicon surfaces have been used to noncovalently or covalently immobilize biomolecules (Wagner, 1998). The most commonly used substrate is muscovite mica because an atomically flat and negatively charged surface is conveniently obtained by peeling the layered mica before sample deposition. Divalent cations, such as Mg^{2+} and Ni^{2+}, can be included in deposition buffer and these divalent cations can function as salt bridges to absorb the negatively charged biomolecules such as DNA onto the mica surface (Hansma and Laney, 1996; Muller *et al.*, 1997). Alternatively, chemical modification of the mica surface can be used to reverse the surface charges to extend its application (Lyubchenko *et al.*, 1993; Shlyakhtenko *et al.*, 2003; Podesta *et al.*, 2004). In addition, lipid bilayers prepared on mica surfaces by the Langmuir–Blodgett (LB) technique can be used as substrates for the reconstitution of membrane proteins (Stahlberg *et al.*, 2001). Finally, cationic lipid bilayers on mica have also been used to strongly anchor double-stranded DNA (dsDNA) to achieve high-resolution images in liquids (Mou *et al.*, 1995).

A

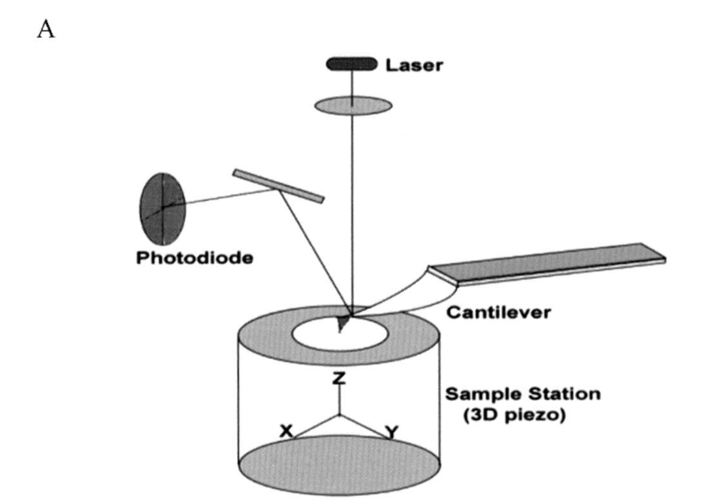

B

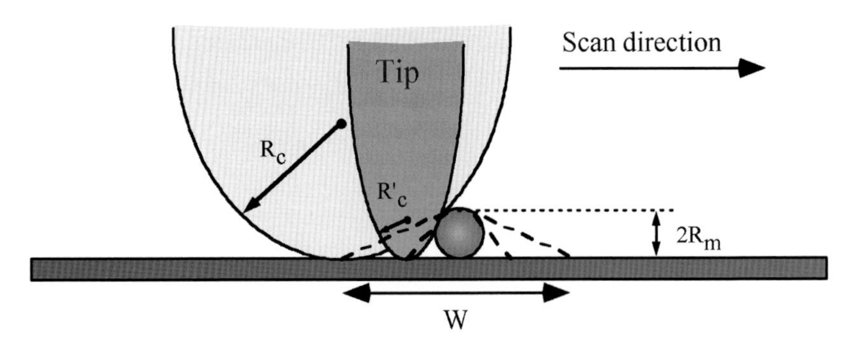

Figure 2.1. Principles of atomic force microscopy (AFM). (A) Schematic of AFM. (B) Illustration of dilation of sample by AFM tips. R_c and R_c' are the tip radii of blunt and sharp AFM tips, respectively. R_m is the half height of the particle. W is the width of the particle in the AFM image.

2.2.3. Imaging in Air

In general, imaging in air is less time consuming than imaging under solution and can provide valuable information about the structural properties of protein–protein complexes in solution and on nucleic acids or matrix proteins. For imaging in air, the sample is first deposited onto a substrate in the buffer of choice. All the water and buffer used for AFM sample storage and deposition should be filtered and screened for absence of small particles by AFM imaging before use. For preparing samples to study DNA–protein complexes, if the concentration of free protein is too high, the DNA–protein complexes can be isolated using a spin-column method before sample deposition (Hoyt *et al.*, 2001).

Samples can be incubated on the surface for up to several minutes before rinsing and drying. An immediate rinse is preferred because longer incubation on the surface increases the chance that the molecules or complexes could be altered by interaction with the surface. After deposition of the sample, the surface is rinsed with filtered distilled deionized water, excess water is wicked off using a piece of filter paper, and the surface is dried with a gentle stream of filtered nitrogen gas. Rinsing the surface is required to remove buffer components, but overrinsing could denature samples and decrease the coverage. Underdrying can reduce AFM resolution because macromolecules can move around on moist surfaces.

For sample deposition, the concentration of macromolecules in solution needs to be sufficiently diluted so that the amount of sample deposited onto the substrate surface is not too crowded. This minimizes the chance of two separate molecules coincidently landing on the same spot on the surface. For DNA, a reasonable coverage on a mica surface can be obtained with DNA concentrations in the range of $1-10\,\mu g/mL$. However, DNA coverage on mica can be dramatically affected by the buffer contents. For example, Mg^{2+} in the buffer can increase DNA coverage on the mica, but monovalent ions will decrease the DNA coverage. For proteins, the required concentration (typically $< 1\,\mu M$) varies depending on the protein. For some proteins, the protein coverage on mica is less dependent on the salt concentration compared with DNA. However, it has been shown that monovalent cations can also inhibit the adsorption of a number of different proteins onto mica (Czajkowsky and Shao, 2003). Kinetic experiments indicated that the transport of DNA molecules from the solution onto the surface is governed solely by diffusion, and analyses of protein and DNA binding to mica indicated that they bind irreversibly over the time scale of deposition (Lee and Belfort, 1989; Rivetti *et al.*, 1996; Gettens *et al.*, 2005). In addition, DNA molecules deposited onto freshly cleaved mica were able to equilibrate on the surface, as in an ideal two-dimensional (2D) solution (Rivetti *et al.*, 1996).

It should be kept in mind that a major assumption when interpreting AFM data is that what observed on the surface is what is in solution, that is, deposition on the surface does not alter the populations or structures. In many cases, this assumption is valid. However, there are exceptions. For example, it is possible that a protein can induce a 3D topology in the DNA such that protein–DNA complex must be distorted to lie flat on the surface. If surface-induced problem is a concern, in some cases, changing to different kinds of substrate for sample deposition can minimize the problem.

2.2.4. Imaging in Liquids

Imaging samples in liquids by AFM offers several advantages over imaging in air. One obvious advantage is the ability to follow the dynamic structural changes of native single molecules, as well as the interactions between macromolecules, in physiologically relevant buffers in real time. For imaging in liquids, a liquid chamber is needed to seal the buffer and allow for buffer exchange. A flow apparatus can be set up to facilitate the switching between different buffers and minimize the thermal drift of the instrument (Guthold *et al.*, 1999b). Accessory proteins, substrates, cofactors, and inhibitors can be injected into the fluid chamber. These procedures permit one to observe dynamic conformational changes of the same single protein or the interactions between macromolecules before and after the addition of these chemical and physical factors (Oberleithner *et al.*, 1996; Osmulski and Gaczynska, 2000). Accordingly, direct correlation between structural and functional states of individual biomolecules can be made. Such information can be elusive using other techniques such as electron microscopy (EM), crystallography, and AFM in air, which take static pictures of macromolecules in nonnative environments. The second major advantage of imaging in liquids is the minimal force that can be applied to the sample during imaging due to the elimination of capillary forces (Drake *et al.*, 1989). Consequently, the deformation of biological samples is reduced relative to imaging in air, which is a prerequisite to high-resolution imaging of soft biological samples. For close-packed macromolecules, such as 2D crystal arrays, contact mode in liquids has generated images with lateral resolutions down to 0.41 nm and vertical resolutions down to 0.10 nm (Muller *et al.*, 1998; Stahlberg *et al.*, 2001; Fotiadis and Engel, 2004). To minimize possible deformation of the biological specimen by the tip, soft cantilevers with spring constants $\sim$0.1 N/m must be used and scanning must be done at minimal tip force ($\sim$100–300 pN) (Fotiadis and Engel, 2004). Oscillating mode in liquids is generally preferable over contact mode for imaging samples with macromolecules loosely attached to the surface. Oscillating mode minimizes lateral forces exerted by the tip and the detachment of the sample from surface during the scan. In addition, for imaging weakly bound individual macromolecules in liquids, the jumping mode, an imaging mode that has not been widely available on commercial instruments, can minimize lateral and vertical forces and has advantages over contact and oscillating modes (Moreno-Herrero *et al.*, 2004). In jumping-mode AFM, at each image point, first the topography of the sample is measured during a feedback phase of a cycle, and then the tip–sample interaction is evaluated in real time as the tip is moved away and toward the sample. As the tip is controlled in such way that it moves laterally to the next point at the maximum tip–sample separation, the lateral forces that can detach the samples from surface and lower the image resolution are greatly minimized.

So far, only a small percentage of the published work done using AFM has been performed in liquids because imaging biomolecules in aqueous solutions remains challenging. First, to watch the dynamic processes in liquids, the right conditions must be identified. Specifically, the samples must bind tightly enough to the surface to allow good imaging but loosely enough to allow the interactions to occur on the surface (Guthold *et al.*, 1999b; Jiao *et al.*, 2001). Second, the scan rates of commercial AFMs are slow. Many biological reactions happen in the order of milliseconds to seconds, but for commercial AFMs, it will take ~30 s to collect a 1 μm × 1 μm image at reasonable resolution (Jiao *et al.*, 2001). In the past several years, however, developments have been made in both the instrumentation and the cantilevers, which have improved the reliability of the instrument and allowed faster scan rates (Han *et al.*, 1996; Ratcliff *et al.*, 1998; Viani *et al.*, 1999; Ando *et al.*, 2001; Rogers *et al.*, 2004).

2.2.5. Force Spectroscopy Mode

The ability to form biomolecular assemblies is fundamentally governed by the long-range and short-range interacting forces between macromolecules. Thermodynamics and dynamics are the traditional tools for us to determine the strength of biomolecular interactions. Although the force between interacting components can be measured directly by some methods, such as the surface force apparatus, these methods lack the spatial resolution to give information at the molecular level (Florin *et al.*, 1994). Recent developments on force spectroscopy mode AFM and optical tweezers opened an exciting area for understanding the strength of interactions at a single-molecule level (Bustamante *et al.*, 2000). Comparing with optical tweezers, in terms of sensitivity of force measurements, conventional AFMs can detect forces in the range of 0.01–100 nN, whereas optical tweezers can exert forces in the range of 1–200 pN (Leckband, 2000). Moreover, subpiconewton forces have been resolved using specialized instrumental developments (Tokunaga *et al.*, 1997). The detection limit of AFM force spectroscopy would meet the requirements needed for detecting the interacting forces in biomolecular assemblies, which are in the piconewton range (Florin *et al.*, 1994; Luckham and Smith, 1998). For example, at saturating nucleoside triphosphate concentrations, RNA polymerase (RNAP) molecules stalled reversibly at a mean applied force estimated to be 14 pN (Yin *et al.*, 1995).

For force measurement, one interacting partner is attached to the AFM tip using techniques such as chemical coating and biological functionalizations (Figure 2.2; Colton *et al.*, 1998). The interaction between AFM tips and surfaces is recorded as force curves when the tip approaches or retracts from surfaces (Zlatanova *et al.*, 2000). The absolute force can be deduced from the spring constant of cantilevers using established force laws [see Heinz and Hoh (1999) for review]. AFM force spectroscopy has been applied in many biological areas, such as protein–DNA, antigen–antibody pairs, protein–ligand, protein–membrane, and protein–cell interactions. A detailed review on these applications is out of the range of this chapter and some excellent reviews are available elsewhere (Clausen-Schaumann *et al.*, 2000; Leckband, 2000).

Although force spectroscopy has powerful capabilities, some limitations exist in current force spectroscopy applied in biological investigations. The force measurements are so sensitive to the sample preparation and the conditions of measurements that it is difficult to compare the absolute forces obtained by different groups. Consequently, monitoring how the force changes with conditions, such as the

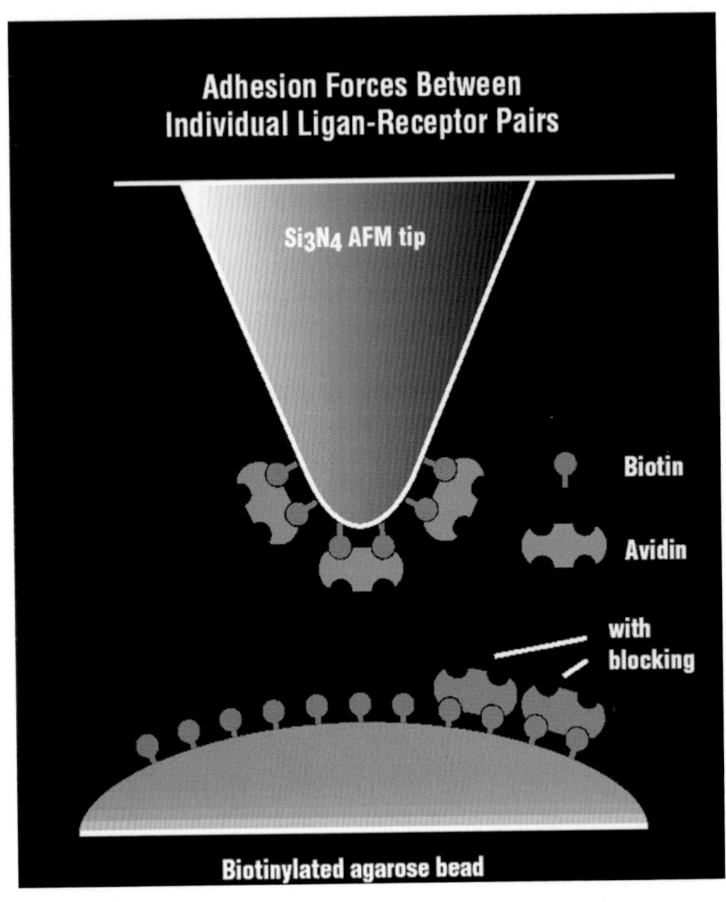

Figure 2.2. A schematic view of force spectroscopy mode. One partner (Avidin) in biomolecular interactions is attached to the atomic force microscopy (AFM) tip. Another partner (Biotin) is immobilized on the surface. The figure is reproduced from Florin *et al.* (1994) with permission.

conditions of the measurement and the buffer, is a more reliable way to understand molecular interactions in cell biology than measuring absolute force values.

2.2.6. Lateral Force Manipulation

Besides stretching the molecules using force spectroscopy mode to learn about the mechanical properties of macromolecular assemblies, AFM can also be used to manipulate sample using the lateral force with a specialized system, called nanoManipulator. The nanoManipulator system integrates the AFM with a virtual–reality interface that gives investigators new ways of interacting with objects at the nanometer scale (Guthold *et al.*, 1999a, 2000; Sincell, 2000).

Using a force-feedback pen, the user can touch the surface and directly manipulate the object. The manipulation is accomplished by exporting the data to a PHANTOM controller (SensAble Technologies, Cambridge, MA). This procedure allows the investigators to precisely locate the objects and features by feel whereas the tip makes the modification. Samples can be manipulated in contact mode and the changes before and after manipulation can be monitored using oscillating mode imaging. Samples can be bent, translated, rotated, and dissected. Mechanical properties of biological samples can be measured directly by recording the vertical and lateral forces during the manipulation process. The rupture forces of fibrin (see discussion in Section 2.3.6 and Figure 2.9) and DNA have been measured using nanoManipulator, and nonspecific binding between adenovirus and silicon surface has been monitored (Guthold *et al.*, 2000, 2004). Forces ranging from a few piconewtons to several micronewtons can be measured using the nanoManipulator, expanding the range of forces ($10^{-9}-10^{-14}$ N) measured by other single-molecule manipulation techniques, such as microneedles, flow field, magnetic field, and optical tweezers (Bustamante *et al.*, 2000). Compared with other single-molecule manipulation techniques, the nanoManipulator has the advantages of easy sample preparation and the ability to monitor the sample before and after the manipulation. One disadvantage of this technique is that the surface interaction may complicate the interpretation of the data. Besides measuring the physical properties of biological samples, in the future, the nanoManipulator maybe used as a tool to push the macromolecules together and watch the interaction in real time in liquids.

2.2.7. Postprocessing of AFM Images and Generating Quantitative Data

Importantly, the conclusion on the properties of protein or protein–protein complexes should be based on the analysis of statistically significant number of images collected from different depositions. Processing and quantitative analysis of AFM images are straightforward. The software-controlling AFM instruments, such as the Nanoscopes of Digital Instruments, Inc. (Santa Barbara, CA), can accomplish diverse tasks. For example, the images can be flattened to smooth the image. In addition, there are analysis tools to measure the contour length and bend angles of molecules such as DNA, as well as the height and volume of molecules. Additional software, such as Image SXM (http://reg.ssci.liv.ac.uk/) and NIH Image, are useful for the quantitative analysis of the size and the shape of molecules and complexes (Ratcliff and Erie, 2001). Image SXM allows raw image files to be opened without losing image information.

One-by-one measurements of DNA contour lengths and bend angles made by tracing each molecule using Nanoscope software is tedious and time consuming. To address this problem, custom software has been developed to automate the process. For example, a custom program written in MATLAB (Mathworks, Natick, MA) has been developed to increase the efficiency of measuring the DNA contour lengths and DNA bend angles induced by protein binding (Wang *et al.*, 2003).

2.3. CHARACTERIZATION OF PROTEIN–PROTEIN COMPLEXES

2.3.1. Characterization of Protein–Protein Interactions Based on Size Information

The heights of proteins as measured by AFM can be affected by various factors, such as the orientation of proteins on the surface and electrostatic interactions between macromolecules and the tip (Muller and Engel, 1997). However, the volumes of proteins in AFM can be used to compare protein size and derive useful information, such as evidence for protein–protein association and protein–protein equilibrium association constants (K_a) (Wyman *et al.*, 1997; Schneider *et al.*, 1998; Ratcliff and Erie, 2001; Yang *et al.*, 2003; Schlacher *et al.*, 2005).

The detailed procedures for volume analysis using Image SXM were described in the supplementary information section of Ratcliff and Erie (2001). Briefly, the first step in image analysis is to determine the height of the surface (S), which is generally nonzero. This surface height must be subtracted from the measured height of each protein before volume determination (Ratcliff and Erie, 2001). After the surface height is measured, each protein is then highlighted individually using the density slice utility in Image SXM. The density slice selects the pixels above the surface that represent the proteins to be analyzed. The image analysis function in Image SXM scans the image and selects all the highlighted proteins within the density slice. Analysis of each protein within the slice is then performed. In addition to height and area information, which permit the calculation of the protein volume, Image SXM can calculate the major and the minor axes by fitting the cross section of each protein to an ellipse. The volume for each protein, V_i, is calculated by multiplying the area, A_i, by the corrected average height (total average height, M_i, minus surface height, S)

$$V_i = A_i(M_i - S).$$

It has been shown using a large number of proteins that there is a quantitative linear dependence of the protein volume measured from AFM images on the molecular mass of proteins (Wyman *et al.*, 1997; Schneider *et al.*, 1998; Ratcliff and Erie, 2001; Figure 2.3). Consequently, volume analysis is a robust and reliable method to obtain the stoichiometries of protein–protein assemblies (Schneider *et al.*, 1998; Ratcliff and Erie, 2001). In addition, it is possible to use this volume analysis to determine homo- or heteroprotein–protein equilibrium association constants (K_a). This methodology has been used to determine the equilibrium association constant for the dimerization of *Escherichia coli* DNA helicase II (UvrD), in which the shifts in the distribution of protein oligomeric states under different protein concentrations were analyzed (Figure 2.4A and B; Ratcliff and Erie, 2001). Recently, analysis of protein volumes from AFM images directly revealed the DNA-independent interaction between pol V and RecA (Figure 2.4C; Schlacher *et al.*, 2005). The conclusion is based on the observation that incubation of RecA and pol V together resulted in the formation of particles with a larger average volume than that seen with either protein alone (Figure 2.4C; Yang *et al.*, 2003; Schlacher *et al.*, 2005). It is worth noting that the observed AFM volume of pol V–RecA complex is consistent with the predicted volume of pol V–RecA protein complex based on the standard curve shown in Figure 2.3. The direct interaction between pol V and RecA observed in AFM is a key observation supporting the idea that the role of RecA is not simply to target pol V to a stalled replication fork. Instead, RecA directly activates pol V, possibly as a subunit of the active pol V holoenzyme complex in translesion DNA synthesis (Schlacher *et al.*, 2005).

AFM volume analysis is also useful for determining the oligomeric state of proteins on DNA (Wyman *et al.*, 1997; Xue *et al.*, 2002; Bao *et al.*, 2003). For example, the effects of phosphorylation and mutation of nitrogen regulatory protein C (NtrC) from *Salmonella typhimurium* on the oligomeric state of NtrC at its specific DNA recognition site have been studied using volume analysis (Wyman *et al.*, 1997). This study provided evidence that large oligomers of NtrC are important for activating transcription (Wyman *et al.*, 1997). In another example, the oligomeric state of hWRN-N, a fragment of human WRN gene product, has been studied using AFM volume analysis (Xue *et al.*, 2002). The study showed that hWRN-N is in a trimer–hexamer equilibrium in the absence

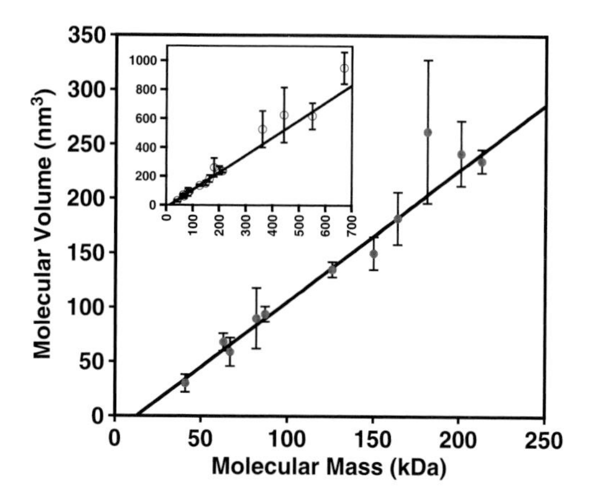

Figure 2.3. Plot of protein volume versus molecular mass. The volumes were determined as described in reference (Ratcliff and Erie, 2001). Data are shown for 15 proteins and protein–protein complexes. In the main plot, there are alcohol dehydrogenase (41 kDa), hWRN-N_{70-240} trimer (63 kDa), bovine serum albumin (67 kDa), UvrD monomer (82 kDa), PCNA trimer (87 kDa), hWRN-N_{70-240} hexamer (126 kDa), hWRN-N_{70-240} trimer, PCNA trimer complex (150 kDa), UvrD dimer (164 kDa), *Taq* MutS dimmer (181 kDa), β-amylose (201 kDa), hWRN-N_{70-240} hexamer, and PCNA trimer complex (213 kDa). The insert plot also includes *Taq* MutS tetramer (362 kDa), apoferritin (443 kDa), RNA polymerase (RNAP) (550 kDa), and thyroglobulin (670 kDa). The line represents the weighted least-square fit of the data, which is described by the equation $V = 1.2(\text{MW}) - 15.5$, where V is atomic force microscopy (AFM) volume and MW is molecular weight ($R^2 = 0.983$). The error bars represent the standard deviation of the distribution for each protein. The data are taken from references (Ratcliff *et al.*, 1998 and Yang *et al.*, 2003) and unpublished results.

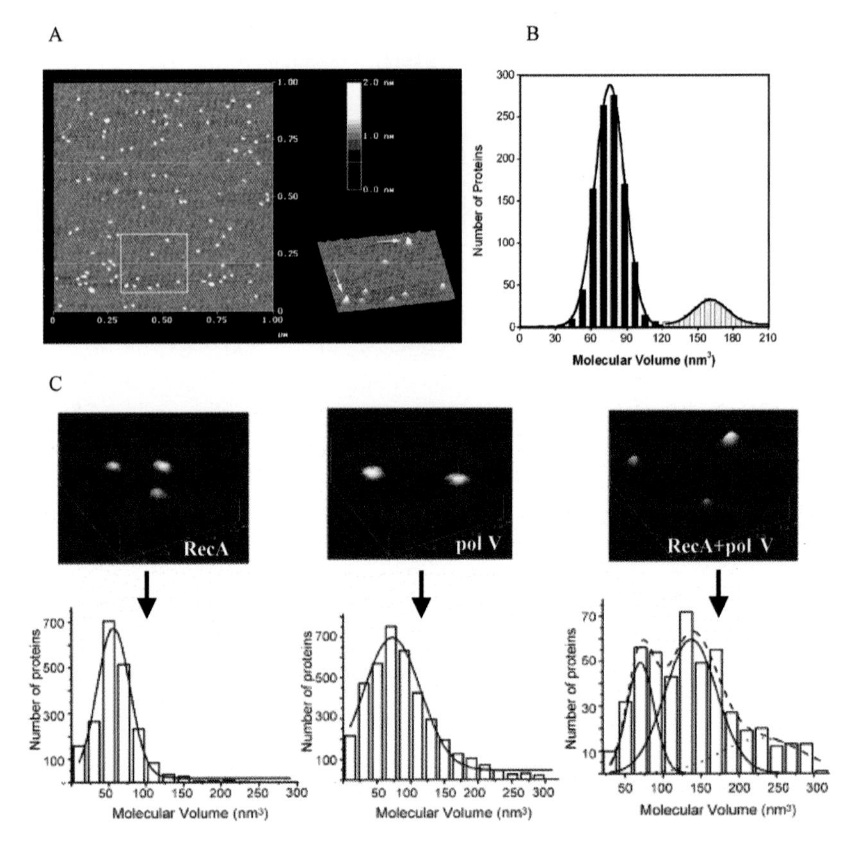

Figure 2.4. Using atomic force microscopy (AFM) volume analysis to investigate oligomeric states of proteins and protein–protein interactions. (A) AFM image of UvrD at 50 nM. The image shows proper surface coverage for volume analysis. The surface plot (inset) represents the rectangle area within the image. Arrows in the inset point to dimers; the other proteins within the inset are monomers. Image is reproduced from Ratcliff and Erie (2001) with permission. (B) Gaussian fits of the volume histogram for UvrD (250 nM). The solid lines are Gaussian fits of the volume data for monomers (solid bars) and dimers (hatched bars). Each species was fit independently. The number of proteins under each curve represents the species' population. Plot is reproduced from Ratcliff and Erie (2001) with permission. (C) AFM images of RecA, pol V, and pol V–RecA complexes. The image size is 100 nm × 100 nm for all images. The arrows point to the volume distributions for each protein. A Gaussian distribution was calculated for each data set and is displayed as a solid black line. For the pol V–RecA complexes, the distinct peaks from the Gaussian function are displayed as solid black lines and the distribution for all peaks is displayed as dashed lines. Images and plots are reproduced from Schlacher *et al.* (2005) with permission.

of DNA, but it is primarily hexamer, the active form for its functions, in the presence of DNA substrates.

AFM volume analysis is useful to study the protein–protein association because it can be complementary to other techniques. For example, analytical ultracentrifugation and isothermal titration calorimetry require high concentrations of samples, which would limit their use in assaying proteins with low solubility or with tight-binding constants. In addition, sample deposition for AFM imaging can be done over a wide range of temperature ranges on a hot plate, which is advantageous for studying thermophilic proteins.

Although volume analysis is a powerful method for determining protein stoichiometries and protein equilibrium association constants, care must be taken when acquiring the images (Ratcliff and Erie, 2001). First, the tip geometry and the strength of the tip–surface interactions need to be consistent in all the experiments, because AFM images are the convolution of tip–surface interactions. Standard curves relating AFM volume and molecular weight of protein may need to be generated in each laboratory using different AFM instruments. It is highly recommended to use a new cantilever each time when collecting final images for volume analysis purpose. Alternatively, another protein with known size close to the protein of interest can be added to the protein sample as a size standard (Verhoeven *et al.*, 2002). In addition, when imaging the DNA–protein complex, the AFM volume of DNA can be used to normalize the AFM volume of proteins (Wang, 2003). It should also be kept in mind that factors that affect the surface deposition of proteins may influence the measured association constant (Ratcliff and Erie, 2001). Two major factors are interaction of the proteins with the surface and diffusional properties of the proteins. If the oligomerized protein diffuses significantly more slowly than the monomer, its population could be underrepresented in an AFM image when very short deposition times are used. This effect would decrease the apparent association constant. Fortunately, diffusional factors should only become important for higher-order oligomers, because the diffusion coefficients of globular proteins depend inversely on the cube root of their molecular weights. Finally, it is worth mentioning that the use of volume analysis is not easily applicable to proteins whose molecular mass is <20 kD because of the limitation of AFM resolution.

2.3.2. Characterization of Architectures of Proteins and Protein–Protein Complexes

Even though AFM imaging cannot provide structural information on protein at atomic level, it can provide a unique view into the architecture of protein and protein–protein complexes. Thus, structural information from AFM images can further our understanding of structure–function relationship of proteins.

The human Rad50–Mre11 complex has important functions in double-strand break repair. AFM imaging revealed a central globular domain from which two long thin structures (arms) extended (Figure 2.5A–F; de Jager *et al.*, 2001). Based on AFM images, along with sedimentation equilibrium data suggesting that the Rad50–Mre11 complex consists of two Rad50 and two Mre11 molecules, an intermolecular coiled-coil structure has been proposed (de Jager *et al.*, 2001). In addition, AFM images of Rad50–Mre11 complexes on DNA showed an important role of the arms of Rad50–Mre11 in bridging two DNA fragments (see more discussion in Section 2.3.5; de Jager *et al.*, 2001).

AFM imaging has also been used to investigate the multisubunit $GABA_A$ receptor, which is the major inhibitory neurotransmitter receptor in the brain (Edwardson and Henderson, 2004). It is known that all these receptors have a common structure, with five subunits arranged around a pseudo fivefold axis. However, there are six possible subunit arrangements in a $GABA_A$ receptor of stoichiometry 2α:2β:γ (Figure 2.5G). The physiological relevant arrangement of the subunits around the receptor rosette has not been established. In one study, a hexahistidine (His_6) tag was attached onto the C-terminus of the α-subunit and the purified receptors containing all three subunits were incubated with an anti-His_6 immunoglobulin G (Edwardson and Henderson, 2004). The resulting receptor–immunoglobulin G complexes were imaged by AFM (Figure 2.5H). AFM images revealed that the most common angle between the two antibody tags was 135° (Figure 2.5I), close to the expected value of 144° if the two α-subunits are separated by a third subunit. This result excludes three (arrangements 3, 4, and 6 in Figure 2.5G) of the six possible arrangements of the subunits around the receptor rosette for majority of the $GABA_A$ receptors. Meanwhile, a small percentage of the complexes has an angle of 75° between the two bound antibodies (Figure 2.5I), suggesting the receptor has a mixed population of configurations. The understanding of the architecture of the $GABA_A$ receptor will help to design drugs with higher specificity (Edwardson and Henderson, 2004).

As mentioned earlier, AFM imaging in air using commercial microfabricated AFM cantilevers has generated useful information on the architecture of some proteins (de Jager *et al.*, 2001; Edwardson and Henderson, 2004). However, expansion of the application of AFM to study the architecture of more proteins depends on higher-resolution images. It is well established now that besides instrumental factors, major factors that can affect the quality of the image include the shape of the tip, tip–sample interactions, stable immobilization of the sample on the surface, as well as the pH and the ionic strength of the buffer used for absorbing and scanning the sample (Mou *et al.*, 1995; Muller and Engel, 1997; Muller *et al.*, 1999; Hafner *et al.*, 2001). At room temperature, easy deformation by the scanning tips and the thermal motion of most macromolecules make it hard to achieve high-resolution images. Close packing of the sample on the surface can reduce this problem to some degree, and subnanometer resolution images have been acquired on these samples by imaging in liquids (Muller *et al.*, 1998, 2001; Stahlberg *et al.*, 2001; Conroy *et al.*, 2004). Cryo-AFM holds the promise for imaging a large variety of biological samples at high resolution, comparable with EM (Shao and Zhang, 1996; Shao *et al.*, 2000; Sheng and Shao, 2002; Sheng *et al.*, 2003). Meanwhile, even with the state-of-art techniques such as cryo-AFM, the resolving power of AFM will not be fully reached without a well-defined ultrasharp tip. Carbon nanotube tips are the most promising candidates for the next generation of ultrasharp AFM probes (Woolley *et al.*, 2000; Hafner *et al.*, 2001). Next, we discuss some examples of high-resolution AFM images achieved on 2D protein crystals and using carbon nanotube tips.

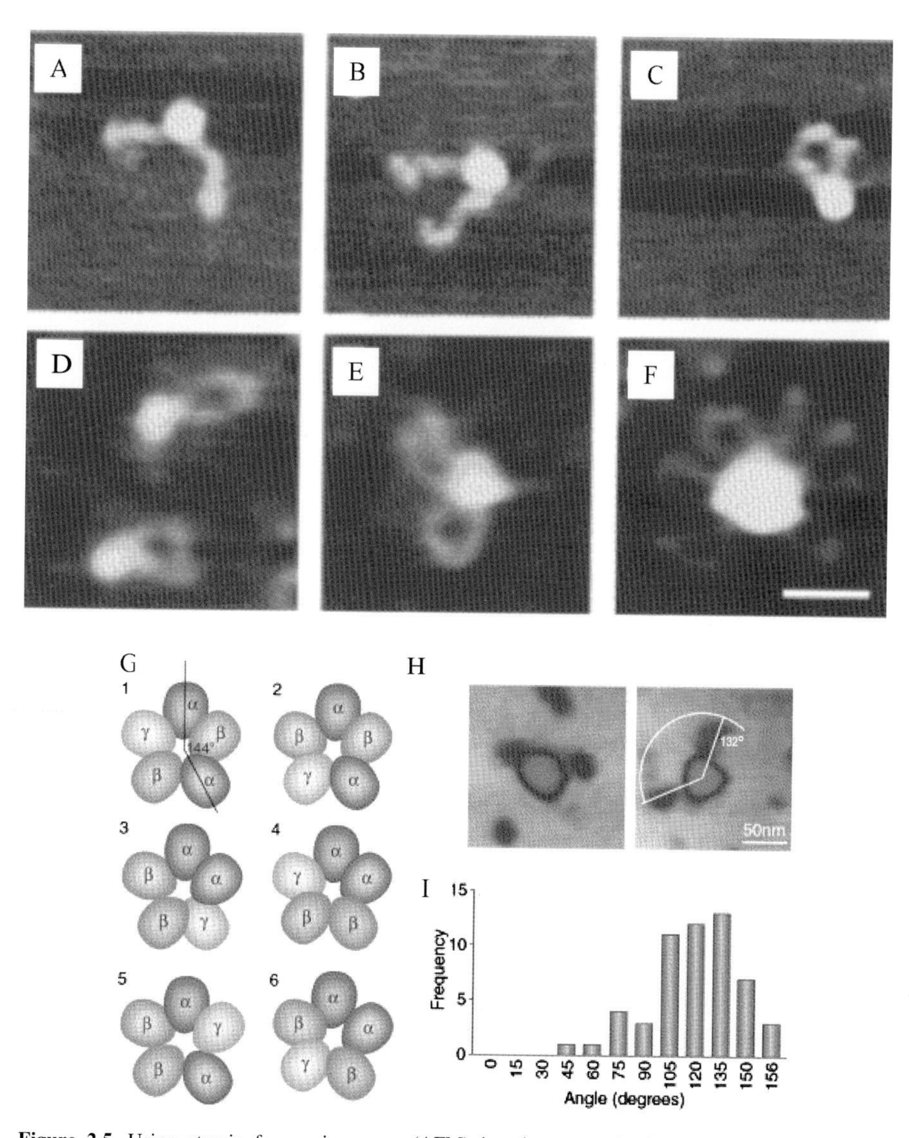

Figure 2.5. Using atomic force microscopy (AFM) imaging to study the architectures of protein complexes. (A–F) Representative AFM images of human Rad50–Mre11 complexes captured by tapping mode in air. Rad50–Mre11 exhibited a distinct architecture with a central globular domain from which arms of 40–50 nm protruded. The arms were observed in a variety of conformations. Images are reproduced from de Jager *et al.* (2001) with permission. (G–I) Analysis of $GABA_A$ receptor architecture by AFM. (G) Possible arrangements of subunits in a $GABA_A$ receptor composed of 2α-, 2β-, and 1γ-subunits. (H) AFM images of complexes between $GABA_A$ receptors, with His_6 tags on their α subunits, and mouse monoclonal anti-His_6 immunoglobulin G molecules. (I) Distribution of angles between antibody molecules in complexes between $GABA_A$ receptors and anti-His_6 immunoglobulin G. Note that the histogram has a major peak at 135° and a possible minor peak at 75°. (G–I) are reproduced from Edwardson and Henderson (2004) with permission.

Due to limited hydrophilic surfaces, membrane proteins do not readily form 3D crystals for X-ray crystallography. However, 2D membrane protein crystals reconstituted in the presence of lipids are more stable, and a large number of membrane proteins have been crystallized in this manner. Electron crystallography of 2D crystals has provided static structural information at atomic resolution (Stahlberg *et al.*, 2001). Imaging in liquids using AFM provides an advantage over EM, in that the native environments and biological activities of these membrane proteins can be preserved throughout sample preparation and scanning. AFM is the only technique that gives insights into the surface structures and the dynamics of membrane proteins at subnanometer resolution. The high resolution is partly due to the elimination of capillary forces (Drake *et al.*, 1989). Müller *et al.* demonstrated that by adjusting the pH and the electrolytes in the buffer, electrostatic double-layer repulsion between the tip and the sample can be reduced, resulting in reduced vertical and lateral forces between the AFM tip and the sample (Muller *et al.*, 1999). In addition, the 2D crystals are strongly anchored to the substrates in liquids, and the force applied to the AFM tips is believed to be distributed over a large sample area on these 2D crystals. The factors mentioned collectively earlier dramatically reduce the sample deformation during scanning. In addition, only the small sharp protrusion at the end of tip is believed to sense the short-range electrostatic repulsion that confers high-resolution structural information. Subnanometer resolution imaging has been demonstrated by the images of purple membrane (which consists of bacteriorhodopsin and lipids) (Muller *et al.*, 1999). A lateral resolution of 0.6 nm (width at half-maximum height) is reproducible in these images.

Carbon nanotubes further extend the power of AFM as a tool to characterize architecture of proteins. Carbon nanotubes have advantage over the microfabricated probes due to their small radii (0.7–5 nm for single-walled nanotubes), high aspect ratio, extremely large Young's modulus (stiffness), and the ability to be elastically buckled under large load. There are several elegant reviews on the fabrication technologies and high-resolution imaging using carbon nanotubes (Stevens *et al.*, 2000; Woolley *et al.*, 2000; Hafner *et al.*, 2001). For example, using carbon nanotubes, the two sides of GroES, a component of the GroEL–GroES chaperonin system involved in protein folding, have been resolved (Hafner *et al.*, 2001). One side was seen as a ring-like structure with an 11-nm outer diameter and the other face looked like a dome with the same diameter. In future, it is possible that carbon nanotube AFM tips will make it routine to obtain images at nanometer resolution.

2.3.3. Characterization of Recognition Specificities of Proteins

The first step for many proteins to participate in DNA transaction events is the recognition of specific sites, such as a DNA replication origin sequence, promoter region, a DNA damage or mismatch. Traditionally, bulk solution measurements, such as electrophoretic mobility-shift assays (EMSA), filter-binding assays, surface plasmon resonance (SPR), and calorimetric assays are used to study the DNA–protein binding affinity and specificity. The limitation with these bulk solution methods is that the observed affinities are the weighted sum of all interactions, including the specific, the nonspecific, and the DNA ends (if a linear fragment is used). These bulk assays can only determine the apparent binding constants to the entire DNA fragment. For example, if a protein has significant binding affinity for the DNA ends, the apparent binding constant to short linear DNA fragment measured by bulk assays would obscure the true specificity to specific sites. Visualization of protein on DNA by AFM not only provides information on the extent of binding of protein to a DNA fragment, but also on where the protein is bound to DNA. In addition, for AFM imaging in air, the DNA–protein complex present in the reaction mixture at the time of deposition is fixed on the surface for imaging. Therefore, it is possible to observe DNA–protein complexes that might dissociate in gel electrophoresis-based assays (de Jager *et al.*, 2001).

From the earlier work using AFM, it is appreciated that indeed AFM can be used to identify the specific recognition site of a protein on DNA or on an extracellular matrix protein (Erie *et al.*, 1994; Allison *et al.*, 1996). This identification is achieved by looking at the position distribution of the protein on the DNA or a matrix protein, which have one or more specific recognition sites at defined locations (Figure 2.6). For example, AFM was used to investigate the binding of Factor IX, a 57-kDa zymogen of a serine protease that participates in blood coagulation, to Collagen IV (Wolberg *et al.*, 1997). Collagen IV is an extracellular matrix protein and a major component of the basement membrane region of endothelial cells. In this study, antibody A-5 was used to increase the apparent size of Factor IX to make it more easily identifiable on Collagen IV. Figure 2.6A shows a representative AFM image of Factor IX–A-5 complex bound to Collagen IV. In addition, Wolberg *et al.* identified two specific binding sites by plotting the position distribution of Factor IX–A-5 complexes on Collagen IV (Figure 2.6B).

In principle, for a linear substrate, the distribution of positions of the protein on the substrate provides a direct measurement of binding affinity to specific and nonspecific sites. Recently, the theoretical basis for rigorous analysis of DNA–protein complexes from AFM images to estimate specific and nonspecific DNA-binding constants and specificities has been worked out (Yang *et al.*, 2005). Using this analysis on the previously published data on human DNA damage recognition protein XPC-HR23B and human 8-oxoguanine DNA glycosylase (hOGG1) demonstrated that for proteins with high specificity, methods based on AFM images yield similar numbers of binding specificity compared with bulk biochemical assays (Chen *et al.*, 2002; Janicijevic *et al.*, 2003b; Yang *et al.*, 2005). However, this new study demonstrated that in the case of protein that has high affinity to DNA ends, compared with bulk solution measurement, single molecule methods based on AFM images can provide more accurate measurement of a binding constant to an individual site (Yang *et al.*, 2005). Based on EMSA, the specificity of *Taq* MutS, which is involved in DNA mismatch recognition, for a T-bulge on a short DNA fragment is ~1700; whereas, it is only ~30 for *E. coli* MutS. AFM imaging and statistical analysis revealed that *E. coli* MutS binds to

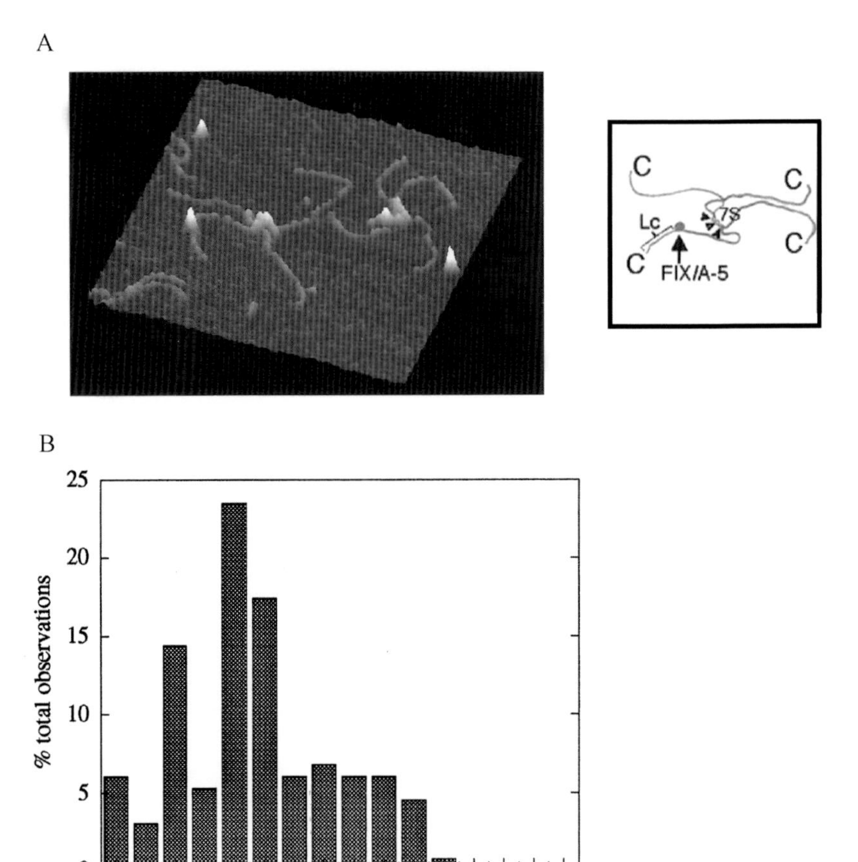

Figure 2.6. Using atomic force microscopy (AFM) imaging to study the binding specificity of proteins. (A) Surface view of Factor IX–A-5 complexes bound to Collagen IV C termini. 7S domains and Factor IX–A-5 complexes are indicated in the diagram on the right-hand side of the image. (B) Distribution of distances of the Factor IX–A-5 complexes from the free (C-terminal) end of the Collagen IV monomers. Both the peaks (located at 98 ± 13 nm and 50 ± 13 nm, respectively) were statistically significant ($p < 0.0004$). The larger peak at 98 nm likely represents a high-affinity, specific Factor IX binding site on the collagenous domain, whereas the smaller peak may represent a secondary, specific Factor IX binding site, which has threefold to fourfold lower affinity for Collagen IV than the primary site. Image and plot are reproduced from Wolberg *et al.* (1997) with permission.

DNA ends with an affinity that is only approximately five times less than that to a GT mismatch. This study suggested that the apparent differences in specificity of *Taq* and *E. coli* MutS for a T-bulge determined from bulk measurements likely result from differences in the extent to which end binding is being detected in bulk assays. In summary, the binding specificities observed from AFM are more accurate. Specifically, AFM provides a direct measure of the relative affinities for the different sites on individual DNA fragments; whereas, bulk assays yield relative affinities of a protein to different DNA fragments. Protein-binding specificity to DNA ends is also of great interest, because DNA ends are common intermediates in genome metabolism and can be caused by endogenous and exogenous DNA-damaging agents. Failure to correctly process these DNA ends can lead to mutations, uncontrolled cell growth, and carcinogenesis. It has been discovered that to prevent these detrimental effects of DNA ends, some proteins can recognize DNA ends and function as DNA damage sensor and recruit other protein factors for further processing of DNA ends. AFM imaging provides direct evidence that some proteins indeed preferentially bind to DNA ends. The examples are Rad50 and Mre11 (R–M) complex, ataxia-telangiectasia gene product ATM, MutS, and DNA-dependent protein kinase (DNA-PK) (Yaneva *et al.*, 1997; Smith *et al.*, 1999; de Jager *et al.*, 2001, 2002).

It is worth mentioning that although both selective and systemic alterations in the DNA occupancy by the surface can affect the absolute constants measured by AFM method, only biased alterations between the occupancy on the specific site and that on nonspecific sites will affect specificities. There are more detailed discussions on this subject in Yang *et al.* (2005).

AFM imaging also makes it possible to observe multiple recognition events on distinct sites on the same DNA fragment or on an extracellular matrix protein (Allison *et al.*, 1996; Gaczynska *et al.*, 2004). For example, in eukaryotes, the initiation of DNA replication depends on the recognition of origin sequence by the origin recognition complex (ORC), a heteromeric six-subunit protein complex. Gaczynska *et al.* used AFM to examine the binding of *Schizosaccharomyces pombe* (*sp*) ORC to the *sp* autonomously replicating sequence 1 (*ars1*) and found two binding sites in *ars1* (Gaczynska *et al.*, 2004). Although one binding site discovered in this study was consistent with previous studies, another binding site was not previously identified. In addition, by using each single subunit of *sp*ORC and mutant proteins, AFM imaging further identified one of the six subunits of *sp*ORC and its structural element that is responsible for binding to *ars1*. This study is a good example showing that AFM can be used as a great tool for studying recognition events involving multiple sites, such as transcription and replication initiation processes.

For mapping the binding sites of protein on circular DNA, the challenge is to add a position marker on circular DNA. Site-specific labeling of covalently closed circular DNA has been achieved by using triple helix-forming oligonucleotides (Zelphati *et al.*, 2000). The binding of avidin or streptavidin to a biotin group on the oligonucleotide probes can serve as a position marker. In addition, two approaches have been described for stably conjugating peptides, proteins, and oligonucleotides onto plasmid DNA (Zelphati *et al.*, 2000). Recently, the Lyubchenko group used an inverted repeat sequence, which forms a cruciform as a position marker to study the binding of the Zα domain of human ADAR1 on a supercoiled plasmid (Lushnikov *et al.*, 2004).

2.3.4. Characterization of Mechanisms of Action of Proteins on DNA

AFM can not only be used to investigate the partnership between proteins, but also to provide information on their mechanisms of action, because each DNA–protein complex in AFM image is a snapshot of protein caught in action on DNA. In addition, using AFM to investigate protein-induced conformational changes in DNA is straightforward because the topographic difference between proteins and DNA is obvious in an AFM image.

Large conformational changes in both proteins and DNA can occur when proteins bind to DNA or exchange between specific sites and nonspecific sites. Sequence-dependent DNA helix deformability is an important component of recognition of specific sequence on DNA, alongside the more generally recognized patterns of hydrogen bonding (Dickerson, 1998). For sequence-specific enzymes, such DNA deformation, may contribute to the correct assembly of active site residues and provide access to specific DNA moieties. In addition, protein-induced DNA bending and wrapping play an important architectural role in assembling the specific DNA–protein complex for DNA replication, regulation of transcription, and condensation of DNA into chromatin.

Two early AFM studies investigated the DNA bending induced by *E. coli* RNAP (Rees *et al.*, 1993) and bacteriophage λ Cro, a small transcription regulatory protein (Erie *et al.*, 1994). In the AFM study of the RNAP, the DNA appeared bent in open promoter complexes containing RNAP bound to the promoter, and more severely bent in elongation complexes in which RNAP has synthesized a 15-nucleotide transcript (Rees *et al.*, 1993). The different bent conformations of DNA induced by RNAP were proposed to be the characteristics of polymerase transiting from the open promoter complexes to the elongation complexes (Rees *et al.*, 1993). In the Cro study, Erie *et al.* analyzed the fundamental roles of protein-induced DNA bending at specific sites as well as at nonspecific sites (Erie *et al.*, 1994). Protein-induced bending at nonspecific sites was observed and it was suggested to be important for protein in searching for specific sites and increasing specificity on the target sites (Erie *et al.*, 1994). AFM has also been used to study specific and nonspecific photolyase–DNA complexes (van Noort *et al.*, 1999). Photolyase is an enzyme that binds to UV-induced thymidine dimers and reverses the cross-linking of adjacent pyrimidines by using the energy of visible light. Contrary to Cro–DNA complexes, nonspecific photolysase–DNA complexes show no significant bending but increased rigidity compared with naked DNA, whereas specific complexes show average DNA bending of 36° and higher flexibility (van Noort *et al.*, 1999).

It is worth pointing out that though other DNA bending assays, such as gel mobility, yield a single or average bend angle, AFM provides the spatial distribution of bending along the DNA and dynamic bending histograms of DNA–protein assemblies bound at the same location; that is, the full distribution of angles is observed. Knowledge of the full distribution of bend angles can provide unique insight into the mechanism of action by proteins, such as in the case of MutS (Wang *et al.*, 2003). The family of MutS proteins initiates DNA mismatch repair (MMR) by the recognition of base–base mismatches and insertion or deletion loops. To gain insight into the mechanism by which MutS discriminates between mismatch and homoduplex DNA, protein-induced DNA bending at specific and nonspecific MutS–DNA complexes has been studied by AFM (Wang *et al.*, 2003; Figure 2.7A). Interestingly, MutS–DNA complexes exhibit a single population of conformations, in which the DNA is bent at homoduplex sites, but two populations of conformations, bent and unbent, at mismatch sites (Figure 2.7B). These results suggest that the specific recognition complex is one in which the DNA is unbent. Combining these results with existing biochemical and crystallographic data led to the proposal that MutS (i) binds to DNA nonspecifically and bends it in search of a mismatch; (ii) on specific recognition of a mismatch, undergoes a conformational change to an initial recognition complex in which the DNA is kinked, with interactions similar to those in the published crystal structures; and (iii) finally undergoes a further conformational change to the ultimate recognition complex in which the DNA is unbent (Wang *et al.*, 2003). The results from this study provide one structural explanation that can contribute to the further understanding of how MutS achieves high MMR specificity.

Interestingly, AFM imaging has shown that simultaneous binding of two cellular transcription factors, nuclear factor I (NFI) and octamer-binding protein (Oct-1), can induce a collective bend in DNA, when bound to the transcription origin (Mysiak *et al.*, 2004a,b). Mysiak *et al.* observed that NFI induced a 60° bend in the origin DNA, whereas Oct-1 induced a 42° bend. Simultaneous binding of NFI and Oct-1 induces an 82° bend. It was suggested that this collective DNA bending can lead to a synergistic enhancement of DNA replication (Mysiak *et al.*, 2004b).

DNA bending induced by other proteins, such as EcoRI and adenine N6 DNA methyltransferases, has also been observed using AFM (Garcia *et al.*, 1996; Allan *et al.*, 1999; van Noort *et al.*, 1999; Mysiak *et al.*, 2004a). In addition to DNA bending, other DNA distortions, such as DNA wrapping, have been observed for protein involved in transcription, DNA repair, replication initiation, and chromatin remodeling. These proteins include RNAP, UvrB, human replication protein A (RPA), chromatin remodeling factor CSB, DNA gyrase, histone protein, and DNA replication ORC (Rivetti *et al.*, 1999; Verhoeven *et al.*, 2001; Lysetska *et al.*, 2002; Kepert *et al.*, 2003; Rivetti *et al.*, 2003; Gaczynska *et al.*, 2004; Heddle *et al.*, 2004). In AFM images, DNA wrapping around a protein is observed as a reduced DNA contour length when comparing DNA length from DNA–protein complexes with free DNA.

AFM has also been used to visualize the unwinding of duplex RNA by DbpA, which is a DEAD box helicase (Henn *et al.*, 2001). The DEAD box protein family catalyzes the hydrolysis of ATP in the presence of RNA. From AFM imaging, Henn *et al.* (2001) observed that DbpA was bound to the end of RNA molecule with a ssRNA overhang, indicating that DbpA requires an ssRNA or moderate fork junction for binding before performing the unwinding activity. The unwinding of

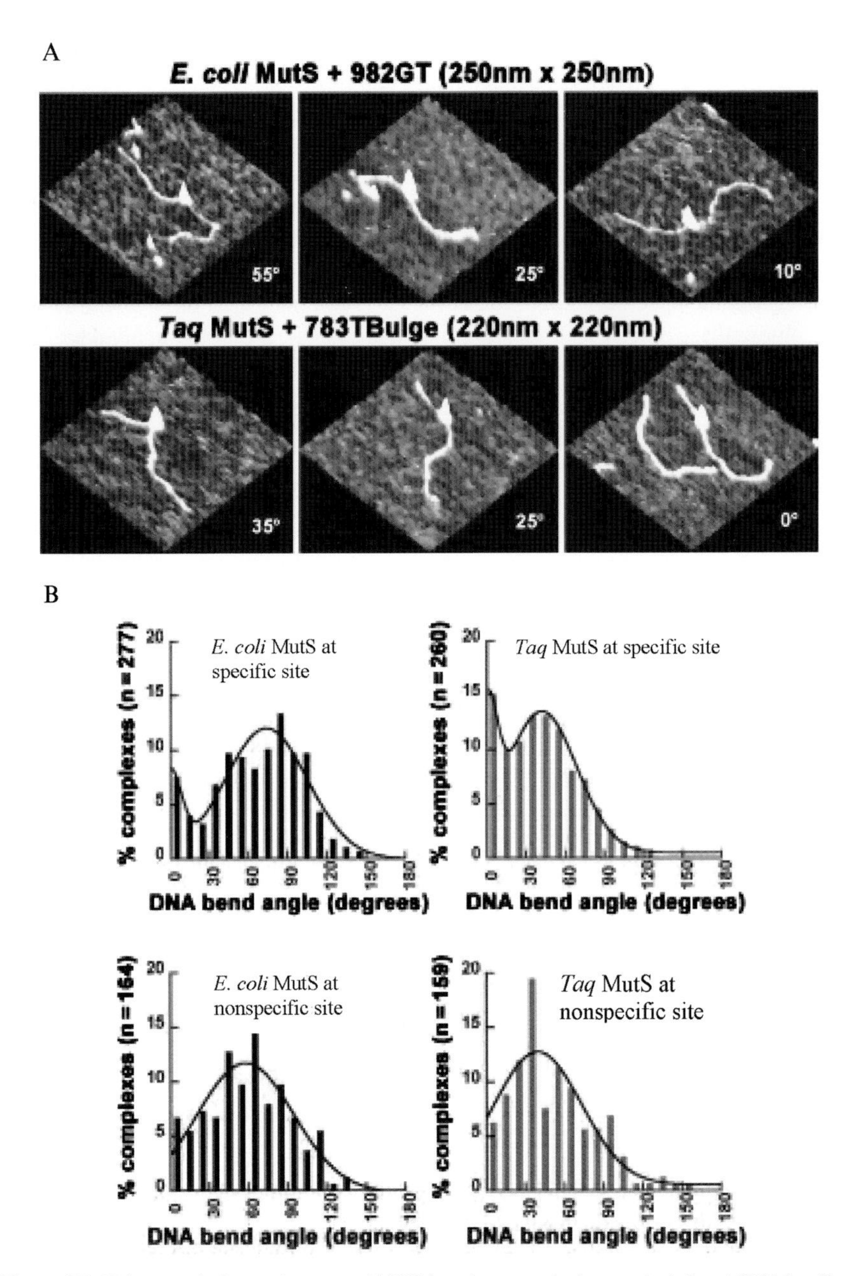

Figure 2.7. Using atomic force microscopy (AFM) imaging to study the protein-induced DNA bending. (A) AFM surface plots of *E. coli* MutS bound to a G–T mismatch and *Taq* MutS bound to a 1T-bulge. MutS-induced DNA bend angles are shown on each image. (B) Histograms of DNA bend angles induced by *E. coli* and *Taq* MutS bound to mismatch (specific complexes, *upper panels*) and homoduplex (nonspecific complexes, *lower panels*) sites. The images and the plots are reproduced from Wang *et al.* (2003) with permission.

duplex RNA was observed as ATP simulated formation of Y-shaped intermediate, representing strand separation.

2.3.5. Characterization of Collective Actions of Protein Assemblies at Multiple Sites on DNA

Although some reactions on DNA are carried out by a single protein at one specific site, many reactions involve multiple proteins interacting with multiple sites that span long distances along the DNA. This section highlights the applications of AFM to study the cooperative protein binding to DNA, protein-mediated DNA condensation, DNA looping, and end joining.

2.3.5.1. Cooperative DNA Binding by Proteins

RecA is a classic allosterically regulated enzyme, and ATP binding leads to a dramatic increase in DNA-binding affinity and a cooperative assembly of RecA subunits to form a helical nucleoprotein filament. RecA filaments on DNA play an important role in promoting recombination between two DNA strands and regulating the SOS response in bacteria (McGrew and Knight, 2003). AFM imaging of RecA filaments using carbon nanotube probes revealed a 10-nm pitch of RecA–dsDNA complex (Umemura *et al.*, 2001), which is consistent with the observation from EM. This study shows that high-resolution AFM images of RecA filaments can provide useful information for constructing a 3D structural model for a RecA–DNA filament. Recently, examination of the formation of RecA–dsDNA complex as a function of time using AFM provided insight into the mechanism of assembly of the RecA filament (Sattin and Goh, 2004). In the study carried by Sattin *et al.*, RecA was incubated with nicked plasmid DNA in the presence of ATPγS (adenosine-5′ (γ-thio)-triphosphate), and aliquots of reaction were taken at different time points and deposited on mica for AFM imaging. AFM imaging revealed that extensive polymerization of the RecA along DNA did not happen until after 15 min, and at later time, the RecA coverage of plasmid DNA was continuous, with no instances of a plasmid with more than one continuous stretch of RecA filament (Figure 2.8A–C). These findings suggested that the nucleation step of RecA binding to DNA is very unfavorable, and it is slow, whereas the polymerization step is fast. In addition, AFM imaging also showed that homologs of RecA, such as archaeal RadA and *Saccharomyces cerevisiae* Dmc1, also form protein filaments on DNA (Seitz *et al.*, 1998; Chang *et al.*, 2005).

MutL and its homologs are essential components of postreplicative DNA MMR systems (Modrich and Lahue, 1996; Buermeyer *et al.*, 1999; Kolodner and Marsischky, 1999). In addition, MutL and its homologs also participate in a variety of other DNA transactions, such as cell-cycle checkpoint control, apoptosis, and regulation of homologous and homeologous recombination (Buermeyer *et al.*, 1999; Bellacosa, 2001). The detailed knowledge of DNA-binding mechanism of MutL homologs is a perquisite to understanding their roles in different DNA transactions. Using AFM, it was demonstrated that yeast MutL homolog, MLH1–PMS1, can form cooperative protein assemblies on either one or two strands of DNA (Figure 2.8D–F), suggesting that it has at least one independent DNA-binding site on each of its subunits (Hall *et al.*, 2001; Drotschmann *et al.*, 2002; Hall *et al.*, 2003). Unlike RecA filaments, there were multiple MLH1–PMS1 protein tracts on one M13 plasmid. These data suggest that for MLH1–PMS1, although the nucleation event is significantly less favorable than the polymerization, it is not so unfavorable as in the case of RecA for which only a single protein tract is seen on each plasmid.

2.3.5.2. Protein-Mediated DNA Condensation

Genomic DNA needs to be packaged to fit into its cellular compartments (Sato *et al.*, 1999; Dame *et al.*, 2000; Brewer *et al.*, 2003; Ceci *et al.*, 2004; Friddle *et al.*, 2004). Mammals and the budding yeast package mtDNA in compact globular

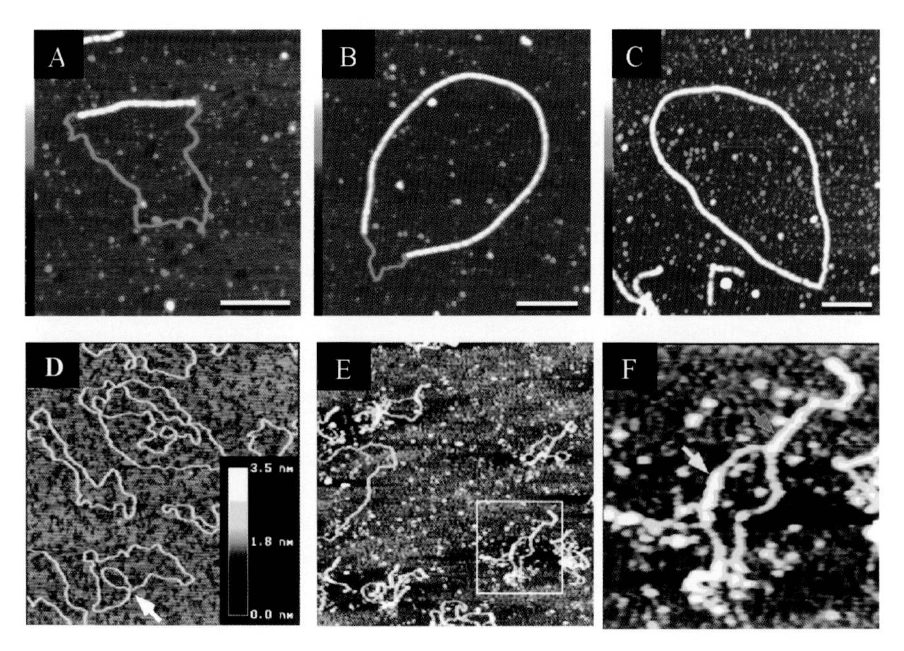

Figure 2.8. Using atomic force microscopy (AFM) imaging to study the cooperative binding of protein on dsDNA. (A–C) Typical RecA–DNA complexes observed at different incubation times at 37°C: (A) After 30 min—∼38% coverage of DNA by RecA; (B) 45 min—∼88% coverage; and (C) 60 min—DNA fully coated by RecA. The images are reproduced from Sattin and Goh (2004) with permission. (D–F) Cooperative binding of yeast MLH1–PMS1 heterodimer to dsDNA: (D) M13mp2 RFI DNA alone. (E) M13 RFI DNA in the presence of MLH1–PMS1. The scan size is 1500 nm for (D) and (E). (F) Zoomed view of the boxed region in (E). The light blue arrow indicates a tract of cooperatively bound MLH1–PMS1 associated with a single dsDNA region. The red arrow indicates a tract of cooperatively bound MLH1–PMS1 associated with two dsDNA regions. The images are reproduced from Hall *et al.* (2001) with permission.

structures are similar to a bacterial nucleoid, and its structure is distinctly different from the packaging of DNA into chromatin in the cell nucleus, which involves wrapping of the DNA around histone proteins. DNA compaction involves collective efforts of multiple copies of proteins over a large distance at discrete sites. Compaction of linear DNA by yeast mitochondrial protein Abf2p was investigated by AFM as a function of protein concentration (Brewer *et al.*, 2003). At low protein concentrations, DNA bending and compaction were observed, whereas at high protein concentrations, round compact objects were observed (Brewer *et al.*, 2003). Based on these observations, Friddle *et al.* (2004) suggested that Abf2p compacts DNA by introducing sharp bends into the DNA backbone. On the other hand, for *E. coli* H-NS protein, DNA bridging mediated by H-NS and highly compacted DNA was observed with co-occurrence of relative high features on the DNA. These data suggest that compaction of DNA is mediated by DNA bridging and extensive oligomerization of bound H-NS molecules (Dame *et al.*, 2000). In addition, at higher ratio of H-NS–DNA, the complexes have a rod-like appearance, indicating high level of condensation through network formation. The protein RdgC, which was suggested to play a role in replication and recombination, has also been implicated in promoting DNA condensation (Tessmer *et al.*, 2005). For the protein RdgC, both RdgC-induced DNA bending and protein–protein mediated strongly interwound dense DNA structures were observed in AFM images (Tessmer *et al.*, 2005).

2.3.5.3. Proteins Bind to Multiple Sites on DNA Leading to DNA Looping and End Joining

Two pathways can lead to DNA looping. First, protein–protein complexes can bind concurrently to two sites forming a loop with fixed size. Second, energy-dependent translocation of one or more subunits along the DNA relative to the other subunit can result in DNA looping with loop size not fixed. Examples are DNA looping by restriction enzymes and during transcription activation process (Rippe *et al.*, 1997; Halford *et al.*, 2004). The restriction enzymes that interact with two sites on the DNA include all the endonucleases from the Type I and III restriction–modification systems and many of the type II restriction enzymes (Halford *et al.*, 2004). Even though people are more familiar with the restriction enzymes that act at a single specific site, these enzymes are indeed a minority compared with those needing two sites. DNA looping on linearized plasmid, through dimerization of proteins bound to two binding sites, has been observed with a type IA restriction enzyme, EcoKI (Berge *et al.*, 2000). It was suggested that the requirement for interaction with two sites for cleavage may increase specificity and prevent the promiscuous cleavage by restriction endonucleases (Halford *et al.*, 2004).

Transcriptional enhancers are *cis*-acting DNA elements that are binding sites for regulatory proteins and function at large distances from promoter elements to stimulate transcription (Xu and Hoover, 2001). Enhancer-like elements have been discovered in eukaryotes and in a wide variety of bacteria. The regulatory proteins that bind to enhancers must contact RNAP to activate transcription. Interactions between enhancer-binding proteins and RNAP can occur by either DNA looping or tracking of the enhancer-binding protein along the DNA. AFM has shown that contact of an activator protein NtrC and sigma (54)-RNAP holoenzyme is mediated by DNA looping (Rippe *et al.*, 1997).

In an earlier section, we mentioned the importance of the correct processing of DNA ends to the integrity of the genome. DNA-PK plays important roles in DNA double-strand break repair and immunoglobulin gene rearrangement. The DNA-PK holoenzyme is composed of three polypeptide subunits: The DNA-binding Ku70/Ku86 heterodimer and a catalytic subunit (DNA-PKcs). AFM has been used to visualize the interaction of Ku and DNA-PK with DNA (Cary *et al.*, 1997). Ku-mediated DNA looping was observed, and the loop formation was independent of DNA sequence and of the presence of DNA-PKcs (Cary *et al.*, 1997). To differentiate whether Ku tethered DNA ends through self-association of the DNA-bound proteins or a capacity for the complex to bind two DNA ends simultaneously, gel filtration of Ku in the absence and the presence of DNA was conducted. Gel filtration showed no evidence of self-association of two Ku70/Ku86 in the absence of DNA but association of two Ku70/Ku86 in the presence of DNA. These observations suggested that Ku binding at DNA double-strand breaks leads to Ku self-association and a physical tethering of the broken DNA strands. AFM imaging directly supports the notion that the Ku70/Ku86 heterodimer is capable of holding or tethering broken ends in place through self-association to ensure physical proximity of DNA ends for correct repair. It is worth mentioning that AFM imaging also demonstrated that telomere repeat binding factor, TRF2, also uses DNA-dependent multimerization to form DNA loops (Yoshimura *et al.*, 2004). It was suggested that this TRF protein-mediated DNA loop might be an important intermediate structure for protection of chromosome ends.

Previously, we discussed that AFM imaging has resolved the globular and arm domain of human Rad50–Mre11 complex and the study of its end-binding specificity (de Jager *et al.*, 2001). AFM imaging has also demonstrated that Rad50–Mre11 formed large oligomeric protein complexes at DNA ends, and the arms protruded from the DNA (de Jager *et al.*, 2001). Based on these observations, it is suggested that Rad50–Mre11 oligomers accumulate at broken DNA ends and keep the ends in close proximity by interaction of the end-bound Rad50–Mre11 oligomers through interaction of their arm domains (de Jager *et al.*, 2001, 2002).

2.3.6. Characterization of Protein Fibers

Most of the proteins we have discussed so far are globular proteins, in which the polypeptide chain folds into a compact shape like a ball with an irregular surface. Fibrous proteins form the cytoskeleton inside the cell. They are also a main component of the gel-like extracellular matrix that is responsible for tissue strength and resilience and promoting cell growth and differentiation. In addition, blood clots, which prevent the loss of blood after injury, mainly consist of fibrin fibers attached to platelets. Understanding the structure and mechanical properties of protein fibers can help us understand the normal function of these fibrous proteins and various disease states that are caused by the defects in the polymerization and degradation of fibrous proteins. The ability of AFM to image in air and liquids, and to manipulate samples makes it a unique tool for studying fibrous proteins.

Recently, Guthold *et al.* demonstrated the ability of AFM to visualize and mechanically manipulate individual fibrin fibers, which are the key structural components of blood clots. Using lateral force manipulation (see Section 2.2.6 and Figure 2.9), Guthold *et al.* determined the rupture force, F_R, as a function of diameter of fibrin fiber, D. They found that the rupture force increased with increasing diameter as $F_R \sim D^{1.30 \pm 0.06}$. This observation suggested that the molecule density (ρ) of fibrin fiber varies as $\rho(D) \sim D^{0.7}$, which means that thinner fibers are denser than thicker fibers. Future comparison studies of the mechanical properties of fibrin fibers formed from precursors obtained from healthy people and people with blood clotting disorder will further our understanding of the disease state.

Abnormal formation of protein fibril by aggregation of normally soluble proteins can also cause various diseases. The amyloid fibril, which has characteristic filamentous structures, is involved in a range of human diseases, such as Alzheimer's disease, type II diabetes, Parkinson's diseases, and Huntington's disease. *In vitro*, the polymerization of many amyloids can be initiated by the formation of a seeding nucleus or protofibril followed by the assembly of soluble protein assemblies with these nuclei to form amyloid fibers (see Figure 2.10A). Therapeutic intervention for these diseases depends on detailed understanding of the molecular mechanisms governing fibril assembly. AFM can recapitulate the fibril morphologies that have been shown by EM (Goldsbury *et al.*, 1999). For example, for amylin fibril, which is the protein component of the pancreatic amyloid deposits in type II diabetes, AFM imaging revealed a distinct 25-nm crossover repeat that is consistent with the observation from EM (Goldsbury *et al.*, 1999). More importantly, time-lapse AFM imaging in liquids has been used to continuously monitor the growth, directionality, and changes in the morphology of individual fibrils (see Figure 2.10B; Goldsbury *et al.*, 1999, 2001; Stolz *et al.*, 2000). This time-lapse AFM imaging in liquids showed that growth of the amylin and β-amyloid protofibrils was bidirectional (see Figure 2.10B; Goldsbury *et al.*, 1999; Blackley *et al.*, 2000). In addition, the outgrowth of protofibrils from a common amyloid core is also observed (Blackley *et al.*, 2000).

AFM has been used to look for the factors, possibly an abnormal metabolite, that can dramatically accelerate the amyloidogenesis, which might explain why wild-type protein only misfolds in a small percentage of human population (Zhang *et al.*, 2004). Using AFM, Zhang *et al.* reported that in the absence of ketoaldehyde or its aldol product Aβ amyloidogenesis was not observed whereas in their presence spherical assemblies appeared (Zhang *et al.*, 2004). On addition of fibrillar seeds, the spherical aggregates are rapidly converted into fibril, whereas without the ketoaldehyde, the process was much slower.

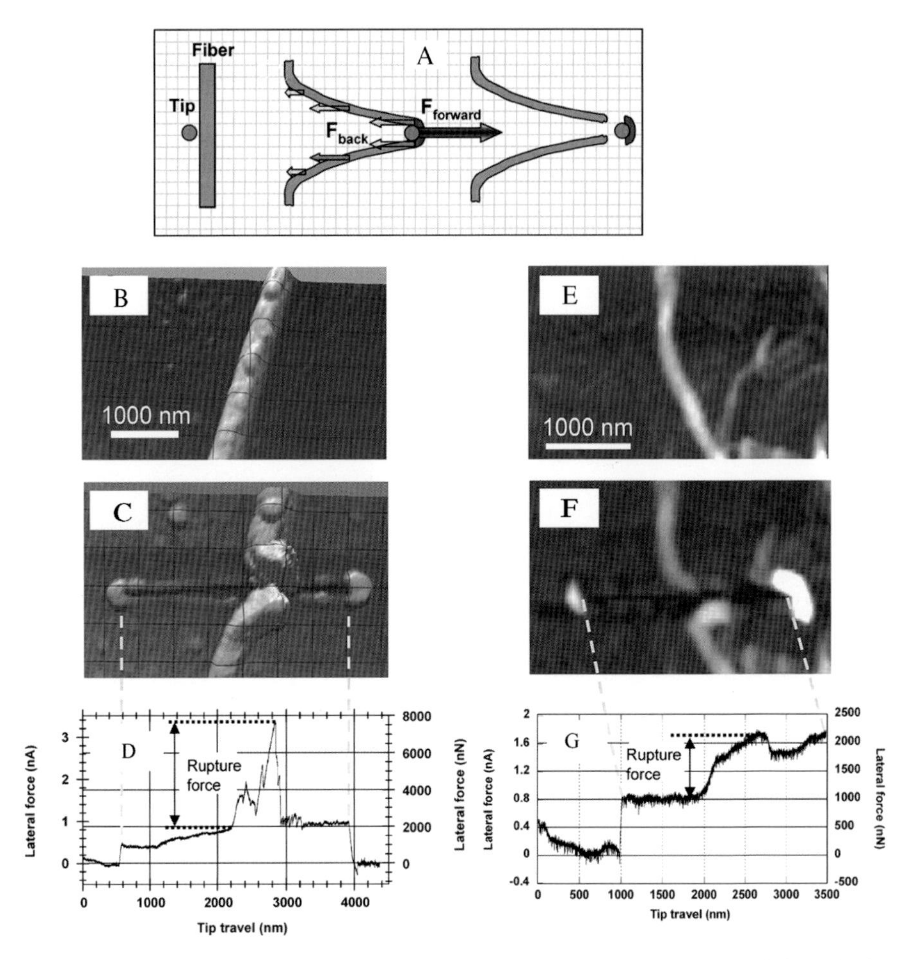

Figure 2.9. Lateral force manipulation of fibrin fiber. (A) Schematics of a fiber manipulation. The tip contacts the fiber, and stretches it until it ruptures. During manipulation, two main forces (a force pair) act on the fiber segment that ruptures (dark red): the backward distributed frictional force and the forward applied tip force. Those two forces balance each other and cause the fiber to deform and to eventually rupture. (B–G) atomic force microscopy (AFM) images of two fibrin fibers before and after being ruptured by the AFM tip, and corresponding lateral force versus tip travel during this manipulation. Dotted vertical lines between C and D, and F and G align scratched trace in images after manipulation with tip travel trace. Images and diagrams are reproduced from Guthold *et al.* (2004) with permission.

AFM has also been used in search for the fibril elongation inhibitors (Li *et al.*, 2004). For example, AFM has been used to monitor the dopamine-induced dissolution of single α-synuclein fibril immobilized on mica in aqueous solution (Li *et al.*, 2004). Dopamine is a key neuromuscular neurotransmitter. This study showed that in the presence of dopamine, the initial fibril was significantly disassembled in 1 h and is completely dissociated in 2 h (see Figure 2.10C). Even though dopamine is

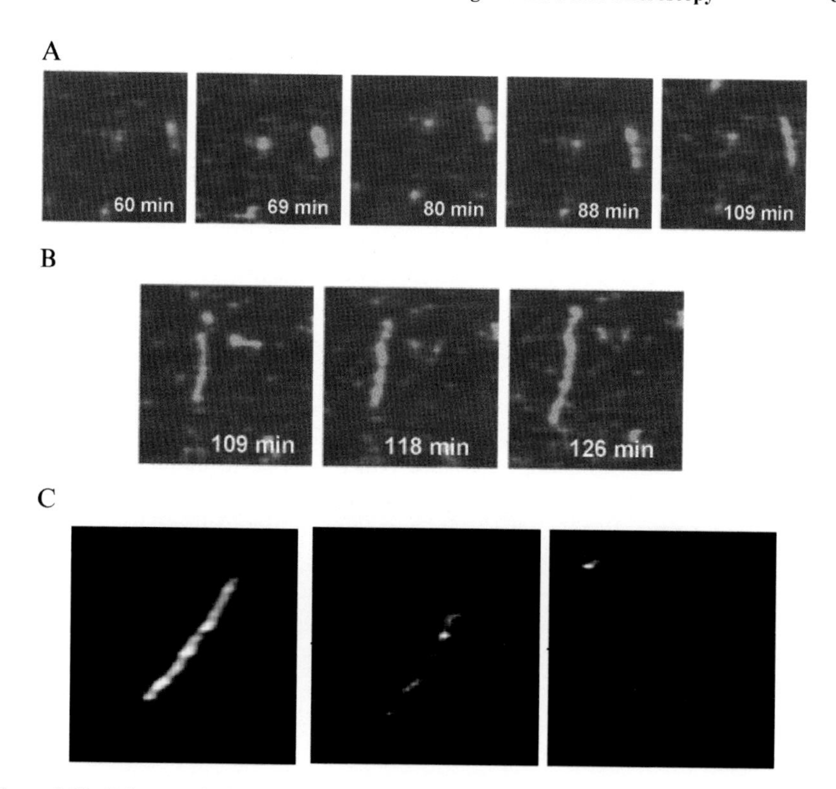

Figure 2.10. Using atomic force microscopy (AFM) to study the structure and assembly of amyloid fibrils. (A) The formation of protofibrils through single aggregate of β-amyloid and subsequent elongation by the addition of further aggregate units of β-amyloid. (B) The bidirectional elongation of a β-amyloid protofibril by the addition of aggregate units of β-amyloid. The images from (A) and (B) panels are reproduced from Blackley *et al.* (2000) with permission. (C) Disassembly of a single fibril of α-synuclein monitored by time-lapse AFM in liquids. Dopamine was added to a solution of preformed α-synuclein fibrils and an aliquot was deposited on mica. Panels from left to right show a specific fibril 0, 1, and 2 h after adding dopamine. The images are reproduced from Li *et al.* (2004) with permission.

not suitable for use as a drug, future AFM studies will open new possibilities for therapeutic intervention of disease caused by protein aggregations.

2.3.7. Following Biological Processes Using Imaging in Liquids

Applications of AFM imaging in liquids to study membrane proteins, fibril formation, and degradation have been mentioned in the earlier sections. This section discusses more applications of imaging in liquids. The biomolecular assemblies and interactions in biological pathways take place in a timed fashion in the cell. Time-lapse AFM imaging in liquids can be used to follow these processes under near-physiological conditions. For example, the DNA-directed synthesis of RNA by *E. coli* RNAP has been observed using AFM solution imaging (Argaman *et al.*, 1997). In addition, the degradation of DNA by the nuclease DNase I was followed by oscillating mode AFM in solution in the presence of nickel ions (Bezanilla *et al.*, 1994). Many quantitative dynamic properties of these processes can be obtained using time-lapse AFM in solution. For example, the rate of the diffusion of *E. coli* RNAP on the DNA was measured using time-lapse AFM. The rate was found to be 1.5 nucleotides per second, which is about three times slower than the speed in solution. This was expected because the surface could hamper the diffusion (Guthold *et al.*, 1999b). Finally, very detailed enzyme kinetics of phospholipase A_2 has been analyzed using AFM imaging in liquids (Balashev *et al.*, 2001).

In addition to the earlier quantitative assays, it is also possible to directly qualitatively correlate structural conformations and functional states of individual biomolecular assemblies using time-lapse AFM imaging (Stolz *et al.*, 2000). For example, the conformational change of nuclear pore complexes modulated by ATP, calcium, and carbon dioxide have been studied using time-lapse AFM (Rakowska *et al.*, 1998; Stoffler *et al.*, 1999; Oberleithner *et al.*, 2000). These studies disclosed that ATP and calcium induce the pore contraction and facilitate the transportation of macromolecules between the nucleus and the cytosol, whereas carbon dioxide induces pore collapse and functions to isolate the nucleus. Although time-lapse AFM has powerful capabilities, limitations on this dynamic approach also exist. First, the resolution limitation due to the noise in time-lapse imaging in liquids limits its use for studying small biomolecular assemblies. Second, limitations of scanning speed of commercial AFM instruments make it impossible to observe many fast biological processes (Ando *et al.*, 2001); however, this limitation is overcome with recent advances in instrumentation.

2.3.8. AFM as a Tool for Proteomics

The accelerating determination of genome sequences and their interpretation (genomics) has brought with it one of the most daunting challenges to modern biological sciences, that is, the concomitant determination of the structure, function, protein–protein interactions, and expression of all the corresponding proteins that are encoded by the genomic DNA. The rapid advance of the field of proteomics depends on traditional biochemical techniques and new innovations. Microarrays have been established as a standard for parallel screening of the nucleic acids profiles. Recently, exciting new technology based on AFM has emerged which has the potential to be used in high-throughput screening of protein–protein interactions (Lynch *et al.*, 2004). A new instrument called the NanoArrayer can mechanically mediate direct deposition of materials on surface with spot size several hundred nanometers to 2 μm in diameter. After adding the second protein, the detection of protein–protein interactions on the protein array can be carried out by fluorescence microscopy (in case the second protein added is fluorescent) or monitoring the height increase by AFM imaging.

2.4. CONCLUDING REMARKS

AFM is a versatile tool, which allows us to investigate the protein–protein interactions from different perspectives. In the past several years, there is an exciting trend of designing new diversified AFM instruments. Besides the Nanomanipulator and Nanoarrayer mentioned earlier, people have combined AFM with fluorescence microscope (Vickery and Dunn, 2001; Kassies *et al.*, 2005). These emerging new instruments will continue to expand our vision and reach into the realm of protein–protein complexes at the molecular level.

ACKNOWLEDGMENTS

We thank Drs Thomas A. Kunkel and Bennett Van Houten for helpful comments on the manuscript. This work was supported in part by National Institutes of Health grants GM54136 and the American Cancer Society (to D.A.E.).

REFERENCES

Abels, J. A., Moreno-Herrero, F., van der Heijden, T., Dekker, C., and Dekker, N. H. (2005). Single-molecule measurements of the persistence length of double-stranded RNA. *Biophys J* 88:2737–2744.

Abu-Lail, N. I. and Camesano, T. A. (2003). Polysaccharide properties probed with atomic force microscopy. *J Microsc* 212:217–238.

Allan, B. W., Garcia, R., Maegley, K., Mort, J., Wong, D., Lindstrom, W., Beechem, J. M., and Reich, N. O. (1999). DNA bending by EcoRI DNA methyltransferase accelerates base flipping but compromises specificity. *J Biol Chem* 274:19269–19275.

Allison, D. P., Kerper, P. S., Doktycz, M. J., Spain, J. A., Modrich, P., Larimer, F. W., Thundat, T., and Warmack, R. J. (1996). Direct atomic force microscope imaging of EcoRI endonuclease site specifically bound to plasmid DNA molecules. *Proc Natl Acad Sci USA* 93:8826–8829.

Ando, T., Kodera, N., Takai, E., Maruyama, D., Saito, K., and Toda, A. (2001). A high-speed atomic force microscope for studying biological macromolecules. *Proc Natl Acad Sci USA* 98:12468–12472.

Argaman, M., Golan, R., Thomson, N. H., and Hansma, H. G. (1997). Phase imaging of moving DNA molecules and DNA molecules replicated in the atomic force microscope. *Nucleic Acids Res* 25:4379–4384.

Balashev, K., Jensen, T. R., Kjaer, K. and Bjornholm, T. (2001). Novel methods for studying lipids and lipases and their mutual interaction at interfaces. Part I. Atomic force microscopy. *Biochimie* 83:387–397.

Bao, K. K., Wang, H., Miller, J. K., Erie, D. A., Skalka, A. M., and Wong, I. (2003). Functional oligomeric state of avian sarcoma virus integrase. *J Biol Chem* 278:1323–1327.

Bellacosa, A. (2001). Functional interactions and signaling properties of mammalian DNA mismatch repair proteins. *Cell Death Differ* 8:1076–1092.

Berge, T., Ellis, D. J., Dryden, D. T., Edwardson, J. M., and Henderson, R. M. (2000). Translocation-independent dimerization of the EcoKI endonuclease visualized by atomic force microscopy. *Biophys J* 79:479–484.

Bezanilla, M., Drake, B., Nudler, E., Kashlev, M., Hansma, P. K., and Hansma, H. G. (1994). Motion and enzymatic degradation of DNA in the atomic force microscope. *Biophys J* 67:2454–2459.

Binnig, G., Quate, C. F., and Gerber, C. (1986). Atomic force microscope. *Phys Rev Lett* 56:930–933.

Blackley, H. K., Sanders, G. H., Davies, M. C., Roberts, C. J., Tendler, S. J., and Wilkinson, M. J. (2000). In-situ atomic force microscopy study of beta-amyloid fibrillization. *J Mol Biol* 298:833–840.

Bockelmann, U. (2004). Single-molecule manipulation of nucleic acids. *Curr Opin Struct Biol* 14: 368–373.

Bonin, M., Oberstrass, J., Lukacs, N., Ewert, K., Oesterschulze, E., Kassing, R., and Nellen, W. (2000). Determination of preferential binding sites for anti-dsRNA antibodies on double-stranded RNA by scanning force microscopy. *RNA* 6:563–570.

Brewer, L. R., Friddle, R., Noy, A., Baldwin, E., Martin, S. S., Corzett, M., Balhorn, R., and Baskin, R. J. (2003). Packaging of single DNA molecules by the yeast mitochondrial protein Abf2p. *Biophys J* 85:2519–2524.

Bucior, I. and Burger, M. M. (2004). Carbohydrate–carbohydrate interactions in cell recognition. *Curr Opin Struct Biol* 14:631–637.

Buermeyer, A. B., Deschenes, S. M., Baker, S. M., and Liskay, R. M. (1999). Mammalian DNA mismatch repair. *Annu Rev Genet* 33:533–564.

Bustamante, C., Macosko, J. C., and Wuite, G. J. (2000). Grabbing the cat by the tail: manipulating molecules one by one. *Nat Rev Mol Cell Biol* 1:130–136.

Bustamante, C. and Rivetti, C. (1996). Visualizing protein-nucleic acid interactions on a large scale with the scanning force microscope. *Annu Rev Biophys Biomol Struct* 25:395–429.

Bustamante, C., Zuccheri, G., Leuba, S. H., Yang, G., and Samori, B. (1997). Visualization and analysis of chromatin by scanning force microscopy. *Methods* 12:73–83.

Carrion-Vazquez, M., Oberhauser, A. F., Fowler, S. B., Marszalek, P. E., Broedel, S. E., Clarke, J., and Fernandez, J. M. (1999). Mechanical and chemical unfolding of a single protein: a comparison. *Proc Natl Acad Sci USA* 96:3694–3699.

Cary, R. B., Peterson, S. R., Wang, J., Bear, D. G., Bradbury, E. M., and Chen, D. J. (1997). DNA looping by Ku and the DNA-dependent protein kinase. *Proc Natl Acad Sci USA* 94:4267–4272.

Ceci, P., Cellai, S., Falvo, E., Rivetti, C., Rossi, G. L., and Chiancone, E. (2004). DNA condensation and self-aggregation of *Escherichia coli* Dps are coupled phenomena related to the properties of the N-terminus. *Nucleic Acids Res* 32:5935–5944.

Chang, Y. C., Lo, Y. H., Lee, M. H., Leng, C. H., Hu, S. M., Chang, C. S., and Wang, T. F. (2005). Molecular visualization of the yeast Dmc1 protein ring and Dmc1-ssDNA nucleoprotein complex. *Biochemistry* 44:6052–6058.

Chen, L., Haushalter, K. A., Lieber, C. M., and Verdine, G. L. (2002). Direct visualization of a DNA glycosylase searching for damage. *Chem Biol* 9:345–350.

Chilkoti, A., Boland, T., Ratner, B. D., and Stayton, P. S. (1995). The relationship between ligand-binding thermodynamics and protein-ligand interaction forces measured by atomic force microscopy. *Biophys J* 69:2125–2130.

Clausen-Schaumann, H., Seitz, M., Krautbauer, R., and Gaub, H. E. (2000). Force spectroscopy with single bio-molecules. *Curr Opin Chem Biol* 4:524–530.

Colton, R. J., Engel, A., Frommer, J. E., Gaub, H. E., Gewirth, A. A., Guckenberger, R., Rabe, J., Heckl, W. M., and Parkinson, B. (1998). *Procedures in Scanning Probe Microscopies.* John Wiley & Sons, London.

Conroy, M. J., Jamieson, S. J., Blakey, D., Kaufmann, T., Engel, A., Fotiadis, D., Merrick, M., and Bullough, P. A. (2004). Electron and atomic force microscopy of the trimeric ammonium transporter AmtB. *EMBO Rep* 5:1153–1158.

Czajkowsky, D. M. and Shao, Z. (2003). Inhibition of protein adsorption to muscovite mica by monovalent cations. *J Microsc* 211:1–7.

Dame, R. T., Wyman, C., and Goosen, N. (2000). H-NS mediated compaction of DNA visualised by atomic force microscopy. *Nucleic Acids Res* 28:3504–3510.

Dammer, U., Popescu, O., Wagner, P., Anselmetti, D., Guntherodt, H. J., and Misevic, G. N. (1995). Binding strength between cell adhesion proteoglycans measured by atomic force microscopy. *Science* 267:1173–1175.

de Jager, M., van Noort, J., van Gent, D. C., Dekker, C., Kanaar, R., and Wyman, C. (2001). Human Rad50/Mre11 is a flexible complex that can tether DNA ends. *Mol Cell* 8:1129–1135.

de Jager, M., Wyman, C., van Gent, D. C., and Kanaar, R. (2002). DNA end-binding specificity of human Rad50/Mre11 is influenced by ATP. *Nucleic Acids Res* 30:4425–4431.

Dickerson, R. E. (1998). DNA bending: the prevalence of kinkiness and the virtues of normality. *Nucleic Acids Res* 26:1906–1926.

Drake, B., Prater, C. B., Weisenhorn, A. L., Gould, S. A., Albrecht, T. R., Quate, C. F., Cannell, D. S., Hansma, H. G., and Hansma, P. K. (1989). Imaging crystals, polymers, and processes in water with the atomic force microscope. *Science* 243:1586–1589.

Drotschmann, K., Hall, M. C., Shcherbakova, P. V., Wang, H., Erie, D. A., Brownewell, F. R., Kool, E. T., and Kunkel, T. A. (2002). DNA binding properties of the yeast Msh2-Msh6 and Mlh1–Pms1 heterodimers. *Biol Chem* 383:969–975.

Edwardson, J. M. and Henderson, R. M. (2004). Atomic force microscopy and drug discovery. *Drug Discov Today* 9:64–71.

Engel, A. and Muller, D. J. (2000). Observing single biomolecules at work with the atomic force microscope. *Nat Struct Biol* 7:715–718.

Erie, D. A., Yang, G., Schultz, H. C., and Bustamante, C. (1994). DNA bending by Cro protein in specific and nonspecific complexes: implications for protein site recognition and specificity. *Science* 266:1562–1566.

Fisher, T. E., Marszalek, P. E., and Fernandez, J. M. (2000). Stretching single molecules into novel conformations using the atomic force microscope. *Nat Struct Biol* 7:719–724.

Florin, E. L., Moy, V. T., and Gaub, H. E. (1994). Adhesion forces between individual ligand-receptor pairs. *Science* 264:415–417.

Fotiadis, D. and Engel, A. (2004). High-resolution imaging of bacteriorhodopsin by atomic force microscopy. *Methods Mol Biol* 242:291–303.

Friddle, R. W., Klare, J. E., Martin, S. S., Corzett, M., Balhorn, R., Baldwin, E. P., Baskin, R. J., and Noy, A. (2004). Mechanism of DNA compaction by yeast mitochondrial protein Abf2p. *Biophys J* 86:1632–1639.

Fritzsche, W., Takac, L., and Henderson, E. (1997). Application of atomic force microscopy to visualization of DNA, chromatin, and chromosomes. *Crit Rev Eukaryot Gene Expr* 7:231–240.

Gaczynska, M., Osmulski, P. A., Jiang, Y., Lee, J. K., Bermudez, V., and Hurwitz, J. (2004). Atomic force microscopic analysis of the binding of the *Schizosaccharomyces pombe* origin recognition complex and the spOrc4 protein with origin DNA. *Proc Natl Acad Sci USA* 101:17952–17957.

Garcia, R. A., Bustamante, C. J., and Reich, N. O. (1996). Sequence-specific recognition of cytosine C5 and adenine N6 DNA methyltransferases requires different deformations of DNA. *Proc Natl Acad Sci USA* 93:7618–7622.

Gettens, R. T., Bai, Z., and Gilbert, J. L. (2005). Quantification of the kinetics and thermodynamics of protein adsorption using atomic force microscopy. *J Biomed Mater Res A* 72:246–257.

Goldsbury, C., Aebi, U., and Frey, P. (2001). Visualizing the growth of Alzheimer's A beta amyloid-like fibrils. *Trends Mol Med* 7:582.

Goldsbury, C., Kistler, J., Aebi, U., Arvinte, T., and Cooper, G. J. (1999). Watching amyloid fibrils grow by time-lapse atomic force microscopy. *J Mol Biol* 285:33–39.

Guthold, M., Falvo, M., Matthews, W. G., Paulson, S., Mullin, J., Lord, S., Erie, D., Washburn, S., Superfine, R., Brooks, F. P. Jr., and Taylor, R. M. II (1999a). Investigation and modification of molecular structures with the nanoManipulator. *J Mol Graph Model* 17:187–197.

Guthold, M., Falvo, M., Matthews, W. G., Paulson, S., Mullin, J., Lord, S., Erie, D., Washburn, S., Superfine, R., Brooks, F. P., and Taylor, R. M. (2000). Investigation and modification of molecular structures with the nanoManipulator. *J Mol Graph Model* 17:187–197.

Guthold, M., Liu, W., Stephens, B., Lord, S. T., Hantgan, R. R., Erie, D. A., Taylor, R. M. Jr., and Superfine, R. (2004). Visualization and mechanical manipulations of individual fibrin fibers suggest that fiber cross section has fractal dimension 1.3. *Biophys J* 87:4226–4236.

Guthold, M., Zhu, X., Rivetti, C., Yang, G., Thomson, N. H., Kasas, S., Hansma, H. G., Smith, B., Hansma, P. K., and Bustamante, C. (1999b). Direct observation of one-dimensional diffusion and transcription by *Escherichia coli* RNA polymerase. *Biophys J* 77:2284–2294.

Gutsmann, T., Fantner, G. E., Kindt, J. H., Venturoni, M., Danielsen, S., and Hansma, P. K. (2004). Force spectroscopy of collagen fibers to investigate their mechanical properties and structural organization. *Biophys J* 86:3186–3193.

Hafner, J. H., Cheung, C. L., Woolley, A. T., and Lieber, C. M. (2001). Structural and functional imaging with carbon nanotube AFM probes. *Prog Biophys Mol Biol* 77:73–110.

Halford, S. E., Welsh, A. J., and Szczelkun, M. D. (2004). Enzyme-mediated DNA looping. *Annu Rev Biophys Biomol Struct* 33:1–24.

Hall, M. C., Shcherbakova, P. V., Fortune, J. M., Borchers, C. H., Dial, J. M., Tomer, K. B., and Kunkel, T. A. (2003). DNA binding by yeast Mlh1 and Pms1: implications for DNA mismatch repair. *Nucleic Acids Res* 31:2025–2034.

Hall, M. C., Wang, H., Erie, D. A., and Kunkel, T. A. (2001). High affinity cooperative DNA binding by the yeast Mlh1–Pms1 heterodimer. *J Mol Biol* 312:637–647.

Han, W., Lindsay, S. M., and Jing, T. (1996). A magnetically driven oscillating probe microscope for operation in fluids. *Appl Phys Lett* 69:4111–4114.

Hansma, H. G. (2001). Surface biology of DNA by atomic force microscopy. *Annu Rev Phys Chem* 52:71–92.

Hansma, H. G., Kasuya, K., and Oroudjev, E. (2004). Atomic force microscopy imaging and pulling of nucleic acids. *Curr Opin Struct Biol* 14:380–385.

Hansma, H. G. and Laney, D. E. (1996). DNA binding to mica correlates with cationic radius: assay by atomic force microscopy. *Biophys J* 70:1933–1939.

Hansma, P. K., Elings, V. B., Marti, O., and Bracker, C. E. (1988). Scanning tunneling microscopy and atomic force microscopy: application to biology and technology. *Science* 242:209–216.

Heddle, J. G., Mitelheiser, S., Maxwell, A., and Thomson, N. H. (2004). Nucleotide binding to DNA gyrase causes loss of DNA wrap. *J Mol Biol* 337:597–610.

Heinz, W. F. and Hoh, J. H. (1999). Spatially resolved force spectroscopy of biological surfaces using the atomic force microscope. *Trends Biotechnol* 17:143–150.

Henderson, R. M., Edwardson, J. M., Geisse, N. A., and Saslowsky, D. E. (2004). Lipid rafts: feeling is believing. *News Physiol Sci* 19:39–43.

Henn, A., Medalia, O., Shi, S. P., Steinberg, M., Franceschi, F., and Sagi, I. (2001). Visualization of unwinding activity of duplex RNA by DbpA, a DEAD box helicase, at single-molecule resolution by atomic force microscopy. *Proc Natl Acad Sci USA* 98:5007–5012.

Hoyt, P. R., Doktycz, M. J., Warmack, R. J., and Allison, D. P. (2001). Spin-column isolation of DNA-protein interactions from complex protein mixtures for AFM imaging. *Ultramicroscopy* 86:139–143.

Ikai, A. and Afrin, R. (2003). Toward mechanical manipulations of cell membranes and membrane proteins using an atomic force microscope: an invited review. *Cell Biochem Biophys* 39:257–277.

Janicijevic, A., Ristic, D., and Wyman, C. (2003a). The molecular machines of DNA repair: scanning force microscopy analysis of their architecture. *J Microsc* 212:264–272.

Janicijevic, A., Sugasawa, K., Shimizu, Y., Hanaoka, F., Wijgers, N., Djurica, M., Hoeijmakers, J. H., and Wyman, C. (2003b). DNA bending by the human damage recognition complex XPC-HR23B. *DNA Repair (Amst)* 2:325–336.

Jena, B. P. (2004). Discovery of the Porosome: revealing the molecular mechanism of secretion and membrane fusion in cells. *J Cell Mol Med* 8:1–21.

Jiao, Y., Cherny, D. I., Heim, G., Jovin, T. M., and Schaffer, T. E. (2001). Dynamic interactions of p53 with DNA in solution by time-lapse atomic force microscopy. *J Mol Biol* 314:233–243.

Kassies, R., van der Werf, K. O., Lenferink, A., Hunter, C. N., Olsen, J. D., Subramaniam, V., and Otto, C. (2005). Combined AFM and confocal fluorescence microscope for applications in bio-nanotechnology. *J Microsc* 217:109–116.

Kepert, J. F., Toth, K. F., Caudron, M., Mucke, N., Langowski, J., and Rippe, K. (2003). Conformation of reconstituted mononucleosomes and effect of linker histone H1 binding studied by scanning force microscopy. *Biophys J* 85:4012–4022.

Kolodner, R. D. and Marsischky, G. T. (1999). Eukaryotic DNA mismatch repair. *Curr Opin Genet Dev* 9:89–96.

Leckband, D. (2000). Measuring the forces that control protein interactions. *Annu Rev Biophys Biomol Struct* 29:1–26.

Lee, C. S. and Belfort, G. (1989). Changing activity of ribonuclease A during adsorption: a molecular explanation. *Proc Natl Acad Sci USA* 86:8392–8396.

Leuba, S. H., Bennink, M. L., and Zlatanova, J. (2004). Single-molecule analysis of chromatin. *Methods Enzymol* 376:73–105.

Li, J., Zhu, M., Manning-Bog, A. B., Di Monte, D. A., and Fink, A. L. (2004). Dopamine and L-dopa disaggregate amyloid fibrils: implications for Parkinson's and Alzheimer's disease. *Faseb J* 18: 962–964.

Liphardt, J., Onoa, B., Smith, S. B., Tinoco, I. J., and Bustamante, C. (2001). Reversible unfolding of single RNA molecules by mechanical force. *Science* 292:733–737.

Luckham, P. F. and Smith, K. (1998). Direct measurement of recognition forces between proteins and membrane receptors. *Faraday Discuss* 307–320; discussion 331–343.

Lushnikov, A. Y., Brown, B. A. II, Oussatcheva, E. A., Potaman, V. N., Sinden, R. R., and Lyubchenko, Y. L. (2004). Interaction of the Zalpha domain of human ADAR1 with a negatively supercoiled plasmid visualized by atomic force microscopy. *Nucleic Acids Res* 32:4704–4712.

Lynch, M., Mosher, C., Huff, J., Nettikadan, S., Johnson, J., and Henderson, E. (2004). Functional protein nanoarrays for biomarker profiling. *Proteomics* 4:1695–1702.

Lysetska, M., Knoll, A., Boehringer, D., Hey, T., Krauss, G., and Krausch, G. (2002). UV light-damaged DNA and its interaction with human replication protein A: an atomic force microscopy study. *Nucleic Acids Res* 30:2686–2691.

Lyubchenko, Y. L., Gall, A. A., Shlyakhtenko, L. S., Harrington, R. E., Jacobs, B. L., Oden, P. I., and Lindsay, S. M. (1992). Atomic force microscopy imaging of double stranded DNA and RNA. *J Biomol Struct Dyn* 10:589–606.

Lyubchenko, Y. L., Jacobs, B. L., Lindsay, S. M., and Stasiak, A. (1995). Atomic force microscopy of nucleoprotein complexes. *Scanning Microsc* 9:705–724; discussion 724–727.

Lyubchenko, Y. L., Oden, P. I., Lampner, D., Lindsay, S. M., and Dunker, K. A. (1993). Atomic force microscopy of DNA and bacteriophage in air, water and propanol: the role of adhesion forces. *Nucleic Acids Res* 21:1117–1123.

Marti, O., Elings, V., Haugan, M., Bracker, C. E., Schneir, J., Drake, B., Gould, S. A., Gurley, J., Hellemans, L., Shaw, K., *et al.* (1988). Scanning probe microscopy of biological samples and other surfaces. *J Microsc* 152 (Pt 3):803–809.

McGrew, D. A., and Knight, K. L. (2003). Molecular design and functional organization of the RecA protein. *Crit Rev Biochem Mol Biol* 38:385–432.

Modrich, P. and Lahue, R. (1996). Mismatch repair in replication fidelity, genetic recombination, and cancer biology. *Annu Rev Biochem* 65:101–133.

Moreno-Herrero, F., Colchero, J., Gomez-Herrero, J., and Baro, A. M. (2004). Atomic force microscopy contact, tapping, and jumping modes for imaging biological samples in liquids. *Phys Rev E Stat Nonlin Soft Matter Phys* 69:031915.

Mou, J., Czajkowsky, D. M., Zhang, Y., and Shao, Z. (1995). High-resolution atomic-force microscopy of DNA: the pitch of the double helix. *FEBS Lett* 371:279–282.

Muller, D. J., Amrein, M., and Engel, A. (1997). Adsorption of biological molecules to a solid support for scanning probe microscopy. *J Struct Biol* 119:172–188.

Muller, D. J., Dencher, N. A., Meier, T., Dimroth, P., Suda, K., Stahlberg, H., Engel, A., Seelert, H., and Matthey, U. (2001). ATP synthase: constrained stoichiometry of the transmembrane rotor. *FEBS Lett* 504:219–222.

Muller, D. J. and Engel, A. (1997). The height of biomolecules measured with the atomic force microscope depends on electrostatic interactions. *Biophys J* 73:1633–1644.

Muller, D. J., Fotiadis, D., and Engel, A. (1998). Mapping flexible protein domains at subnanometer resolution with the atomic force microscope. *FEBS Lett* 430:105–111.

Muller, D. J., Fotiadis, D., Scheuring, S., Muller, S. A., and Engel, A. (1999). Electrostatically balanced subnanometer imaging of biological specimens by atomic force microscope. *Biophys J* 76:1101–1111.

Mysiak, M. E., Bleijenberg, M. H., Wyman, C., Holthuizen, P. E., and van der Vliet, P. C. (2004a). Bending of adenovirus origin DNA by nuclear factor I as shown by scanning force microscopy is required for optimal DNA replication. *J Virol* 78:1928–1935.

Mysiak, M. E., Wyman, C., Holthuizen, P. E., and van der Vliet, P. C. (2004b). NFI and Oct-1 bend the Ad5 origin in the same direction leading to optimal DNA replication. *Nucleic Acids Res* 32:6218–6225.

Oberleithner, H., Schillers, H., Wilhelmi, M., Butzke, D., and Danker, T. (2000). Nuclear pores collapse in response to CO2 imaged with atomic force microscopy. *Pflugers Arch* 439:251–255.

Oberleithner, H., Schneider, S., and Bustamante, J. O. (1996). Atomic force microscopy visualizes ATP-dependent dissociation of multimeric TATA-binding protein before translocation into the cell nucleus. *Pflugers Arch* 432:839–844.

Ohnesorge, F. M., Horber, J. K., Haberle, W., Czerny, C. P., Smith, D. P., and Binnig, G. (1997). AFM review study on pox viruses and living cells. *Biophys J* 73:2183–2194.

Osmulski, P. A. and Gaczynska, M. (2000). Atomic force microscopy reveals two conformations of the 20 S proteasome from fission yeast. *J Biol Chem* 275:13171–13174.

Podesta, A., Imperadori, L., Colnaghi, W., Finzi, L., Milani, P., and Dunlap, D. (2004). Atomic force microscopy study of DNA deposited on poly L-ornithine-coated mica. *J Microsc* 215:236–240.

Radmacher, M., Tillamnn, R. W., Fritz, M., and Gaub, H. E. (1992). From molecules to cells: imaging soft samples with the atomic force microscope. *Science* 257:1900–1905.

Rakowska, A., Danker, T., Schneider, S. W., and Oberleithner, H. (1998). ATP-Induced shape change of nuclear pores visualized with the atomic force microscope. *J Membr Biol* 163:129–136.

Ratcliff, G. C. and Erie, D. A. (2001). A novel single-molecule study to determine protein–protein association constants. *J Am Chem Soc* 123:5632–5635.

Ratcliff, G. C., Erie, D. A., and Superfine, R. (1998). Photothermal modulation for oscillating mode atomic force microscopy in solution. *Appl Phys Lett* 72:1911–1913.

Rees, W. A., Keller, R. W., Vesenka, J. P., Yang, G., and Bustamante, C. (1993). Evidence of DNA bending in transcription complexes imaged by scanning force microscopy. *Science* 260: 1646–1649.

Rippe, K., Guthold, M., von Hippel, P. H., and Bustamante, C. (1997). Transcriptional activation via DNA-looping: visualization of intermediates in the activation pathway of *E. coli* RNA polymerase x sigma 54 holoenzyme by scanning force microscopy. *J Mol Biol* 270:125–138.

Rivetti, C., Codeluppi, S., Dieci, G., and Bustamante, C. (2003). Visualizing RNA extrusion and DNA wrapping in transcription elongation complexes of bacterial and eukaryotic RNA polymerases. *J Mol Biol* 326:1413–1426.

Rivetti, C., Guthold, M., and Bustamante, C. (1996). Scanning force microscopy of DNA deposited onto mica: equilibration versus kinetic trapping studied by statistical polymer chain analysis. *J Mol Biol* 264:919–932.

Rivetti, C., Guthold, M., and Bustamante, C. (1999). Wrapping of DNA around the *E. coli* RNA polymerase open promoter complex. *EMBO J* 18:4464–4475.

Rogers, B., Manning, L., Sulchek, T., and Adams, J. D. (2004). Improving tapping mode atomic force microscopy with piezoelectric cantilevers. *Ultramicroscopy* 100:267–276.

Ros, R., Eckel, R., Bartels, F., Sischka, A., Baumgarth, B., Wilking, S. D., Puhler, A., Sewald, N., Becker, A., and Anselmetti, D. (2004). Single molecule force spectroscopy on ligand-DNA complexes: from molecular binding mechanisms to biosensor applications. *J Biotechnol* 112:5–12.

Rounsevell, R., Forman, J. R., and Clarke, J. (2004). Atomic force microscopy: mechanical unfolding of proteins. *Methods* 34:100–111.

Safinya, C. R. (2001). Structures of lipid-DNA complexes: supramolecular assembly and gene delivery. *Curr Opin Struct Biol* 11:440–448.

Sato, M. H., Ura, K., Hohmura, K. I., Tokumasu, F., Yoshimura, S. H., Hanaoka, F., and Takeyasu, K. (1999). Atomic force microscopy sees nucleosome positioning and histone H1-induced compaction in reconstituted chromatin. *FEBS Lett* 452:267–271.

Sattin, B. D. and Goh, M. C. (2004). Direct observation of the assembly of RecA/DNA complexes by atomic force microscopy. *Biophys J* 87:3430–3436.

Schlacher, K., Leslie, K., Wyman, C., Woodgate, R., Cox, M. M., and Goodman, M. F. (2005). DNA polymerase V and RecA protein, a minimal mutasome. *Mol Cell* 17:561–572.

Schneider, S. W., Larmer, J., Henderson, R. M., and Oberleithner, H. (1998). Molecular weights of individual proteins correlate with molecular volumes measured by atomic force microscopy. *Pflugers Arch* 435:362–367.

Seitz, E. M., Brockman, J. P., Sandler, S. J., Clark, A. J., and Kowalczykowski, S. C. (1998). RadA protein is an archaeal RecA protein homolog that catalyzes DNA strand exchange. *Genes Dev* 12:1248–1253.

Shao, Z., Shi, D., and Somlyo, A. V. (2000). Cryoatomic force microscopy of filamentous actin. *Biophys J* 78:950–958.

Shao, Z. and Zhang, Y. (1996). Biological cryo atomic force microscopy: a brief review. *Ultramicroscopy* 66:141–152.

Sheng, S., Gao, Y., Khromov, A. S., Somlyo, A. V., Somlyo, A. P., and Shao, Z. (2003). Cryo-atomic force microscopy of unphosphorylated and thiophosphorylated single smooth muscle myosin molecules. *J Biol Chem* 278:39892–39896.

Sheng, S. and Shao, Z. (2002). Cryo-atomic force microscopy. *Methods Cell Biol* 68:243–256.

Shlyakhtenko, L. S., Gall, A. A., Filonov, A., Cerovac, Z., Lushnikov, A., and Lyubchenko, Y. L. (2003). Silatrane-based surface chemistry for immobilization of DNA, protein-DNA complexes and other biological materials. *Ultramicroscopy* 97:279–287.

Sincell, M. (2000). NanoManipulator lets chemists go Mano to Nano with molecules. *Science* 290: 1530.

Smith, G. C., Cary, R. B., Lakin, N. D., Hann, B. C., Teo, S. H., Chen, D. J., and Jackson, S. P. (1999). Purification and DNA binding properties of the ataxia-telangiectasia gene product ATM. *Proc Natl Acad Sci USA* 96:11134–11139.

Stahlberg, H., Fotiadis, D., Scheuring, S., Remigy, H., Braun, T., Mitsuoka, K., Fujiyoshi, Y., and Engel, A. (2001). Two-dimensional crystals: a powerful approach to assess structure, function and dynamics of membrane proteins. *FEBS Lett* 504:166–172.

Stevens, R. M. D., Frederick, N. A., Smith, B. L., Morse, D. E., Stucky, G. D., and Hansma, P. K. (2000). Carbon nanotubes as probes for atomic force microscopy. *Nanotechnology* 11:1–5.

Stoffler, D., Goldie, K. N., Feja, B., and Aebi, U. (1999). Calcium-mediated structural changes of native nuclear pore complexes monitored by time-lapse atomic force microscopy. *J Mol Biol* 287:741–752.

Stolz, M., Stoffler, D., Aebi, U., and Goldsbury, C. (2000). Monitoring biomolecular interactions by time-lapse atomic force microscopy. *J Struct Biol* 131:171–180.

Tamayo, J. (2003a). Structure of human chromosomes studied by atomic force microscopy. *J Struct Biol* 141:198–207.

Tamayo, J. (2003b). Structure of human chromosomes studied by atomic force microscopy. Part II. Relationship between structure and cytogenetic bands. *J Struct Biol* 141:189–197.

Tessmer, I., Moore, T., Lloyd, R. G., Wilson, A., Erie, D. A., Allen, S., and Tendler, S. J. (2005). AFM studies on the role of the protein RdgC in bacterial DNA recombination. *J Mol Biol* 350:254–262.

Tokunaga, M., Aoki, T., Hiroshima, M., Kitamura, K., and Yanagida, T. (1997). Subpiconewton intermolecular force microscopy. *Biochem Biophys Res Commun* 231:566–569.

Umemura, K., Komatsu, J., Uchihashi, T., Choi, N., Ikawa, S., Nishinaka, T., Shibata, T., Nakayama, Y., Katsura, S., Mizuno, A., Tokumoto, H., Ishikawa, M., and Kuroda, R. (2001). Atomic force microscopy of RecA–DNA complexes using a carbon nanotube tip. *Biochem Biophys Res Commun* 281:390–395.

van Noort, J., Orsini, F., Eker, A., Wyman, C., de Grooth, B., and Greve, J. (1999). DNA bending by photolyase in specific and non-specific complexes studied by atomic force microscopy. *Nucleic Acids Res* 27:3875–3880.

Verhoeven, E. E., Wyman, C., Moolenaar, G. F., and Goosen, N. (2002). The presence of two UvrB subunits in the UvrAB complex ensures damage detection in both DNA strands. *EMBO J* 21:4196–4205.

Verhoeven, E. E., Wyman, C., Moolenaar, G. F., Hoeijmakers, J. H., and Goosen, N. (2001). Architecture of nucleotide excision repair complexes: DNA is wrapped by UvrB before and after damage recognition. *EMBO J* 20:601–611.

Viani, M. B., Schaffer, T. E., Chand, A., Rief, M., Gaub, H. E., and Hansma, P. K. (1999). Small cantilevers for force spectroscopy of single molecules. *J Appl Phys* 86:2258–2262.

Vickery, S. A. and Dunn, R. C. (2001). Combining AFM and FRET for high resolution fluorescence microscopy. *J Microsc* 202:408–412.

Wagner, P. (1998). Immobilization strategies for biological scanning probe microscopy. *FEBS Lett* 430:112–115.

Wang, H. (2003). Atomic force microscopy studies of initiation events in DNA mismatch repair: In *Matls Sci*, University of North Carolina at Chapel Hill, Chapel Hill.

Wang, H., Yang, Y., Schofield, M. J., Du, C., Fridman, Y., Lee, S. D., Larson, E. D., Drummond, J. T., Alani, E., Hsieh, P., and Erie, D. A. (2003). DNA bending and unbending by MutS govern mismatch recognition and specificity. *Proc Natl Acad Sci USA* 100:14822–14827.

Willemsen, O. H., Snel, M. M., Cambi, A., Greve, J., De Grooth, B. G., and Figdor, C. G. (2000). Biomolecular interactions measured by atomic force microscopy. *Biophys J* 79:3267–3281.

Wolberg, A. S., Stafford, D. W., and Erie, D. A. (1997). Human factor IX binds to specific sites on the collagenous domain of collagen IV. *J Biol Chem* 272:16717–16720.

Woolley, A. T., Cheung, C. L., Hafner, J. H., and Lieber, C. M. (2000). Structural biology with carbon nanotube AFM probes. *Chem Biol* 7:193–204.

Wyman, C., Rombel, I., North, A. K., Bustamante, C., and Kustu, S. (1997). Unusual oligomerization required for activity of NtrC, a bacterial enhancer-binding protein. *Science* 275:1658–1661.

Xu, H. and Hoover, T. R. (2001). Transcriptional regulation at a distance in bacteria. *Curr Opin Microbiol* 4:138–144.

Xue, Y., Ratcliff, G. C., Wang, H., Davis-Searles, P. R., Gray, M. D., Erie, D. A., and Redinbo, M. R. (2002). A minimal exonuclease domain of WRN forms a hexamer on DNA and possesses both 3′– 5′ exonuclease and 5′-protruding strand endonuclease activities. *Biochemistry* 41:2901–2912.

Yaneva, M., Kowalewski, T., and Lieber, M. R. (1997). Interaction of DNA-dependent protein kinase with DNA and with Ku: biochemical and atomic-force microscopy studies. *EMBO J* 16:5098–5112.

Yang, Y., Sass, L. E., Du, C., Hsieh, P., and Erie, D. A. (2005). Determination of protein-DNA binding constants and specificities from statistical analyses of single molecules: MutS-DNA interactions. *Nucleic Acids Res* 33:4322–4334.

Yang, Y., Wang, H., and Erie, D. A. (2003). Quantitative characterization of biomolecular assemblies and interactions using atomic force microscopy. *Methods* 29:175–187.

Yin, H., Wang, M. D., Svoboda, K., Landick, R., Block, S. M., and Gelles, J. (1995). Transcription against an applied force. *Science* 270:1653–1657.

Yoshimura, S. H., Maruyama, H., Ishikawa, F., Ohki, R., and Takeyasu, K. (2004). Molecular mechanisms of DNA end-loop formation by TRF2. *Genes Cells* 9:205–218.

Zelphati, O., Liang, X., Nguyen, C., Barlow, S., Sheng, S., Shao, Z., and Felgner, P. L. (2000). PNA-dependent gene chemistry: stable coupling of peptides and oligonucleotides to plasmid DNA. *Biotechniques* 28:304–310, 312–314, 316.

Zhang, Q., Powers, E. T., Nieva, J., Huff, M. E., Dendle, M. A., Bieschke, J., Glabe, C. G., Eschenmoser, A., Wentworth, P., Jr., Lerner, R. A., and Kelly, J. W. (2004). Metabolite-initiated protein misfolding may trigger Alzheimer's disease. *Proc Natl Acad Sci USA* 101:4752–4257.

Zhuang, X. and Rief, M. (2003). Single-molecule folding. *Curr Opin Struct Biol* 13:88–97.

Zlatanova, J. and Leuba, S. H. (2003). Chromatin fibers, one-at-a-time. *J Mol Biol* 331:1–19.

Zlatanova, J., Lindsay, S. M., and Leuba, S. H. (2000). Single molecule force spectroscopy in biology using the atomic force microscope. *Prog Biophys Mol Biol* 74:37–61.

3

Combined Solid-Phase Detection Techniques for Dissecting Multiprotein Interactions on Membranes

Jacob Piehler

3.1. INTRODUCTION

After the completion of numerous genome-sequencing projects, analytical tools for characterizing protein functions are increasingly in the demand. One key challenge is the mechanistic dissection of the assembling of multiprotein complexes, which are involved in numerous cellular functions. Among these, protein–protein interactions at membranes have been particularly elusive, because binding to the membrane and simultaneous interaction in plane of the membrane are extremely difficult to deconvolute, requiring both biochemically and spectroscopically highly demanding techniques. Very diverse approaches for measuring protein–protein interactions are established, among which solid-phase techniques have tremendously gained importance during recent years. Monitoring protein interactions on surfaces has particular advantages compared with the techniques detecting them in solution:

(i) Sample handling is much more versatile in solid-phase assay formats, allowing for exchanging the sample in a flow-through system within a few hundred milliseconds. Thus, the kinetics of association and dissociation can be monitored up to rate constant of several seconds. For the

J. PIEHLER • Institute of Biochemistry, Biocenter N210, Johann Wolfgang Goethe-University, Max-von-Laue-Straße 9, 60438 Frankfurt am Main, Germany. Tel: +49 (0)69 79829468; fax: +49 (0)69 798294695; e-mail: j.piehler@em.uni-frankfurt.de

analysis of cooperative interactions in multiprotein complex formation, the possibility for complex multistep injection schemes is particularly important and powerful. In case of protein interaction on membranes, lipid layers assembling and reconstitution of proteins can be carried out in a stepwise fashion using suitable injection sequences. Furthermore, flow-through sample handling is readily automated, thus increasing the experimental throughput and reproducibility.

(ii) Protein interactions at membrane are *per se* interactions at interfaces. Solid-supported lipid bilayers (Brian and McConnell, 1984) and polymer-supported lipid bilayers (Sackmann, 1996; Sackmann and Tanaka, 2000; Sinner and Knoll, 2001; Tanaka and Sackmann, 2005) provide powerful means for reconstituting membrane proteins on solid supports under well-defined conditions.

(iii) Most importantly, numerous spectroscopic techniques are available for selectively probing protein interactions on surfaces with high sensitivity. Next to fluorescence-based detection, surface-sensitive techniques, which detect intrinsic properties of biomolecules such as mass or polarizability, are more promising. These label-free techniques and their combination with surface-sensitive fluorescence detection are discussed in more detail in the following sections. The application of simultaneous label-free and fluorescence detection for dissecting multiprotein complex formation on membranes is also presented.

3.2. LABEL-FREE DETECTION TECHNIQUES

3.2.1. Overview

In the past 15 years, numerous techniques have been established which enable to detect biomolecular interaction on surfaces without involving additional labels. Both optical and acoustic interrogation approaches have proved capable for monitoring mass deposition of a few pg/mm2 in real time (Cooper, 2003; Homola, 2003). Acoustic techniques such as the quartz crystal microbalance (QCM) or surface acoustic wave (SAW) directly sense the mass of the adsorbed proteins, which typically include the water shell bound to the biomolecules (Marx, 2003). Furthermore, the changes in the viscoelastic properties of the attached layers can be detected by monitoring the heat dissipation (QCM-D) (Hook *et al.*, 2001). QCM-D is not ideal for absolutely quantifying adsorbed mass on surfaces, but provides a qualitative readout for interfacial processes other than mass adsorption. In combination with optical techniques such as surface plasmon resonance (SPR) (see later), QCM-D has been used to study the changes in morphology and hydration of surface layers, enabling detecting and monitoring conformational changes during ligand binding (Reimhult *et al.*, 2004).

Label-free optical detection techniques (Haake *et al.*, 2000; Gauglitz, 2005) are based on the fact that the refractive index of biomolecules is higher than that of water.

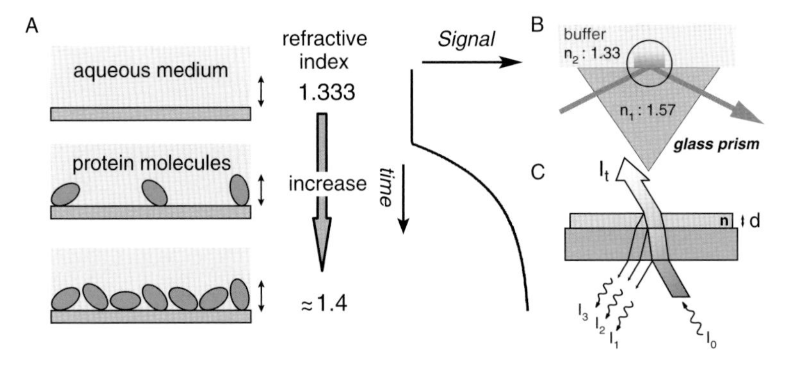

Figure 3.1. Principles of optical label-free detection. (A) On binding of proteins, the refractive index adjacent to the surface is increased, which can be detected in real time by directional reflection or evanescent field interrogation. (B) Changes in refractive index can be probed in a surface-sensitive manner by the evanescent field, which penetrates several hundred nanometers into the superstrate on total internal reflection of a light beam. (C) Directional reflection at transparent layer systems splits the light beam into reflected and transmitted light beams. Phase and amplitude of these light beams carry information about the refractive indices and the thickness of the layer system.

On protein binding to a surface, the refractive index of the interfacial layer is continuously increasing because of the replacement of water (Figure 3.1A). These changes in refractive index can be probed in a surface-sensitive manner either by evanescent field interrogation (Figure 3.1B) or by directional reflection (Figure 3.1C).

3.2.2. Evanescent Field Techniques

An evanescent field is generated on total internal reflection of a light beam at the interface with decreasing refractive index. Owing to the exponential decay of the evanescent field with a penetration depth of a few hundred nanometers, light propagation is affected by changes in the refractive index in the vicinity of the surface. Several approaches to monitor these changes in real time have been implemented. Among these, SPR has proved to be particularly powerful for the detection of biomolecular interaction, and numerous implements of this technique have been described (Lofas *et al.*, 1991; Harris *et al.*, 1999; Malmqvist, 1999; Haes and Van Duyne, 2002). The principle and the application of SPR for dissecting multiprotein complexes are elaborated in detail in Chapter 4, and therefore is not discussed further in this chapter. In contrast to classic SPR detection, the more recently established technique of coupled plasmon-waveguide resonance (CPWR) spectroscopy provides more information on the optical properties of the surface bound layers, as it employs parallel and perpendicular polarized light for plasmon excitation (Salamon *et al.*, 1997). This can be used for characterizing the properties of the attached layer more precisely. Thus, not only binding to the surface, but also conformational changes or phase segregation during interaction with membrane proteins can be studied (Tollin *et al.*, 2003; Salamon *et al.*, 2005).

Other important evanescent field techniques are based on grating couplers (GC) (Lukosz *et al.*, 1991) or on frustrated total internal reflection (resonant mirror, RM) (Edwards *et al.*, 1995). In contrast to the SPR transducer, which is based on a thin gold layer, transducers for GC and RM employ layers of transparent metal oxide layers. Both GC and RM detect the angle required for coupling of a light beam into a waveguide, which depends on the refractive index on the surface of the waveguide. Although these techniques have not proved competitive to SPR in the field of label-free detection, in principle, they offer very elegant means for combination with total internal reflection fluorescence spectroscopy (TIRFS), because their nonmetallic substrates are well compatible with fluorescence detection. So far, however, this possibility has not been exploited.

3.2.3. Techniques Based on Directional Reflection

Optical approaches based on directional reflection have also been used successfully for detecting protein binding to surfaces in real time. Among these, the classic technique is ellipsometry, which detects the changes in phase and amplitude of polarized light on reflection at a surface. Thus, both the refractive index and the thickness of a layer can be monitored, providing detailed information about surface-bound protein layers. However, the experimental setup for ellipsometry is rather elaborate, and the combination of this technique with a flow-through system and the evaluation of data are not straightforward. In contrast, a related technique based on reflectance interference (RIf) detection is much more readily implemented (Brecht *et al.*, 1993; Schmitt *et al.*, 1997; Gauglitz, 2005). This method detects the change in optical thickness of a thin layer by the shift of the interference spectrum (Figure 3.2). Spectral RIf detection (RIfS) has proven rugged and powerful for label-free detection of protein–protein interactions on surfaces and solid-supported membranes (Piehler and Schreiber, 2001; Lamken *et al.*, 2004; Lata and Piehler, 2005). Detection limits of a few pg/mm^2 have been reached by this technique (Piehler *et al.*, 1996; Hanel and Gauglitz, 2002). RIfS selectively probes the layer of attached proteins, and is, therefore, much less sensitive to background changes in refractive index, which is also probed by evanescent field techniques. Thus, referencing of background signals is not required. Recently, its potential for parallel detection has been demonstrated (Birkert and Gauglitz, 2002; Kroger *et al.*, 2002).

3.3. SIMULTANEOUS LABEL-FREE AND FLUORESCENCE DETECTION

Label-free detection techniques are powerful tools for monitoring and absolutely quantifying surface mass deposition in real time with detection limits of a few pg/mm^2. Owing to their mass sensitivity, however, label-free detection techniques can neither discriminate between different types of molecules, nor provide direct information on interactions in plane of the transducer surface. Furthermore, the

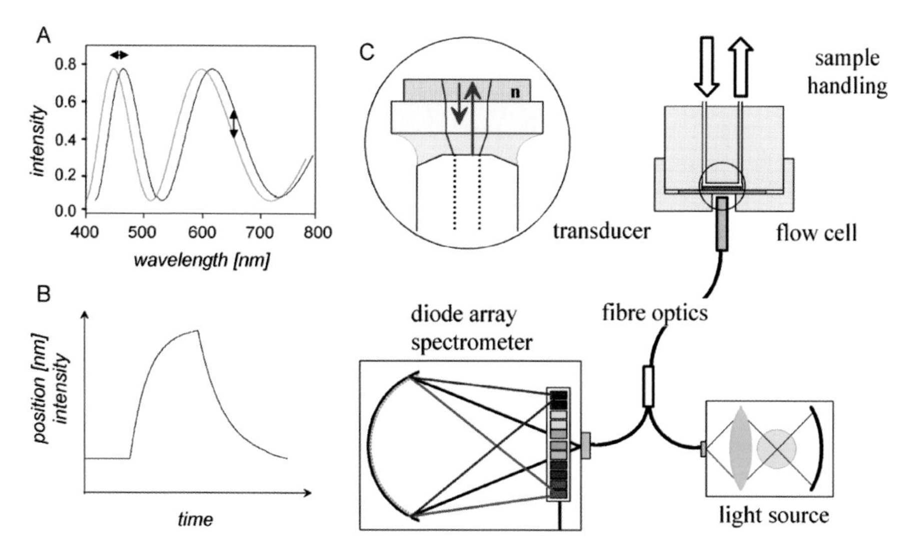

Figure 3.2. Principle of RIf. (A) White light reflected at a thin transparent layer gives a spectral interference pattern, which is shifted on protein binding to the surface. (B) Monitoring the shift of the interference pattern or the intensity in the inflection point (indicated by the arrows in panel A) with time yields a binding curve. (C) Typical setup for spectral RIf. White light from a tungsten lamp is guided perpendicular onto the transducer, and the reflected light is collected into the same fiber, and spectrally detected in a diode array spectrometer.

generic nature of label-free detection implicates high sensitivity to background signals such as bulk refractive index changes and nonspecific binding. In contrast, fluorescence detection is very selective and sensitive, and provides the possibility of discriminating multiple species within a complex by using different fluorescent dyes. Furthermore, fluorescence quenching and resonance energy transfer enable for selectively probing individual interactions within complexes. Surface-sensitive fluorescence detection by TIRFS is a powerful tool for monitoring ligand interactions at surfaces and solid-supported membranes (Axelrod *et al.*, 1984; Thompson *et al.*, 1993, 1997; Thompson and Lagerholm, 1997; Schmid *et al.*, 1998). Here, the evanescent field is used for selectively exciting fluorescence at the transducer surface. For their complementary properties, combination of TIRFS with label-free, mass-sensitive detection appears highly attractive. Combination of TIRFS with SPR (surface plasmon field enhanced fluorescence spectroscopy) has been shown to be a powerful tool for characterizing processes at interfaces (Liebermann and Knoll, 2000; Neumann *et al.*, 2002). This technique, however, uses the same light source for fluorescence excitation as for SPR detection, thus limiting the flexibility of each technique. The thin noble metal layers required for SPR are very useful for surface modification by the self-assembling of alkyl thiols, but can be problematic for fluorescence detection, because they promote radiationless deactivation by energy transfer in a surface distance-dependent manner.

Recently, a novel combination of TIRFS with reflectance interferometry at a thin silica layer was described for studying lateral interactions at supported lipid bilayers (Gavutis *et al.*, 2005). To spectroscopically separate the two detection techniques, single wavelength RIf detection in the NIR was implemented. Thus, the full visible range can be used for fluorescence detection. The experimental setup is depicted in Figure 3.3. Fluorescence excitation and emission were kept independent of RIf illumination by implementing monochromatic RIf detection in the near-IR region (Figure 3.3C). Thus, a simple and rugged setup for simultaneous mass-sensitive and fluorescence detection was implemented without compromising the flexibility of either technique. Complete spectral separation of the two techniques is more important as high-power illumination for optimum RIf detection could be applied without photobleaching the fluorophores absorbed in the visible region. Fluorescence excitation is possible with different excitation sources and different excitation powers without compromising RIf detection. Vice versa, the performance of the TIRFS setup is completely independent of RIf illumination.

Combination of label-free detection with TIRFS adds several important features: (i) fluorescence labeling of several components can be avoided, reducing signal cross talk and possible effects on protein function; (ii) surface concentrations are directly quantified and their changes are monitored, which is extremely important in case of sensitive multicomponent surface architectures; and (iii) straightforward calibration of fluorescence signals with respect to surface coverage. The RIf transducer element—a silica layer on top of a glass substrate—is fully transparent

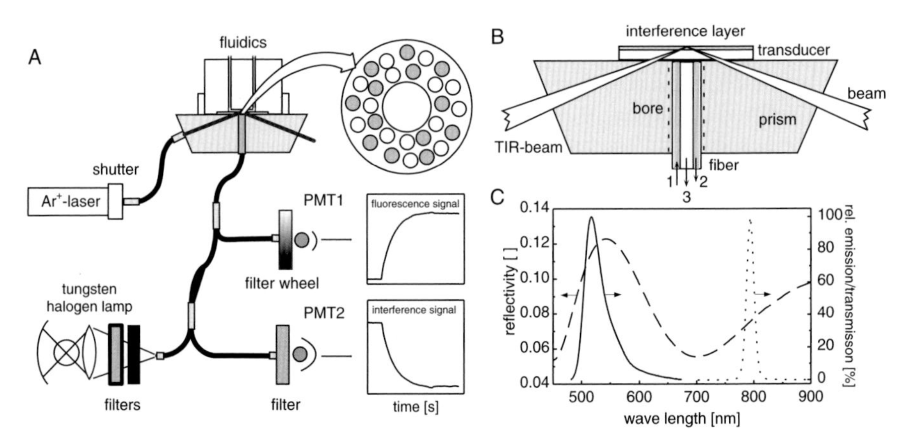

Figure 3.3. Simultaneous TIRFS–RIf detection. (A) Schematic of the setup, the cross section of the fiber at the interface to the transducer is shown in the inset. (B) Enlarged view of the coupling of the light beam for fluorescence excitation into the RIf transducer, and the fibers for RIf illumination (1), RIf detection (2), and fluorescence detection (3). (C) Spectral separation of RIf and TIRFS detection: Reflectivity of the RIf transducer at perpendicular illumination (– – –) and transmission of the interference filter used for RIf detection (……), in comparison with the fluorescence emission spectrum of AF488 (——). Reproduced from Gavutis *et al.* (2005) with permission.

and does not quench surface-proximal fluorophores, unlike noble metal surfaces required for detection by SPR, and therefore is ideally compatible with TIRFS.

Simultaneous TIRFS-RIf detection was applied for dissecting ligand-induced cross-linking of the type-I interferon (IFN) receptor *in vitro* (Lamken *et al.*, 2004). The extracellular domains of the receptor subunits ifnar1 (ifnar1-EC) and ifnar2 (ifnar2-EC) were tethered onto solid-supported, fluid lipid bilayers to mimic the anchoring of the receptor subunits on the plasma membrane (Figure 3.4). For stable anchoring of the proteins to the membrane, a lipid-like molecule carrying a *bis*-NTA headgroup (Lata *et al.*, 2006a) was incorporated into the membrane (Figure 3.4A and B). Binding of the ligand IFNα2 and various mutants with different affinities toward ifnar1-EC and ifnar2-EC was studied using site-specific fluorescence labeling with Oregon Green 488 (OG488) and Alexa Fluor 488 (AF488) dyes. A typical binding assay including assembling of the lipid bilayer, tethering of ifnar2-EC, and the interaction with the fluorescence-labeled ligand IFNα2 is shown in Figure 3.4C. The advantages of simultaneous detection are

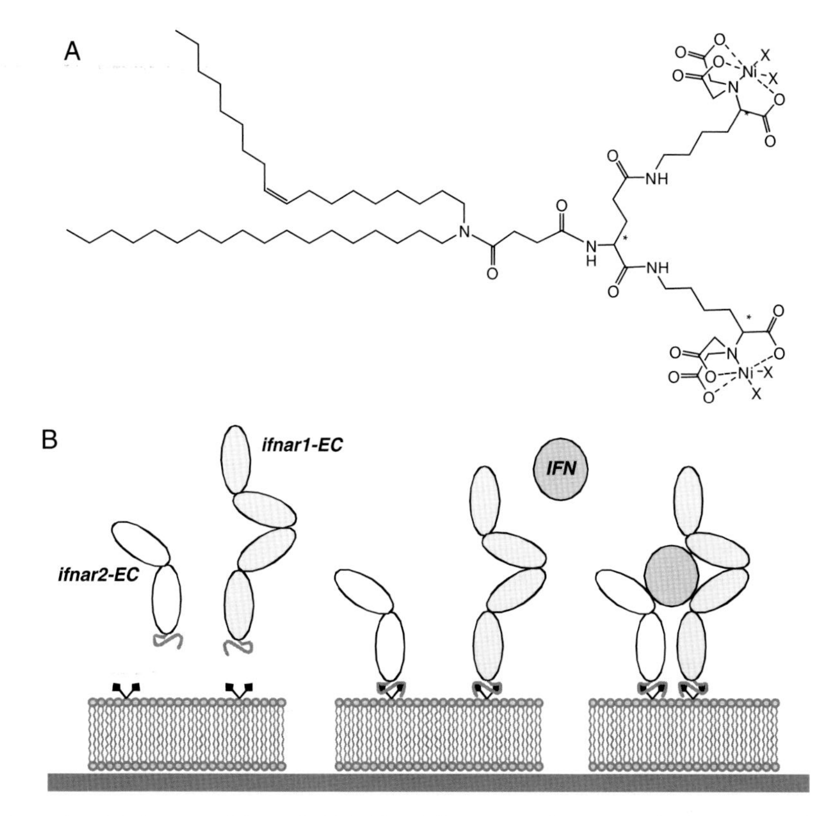

Figure 3.4. (A) Structure of the *bis*-NTA lipid used for tethering the extracellular receptor domains on supported lipid bilayer in a stable, yet reversible manner (B).

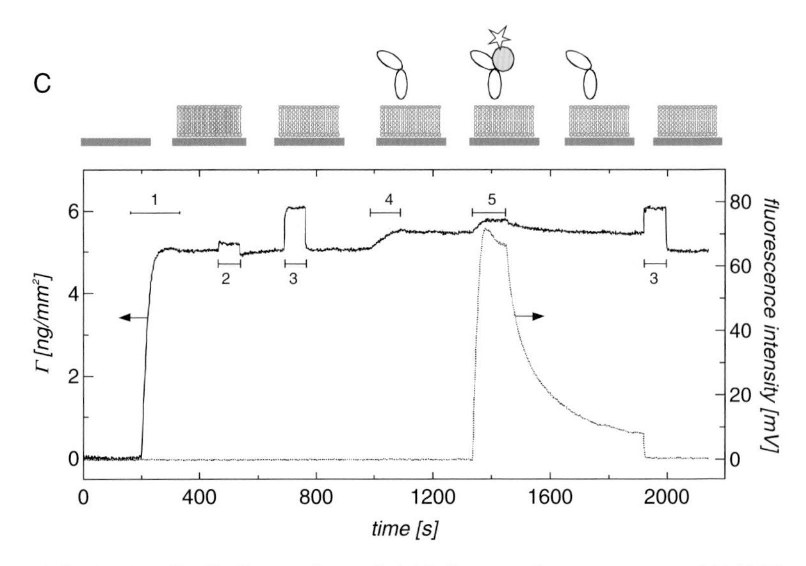

Figure 3.4. (*Continued*) (C) Course of a typical binding experiment on supported lipid bilayers: Injection of (1) SOPC SUVs, (2) 10 mM nickel(II)chloride, (3) 200 mM imidazole, (4) 300 nM ifnar2-H10, and (5) 100 nM ^{AF488}IFNα2. Reproduced from Gavutis *et al.* (2005) with permission.

well demonstrated by this experiment: All steps of membrane assembling and protein tethering are monitored and controlled by RIf detection in a quantitative manner, whereas binding of the ligand is detected with a much higher signal-to-noise ratio by TIRFS. However, simultaneous mass-sensitive and fluorescence detection provides powerful means for dissecting interactions involved in multi-protein complex formation, some of which is discussed in more detail in the later sections.

3.3.1. Cooperativity in Multiprotein Complexes

One important feature of multiprotein complex formation is the cooperative interaction between several partners (see also Chapter 4). In case of simultaneous interaction of the ligand IFNα2 with the two receptor subunits ifnar1 and ifnar2, cooperative interaction by contact between the receptor subunits or conformational changes of the ligand is very likely, and has been observed for several other members of the cytokine receptor family. A typical ''sandwich'' assay for identifying and characterizing cooperativity by simultaneous RIf and dual-color TIRFS detection is shown in Figure 3.5. Here, the ligand was site-specifically labeled with OG488, whereas ifnar1-H10 was through its histidine tag with an Cy5-analog dy (FEW646) (Lata *et al.*, 2006b). Because all species and the total mass were monitored independently, highly robust fitting of these binding curves was possible. Thus, noncooperative interaction of ifnar1-EC and ifnar2-EC with IFNα2 was confirmed by this assay.

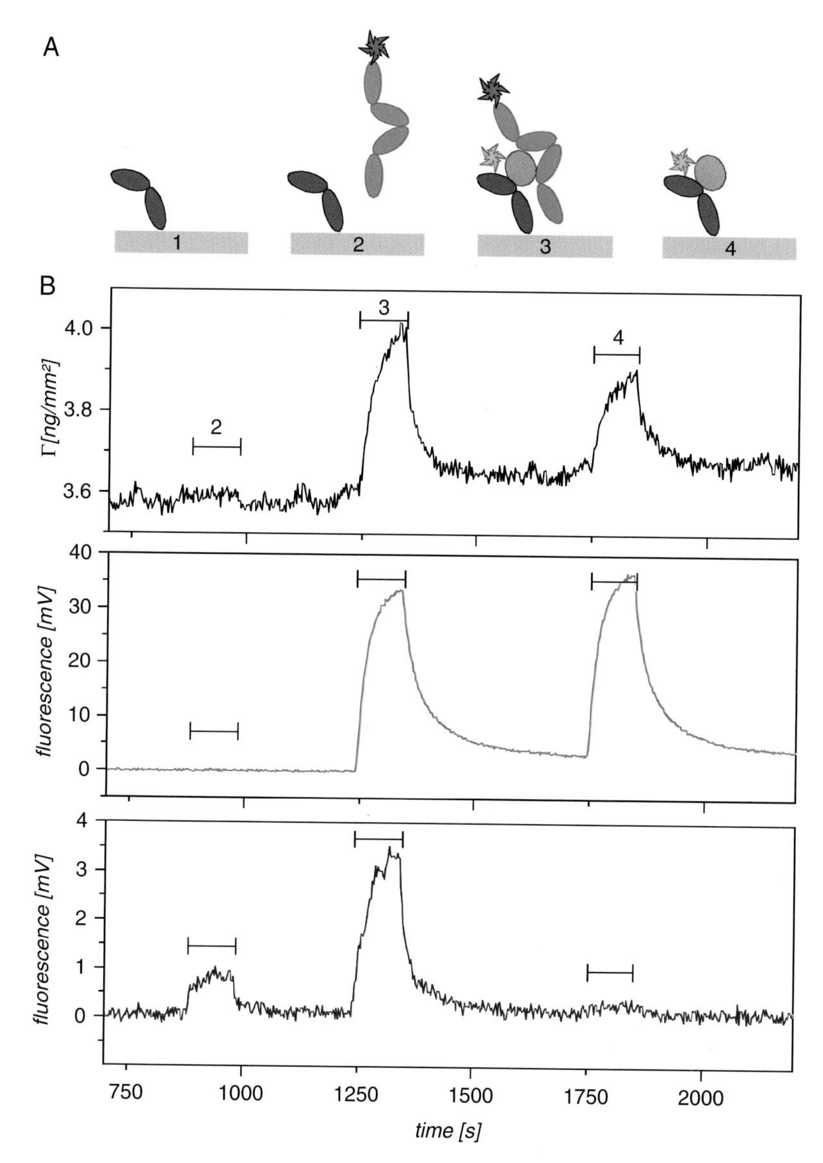

Figure 3.5. Dissection of multiprotein complex formation by surface-sensitive detection. (A) Schematic of the binding assays: After immobilization of ifnar2-H10 (1), first 200 nM ^{FEW646}ifnar1-H10 with 800 nM was injected (2), then the complex of ^{FEW646}ifnar1-H10 with 100 nM ^{OG488}IFNα2 (3), and then 100 nM ^{OG488}IFNα2 alone (4). (B) Typical binding curves as detected by simultaneous TIRFS-RIf detection: Mass signal (black), OG488 fluorescence (green), and FEW646 fluorescence (red).

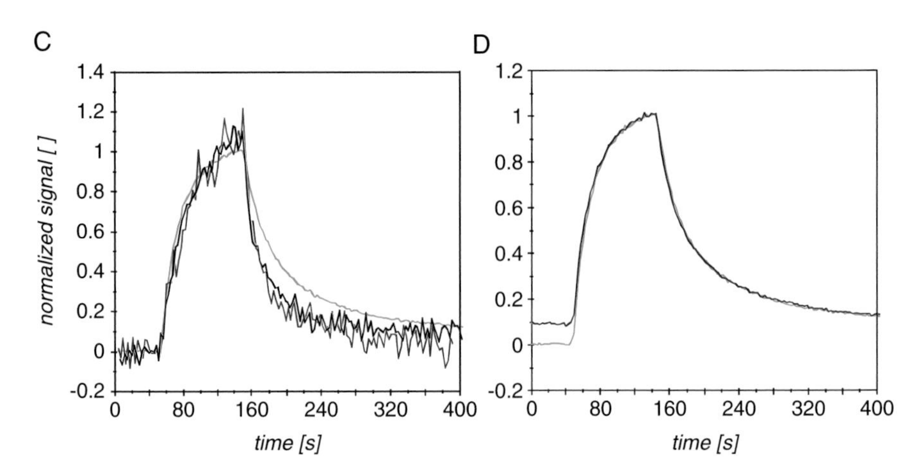

Figure 3.5. (*Continued*) (C) Overlay of the three signals during injection 2 normalized to the maximum signal (same color coding as B). (D) Overlay of the green fluorescence signal during injection 3 (green) and injection 4 (blue), normalized to the maximum signal.

3.3.2. Cross-Linking of Receptor Subunits on Membranes

Simultaneous interaction and cross-linking of two or more receptor subunits on the plasma membrane is the basis of activation of cytokine receptors and receptor tyrosine kinases. However, preassociated receptor subunits have been proposed, and the role of membrane anchoring for these mechanisms has been discussed controversially (Sebald and Mueller, 2003; Stroud and Wells, 2004). To study such processes *in vitro*, ligand-induced receptor cross-linking was mimicked by tethering ifnar1-EC and ifnar2-EC onto solid-supported lipid bilayers in an oriented fashion. Based on simultaneous fluorescence and mass-sensitive detection, the mechanism of ligand-induced receptor assembling on the membrane was deconvoluted (Gavutis *et al.*, 2005). A key parameter in these studies is the control and absolute quantification of the surface concentrations of the receptor subunits. This is demonstrated in Figure 3.6A–D, where the binding and the dissociation of fluorescence-labeled IFNα2 to the receptor subunits ifnar1-EC and ifnar2-EC tethered onto solid-supported membranes in different surface concentrations were monitored. Strikingly, the ligand dissociation kinetics was affected by the absolute and relative concentrations of the receptor subunits. These experiments are important for determining the stoichiometry of the receptor–ligand complex on the membrane and to identify preassembled receptors. In case of the type-I IFN receptor, a dynamic equilibrium between binary and ternary complexes was concluded from these measurements.

Based on these results and further studies by FRAP and FRET, two step of the ternary complex was shown. However, the assembling pathway depends on the surface concentrations and the relative association rate constants of the interaction

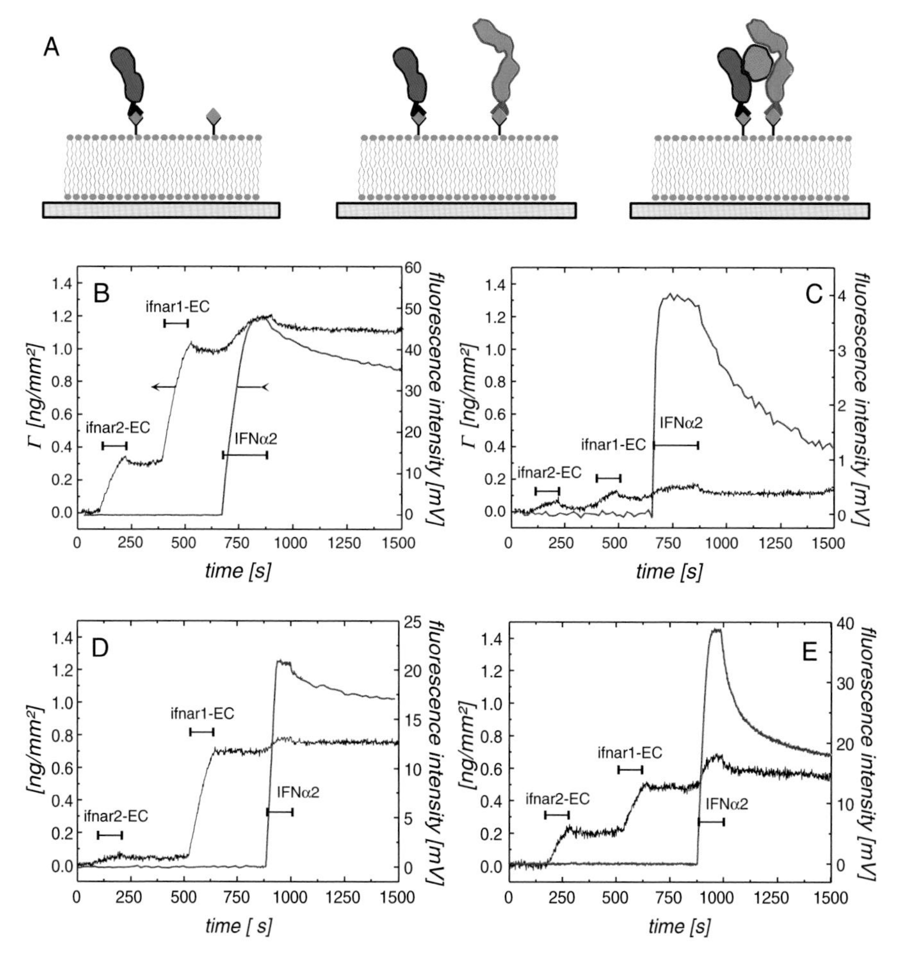

Figure 3.6. Interaction of fluorescence-labeled IFNα2 with ifnar1-EC and ifnar2-EC coimmobilized onto supported lipid bilayers at different absolute and relative amounts. (A) Schematic of the binding assay. (B–E) Signals detected by TIRFS (.....) and by RIf (——) during sequential tethering of ifnar2-EC and ifnar1-EC followed by injection of the ligand. (B) 12 fmol/mm^2 of both ifnar2-EC (25 kDa) and ifnar1-EC (57 kDa). (C) 2 fmol/mm^2 of both ifnar2-EC and ifnar1-EC. (D) 2 fmol/mm^2 of ifnar2-EC and 10 fmol/mm^2 of ifnar1-EC. (E) 8 fmol/mm^2 of ifnar2-EC and 5 fmol/mm^2 of ifnar1-EC. Reproduced from Gavutis *et al.* (2005) with permission.

with the receptor subunits (Figure 3.7A). Owing to the tenfold faster association of wild-type IFNα2 to ifnar2-EC, pathway 1 is dominated at stoichiometric surface concentrations of the receptor subunits (Gavutis *et al.*, 2005). Thus, the equilibrium dissociation constant of the two-dimensional interaction of ifnar1-EC with IFNα2 bound to ifnar2-EC on the lipid bilayer surface on the membrane K_2 determines the

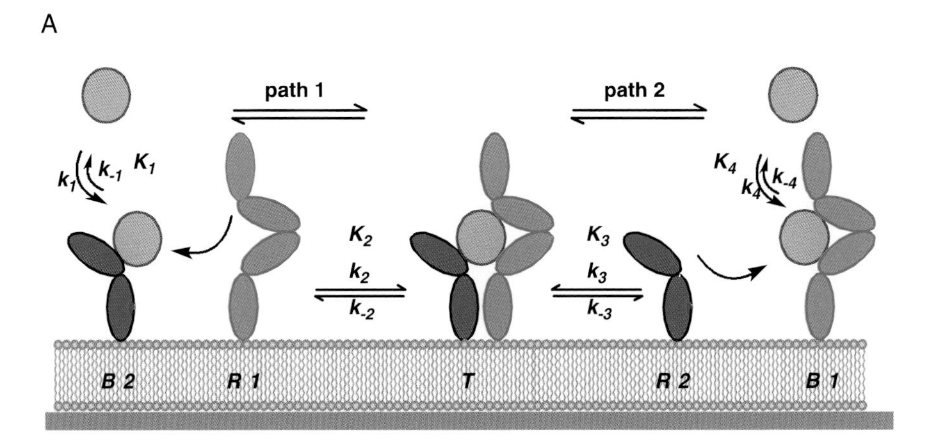

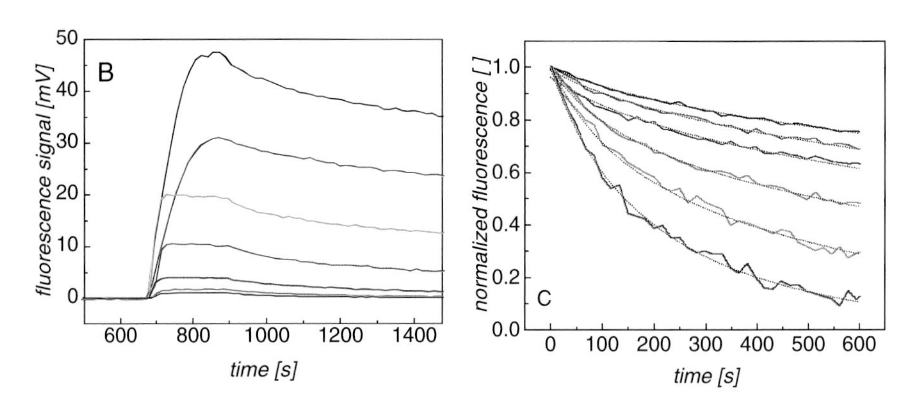

Figure 3.7. Determination of two-dimensional equilibrium dissociation constants from ligand dissociation kinetics. (A) Possible mechanisms of ternary complex assembling and dissociation. (B) Ligand-binding curves at different receptor surface concentrations. (C) Normalized dissociation curves, and fit of a model assuming that only path 1 contributes to ligand dissociation. Reproduced from Gavutis *et al.* (2005) with permission.

equilibrium between binary and ternary complexes. Because ligand dissociation is much faster than the binary complex (at least 200-fold), this equilibrium can be probed by monitoring ligand dissociation kinetics at different surface concentrations of the receptor subunits (Figure 3.7B and C). A particular important parameter for the fitting was the precise parameterization of the surface concentrations of the receptor subunits, which was possible due to the mass-sensitive detection by RIf. Strikingly, these dissociation curves were consistently fitted by a two-step dissociation model according to pathway 1 (Figure 3.7A), yielding a two-dimensional equilibrium constant of ~ 40 molecules/μm^2 (Gavutis *et al.*, 2005).

3.3.3. Two-Dimensional Interaction Kinetics on Membranes

A key problem of protein–protein interaction on membranes is the direct measurement of interaction rate constants, because this requires a rapid change of concentrations. Although this is very difficult to achieve for proteins integrated into the membrane, it is well possible for proteins tethered to the membrane through a tag (Lata *et al.*, 2006a). Using this approach, the two-dimensional dissociation kinetics was monitored by an elegant FRET assay shown in Figure 3.8 (Gavutis *et al.*, 2006). Donor-labeled ifnar2-H10 and unlabeled ifnar1-H10 were tethered onto the membrane in stoichiometric concentrations. On binding of the ligand labeled with an acceptor dye (Alexa Fluor 568, AF568), FRET was detected both on the donor and the acceptor channels (Figure 3.8B). Slow dissociation of the ligand was observed, as expected at these surface concentrations. After a second injection of

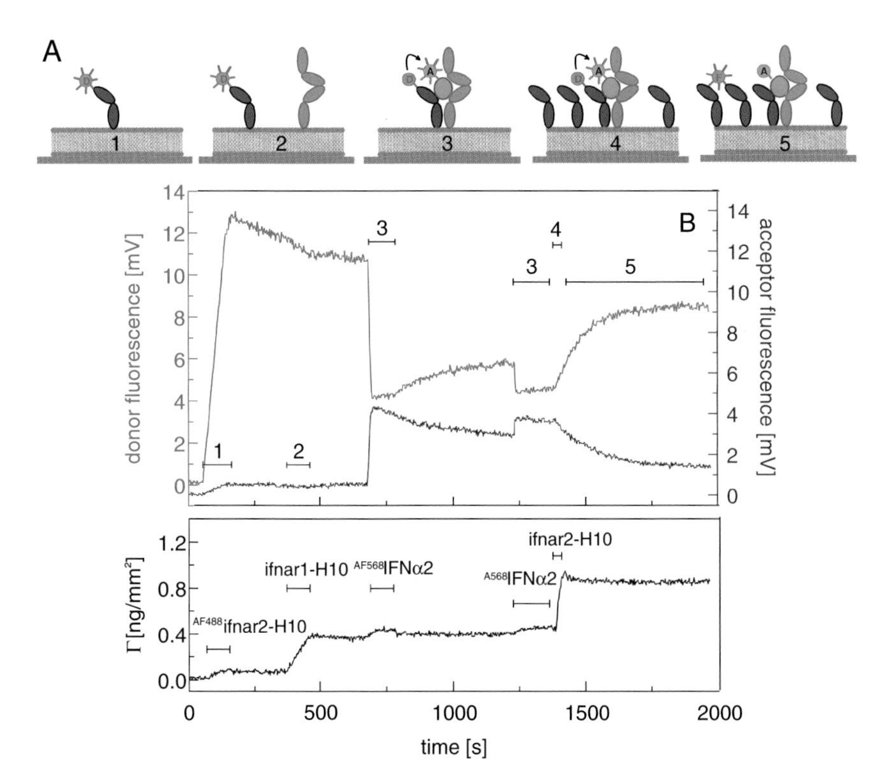

Figure 3.8. Monitoring two-dimensional dissociation kinetics by pulse chasing of a ternary complex. (A) Principle of surface dissociation rate constant determination as detected by FRET: The ternary complex on fluid lipid membrane is formed by sequential injection of ^{AF488}ifnar2-H10 (1), ifnar1-H10 (2), and ^{AF568}IFNα2 (3). Equilibrium was then perturbed by rapidly tethering an excess of nonlabeled ifnar2-H10 onto the membrane (4), which exchanges the labeled ifnar2-H10 in the ternary complex (5). (B) Course of a typical experiment monitoring donor fluorescence (green) and acceptor (red trace) fluorescence by TIRFS and the mass loading by RIf (black).

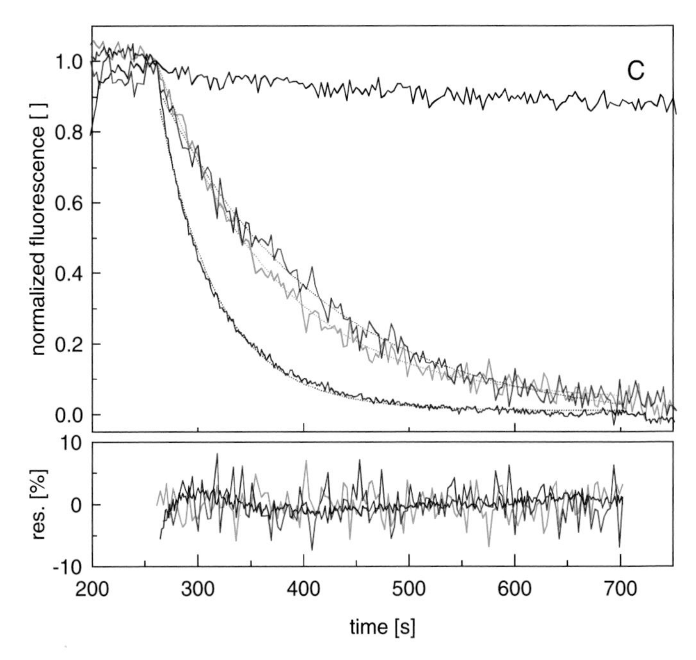

Figure 3.8. (*Continued*) (C) Comparison of the surface dissociation rates from donor (green) and acceptor (red) channels with the dissociation of ^{AF568}IFNα2 from ifnar2-H10 alone (blue). A control experiment was carried out in the same way, but with unlabeled ifnar2-H10 in (1) and with direct excitation of ^{AF568}IFNα2 confirmed negligible ligand dissociation from the surface (black). The residuals from monoexponential curve fits are shown at the bottom. Reproduced from Gavutis *et al.* (2006) with permission.

the ligand, however, the resulting ternary complex was chased by rapidly tethering a large excess of unlabeled ifnar2-H10 onto the membrane. Much faster recovery of the donor fluorescence and very similar decay of the sensitized fluorescence were observed. However, this was not due to ligand dissociation, which was hardly detectable under these conditions (Figure 3.8C), but to the exchange of labeled versus unlabeled ifnar2-EC in the ternary complex. Thus, the two-dimensional dissociation kinetics of the ifnar2-H10/IFNα2 interaction as the rate-limiting step of this process was probed. Unexpectedly, slower dissociation in plane of the membrane was observed, suggesting a more complicated dissociation mechanism.

Chasing with unlabeled ligand also turned out to be a valuable approach to determine two-dimensional rate constants (Gavutis *et al.*, 2006). Within a single experimental cycle, two-dimensional association and dissociation kinetics can be determined. This is demonstrated in Figure 3.9 for the interaction between ifnar1-H10 and IFNα2. Here, a substantial stoichiometric excess of ifnar2-H10 I47A was immobilized, which binds IFNα2 with a complex lifetime of $\sim$5 s. The IFNα2 mutant HEQ was used as the ligand, which binds ifnar1-EC with a complex lifetime of $\sim$ 20 s (Jaitin *et al.*, 2005). After formation of the ternary complex, ligand

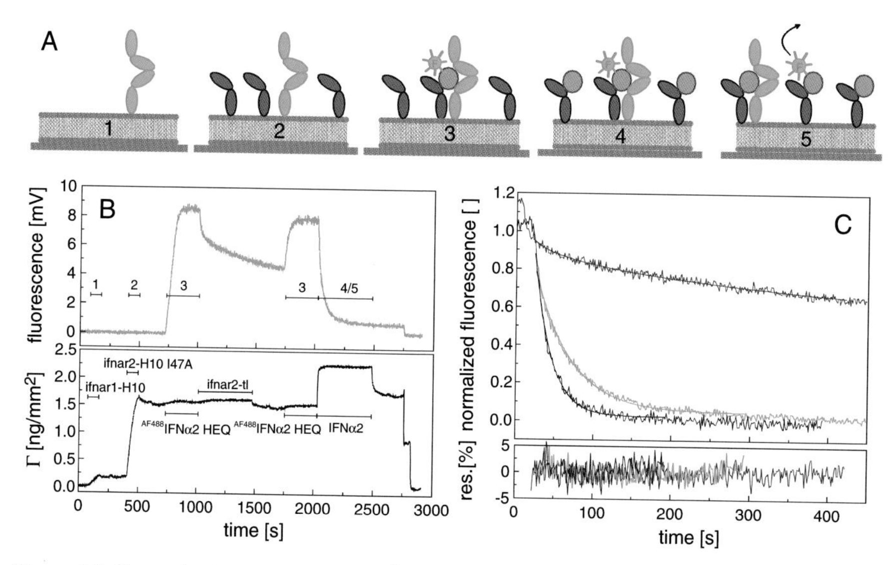

Figure 3.9. Determination of two-dimensional rate constants for pathway 1. (A) Schematic of the assay: Ternary complex on fluid lipid membrane was formed by sequential injection of ifnar1-H10 (1), excess ifnar2-H10 I47A (2), and ^{AF488}IFNα2 HEQ (3). On loading the excess binding sites of ifnar2-H10 with unlabeled IFNα2 (4), labeled IFNα2 in the ternary complex was exchanged (5). (B) Course of a typical experiment as detected by TIRFS (green) and RIf (black). During spontaneous ligand dissociation, 2 μM ifnar2-tl was maintained in the background to eliminate rebinding. After the second injection of ^{AF488}IFNα2 HEQ, 1 μM unlabeled IFNα2 wild type was injected. (C) Overlay of the normalized dissociation curves: Spontaneous dissociation from the ternary complex (red) and ligand exchange kinetics washing with 1 μM IFNα2 HEQ (green). Dissociation from Ifnar1-H10 alone is shown for comparison (blue). The residuals from the curve fits are shown at the bottom. Reproduced from Gavutis *et al.* (2006) with permission.

bound to excess ifnar2-EC dissociated within a few ten seconds. The second slow phase is the spontaneous ligand dissociation from the ternary complex through path 1 (cf. Figure 3.7A), which can be used for determining the two-dimensional dissociation constant of the ifnar1–EC IFNα2 complex (K_2 in Figure 3.7A). After reloading the ligand, nonlabeled ligand was injected, and the exchange of labeled versus unlabeled ligand was monitored. Here, the rate-limiting step is the two-dimensional dissociation of IFNα2 from ifnar1-EC (k_{-2} in Figure 3.7A). Thus, this two-dimensional interaction can be fully characterized in a single experimental cycle, and under the same set of conditions with respect to receptor surface concentrations.

3.4. CONCLUSIONS

Label-free detection techniques are compelling tools for monitoring protein–protein interactions in real time. They do not require labeling of the interacting compounds, and provide quantitative information not only about interaction rates and affinities, but also on the stoichiometry of an interaction. Because all mass exchange on the surface is monitored, building up more complex surface architectures can be controlled quantitatively. For studying multiprotein complex formation, TIRFS offers versatile means for discriminating different species and for selectively probing interactions by FRET. Owing to the complementary features of the detection techniques, the combination of label-free and TIRFS detection is ideal for dissecting complex protein–protein interaction mechanism, and for characterizing individual interactions within multiprotein complex formation. Surface-sensitive techniques are more suitable for studying protein complex formations at solid-supported membranes. Here, the possibility to monitor membrane assembling and protein reconstitution, and to absolutely quantify surface concentrations of tethered proteins by label-free detection is a key advantage. Membrane protein complexes play a key role in processes such as signaling, transport, and energy metabolism; yet the mechanisms and the principles governing their assembling have hardly been studied. Combinations of solid-phase techniques will substantially contribute to understand the molecular mechanisms of these processes.

REFERENCES

Axelrod, D., Burghardt, T. P., and Thompson, N. L. (1984). Total internal reflection fluorescence. *Annu Rev Biophys Bioeng* 13:247–268.

Birkert, O. and Gauglitz, G. (2002). Development of an assay for label-free high-throughput screening of thrombin inhibitors by use of reflectometric interference spectroscopy. *Anal Bioanal Chem* 372:141–147.

Brecht, A., Gauglitz, G., and Polster, J. (1993). Interferometric immunoassay in a FIA-system: a sensitive and rapid approach in label-free immunosensing. *Biosens Bioelectron* 8:387–392.

Brian, A. A. and McConnell, H. M. (1984). Allogeneic stimulation of cytotoxic T cells by supported planar membranes. *Proc Natl Acad Sci USA* 81:6159–6163.

Cooper, M. A. (2003). Label-free screening of bio-molecular interactions. *Anal Bioanal Chem* 377: 834–842.

Edwards, P. R., Gill, A., Pollard-Knight, D. V., Hoare, M., Buckle, P. E., Lowe, P. A., and Leatherbarrow, R. J. (1995). Kinetics of protein–protein interactions at the surface of an optical biosensor. *Anal Biochem* 231:210–217.

Gauglitz, G. (2005). Direct optical sensors: principles and selected applications. *Anal Bioanal Chem* 381:141–155.

Gavutis, M., Jaks, E., Lamken, P., and Piehler, J. (2006). Determination of the 2-dimensional interaction rate constants of a cytokine receptor complex. *Biophys J* 90:3345–3355.

Gavutis, M., Lata, S., Lamken, P., Müller, P., and Piehler, J. (2005). Lateral ligand-receptor interactions on membranes probed by simultaneous fluorescence-interference detection. *Biophys J* 88: 4289–4302.

Haake, H. M., Schutz, A., and Gauglitz, G. (2000). Label-free detection of biomolecular interaction by optical sensors. *Fresenius J Anal Chem* 366:576–585.

Haes, A. J. and Van Duyne, R. P. (2002). A nanoscale optical biosensor: sensitivity and selectivity of an approach based on the localized surface plasmon resonance spectroscopy of triangular silver nanoparticles. *J Am Chem Soc* 124:10596–10604.

Hanel, C. and Gauglitz, G. (2002). Comparison of reflectometric interference spectroscopy with other instruments for label-free optical detection. *Anal Bioanal Chem* 372:91–100.

Harris, R. D., Luff, B. J., Wilkinson, J. S., Piehler, J., Brecht, A., Gauglitz, G., and Abuknesha, R. A. (1999). Integrated optical surface plasmon resonance immunoprobe for simazine detection. *Biosens Bioelectron* 14:377–386.

Homola, J. (2003). Present and future of surface plasmon resonance biosensors. *Anal Bioanal Chem* 377:528–539.

Hook, F., Kasemo, B., Nylander, T., Fant, C., Sott, K., and Elwing, H. (2001). Variations in coupled water, viscoelastic properties, and film thickness of a Mefp-1 protein film during adsorption and cross-linking: a quartz crystal microbalance with dissipation monitoring, ellipsometry, and surface plasmon resonance study. *Anal Chem* 73:5796–5804.

Jaitin, D., Roisman, L. C., Jaks, E., Gavutis, M., Piehler, J., Van der Heyden, J., Uze, G., and Schreiber, G. (2005). Inquiring into the differential actions of interferons: an IFNa2 mutant endowed with enhanced binding affinity to IFNAR1 is functionally similar to IFNbeta. *Mol Cell Biol* 26: 1888–1897.

Kroger, K., Bauer, J., Fleckenstein, B., Rademann, J., Jung, G., and Gauglitz, G. (2002). Epitope-mapping of transglutaminase with parallel label-free optical detection. *Biosens Bioelectron* 17:937–944.

Lamken, P., Lata, S., Gavutis, M., and Piehler, J. (2004). Ligand-induced assembling of the type I interferon receptor on supported lipid bilayers. *J Mol Biol* 341:303–318.

Lata, S., Gavutis, M., and Piehler, J. (2006a). Monitoring the dynamics of ligand-receptor complexes on model membranes. *J Am Chem Soc* 128:6–7.

Lata, S., Gavutis, M., Tampé, R., and Piehler, J. (2006b). Specific and stable fluorescence labeling of histidine-tagged proteins for dissecting multi-protein complex formation. *J Am Chem Soc* 128:2365–2372.

Lata, S. and Piehler, J. (2005). Stable and functional immobilization of histidine-tagged proteins via multivalent chelator head-groups on a molecular poly(ethylene glycol) brush. *Anal Chem* 77:1096–1105.

Liebermann, T. and Knoll, W. (2000). Surface-plasmon field-enhanced fluorescence spectroscopy. *Colloids Surf A-Physicochem Eng Asp* 171:115–130.

Lofas, S., Malmqvist, M., Ronnberg, I., Stenberg, E., Liedberg, B., and Lundstrom, I. (1991). Bioanalysis with surface-plasmon resonance. *Sens Actuators B Chem* 5:79–84.

Lukosz, W., Clerc, D., Nellen, P. M., Stamm, C., and Weiss, P. (1991). Output grating couplers on planar optical waveguides as direct immunosensors. *Biosens Bioelectron* 6:227–232.

Malmqvist, M. (1999). BIACORE: an affinity biosensor system for characterization of biomolecular interactions. *Biochem Soc Trans* 27:335–340.

Marx, K. A. (2003). Quartz crystal microbalance: a useful tool for studying thin polymer films and complex biomolecular systems at the solution-surface interface. *Biomacromolecules* 4:1099–1120.

Neumann, T., Johansson, M. L., Kambhampati, D., and Knoll, W. (2002). Surface-plasmon fluorescence spectroscopy. *Adv Funct Mater* 12:575–586.

Piehler, J., Brecht, A., and Gauglitz, G. (1996). Affinity detection of low molecular weight analytes. *Anal Chem* 68:139–143.

Piehler, J. and Schreiber, G. (2001). Fast transient cytokine-receptor interactions monitored in real time by reflectometric interference spectroscopy. *Anal Biochem* 289:173–186.

Reimhult, E., Larsson, C., Kasemo, B., and Hook, F. (2004). Simultaneous surface plasmon resonance and quartz crystal microbalance with dissipation monitoring measurements of biomolecular adsorption events involving structural transformations and variations in coupled water. *Anal Chem* 76:7211–7220.

Sackmann, E. (1996). Supported membranes: scientific and practical applications. *Science* 271:43–48.

Sackmann, E. and Tanaka, M. (2000). Supported membranes on soft polymer cushions: fabrication, characterization and applications. *Trends Biotechnol* 18:58–64.

Salamon, Z., Devanathan, S., Alves, I. D., and Tollin, G. (2005). Plasmon-waveguide resonance studies of lateral segregation of lipids and proteins into microdomains (rafts) in solid-supported bilayers. *J Biol Chem* 280:11175–11184.

Salamon, Z., Macleod, H. A., and Tollin, G. (1997). Coupled plasmon-waveguide resonators: a new spectroscopic tool for probing proteolipid film structure and properties. *Biophys J* 73:2791–2797.

Schmid, E. L., Tairi, A. P., Hovius, R., and Vogel, H. (1998). Screening ligands for membrane protein receptors by total internal reflection fluorescence: the 5-HT3 serotonin receptor. *Anal Chem* 70:1331–1338.

Schmitt, H. M., Brecht, A., Piehler, J., and Gauglitz, G. (1997). An integrated system for optical biomolecular interaction analysis. *Biosens Bioelectron* 12:809–816.

Sebald, W. and Mueller, T. D. (2003). The interaction of BMP-7 and ActRII implicates a new mode of receptor assembly. *Trends Biochem Sci* 28:518–521.

Sinner, E. K. and Knoll, W. (2001). Functional tethered membranes. *Curr Opin Chem Biol* 5:705–711.

Stroud, R. M. and Wells, J. A. (2004). Mechanistic diversity of cytokine receptor signaling across cell membranes. *Sci STKE*, re7.

Tanaka, M. and Sackmann, E. (2005). Polymer-supported membranes as models of the cell surface. *Nature* 437:656–663.

Thompson, N. L., Drake, A. W., Chen, L., and Vanden Broek, W. (1997). Equilibrium, kinetics, diffusion and self-association of proteins at membrane surfaces: measurement by total internal reflection fluorescence microscopy. *Photochem Photobiol* 65:39–46.

Thompson, N. L. and Lagerholm, B. C. (1997). Total internal reflection fluorescence: applications in cellular biophysics. *Curr Opin Biotechnol* 8:58–64.

Thompson, N. L., Pearce, K. H., and Hsieh, H. V. (1993). Total internal reflection fluorescence microscopy: application to substrate-supported planar membranes. *Eur Biophys J* 22:367–378.

Tollin, G., Salamon, Z., Cowell, S., and Hruby, V. J. (2003). Plasmon-waveguide resonance spectroscopy: a new tool for investigating signal transduction by G-protein coupled receptors. *Life Sci* 73:3307–3311.

4

Surface Plasmon Resonance Biosensing in the Study of Ternary Systems of Interacting Proteins

Eric J. Sundberg, Peter S. Andersen, Inna I. Gorshkova, and Peter Schuck*

4.1. INTRODUCTION

Surface plasmon resonance (SPR) biosensors are optical evanescent wave sensors, where light in total internal reflection is used to probe properties of the solution adjacent to the surface. SPR occurs through the interaction of light with a thin metal film, which can be used to measure the refractive index of the solution close to the surface with high sensitivity (Kretschmann and Raether, 1968; Knoll, 1998). This is exploited in the study of protein interactions by immobilizing one binding partner to the surface, and observing the change in local refractive index during the interaction with a label-free soluble-binding partner.

SPR technology became a popular tool for studying protein interactions in the 1990s with the introduction of commercial instruments. It is closely related in many ways to the optical biosensor techniques described in Chapter 3. Nevertheless, the widespread availability, the versatility of this approach, and the many published applications of qualitative and quantitative binding studies warrant a separate chapter on the potential of SPR biosensors in the study of complex systems of interacting proteins. This chapter is not intended to be an exhaustive review, but is

E. J. SUNDBERG • Boston Biomedical Research Institute, Watertown, MA, USA. P. S. ANDERSEN • Symphogen A/S, Copenhagen, Denmark. I. I. GORSHKOVA • National Institute for Biomedical Imaging and Bioengineering, National Institutes of Health, Bethesda, MD, USA. P. SCHUCK • National Institute for Biomedical Imaging and Bioengineering, National Institutes of Health, 13 South Drive, Bethesda, MD 20892 and *Corresponding author: Tel: +1 301 4351950; fax: +1 301 4801242; e-mail: pschuck@helix.nih.gov

instead meant to critically highlight only certain features and selected approaches for the analysis of multiprotein complexes. For more general information, practical protocols, or literature reviews highlighting other topics, the reader is referred to many published reviews (e.g., O'Shannessy, 1994; Malmborg and Borrebaeck, 1995; van der Merwe and Barclay, 1996; Schuck, 1997b; Huber *et al.*, 1999; Hall, 2001; Cooper, 2002; Alves *et al.*, 2005; Lee *et al.*, 2005; Pattnaik, 2005).

It has been generally recognized that, when SPR biosensors were first commercially introduced, the interpretation of the surface-binding kinetic traces was frequently too simplistic, and based in many cases on unwarranted assumptions regarding the ideality of the measurement and the information content of the recorded traces (Schuck, 1997a). However, as many elegant and well-controlled SRP studies in the literature have demonstrated, this should not detract from the possibility to use SPR technology as a reliable biophysical research tool, since knowledge about experimental and analytical aspects available have evolved significantly since the initial introduction. SPR is extremely flexible in the experimental setup, and very useful for studying many facets of multiprotein complex formation. In particular, in the context of a combination with other biophysical tools, it can provide unique and reliable information.

From the perspective of studying reversibly associating proteins, there are several essential aspects of SPR biosensing that should be noted in comparison with solution-based methods. First, for characterizing reversible interactions, the surface immobilization eliminates the need to work with both proteins at concentrations close to the order of magnitude of the binding constant (K_D). Instead, the surface concentration of the immobilized protein can be chosen to optimize the magnitude of the signal, whereas the soluble-binding partner is typically in the range of 0.1–10-fold the K_D. Only the concentration of the soluble species governs the fractional saturation of sites. This permits the study of reversible, very high affinity interactions without running into problems of either too low signal or entirely stoichiometric binding. In principle, interactions with affinities spanning a range of at least 10^4–$10^{10}\ M^{-1}$ can be characterized by SPR. Of course, surface immobilization may introduce problems itself because of chemical modification and conformational constraints. Several commonly used approaches will be described later. In addition, the properties of the surface itself need to be considered when interpreting the binding experiments. Different strategies addressing this problem, such as the use of reference surfaces and competition-binding approaches, will be discussed.

A second essential feature of SPR biosensing is the ability to observe in real time the kinetics of the surface-binding process. When the soluble-binding partner is injected into the flow across the sensor surface, its time course of accumulation at the surface sites and the attainment of a steady state can usually be readily observed, as well as the dissociation of the complex after chasing the surface with a flow of buffer. This has great potential for the elucidation of the binding kinetics of interacting proteins, for which chemical association rates of up to 10^5–$10^6\ M^{-1}s^{-1}$ and dissociation rates of between 10^{-5} and $10^{-1}\ s^{-1}$ are generally accessible.

In principle, this can also permit the estimation of equilibrium constants even in the absence of reaching true steady state, even though there are special considerations and precautions needed for this extrapolation because of the macroscopic transport of soluble analyte from the flow to the sensor surface. This can be particularly advantageous in situations in which protein quantities are limited, or where the equilibration time is impractically long.

Third, the biosensor surface can act effectively as a miniaturized affinity chromatography matrix, allowing for the separation of unbound soluble protein, or any other contaminating or nonparticipating molecules, from the surface-bound complex. For systems with slow dissociation rates, this has preparative implications, as the amount of material captured can be compatible with mass spectroscopy. It also has important analytical consequences, since the preformed surface-bound complex is available for further investigation by probing binding with a third protein species, which can be a powerful tool for qualitative or quantitative studies of multiprotein complexes.

In this chapter, we will (1) describe the basic principles of SPR detection, a variety of strategies for functional attachment of proteins to the sensor surface, and summarize some standard kinetic and thermodynamic experiments for binary protein interactions; (2) outline a set of experimental restrictions and controls that are aimed at ensuring the absence of characteristic artifacts arising from using surface binding as a probe for protein interactions; (3) illustrate the capacity of SPR biosensing in multiprotein interactions or interactions with multiple conformations and cooperativity; and (4) provide a selective literature review of practical applications to different types of interacting protein systems, focusing largely on the ternary complex formation between bacterial superantigen (SAG), T-cell receptor (TCR), and major histocompatibility complex (MHC) molecules (Andersen *et al.*, 2002).

4.2. SURFACE PLASMON RESONANCE BASICS

4.2.1. Physical Principle

The SPR biosensor experiment exploits surface-confined electromagnetic fields for real-time measurement of the refractive index of the medium in the immediate vicinity of the sensor surface (Kretschmann and Raether, 1968; Raether, 1977; Lukosz, 1991; Garland, 1996; Knoll, 1998). An evanescent wave is created when light strikes the interface between the glass of a sensor surface and the assay buffer at an angle in total internal reflection. The typical decay length of the evanescent field for visible light is on the order of a few hundred nanometers. If a thin metal layer is located at the interface, the light can cause surface charge density waves (surface plasmons) in the free electrons on the metal film. For a specific angle of incidence of the light, the wave vector of the reflected light and the surface plasmons are in phase, and resonant energy transfer can take place, which diminishes the intensity of the reflected light. This is called SPR. Since the angle of minimum reflectivity depends strongly on the refractive index of the solution in the vicinity of the sensor surface, the analysis of the angular-dependent reflectivity can be used to determine this refractive index with a precision of up to 6–7 decimal places. Thus, even though the refractive index increment of proteins is only 0.15–0.19 ml/g, and thus not dramatically different from water and not a highly sensitive parameter, protein accumulation on the sensor surface can be followed with high precision via the concurrent local increase of refractive index.

Other optical evanescent wave sensors are based on waveguide principles, where a thin high-refractive index layer is deposited on the interface. This permits the detection of phase shifts of the guided light in two polarizations, both having different decay lengths of the evanescent field, which in combination gives information about the effective average layer thickness of a protein film in addition to the refractive index (Lukosz, 1991). Even though these approaches yield a richer data basis on protein interactions, and can provide information on conformational changes (Salamon *et al.*, 1994, 1998), they so far did not have as widespread applications as SPR (for a more detailed review, see Alves *et al.*, 2005). Several other label-free optical techniques for assessing surface-bound protein have been developed (see Brecht and Gauglitz (1995), Chapter 3, and references cited therein). This also includes high-throughput formats and imaging techniques. The following discussion is restricted to SPR, but many aspects apply directly to other evanescent wave biosensors, as well.

Figure 4.1A shows the basic idea of a surface-binding experiment. In the SPR instrument from Biacore AB (www.biacore.com), sample is supplied in microfluidic channels containing an HPLC-like injection loop (Sjölander and Urbaniczky, 1991). Alternative approaches eliminating the constraints from the finite volume of the sample plug (and the resulting limitation in contact time and flow rate of the sample with the sensor surface) have been reported using cuvette-type systems (such as the IAsys instrument from Thermo, UK), circulating sample (Schuck *et al.*, 1998), and oscillating flow (Abrantes *et al.*, 2001), the latter combining a significant reduction of sample volume with high flow rates and extended observation time (Abrantes *et al.*, 2001). In any case, the change in reflectivity, caused by the binding or dissociation of molecules from the sensor surface, is in first approximation proportional to the mass of bound material and is recorded in a so-called sensorgram (Figure 4.1B). After (usually covalent) immobilization of one binding partner to the surface, the sensorgram permits following in real time the increasing response as molecules dissolved in the sample flowing across the surface interact with the surface sites. The response remains constant if the interaction attains a steady state. When sample flow is replaced by buffer, the response decreases as the interacting-binding partners dissociate and the soluble molecule is released from the surface. The signal from bulk refractive changes is measured separately on a similarly treated but nonfunctionalized surface, and subtracted from the traces to be analyzed (Ober and Ward, 1999a; Karp *et al.*, 2005). The analysis of the net binding traces can be used to derive information on kinetic rate constants and equilibrium

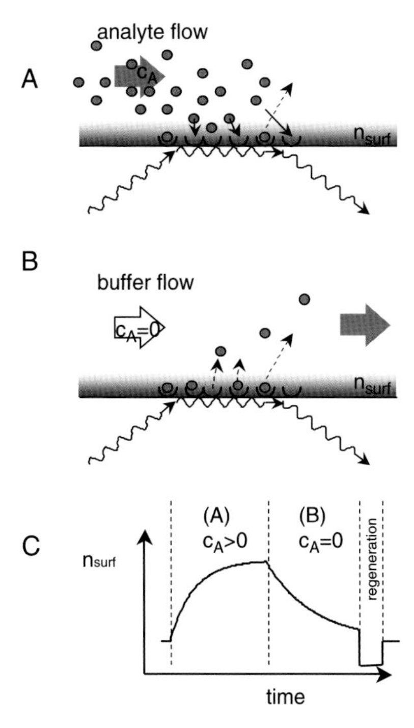

Figure 4.1. Schematic presentation of a typical optical biosensor experiment. Light is coupled into a structure that allows generation of surface-confined electromagnetic waves, which are sensitive to the refractive index of the solution close to the surface, n_{surf}, in the range of the evanescent field. Typical penetration depths of the sensitive volume into the solution are in the order of 100 nm. Ligands are attached to the sensor surface, as indicated by half-circles. (A) When analytes (full circles) are introduced into the solution above the surface, reversible interactions with the ligand leads to binding and dissociation events. (B) When the surface is washed by running buffer in the absence of analyte, only dissociation events take place. (C) Signal obtained from probing the refractive index n_{surf} during the sequential application of the configurations depicted in A and B, given in arbitrary units. Following the association phase (A) and the dissociation phase (B), usually a regeneration procedure is applied for removing the remaining analyte from the surface before a new experimental cycle takes place.

constants of the interaction. The following section will outline some of the fundamental aspects of such experiments.

4.2.2. Immobilization

The goal of immobilization is the stable coupling of the ligand to the sensor surface in its active form. To prevent irreversible adsorption of protein to the metal film, the surface can be coated with self-assembled monolayer of alkyl-thiols (Löfås and Johnsson, 1990). Further, in many cases, a sensor surface is chosen that is coated with a polymer, such as carboxymethyl dextran, in order to serve as an immobilization matrix and to suppress nonspecific interactions (Löfås and

Johnsson, 1990). Several methods exist for immobilization of ligand to the sensor chip surface (O'Shannessy *et al.*, 1992). The strategy of immobilization is related to the choice of the regeneration procedure for the surface, which is frequently required to strip remaining bound protein from the surface before starting a new cycle of association and dissociation. For an overview and practical protocols, see Schuck *et al.* (1999).

The most commonly used procedure is to covalently couple molecules to a carboxymethyl dextran-coated sensor surface via amine, thiol, aldehyde, or carboxyl groups. It should be noted that these immobilization methods differ in their exposure of the ligand to relatively harsh conditions. Electrostatic preconcentration of protein below its p*I* under low salt conditions in the negatively charged carboxymethyl dextran can provide highly effective immobilization requiring only microgram amounts of protein (Johnsson *et al.*, 1991).

Other approaches rely on capturing the ligand noncovalently with another (covalently immobilized) protein, such as an antibody, protein A, or streptavidin, or capture of polyhistidine tags (Lata and Piehler, 2005). An elegant capture method for glycosylated proteins can be the use of lectins, such as concanavalin A, which recognize polysaccharides. This has been applied, for example, to the surface attachment of detergent-solubilized rhodopsin (Rebois *et al.*, 2002; Northup, 2004).

For membrane receptors, techniques have been developed for creating supported or tethered planar lipid bilayers on the sensor surface (Atanasov *et al.*, 2005; Tanaka and Sackmann, 2005). Proteins may be inserted into the bilayers or cross-linked to the lipid head groups (Cornell *et al.*, 2001). This can provide lateral mobility and establish rotational preorientation of the molecules, which may profoundly influence the binding kinetics and thermodynamics. An example for a study of ligand interactions with extracellular receptor domains attached to supported lipid bilayers can be found in Chapter 3 on optical biosensors. Alternatively, hybrid bilayers consisting of surface-attached alkanethiol monolayers and phospholipid have been developed (Plant *et al.*, 1995). Establishing methods for functional reconstitution of integral membrane proteins in supported lipid bilayers is a subject of intense current research. Techniques for the deposition of whole membrane fragments have been described (Rao *et al.*, 1997).

Some of the chemical methods will result in cross-links to the surface in many different orientations. The site-specific capture or cross-link is usually preferable, even though it may be unclear if the physical orientation at the surface is uniform, for example, due to random immobilization applied to the capturing molecule.

Clearly, the choice of immobilization methods is specific to the molecular system studied. Regardless of which method is chosen, it is of paramount importance to control the number of available surface-binding sites, or the average immobilized ligand density, as numerous surface-related artifacts may be invoked if the surface density is too high (one is described Section 4.2). This density of immobilized ligand can be measured as the difference in the SPR signal before and after the immobilization procedure, which also provides an estimation of the maximal analyte-binding capacity. For kinetic experiments, the immobilized ligand density should be maintained at a relatively low level, with the maximal binding capacity on the order of 50–200-fold the instrument noise. Another reason for maintaining low surface density of immobilized proteins may be the concern for some systems that the relatively high local concentration in the immobilization matrix may promote oligomerization or aggregation. In this case, comparison with binding studies in the opposite orientation, or with solution competition assays (see later), may be useful diagnostic tools.

As is apparent from the similarity of a flow-based biosensor surface with an affinity chromatographic matrix (Winzor, 2000), the largest difference being in the scale and the ability of real-time detection of bound analyte, the biosensor surface can serve an additional preparative function. The amounts of surface-bound material can be compatible with mass spectroscopic detection. Practical approaches have been developed to achieve recovery of the captured analyte and delivery to a mass spectrometer from SPR and other detectors, with different efficiency and sensitivity (Krone *et al.*, 1997; Sonksen *et al.*, 1998; Natsume *et al.*, 2000, 2002; Gilligan *et al.*, 2002; Mehlmann *et al.*, 2005). The use of biosensors in affinity purification step can be superior to immunoprecipitation and conventional affinity chromatography with regard to the quality of the capture process (Williams and Addona, 2000). This can permit the identification of the soluble-binding partners from a mixture exposed to the sensor surface, and determine possible protein modifications essential for binding.

4.2.3. Binding Analysis

Although the interpretation may require consideration of different processes, the typical SPR experiments produce data that are highly quantitative and reproducible. The following describes a bimolecular reaction of a soluble analyte with the surface-immobilized ligand. If the analyte concentration c is held constant, for example, due to a replenishing with a flow or due to a negligible number of surface-bound analyte molecules, the binding progress $s(t)$ follows the rate equation

$$\frac{\mathrm{d}s}{\mathrm{d}t} = k_{\mathrm{on}}c(s_{\max} - s) - k_{\mathrm{off}}s, \tag{4.1}$$

where $s_{\max}$ denotes the signal at full saturation, and k_{on} and k_{off} the chemical on- and off-rate constants, with $K_{\mathrm{A}} = k_{\mathrm{on}}/k_{\mathrm{off}}$. If we apply the analyte at time t_0 for a contact time t_{c}, we can integrate the rate equation and arrive at the binding progress in the association phase

$$s_{\mathrm{a}}(c, t) = s_{\mathrm{eq}}(c)\left(1 - \mathrm{e}^{-(k_{\mathrm{on}}c + k_{\mathrm{off}})(t - t_0)}\right) \tag{4.2}$$

with the steady-state response

$$s_{\mathrm{eq}}(c) = \frac{s_{\max}}{1 + (K_{\mathrm{A}}c)^{-1}} \tag{4.3}$$

(Langmuir, 1918). After the analyte is removed, we see dissociation of the bound analyte from the surface with

$$s_d(c,t) = s_a(c,t_c)e^{-k_{off}(t-t_0+t_c)}. \tag{4.4}$$

Both association and dissociation are proportional to s_{max}, and are single exponential, ascending or descending traces. Thus, information content on the kinetic rate constants is present mostly in the curvature of the sensorgrams (corresponding, e.g., to the analysis of molar masses in sedimentation equilibrium, Chapter 10), and sufficiently long observation times are required. With the commercial SPR systems, baseline stability is usually sufficiently high for experiments over several hours or more. Typically, a nonlinear regression to globally fit the kinetic-binding curves obtained at different loading concentrations results in the most robust estimates of rate constants (O'Shannessy *et al.*, 1993; Morton *et al.*, 1995). Ober and Ward (1999b) have determined the limitations of accuracy of the kinetic constants that can be obtained from noisy SPR data. They have also demonstrated the application of a subspace algorithm, a new noniterative technique to directly fit exponential data, to SPR analysis (Ober *et al.*, 2003).

Unfortunately, for reasons outlined later, the observation of binding kinetic data—with injection times and binding capacity appropriate for detailed kinetic modeling—that strictly follows Eqs. (4.3) and (4.4) is very rare (Karlsson *et al.*, 1994; Schuck *et al.*, 1998), despite the common uncritical use of this model in the past literature (Figure 4.2A). Nevertheless, the single-exponential binding is very important as a theoretically ''ideal'' case, against which experimental curves can be compared.

Alternatively, the equilibrium-binding constant can be derived from the concentration dependence of the steady-state signal, $s_{eq}(c)$, which follows a Langmuir isotherm (Eq. (4.3) and Figure 4.2B). This approach has the fundamental advantage that it is completely independent of the kinetic pathway, simplifying the analysis considerably. This consideration can be important in the study of more complex interactions (see later). In addition, the surface can be titrated in a configuration not requiring surface regeneration (Schuck *et al.*, 1998; Abrantes *et al.*, 2001).

A third very basic and highly useful type of experiment is the competition of surface binding with soluble forms of the surface-immobilized molecule. This approach uses the fact that the solution interaction leaves only a fraction of molecules free to interact with the surface sites. In the flow system, the contact time of the soluble mixture with the surface is typically too short, and the number of surface-binding molecules is typically too low, to warrant corrections in the equilibrium of the soluble molecules due to depletion caused by surface binding, even though numerical corrections can be applied. For a 1:1 interaction, the concentration of unbound protein can be calculated as

$$[A]_{free} = [A]_0 - 0.5\left\{[B] + [A]_0 + \frac{1}{K_{AB}} - \left(\left([B] + [A]_0 + \frac{1}{K_{AB}}\right)^2 - 4[B][A]_0\right)^{0.5}\right\} \tag{4.5}$$

with $[A]_{free}$ and $[A]_0$ denoting the free and the total of a soluble protein ''A'', interacting in equilibrium with a soluble protein ''B'' at concentration [B], and the

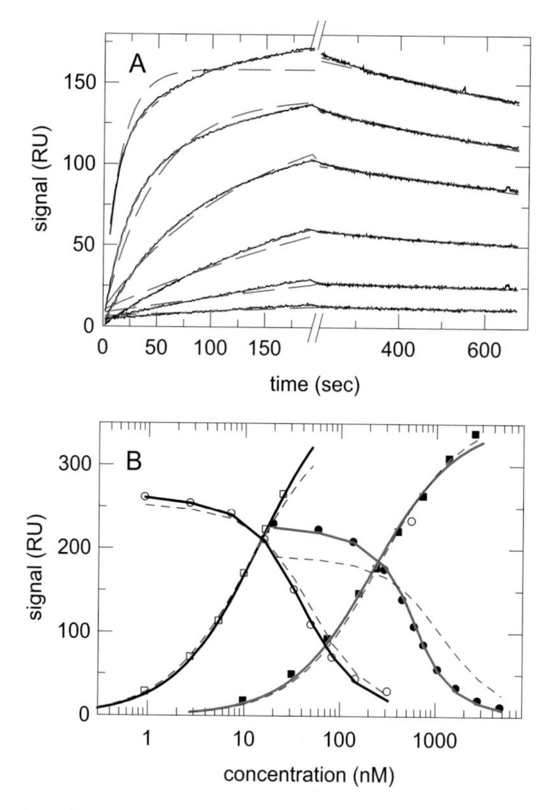

Figure 4.2. (A) Kinetic surface-binding traces of myoglobin binding to a surface-immobilized antibody, at concentrations between 4.6 and 990 nM (black lines), and global fit using a two-site model (red dotted line). For this interaction, in contrast to the data shown in this figure, Roden and Myszka (1996) presented a data set comprising of shorter contact times (<50 s) and a limited concentration range (maximum concentration 330 nM) and concluded that the interaction follows an ideal 1:1 reaction (Roden and Myszka, 1996). While such truncated data could be well-reproduced with the present surface, the best-fit single-site model to the present, more comprehensive, data set, however, is shown as blue dashed lines, clearly incompatible with the data. This highlights the necessity of collecting data with high information content (from sufficient signal/noise ratio and traces with significant curvature) to permit discrimination between different models for the surface-binding process and to arrive at a qualitatively and quantitatively reliable interpretation. For details, see Schuck *et al.* (1998). (B) Equilibrium surface-binding signals (squares) from equilibrium titration (Schuck *et al.*, 1998) for the interaction of NC10 Fab to immobilized whale neuraminidase (filled symbols) and immobilized tern neuraminidase (open symbols). The solution competition isotherm using a fixed concentration of NC10 and variable soluble neuraminidase competing with the surface sites is shown as circles. The best-fit analysis allowing for different binding constants in solution and at the surface is shown as solid lines, and the global fit assuming the affinity to be identical in the solution and at the surface is shown as dotted lines. For whale neuraminidase, this suggests the solution interaction to be ~five-fold stronger, while for immobilized tern neuraminidase the surface- and the solution-binding constants are virtually identical. For details, see Schuck *et al.* (1998).

isotherm from the concentration dependence of the free soluble analyte binding to the surface sites can be quantitatively analyzed (Figure 4.2B). Further, the surface-binding properties of the soluble analyte can be empirically characterized (e.g., using a calibration curve for either the steady-state binding or the initial slopes), and the interaction with the immobilized surface site can be understood solely as a means to generate a concentration-dependent signal. In this way, this approach allows the probing of protein interactions in a manner entirely free from potential artifacts due to surface immobilization. A combination of both approaches permits probing whether the affinity of the surface sites is affected because of conformational constraints or the microenvironment at the surface (Figure 4.2B).

4.3. LIMITATIONS OF USING SURFACE-IMMOBILIZED SITES TO STUDY PROTEIN–PROTEIN INTERACTIONS

Besides the advantageous features of using biosensors for studying protein interactions outlined earlier, they also impose certain restrictions and difficulties that arise fundamentally from the fact that the sites are localized to the surface, beyond the potential problem of functional protein immobilization. These include mass transport limitations and the analysis of multivalent analytes. Although both the role of ligand diffusion to a cell-surface receptor, lateral receptor interactions in the plane of the membrane, as well as multivalent binding may be very important factors, or perhaps sometimes even govern, the process to effect a biological function *in vivo*, the correspondence of a cell surface *in vivo*, and the SPR biosensor surface is very difficult to establish. For example, the precise replication of the surface density, the lateral distribution and mobility in the plane of the surface is not possible, although they greatly influence the effects of multivalent attachment and mass transport. Likewise, the detailed physicochemical properties of the surface and the adjacent environment can be expected to exert a great influence on mass transport. Therefore, unless a well-controlled biophysical model system can be established, for example, such as described in Chapter 3, it seems usually advantageous to attempt to use the SPR sensor as a tool to characterize the interactions in solution, and to conduct the experiment under conditions free of surface-related effects, governed only by the association and dissociation events arising from the forces between protein interfaces and not from their spatial location. Understanding the dynamics of these processes will shed light on the properties of the proteins, and serve as a basis for studying their function in the biological context. This is obviously the goal also for the study of protein interactions that take place between soluble molecules away from any surface.

4.3.1. Multivalent Analytes

A classic example of a multivalent analyte is a full-length I_gG antibody. If we consider a bivalent antibody interacting with antigen immobilized to the sensor

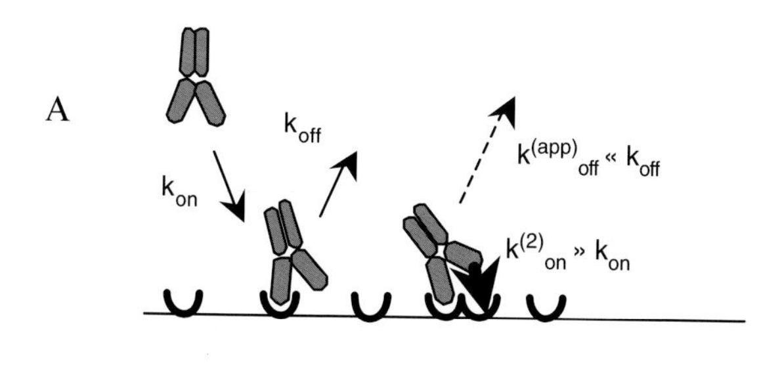

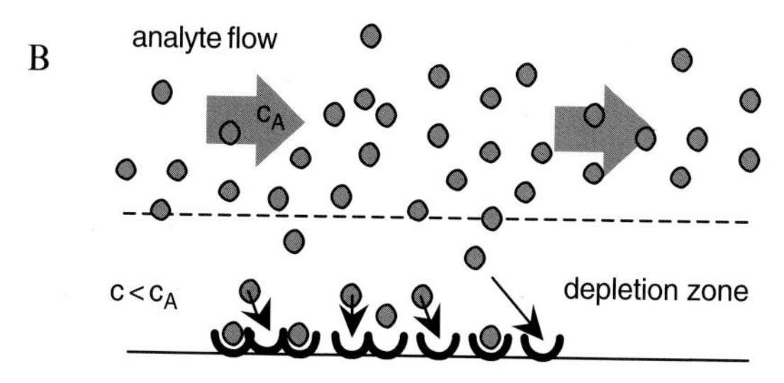

Figure 4.3. Experimental configurations of binding studies with surface-immobilized proteins exhibiting fundamental difficulties include (A) multivalent soluble analytes and (B) binding with rapid chemical kinetics and slow transport rate constants.

surface (Figure 4.3A), it may be bound either through one or through both combining sites, and since the SPR signal is sensitive essentially only to the total protein mass residing within the evanescent field, both states are indistinguishable. For an antibody molecule that has formed only a single antibody–antigen interface, but where a free antigen molecule is within accessible radius of the second combining site, the probability of making the second interaction is greatly enhanced over the probability of making the first interaction. At the same time, for the doubly bound molecule to be released from the surface requires simultaneous dissociation of both bonds, which is a statistically rare event compared with the dissociation of only one bond. Both effects strongly stabilize the surface-bound molecule, and, as a consequence, the apparent off-rate constant of whole antibody molecules dissociating from the antigen surface is frequently many orders of magnitudes slower than that of the individual antibody–antigen interfaces. The apparent K_D values may be many orders of magnitude lower than the true K_D (Ladbury *et al.*, 1995).

The computational modeling of this effect has proven very difficult due to the dependence on the spatial distribution of the surface-immobilized antigen, and this is compounded by the fact that the use of a flexible matrix for surface immobilization introduces many additional variables (Nieba *et al.*, 1996; Muller *et al.*, 1998). A possible experimental approach is the use of very low density surface immobilization, to minimize the probability of simultaneous binding of both antibody-combining sites, but it seems problematic to establish to what degree this approach is successful, since it would require spatially uniform immobilization (Nieba *et al.*, 1996). While the total amount of surface-immobilized antigen that can be controlled, it is notoriously difficult to control or even measure the spatial distribution, in particular when using an extended polymer, such as the popular carboxymethyl dextran, as an immobilization matrix.

As a consequence, the best method to reliably avoid this artifact and to permit a quantitative thermodynamic or kinetic analysis of the interaction is to immobilize the multivalent species and use soluble analytes that are only monovalent with regard to the surface-immobilized species. This also rules out the use of self-associating species as a soluble analyte, and makes very difficult the quantitative study of homogeneous protein self-association processes. However, it may be used, for example, in some configurations to characterize the process of subunit dissociation after selective immobilization of oligomeric complexes, or in fibrillation processes for the association and the dissociation of monomers to and from surface-immobilized nuclei or fibers (Hasegawa *et al.*, 2002).

The same restriction also imposes stringent requirements for sample preparation. Even at low levels, soluble oligomeric aggregates can accumulate at the sensor surface, and have been found to have considerable influence on the apparent association and dissociation kinetics. If unrecognized oligomeric impurities are present, this may lead to biphasic binding and dissociation curves, and cause significant deviations of binding constants. Depending on the size of the analyte, size-exclusion chromatography may be used to ensure the absence of relatively stable oligomeric aggregates (Davis *et al.*, 1998; Andersen *et al.*, 1999; Schuck *et al.*, 1999). As an analytical tool, sedimentation velocity analytical ultracentrifugation (see Chapter 16) is playing an important role in the highly sensitive detection of oligomers and larger aggregates below the 1% level. These may be a result of imperfect refolding of proteins, protein degradation, or low-level aggregation after protein freeze–thaw cycles. Sedimentation velocity is also a reliable tool for the detection and the characterization of reversible oligomerization.

4.3.2. Mass Transport Limitation

Transport limitation arises as a consequence of the experimental configuration in which the soluble analyte is initially not mixed well with the immobilized sites at the start of the binding experiments, and of the requirement to hold the analyte concentration constant during the experiment. If the macroscopic supply of analyte to the sensor surface is not sufficiently fast compared with the chemical reaction, the observed surface-binding kinetics reflects the characteristics of this transport step rather than the molecular binding parameters. An example for a methodology to study surface-binding kinetics that overcomes this problem is total internal reflection fluorescence correlation spectroscopy, where the kinetic information is extracted from the equilibrium fluctuations (Lieto *et al.*, 2003). In SPR, the range of on-rate constants where one may observe the onset of transport limitation is at 10^5–10^6 M^{-1} s^{-1} (Jönsson and Malmqvist, 1992; Schuck, 1997b), but this is strongly dependent on size, charge, and nonspecific binding of the analyte, as well as on the surface density of the immobilized sites (Karlsson, 1994; Schuck, 1996). Mass transport is a practical problem in many (or perhaps most) kinetic SPR-binding experiments (Karlsson *et al.*, 1994), and if disregarded, the apparent binding parameters may be up to several orders of magnitude in error. Unfortunately, the procedure to account for mass transport initially proposed by the manufacturer (Biacore) (Karlsson *et al.*, 1991) was incorrect (Glaser, 1993; Schuck and Minton, 1996). However, since then, the transport process has been well studied, and analytical and diagnostic methods have been developed.

One can examine the transport process on three different levels. First, the rate of macroscopic buffer exchange in the flow cell has been considered (Glaser, 1993; van der Merwe *et al.*, 1994; Hall *et al.*, 1996), which clearly provides estimates for an upper limit of the detectable rate constants. A more detailed view is possible considering the diffusion of the analyte through the laminar flow across the surface (Glaser, 1993; Karlsson *et al.*, 1994; Yarmush *et al.*, 1996). This predicts that the transport rate increases only with the cube root of the flow rate. Finally, an additional transport step is the diffusion through the array of binding sites within the immobilization matrix, if such a matrix is used. Even though the matrix is generally very thin (estimates for some of the commercially available surfaces range between 100 and 400 nm (Stenberg *et al.*, 1991; Yeung *et al.*, 1995)), it has been proposed that it may under some conditions be the rate-limiting step of transport (Schuck, 1996). According to Wofsy and Goldstein (2002), the latter requires the diffusion coefficient in the matrix to be substantially lower (one or two orders of magnitude) than that in the bulk. Such conditions may exist for highly functionalized matrices, where steric hindrance and electrostatic interactions with the charged polymer lead to restricted diffusion (Schuck, 1996). Further, nonspecific binding to the sensor surface can reduce the analyte mobility very substantially (Crank, 1975), even at a level where nonspecific binding causes signals only on the order of common bulk refractive index offsets (Schuck, 1997a). This question was addressed experimentally in studies with different model systems, where the presence of the dextran matrix did (Piehler *et al.*, 1999; Fong *et al.*, 2002) and did not (Karlsson and Fält, 1997; Parsons and Stockley, 1997) have an influence on the surface-binding kinetics.

A hallmark of reaction, diffusion, and convection processes in general is the formation of spatial gradients and reaction fronts. Computer simulations for the binding process in the SPR biosensor showed that under transport-limited conditions, spatial gradients within the sensing volume can form, which may generate characteristic artifacts in the measured signal due to the spatially inhomogeneous sensitivity of detection. Spatially inhomogeneous binding progress in a direction perpendicular to the surface was experimentally observed by dual-color SPR during the immobilization of streptavidin into a dextran hydrogel (Zacher and Wischerhoff, 2002). It has been proposed that spatial inhomogeneities may explain some experimentally reproducibly observed artifacts in strongly transport-limited-binding experiments, which show an increasing slope in the association phase and an increasing signal in the dissociation phase, if it follows only partial saturation of the surface sites (Schuck, 1996). In particular, the increasing signal in the dissociation phase cannot be explained from chemical kinetics alone, without invoking spatial gradients and considering the spatially nonuniform sensitivity. Besides the obvious problem is that when transport is the rate-limiting step of binding, relatively less information on the chemical kinetics can be contained in the surface-binding traces, such effects from spatial parameters governing transport and the detection of binding considerably constrain the reliability of kinetically modeling strongly mass transport-limited surface binding (see later).

The simplest form of addressing gradients in the analyte concentration is a two-compartment model. In this highly simplified model, the transport is considered in a framework similar to chemical kinetics, as an abstract-partitioning step from a well-mixed compartment into another well-mixed compartment close to the sensor surface containing the binding sites

$$\begin{aligned}\frac{\mathrm{d}c_{\mathrm{surf}}}{\mathrm{d}t} &= k_{\mathrm{tr}}(c_{\mathrm{bulk}}-c_{\mathrm{surf}})-k_{\mathrm{on}}(b_{\mathrm{tot}}-b)c_{\mathrm{surf}}+k_{\mathrm{off}}b \\ \frac{\mathrm{d}b}{\mathrm{d}t} &= k_{\mathrm{on}}(b_{\mathrm{tot}}-b)c_{\mathrm{surf}}-k_{\mathrm{off}}b\end{aligned} \tag{4.6}$$

with c_{surf} and c_{bulk} denoting the concentration of the soluble analyte close to the surface and in the bulk, b and b_{tot} the bound analyte and the surface-binding capacity, respectively, and k_{tr} an effective transport rate constant. Even though limitations of this model are obvious, since it implies the existence of well-mixed surface and flow compartments, it can serve as a first-order approximation of transport contributions to the binding kinetics for the range of chemical binding reactions at the onset of transport limitation, that is, where the molecular on-rate constants are slow enough to make the overall surface-binding kinetics essentially reaction controlled and only slightly influenced by the transport step (Schuck and Minton, 1996).

The basic features of transport limitation are reduced rate constants of both surface binding and dissociation. Close to the surface, the capture of soluble analyte by the immobilized sites proceeds at a higher rate constant than the analyte supply from the bulk flow, such that a depletion zone is established where the soluble analyte concentration is reduced (Figure 4.3B). *Vice versa*, in the dissociation phase the rate constant of dissociation is faster than the analyte can be rinsed out, such that a zone of nonvanishing analyte concentration is maintained, which can be subjected to rebinding to the empty surface sites. In both cases, the extent of transport influence is governed by the ratio of the surface-binding flux $k_{\mathrm{on}}b_{\mathrm{free}}c$ (with b_{free} denoting the concentration of free surface sites) relative to the transport flux $k_{\mathrm{tr}}c$, that is, the transport influence is governed by the chemical on-rate constant. In the association phase, the depletion zone causes a reduction in the curvature of the surface-binding progress, which, at analyte concentrations greater than K_{D}, is distinct from the exponential binding progress expected in the absence of transport limitation. If the dissociation is started from close to saturation, relatively fewer empty surface sites are available for rebinding as compared with dissociation at lower binding levels, and the resulting time course of dissociation can be empirically fit very well with a double exponential, where both apparent rate constants are below (and possibly far from) the true molecular off-rate constant (Figure 4.4). Unfortunately, at concentrations smaller than K_D, the transport-limited traces still closely follow single exponential shape, mimicking ideal binding at reduced rate constants.

For the onset of transport influence on the surface-binding kinetics, the two-compartment model predicts (under steady-state conditions for the free analyte close to the sensor surface) the following influence on the apparent rate constants

$$\frac{k_{\text{on,app}}}{k_{\text{on,true}}} = \frac{k_{\text{off,app}}}{k_{\text{off,true}}} = \frac{1}{1 + b_{\text{free}}(t)k_{\text{on,true}}/k_{\text{tr}}}, \tag{4.7}$$

where the subscripts app and true refer to the apparent and true on- and off-rate constants, and k_{tr} denotes the transport rate constant. Eq. (4.7) shows that the transport effect scales directly with the number of available surface sites. Therefore, limiting the immobilized density of sites is an important strategy to minimize mass

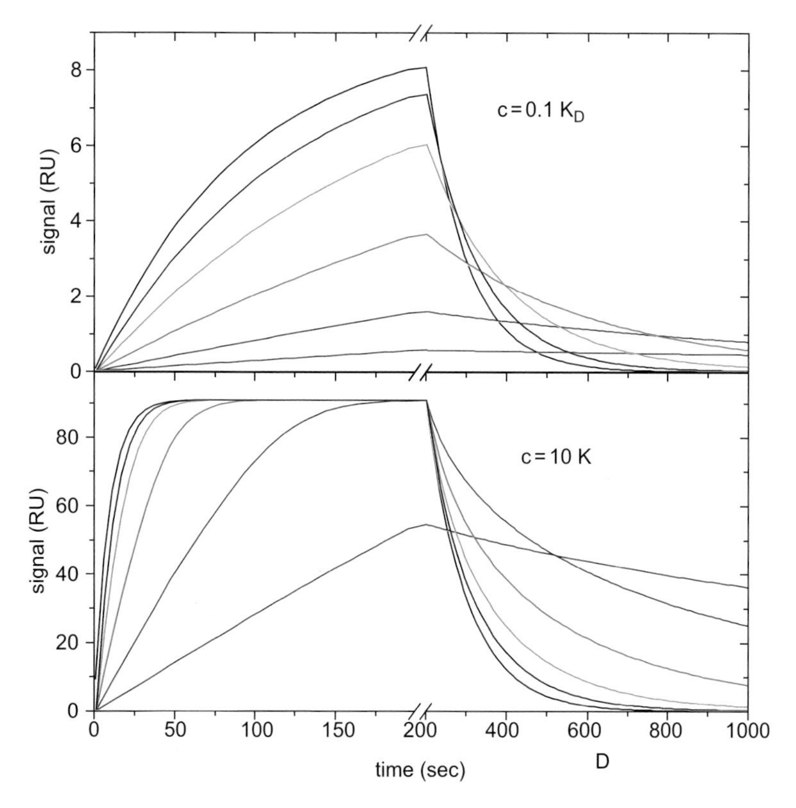

Figure 4.4. Transport-limited-binding kinetics for different analyte concentrations. Simulated with $k_{on}{}^{*}r_{max}/k_{tr} = 0$ (black), 0.33 (blue), 1 (cyan), 3.3 (green), 10 (magenta), and 33 (red). At low concentrations ($c < K_D$), traces are close to single exponentials, whereas at high concentrations ($c > K_D$) traces show a linearized association and a double exponential dissociation phase.

transport, and the variation of surface density of the immobilized sites is an excellent test for the presence of mass transport influence. In the context of a global fit of binding progress to the different surfaces, this may allow the derivation of estimates of the magnitude of k_{tr}. Another excellent qualitative test is the injection of a soluble form of the immobilized binding partner during the dissociation phase, which can serve as a competitor and prevent rebinding to the surface sites and lead to an accelerated dissociation if transport limitation is present.

4.4. STUDYING INTERACTING SYSTEMS WITH MULTIPLE COMPONENTS, BINDING SITES, OR CONFORMATIONAL STATES

Most of the fundamentals of cellular processes rely on the formation of complexes of multiple proteins, as they exceed the mechanistic limitations of simple binary binding reactions. The identification of vast networks of interacting proteins within the cell (McCraith *et al.*, 2000; Uetz *et al.*, 2000; Walhout *et al.*, 2000; Ito *et al.*, 2001; Rain *et al.*, 2001; Gavin *et al.*, 2002; Ho *et al.*, 2002; Giot *et al.*, 2003; Li *et al.*, 2004) has highlighted the interdependency of many of these processes and their reliance on such multiprotein complexes. Analysis of the associations of multiple proteins will likely become only more important in the post-genomic era as focus shifts from genomes to interactomes, the networks of protein–protein interactions encoded by whole genomes. While the rules that govern the interaction between two individual proteins in forming a bimolecular complex are yet to be fully elucidated (Bogan and Thorn, 1998; Lo Conte *et al.*, 1999; Sundberg and Mariuzza, 2000, 2002; Ma *et al.*, 2001), the association of more than two binding partners in a single multiprotein complex introduces further levels of complexity to the binding reaction that must also be addressed.

One hallmark of multiprotein complexes is cooperativity. As the affinity of proteins for their ligands is a fundamental property that determines the dynamic range within which they operate, binding affinity either gained or lost via cooperative interactions makes important contributions to the functionality of the resulting multiprotein complex (Germain and Stefanova, 1999; Courey, 2001). To understand fully the role of cooperative binding in protein function, it is necessary to describe quantitatively each binary reaction that together comprises the multiprotein association and how each of these reactions is affected energetically by others in the overall complex. Studying the binding mechanisms of multiprotein complexes, however, is complicated by the intricacy of the reaction schemes, as well as the need for structural information and highly homogeneous sources of purified protein. While the rate at which atomic structures of multiprotein complexes are described continues to increase, the energetic analysis of these higher-order molecular interactions, in which cooperativity is likely to be a frequent attribute, has lagged well behind.

When studying systems of more than two binding partners or with multiple sites or conformational states, a variety of experimental configurations are possible in the SPR biosensor. To illustrate the potential of the technique, we consider three systems: (1) binding of A and B where A can undergo a conformational change to A^*; (2) a molecule A with multiple binding sites for B; and (3) a molecule A that has separate sites for two molecules B and C.

4.4.1. Multiple Conformational States

Several authors have considered the presence of different conformational states and binding-induced conformational changes in the interaction proteins, including quantitative studies of antibody–antigen interactions (Glaser and Hausdorf, 1996; Lipschultz *et al.*, 2000) and other protein–protein interactions (De Crescenzo *et al.*, 2000; Honjo *et al.*, 2002; Khursigara *et al.*, 2005), as well as mapping of binding-induced conformational changes in proteins with conformationally specific antibodies (e.g., Dubs *et al.*, 1992; Cohen *et al.*, 1995; Fischer *et al.*, 1996).

For the reactions $A \overset{K_C}{\longleftrightarrow} A^*$, $A+B \overset{K}{\longleftrightarrow} (AB)$, and $A^*+B \overset{K^*}{\longleftrightarrow} (AB)^*$, let us denote K_C as the equilibrium constant for the conformational transition, and K and K^* the equilibrium constants for the formation of the complexes (AB) and $(AB)^*$, respectively. Since the SPR signal is proportional to the mass and refractive index increment, and in practice essentially insensitive to conformations (sensitivity to conformation is possible with waveguide sensors that use two evanescent fields with different polarization and decay length, thus obtaining more information on the geometric orientation of the protein at the surface (Salamon *et al.*, 1999; Alves *et al.*, 2004)), the experimentally measured surface-bound complex is $([ab]+[ab]^*)$ (with the lower case letters denoting species concentrations). The mass action law for the formation of the experimentally measurable complex $([ab]+[ab]^*)$ from $(a+a^*)$ and b follows an equilibrium constant $K_{app} = K(1 + K_C K^*/K)(1 + K_C)$; thus, not surprisingly, it cannot be inferred from an equilibrium experiment that we have different conformations of A, or if there is preferential binding of B to either one.

This situation is different in kinetic experiments, if the protein with the multiple conformational states is immobilized to the sensor surface (here one assumes that the immobilization procedure does not affect the ability of the protein to adopt the entire range of conformations that it does in solution). In this case, binding progress follows the rate equations

$$\begin{aligned}
\frac{da}{dt} &= -k_{f,a}a + k_{r,a}a^* - k_{on}ab + k_{off}c \\
\frac{da^*}{dt} &= +k_{f,a}a - k_{r,a}a^* - k_{on}^* a^* b + k_{off}^* c^* \\
\frac{dc}{dt} &= -k_{f,c}c + k_{r,c}c^* + k_{on}ab - k_{off}c \\
\frac{dc^*}{dt} &= +k_{f,c}c - k_{r,c}c^* + k_{on}^* a^* b - k_{off}^* c^* \\
S(t) &\sim c(t) + c^*(t)
\end{aligned} \tag{4.8}$$

with a and c denoting the surface concentration of the free surface-immobilized protein and the complex, respectively, b the concentration of the soluble analyte, the asterisk referring to the alternate conformation, $k_{f,a}$ and $k_{r,a}$ denoting the forward and the reverse rate constants for conformational transition of the unliganded molecule A, and the subscript a and c indicating the analogous rate constants for conformational change of the free molecules and the complex, respectively. The most interesting feature of this situation occurs if the two conformations result in different complex stabilities, as illustrated in Figure 4.5. Initial rapid binding to the low-affinity (high k_{off}) sites is followed by continuous conversion of the low affinity with the more slowly dissociating (generally high-affinity) sites (Figure 4.5, red and magenta lines). This introduces a history dependence of the dissociation process, which is indicative of more complex interactions. At an equivalent binding level, the dissociation is slower for configurations with longer contact times preceding the dissociation (providing more time for conversion of the sites). Such a history dependence cannot be explained by the existence of multiple independent classes of surface sites. The analysis of binding with conformational changes is reviewed by Lipschultz *et al.* (2000).

Unfortunately, this history-dependent behavior is not unique to systems with conformational changes. Very similar curves are observed for binding to a single class of surface sites, if the analyte is heterogeneous and consists of multiple, competitively binding species with different binding properties (circles in Figure 4.5) (Svitel *et al.*, 2003). This situation may arise if the soluble protein exists in different conformations with different binding properties (in the flow system, the time which the analyte spends close to the sensor surface is typically too short to allow conformational transitions of the soluble analyte to be relevant before it is being replenished). Alternatively, the analyte may exist in mixtures of monomeric and multimeric forms that differ in their binding properties, for example, due to avidity effects (see earlier).

It can be difficult to distinguish between these cases on the basis of the SPR-binding data alone. This has generally been observed for models with complex kinetic-binding schemes (Glaser and Hausdorf, 1996; Karlsson and Fält, 1997; Schuck, 1997a; De Crescenzo *et al.*, 2000). This is rooted in the well-known problems of fitting noisy exponentials, reminiscent of data analysis problems in many other biophysical disciplines, some of which are illustrated in other chapters. The ambiguity can be resolved, however, if independent information is available from other biophysical methods that can report on conformational changes. For example, De Crescenzo and colleagues (2000) discuss the consistency of the conformational change model for the interaction of transforming growth factor α with epidermal growth factor receptor extracellular domain with the results from circular dichroism (Greenfield *et al.*, 1989). Other techniques to study conformational changes upon binding are described in other chapters, including hydrodynamic, spectroscopic, and mass spectrometry approaches.

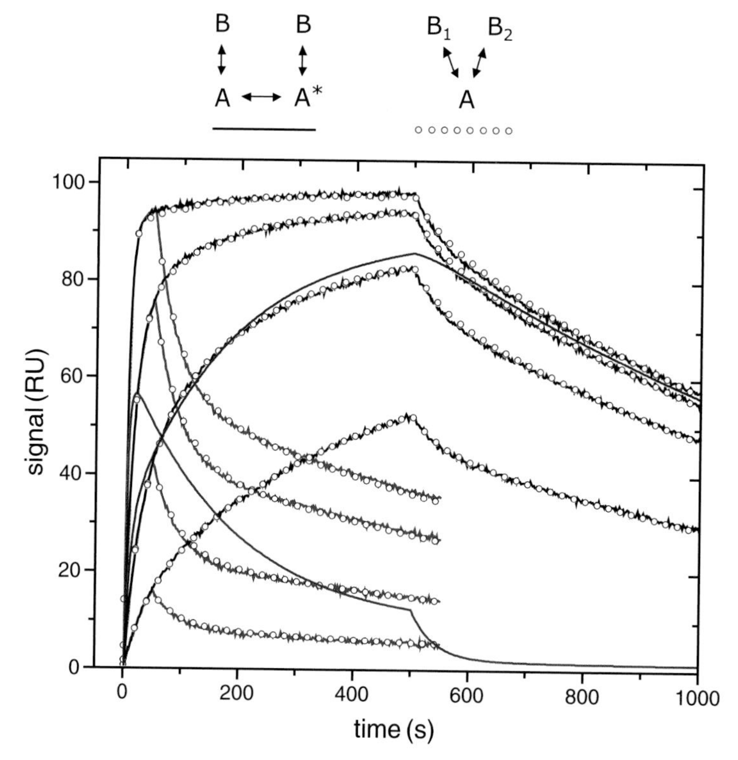

Figure 4.5. Illustration of surface-binding kinetics with surface sites undergoing a conformational change. Binding curves are simulated for a low-affinity conformation ($K_{D,1} = 100\,\text{nM}$, $k_{dis,1} = 2 \times 10^{-2}\ \text{s}^{-1}$) and a high-affinity conformation ($K_{D,2} = 5\,\text{nM}$, $k_{dis,2} = 5 \times 10^{-4}\ \text{s}^{-1}$) with slow conformational transition ($k_{cf,12} = 1 \times 10^{-2}\ \text{s}^{-1}$) and with initially two-thirds of the unliganded surface sites in the low-affinity state, and with a shift of the conformational equilibrium toward the high-affinity site due to stabilization by the ligand (tenfold excess of high-affinity sites in equilibrium). Surface-binding curves are for soluble analyte concentrations of 30, 100, 300, and 1,000 nM, for 50 and 500 s contact time (solid blue and black lines, respectively). It should be noted that the dissociation after long contact times is much slower than that after short contact times. This is due to the slow conformational transition of the bound and the free surface sites. For illustration, the conversion of the complex with low affinity into high-affinity conformation is indicated for the highest analyte concentration by solid red and magenta lines, respectively. The dependence of the off rate on the contact time is qualitatively incompatible with the existence of multiple conformationally stable surface sites. However, very similar binding progress can be obtained for single, stable surface sites, if the analyte is heterogeneous in its binding properties. In this case, lower-affinity analyte is replaced during the contact time by higher-affinity, slower dissociating analyte. The best-fit traces for a mixture of two analytes binding competitively to the same surface site are indicated as circles. In the case shown, the binding is calculated for a low-affinity analyte with $K_{D,1} = 133\,\text{nM}$ and $k_{dis,1} = 2.2 \times 10^{-2}\ \text{s}^{-1}$ and a high-affinity analyte with $K_{D,2} = 7\,\text{nM}$, $k_{diss,2} = 9.5 \times 10^{-4}\ \text{s}^{-1}$, with the second site at a constant 46% of the total analyte concentration. The root-mean-square (rms) deviation between the model of two surface sites with conformational equilibrium and the model of competitively binding analyte mixtures is 0.7 RU, which is experimentally virtually indistinguishable.

4.4.2. Multiple Binding Sites

The second system we consider is a soluble protein B binding to multiple independent surface sites $A_1, A_2, \ldots A_n$, each with different binding properties. This model can describe binding to a surface-immobilized protein with different, independent epitopes, binding to an intrinsically heterogeneous ensemble of proteins at the surface (e.g., due to sequence variability or variation in glycosylation), or it may be used to account for partially inactivated sites from surface immobilization or from constraints imposed by heterogeneity in the microenvironment of the surface sites. Frequently, the number of existing classes of surface sites may not be known *a priori*, or the binding properties may vary quasicontinuously for different subclasses of sites. A useful approach is the description of the surface sites as a continuous distribution $c(k_{\text{off}}, K_{\text{D}})$, where the differential $c(k_{\text{off}}, K_{\text{D}})\text{d}k_{\text{off}}\,\text{d}K_{\text{D}}$ is the surface concentration of sites with off-rate constants between k_{off} and $k_{\text{off}} + \text{d}k_{\text{off}}$ and with equilibrium constant between K_{D} and $K_{\text{D}} + \text{d}K_{\text{D}}$ (Svitel *et al.*, 2003). The experimentally measured kinetic-binding curves can be modeled as

$$S(b,t) \cong \int_{K_{\text{D,min}}}^{K_{\text{D,max}}} \int_{k_{\text{off,min}}}^{k_{\text{off,max}}} c(k_{\text{off}}, K_{\text{D}})\, s\,(k_{\text{off}}, K_{\text{D}}, b, t)\, \text{d}k_{\text{off}}\, \text{d}K_{\text{D}}, \tag{4.9}$$

where $S(b, t)$ are binding signals at times t and concentrations of soluble analyte b, and with $s(k_{\text{off}}, K_{\text{D}}, b, t)$ denoting the standard pseudo first-order binding kinetics Eqs (4.2–4.4). Equation (4.9) can be modified with a compartment-like transport step to account for the binding kinetics at the onset of transport-limited binding (Svitel *et al.*, submitted). This represents a data transformation from the space of surface-binding signals to a space of binding constants, reminiscent to a Laplace transform of exponentials. Although the numerical solution of Eq. (4.9) is unstable for high discretization of k_{off} and K_{D}, it can be combined with maximum entropy or Tikhonov regularization approaches that will determine the simplest distribution consistent with the experimental data (at the given signal and noise level) (Hansen, 1998). This regularization strategy has been introduced by Provencher (1982) very successfully to the analysis of dynamic light-scattering data in the software CONTIN. The approach in Eq. (4.9) is similar to the concept of using sedimentation coefficient distributions $c(s)$ in analytical ultracentrifugation (Schuck, 2000) to account for all sedimentable material (chapter 16). Likewise, the surface site distribution allows experimental imperfections, such as arising from immobilization, baseline drifts and nonspecific binding, as well as multiexponential binding from many specific sites, to be transformed from the original data space into the space of binding constants, where they can be identified and excluded from further consideration, without biasing or oversimplifying the analysis of the sites of interest.

Figure 4.6 shows the application to different antibody–antigen interactions, and illustrates the gain in quality of fit if the heterogeneity of the surface sites is

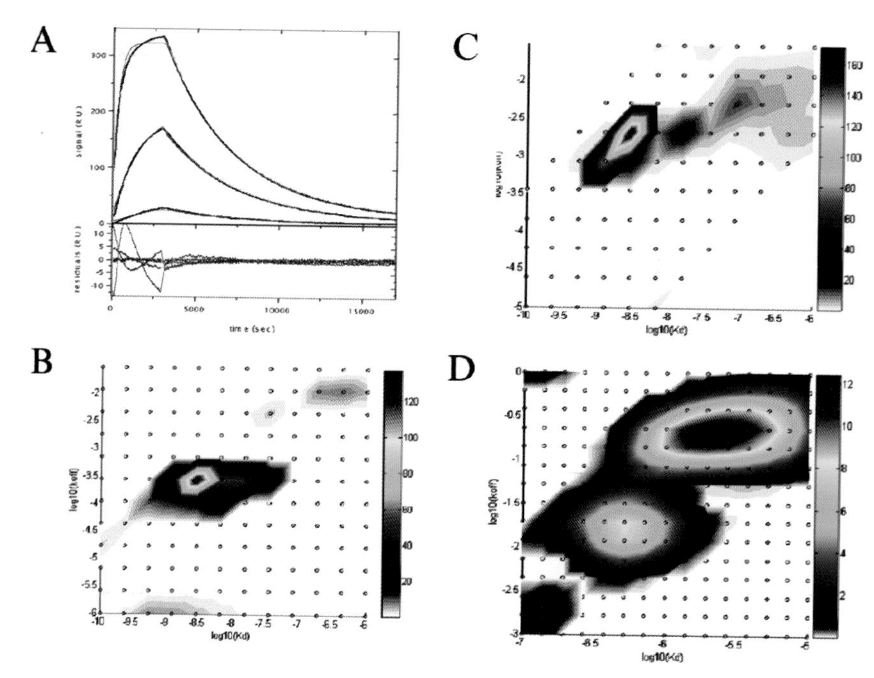

Figure 4.6. Distributions of surface sites calculated from families of experimental kinetic-binding curves. (A) (*top*) Binding traces for the antigen B5R (0.5, 5, and 50 nM) binding to antibody 19C2 (Schmelz *et al.*, 1994) (a kind gift from Dr Berhard Moss), immobilized on a long-chain carboxymethyl dextran (Biacore CM5) (Chen *et al.*, 2006). Experimental traces (black) and best-fit assuming a single site model (blue lines). (A) (*bottom*) Residuals of the fit with a single site model (blue lines, rms deviation (rmsd) = 3.4 RU) and with the distribution model (red lines, rmsd = 0.43 RU). (B) Distribution reveals a low level of heterogeneity, indicating a small subpopulation of slightly lower-affinity sites over a narrow K_D range. The circles indicate the gridpoints for the numerical solution of Eq. (4.9). (C) Distribution of surface sites for transport-limited binding of β_2-microglobulin to an immobilized monoclonal antibody (Svitel *et al.*, submitted). (D) The distribution of sites for binding of a soluble Fab fragment of a variant of mAb CC49 to immobilized bovine mucin (Svitel *et al.*, 2003). The CC49 antibody recognizes the tumor-associated glycoprotein TAG-72 via epitopes. They are the trisaccharide Galβ(1–3)[NeuAcα(2–6)]GalNAc and the disaccharide structure [NeuNAcα(2–6)]GalNAc, linked to serine or threonine side chains (Hanisch *et al.*, 1989). Both the disaccharide and the trisaccharide structures are also present in bovine and ovine submaxillary mucins (Reddish *et al.*, 1997). Therefore, the interaction of the Fab with immobilized mucin is an example where intrinsically multiple classes of ligand are present. Accordingly, the two main peaks of the distribution have been attributed to the trisaccharide and the disaccharide structures.

accounted for. This model has been applied, for example, in a study of the ligand recognition of $\alpha_x\beta_2$ integrin (Vorup-Jensen *et al.*, 2005), and in the characterization of antibody–antigen interactions (Chen *et al.*, submitted). An analogous (one-dimensional) distribution of affinity constants can be applied to the isotherm analysis of steady-state binding data, as applied to the analysis of anthrax toxin residues interacting with the neutralizing antibody (Rosovitz *et al.*, 2003).

4.4.3. Trimolecular Interactions

Perhaps the most interesting system in the present context is a ternary interaction of protein A binding two proteins B and C on separate sites, forming complexes [AB], [AC], and [ABC]. If A is immobilized to the sensor surface, binding can be probed with B, C, and mixtures of B and C. Mutual interactions of B and C may be detected through deviations of the binding signal of the mixture from being a linear superposition of the predetermined binding curves of B and C alone. The steady-state binding signal S as a function of solution concentrations b and c follows the isotherm

$$S(b,\, c) = \frac{s_{\mathrm{A,active}}}{M_{\mathrm{A}}} \frac{M_{\mathrm{B}} b K_{\mathrm{AB}} + M_{\mathrm{C}} c K_{\mathrm{AC}} + (M_{\mathrm{B}} + M_{\mathrm{C}}) ab\, \alpha K_{\mathrm{AB}} K_{\mathrm{AC}}}{1 + bK_{\mathrm{AB}} + cK_{\mathrm{AC}} + ab\, \alpha K_{\mathrm{AB}} K_{\mathrm{AC}}} \tag{4.10}$$

with $s_{\mathrm{A,active}}$ denoting the surface concentration, in signal units, of the active fraction of the immobilized species A, and the factors M_{A}, M_{B}, and M_{C} are the molecular weights accounting for the signal contributions due to the different size of the binding partners (a further correction should be applied for differences in refractive index of the components). The constant α indicates positive ($\alpha > 1$) or negative ($0 < \alpha < 1$) cooperativity.

Figure 4.7 shows binding isotherms expected for the steady-state signal of equimolar mixtures of B and C. For positive cooperativity, the isotherm becomes slightly steeper and exhibits a midpoint at significantly lower concentrations than the K_{D} (for simplicity, it is assumed in this illustration that $K_{\mathrm{AB}} = K_{\mathrm{AC}}$, but similar conclusions can be drawn in the general case). In contrast, in the presence of negative cooperativity, the binding at low concentration is less affected since both molecules B and C can bind to different molecules of A available at the surface. However, the isotherm becomes broader and biphasic, as it requires much higher concentrations to exceed half-saturation of the surface sites.

In some regard, the possibility at lower concentration of parallel binding to different surface sites may represent a practical limitation of this configuration. Therefore, we next consider an arrangement where B is immobilized to the surface, and both A and C are soluble. If C does not interact with B, it can bind to the surface only via A as a bridge. If we assume that A and C are applied to the sensor surface as an equilibrium mixture, the concentrations of free A and of the complex [AC] in solution can be calculated from the solution interaction isotherm [Eq. (4.5)], and the total surface-binding isotherm then follows

$$S(a_{\mathrm{tot}}, c_{\mathrm{tot}}) = \frac{s_{\mathrm{B,active}}}{M_{\mathrm{B}}} \frac{M_{\mathrm{A}} K_{\mathrm{AB}} a_{\mathrm{free}} + (M_{\mathrm{A}} + M_{\mathrm{C}}) \alpha K_{\mathrm{AB}} [\mathrm{AC}]}{1 + K_{\mathrm{AB}} a_{\mathrm{free}} + \alpha K_{\mathrm{AB}} [\mathrm{AC}]}. \tag{4.11}$$

The steady-state surface-binding signal dependent on the total concentrations of A and C exhibits several characteristic features (Figure 4.8). First, at any given concentration of A, the coinjection with C to form soluble complex [AB] that can bind to the surface will increase the signal, as reflected by the lines in Figure 4.8A at

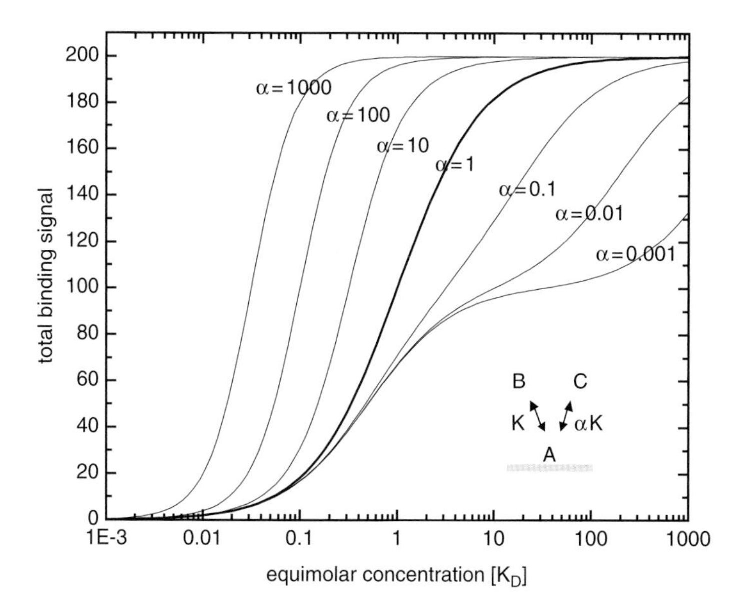

Figure 4.7. Steady-state binding curves for two soluble ligands B and C binding to the same surface-immobilized protein A, with different cooperativity factors α in Eq. (4.10). It is assumed that B and C are in equimolar concentration, have the same signal contribution, and that both have the same binding constant to A ($K_{AB} = K_{AC}$).

constant a_{tot} and varying c_{tot}. Interestingly, the perpendicular lines, showing the signal for a given concentration of soluble ligand c_{tot} and varying concentration a_{tot} will exhibit a maximum. This occurs because as we are increasing the molar excess of A over C, free A will increasingly compete with [AC] for the available surface-binding sites, and due to the smaller mass of A compared with [AC], this will result in a decrease of the measured signal. The effect of cooperativity is that of modulating the affinity of [AC] to the surface relative to free A. Positive cooperativity will lead to a stronger enhancement of surface binding, for example, visible in the higher slope of lines at constant a_{tot} and increasing c_{tot} (Figure 4.8B). For negative cooperativity, the same lines of constant a_{tot} can exhibit a qualitatively different feature, in the form of a negative slope (Figure 4.8C). This is due to competition for A between soluble C and surface-immobilized B, and can be clearly visible even at relative low concentrations of A. An application of this model will be described later for the study of ternary interactions of a bacterial SAG, MHC molecules, and TCR fragments (see Section 4.4). For the cooperative multistep association of RANTES with heparin, a related, but more complex, binding scheme was proposed by Vivès *et al.* (2002), which was described with a model for the equilibrium isotherm binding.

This approach of the coinjection of mixtures of soluble molecules is suitable also for kinetic experiments. Examples are the kinetic study of the ternary

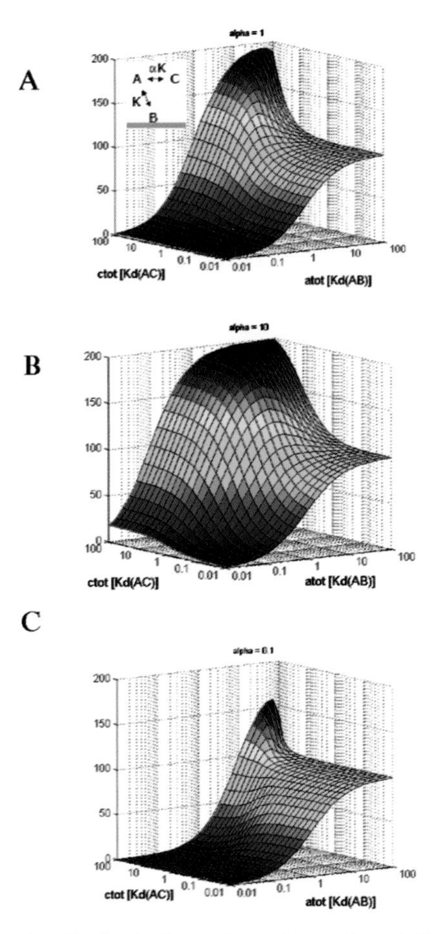

Figure 4.8. Steady-state surface-binding isotherms for two interacting soluble proteins A and C, which can form a reversible soluble complex [AC], binding to the surface-immobilized protein B forming surface-bound complexes [AB] and [ABC]. The effect of different cooperativity factors α from Eq. (4.11) is shown. It is assumed that A and C have the same signal contributions. The total soluble concentrations of A and C are expressed in units of the equilibrium dissociation constants. The cooperativity factors are 1 (A), 10 (B) and 0.1 (C), respectively.

interaction of TCR, MHC class II, and SAG SEA (Redpath *et al.*, 1999) in the interaction of fibroblast growth factor, its receptor, and heparin (Ibrahimi *et al.*, 2004), and the interactions of insulin-like growth factor-binding proteins (Beattie *et al.*, 2005).

For systems with suitably slow dissociation, the kinetic experiments allow for an even more direct visualization of the second binding site for A, if sequential injections of A and C are used, such that C comes in contact only with the preformed surface-bound complex AB (Figure 4.9). This configuration exploits the ability of the flow system to separate soluble A from the complex captured

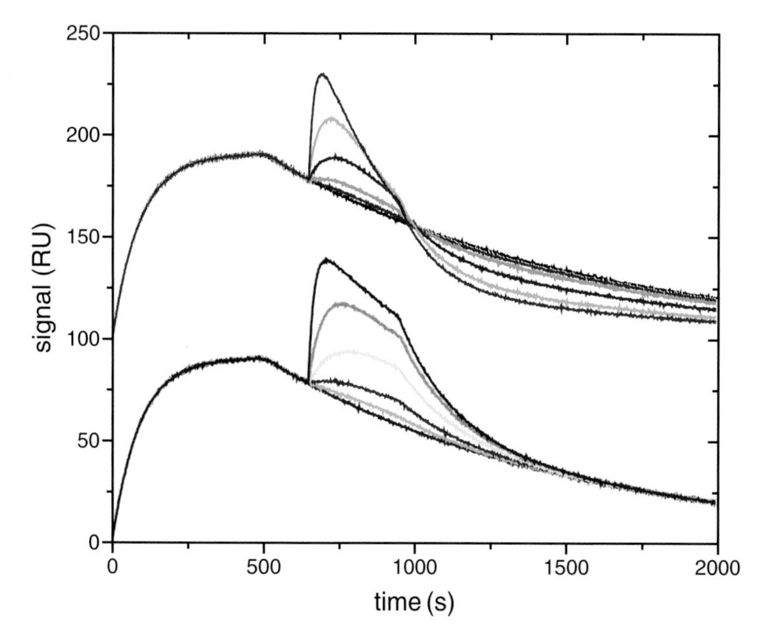

Figure 4.9. Examples of kinetic traces of triple protein interactions. Simulated binding curves of a soluble protein A binding to surface-immobilized surface sites B, followed by binding of protein C to the preformed surface-bound complex [AB]. Binding curves are calculated for binding of A with an equilibrium constant $K_{D,AB} = 10\,nM$, an off-rate constant of $k_{off,AB} = 1 \times 10^{-3}\ s^{-1}$, and with maximal capacity of 100 RU units. Contact time of A with the surface-immobilized B is 500 s. After 200 s of dissociation in a buffer wash, protein C is introduced into the flow for 300 s at concentrations of 10 (red), 30 (green), 100 (blue), 300 (cyan), or 1,000 nM (magenta), followed by rinsing with buffer. C is assumed to have the same mass as A, resulting in the same signal contribution. Binding of C to preformed [AB] proceeds with $K_{D,C[AB]} = 100\,nM$ and $k_{off,AB} = 5 \times 10^{-3}\ s^{-1}$. For reference, the undisturbed dissociation of A without introduction of C is indicated by the black line. The lower curves assume no cooperativity of C on the interaction of A–B. In contrast, the upper curves (offset by 100 for clarity) illustrate the case where C destabilizes the interaction of A–B, with a threefold enhancement of dissociation [CA]–B over A–B.

at the sensor surface. Since the complex will continuously dissociate, these experiments can generally only be conducted in conjunction with a global kinetic analysis of the binding curves. The rate equations during the dissociation follow

$$\begin{aligned}
\frac{d(ab)}{dt} &= -k_{off,a}(ab) - k_{on,C}(ab)c_{sol} + k_{off,C}(abc) \\
\frac{d(abc)}{dt} &= k_{on,C}(ab)c_{sol} - (k_{off,AC} + k_{off,C})(abc) \qquad (4.12) \\
S(t) &\sim M_A(ab)(t) + (M_A + M_B)(abc)(t)
\end{aligned}$$

with ab and abc denoting the surface concentrations of the complexes [AB] and [ABC], respectively, c_{sol} the solution concentration of C, $k_{off,a}$ the off-rate constant of A from B, $k_{on,C}$ and $k_{off,C}$ the on- and off-rate constants of C binding to A, and $k_{off,AC}$ the off-rate constant of the complex [AC] dissociating from B. Families of curves with different c_{sol} are shown in Figure 4.9. If binding of C to the preformed complex [AB] destabilizes the interface of A and B, a characteristic acceleration of overall surface dissociation may be discerned. This can be taken as a qualitative sign for such a destabilizing effect of C. As with the experimental configurations described earlier for studying this system, it cannot be deduced from the data in Figure 4.9 if the observed cooperativity between the ligands B and C stems from conformational changes propagated through A, or from the proximity of B and C on A leading to repulsive or attractive interactions between the two ligands. As will be shown later, this question may be addressed with control experiments analyzing interactions of B and C, and in combination with a structural technique. In addition, circular dichroism can be useful to detect conformational changes in the triple complex (Arthos *et al.*, manuscript in preparation).

The kinetic analysis in this configuration is described in more detail by Joss *et al.* (1998). It has been applied, for example, by Walsh *et al.* (2003) to the study of human growth hormone (hGH) receptor–ligand interactions. hGH forms a 1:2 ligand–receptor complex involving two distinct receptor sites. In this study, the extracellular domain containing site 1 of the receptor was immobilized on a dextran matrix, saturated with soluble hGH, and the binding kinetics of soluble receptor with the preformed 1:1 complex was studied. Alanine mutations in hGH site 2 were inserted to study the role of side chains contributing to the overall binding energy of site 2 and the cooperativity of their contributions (Walsh *et al.*, 2003). In a different configuration, Bernat *et al.* (2003) employed a dextran-free carboxylated surface for the immobilization of the hGH receptor, such that no lateral receptor interactions are permitted and upon saturation with hGH only hGH/receptors site 1 complexes can be formed, which, in turn, can capture soluble receptor to form the 1:2 complex, making the latter reaction accessible for study.

Sequential binding studies can be carried out with systems of greater complexity than those discussed so far. For example, Schuster *et al.* (1993) have exploited a configuration of sequential injection of proteins participating in a quaternary chemotactic signal transduction complex. For qualitative information, even higher-order multiprotein complexes have been studied by sequential binding of soluble ligands, such as the DNA polymerase III holoenzyme complexes (Dallmann and McHenry, 1995) and multi-antibody–antigen complexes in context of antibody epitope mapping (Fägerstam *et al.*, 1990).

For the study of trimolecular interactions, a third experimental configuration is possible; the coupling of two different molecules to the same surface (in the terminology used earlier, it would represent the surface immobilization of both B and C, to probe the reversible interaction of soluble A). This seems attractive when studying interactions with multiple copies of cell-surface receptor molecules. However, when using a polymeric immobilization matrix, a possible concern for the quantitative interpretation of binding in this orientation appears to be the unknown relative mobility of the immobilized molecules within the matrix and the possibility of energetic constraints from the microenvironment influencing their ability to interact with each other.

In different situations, this approach has been applied in the literature to study receptor–ligand interactions (Cunningham and Wells, 1993; Myszka *et al.*, 1996; Stokes *et al.*, 2005). In the study of hGH receptor–ligand interactions, Cunningham and Wells (1993) established that by introducing mutations in different residues of the receptor, the mode of soluble ligand binding to immobilized receptor could be controlled, allowing either only 1:1 complexes via site 1 of the receptor, or the formation of ligand-induced receptor dimerization via both sites 1 and 2. This demonstrated sufficient mobility of the immobilized hGH receptor in the dextran matrix to permit formation of ternary complexes. Changes in the binding properties induced by a series of alanine mutations of residues involved in site 1 binding of hGH were measured and the relative rate constants for formation of the 1:1 complex determined, which revealed the energetic contributions of the hGH side chains to the overall binding energy of the binary complex of site 1. A quantitative interpretation of ternary ligand–receptor interactions involving mixed receptor surfaces was attempted by Myszka *et al.* (1996). Both α and β subunits of the interleukin-2 (IL-2) receptor were sequentially immobilized, and binding of soluble IL-2 was observed. Slightly faster surface binding and significantly slower dissociation was measured for the mixed surface compared with the surfaces with each of the individual receptor subunits immobilized alone. However, the binding kinetics was obscured by strong mass transport limitation, and for the reasons outlined earlier, it is an open question how much the immobilization of two subunits into the dextran matrix would affect the energetics for triple complex formation (Myszka *et al.*, 1996). In a different system, Stokes *et al.* (2005) created mixed surfaces containing meningococcal transferrin-binding proteins A and B, integral membrane proteins of the outer membrane, using supported dimyristoyl phosphatidylcholine (DMPC) monolayers in the absence of a dextran matrix. Human transferrin binding to the mixed surface was more stable than binding to surfaces containing either one of the receptors alone (Stokes *et al.*, 2005). A powerful approach for studying receptor subunits attached to supported lipid bilayers, the interactions in the plane of the membrane, and their role in ligand binding was developed by Piehler (see Chapter 3).

The above-mentioned three cases may serve as examples for the experimental configurations that can be encountered in the analysis of complex interacting systems with SPR biosensing. Extensions to more complex systems may be possible, although it should be noted that the information content from SPR-binding traces is frequently limited due to the problem that the experimentally measured exponential kinetic-binding curves and sigmoid steady-state isotherms may be fitted equally well with many different models. Further, more complex models can be significantly more susceptible to experimental imperfections in the sample purity, aggregation state, and detection. SPR shares these difficulties with many other techniques. Therefore, it is frequently crucial to combine the SPR data with other complementary techniques to alleviate the problem of amibuity in the modeling.

4.5. A PRACTICAL APPLICATION TO THE STUDY OF MULTIPROTEIN COMPLEXES

Instead of compiling a comprehensive literature review, we will highlight in the following the potential of SPR biosensing for the quantitative analysis of the thermodynamics of ternary complex formation by providing a more in-depth description of the application of SPR biosensing to the simultaneous interaction of a SAG with MHC and TCR molecules. This interaction exhibits both positive and negative cooperativity, and the interpretation of the SPR data is aided by the availability of structural models derived from X-ray crystallographic studies. It is likely that similar analyses are possible for other ternary interactions.

4.5.1. Superantigen–Major Histocompatibility Complex–T-Cell Receptor Interactions

SAGs are immunostimulatory and disease-associated proteins of bacterial or viral origin that bind simultaneously to class II MHC and TCR molecules on the surfaces of antigen-presenting cells and T lymphocytes, respectively (Figure 4.10A) (Sundberg *et al.*, 2002). Contrary to the presentation of processed antigenic peptides by MHC to TCR, SAGs bind to MHC molecules outside of the peptide-binding groove and interact with TCR Vβ domains, resulting in the stimulation of a large fraction (up to 5–20%) of the T-cell population. In this way, SAGs are able to circumvent the normal mechanism of T-cell activation associated with antigenic peptide–MHC complexes, leading not only to massive T-lymphocyte proliferation, but also T-cell anergy and death. Accordingly, SAGs have been implicated in the pathogenesis of a number of human diseases, including toxic shock syndrome (Bohach *et al.*, 1990), food poisoning (Kotzin *et al.*, 1993), and several autoimmune disorders (Renno and Acha-Orbea, 1996), and are classified as bioterror reagents.

We studied the bacterial SAG, *Staphylococcus* enterotoxin C3 (SEC3), in complex with the human MHC class II molecules human lymphocyte antigen (HLA)-DR1 and a murine TCR (Andersen *et al.*, 2002). In this system, the MHC molecule and the TCR do not form a specific binary complex in the absence of SEC3. Structures of each of the relevant binary complexes that comprise the MHC–SEC3–TCR ternary complex, including the SEC3–MHC class II (Sundberg *et al.*, 2003), αβTCR heterodimer (Garcia *et al.*, 1996), and SEC3–TCRβ-chain (Fields *et al.*, 1996) complexes, have made it possible to assemble a trimolecular structural model. As shown in Figure 4.10A, this model predicts that the complex is stabilized through three distinct interfaces: SEC3–MHCα subunit, SEC3–TCRβ-chain, and TCRα-chain–MHCβ subunit.

The direct MHC–TCR interaction in this supramolecular complex has been verified by biochemical studies (Andersen *et al.*, 1999). The binding energy derived from the direct contacts between the TCR and the MHC is accounted for in terms of cooperativity in the triple complex, since these molecules would not interact in the

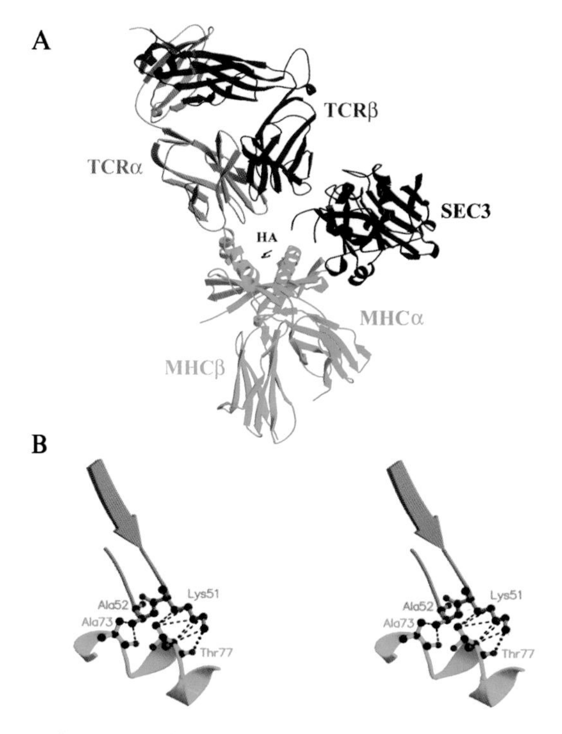

Figure 4.10. Structural model of the T-cell receptor–*Staphylococcus* enterotoxin C3–DR1 (TCR–SEC3–DR1) complex. (A) Model of the major histocompatibility complex–SEC3–TCRαβ (MHC–SEC3–TCRαβ) heterodimer complex produced by superimposing the human lymphocyte antigen–DR1–SEC3 (HLA–DR1 (HA 306–318)–SEC3) complex (Andersen *et al.*, 2002), the SEC3–14.3.d TCRβ-chain complex (Fields *et al.*, 1996), and the 2C TCRαβ heterodimer complex (Garcia *et al.*, 1996). Overlapping SEC3 and TCRβ molecules have been removed for clarity. Colors are as follows: SEC3, blue; MHCα subunit, green; MHCβ subunit, cyan; HA (306–318) antigenic peptide, magenta; TCRα-chain, orange; and TCRβ-chain, red. (B) Molecular modeling of the DR1β subunit–TCRα-chain interface. Interactions between residues Lys51 and Ala52 from the CDR2 loop of the Vα and residues Ala73 and Thr77 from the DR1β subunit. Van der Waals interactions are indicated by dashed lines and a potential hydrogen bond between the N^{ζ} atom of Lys51 and the main chain carbonyl O of Thr77 is indicated by a dotted line. Intermolecular contacts were defined by atomic pair distances (in Å) less than or equal to the following: C–C, 4.1; C–N, 3.8; C–O, 3.7; N–N, 3.4; N–O, 3.4; O–O, 3.3. Colors are as follows: MHCβ subunit, cyan; TCRα-chain, orange; carbon atoms, black; nitrogen atoms, blue; and oxygen atoms, red. Reproduced from Andersen *et al.* (2002) with permission.

absence of SEC3. Molecular modeling of this binding site (Figure 4.10B) indicates a relatively small protein–protein interface (with a buried surface area of only $\sim 400\,\text{Å}^2$) consisting of numerous Van der Waals interactions and a single hydrogen bond. To assess the cooperative energetics involved in the formation of the ternary complex, the contribution of the TCRα-chain–MHCβ subunit contacts must be assessed directly by comparing trimolecular complex formation in both the presence and the absence of the TCRα-chain. This is possible with this system because biochemical and crystallographic studies have demonstrated that the isolated TCRβ-chain can interact functionally with SEC3 (Gascoigne and Ames, 1991; Malchiodi *et al.*, 1995; Fields *et al.*, 1996).

Figure 4.11E and F shows the response of a fixed concentration of soluble SEC3 binding to immobilized αβTCR heterodimer and the TCRβ-chain, respectively, in the presence of increasing concentrations of MHC. It is evident that increasing concentrations of MHC caused increasing accumulation of mass in the TCRαβ-coupled surface (Figure 4.11E), consistent with the binding of preformed SEC3–MHC complexes to the TCR (note that without SEC3, MHC does not bind TCR). In addition, the kinetics became significant slower, indicating enhanced stability of the trimolecular complex. In contrast, the TCRβ-chain-coupled surface responded with a lower signal when the concentration of MHC was increased (Figure 4.11F). This demonstrates that SEC3–MHC complexes have lower affinity to the TCRβ-chain than does SEC3 alone, representing negative cooperativity.

As with all SPR analyses, the experimental approach is necessarily system specific. This is especially true when studying ternary complexes. In principle, several different approaches could have been taken for analyzing the MHC–SEC3–TCR interaction (Figure 4.11A–D). Due to the relative immobilization capacities and solubilities of the constituent molecules in the supramolecular complex, the experimental design possibilities were, in reality, limited. The MHC molecule consists of three noncovalently associated polypeptides and is characterized by a relatively low isoelectric point, and thus, is intolerant to the chemical procedures necessary for amine coupling to the biosensor surface. This excluded the experimental setup depicted in Figure 4.11B. The SEC3 molecule, in contrast, is very stable and has a neutral p*I*, and therefore is not restricted in its immobilization. Thus, the experimental setup of Figure 4.11A could be used. The full-length TCR, expressed as a soluble αβ heterodimer in insect cells, had a strong tendency to aggregate. Thus, only few measurements could be made with soluble TCR immediately after size-exclusion chromatography purification, as it would not remain stable throughout the data collection period (24–48 h at ambient temperature). The αβTCR heterodimer could be immobilized (Figure 4.11C). Furthermore, it was also possible to immobilize the TCRβ-chain alone in a functional state, thereby allowing for direct comparison of complex formation in the absence of presence of the cooperative binding site on the TCRα-chain (Figure 4.11D).

Initially, each of the relevant bimolecular interactions was analyzed. The binding of SEC3 to surface-immobilized αβTCR heterodimer was evaluated by injecting increasing concentrations of SEC3 (Figure 4.12, lower insert). For the MHC–SEC3 interaction, the binding constant in solution was determined by a solution competition assay. First, SEC3 was immobilized and serial dilutions of MHC were applied to yield the K_D to the immobilized SEC3 (Figure 4.12, upper insert, in blue). Second, the binding between SEC3 and MHC in solution was determined by mixing a constant DR1 concentration with serial dilutions of SEC3 and passing these mixtures over a SEC3-coupled surface. The solution K_D was then

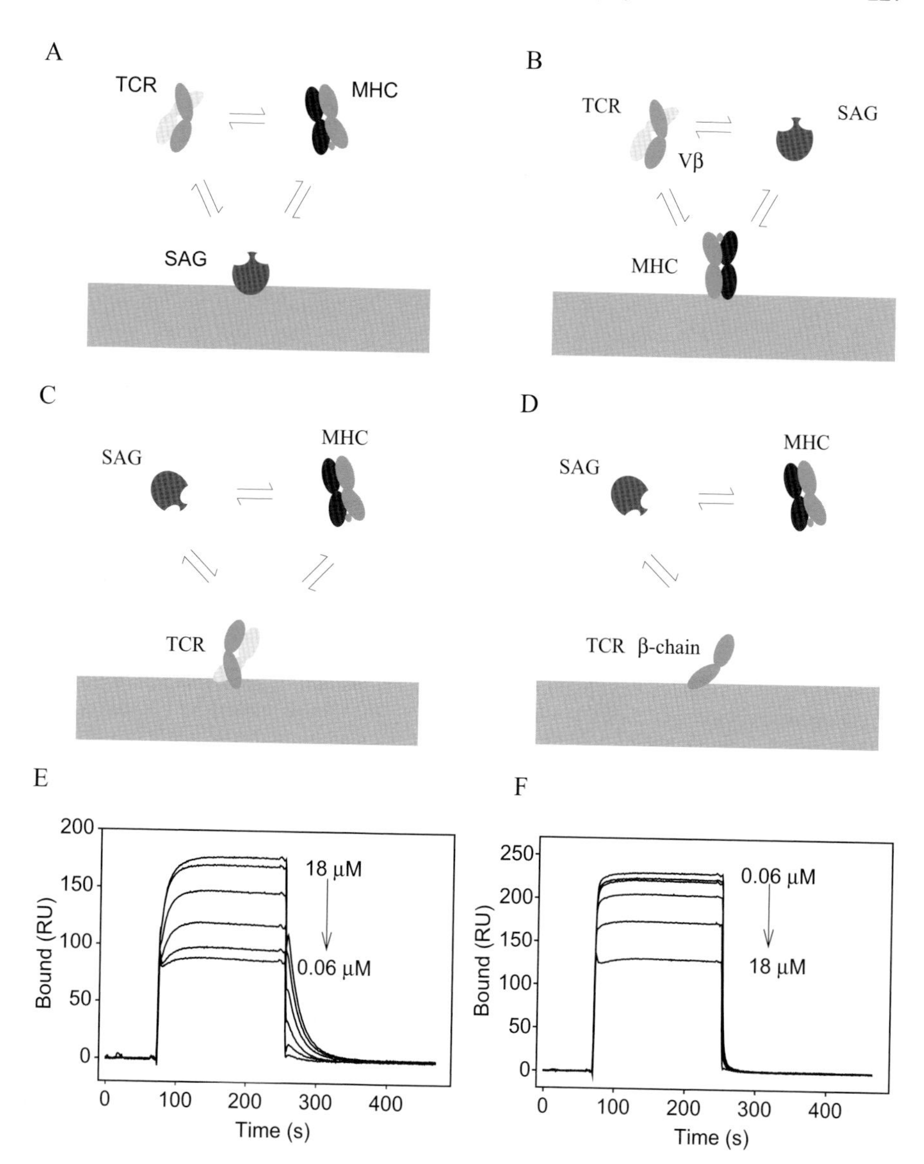

Figure 4.11. (A–D): Possible orientations for studying the interaction between T-cell receptor (TCR) or TCRβ-chain, superantigen (SAG), and major histocompatibility complex (MHC) molecules. (E, F) Positive and negative cooperative effects are involved in ternary complex formation. (E) Overlay sensorgrams of soluble *Staphylococcus* enterotoxin C3 (SEC3) (3.8 μM) binding to immobilized TCRαβ heterodimer in the presence of increasing concentrations of DR1 (0.06, 0.18, 0.60, 1.8, 6.0, and 18 μM as indicated). (B) SEC3 and DR1 binding to immobilized TCRβ-chain. Concentrations of SEC3 and DR1 are as in (E).

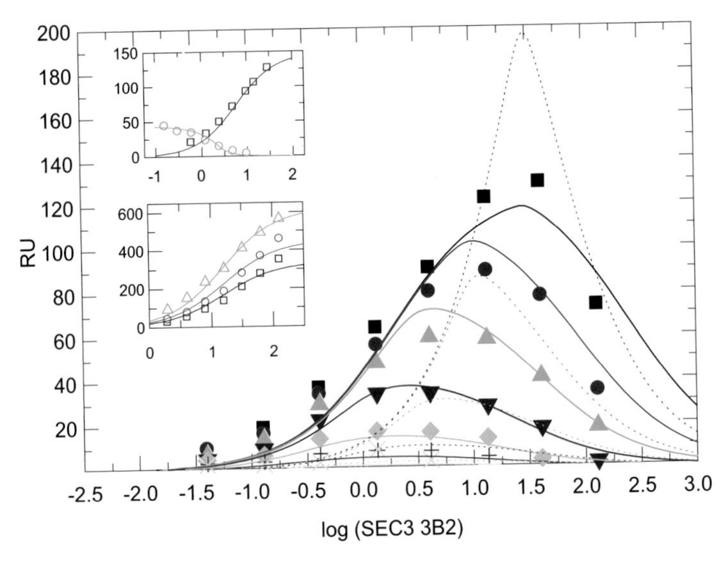

Figure 4.12. Fitting of ternary complex binding results. Representative plots of the global fitting procedure. Solid lines are calculated best-fit isotherms for a model including cooperative binding. For comparison, dotted lines represent the best-fit isotherms calculated in the absence of cooperative binding. Amount of ternary complex formed on immobilized TCRαβ heterodimer using varying concentrations of *Staphylococcus* enterotoxin C3 (SEC3) mixed with DR1 at 30 μM (black squares), 9.5 μM (red circles), 3.0 μM (green triangles), 1.0 μM (blue trangles), 0.3 μM (light blue diamonds), 0.1 μM (purple crosses), and 0.03 μM (yellow crosses). These isotherms reflect the ratio of free SEC3 to SEC3–DR1 complex in solution offered to the immobilized TCRαβ heterodimer. At a fixed concentration of DR1, increasing SEC3 leads to higher SEC3–DR1 complex concentrations available for ternary surface complex formation. At SEC3 concentrations higher than an optimal value, most surface sites will be occupied with unliganded SEC3. The data shown in the insets contain the information on the binary SEC3–T-cell receptor (TCR) and SEC3–DR1 interactions. Upper insert shows dilutions of DR1 binding to immobilized SEC3 (blue squares) and dilutions of SEC3 competing out binding of DR1 to immobilized SEC3 (green circles). Lower insert shows binding of SEC3 to two TCRαβ heterodimer coupled surfaces (red circles and black squares) and one TCRβ surface (green triangles). Reproduced from Andersen *et al.* (2002) with permission.

determined by fitting equilibrium-binding data to a competition model (Figure 4.12, upper insert, in green), using Eq. (4.5). Determination of their solution-binding constant enables the confident prediction of the relative abundance of soluble SEC3–DR1 complexes in the samples used for the study of the ternary complex formation.

The ternary complex formation was explored by acquiring binding data of mixtures of soluble SEC3 and soluble MHC interacting with either surface-immobilized TCRαβ heterodimers, surface-immobilized TCRβ-chains, or blank sensor surfaces (for correcting the signals for contributions arising from the refractive index difference of the sample plugs from the running buffer). SEC3–DR1 mixtures were applied covering a range of concentrations and molar ratios.

To quantify the interaction, a two-dimensional binding isotherm was constructed from the binding data (Figure 4.12). Because the dissociation of SEC3 from TCRαβ heterodimers (in the absence of DR1) was much faster than the dissociation of the ternary complex, and because the DR1 did not bind to the TCR surfaces, the amount of ternary complex formed on TCRαβ heterodimer-coupled surfaces could be estimated from the dissociation phases by extrapolation to the start of the dissociation phase using the known half-life of the ternary complex of 22 s. In contrast, for the surface with the TCRβ-chain, the ternary complex dissociated much faster (Andersen *et al.*, 1999) and such a selective observation as for triple protein complex was not possible. Instead, the total surface binding at equilibrium was recorded at the end of the association phase for each mixture of SEC3 and DR1.

The final data sets consisted of 199 and 135 unique data points from the binary- and the ternary-binding isotherms, respectively. These data were subjected to global analysis, using the formulas for the binary surface binding and solution competition isotherms, Eqs (4.3) and (4.5), respectively, and the triple complex isotherm introduced earlier in Eq. (4.11). The parameters included the equilibrium association constant $K_{D,S(surf)}$ for the titration of immobilized SEC3 with soluble DR1, the solution interaction constant $K_{D,S}$ between soluble SEC3 and DR1 in the competition experiment, the binding constant $K_{S,T}$ of soluble SEC3 with immobilized TCR, and the binding constant $K_{DS,T}$ of the preformed soluble DR1–SEC3 complex to immobilized TCR.

Results are presented in Table 4.1 for both αβTCR heterodimers and TCRβ-chains. We found the overall free energy change due to cooperative binding ($\Delta G_{\alpha\beta}$)

Table 4.1. Summary of binding constants

	Unit	SEC3	SEC3
Previous estimates			
$K_{D(SEC3/TCR)}$	μM	22*	–
$K_{D(imm.\ SEC3\ or\ SEC3/sol.\ DR1)}$	μM	270*	4.6*
Independent fits—bimolecular interactions			
$K_{D(SEC3/TCR)}$	μM	–	18 ± 2
$K_{D(imm\ SEC3/sol.\ DR1)}$	μM	–	4.3 ± 1.4
$K_{D(sol.SEC3/sol.\ DR1)}$	μM	–	0.26 ± 0.07
Global fits of ternary complex—bimolecular interactions			
$K_{D(SEC3/TCR)}$	μM	–	22 ± 7
$K_{D(imm\ SEC3/sol.\ DR1)}$	μM	–	4.3 ± 1.4
$K_{D(sol.\ SEC3/sol.\ DR1)}$	μM	–	1.25 ± 1.0
Global fits of ternary complex–cooperative interactions			
$\Delta G_{\alpha\beta}$	kcal mol^{-1}	–	−1.6 ± 0.3
$\Delta G_{v\beta}$	kcal mol^{-1}	–	0.8 ± 0.3
$\Delta G_{v\alpha}$	kcal mol^{-1}	–	−2.4 ± 0.1

*Data from Andersen *et al.* (2002).

to be $-1.6 \pm 0.3\,\text{kcal mol}^{-1}$. By removing the TCRα-chain, and thereby disrupting the cooperative basis for ternary complex formation, we measure a significant unfavorable cooperative free energy ($\Delta G_{V\beta}$) of $0.8 \pm 0.3\,\text{kcal mol}^{-1}$, which indicates that simultaneous binding of TCR and DR1 to SEC3 has significant energetic costs. Finally, the energetic contribution of the DR1β subunit–TCRα-chain interactions ($\Delta G_{V\alpha} = \Delta G_{\alpha\beta} - \Delta G_{V\beta}$) was estimated to be $-2.4 \pm 0.1\,\text{kcal mol}^{-1}$. This translates into an $\sim$50-fold increase in SEC3 affinity for TCR as a consequence of favorable cooperative interactions and illustrates the significance of the TCR Vα domain in maintaining the biological activity of SEC3. The higher stability of the ternary complex leads to a lower minimal SEC3 concentration at which the ternary complex can be formed, as well as higher SEC3 concentrations at which self-inhibition (i.e., the competition of free SEC3 with SEC3–MHC complexes for TCR binding) takes place. Thus, as a consequence of cooperative binding, the concentration range over which SEC3 can perform its biological function, that is, cross-linking TCR and MHC molecules on apposing cell surfaces, is expanded.

4.5.2. Temperature-Dependent Binding Analysis

For binary interactions, a full set of thermodynamic parameters may be derived by measuring association and dissociation rates at multiple temperatures. This was shown elegantly using SPR for pMHC–TCR interactions (Boniface *et al.*, 1999; Willcox *et al.*, 1999; Garcia *et al.*, 2001), and has since been used for SAG–TCR interactions (Yang *et al.*, 2003). Estimates for the enthalpy, entropy, and heat capacity of binding can be derived from the van't Hoff analysis of affinity as a function of temperature. Further, Arrhenius plots of $\ln k_a$ or $\ln k_d$ versus $1/T$ determines the slope of the curve from which the enthalpies of association ($\Delta H_a^{\ddagger}$) and dissociation ($\Delta H_d^{\ddagger}$), respectively, can be determined. These, in turn, provide the activation energies of the interaction, where $\Delta H_a^{\ddagger} = -E_a^{(ass)} - RT$, $\Delta H_d^{\ddagger} = -E_a^{(diss)} - RT$, and $\Delta H^{\ddagger} = \Delta H_a^{\ddagger} - \Delta H_d^{\ddagger}$. High activation energy values indicate that association and dissociation of the complex are impeded by significant energetic barriers. Large, unfavorable entropic changes upon binding can be explained by conformational changes that result in a decrease in mobility of flexibility of the binding sites (i.e., local folding) or from changes in solvent (i.e., incorporation of ordered water molecules in the molecular interface).

When multiple temperature SPR experiments are performed with great care (i.e., to eliminate any aggregates before injection of the analyte), any nonlinearity of $\ln k_a$ or $\ln k_d$ versus T plots that is observed is indicative of significant changes in heat capacity. For rigid-body associations, changes in heat capacity can often be predicted accurately based on changes in buried surface area upon complex formation (Spolar and Record, 1994). Large discrepancies between calculated and observed heat capacities indicate that the association likely involves conformational changes. Induced-fit mechanisms may be further supported through dissection of the total entropy changes. ΔS_{assoc} for protein–ligand interactions includes several contributions and can be expressed as $\Delta S_{\text{assoc}} = \Delta S_{\text{HE}} + \Delta S_{\text{tr}} + \Delta S_{\text{other}}$, where ΔS_{HE} represents entropy changes from hydrophobic effect, ΔS_{tr} entropy changes from rotational and translational entropies of the molecules, and ΔS_{other} any other entropic effects such as vibrational or conformational entropies. For rigid-body interactions, ΔS_{other} is negligible; for induced-fit mechanisms ΔS_{other} is significantly negative. ΔS_{other} values calculated for processes known to couple local or long-range folding to site-specific binding have been reported, and the number of residues that change conformations upon binding can be estimated (Spolar and Record, 1994). This topic is further explored in Chapter 8 on the extended interface.

Even in the absence of rigorous quantitative description of the interaction, this type of analysis may be used to provide qualitative information on the role of protein flexibility in binding. Thus, to explore if molecular flexibility might be involved in the SEC3-mediated cross-linking of the αβTCR heterodimer and MHC, the binding characteristics of various combinations of these three proteins were analyzed at temperatures ranging from 10°C to 30°C at 4°C intervals (Figure 4.13). Binding of MHC to immobilized SEC3 demonstrated temperature-dependent reaction kinetics, with binding kinetics slowing significantly as the temperature was decreased (Figure 4.13A). The most likely mediators of temperature dependence in the SEC3–MHC interaction are the N-terminal residues of the SEC3 disulfide loop, known to be highly flexible, which align to, and make contact with, an α-helix of the MHC molecule (Sundberg *et al.*, 2003). In contrast, the kinetics of SEC3 binding to the αβTCR heterodimer were significantly less temperature dependent (Figure 4.13B). Combining SEC3 and MHC in solution and monitoring for binding to immobilized αβTCR heterodimer (Figure 4.13C) showed a significant change in temperature effects relative to SEC3 binding in the absence of MHC, with the kinetics of this reaction being slower and more highly temperature dependent. This suggests that modes of conformational flexibility still present in the binary complexes may be lost when forming a ternary complex. The increase in temperature dependence of SEC3 binding to the αβTCR heterodimer in the presence of DR1, may be the result of flexibility contributed by the MHCβ subunit–TCRα-chain interface (Figure 4.10C) in the ternary complex. Accordingly, no increase in temperature dependence on binding kinetics is seen in complexes involving only the TCRβ-chain, indicating that SEC3 alone likely does not undergo substantial conformational change upon simultaneous binding to TCR and MHC.

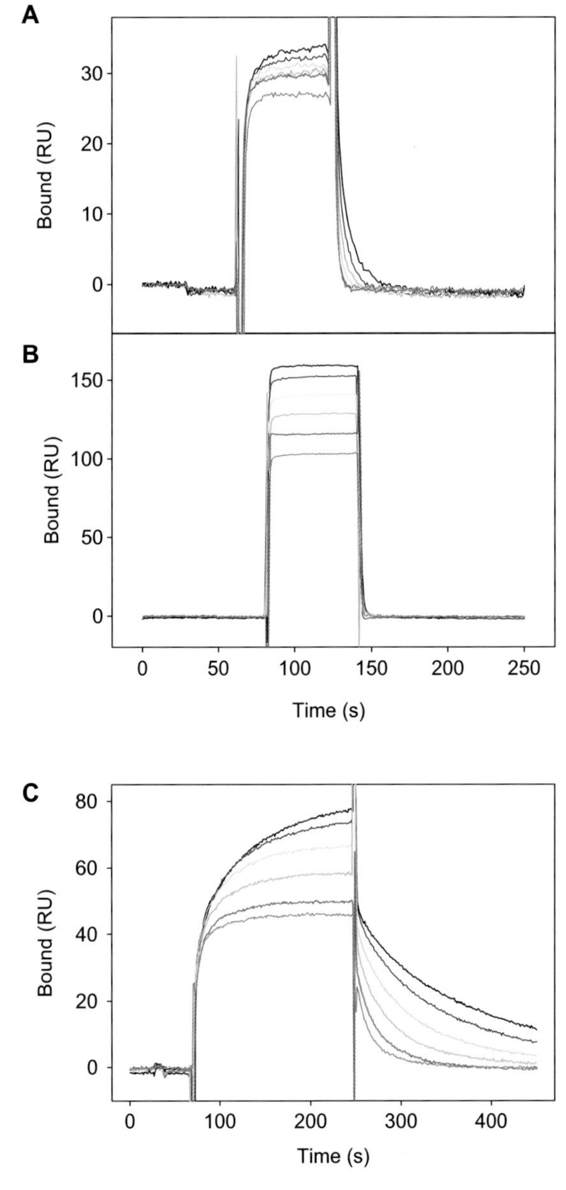

Figure 4.13. The effect of temperature on *Staphylococcus* enterotoxin C3 (SEC3) binding. Overlay sensorgrams of binding reactions at selected equidistant temperatures ranging from 10°C to 30°C. (A) Binding of 8 μM DR1 to immobilized SEC3. (B) Binding of 20 μM SEC3 to immobilized TCRαβ heterodimer. (C) Mixture of 4 μM SEC3 and 3 μM DR1 binding to immobilized TCRαβ heterodimer. Colors of sensorgram traces are as follows: 10°C, black; 14°C, red; 18°C, yellow; 22°C, green; 26°C, blue; and 30°C, magenta. Reproduced from Andersen *et al.* (2002) with permission (we correct the color label of the legend).

4.6. CONCLUSIONS

SPR biosensing can be a very powerful tool in the study of multiprotein complexes and protein interactions involving multiple conformational states. This is amply demonstrated by the numerous examples in the literature of well-controlled studies, where SPR biosensing provided unique insights into the kinetics and the thermodynamics of triple and higher-order protein complexes, revealing cooperativity of binding and details such as the contribution of protein flexibility to the binding process.

SPR biosensing is an extremely versatile approach. Due to the diversity of systems of interacting proteins and the configurations used in their study as described in the literature, we have not provided a systematic account, but tried only to highlight promising opportunities for study. However, as illustrated in the in-depth application to the TCR–MHC–SAG interaction, only particular experimental approaches may appear practical for a given system. For multiprotein interactions, particularly attractive are the configurations exploiting the temporary purification of binary complex through reversible analyte capture, which provide important opportunities for quantitative study. Another advantageous feature of SPR biosensing is the comparatively low amount of material required, even though a well-controlled study may frequently require auxiliary application of other biophysical and biochemical techniques (such as size-exclusion chromatography) with higher protein consumption. Like most biophysical techniques, SPR biosensing also gains in strength considerably when conducted in parallel with other techniques, in particular for elucidating complex-binding reactions. For example, this includes combinations with analytical ultracentrifugation (to study oligomeric state of analyte proteins, and to provide complementary information on the number and stoichiometry of complexes, as well as solution-binding constants for moderate to weak interactions), isothermal titration calorimetry (providing orthogonal thermodynamic information in free solution), and CD (for the study of conformational changes upon binding).

With the technology maturing, critical knowledge has been accumulated about how to conduct reliable studies, including the requirements to generate data with sufficient information content, the nature of potential artifacts, and the potential pitfalls of unquestioned idealizing assumptions, all factors that may lead to quantitatively and qualitatively wrong conclusions if unrecognized. It also has become apparent that SPR shares analytical problems with many other techniques that deal with the analysis of smoothly increasing or decreasing exponential data sets. As a consequence, configurations appear advantageous when different qualitative features of the protein interaction correspond to qualitative features of the recorded binding traces, rather than details in the quantitative analysis (since even though the latter may be very reproducible, this does not necessarily imply that they can be reliably interpreted).

Many of the current limitations are related to the frequently unknown microenvironment of the protein at the surface, and to the fact that a single type of binding signal is the only source of information to probe the immobilized protein and the interaction. New developments may enable the extension to technology with multiple signals reporting with multiple evanescent fields (and sensitivity). This includes, for example, dual-color SPR (Zacher and Wischerhoff, 2002), use of waveguides (Lukosz, 1991; Salamon *et al.*, 1998; Cross *et al.*, 2003; Alves *et al.*, 2004), combination with fluorescence detection (Liebermann and Knoll, 2000), and many other optical detection approaches (Gauglitz, 2005). It may also be combined with lateral resolution, such as in SPR microscopy (Rothenhäusler and Knoll, 1988; Shumaker-Parry and Campbell, 2004). That a large fraction or even a majority of binding progress is transport limited shows that the sensing technology has to be accompanied with an efficient mass transfer mechanism to further optimize the utility of this method in the study of proteins interactions. An example for a related biosensor approach exploiting multiple detection methods can be found in Chapter 3.

ACKNOWLEDGMENT

We thank Dr Raimund Ober for critically reading the manuscript.

REFERENCES

Abrantes, M., Magone, M. T., Boyd, L. F., and Schuck, P. (2001). Adaptation of the Biacore X surface plasmon resonance biosensor for use with small sample volumes and long contact times. *Anal Chem* 73:2828–2835.

Alves, I. D., Cowell, S. M., Salamon, Z., Devanathan, S., Tollin, G., and Hruby, V. J. (2004). Different structural states of the proteolipid membrane are produced by ligand binding to the human delta-opioid receptor as shown by plasmon-waveguide resonance spectroscopy. *Mol Pharmacol* 65:1248–1257.

Alves, I. D., Park, C. K., and Hruby, V. J. (2005). Plasmon resonance methods in GPCR signaling and other membrane events. *Curr Protein Pept Sci* 6:293–312.

Andersen, P. S., Lavoie, P. M., Sekaly, R. P., Churchill, H., Kranz, D. M., Schlievert, P. M., Karjalainen, K., and Mariuzza, R. A. (1999). Role of the T cell receptor alpha chain in stabilizing TCR–superantigen–MHC class II complexes. *Immunity* 10:473–483.

Andersen, P. S., Schuck, P., Sundberg, E. J., Geisler, C., Karjalainen, K., and Mariuzza, R. A. (2002). Quantifying the energetics of cooperativity in a ternary protein complex. *Biochemistry* 41:5177–5184.

Atanasov, V., Knorr, N., Duran, R. S., Ingebrandt, S., Offenhausser, A., Knoll, W., and Koper, I. (2005). Membrane on a chip: a functional tethered lipid bilayer membrane on silicon oxide surfaces. *Biophys J* 89:1780–1788.

Beattie, J., Phillips, K., Shand, J. H., Szymanowska, M., Flint, D. J., and Allan, G. J. (2005). Molecular recognition characteristics in the insulin-like growth factor (IGF)-insulin-like growth factor binding protein-3/5 (IGFBP-3/5) heparin axis. *J Mol Endocrinol* 34:163–175.

Bernat, B., Pal, G., Sun, M., and Kossiakoff, A. A. (2003). Determination of the energetics governing the regulatory step in growth hormone-induced receptor homodimerization. *Proc Natl Acad Sci USA* 100:952–957.

Bogan, A. A. and Thorn, K. S. (1998). Anatomy of hot spots in protein interfaces. *J Mol Biol* 280:1–9.

Bohach, G. A., Fast, D. J., Nelson, R. D., and Schlievert, P. M. (1990). Staphylococcal and streptococcal pyrogenic toxins involved in toxic shock syndrome and related illnesses. *Crit Rev Microbiol* 17:251–272.

Boniface, J. J., Reich, Z., Lyons, D. S., and Davis, M. M. (1999). Thermodynamics of T cell receptor binding to peptide-MHC: evidence for a general mechanism of molecular scanning. *Proc Natl Acad Sci U S A* 96:11446–11451.

Brecht, A. and Gauglitz, G. (1995). Optical probes and transducers. *Biosens Bioelectron* 10:923–936.

Chen, Z., Earl, P., Damon, I., Zhou, Y.-H., Yu, F., Sebrell, A., Emerson, S., Americo, J., Cohen, G. H., Svitel, J., Schuck, P., Satterfield, W., Moss, B., and Purcell, R. (2006). A chimpanzee monoclonal antibody to vaccinia virus B5R protein protects mice against virulent vaccinia virus and neutralizes vaccinia and smallpox viruses in vitro. *J Infect Dis* 193:625–633.

Cohen, P., Simon, D., Badouaille, G., Mani, J. C., Portefaix, J. M., and Pau, B. (1995). New monoclonal antibodies directed against the propart segment of human prorenin as a tool for the exploration of prorenin conformation. *J Immunol Methods* 184:91–100.

Cooper, M. A. (2002). Optical biosensors in drug discovery. *Nat Rev Drug Discov* 1:515–528.

Cornell, B. A., Krishna, G., Osman, P. D., Pace, R. D., and Wieczorek, L. (2001). Tethered-bilayer lipid membranes as a support for membrane-active peptides. *Biochem Soc Trans* 29:613–617.

Courey, A. J. (2001). Cooperativity in transcriptional control. *Curr Biol* 11:R250–R252.

Crank, J. (1975). *The Mathematics of Diffusion*. Clarendon Press, Oxford.

Cross, G. H., Reeves, A. A., Brand, S., Popplewell, J. F., Peel, L. L., Swann, M. J., and Freeman, N. J. (2003). A new quantitative optical biosensor for protein characterisation. *Biosens Bioelectron* 19:383–390.

Cunningham, B. C. and Wells, J. A. (1993). Comparison of a structural and a functional epitope. *J Mol Biol* 234.

Dallmann, H. G. and McHenry, C. S. (1995). DnaX complex of *Escherichia coli* DNA polymerase III holoenzyme. *J Biol Chem* 270:29563–29569.

Davis, S. J., Ikemizu, S., Wild, M. K., and van der Merwe, P. A. (1998). CD2 and the nature of protein interactions mediating cell–cell recognition. *Immunol Rev* 163:217–236.

De Crescenzo, G., Grothe, S., Lortie, R., Debanne, M. T., and O'Connor-McCourt, M. (2000). Real-time kinetic studies on the interaction of transforming growth factor alpha with the epidermal growth factor receptor extracellular domain reveal a conformational change model. *Biochemistry* 39:9466–9476.

Dubs, M.-C., Altschuh, D., and VanRegenmortel, M. H. V. (1992). Mapping of viral epitopes with conformationally specific monoclonal antibodies using biosensor technology. *J Chromatogr* 597:391–396.

Fägerstam, L. G., Frostell, Å., Karlsson, R., Kullman, M., Larsson, A., Malmqvist, M., and Butt, H. (1990). Detection of antigen–antibody interactions by surface plasmon resonance. Application to epitope mapping. *J Mol Recognit* 3:208–214.

Fields, B. A., Malchiodi, E. L., Li, H., Ysern, X., Stauffacher, C. V., Schlievert, P. M., Karjalainen, K., and Mariuzza, R. A. (1996). Crystal structure of a T-cell receptor beta-chain complexed with a superantigen. *Nature* 384:188–192.

Fischer, P. B., Karlsson, G. B., Butters, T. D., Dwek, R. A., and Platt, F. M. (1996). *N*-butyldeoxynojirimycin-mediated inhibition of human immunodeficiency virus entry correlates with changes in antibody recognition of the V1/V2 region of gp120. *J Virol* 70:7143–7152.

Fong, C.-C., Wong, M.-S., Fong, W.-F., and Yang, M. (2002). Effect on hydrogel matrix on binding kinetics of protein–protein interactions on the sensor surface. *Anal Chim Acta* 456:201–208.

Garcia, K. C., Degano, M., Stanfield, R. L., Brunmark, A., Jackson, M. R., Peterson, P. A., Teyton, L., and Wilson, I. A. (1996). An alpha beta T cell receptor structure at 2.5 A and its orientation in the TCR–MHC complex. *Science* 274:209–219.

Garcia, K. C., Radu, C. G., Ho, J., Ober, R. J., and Ward, E. S. (2001). Kinetics and thermodynamics of T cell receptor–autoantigen interactions in murine experimental autoimmune encephalomyelitis. *Proc Natl Acad Sci USA* 98:6818–6823.

Garland, P. B. (1996). Optical evanescent wave methods for the study of biomolecular interactions. *Q Rev Biophys* 29:91–117.

Gascoigne, N. R. and Ames, K. T. (1991). Direct binding of secreted T-cell receptor beta chain to superantigen associated with class II major histocompatibility complex protein. *Proc Natl Acad Sci USA* 88:613–616.

Gauglitz, G. (2005). Direct optical sensors: principles and selected applications. *Anal Bioanal Chem* 381:141–155.

Gavin, A. C., Bosche, M., Krause, R., Grandi, P., Marzioch, M., Bauer, A., Schultz, J., Rick, J. M., Michon, A. M., Cruciat, C. M., Remor, M., Hofert, C., Schelder, M., Brajenovic, M., Ruffner, H., Merino, A., Klein, K., Hudak, M., Dickson, D., Rudi, T., Gnau, V., Bauch, A., Bastuck, S., Huhse, B., Leutwein, C., Heurtier, M. A., Copley, R. R., Edelmann, A., Querfurth, E., Rybin, V., Drewes, G., Raida, M., Bouwmeester, T., Bork, P., Seraphin, B., Kuster, B., Neubauer, G., and Superti-Furga, G. (2002). Functional organization of the yeast proteome by systematic analysis of protein complexes. *Nature* 415:141–147.

Germain, R. N. and Stefanova, I. (1999). The dynamics of T cell receptor signaling: complex orchestration and the key roles of tempo and cooperation. *Annu Rev Immunol* 17:467–522.

Gilligan, J. J., Schuck, P., and Yergey, A. L. (2002). Mass spectrometry after capture and small-volume elution of analyte from a surface plasmon resonance biosensor. *Anal Chem* 74:2041–2047.

Giot, L., Bader, J. S., Brouwer, C., Chaudhuri, A., Kuang, B., Li, Y., Hao, Y. L., Ooi, C. E., Godwin, B., Vitols, E., Vijayadamodar, G., Pochart, P., Machineni, H., Welsh, M., Kong, Y., Zerhusen, B., Malcolm, R., Varrone, Z., Collis, A., Minto, M., Burgess, S., McDaniel, L., Stimpson, E., Spriggs, F., Williams, J., Neurath, K., Ioime, N., Agee, M., Voss, E., Furtak, K., Renzulli, R., Aanensen, N., Carrolla, S., Bickelhaupt, E., Lazovatsky, Y., DaSilva, A., Zhong, J., Stanyon, C. A., Finley, R. L. Jr., White, K. P., Braverman, M., Jarvie, T., Gold, S., Leach, M., Knight, J., Shimkets, R. A., McKenna, M. P., Chant, J., and Rothberg, J. M. (2003). A protein interaction map of *Drosophila melanogaster*. *Science* 302:1727–1736.

Glaser, R. W. (1993). Antigen–antibody binding and mass transport by convection and diffusion to a surface: a two-dimensional computer model of binding and dissociation kinetics. *Anal Biochem* 213:152–161.

Glaser, R. W. and Hausdorf, G. (1996). Binding kinetics of an antibody against HIV p24 core protein measured with real-time biomolecular interaction analysis suggest a slow conformational change in antigen p24. *J Immunol Methods* 189:1–14.

Greenfield, C., Hiles, I., Waterfield, M. D., Federwisch, M., Wollmer, A., Blundell, T. L., and McDonald, N. (1989). Epidermal growth factor binding induces a conformational change in the external domain of its receptor. *Embo J* 8:4115–4123.

Hall, D. (2001). Use of optical biosensors for the study of mechanistically concerted surface adsorption processes. *Anal Biochem* 288:109–125.

Hall, D. R., Cann, J. R., and Winzor, D. J. (1996). Demonstration of an upper limit to the range of association rate constants amenable to study by biosensor technology based on surface plasmon resonance. *Anal Biochem* 235:175–184.

Hanisch, F. G., Uhlenbruck, G., Egge, H., and Peter-Katalinic, J. (1989). A B72.3 second-generation-monoclonal antibody (CC49) defines the mucin-carried carbohydrate epitope Gal beta(1–3) [NeuAc alpha(2–6)]GalNAc. *Biol Chem Hoppe Seyler* 370:21–26.

Hansen, P. C. (1998). *Rank-Deficient and Discrete Ill-Posed Problems: Numerical Aspects of Linear Inversion*. SIAM, Philadelphia.

Hasegawa, K., Ono, K., Yamada, M., and Naiki, H. (2002). Kinetic modeling and determination of reaction constants of Alzheimer's beta-amyloid fibril extension and dissociation using surface plasmon resonance. *Biochemistry* 41:13489–13498.

Ho, Y., Gruhler, A., Heilbut, A., Bader, G. D., Moore, L., Adams, S. L., Millar, A., Taylor, P., Bennett, K., Boutilier, K., Yang, L., Wolting, C., Donaldson, I., Schandorff, S., Shewnarane, J., Vo, M., Taggart, J., Goudreault, M., Muskat, B., Alfarano, C., Dewar, D., Lin, Z., Michalickova, K., Willems, A. R., Sassi, H., Nielsen, P. A., Rasmussen, K. J., Andersen, J. R., Johansen, L. E.,

Hansen, L. H., Jespersen, H., Podtelejnikov, A., Nielsen, E., Crawford, J., Poulsen, V., Sorensen, B. D., Matthiesen, J., Hendrickson, R. C., Gleeson, F., Pawson, T., Moran, M. F., Durocher, D., Mann, M., Hogue, C. W., Figeys, D., and Tyers, M. (2002). Systematic identification of protein complexes in *Saccharomyces cerevisiae* by mass spectrometry. *Nature* 415:180–183.

Honjo, E., Watanabe, K., and Tsukamoto, T. (2002). Real-time kinetic analyses of the interaction of ricin toxin A-chain with ribosomes prove a conformational change involved in complex formation. *J Biochem* (*Tokyo*) 131:267–275.

Huber, A., Demartis, S., and Neri, D. (1999). The use of biosensor technology for the engineering of antibodies and enzymes. *J Mol Recognit* 12:198–216.

Ibrahimi, O. A., Zhang, F., Hrstka, S. C., Mohammadi, M., and Linhardt, R. J. (2004). Kinetic model for FGF, FGFR, and proteoglycan signal transduction complex assembly. *Biochemistry* 43:4724–4730.

Ito, T., Chiba, T., Ozawa, R., Yoshida, M., Hattori, M., and Sakaki, Y. (2001). A comprehensive two-hybrid analysis to explore the yeast protein interactome. *Proc Natl Acad Sci USA* 98:4569–4574.

Johnsson, B., Löfås, S., and Lindquist, G. (1991). Immobilization of proteins to a carboxymethyldextran-modified gold surface for biospecific interaction analysis in surface plasmon resonance sensors. *Anal Biochem* 198:268–277.

Jönsson, U. and Malmqvist, M. (1992). Real time biospecific interaction analysis. *Adv Biosens* 2:291–336.

Joss, L., Morton, T. A., Doyle, M. L., and Myszka, D. G. (1998). Interpreting kinetic rate constants from optical biosensor data recorded on a decaying surface. *Anal Biochem* 261:203–210.

Karlsson, R. (1994). Real-time competitive kinetic analysis of interactions between low-molecular-weight ligands in solution and surface-immobilized receptors. *Anal Biochem* 221:142–151.

Karlsson, R. and Fält, A. (1997). Experimental design for kinetic analysis of protein–protein interactions with surface plasmon resonance biosensors. *J Immunol Methods* 200:121–133.

Karlsson, R., Michaelsson, A., and Mattson, L. (1991). Kinetic analysis of monoclonal antibody–antigen interactions with a new biosensor based analytical system. *J Immunol Methods* 145:229–240.

Karlsson, R., Roos, H., Fägerstam, L., and Persson, B. (1994). Kinetic and concentration analysis using BIA technology. *Methods: A Companion to Methods Enzymol* 6:99–110.

Karp, N. A., Edwards, P. R., and Leatherbarrow, R. J. (2005). Analysis of calibration methodologies for solvent effects in drug discovery studies using evanescent wave biosensors. *Biosens Bioelectron* 21:128–134.

Khursigara, C. M., De Crescenzo, G., Pawelek, P. D., and Coulton, J. W. (2005). Kinetic analyses reveal multiple steps in forming TonB–FhuA complexes from *Escherichia coli*. *Biochemistry* 44:3441–3453.

Knoll, W. (1998). Interfaces and thin films as seen by bound electromagnetic waves. *Annu Rev Phys Chem* 49:569–638.

Kotzin, B. L., Leung, D. Y., Kappler, J., and Marrack, P. (1993). Superantigens and their potential role in human disease. *Adv Immunol* 54:99–166.

Kretschmann, E. and Raether, H. (1968). Radiative decay of non-radiative surface plasmons excited by light. *Z Naturforsch* 23a:2135–2136.

Krone, J. R., Nelson, R. W., Dogruel, D., Williams, P., and Granzow, R. (1997). BIA/MS: interfacing biomolecular interaction analysis with mass spectrometry. *Anal Biochem* 244:124–132.

Ladbury, J. E., Lemmon, M. A., Zhou, M., Green, J., Botfield, M. C., and Schlesinger, J. (1995). Measurement of the binding of tyrosyl phosphopeptides to SH2 domains: a reappraisal. *Proc Natl Acad Sci USA* 92:3199–3203.

Langmuir, I. (1918). The adsorption of gases on plane surfaces of glass, mica and platinum. *J Am Chem Soc* 40:1361–1403.

Lata, S. and Piehler, J. (2005). Stable and functional immobilization of histidine-tagged proteins via multivalent chelator head groups on a molecular poly(ethylene glycol) brush. *Anal Chem* 77:1096–1105.

Lee, H. J., Yan, Y., Marriott, G., and Corn, R. M. (2005). Quantitative functional analysis of protein complexes on surfaces. *J Physiol* 563:61–71.

Li, S., Armstrong, C. M., Bertin, N., Ge, H., Milstein, S., Boxem, M., Vidalain, P. O., Han, J. D., Chesneau, A., Hao, T., Goldberg, D. S., Li, N., Martinez, M., Rual, J. F., Lamesch, P., Xu, L., Tewari, M., Wong, S. L., Zhang, L. V., Berriz, G. F., Jacotot, L., Vaglio, P., Reboul, J., Hirozane-Kishikawa, T., Li, Q., Gabel, H. W., Elewa, A., Baumgartner, B., Rose, D. J., Yu, H., Bosak, S., Sequerra, R., Fraser, A., Mango, S. E., Saxton, W. M., Strome, S., Van Den Heuvel, S., Piano, F., Vandenhaute, J., Sardet, C., Gerstein, M., Doucette-Stamm, L., Gunsalus, K. C., Harper, J. W., Cusick, M. E., Roth, F. P., Hill, D. E., and Vidal, M. (2004). A map of the interactome network of the metazoan *C. elegans*. *Science* 303:540–543.

Liebermann, T. and Knoll, W. (2000). Surface-plasmon filed-enhanced fluorescence spectroscopy. *Colloids Surf A* 171:115–130.

Lieto, A. M., Cush, R. C., and Thompson, N. L. (2003). Ligand–receptor kinetics measured by total internal reflection with fluorescence correlation spectroscopy. *Biophys J* 85:3294–3302.

Lipschultz, C. A., Li, Y., and Smith-Gill, S. (2000). Experimental design for analysis of complex kinetics using surface plasmon resonance. *Methods* 20:310–318.

Lo Conte, L., Chothia, C., and Janin, J. (1999). The atomic structure of protein–protein recognition sites. *J Mol Biol* 285:2177–2198.

Löfås, S. and Johnsson, B. (1990). A novel hydrogel matrix on gold surfaces in surface plasmon resonance sensors for fast and efficient covalent immobilization of ligands. *J Chem Soc, Chem Commun* 21:1526–1528.

Lukosz, W. (1991). Principles and sensitivities of integrated optical and surface plasmon sensors for direct affinity sensing and immunosensing. *Biosens Bioelectron* 6:215–225.

Ma, B., Wolfson, H. J., and Nussinov, R. (2001). Protein functional epitopes: hot spots, dynamics and combinatorial libraries. *Curr Opin Struct Biol* 11:364–369.

Malchiodi, E. L., Eisenstein, E., Fields, B. A., Ohlendorf, D. H., Schlievert, P. M., Karjalainen, K., and Mariuzza, R. A. (1995). Superantigen binding to a T cell receptor beta chain of known three-dimensional structure. *J Exp Med* 182:1833–1845.

Malmborg, A. C. and Borrebaeck, C. A. (1995). BIAcore as a tool in antibody engineering. *J Immunol Methods* 183:7–13.

McCraith, S., Holtzman, T., Moss, B., and Fields, S. (2000). Genome-wide analysis of vaccinia virus protein–protein interactions. *Proc Natl Acad Sci USA* 97:4879–4884.

Mehlmann, M., Garvin, A. M., Steinwand, M., and Gauglitz, G. (2005). Reflectometric interference spectroscopy combined with MALDI-TOF mass spectrometry to determine quantitative and qualitative binding of mixtures of vancomycin derivatives. *Anal Bioanal Chem* 382:1942–1948.

Morton, T. A., Myszka, D. G., and Chaiken, I. M. (1995). Interpreting complex binding kinetics from optical biosensors: a comparison of analysis by linearization, the integrated rate equation, and numerical integration. *Anal Biochem* 227:176–185.

Muller, K. M., Arndt, K. M., and Pluckthun, A. (1998). Model and simulation of multivalent binding to fixed ligands. *Anal Biochem* 261:149–158.

Myszka, D., Arulanantham, P. R., Sana, T., Wu, Z., Morton, T. A., and Ciardelli, T. L. (1996). Kinetic analysis of ligand binding to interleukin-2 receptor complexes created on an optical biosensor surface. *Protein Sci* 5:2468–2478.

Natsume, T., Nakayama, H., Jansson, O., Isobe, T., Takio, K., and Mikoshiba, K. (2000). Combination of biomolecular interaction analysis and mass spectrometric amino acid sequencing. *Anal Chem* 72:4193–4198.

Natsume, T., Yamauchi, Y., Nakayama, H., Shinkawa, T., Yanagida, M., Takahashi, N., and Isobe, T. (2002). A direct nanoflow liquid chromatography-tandem mass spectrometry system for interaction proteomics. *Anal Chem* 74:4725–4733.

Nieba, L., Krebber, A., and Plückthun, A. (1996). Competition BIAcore for measuring true affinities: large differences from values determined from binding kinetics. *Anal Biochem* 234:155–165.

Northup, J. (2004). Measuring rhodopsin-G-protein interactions by surface plasmon resonance. *Methods Mol Biol* 261:93–112.

O'Shannessy, D. J. (1994). Determination of kinetic rate and equilibrium binding constants for macromolecular interactions: a critique of the surface plasmon resonance literature. *Curr Opin Biotechnol* 5:65–71.

O'Shannessy, D. J., Brigham-Burke, M., and Peck, K. (1992). Immobilization chemistries suitable for use in the BIAcore surface plasmon resonance detector. *Anal Biochem* 205:132–136.

O'Shannessy, D. J., Brigham-Burke, M., Soneson, K. K., Hensley, P., and Brooks, I. (1993). Determination of rate and equilibrium binding constants for macromolecular interactions using surface plasmon resonance: use of nonlinear least squares analysis methods. *Anal Biochem* 212:457–468.

Ober, R. J., Caves, J., and Ward, E. S. (2003). Analysis of exponential data using a noniterative technique: application to surface plasmon experiments. *Anal Biochem* 312:57–65.

Ober, R. J. and Ward, E. S. (1999a). The choice of reference cell in the analysis of kinetic data using BIAcore. *Anal Biochem* 271:70–80.

Ober, R. J. and Ward, E. S. (1999b). The influence of signal noise on the accuracy of kinetic constants measured by surface plasmon resonance experiments. *Anal Biochem* 273:49–59.

Parsons, I. D. and Stockley, P. G. (1997). Quantitation of the *Escherichia coli* methionine repressor–operator interaction by surface plasmon resonance is not affected by the presence of a dextran matrix. *Anal Biochem* 254:82–87.

Pattnaik, P. (2005). Surface plasmon resonance: applications in understanding receptor–ligand interaction. *Appl Biochem Biotechnol* 126:79–92.

Piehler, J., Brecht, A., Hehl, K., and Gauglitz, G. (1999). Protein interactions in covalently attached dextran layers. *Colloids Surf B Biointerfaces* 13:325–336.

Plant, A. L., Brigham-Burke, M., Petrella, E. C., and O'Shannessy, D. J. (1995). Phospholipid/alkanethiol bilayers for cell-surface receptor studies by surface plasmon resonance. *Anal Biochem* 226:342–348.

Provencher, S. W. (1982). CONTIN: a general purpose constrained regularization program for inverting noisy linear algebraic and integral equations. *Comput Phys Commun* 27:229–242.

Raether, H. (1977). Surface plasma oscillations and their applications. In: Hass, G., Francombe, M. H., and Hoffman, R. W. (eds), *Physics of Thin Films*, volume 9. Academic Press, New York, pp. 145–261.

Rain, J. C., Selig, L., De Reuse, H., Battaglia, V., Reverdy, C., Simon, S., Lenzen, G., Petel, F., Wojcik, J., Schachter, V., Chemama, Y., Labigne, A., and Legrain, P. (2001). The protein–protein interaction map of *Helicobacter pylori*. *Nature* 409:211–215.

Rao, N. M., Plant, A. L., Silin, V., Wight, S., and Hui, S. W. (1997). Characterization of biomimetic surfaces formed from cell membranes. *Biophys J* 73:3066–3077.

Rebois, R. V., Schuck, P., and Northup, J. K. (2002). Elucidating kinetic and thermodynamic constants for interaction of G protein subunits and receptors by surface plasmon resonance spectroscopy. *Methods Enzymol* 344:15–42.

Reddish, M. A., Jackson, L., Koganty, R. R., Qiu, D., Hong, W., and Longenecker, B. M. (1997). Specificities of anti-sialyl-Tn and anti-Tn monoclonal antibodies generated using novel clustered synthetic glycopeptide epitopes. *Glycoconj J* 14:549–560.

Redpath, S., Alam, S. M., Lin, C. M., O'Rourke, A. M., and Gascoigne, N. R. (1999). Cutting edge: trimolecular interaction of TCR with MHC class II and bacterial superantigen shows a similar affinity to MHC:peptide ligands. *J Immunol* 163:6–10.

Renno, T. and Acha-Orbea, H. (1996). Superantigens in autoimmune diseases: still more shades of gray. *Immunol Rev* 154:175–191.

Roden, L. D. and Myszka, D. G. (1996). Global analysis of a macromolecular interaction measured on BIAcore. *Biochem Biophys Res Commun* 225:1073–1077.

Rosovitz, M. J., Schuck, P., Varughese, M., Chopra, A. P., Mehra, V., Singh, Y., McGinnis, L. M., and Leppla, S. H. (2003). Alanine-scanning mutations in domain 4 of anthrax toxin protective antigen reveal residues important for binding to the cellular receptor and to a neutralizing monoclonal antibody. *J Biol Chem* 278:30936–30944.

Rothenhäusler, B. and Knoll, W. (1988). Surface-plasmon microscopy. *Nature* 332:615–617.

Salamon, Z., Brown, M. F., and Tollin, G. (1999). Plasmon resonance spectroscopy: probing molecular interactions within membranes. *Trends Biochem Sci* 24:213–219.

Salamon, Z., Huang, D., Cramer, W. A., and Tollin, G. (1998). Coupled plasmon-waveguide resonance spectroscopy studies of the cytochrome b6f/plastocyanin system in supported lipid bilayer membranes. *Biophys J* 75:1874–1885.

Salamon, Z., Wang, Y., Brown, M. F., Macleod, H. A., and Tollin, G. (1994). Conformational changes in rhodopsin probed by surface plasmon resonance spectroscopy. *Biochemistry* 33:13706–13711.

Schmelz, M., Sodeik, B., Ericsson, M., Wolffe, E. J., Shida, H., Hiller, G., and Griffiths, G. (1994). Assembly of vaccinia virus: the second wrapping cisterna is derived from the trans Golgi network. *J Virol* 68:130–147.

Schuck, P. (1996). Kinetics of ligand binding to receptor immobilized in a polymer matrix, as detected with an evanescent wave biosensor. I. A computer simulation of the influence of mass transport. *Biophys J* 70:1230–1249.

Schuck, P. (1997a). Reliable determination of binding affinity and kinetics using surface plasmon resonance biosensors. *Curr Opin Biotechnol* 8:498–502.

Schuck, P. (1997b). Use of surface plasmon resonance to probe the equilibrium and dynamic aspects of interactions between biological macromolecules. *Ann Rev Biophys Biomol Struct* 26:541–566.

Schuck, P. (2000). Size distribution analysis of macromolecules by sedimentation velocity ultracentrifugation and Lamm equation modeling. *Biophys J* 78:1606–1619.

Schuck, P., Boyd, L. F., and Andersen, P. S. (1999). Measuring protein interactions by optical biosensors. In: Coligan, J. E., Dunn, B. M., Ploegh, H. L., Speicher, D. W., and Wingfield, P. T. (eds), *Current Protocols in Protein Science*, volume 2. Wiley, New York, pp. 20.2.1–20.2.21.

Schuck, P., Millar, D. B., and Kortt, A. A. (1998). Determination of binding constants by equilibrium titration with circulating sample in a surface plasmon resonance biosensor. *Anal Biochem* 265:79–91.

Schuck, P. and Minton, A. P. (1996). Analysis of mass transport limited binding kinetics in evanescent wave biosensors. *Anal Biochem* 240:262–272.

Schuster, S. C., Swanson, R. V., Alex, L. A., Bourret, R. B., and Simon, M. I. (1993). Assembly and function of a quaternary signal transduction complex monitored by surface plasmon resonance. *Nature* 365:343–347.

Shumaker-Parry, J. S. and Campbell, C. T. (2004). Quantitative methods for spatially resolved adsorption/desorption measurements in real time by surface plasmon resonance microscopy. *Anal Chem* 76:907–917.

Sjölander, S. and Urbaniczky, C. (1991). Integrated fluid handling system for biomolecular interaction analysis. *Anal Chem* 63:2338–2345.

Sonksen, C. P., Nordhoff, E., Jansson, O., Malmqvist, M., and Roepstorff, P. (1998). Combining MALDI mass spectrometry and biomolecular interaction analysis using a biomolecular interaction analysis instrument. *Anal Chem* 70:2731–2736.

Spolar, R. S. and Record, M. T. Jr. (1994). Coupling of local folding to site-specific binding of proteins to DNA. *Science* 263:777–784.

Stenberg, E., Persson, B., Roos, H., and Urbaniczky, C. (1991). Quantitative determination of surface concentration of protein with surface plasmon resonance using radiolabeled proteins. *J Coll Interface Sci* 143:513–526.

Stokes, R. H., Oakhill, J. S., Joannou, C. L., Gorringe, A. R., and Evans, R. W. (2005). Meningococcal transferrin-binding proteins A and B show cooperation in their binding kinetics for human transferrin. *Infect Immun* 73:944–952.

Sundberg, E. J., Andersen, P. S., Schlievert, P. M., Karjalainen, K., and Mariuzza, R. A. (2003). Structural, energetic and functional analysis of a protein–protein interface at distinct stages of affinity maturation. *Structure* 11:1151–1161.

Sundberg, E. J., Li, Y., and Mariuzza, R. A. (2002). So many ways of getting in the way: diversity in the molecular architecture of superantigen-dependent T-cell signaling complexes. *Curr Opin Immunol* 14:36–44.

Sundberg, E. J. and Mariuzza, R. A. (2000). Luxury accommodations: the expanding role of structural plasticity in protein–protein interactions. *Structure Fold Des* 8:R137–R142.

Sundberg, E. J. and Mariuzza, R. A. (2002). Molecular recognition in antigen–antibody complexes. *Adv Protein Chem* 61:119–160.

Svitel, J., Balbo, A., Mariuzza, R. A., Gonzales, N. R., and Schuck, P. (2003). Combined affinity and rate constant distributions of analyte or ligand populations from experimental surface binding and kinetics and equilibria. *Biophys J* 84:4062–4077.

Svitel, J., Boukari, H., Gorshkova, I. I., Sackett, D. L., and Schuck, P. (submitted). Probing the functional heterogeneity of surface binding sites.

Tanaka, M. and Sackmann, E. (2005). Polymer-supported membranes as models of the cell surface. *Nature* 437:656–663.

Uetz, P., Giot, L., Cagney, G., Mansfield, T. A., Judson, R. S., Knight, J. R., Lockshon, D., Narayan, V., Srinivasan, M., Pochart, P., Qureshi-Emili, A., Li, Y., Godwin, B., Conover, D., Kalbfleisch, T., Vijayadamodar, G., Yang, M., Johnston, M., Fields, S., and Rothberg, J. M. (2000). A comprehensive analysis of protein–protein interactions in *Saccharomyces cerevisiae*. *Nature* 403:623–627.

van der Merwe, P. A. and Barclay, A. N. (1996). Analysis of cell-adhesion molecule interactions using surface plasmon resonance. *Curr Opin Immunol* 8:257–261.

van der Merwe, P. A., Barclay, A. N., Mason, D. W., Davies, E. A., Morgan, B. P., Tone, M., Krishnam, A. K. C., Ianelli, C., and Davis, S. J. (1994). Human cell-adhesion molecule CD2 binds CD58 (LFA-3) with a very low affinity and an extremely fast dissociation rate but does not bind CD48 or CD59. *Biochemistry* 33:10149–10160.

Vives, R. R., Sadir, R., Imberty, A., Rencurosi, A., and Lortat-Jacob, H. (2002). A kinetics and modeling study of RANTES(9–68) binding to heparin reveals a mechanism of cooperative oligomerization. *Biochemistry* 41:14779–14789.

Vorup-Jensen, T., Carman, C. V., Shimaoka, M., Schuck, P., Svitel, J., and Springer, T. A. (2005). Exposure of acidic residues as a danger signal for recognition of fibrinogen and other macromolecules by integrin alphaXbeta2. *Proc Natl Acad Sci USA* 102:1614–1619.

Walhout, A. J., Sordella, R., Lu, X., Hartley, J. L., Temple, G. F., Brasch, M. A., Thierry-Mieg, N., and Vidal, M. (2000). Protein interaction mapping in *C. elegans* using proteins involved in vulval development. *Science* 287:116–122.

Walsh, S. T., Jevitts, L. M., Sylvester, J. E., and Kossiakoff, A. A. (2003). Site 2 binding energetics of the regulatory step of growth hormone-induced receptor homodimerization. *Protein Sci* 12:1960–1970.

Willcox, B. E., Gao, G. F., Wyer, J. R., Ladbury, J. E., Bell, J. I., Jakobsen, B. K., and van der Merwe, P. A. (1999). TCR binding to peptide-MHC stabilizes a flexible recognition interface. *Immunity* 10:357–365.

Williams, C. and Addona, T. A. (2000). The integration of SPR biosensors with mass spectrometry: possible applications for proteome analysis. *Trends Biotechnol* 18:45–48.

Winzor, D. J. (2000). From gel filtration to biosensor technology: the development of chromatography for the characterization of protein interactions. *J Mol Recognit* 13:279–298.

Wofsy, C. and Goldstein, B. (2002). Effective rate models for receptors distributed in a layer above a surface: application to cells and Biacore. *Biophys J* 82:1743–1755.

Yang, J., Swaminathan, C. P., Huang, Y., Guan, R., Cho, S., Kieke, M. C., Kranz, D. M., Mariuzza, R. A., and Sundberg, E. J. (2003). Dissecting cooperative and additive binding energetics in the affinity maturation pathway of a protein–protein interface. *J Biol Chem* 278:50412–50421.

Yarmush, M. L., Patankar, D. B., and Yarmush, D. M. (1996). An analysis of transport resistances in the operation of BIAcore; implications for kinetic studies of biospecific interactions. *Mol Immunol* 33:1203–1214.

Yeung, D., Gill, A., Maule, C. H., and Davies, R. J. (1995). Detection and quantification of biomolecular interactions with optical biosensors. *Trends Anal Chem* 14:49–56.

Zacher, T. and Wischerhoff, E. (2002). Real-time two-wavelength surface plasmon resonance as a tool for the vertical resolution of binding processes in biosensing hydrogels. *Langmuir* 18:1748–1759.

5

Mass Spectrometry for Studying Protein Modifications and for Discovery of Protein Interactions

Peter S. Backlund Jr. and Alfred L. Yergey*

5.1. INTRODUCTION

Mass spectrometry is a tool that has been widely used for investigations of topics in the realm of biophysics. It has been said that, in general, the reach of mass spectrometry's applications is limited only by an investigator's ability to imagine novel uses. Biophysically relevant applications range from studies of the thermodynamics of the hydration spheres of isolated small molecules and the characterization of lipid compositions of organ systems to the modeling of interaction surfaces of protein complexes and the representation of the dimensions of very large, multiple protein complexes. The first two of these applications are out of the scope of this volume but other chapters describe the latter two applications in detail, specifically the application of H/D exchange (Chapter 6) and electrospray ionization (ESI) (Chapter 15) to the analysis of intact protein complexes. The purpose of this chapter is to describe a widely used set of biochemical and mass spectrometric tools that are more or less routinely applied to the characterization of proteins at present.

These protein characterizations can be thought of in three broad groups:

1. Verification of molecular weight
2. Identification of ''unknown'' proteins
3. Mapping posttranslational modifications of specific proteins.

P. S. BACKLUND JR. and A. L. YERGEY • Section on Mass Spectrometry and Metabolism, LCMB, NICHD, NIH, Bethesda, MD 20892 and *Corresponding author: Ph: 301-496-5531; fax: 301-480-5793; e-mail: aly@helix.nih.gov.

In each case of these three rather broad areas, the combination of biochemical isolation tools and mass spectrometric methods must be used appropriately in order for an investigator to achieve satisfactory results. We discuss a number of the most widely used techniques and illustrate them with examples while referring the reader to several of the large number of reviews currently available (Graves and Haystead, 2002; Steen and Mann, 2004).

5.2. MASS SPECTROMETRY

Mass spectrometry requires that for any molecule to be analyzed, it must first be converted into a gas phase ion. This production of ions, with either positive or negative charges, can be viewed quite independently from their analysis. That is, within very wide margins, any ionization technique can be coupled to any type of mass analyzer. Since mass analyzers in reality are analyzers of the ratio of mass to charge, m/z, the number of charges present in an ion may be single or multiple. Having formed a set of gas phase ions, mass analyzers use a variety of techniques to manipulate the ions ranging from drift times of ions at high velocity to the stability of ion paths in radio-frequency fields to determine m/z values of the species introduced into them.

For example, consider a protein of molecular weight 50 kDa.[1] On the one hand, if this protein is analyzed as a singly charged ion, typically formed by the addition of a proton as described later, then it would appear at m/z of 5.0×10^4. If on the other hand, it were ionized by another widely used method, the protein would appear as a group of ions, each with multiple charges from multiple proton additions; in such a case, the mass spectrum might have signals at m/z of 981.392, 1,001.000, and 1,021.408, corresponding to charge states of 51, 50, and 49, respectively.

5.2.1. Ionization

The modern era of mass spectrometric protein characterization employs two principal types of ionization: ESI (Yamashita and Fenn, 1984; Fenn *et al.*, 1989) and matrix-assisted laser desorption ionization (MALDI)[2] (Karas *et al.*, 1987; Karas and Hillenkamp, 1988; Tanaka *et al.*, 1988). Both of these types of ionization have been the subject of a great number of studies and reviews (Graves and Haystead, 2002; Steen and Mann, 2004), and, therefore, only a brief overview of the basic aspects of their operation is presented in the following sections.

5.2.1.1. Electrospray Ionization

The operation of an ESI source is illustrated in Figure 5.1A. Liquid, typically effluent from a reverse-phase liquid chromatographic separation, passes through a

[1] The Dalton (Da) is the unit of atomic mass and is equivalent to the weight of $\frac{1}{12}$ of a carbon atom.

[2] A very recently developed method, electron capture dissociation (ECD) will not be discussed in this chapter.

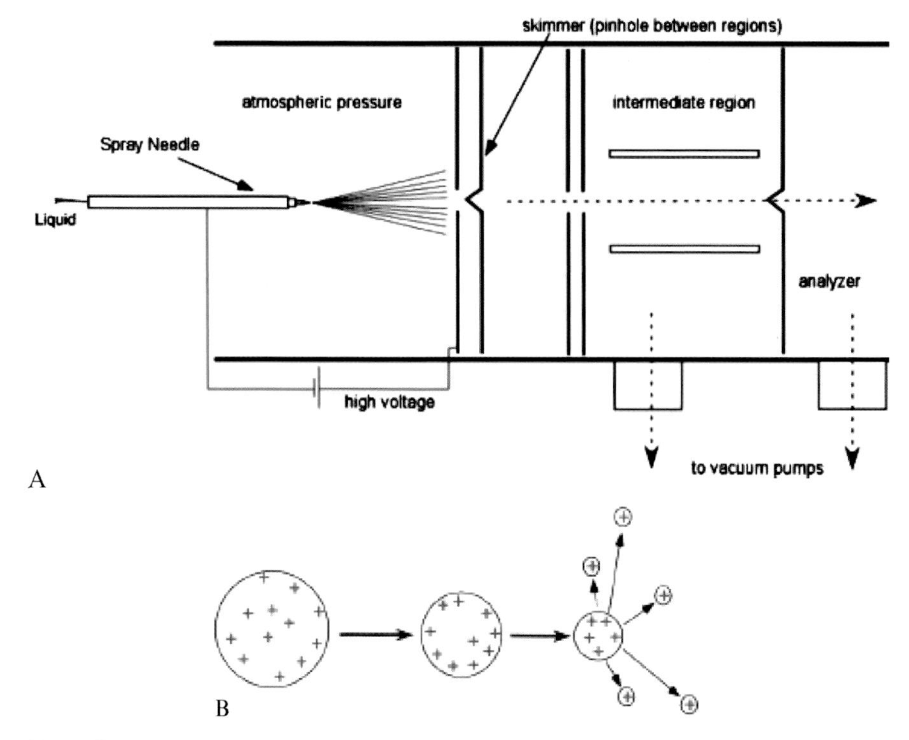

Figure 5.1. (A) Schematic diagram of electrospray ionization (ESI) source. (B) Schematic of droplet breakup in electrospray ionization (ESI).

narrow bore tube, often a section of fused silica capillary, at high voltage. A Taylor cone of the charged liquid forms at the exit of the capillary and subsequently disperses as a sheath of charged droplets that is directed toward a counter electrode, the entrance orifice of the mass spectrometer. At this point, a reduction in pressure from atmospheric to the high vacuum conditions, $\leq 10^{-5}$ torr, necessary for mass analysis, is carried out. A common configuration of this reduction process is the two-stage configuration shown in the figure; it consists of a nozzle–skimmer pair followed by an intermediate region maintained at $\sim 10^{-4}$ torr.

There are several interesting phenomena that occur in the process of delivering unsolvated ions to the analyzer. The first of these is the mechanism of droplet breakup following dispersion of the Taylor cone into a sheath. These processes are illustrated schematically in Figure 5.1B. Droplets in the sheath are charged positively or negatively, depending on the polarity of the high voltage applied to the liquid, with excess charges distributed on the surface of the droplets. In the course of traversing the distance between the high-voltage capillary and the ion-sampling orifice, the solvent evaporates from the droplets. As the solvent evaporates, the surface charge density increases until it exceeds the Rayleigh limit (Rayleigh, 1882), a function of droplet radius, the number of charges present, the liquid density, and the permittivity of the medium. On exceeding this limit, the droplet explodes and releases smaller droplets; this process continues until essentially bare charged molecules are left. In the case of positive ions, the charges are present as protons, added to the analyte molecules in the original solution; this is generally accomplished by insuring an excess of protons in the solution by using a volatile acid, acetic, formic, and to a lesser extent trifluoroacetic acid (TFA), in preparing the liquid phases used in ESI. In the case of proteins and peptides, the presence of basic sites on the molecules, coupled with excess protons in solution, leads to multiply charged ions.

There are a variety of ways of achieving the goal of delivering bare ions to a mass analyzer, but a general rule of thumb is that solution flow rates, voltage, and distance between capillary and sampling orifice must be balanced so that an essentially "dry" beam is present at the orifice. At high solution flow rates, $>1\ \mu L/min$, voltages applied to the solution are often in excess of 2 kV and the drying process is often assisted by the presence of a drying gas flowing countercurrent to the spray. As flow rates are lowered, the need for drying gas is eliminated, the capillary can be placed very close to the sampling orifice and voltages reduced to ~ 1 kV. An additional advantage accrues to the use of low, $\sim 200\ nL/min$, and very low, $\sim 5\ nL/min$, flow rates. As Figure 5.1A shows, the sheath of charged droplets includes a large solid angle. At high flow rates, only a small fraction of this solid angle can be sampled by the orifice while at low and very low flows, virtually the entire sheath can be sampled by the orifice. This has the clear advantage of collecting a larger fraction of the ions produced for subsequent mass analysis and gives rise to an apparent increase in sensitivity at the lower flow rates. In fact, this is only partially true. ESI is a so-called "concentration-dependent" method of ionization. This means that effective production of ion signals requires an analyte concentration in the order of micromolar, e.g., $1\ pmol/\mu L$, $1\ fmol/nL$, and so on, in order to give useful responses. In other words, while the absolute number of ions is reduced, they must be present at more or less fixed concentrations.

A second phenomenon of significance in the formation of electrosprayed ions is a result of the free jet expansion that occurs with the passage of gas and ions through the sampling orifice. This adiabatic expansion results in the enthalpy of the gas being converted into bulk flow kinetic energy and leads to a decrease in local gas temperature (Schneider and Chen, 2000; Fenn, 2002), typically from 295 to ~ 10 K close to the skimmer. Simultaneously, within the supersonic jet, both the vibrational and the rotational modes of the molecules and the ions are frozen as a consequence of a near absence of collisions in the jet. The implications for mass spectrometry are substantial because these processes lead to molecular ions that are formed under "soft" conditions, which almost inevitably produce intact multiply charged molecular ions, a major accomplishment when the possible decomposition pathways would be available to a large vibrationally excited gas phase ion. An electrospray mass spectrum of myoglobin, $M_r = 16{,}951$ Da, is shown in Figure 5.2. The envelope of peaks illustrates the extent of multiple charging; since each charge state represents an independent measurement of molecular weight, when the charge states are

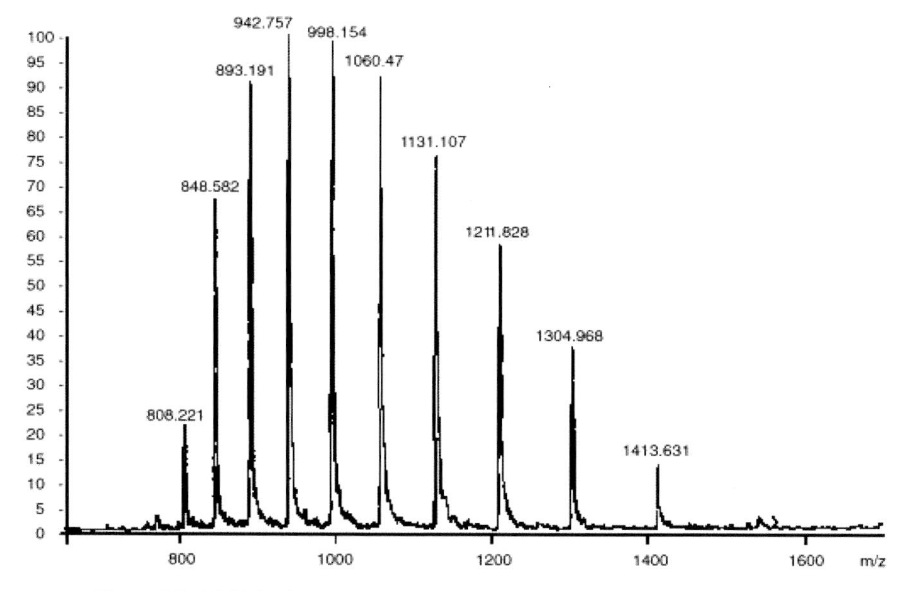

Figure 5.2. Multiple charge states for protein ions from electrospray ionization (ESI).

assigned correctly, the determination of M_r typically has a relative standard deviation of $< 0.1\%$. A detailed description of the application of ESI mass spectrometry in the study of intact protein complexes is presented in Chapter 15.

5.2.1.2. Matrix-Assisted Laser Desorption Ionization

The operating principle of a MALDI source is shown schematically in Figure 5.3. The mound of material shown in the center of the sample plate is a preparation that has been air dried, typically by applying analyte first and then the matrix, often after the sample has dried. Analyte molecules such as peptides or proteins, shown as dark ovals in the mound, are mixed with an excess of an organic acid. These acids are the matrix that provides part of the name of the method; the matrices are chosen so that they have an absorption band centered approximately at the frequency of the laser used in an instrument, typically near 235 nm. MALDI preparations are generally made from strongly acidic solutions, typically 0.1% TFA, and gives rise to an excess of free protons in the dried sample or matrix preparation; these are illustrated by the small dark dots in Figure 5.3. Cations such as sodium and potassium are also often present and appear in mass spectra as adduction products rather than protons; careful sample preparation can usually eliminate these adducts, but if one is attempting to obtain spectra from carbohydrates or nucleotides, it is important to have them present.

The typical operating sequence of events in MALDI is for the laser to fire, initiating a timing sequence and heating the surface of the sample or matrix

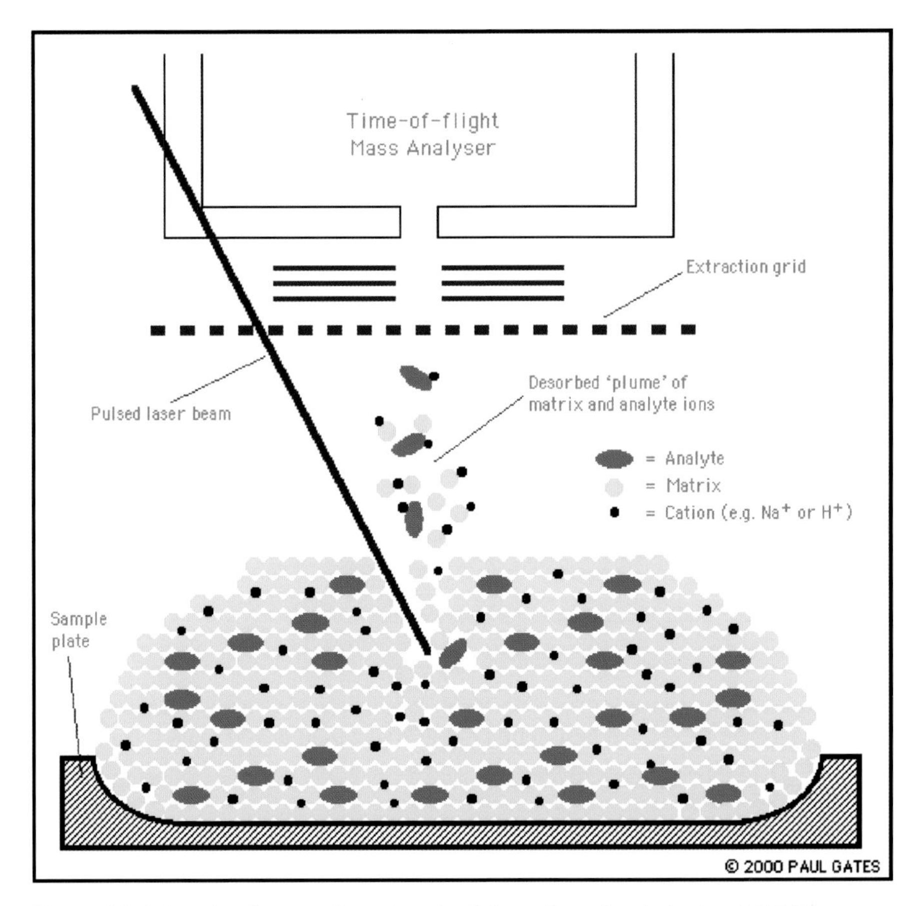

Figure 5.3. Schematic diagram of matrix-assisted laser desorption ionization (MALDI) source. (Copyright 2000, Paul Gates.)

preparation. Within a few nanoseconds of the termination of the laser pulse, a hot plume of neutral molecules and ions forms above the surface of the mound. Investigations of this plume have characterized its initial velocities and angular distribution, but from the perspective of analytical utility, the plume is typically directed away from the surface and toward an electrostatic lens that causes ions in the plume to move toward the mass analyzer. The first element of the lens system is set to pass ions after a fixed delay time that is set to allow for an initial settling out of the plume, typically several hundred nanoseconds following the laser pulse.

Commonly used matrices are shown in Figure 5.4. α-Cyano-4-hydroxy cinnamic acid (CHCA) is typically used in applications involving peptides, sinapinic acid (SA) is used for applications involving intact protein molecular weight determinations and for many applications involving negative ions, and 2,4-dihydroxy benzoic acid (DHB) is used in situations where the others do not work well.

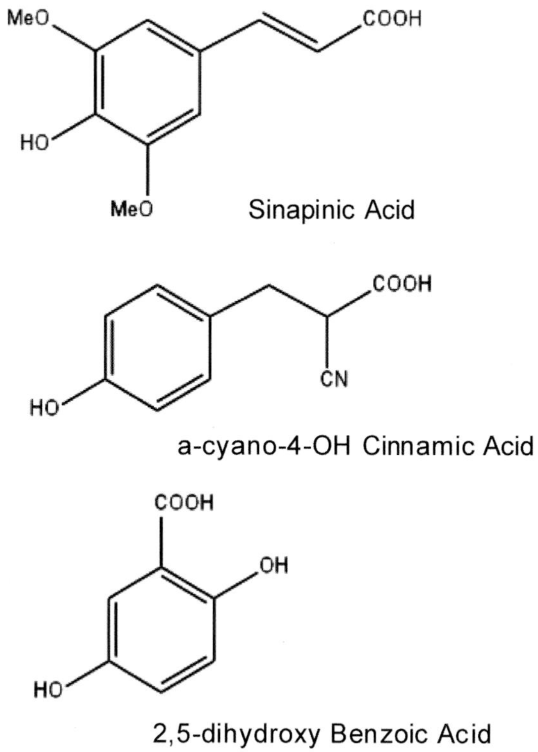

Figure 5.4. Common matrix compounds used for matrix-assisted laser desorption ionization (MALDI) mass spectrometry.

In many respects, there is less fundamental understanding of the physical processes involved in MALDI than is the case for electrospray. Recent experimental work continues to address these fundamental issues and to uncover some very interesting characteristics of these processes (Campbell *et al.*, 2006). This is probably because of the overall greater complexity of these processes and a somewhat less directly accessible system for experimental observation. Among the studies that have been conducted are some on the basis of ion formation, pulse duration and desorption, and in plume processes (Menzel *et al.*, 2002; Dreisewerd, 2003; Karas and Kruger, 2003; Knochenmuss and Zenobi, 2003), as well as theoretical studies by several groups (Bencsura *et al.*, 1997; Zhigilei *et al.*, 1997).

Despite these comments, MALDI is an extremely valuable technique for analytical characterization of proteins and peptides. The technique yields singly charged ions almost exclusively, and thus for proteins requires an analyzer with a high mass range; most commonly, time-of-flight (TOF) mass analysis is coupled with MALDI for protein measurements, but analyzers with lower mass ranges such as quadrupoles and ion traps have been used with MALDI for peptide analyses.

The production of singly charged ions of intact peptides and proteins is at first glance something of an unexpected result considering the amount of energy coupled into a molecule by the laser desorption process; MALDI, unlike ESI, is not a soft ionization technique, but one that yields useful ion signals despite high energies input to the system. There are two likely reasons that these intact ions are observed. First, the plume following the laser flash consists principally of matrix-related species and, while initially at very high temperatures, cools rapidly and by multiple collisions reduces vibrational excitation of species within it. This collisional cooling is not completely effective in that there is much molecular degradation within the plume, as evidenced by the very high levels of chemical noise that are always present at low molecular weight regions of a MALDI spectrum. There is a definite balance among laser intensity, matrix compound, and useful ion signal, and best results are usually obtained at the minimum laser intensity that gives rise to an ion signal. The second reason that ions are observed in MALDI is partially a consequence of ion behavior in TOF analyzers. Once ions leave the source region of a TOF instrument, they enter a field-free region until they reach the detector.[3] Some ions that enter this region have not been completely cooled and, therefore, have sufficient internal energy to fragment, given sufficient time; these ions are called ''metastable'' since they are intact when leaving the source, but break up in transiting the TOF analyzer. Since the metastable ions of a particular m/z received the same energy as all other ions, they reach the detector at a flight time equal to those of the correct m/z that have not fragmented; they are detected at the proper m/z despite having fallen apart. This phenomenon can be exploited in the case of peptides to obtain sequence information, as discussed in the following section.

5.2.2. Mass Analyzers

There are at least six different types of mass analyzers that have been used in protein characterizations, but only three of them, quadrupoles, ion traps, and TOF analyzers, play a significant role at this point. Magnetic sector instruments, while of great importance in other aspects of mass spectrometry, have a rather limited role at this point in time. Fourier-transform mass analyzers (FTMS) (Marshall *et al.*, 1998; Hendrickson and Emmett, 1999) and the recently introduced Orbitrap analyzer (Hu *et al.*, 2005) may have larger roles in the future, but are limited at present, the former due to its high cost and the latter due to its novelty; the promise of Orbitrap is such that it may rapidly supplant the lower-performance FTMS instruments.

In addition to the stand-alone capabilities of the three widely used instrument types, the field of mass analyzer development has given rise to classes of instruments that are combinations of discreet analyzers, e.g., triple quadrupoles and tandem TOFs, and to what are termed hybrid instruments such as quadrupole TOFs and quadrupole traps. Tandem and hybrid instruments play a major role in

[3] This is true for TOF instruments that do not use an electrostatic mirror on ''reflectron'' as discussed below in mass analyzers.

mass spectrometric characterization of proteins. This discussion is limited to brief descriptions of the basics of each of the three major analyzer types as well as an introduction to tandem and hybrid instruments.

5.2.2.1. Quadrupoles and Ion Traps

These devices are two examples of the class of path stability mass analyzers, which also includes the FTMS and Orbitrap instruments; more specifically, they are both mass filters,[4] and are the results of Wolfgang Paul's work (Hill, 2005). A quadrupole mass filter consists of four metal rods, theoretically of hyperbolic cross section but cylindrical in most common practice. They are arranged in parallel and mutually perpendicular, as shown in Figure 5.5. The rods have potentials of $\pm(U + V\cos(\omega t))$ applied pairwise on opposite rods, where U is a DC voltage and $V\cos(\omega t)$ is an RF voltage. The voltages modulate the trajectory of ions entering the quadrupole fields, but not their velocity along the rod axis; these trajectories can be expressed as motion in either of the two mutually perpendicular planes formed between the rod pairs and are described by solutions of the Mathieu second order differential equation. Trajectories leading to ion transmission through the quadrupole field fall into the class of stable solutions to the Mathieu equation and those colliding with the rods undergo unstable oscillations. There are a number of regions of stable oscillation that emerge in solving the Mathieu equation for motion in the two planes of the quadrupoles motion. The stable regions can be expressed in terms of magnitudes of the DC and RF potentials; when the solutions for the two polarities are plotted together, regions of simultaneous stability emerge. Quadrupole mass filters are operated in what is known as the first region of stability, and a plot of this region, known as a stability diagram, is given in Figure 5.6. A mass spectrum is

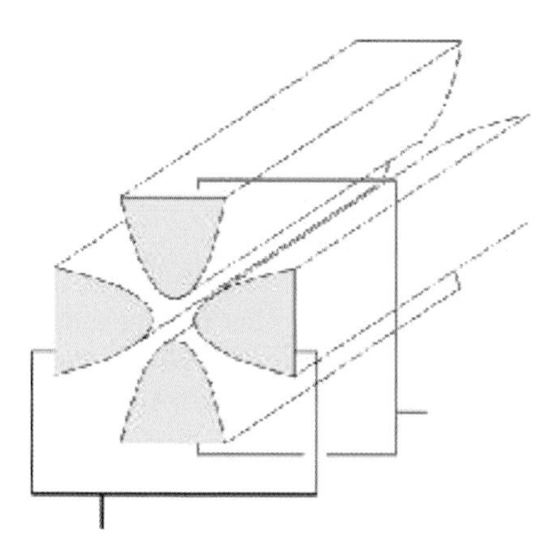

Figure 5.5. Schematic of rod arrangement and connections for quadrupole mass filters.

[4] To be completely correct, only magnetic sector instruments are truly mass ''spectrometers'' since these devices are the only type of mass analyzer in which an ion beam is brought to a focal point at a detector.

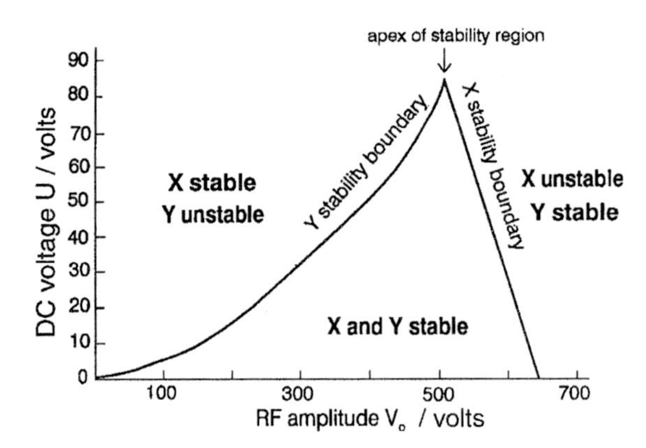

Figure 5.6. Stability diagram for quadrupole mass filters showing axes in voltage units to emphasize scale of operating conditions.

obtained by causing ions of increasing m/z to pass successively through the apex of the stability diagram. This is accomplished in practice by choosing values of DC and RF potentials that correspond to the tip of the stability diagram and then increasing the magnitude of both voltages while maintaining the ratio constant. An excellent discussion of simulating ion trajectories in quadrupole fields can be found in the work of Steel and Henchman (1998), and a somewhat simpler explanation of quadrupole motion can be found in Hill (2005).

There is another application of quadrupoles that is very significant in terms of the instruments used in protein characterization—their use as highly efficient ion transmission devices. In this ion transmission mode, the quadrupole rods have only an RF potential applied to them, still pairwise as in the mass filter application, with the opposite pairs of rods with RF voltages 180° out of phase with one another, as in the mass filter application. This only application of RF is particularly important in transmitting ions between sources and mass analyzers and in a variety of similar roles in tandem instruments.

5.2.2.2. Quadrupole Ion Trap

This type of mass analyzer is usually simply called an ion trap. It is functionally related to the quadrupole mass filter, but is a three-dimensional system rather than two dimensional. An ion trap consists of three hyperbolic electrodes, a ring and two end caps shown schematically in Figure 5.7. Unlike two-dimensional quadrupoles, ion trap instruments are operated principally in an RF voltage mode only, but due to the differences in geometry, ion trajectories tend to stabilize in the center of the trap volume. That is, in order to function, ion trajectories must be stable in both the axial (toward the end caps) and the radial (toward the ring electrode) dimensions. This is accomplished by application of an RF voltage to the ring electrode. When the amplitude of the ring electrode RF is relatively low, all ions above a minimum m/z are trapped, as shown in Figure 5.8, by the presence of three ions of

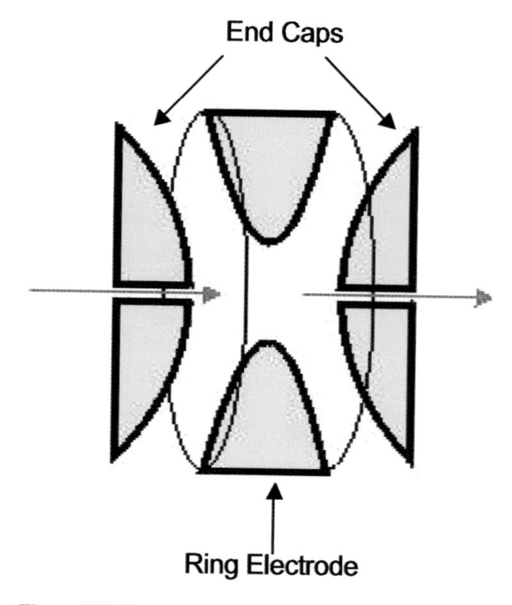

Figure 5.7. Schematic of quadrupole ion trap electrodes.

different mass residents simultaneously in the stability region. Lower m/z ions are trapped more effectively than higher ones. In order to accomplish this without having some trajectories becoming unstable, it was discovered that low pressure He had to be added to the trap (Louris *et al.*, 1987). Masses can be scanned by increasing the amplitude of the applied RF, making low masses progressively

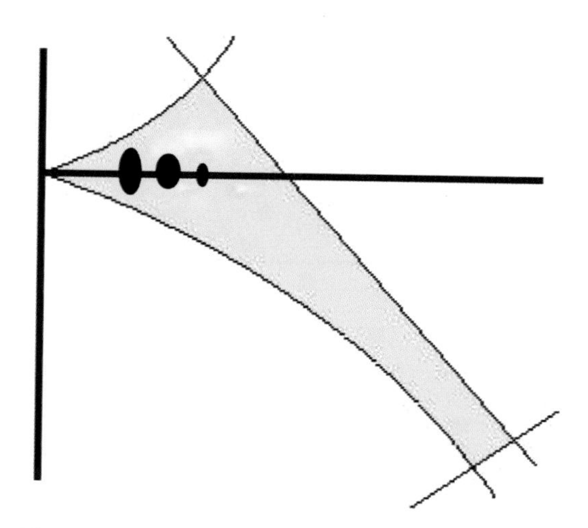

Figure 5.8. Stability diagram for quadrupole ion trap devices.

unstable, causing them to exit from an orifice in either end cap; the portion of ions ejected from the detector end cap is directed to a multiplier detector.

In a practical sense, the maximum RF amplitude is limited, and this would limit the scannable mass range. It has been found, however, that an additional RF voltage can be applied to the end caps to effect a resonance ejection from the trap (Stafford *et al.*, 1984); this resonance ejection effectively generates a region of instability within the normally stable region of operation and has the effect of increasing the useful mass range of the instrument. In addition, this resonance ejection strategy can be used to isolate ions of a particular m/z in the trap by alternatively ejecting m/z values above and below those desired. Finally, this same approach can be used at lower amplitudes of the resonance voltage to excite the selectively trapped ions, causing them to fragment. It is this methodology, isolating ions of a particular m/z and then exciting them to fragmentation, followed by ejection of the fragments, that is used to produce peptide fragments in the trap, as part of the methodology of protein identification discussed in Section 5.3.

5.2.2.3. Time-of-Flight Analyzer

TOF analyzers were introduced in the early 1950s, but it was not until the development of modern detectors and computer-based solid-state timing electronics that these analyzers have become widely used. On the one hand, TOFs are conceptually the simplest mass analyzers, yet in their modern implementation they are capable of extraordinary performance in terms of resolution and mass accuracy; in principle, TOFs have an unlimited mass range. The basis of a TOF analyzer's operation is shown schematically in Figure 5.9. Ions are formed in a well-defined planar region in conjunction with some well-characterized timing event, e.g., a laser flash in MALDI; this signal also begins the operation of a timing clock.

Following the ion formation event, ions are accelerated away from the source by application of a potential to the ion withdrawal lens element. Following this initial acceleration, the ions pass through a second lens element that defines the entrance to the analyzer drift tube, a field-free region. Since all ions fall through the same

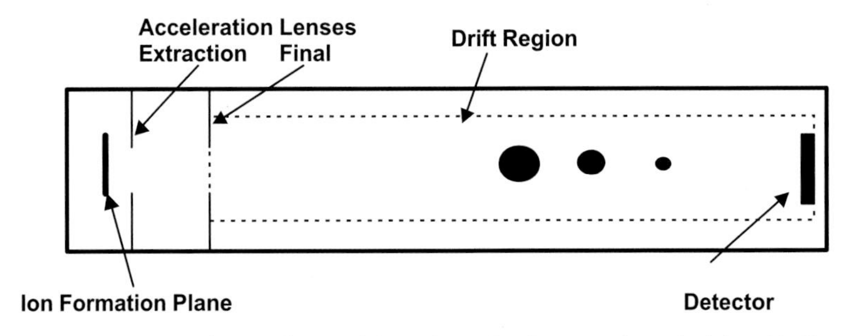

Figure 5.9. Schematic of time-of-flight (TOF) analyzer showing operating principle with ''ions'' of increasing mass shown as progressively larger ellipses.

potential, they all have the same kinetic energy, in addition to any energy distribution present as a consequence of ion formation. Since all ions have the same energy, they have velocities, v inversely proportional to the square roots of their mass, i.e.,

$$v = \left(\frac{2eE}{m}\right)^{1/2},$$

where e is the unit of charge, E is the electrostatic potential, and m is the ionic mass; this is shown schematically with three ions in Figure 5.9.

There are two main refinements of this simple figure that have led to the modern high-performance TOF analyzers. First, the incorporation of a time delay in the extraction of ions from the source region (Medzihradszky *et al.*, 2000) is routinely applied as a first-order correction for energy dispersion of ions that arises as a consequence of effects associated with ion production. This delayed extraction led to a remarkable improvement in ion peak shape and resolution. The second refinement has been the incorporation of an electrostatic mirror or ''reflectron.'' This device serves as a second-order correction for ion energy dispersion and serves to improve resolution in the m/z range up to ~10 kDa. It functions as an energy mirror in that ions of a given m/z, but having a range of kinetic energies, penetrate the mirror to a depth proportional to their energy and are then turned around when reaching zero velocity, exiting the mirror from the end at which they entered. Since the packet of ions of this m/z had a range of energies, different extent of penetration occurred, but all were turned upon reaching zero velocity and all exit the reflector having the same energy; the dispersion is eliminated, and resolution is improved.

5.3. PROTEIN IDENTIFICATION AND CHARACTERIZATION BY MASS SPECTROMETRY

5.3.1. Protein Identification by Mass Spectrometry

The development of ESI and MALDI, as mentioned in Sections 5.2.1.1 and 5.2.1.2, has made it possible to ionize efficiently a wide variety of protein samples for analysis by mass spectrometry. A common experimental question for biochemists and cell biologists is the identification of an unknown protein or proteins present in a biological sample. In the past few years, a relatively standard approach to protein identification by mass spectrometry has been developed, as shown in Figure 5.10. Most biochemical laboratories routinely purify and analyze protein samples using a variety of liquid chromatography or gel electrophoresis methods. However, these methods often use salts or detergents that interfere with ionization, and these contaminants must be removed before mass spectrometry is attempted. For protein identification, the final step in protein preparation usually involves protein separation by either one- or two-dimensional polyacrylamide gel electrophoresis (PAGE). Most PAGE systems use some modification of a discontinuous buffer system in the presence of the detergent sodium dodecyl sulfate (SDS) (Laemmli,

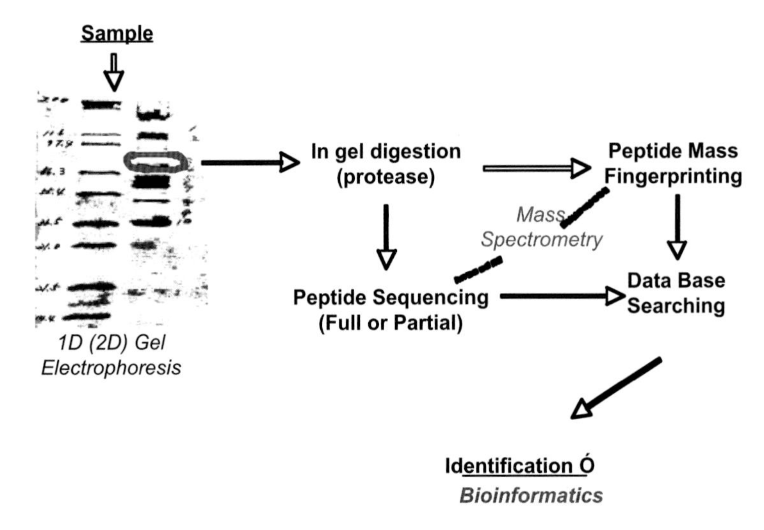

Figure 5.10. Standard approach for protein identification by mass spectrometry.

1970). The samples are treated with a reducing agent to remove peptide cross-linking through disulfide bonds and the protein subunits separate by relative size, assuming a constant ratio of SDS binding per unit mass of protein. This method works well for a wide variety of proteins, since most proteins are well denatured and soluble in the presence of SDS. For complex samples, often a single band on a one-dimensional gel contains multiple comigrating proteins. To obtain a higher resolution of protein species, a two-dimensional PAGE system is commonly used (O'Farrell, 1975), in which proteins are separated on the basis of two independent properties, their isoelectric point in the first dimension, followed by separation by relative size in the SDS–PAGE second dimension. For some studies, the determination of the molecular weight of the protein is desired. However, extraction of intact proteins from gels after SDS–PAGE and removal of contaminants can be difficult, so liquid chromatographic methods are most commonly used when the intact molecular weight is desired.

For protein identification studies, the protein bands or spots in a gel are visualized by some type of stain and then digested with a protease, usually trypsin, to produce peptides that can be extracted from the gel (Shevchenko *et al.*, 1996; Wilm *et al.*, 1996). This approach produces peptides that can then be routinely analyzed by either electrospray or MALDI mass spectrometry when femtomole or higher levels of protein are present in a gel band. The mass spectrometric analysis of peptide mixtures from in-gel digestions can be divided into two basic approaches. In the first, the m/z values of the ion peaks can be used to determine the masses for all the peptide ions produced from the digest. This is most easily performed using MALDI mass spectrometry, since all the ions produced are singly charged, and the TOF analyzer produces a single spectrum from which all of the peptide masses can be calculated directly. This type of data is often referred to as a ''peptide mass fingerprint'' of a protein. If the identity of a protein is known, the peptide mass fingerprint can be used to verify its sequence, and peaks with masses that differ from those predicted may indicate either the presence of mutations or posttranslational modifications of the protein.

For an unknown protein, the protein identity can often be determined from the analysis of the peptide mass fingerprint. This method relies on a database of predicted amino acid sequences of proteins from a given organism, which is used to generate a list of theoretical masses for peptides produced by cleavage of the protein at specific sites by a protease. The experimental peptide mass list is then compared with the list of theoretical peptide masses of each protein in the database, and the protein that best matches the experimental data is determined. In an early demonstration of this method using yeast proteins separated by two-dimensional gels, Shevchenko *et al.* (1996) reported that for 150 gel spots identified, 90% could be identified using this automated method. The remaining proteins required electrospray tandem mass spectrometry to produce peptide sequence tags, followed by database searching for their identification. While protein identification by peptide mass fingerprinting works with many samples, the reliability of this method comes into question in cases where protein mixtures are present or if there is a high degree of posttranslational modification. With the increased availability of mass spectrometers that have the capability to perform tandem mass spectrometry, the preferred method for protein identification has become peptide sequencing using tandem mass spectrometry, as described in the following section.

5.3.2. Peptide Sequencing by MS/MS

Amino acid sequence determination has long been an essential element in the biochemical characterization of a protein. Twenty years ago, the standard method for peptide sequencing was automated Edman sequencing (Hunkapiller *et al.*, 1984). This method involves the sequential removal of amino acids from the amino terminal of the peptide and analysis of the phenylthiohydantoin (PTH)-modified amino acid that is released by reversed-phase HPLC. As the cycle of reactions is repeated, the next amino acid residue from the amino terminal is removed and a sequence ladder is developed from the amino to the carboxyl terminal of the peptide until the end of the peptide is reached, or the released amino acid is below the detection limit of the instrument. Until 1998, automated Edman sequencing was still the standard method of peptide sequencing used in protein structure laboratories surveyed by the Association of Biomolecular Resource Facilities (Henzel *et al.*, 2000). However, several groups reported using the peptide mass, determined by mass spectrometry, to improve on the reliability of their Edman sequencing results. Although Edman sequencing is still used for a variety of sequencing purposes, the increased speed, sensitivity, and ability to deal with peptide mixtures make mass spectrometry an advantageous method for protein identification and analysis (Aebersold and Mann, 2003).

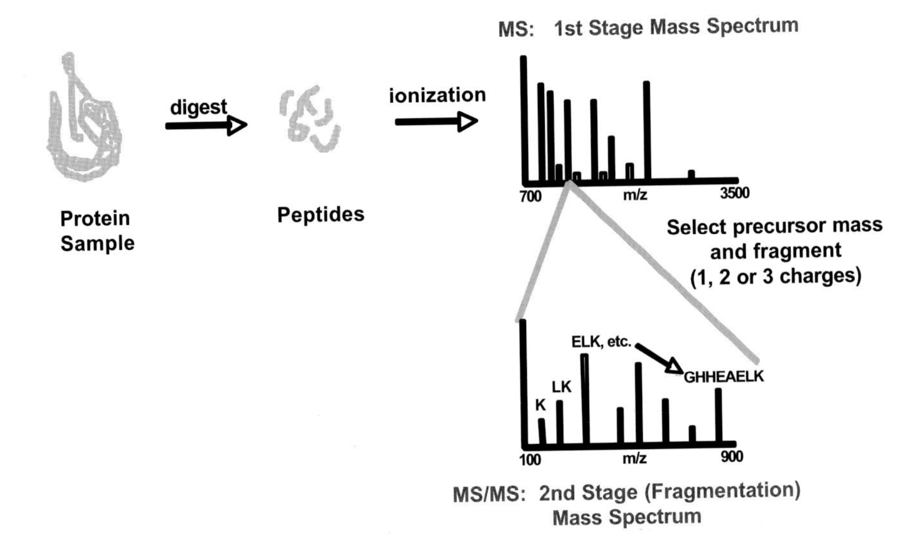

Figure 5.11. Tandem mass spectrometry for peptide sequencing.

For peptide sequencing by tandem mass spectrometry, a peptide ion is mass selected and fragmented, and the masses of the resulting fragment ions are determined (Figure 5.11). The method most commonly used for fragmentation is either metastable decay or collisionally induced decomposition (CID). In either case, sequence information is derived from fragmentation of the ion at the peptide bonds (Figure 5.12), producing either a ladder of y-ions, if the charge is on the C-terminal fragment, or b-ions, if the charge is on the N-terminal fragment. A detailed description of peptide sequencing by mass spectrometry is found in the review by Steen and Mann (2004).

Due to the chemical nature of peptide ion fragmentation, the MS/MS spectra can be quite complex. Complete ladders of either y-ions or b-ions are seldom obtained, making a complete amino acid sequence determination from the mass spectra data alone a difficult task. In addition to y-ion and b-ion formation, bond cleavage at other sites also occurs, generating additional fragments that complicate the interpretation. Peptide sequencing from MS/MS spectra has become a relatively rapid and straightforward process as a result of the development of database search engines that make use of protein sequences databases obtained from translation of cDNA and genomic DNA sequence. Examples of these programs include Sequest (Eng *et al.*, 1994) and Mascot (Perkins *et al.*, 1999), among others. These programs generate tables of theoretical peptide and fragment ion masses for sequences in the database. The programs then search the database sequences for peptides of a given mass, within a range of mass tolerances compatible with the mass spectrometry data, and then test for the best match of the predicted fragment masses to the peaks observed in the MS/MS data. Therefore, even if the MS/MS spectra do not contain

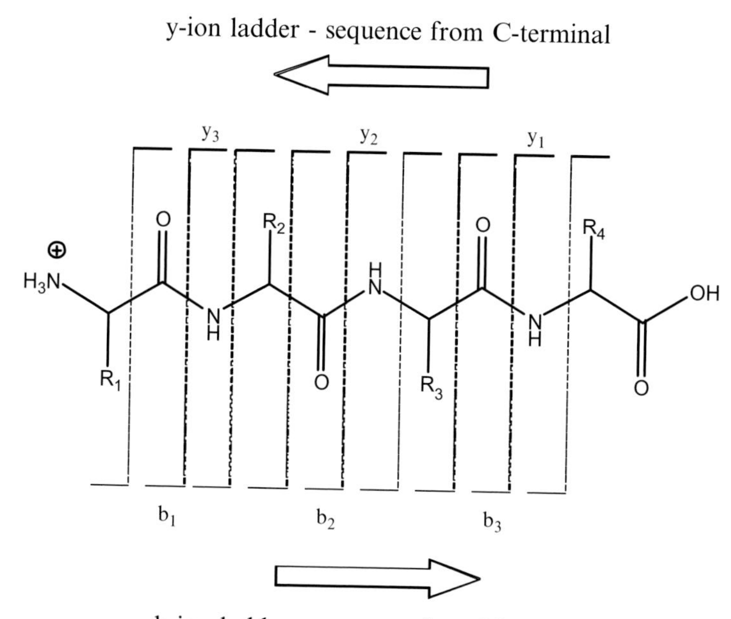

Figure 5.12. Peptide fragment ions.

enough fragment information to determine a unique sequence by *de novo* methods, there is often enough information to match a unique sequence in the database.

There are several factors that may influence sequence identifications using these methods. The first is the extent of fragmentation for the given peptide ion. For ions that do not fragment well, there may not be enough information to determine a unique match. A second factor is noise in the spectra, which may result in enough peaks being matched to an incorrect sequence to generate a false positive match. An additional requirement for this method is that the exact peptide sequence must be present in the database in order to obtain a match. As sequence databases improve, this becomes less of a problem, but there are currently many biologically important species for which the complete genomic DNA sequences are not available.

5.4. ANALYSIS OF PROTEIN COMPLEXES BY MASS SPECTROMETRY

5.4.1. Isolation and Identification of Novel Proteins

The isolation of protein complexes for biochemical analysis presents a significant challenge due to the complexity of cellular protein mixtures, the diversity of protein structures, their chemical reactivity, and the dynamic range of proteins in biological samples. The separation techniques must be gentle enough to preserve specific interactions of the complex, while also being stringent enough to disrupt those interactions that result in nonspecific binding of contaminants. In addition, the dissociation constants for proteins in the complex must be slow enough for the proteins to remain associated throughout the purification steps (or, conversely, the purification method must be rapid enough to preserve the complex when the dissociation rate is high).

For stable protein complexes expressed at fairly high levels in the cell, a variety of chromatographic and sedimentation techniques have been successfully used to isolate these complexes. Examples of complexes isolated using these techniques include the 30S ribosome from *Escherichia coli* (Hardy *et al.*, 1969), clathrin complexes (Pearse, 1976; Prasad and Lippoldt, 1988), and yeast nuclear pore complexes (NPC) (Rout and Blobel, 1993). One of the first large-scale applications of mass spectrometry to the identification of novel proteins in large protein complexes was reported for the yeast NPC (Rout *et al.*, 2000). In this example, the yeast NPCs were isolated by conventional biochemical separations, and the proteins were then further separated by liquid chromatography, using either a hydroxyapatite column or C_4 reversed-phase chromatography in the presence of TFA. Fractions from both of these chromatographic methods were then separated by SDS-PAGE. All of the protein bands present were then subjected to in-gel digestion and the peptides were analyzed by MALDI–TOF mass spectrometry or tandem mass analysis using a MALDI ion-trap mass spectrometer. This study included the analysis by mass spectrometry of 642 gel bands, resulting in the identification of 174 proteins. Of the 174 proteins identified, 40 were confirmed to be associated specifically with NPCs. The success of this method for identifying novel proteins was demonstrated by the finding that only four of the identified proteins were nucleoporin proteins which had already been known to be part of the NPC. The other 36 NPC-associated proteins identified were those expressed from open reading frames for which no protein function had been previously identified. Overall, 29 nucleoporins and 11 transfer factor proteins and NPC-associated proteins were identified using this technique.

Many cellular protein complexes cannot be isolated by standard techniques for biochemical analysis. In efforts to isolate and analyze these protein complexes, a variety of related affinity purification approaches have been developed. While these methods have been successful for specific protein isolations, they each have their individual strengths and limitations.

5.4.2. Immunoaffinity Purification

Due to the high affinity of antibody–ligand interactions, antibodies raised to specific proteins have been widely used to purify protein complexes by immunoprecipitation or immunoaffinity chromatography. For this approach to succeed, the antibody must have a high affinity for the native protein and little cross-reactivity to other related proteins. A drawback of this approach is that the high-quality antibodies suitable for this method can take a lot of time to generate, and they are specific for only one protein. In addition, antibody production requires either purified protein or specific synthetic peptides for immunization of the animals. Advantages of this approach include the high-affinity binding and that the isolated complex proteins have not been artificially modified in their level of expression or structure.

A modification of the immunopurification approach involves attaching the antibody to the surface of a plasmon resonance biosensor chip to produce an affinity chromatography matrix. The protein sample is then passed over the chip using oscillating microfluidics until equilibrium is achieved, and the protein is eluted and collected for further analysis by mass spectrometry or other techniques. Using this approach, the binding and the elution constants can be monitored in real time. Therefore, binding, washing, and elution conditions can be optimized to achieve the best purification and recovery of the protein. Using this BIA–MS method, with an anti-lysozyme antibody, 350 fmol of lysozyme was specifically bound and eluted from a mixture of cytosol proteins (Figure 5.13), and the protein could be identified from MALDI-MS of a tryptic digest of the protein (Gilligan *et al.*, 2002). The current design of biosensor chips limits the amount of antibody that can be coupled, and therefore presents a challenge to binding enough protein for mass spectrometric analysis. However, it seems that modification of the chip design to create a larger surface area for binding would make the method more generally applicable for analysis by mass spectrometry.

5.4.3. Epitope Tags and Tandem Affinity Purification

Another affinity purification approach that has been recently coupled with mass spectrometric analysis is the expression of a protein of interest with a specific epitope tag, which can be used to affinity purify the protein as a complex with other endogenous proteins in the cell. A variety of affinity tags, including polyhistidine, myc, FLAG-epitope, and HA, have been developed to assist in the specific purification of recombinant proteins from cells (Ford *et al.*, 1991). The binding protein to be studied is expressed as a tagged recombinant protein, and the cell lysate is then incubated with the antitag antibody covalently coupled to agarose beads. The beads are then washed and the bound protein is eluted from the column for further analysis (Figeys *et al.*, 2001). Even with high-affinity epitope tags, nonspecific contaminants can still frequently be observed, especially if the complex is expressed at low levels in the cell.

To overcome this problem, a modification of this approach has been developed using two or more distinct affinity tags attached to the protein, or tandem affinity purification (TAP)-tagged method, first described by Rigaut *et al.* (1999). In this method, a vector is constructed to express the protein of interest fused to both calmodulin-binding peptide and IgG-binding domains of protein A. In addition, a TEV protease site is included between the two tags to allow the IgG-binding domains to be cleaved from the expressed protein under mild conditions. Using this method, the cell extract expressing the TAP-tagged protein complex is first bound to an IgG affinity column. After binding and washing, the specifically bound

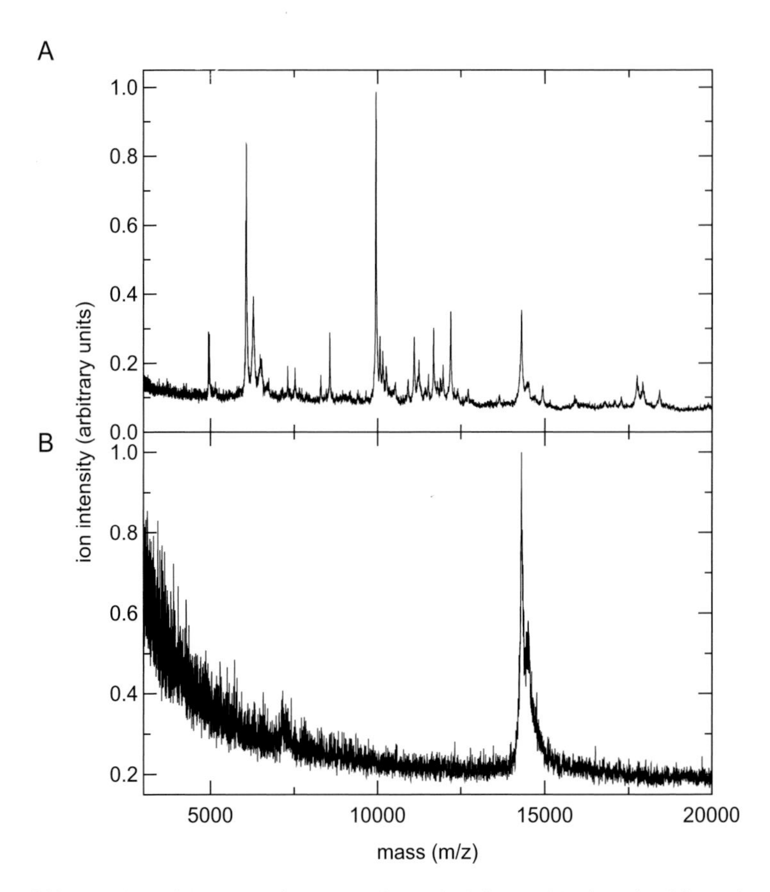

Figure 5.13. Isolation of lysozyme from cytosol protein Mixture. Matrix-assisted laser desorption ionization time-of-flight (MALDI–TOF) mass spectra of 5 μL sample containing 70 nM lysozyme in the presence of cytosol protein mixture: (A) sample spotted directly on the MALDI plate and (B) sample recovered after affinity purification by biosensor capture.

protein is cleaved using TEV protease and eluted from the column. The protein is then bound in the presence of calcium to a second-affinity column containing covalently attached calmodulin. Again, contaminants are washed away, and purified complex is specifically removed from the beads by removing the calcium.

The utility of the TAP method, combined with mass spectrometry for identifying specific protein-binding partners, has been demonstrated for a number of protein complexes. One example of this method is identifying the proteins associated with the yeast *YKL155C* gene (Saveanu *et al.*, 2001). A previous report using a yeast two-hybrid screen to detect protein-binding partners had found an interaction between Ykl155c and the Prp11 splicing factor (Fromont-Racine *et al.*, 1997). However, when the TAP-tagged Ykl155c protein was used to isolate protein complexes, it was clearly shown that Ykl155c was part of the mitochondrial small ribosomal subunit. To confirm the identity of the localization, a complementary TAP-tag experiment was carried out, using the yeast mitochondrial ribosomal protein Mrp1 as the TAP-tagged bait protein. Again, in this experiment, Ykl155c protein was demonstrated to be part of a complex with Mrp1. In addition to identifying Ykl155c as a part of this mitochondrial ribosomal protein complex, 12 novel proteins were also identified. It was further shown that these proteins were required for respiratory competency of the yeast cells.

While the TAP-tag method is considered a generic method for protein complex isolation, a specific target protein must still be cloned into the TAP constructs and the cells must be transfected in order to express the protein of interest. In addition, specific plasmid vector systems must be developed for the expression of the TAP-tagged protein in other organisms, and a number of other vectors for expression of TAP-tagged proteins in other organisms, including *Schizosaccharomyces pombe* (Tasto *et al.*, 2001), *Cenorhabditis elegans* (Polanowska *et al.*, 2004), and mammalian cells (Knuesel *et al.*, 2003) have been described.

5.4.4. Global Screening for Protein Interactions

In addition to using mass spectrometry to identify novel proteins in previously known protein complexes, recent attempts have been made in the use of mass spectrometry to identify proteins in complexes that have not been previously characterized. These may be complexes that are transient in nature or comprised of proteins for which the biological functions have not yet been identified. Two independent groups (Gavin *et al.*, 2002; Ho *et al.*, 2002) reported on using the expression of epitope-tagged bait proteins to pull out a large number of protein complexes for analysis of the binding partners by mass spectrometry. Ho *et al.* (2002) transiently expressed 725 FLAG-tagged proteins from different functional classes to pull out potential binding partners. From these experiments, 3,617 associated proteins were detected, covering 25% of the yeast proteome. Gavin *et al.* (2002) used TAP-tagged proteins expressed in gene-specific cassettes inserted by homologous recombination. This method results in control of the expression of the TAP-tagged proteins by the normal gene promoters for each protein. They reported the results of expressing 1,167 genes, from which 589 protein complexes were purified and analyzed by mass spectrometry. From these purifications, 232 complexes were identified. The size of the protein complexes isolated ranged from 2 to 83 components, with an average of 12 components/complex.

In addition to screening for novel complexes, the isolation of proteins bound to a TAP-tagged bait protein can also assist in assigning biological functions to proteins encoded by uncharacterized open reading frames. This approach has been applied to the characterization of a number of yeast proteins with unknown functions (Hazbun *et al.*, 2003). The first part of their study was to express a large number of unknown yeast proteins with a TAP-tag and to identify those proteins that specifically formed complexes with each of the unknown proteins. Since many of the bound proteins identified had functions that were already known, it was possible to then assign the same type of biochemical function for the TAP-tagged bait proteins. Using this method, they reported the identification of binding proteins for 29 of the TAP-tagged unknown proteins. The identity of the binding partners to these proteins made it possible to ascribe specific biological processes to 27 of these proteins, including involvement in chromosome segregation, DNA repair, mRNA processing, protein biosynthesis, and others.

5.4.5. Characterization of Posttranslational Modifications in Protein Complexes

In addition to identifying the protein components that make up a complex, mass spectrometry has been extremely useful in identifying how these proteins are posttranslationally modified. Changes in posttranslational modifications, such as phosphorylation, are known to be involved in regulating various biochemical signaling pathways, and a great deal of effort has been devoted to studying how these changes may alter the formation of specific protein complexes in the cell. In contrast to protein identification, identification of specific posttranslational modifications can be more demanding. In order to identify a protein by mass spectrometry using database matching, it is usually sufficient to match peptides that make up only a fraction of the total protein sequence. However, for peptide modifications, ions corresponding to the specific modified peptide must be observed. For complex protein samples, the specific ions corresponding to a modified peptide may be difficult to observe against the large background of unmodified peptides, so additional fractionation of the sample is frequently required.

One example of a posttranslational modification associated with protein complex formation is N-terminal acetylation in targeting an Arf-like GTPase, Arl3p, to the Golgi membrane (Behnia *et al.*, 2004). This targeting required binding of the modified Arl3p to another protein, Sys1p, and the specific modification of the amino terminal by acetylation. This modification was demonstrated by affinity purifying an affinity-tagged version of Arl3p and MALDI mass spectrometry, after digestion with either trypsin or GluC, was used to identify the acetylated amino terminal peptide.

For the detection of phosphopeptide-specific binding interactions, a screening method has been described using phosphorylated and unphosphorylated versions of a synthetic peptide to affinity purify proteins that specifically bind to the phosphorylated peptide (Schulze and Mann, 2004). This method was demonstrated using a tyrosine-phosphorylated peptide of the epidermal growth factor (EGF) receptor to specifically bind the Grb2 protein. In these experiments, stable isotope labeling of two cell populations with either ^{12}C or ^{13}C arginine allowed the specific binding to the phosphorylated peptide bait to be determined by the ratio of heavy to light mass peak intensities in MALDI mass spectrometry of the tryptic digests from the bound proteins. In a similar manner, the EFG receptor pathway was investigated using stable isotope labeling of cell proteins, combined with specific affinity purification of the phosphorylated EGF receptor (Blagoev *et al.*, 2003). Many signal-transducing proteins form complexes with the activated EGF receptor, and this method measured changes in the EGF receptor complex on stimulation of the cells with EGF.

REFERENCES

Aebersold, R. and Mann, M. (2003). Mass spectrometry-based proteomics. *Nature* 422:198–207.

Behnia, R., Panic, B., Whyte, J. R., and Munro, S. (2004). Targeting of the Arf-like GTPase Arl3p to the Golgi requires N-terminal acetylation and the membrane protein Sys1p [see comment]. *Nat Cell Biol* 6:405–413.

Bencsura, A., Navale, V., Sadeghi, M., and Vertes, A. (1997). Matrix–guest energy transfer in matrix-assisted laser desorption. *Rapid Commun Mass Spectrom* 11:679–682.

Blagoev, B., Kratchmarova, I., Ong, S. E., Nielsen, M., Foster, L. J., and Mann, M. (2003). A proteomics strategy to elucidate functional protein–protein interactions applied to EGF signaling. *Nat Biotechnol* 21:315–328.

Campbell, J. M., Stein, S. E., Blank, P. S., Epstein, J. A., Vestal, M. E., Yergey, A. L. Fragmentation Leucine Enkephalin as a Function of Lazer Fluence in a MALDI-TOF. *J. Am Soc. Mass Spectrom.* (In Press, 2006).

Dreisewerd, K. (2003). The desorption process in MALDI. *Chem Rev* 103:395–425.

Eng, J. K., McCormack, A. L., and Yates, J. R., III. (1994). An approach to correlate tandem mass spectral data of peptides with amino acid sequences in a protein database. *J Am Soc Mass Spectrom* 5:976–989.

Fenn, J. (2002). Electrospray wings for molecular elephants. *Nobel Lect* 2002:154–184.

Fenn, J., Mann, M., Meng, C., Wong, S., and Whitehouse, C. (1989). Electrospray ionization for mass spectrometry of large biomolecules. *Science* 246:64–71.

Figeys, D., McBroom, L. D., and Moran, M. F. (2001). Mass spectrometry for the study of protein–protein interactions. *Methods (Duluth)* 24:230–239.

Ford, C. F., Suominen, I., and Glatz, C. E. (1991). Fusion tails for the recovery and purification of recombinant proteins. *Protein Expr Purif* 2:95–107.

Fromont-Racine, M., Rain, J. C., and Legrain, P. (1997). Toward a functional analysis of the yeast genome through exhaustive two-hybrid screens [see comment]. *Nat Genet* 16:277–282.

Gavin, A. C., Bosche, M., Krause, R., Grandi, P., Marzioch, M., Bauer, A., Schultz, J., Rick, J. M., Michon, A. M., Cruciat, C. M., Remor, M., Hofert, C., Schelder, M., Brajenovic, M., Ruffner, H., Merino, A., Klein, K., Hudak, M., Dickson, D., Rudi, T., Gnau, V., Bauch, A., Bastuck, S., Huhse, B., Leutwein, C., Heurtier, M. A., Copley, R. R., Edelmann, A., Querfurth, E., Rybin, V., Drewes, G., Raida, M., Bouwmeester, T., Bork, P., Seraphin, B., Kuster, B., Neubauer, G., and Superti-Furga, G. (2002). Functional organization of the yeast proteome by systematic analysis of protein complexes [see comment]. *Nature* 415:141–147.

Gilligan, J. J., Schuck, P., and Yergey, A. L. (2002). Mass spectrometry after capture and small-volume elution of analyte from a surface plasmon resonance biosensor. *Anal Chem* 74:2041–2047.

Graves, P. and Haystead, T. (2002). Molecular biologist's guide to proteomics. *Microbiol Mol Biol Rev* 66:39–63.

Hardy, S. J., Kurland, C. G., Voynow, P., and Mora, G. (1969). The ribosomal proteins of *Escherichia coli*. I. Purification of the 30S ribosomal proteins. *Biochemistry* 8:2897–2905.

Hazbun, T. R., Malmstrom, L., Anderson, S., Graczyk, B. J., Fox, B., Riffle, M., Sundin, B. A., Aranda, J. D., McDonald, W. H., Chiu, C. H., Snydsman, B. E., Bradley, P., Muller, E. G., Fields, S., Baker, D., Yates, J. R., III, and Davis, T. N. (2003). Assigning function to yeast proteins by integration of technologies [see comment]. *Mol Cell* 12:1353–1365.

Hendrickson, C. L. and Emmett, M. R. (1999). Electrospray ionization Fourier transform ion cyclotron resonance mass spectrometry. *Annu Rev Phys Chem* 50:517–536.

Henzel, W. J., Admon, A., Carr, S., Davis, G., DeJongh, K., Lane, W., Rohde, M., and Steinke, L. (2000). ABRF-99SEQ: evaluation of peptide sequencing at high sensitivity. *J Biomol Tech* 11:92–99.

Hill, L. (2005). How a quadrupole mass filter works. http://www.jic.bbsrc.ac.uk/SERVICES/metabolomics/lcms/single1.htm.

Ho, Y., Gruhler, A., Heilbut, A., Bader, G. D., Moore, L., Adams, S. L., Millar, A., Taylor, P., Bennett, K., Boutilier, K., Yang, L., Wolting, C., Donaldson, I., Schandorff, S., Shewnarane, J., Vo, M., Taggart, J., Goudreault, M., Muskat, B., Alfarano, C., Dewar, D., Lin, Z., Michalickova, K., Willems, A. R., Sassi, H., Nielsen, P. A., Rasmussen, K. J., Andersen, J. R., Johansen, L. E., Hansen, L. H., Jespersen, H., Podtelejnikov, A., Nielsen, E., Crawford, J., Poulsen, V., Sorensen, B. D., Matthiesen, J., Hendrickson, R. C., Gleeson, F., Pawson, T., Moran, M. F., Durocher, D., Mann, M., Hogue, C. W., Figeys, D., and Tyers, M. (2002). Systematic identification of protein complexes in *Saccharomyces cerevisiae* by mass spectrometry [see comment]. *Nature* 415:180–193.

Hu, Q., Noll, R., Li, H., Makarov, A., Hardman, M., and Cooks, R. (2005). The Oribitrap: a new mass spectrometer. *J Mass Spectrom* 40:430–443.

Hunkapiller, M. W., Strickler, J. E., and Wilson, K. J. (1984). Contemporary methodology for protein-structure determination. *Science* 226:304–311.

Karas, M., Bachman, D., Bahr, U., and Hillenkamp, F. (1987). Matrix-assisted ultraviolet-laser desorption of nonvolatile compounds. *Int J Mass Spectrom Ion Process* 78:53–68.

Karas, M. and Hillenkamp, F. (1988). Laser desorption ionization of proteins with molecular masses exceeding 10,000 Da. *Anal Chem* 60:2299–2301.

Karas, M. and Kruger, R. (2003). Ion formation in MALDI: the cluster ionization mechanism. *Chem Rev* 103:427–440.

Knochenmuss, R. and Zenobi, R. (2003). MALDI ionization: the role of in-plume processes. *Chem Rev* 103:441–452.

Knuesel, M., Wan, Y., Xiao, Z., Holinger, E., Lowe, N., Wang, W., and Liu, X. D. (2003). Identification of novel protein–protein interactions using a versatile mammalian tandem affinity purification expression system. *Mol Cell Proteom* 2:1225–1233.

Laemmli, U. K. (1970). Cleavage of structural proteins during assembly of head of bacteriophage-T4. *Nature* 227:680–685.

Louris, J., Cooks, R., Syka, J., Kelley, P., and Stafford, G. (1987). Instrumentation, applications, and energy deposition in quadrupole ion-trap tandem mass spectrometry. *Anal Chem* 59:1677–1685.

Marshall, A. G., Hendrickson, C. L., and Jackson, G. S. (1998). Fourier transform ion cyclotron resonance mass spectrometry: a primer. *Mass Spectrom Rev* 17:1–35.

Medzihradszky, K., Campbell, J., Baldwin, M., Juhasz, P., Vestal, M., and Burlingame, A. (2000). The characteristics of peptide collision-induced dissociation using a high-performance MALDI-TOF/TOF tandem mass spectrometer. *Anal Chem* 72:552–558.

Menzel, C., Dreisewerd, K., Berkenkamp, S., and Hillenkamp, F. (2002). The role of the laser pulse duration in infrared matrix-assisted laser desorption/ionization mass spectrometry. *J Am Soc Mass Spectrom* 13:975–984.

O'Farrell, P. H. (1975). High resolution two-dimensional electrophoresis of proteins. *J Biol Chem* 250:4007–4021.

Pearse, B. M. F. (1976). Clathrin—unique protein associated with intracellular transfer of membrane by coated vesicles. *Proc Natl Acad Sci USA* 73:1255–1259.

Perkins, D. N., Pappin, D. J., Creasy, D. M., and Cottrell, J. S. (1999). Probability-based protein identification by searching sequence databases using mass spectrometry data. *Electrophoresis* 20:3551–3567.

Polanowska, J., Martin, J. S., Fisher, R., Scopa, T., Rae, I., and Boulton, S. J. (2004). Tandem immunoaffinity purification of protein complexes from *Caenorhabditis elegans*. *Biotechniques* 36:778–780.

Prasad, K. and Lippoldt, R. E. (1988). Molecular characterization of the AP180 coated vesicle assembly protein. *Biochemistry* 27:6098–6104.

Rayleigh (1882). *Phil. Mag.*184–186.

Rigaut, G., Shevchenko, A., Rutz, B., Wilm, M., Mann, M., and Seraphin, B. (1999). A generic protein purification method for protein complex characterization and proteome exploration. *Nat Biotechnol* 17:1030–1032.

Rout, M. P., Aitchison, J. D., Suprapto, A., Hjertaas, K., Zhao, Y., and Chait, B. T. (2000). The yeast nuclear pore complex: composition, architecture, and transport mechanism. *J Cell Biol* 148: 635–651.

Rout, M. P. and Blobel, G. (1993). Isolation of the yeast nuclear pore complex. *J Cell Biol* 123: 771–783.

Saveanu, C., Fromont-Racine, M., Harington, A., Ricard, F., Namane, A., and Jacquier, A. (2001). Identification of 12 new yeast mitochondrial ribosomal proteins including 6 that have no prokaryotic homologues. *J Biol Chem* 276:15861–15867.

Schneider, B. and Chen, D. (2000). Collision-induced dissociation of ions within the orifice-skimmer region of an electrospray mass spectrometer. *Anal Chem* 72:791–799.

Schulze, W. X. and Mann, M. (2004). A novel proteomic screen for peptide–protein interactions. *J Biol Chem* 279:10756–10764.

Shevchenko, A., Jensen, O. N., Podtelejnikov, A. V., Sagliocco, F., Wilm, M., Vorm, O., Mortensen, P., Shevchenko, A., Boucherie, H., and Mann, M. (1996). Linking genome and proteome by mass spectrometry: large-scale identification of yeast proteins from two dimensional gels. *Proc Natl Acad Sci USA* 93:14440–14445.

Stafford, G., Kelley, P., Syka, J., Reynolds, W., and Todd, J. (1984). Recent improvements in and application of advanced ion trap technology. *Int J Mass Spectrom Ion Process* 60:85–98.

Steel, C. and Henchman, M. (1998). Understanding the quadrupole mass filter through computer simulation. *J Chem Edu* 75:1049–1054.

Steen, H. and Mann, M. (2004). The ABCs (and XYZs) of peptide sequencing. *Nat Rev* 5:699–711.

Tanaka, K., Waki, H., Ido, Y., Akita, S., and Yoshida, Y. (1988). *Rapid Commun Mass Spectrom* 2:151–153.

Tasto, J. J., Carnahan, R. H., McDonald, W. H., and Gould, K. L. (2001). Vectors and gene targeting modules for tandem affinity purification in *Schizosaccharomyces pombe*. *Yeast* 18:657–662.

Wilm, M., Shevchenko, A., Houthaeve, T., Breit, S., Schweigerer, L., Fotsis, T., and Mann, M. (1996). Femtomole sequencing of proteins from polyacrylamide gels by nano-electrospray mass spectrometry. *Nature* (*London*) 379:466–469.

Yamashita, M. and Fenn, J. (1984). Electrospray ion source. Another variation on the free-jet theme. *J Phys Chem* 88:4451–4459.

Zhigilei, L., Kodali, P., and Garrison, B. (1997). On the threshold behavior in laser ablation of organic solids. *Chem Phys Lett* 276:269–273.

6

$H/^2H$ Exchange Mass Spectrometry of Protein Complexes

Elizabeth A. Komives

6.1. INTRODUCTION

Amide proton exchange is one of the oldest methods for probing the structure of proteins in solution (Linderstrom-Lang, 1955). Initially, experiments involved the monitoring of tritium incorporation into whole proteins. Later, NMR spectroscopists developed methods for monitoring the exchange of deuterons (NMR-silent nuclei) for protons (NMR-active nuclei) and NMR has been widely used for monitoring amide proton exchange to study protein folding. The first attempt to study a protein–protein interaction by monitoring amide proton exchange using NMR was reported in 1990 by Paterson *et al.* After this, many such studies were performed, but it became clear that it was difficult to find the protein–protein interface by NMR. Standard NMR experiment sample amides that exchange relatively slowly (with half-lives of hours) and the protein–protein interface is generally on the surface, so these amides exchange more quickly (with half-lives of seconds to minutes) (Orban *et al.*, 1992) (Figure 6.1).

In order to observe the rapidly exchanging amides in unstructured peptides and on protein surfaces, it is necessary to deuterate for short periods of time and then quench the reaction. The amide protons exchange by both an acid-catalyzed reaction and a base-catalyzed reaction such that the rate of exchange by both processes is minimal at pH 2.5 (Bai *et al.*, 1993) (Figure 6.2).

Thus, amide proton exchange reactions can be carried out at physiological buffer and temperature and then the exchange can be quenched by chilling the sample and reducing the pH to 2.5. Under "quench" conditions, amide protons

E. A. KOMIVES • Department of Chemistry and Biochemistry, University of California, San Diego, La Jolla, CA 92093-0378, USA. Tel.: +1 858 534 3058; fax: +1 858 534 6174; e-mail: ekomives@ucsd.edu

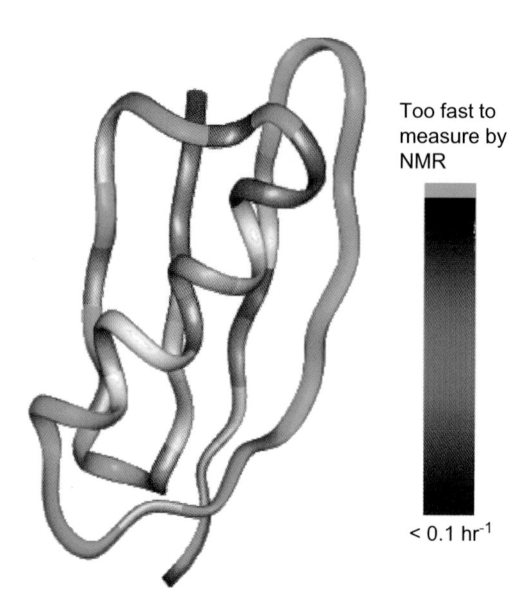

Figure 6.1. Summary of the amide proton exchange data from NMR experiments on the streptococcal protein G B2-domain (Orban *et al.*, 1992). The backbone of the protein is colored according to a scale from blue to red where blue is the slowest measurable rates and red is the fastest measurable rates. Those regions of the protein in which the amides exchanged too quickly to be measured by deuterium exchange experiments are colored green. Most of the surface of the protein exchanges too quickly to be measured by standard NMR experiments.

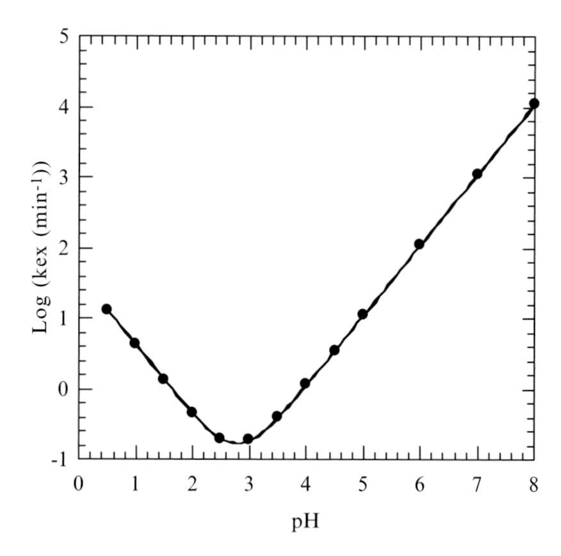

Figure 6.2. Plot of the log of the amide proton exchange rate versus pH extended from the data presented by Bai *et al.* (1993). Both acid-catalyzed and base-catalyzed processes occur, and the rates of these two processes are minimal at approximately pH 2.5.

have a half-life of approximately 20 min and measurements are, therefore, made under these conditions (Bai *et al.*, 1993).

Amide exchange experiments have traditionally been performed to understand protein folding and unfolding where the amide exchange rates relate to hydrogen bond formation (Englander *et al.*, 1997). In the case of protein–protein interfaces, hydrogen bonds are not usually formed across the interface, rather it is a case of decreased solvent accessibility that can arise because of a number of factors, including side chain interactions, decreased loop mobility, and so on. As it is known that the amides in the interior of a protein exchange more slowly than those on the surface, it is nearly always the more rapidly exchanging surface amides that require monitoring for protein–protein interface studies. Typically, amides on the surface of a protein exchange within seconds to minutes. This exchange rate is slower than that for unstructured peptides (Bai *et al.*, 1993), but faster than amides in the core of the protein (Dharmasiri and Smith, 1996).

6.2. TYPES OF AMIDE EXCHANGE EXPERIMENTS ON PROTEIN COMPLEXES

6.2.1. Measurement of Amide Proton Exchange of Whole Proteins in Complex

By comparing amide proton exchange in the separated proteins with that in the protein–protein complex, it is possible to measure the degree to which protein interaction protects one of the proteins from exchange. Essentially, the proteins, either alone or in complex, are allowed to deuterate for varying lengths of time and then the exchange reaction is quenched. The whole protein is sprayed into a quadrupole mass spectrometer and analyzed (Chapter 5). In these sorts of experiments, the protein takes on different numbers of charges and, therefore, shows up as several peaks at different m/z (mass to charge ratio) values. The spectrum can be deconvoluted to yield a single envelope with only one charge and in this way, the mass is accurately determined. The masses of the individually deuterated proteins are expected to be greater than the proteins that were deuterated in the complex. The difference is interpreted as protection from exchange due to the presence of the protein-binding partner. This type of experiment, in which the total difference in amides deuterated in the control versus the complex allowed Redford's group to determine how much the chaperone GroEL protected the folded state of dihydrofolate reductase (Gross *et al.*, 1996). This type of experiment, which measures amide exchange into the entire protein as a whole, has been termed as ''global'' amide exchange experiment. They are most often used in protein-folding studies, but can also be used to confirm whether protection occurrs in a protein–protein complex (see later).

Using matrix assisted laser desorption ionization time-of-flight (MALDI-TOF) mass spectrometry to monitor deuterium incorporation into whole proteins, Fitzgerald's group has pioneered the technique of SUPREX (Powell *et al.*, 2002).

This approach assumes that the ligand binding stabilizes the protein to denaturation and, therefore, less amide exchange occurs at the same concentration of denaturant in the ligand-bound state. MALDI-TOF analysis of full-length proteins does not yield the same mass accuracy as electrospray (ESI), but it is much more tolerant to the presence of denaturants, and thus, the method can be used as a high-throughput screen for ligand binding (Powell and Fitzgerald, 2004).

6.2.2. Measurement of Amide Proton Exchange in Regions of Proteins by Pepsin Digestion of the Exchanged Sample

In the 1970s, Richards' group found that pepsin readily cleaves proteins at pH 2.5 and 0°C, the ''quench'' conditions for amide proton exchange (Rosa and Richards, 1979). This allows for the localization of exchanged amides to segments of the protein defined by the pepsin cleavage sites. David Smith was the first to take advantage of pepsin digestion coupled to mass spectrometry to quantify the amount of deuteration in segments of a protein (Zhang and Smith, 1993). Pepsin is the only robust protease that cleaves most proteins into fragments of approximately 8–20 amino acids under ''quench'' conditions. Pepsin is a ''nonspecific'' protease meaning that its cleavage sites cannot be predicted and, therefore, the peptide fragments must be sequenced in order to be properly identified. However, pepsin cleaves a particular protein the same way every time, so once the pepsin digest products are identified, many experiments can be done without the need for further sequencing (Figure 6.3).

Since this first report, many groups have found that mass spectrometry is a rapid and powerful way to ascertain the extent of amide proton exchange in proteins. In all amide exchange experiments in which pepsin digestion is followed by mass spectrometry, some back-exchange occurs. This is the term that refers to the loss of deuteration after quench, because the sample is now in protiated solvent for some length of time before measurement. Naturally, the more quickly one can process the sample and the more carefully it can be kept at 0°C, the less back-exchange occur. Another way to minimize back-exchange is to carry out the quench and pepsin digestion in deuterated solution. This is not recommended because it can introduce artifacts. Both ESI mass spectrometry (Zhang and Smith, 1993) and MALDI-TOF mass spectrometry (Mandell *et al.*, 1998b) can be used to measure the amount of deuteration on the peptides derived from pepsin digestion of the deuterated proteins (Chapters 5 and 15). There are advantages and disadvantages in each mass spectrometry method. Briefly, ESI requires up-front LC separation of the peptides that must be carried out quickly and at 0°C. Because the peptides elute at different times, the amount of deuterium that is lost due to back-exchange during the analysis must be corrected for each peptide separately. On the other hand, the fact that the peptides are separated can yield more complete coverage of the protein sequence by the peptides that can be analyzed. MALDI-TOF has a larger dynamic range, and therefore no LC separation is required before analysis. This means that all of the peptides have experienced the same time of back-exchange making correction easier. However, because no separation is done, complete coverage of

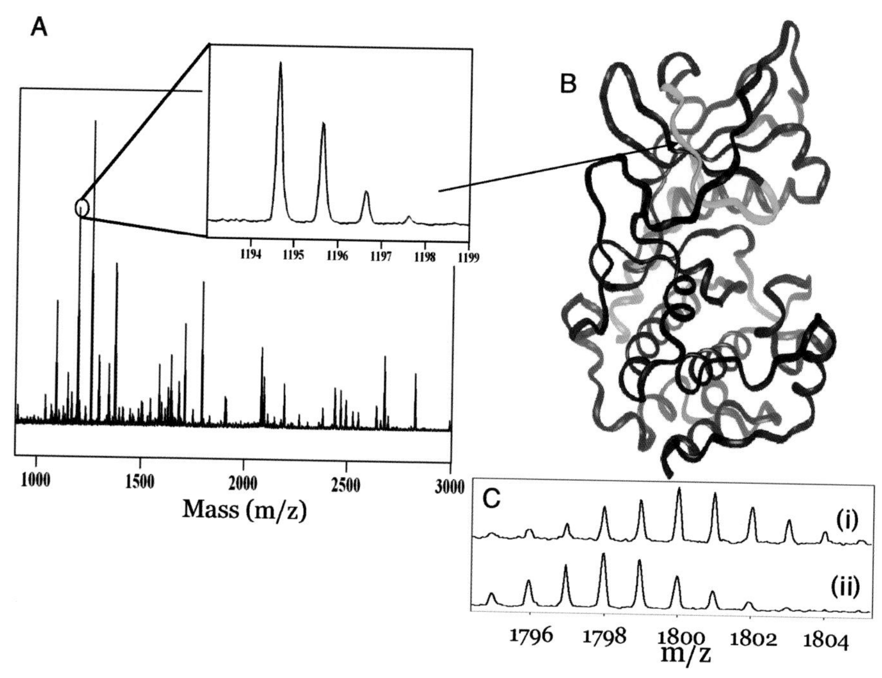

Figure 6.3. Localization of regions of amide exchange in proteins requires digestion with pepsin under quench conditions. Although pepsin is a ''nonspecific'' protease, meaning its cleavages cannot be predicted with high accuracy, it cleaves a particular protein the same way every time. (A) In a matrix assisted laser desorption ionization time-of-flight (MALDI-TOF) mass spectrometry-based amide exchange experiment, information about the entire protein is contained in a single MALDI-TOF mass spectrum, which contains many peptides that cover much of the sequence of the entire protein. The mass envelope of one of the peptides in the spectrum is shown in the inset. (B) After sequencing each of the pepsin digest products using tandem mass spectrometry, it can be mapped back onto the structure of the protein. (C) In an amide proton exchange experiment to measure a protein–ligand interaction, if a region of the protein is protected from exchange by the ligand, this can be seen as less deuterium incorporation into the peptide envelope (ii) as compared with the protein alone (i).

the protein sequence can be more difficult to obtain. When the quench is carried out in protiated solutions, it can be shown that LC-ESI and MALDI analyses require similar time, and result in equivalent back-exchange losses. High resolution data that can be obtained from newer mass spectrometers allow for visualization of differences in amide exchange into peptides in the raw mass spectra (Figure 6.3C).

Although NMR spectroscopists recognized that the interface between two proteins should be protected from exchange, it took some time before mass spectrometrists realized this phenomenon could be measured with high accuracy. The first report of localized interface protection was in the obligate oligomeric protein, rabbit muscle aldolase (Zhang and Smith, 1996). These researchers measured rates of exchange into the peptide segments after ESI and fast atom bombardment MS analysis of each peptide after deuteration times of 2.5 min to 44 h. On the whole, exchange rates correlated with intramolecular hydrogen bonding, except at the subunit interface where few hydrogen bonds were found, but where exchange rates were some of the slowest in the protein. It is important to note that the rates measured were not the rates of exchange of individual amides as can be obtained from NMR experiments, but rather the rates of exchange into each segment of the protein isolated after pepsin digestion.

Measurement of amide proton exchange in protein–protein complexes is becoming increasingly popular, and it would be impossible to cite all of the interesting results that have been forthcoming as a result. The protein interactions within viral coat proteins have been mapped by several investigators using ESI, FTICR, and MALDI-TOF mass spectral methods (Tuma *et al.*, 2001; Lanman *et al.*, 2003; Wang and Smith, 2005). Inter and intramolecular protein–protein and protein–ligand interactions in kinases have also been widely studied (Mandell *et al.*, 1998a; Gmeiner *et al.*, 2001; Nazabal *et al.*, 2003; Lee *et al.*, 2004).

6.2.3. Localization of Small Molecule Binding Sites on Proteins

The first reports of amide exchange being used to probe small molecules binding to proteins investigated NADH and substrate binding to dehydrogenases (Wang *et al.*, 1997, 1998). The dehydrogenase enzymes showed protection at active site segments when either NADH, substrate, or both are bound. Both global amide proton exchange experiments and experiments using pepsin digestion of the quenched sample were carried out and analyzed by ESI triple quadrupole MS. Using "global" exchange experiments, these researchers first showed that approximately 10% of the amides were protected from exchange in the two binary complexes with either the substrate, diaminopimelate, or the cofactor, NADH, bound, whereas 17% of the amides were protected in the ternary complex when both substrate and NADH were bound. They were also able to locate the dimer interface of the enzyme (Wang *et al.*, 1998). Similar methods were used to study the immucillin-H binding to purine nucleoside phosphorylase (Wang *et al.*, 2000). Decreases in amide proton exchange on binding of a transition-state analog suggested reduced protein dynamics in hypoxanthine-guanine phosphoribosyltransferase (Wang *et al.*, 2001). Kaltashov's lab has also done some creative work on small molecule binding to proteins (Xiao *et al.*, 2003). As is discussed later, all of these studies used an on-exchange approach in which the amide proton exchange into the free protein is compared with the exchange into the ligand-bound protein. This approach gives a general idea of regions of the protein that experience altered amide proton exchange on ligand binding, but cannot distinguish conformational changes from direct interface protection.

6.2.4. Combining Amide Proton Exchange with Molecular Docking

The generation of accurate structures of protein–protein complexes by computational docking remains a challenging problem. Often, structures of the individual proteins are known, but the structure of the complex is not, and amide exchange

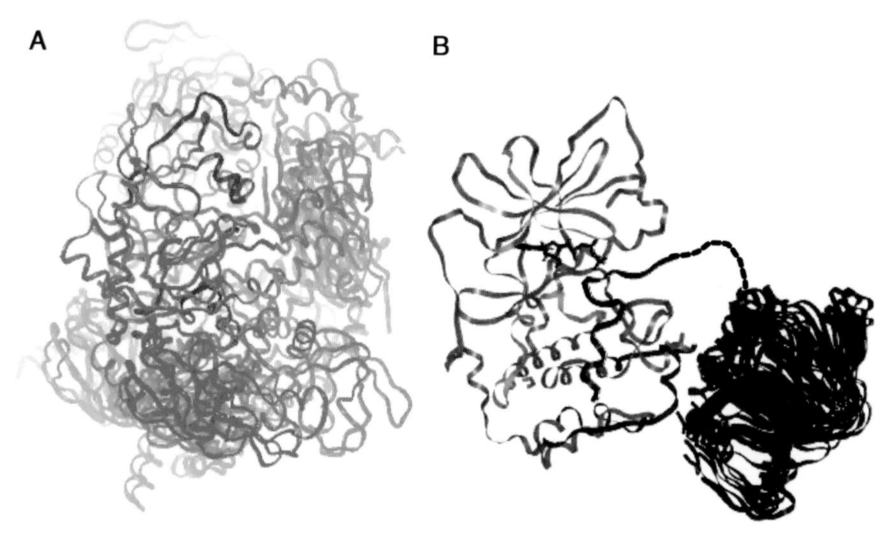

Figure 6.4. Use of amide exchange data to filter results from computational docking can significantly improve the results. On the left is shown the top 50 results from computational docking of the cAMP-binding A-domain of the regulatory subunit of protein kinase A (green) with the catalytic subunit (blue). On the right is shown the final structures after the top 100,000 structures were filtered for consistency with the amide exchange data. The catalytic subunit is gray, the regulatory subunit cAMP-binding domain is blue, and the pseudosubstrate region is black. Within the catalytic subunit, the peptic peptides that were protected when the pseudosubstrate region was bound are colored yellow, and those that were only protected when the cAMP-binding domain of the regulatory subunit was bound are colored red. Within the regulatory subunit, the peptic peptides that were protected when the catalytic subunit was bound are also colored red.

data can be used to localize the binding interface. For protein kinase A, computational docking studies failed to locate the interface between the catalytic and regulatory subunits (Figure 6.4A).

This is probably because the binding is bipartite, involving a pseudosubstrate region (residues 92–99 of the regulatory subunit) and a region of the first cAMP-binding domain also in the regulatory subunit. The structure of this cAMP-binding domain was known, as was the structure of the catalytic subunit with the pseudosubstrate region bound at the active site, but the interface between the catalytic subunit and the cAMP-binding domain was not known. We were able to obtain protection data for both the catalytic and regulatory subunits (Table 6.1). The protection data were used to ''filter'' the computational docking solutions to find the correct interface (Table 6.2; Figure 6.4B; Anand *et al.*, 2003). These data relied on the use of on-exchange experimental data, and when the crystal structure of the protein kinase A holoenzyme structure was solved, it was discovered that the C-subunit segment corresponding to residues 278–289, which showed protection of one amide, was not in the interface. This protection, which was clearly observed, is now thought to have arisen from a conformational change (Kim *et al.*, 2005).

Table 6.1. Peptides that showed protection upon complex formation for the protein kinase A holoenzyme

Catalytic subunit				
Peptide (m/z)	Amides	C-subunit alone	Holoenzyme	Full-length holoenzyme
212–221 (1167.58)	9	6.18 ± 0.25	3.38 ± 0.10	3.37 ± 0.00
247–261 (1793.97)	5	10.32 ± 0.15	7.53 ± 0.02	7.74 ± 0.42
278–289 (1347.75)	11	6.09 ± 0.01	NA	4.99 ± 0.05
Regulatory subunit (cAMP-binding domain)				
Peptide (m/z)	Amides	R-subunit alone	Holoenzyme	Full-length holoenzyme
136–148 (1594.73)	12	5.31 ± 0.41	3.88 ± 0.14	5.81 ± 0.17
222–229 (1011.46)	7	1.56 ± 0.19	0.91 ± 0.03	1.55 ± 0.05
230–238 (1046.61)	8	4.23 ± 0.2	3.38 ± 0.17	4.40 ± 0.13

Table 6.2. Number of docked structures remaining after filtering based on 10 Å Cα–Cα distance

Fragment	10 Å Cα–Cα Filter[a]	10 Å Cα–Cα Filter (# amides protected)[b]	7 Å heavy atom Filter (# amides protected)[c]
212–221 (C)	17,466	9,075 (3)	11,026 (3)
247–261 (C)	6,015	1,609 (3)	1,981 (3)
278–289 (C)	2,856	1,087 (1)	1,414 (1)
136–148 (R)	615	158 (2)	356 (2)
222–229 (R)	103	23 (1)	61 (1)
230–238 (R)	96	23 (1)	60 (1)

[a] The filter required that at least one Cα from the fragment be within 10 Å of any Cα in the binding partner.
[b] The filter required that at least the number of Cα in parentheses from the fragment be within 10 Å of any Cα in the binding partner.
[c] The filter required that at least the number of heavy atoms in parentheses from different residues in the fragment be within 7 Å of any heavy atom in the binding partner.

6.3. THE BASIC EXPERIMENTAL METHOD

The basic experimental scheme that is followed in these experiments involves incubation of the proteins in deuterated buffer for varying lengths of time at physiological pH and temperature. After the incubation is complete, the amide proton exchange reaction is quenched by chilling the sample to 0°C and lowering the pH to 2.5, the sample is digested with pepsin, and then analyzed by mass spectrometry (Figure 6.5). The mass spectrometry can either be carried out by combined LC-ESI or by MALDI-TOF with equivalent results. The first part of this experiment is referred to as an on-exchange experiment because it measures the exchange of deuterons onto the amides of the protein. For analyzing protein–protein complexes, it is often useful to perform an off-exchange experiment in addition.

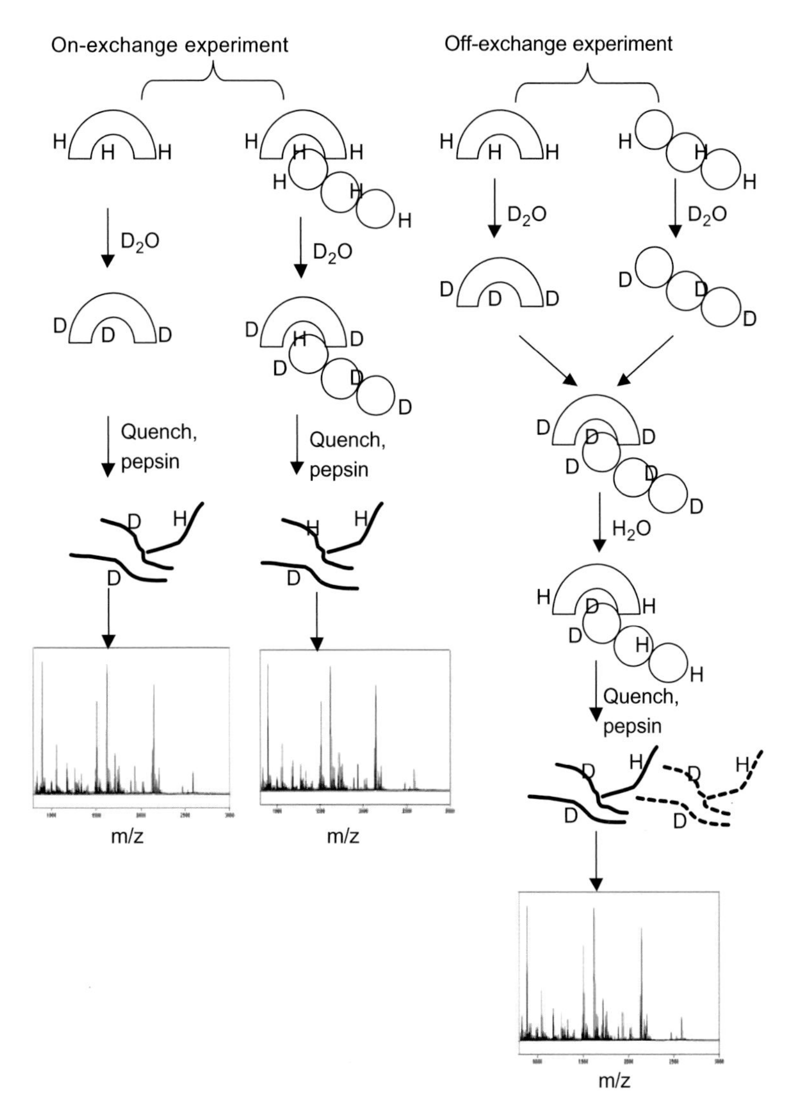

Figure 6.5. Flow chart of the two types of amide exchange experiments used to study protein–ligand interactions. In the on-exchange experiment, the protein–ligand complex interface generally corresponds to the regions in which less deuterium is incorporated compared with control experiments using each protein alone. In the off-exchange experiment, the protein and ligand are first allowed to incorporate deuterium, then the complex is formed and finally deuterons are off-exchanged by dilution back into H_2O. In the off-exchange experiment, the protein–ligand complex interface generally corresponds to the regions in which more deuterons are retained compared with control experiments using each protein alone.

This experiment involves choosing a suitable set of on-exchange times, and then after each protein is subjected to on-exchange, the protein complex is formed and the complex is diluted into H_2O buffer for the off-exchange reaction. The combination of on- and off-exchange experiments gives added information regarding the solvent accessibility of populations of amides at the interface as is explained more fully in the following sections.

6.3.1. Theoretical Considerations for Measurement of Solvent Accessibility at Protein Interfaces

The analysis of solvent accessibility at protein–protein interfaces requires some prior knowledge of the protein–protein binding equilibrium constant. We have previously shown that if the binding constant is weak (>10 nM), then an excess of one binding partner should be used in order to observe interface protection of surface amides (Figure 6.6) (Mandell *et al.*, 2001). This is because typically association rate constants are on the order of 10^4–10^7 $M^{-1}s^{-1}$ and dissociation rate constants are on the order of $0.1 - 0.0001\ s^{-1}$ so that above a K_D of 10 nM, the dissociation rate constant must necessarily be rapid given the diffusion limit of the association rate constant. That said, there are slowly associating complexes, and

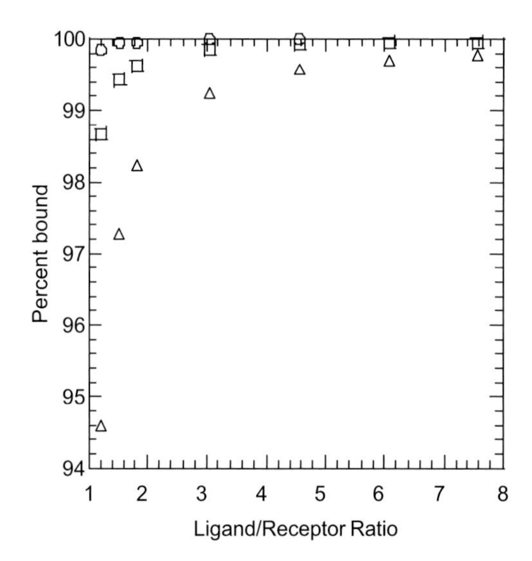

Figure 6.6. Plot of the relationship between the concentration of ligand and the percent of protein bound to the ligand in a typical amide exchange experiment. Theoretical curves for $K_D = 0.1$ and 1 nM were essentially indistinguishable (○). To achieve 100% bound, ligand:receptor ratios greater than 1:1 are required for $K_D = 10$ nM (□) and higher. The curve for $K_D = 50$ nM (△) shows that for these weaker binding affinities, ratios greater than 5:1 are required to achieve 100% bound. Actual experimental data showing the amount of deuterium retained at the interface for the thrombin–thrombomodulin interaction, which has a K_D close to 100 nM, fall closest to the 50 nM curve.

the best way to ensure that the complex stays bound in the experiment is to measure both the association and the dissociation rates directly using SPR, fluorescence, or sedimentation velocity methods (Chapters 4, 13, and 16). If the half-life of the complex is on the order of the time of deuteration, no "protection" is observed, because the amides have time to exchange when the proteins are unbound. It is actually the dissociation rate that determines the half-life of the complex during amide exchange experiments, and the time course of the experiment should reflect knowledge of the dissociation rate constant so that during the amide exchange, the binding partners remain essentially 100% bound. For weakly bound complexes, if the association kinetics are rapid, it is possible to include an excess of the weakly bound ligand and to create a pseudo-100% bound situation. It is usually possible, therefore, to adjust the concentration of one of the binding partners so as to force the equilibrium to essentially 100% bound as shown in Figure 6.6.

Once the 100% bound conditions are achieved, it is interesting to observe which parts of the interface are completely excluded from solvent and which are still partly accessible even in the bound state. Several different groups have observed amides at the protein–protein interface that are completely solvent-excluded as long as the two proteins remain bound (Zhang and Smith, 1996; Nagashima *et al.*, 2000; Mandell *et al.*, 2001; Baerga-Ortiz *et al.*, 2004). For these completely solvent excluded amides, the observed hydrogen/deuterium ($H/^2H$) exchange rate is controlled by the protein–protein dissociation and association rates as well as the intrinsic rate of hydrogen exchange (k_{ex})

$$[R_D L] \underset{k_a}{\overset{k_d}{\rightleftharpoons}} R_D + L \xrightarrow{k_{ex}} R_H, \qquad (6.1)$$

where R_H is the protonated receptor, R_D is the deuterated receptor, L is ligand, k_{ex} is the intrinsic amide exchange rate (min^{-1}) for amides in the uncomplexed receptor, k_d is the rate of dissociation of the complex (min^{-1}), and k_a is the rate of association of the proteins undergoing complexation ($M^{-1}\,\text{min}^{-1}$).

An important consideration is that when measuring amide exchange by mass spectrometry, rates of exchange at individual amides cannot in general be obtained, instead, an ensemble of rates for all the amides in a particular pepsin cleavage product (typically peptides of 7–20 amino acids) is obtained. No matter how many data points are obtained, the exchange rate curve for such peptides can typically be fit to two or three exponentials. Therefore, one can only say that there are so many amides that are exchanging quickly, at an intermediate rate, or slowly in a given segment, but it is impossible to assign a specific exchange rate to a specific amide. When comparing the exchange rate data of the bound protein complex with that of the free protein, one can only say that so many amides moved categories, for example, there are three fewer rapidly exchanging amides, and three more slowly exchanging amides in the complex. Given certain likely assumptions, this information can be used to discover amides that become solvent inaccessible in the complex.

6.3.2. Measurement of Solvent Accessibility of Protein Interfaces

The actual measurement of solvent accessibility at protein–protein interfaces requires knowledge of the amide exchange rates in both the unbound and bound states. Thus, a typical experiment involves on- and off-exchange experiments on both the bound and unbound states (Figure 6.5).

Measurement of the on-exchange of deuterium into each of the proteins involved in the interaction gives important information about the likelihood that a particular segment of the protein is on the surface. This assertion is based on the likely assumption that the interface is at the surface of the protein, and the surface amides ought to exchange fairly rapidly. After on-exchange is measured, a suitable set of times of on-exchange is chosen and these samples are then allowed to form a complex, and are diluted into H_2O for the off-exchange part of the experiment. Solvent inaccessible amides are those that are rapidly deuterated in the free protein, and then are completely protected, i.e., remain deuterated, in the protein–ligand complex. The solvent inaccessible amides are the best indicators of where the protein–protein interface is.

A secondary control that is very useful is to measure the pH dependence of the exchange rates, knowing that near physiological pH, the intrinsic exchange rate is log-linear with [OH-] (Hvidt and Nielsen, 1966). Those amides that are completely protected from exchange except when the proteins dissociate do not show increased exchange at higher pH because their exchange rate is completely dependent on the dissociation rate constant. These two results together are the strongest evidence that the protein region identified as the interface was surface exposed before ligand binding, and that it was highly protected from solvent in the complex. Thus, on-exchange coupled with off-exchange in at least two different pHs results in the highest certainty of identifying regions of the protein that are actually at the protein–ligand interface. Wells and colleagues used site-directed mutagenesis to demonstrate that although protein–protein interfaces appear large in structures of complexes, only a few of the residues at the interface are essential for the interaction (termed the ''hot spot'' of the interaction) (Wells, 1996). For the thrombin–thrombomodulin interaction, we found that either of these two approaches identifies the same region of thrombin as the thrombomodulin-binding site (Mandell *et al.*, 2001) (Figure 6.7). Two surface loops were found to contain solvent-inaccessible amides when thrombomodulin was bound to thrombin; residues 96–112 and residues 54–61. For the larger surface loop, two peptides covered the region, and comparison of the off-exchange kinetics revealed that the longer peptide (residues 97–117) had some amides that off-exchanged at an intermediate rate in the complex which were not present in the shorter peptide (residues 96–112). This allowed us to assign the ''core'' region, containing the solvent inaccessible amides, to residues 96–112 and the ''peripheral'' region to residues 112–117 (Figure 6.7C).

In other more speculative work, we compared the binding kinetics and thermodynamics of thrombomodulin binding to thrombin with those of a monoclonal

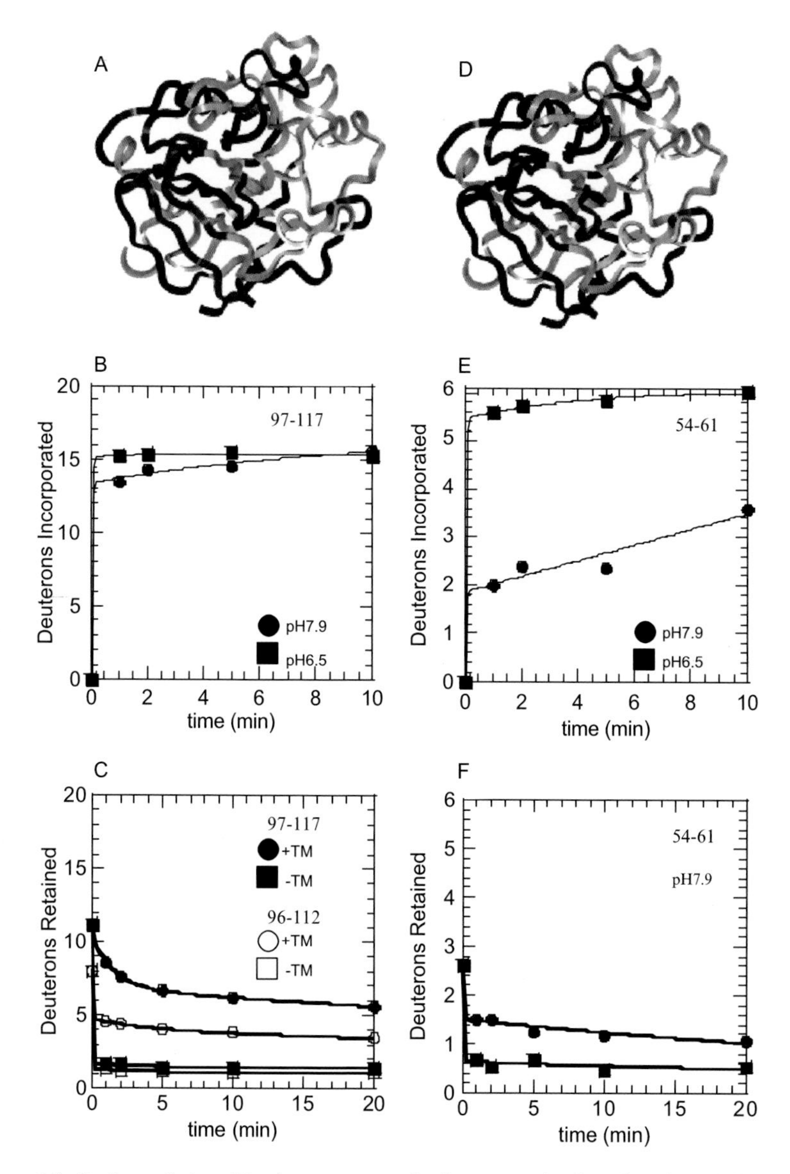

Figure 6.7. Regions of thrombin that were completely protected when thrombomodulin bound. (A) Structure of thrombin showing one surface loop for which we had two overlapping peptides, residues 97–117 and residues 96–112. By comparing the kinetics of on- and off-exchange it was possible to ascertain that residues 112–117 were in the periphery of the interface. (B) Kinetics of on-exchange of residues 97–117. (C) Kinetics of off-exchange of residues 97–117. The first part of the curve shows those amides that were only slightly protected by thrombomodulin and that were not observed in the shorter segment. (D) Structure of thrombin showing the second surface loop, residues 54–61 that also contained a solvent in accessible amide. (E) The second surface loop that showed protection of amide exchange was residues 54–61. (F) Kinetics of off-exchange from thrombin residues 54–61 in the presence and absence of thrombomodulin.

antibody binding to thrombin. These two ligands compete for binding to thrombin but show very different kinetics and thermodynamics (Chapter 8). The antibody bound with slow association and dissociation kinetics and enthalpy-driven thermodynamics. Thrombomodulin bound with rapid association and dissociation kinetics and entropy-driven thermodynamics. Amide proton exchange experiments on the two interactions revealed that many more surface amides were completely excluded from solvent in the entropically driven interaction suggesting that water release from the protein surface on complex formation may contribute to the thermodynamics (Baerga-Ortiz *et al.*, 2004).

Other peripheral regions of thrombin were also protected from exchange when thrombomodulin bound, but these regions were not rendered completely solvent inaccessible in the complex as ascertained by the pH dependence of the off-exchange. Some of these partially excluded regions surrounded the binding site and this is consistent with computational analyses of docked complexes. It is thought that approximately 65% of typical protein–protein interfaces are still partly accessible to H_2O molecules (Lo Conte *et al.*, 1999). One such region is colored pink in Figure 6.7A. Furthermore, other regions of thrombin showed differences in amide proton exchange when the free thrombin was compared with thrombomodulin-bound thrombin. When displayed on the structure of thrombin (and later on the structure of the thrombin–thrombomodulin complex), it was clear that these regions were far from the protein–protein binding site. These regions showed changes in amide proton exchange kinetics that suggested they were conformational changes on binding as is discussed further in the next section.

6.4. CONFORMATIONAL CHANGES ON PROTEIN–PROTEIN INTERACTION

Several groups have used amide exchange detected by mass spectrometry to discover subtle conformational changes that may occur on protein–ligand interaction. Natalie Ahn's group showed that on phosphorylation near the active site cleft, the mitogen-activated protein kinase, ERK2, showed both regions of increased solvent accessibility and regions of decreased solvent accessibility. The residues undergoing these changes were located more than 10 Å from the site of phosphorylation (Hoofnagle *et al.*, 2001). The two-component demethylase CheB undergoes subtle opening of the interface between the regulatory and catalytic subunits on phosphorylation of a critical aspartic acid residue in the regulatory domain (Hughes *et al.*, 2001).

In collaboration with Susan Taylor, we have used amide exchange to observe changes in the regulatory subunit of protein kinase A on binding of cAMP or binding of the catalytic subunit. As expected, cAMP binding caused decreased solvent accessibility at the cAMP-binding site. It was completely unexpected that cAMP would also cause increased solvent accessibility at the catalytic subunit binding site (Figure 6.8). The converse was also true that binding of the catalytic

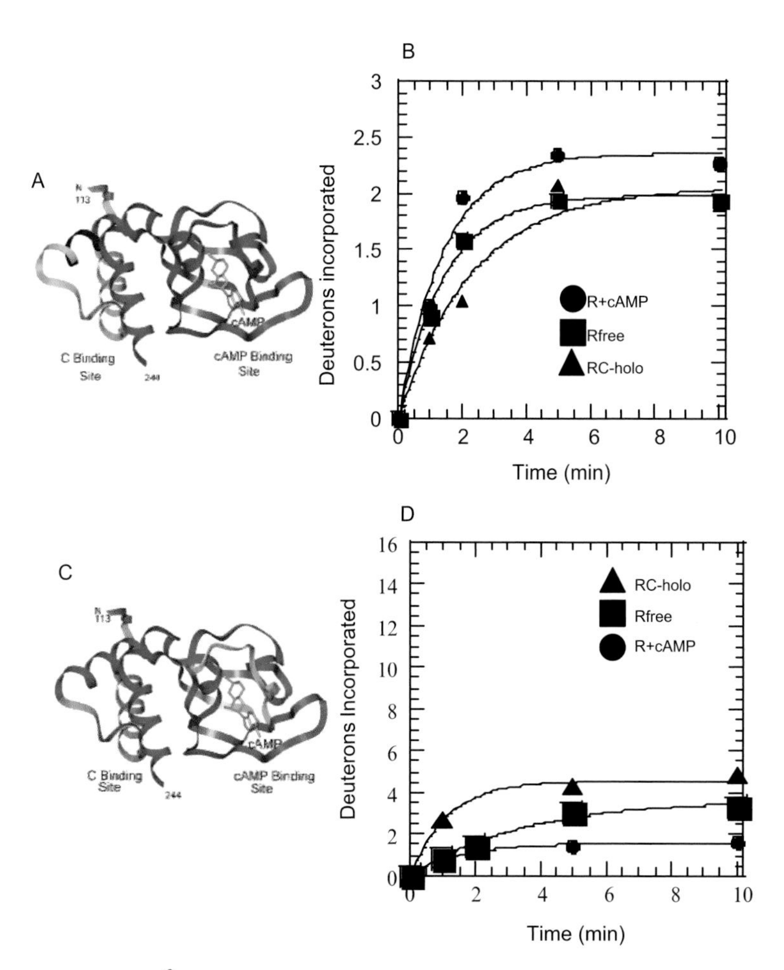

Figure 6.8. Amide $H/^2H$ exchange experiments revealed conformational changes occurring in the regulatory subunit of protein kinase A. (A) Amide exchange in a fragment of the regulatory domain corresponding to the binding site for the catalytic subunit (gold). (B) The kinetics show that when cAMP binds, the binding site for the catalytic subunit becomes more solvent accessible. (C) Amide exchange in a fragment of the regulatory subunit corresponding to the cAMP-binding site (gold). (D) The kinetics show that when the catalytic subunit binds, the cAMP-binding site becomes more solvent accessible.

subunit caused increased solvent accessibility at the cAMP-binding site. These results provided the first clue as to how the regulatory subunit undergoes exclusive binding of only one or the other of its two ligands (Anand *et al.*, 2002). The crystal structure of the holoenzyme does, indeed, show that the conformational changes occur in the regulatory subunit on binding that are consistent with the changes observed in the amide proton exchange (Kim *et al.*, 2005).

Our group has also studied conformational changes that occur on substrate binding in thrombin. Thrombin is thought to be allosterically regulated by

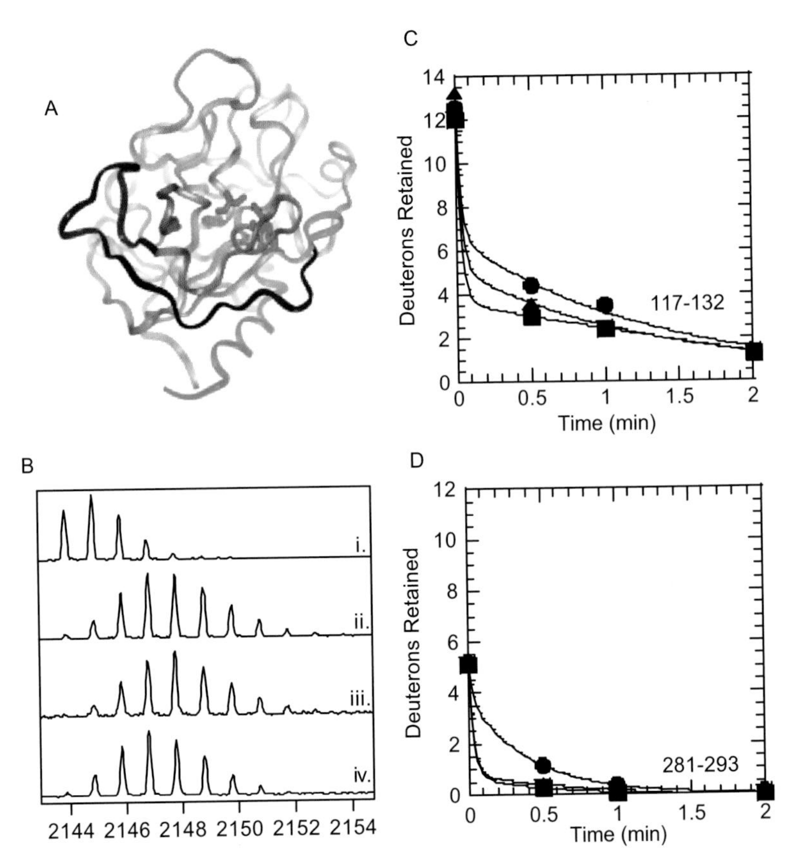

Figure 6.9. Subtle changes in amide proton exchange were observed in thrombin upon thrombomodulin binding. (A) Structure of thrombin showing the thrombomodulin binding loop (red), the ''peripheral'' binding site (pink), the active site loop represented by peptide 117–132 (cyan), and the C-terminal helix also near the active site represented by peptide 281–293 (green). (B) Expansion of mass spectra showing the peptide mass envelope for the thrombin peptide corresponding to residues 117–132 (i) before deuteration; (ii) after 0.5 min off-exchange; (iii) after 0.5 min off-exchange when bound to a cofactor-active thrombomodulin fragment; (iv) after 0.5 min off-exchange when bound to a cofactor-inactive thrombomodulin fragment. (C) Off-exchange kinetics for the same peptide shown in (B) from thrombin alone (▲), thrombin bound to a cofactor-active thrombomodulin fragment (●) and thrombin bound to a cofactor-inactive thrombomodulin fragment (■). (D) Off-exchange kinetics for the peptide corresponding to the C-terminal helix (same symbols as in C).

thrombomodulin, which binds at a site far away from the active site of thrombin (Fuentes-Prior *et al.*, 2000). The potential for allosteric changes was first observed as decreases in solvent accessibility in loops surrounding the thrombin active site when thrombomodulin was bound (Mandell *et al.*, 2001). Compared with the protein kinase A results, this case was more difficult. In protein kinase A, conformational changes were manifested in *increases* in amide exchange. In the case of thrombin, the exchange at the active site loops actually decreased.

The changes were assigned to conformational changes mainly because they were subtle changes in the kinetics of exchange, and were clearly not resulting in solvent-inaccessible amides as would be expected at an interface. Moreover, it was not possible to see how the thrombomodulin could wrap itself around thrombin to form such a large interface. We have now shown that binding of substrates at the active site causes decreased solvent accessibility at the thrombomodulin-binding site and vice versa (Croy *et al.*, 2004; Koeppe *et al.*, 2005). The differences in exchange at the distant sites are real, but are subtle, and it is readily seen from the kinetics that they most likely correspond to redistributions of conformational ensembles and not interface protection or large conformational changes (Figure 6.9). No differences could be seen in thrombin comparing the crystal structures of thrombin alone with thrombin bound to thrombomodulin. These results suggest that amide exchange may uniquely allow a view into the subtle changes in the distribution of the ensemble of states on protein–protein interaction.

REFERENCES

Anand, G. S., Hughes, C. A., Jones, J. M., Taylor, S. S., and Komives, E. A. (2002). Amide $H/^2H$ exchange reveals communication between the cAMP- and catalytic subunit-binding sites in the regulatory subunit of protein kinase A. *J Mol Bio* 323:377–386.

Anand, G. S., Law, D., Mandell, J. G., Snead, A. N., Tsigelny, I., Taylor, S. S., Ten Eyck, L., and Komives, E. A. (2003). Identification of the protein kinase A regulatory RI-catalytic subunit interface by amide $H/^2H$ exchange and protein docking. *Proc Nat Acad Sci USA* 100: 13264–13269.

Baerga-Ortiz, A., Bergqvist, S. P., Mandell, J. G., and Komives, E. A. (2004). Two different proteins that compete for binding to thrombin have opposite kinetic and thermodynamic profiles. *Protein Sci* 13:166–176.

Bai, Y., Milne, J. S., Mayne, L., and Englander, S. W. (1993). Primary structure effects on peptide group hydrogen exchange. *Proteins* 17:75–86.

Croy, C. H., Koeppe, J. R., Bergqvist, S., and Komives, E. A. (2004). Allosteric changes in solvent accessibility observed in thrombin upon active site occupation. *Biochemistry* 43:5246–5255.

Dharmasiri, K. and Smith, D. L. (1996). Mass spectrometric determination of isotopic exchange rates of amide hydrogens located on the surfaces of proteins. *Anal Chem* 68:2340–2344.

Englander, S., Mayne, L., Bai, Y., and Sosnick, T. (1997). Hydrogen exchange: the modern legacy of Linderstrøm-Lang. *Protein Sci* 6:1101–1109.

Fuentes-Prior, P., Iwanaga, Y., Huber, R., Pagila, R., Rumennik, G., Seto, M., Morser, J., Light, D. R., and Bode, W. (2000). Structural basis for the anticoagulant activity of the thrombin–thrombomodulin complex. *Nature* 404:518–525.

Gmeiner, W. H., Xu, I., Horita, D. A., Smithgall, T. E., Engen, J. R., Smith, D. L., and Byrd, R. A. (2001). Intramolecular binding of a proximal PPII helix to an SH3 domain in the fusion protein SH3Hck: PPIIhGAP. *Cell Biochem Biophys* 35:115–126.

Gross, M., Robinson, C. V., Mayhew, M., Hartl, F. U., and Radford, S. E. (1996). Significant hydrogen exchange protection in GroEL-bound DHFR is maintained during iterative rounds of substrate cycling. *Protein Sci* 5:2506–2513.

Hoofnagle, A. N., Resing, K. A., Goldsmith, E. J., and Ahn, N. G. (2001). Changes in protein conformational mobility upon activation of extracellular regulated protein kinase-2 as detected by hydrogen exchange. *Proc Natl Acad Sci USA* 98:956–961.

Hughes, C. A., Mandell, J. G., Anand, G. S., Stock, A. M., and Komives, E. A. (2001). Phosphorylation causes subtle changes in solvent accessibility at the interdomain interface of methylesterase CheB. *J Mol Biol* 307:967–976.

Hvidt, A. and Linderstrom-Lang, K. (1954). Exchange of hydrogen atoms in insulin with deuterium atoms in aqueous solutions. *Biochim Biophys Acta* 14:574–575.

Hvidt, A. and Nielsen, S. O. (1966). Hydrogen exchange in proteins. *Adv Protein Chem* 21:287–386.

Kim, C., Xuong, N. H., and Taylor, S. S. (2005). Crystal structure of a complex between the catalytic and regulatory (RIalpha) subunits of PKA. *Science* 307:690–696.

Koeppe, J. R., Seitova, A., Mather, T., and Komives, E. A. (2005). Thrombomodulin tightens the thrombin active site loops to promote protein C activation. *Biochemistry* 44:14784–14791.

Lanman, J., Lam, T. T., Barnes, S., Sakalian, M., Emmett, M. R., Marshall, A.G., and Prevelige, P. E. Jr. (2003). Identification of novel interactions in HIV-1 capsid protein assembly by high-resolution mass spectrometry. *J Mol Biol* 325:759–772.

Lee, T., Hoofnagle, A. N., Kabuyama, Y., Stroud, J., Min, X., Goldsmith, E. J., Chen, L., Resing, K. A., and Ahn, N. G. (2004). Docking motif interactions in MAP kinases revealed by hydrogen exchange mass spectrometry. *Mol Cell* 14:43–55.

Linderstrom-Lang, K. U. (1955). *Chem Soc Spec Publ* 2:1–20.

Lo Conte, L., Chothia, C., and Janin, J. (1999). The atomic structure of protein–protein recognition sites. *J Mol Biol* 285:2177–2198.

Mandell, J. G., Baerga-Ortiz, A., Akashi, S., Takio, K., and Komives, E. A. (2001). Solvent accessibility of the thrombin–thrombomodulin interface. *J Mol Biol* 306:575–589.

Mandell, J. G., Falick, A. M., and Komives, E. A. (1998a). Identification of protein–protein interfaces by decreased amide proton solvent accessibility. *Proc Natl Acad Sci USA* 95:14705–14710.

Mandell, J. G., Falick, A. M., and Komives, E. A. (1998b). Measurement of amide hydrogen exchange by MALDI-TOF mass spectrometry. *Anal Chem* 70:3987–3995.

Nagashima, M., Werner, M., Wang, M., Zhao, L., Light, D. R., Pagila, R., Morser, J., and Verhallen, P. (2000). An inhibitor of activated thrombin-activatable fibrinolysis inhibitor potentiates tissue-type plasminogen activator-induced thrombolysis in a rabbit jugular vein thrombolysis model. *Thromb Res* 98:333–342.

Nazabal, A., Laguerre, M., Schmitter, J. M., Vaillier, J., Chaignepain, S., and Velours, J. (2003). Hydrogen/deuterium exchange on yeast ATPase supramolecular protein complex analyzed at high sensitivity by MALDI mass spectrometry. *J Am Soc Mass Spectrom* 14:471–481.

Orban, J., Alexander, P., and Bryan, P. (1992). Sequence-specific ^{1}HNMR assignments and secondary structure of the streptococcal protein G B2-domain. *Biochemistry* 31:3604–3611.

Paterson, Y., Englander, S. W., and Roder, H. (1990). An antibody binding site on cytochrome c defined by hydrogen exchange and two-dimensional NMR. *Science* 249:755–759.

Powell, K. D. and Fitzgerald, M. C. (2004). High-throughput screening assay for the tunable selection of protein ligands. *J Comb Chem* 6:262–269.

Powell, K. D., Ghaemmaghami, S., Wang, M. Z., Ma, L., Oas, T. G., and Fitzgerald, M. C. (2002). A general mass spectrometry-based assay for the quantitation of protein–ligand binding interactions in solution pp. *J Am Chem Soc* 124:10256–10257.

Rosa, J. J. and Richards, F. M. (1979). Hydrogen exchange from identified regions of the S-Protein component of ribonuclease as a function of temperature, pH, and the binding of S-peptide. *J Mol Biol* 133:399–416.

Tuma, R., Coward, L. U., Kirk, M. C., Barnes, S., and Prevelige, P. E. (2001). Hydrogen–deuterium exchange as a probe of folding and assembly in viral capsids. *J Mol Biol* 306:389–396.

Wang, F., Blanchard, J. S., and Tang, X. J. (1997). Hydrogen exchange/electrospray ionization mass spectrometry studies of substrate and inhibitor binding and conformational changes of *Escherichia coli* dihydrodipicolinate reductase. *Biochemistry* 36:3755–3759.

Wang, F., Miles, R., Kicska, G., Nieves, E., Schramm, V., and Angeletti, R. (2000). Immucillin-H binding to purine nucleoside phosphorylase reduces dynamic solvent exchange. *Protein Sci* 9: 1660–1668.

Wang, F., Scapin, G., Blanchard, J. S., and Angeletti, R. H. (1998). Substrate binding and conformational changes of *Clostridium glutamicum* diaminopimelate dehydrogenase revealed by hydrogen/deuterium exchange and electrospray mass spectrometry. *Protein Sci* 7:293–299.

Wang, F., Shi, W., Nieves, E., Angeletti, R., Schramm, V., and Grubmeyer, C. (2001). A transition-state analogue reduces protein dynamics in hypoxanthine-guanine phosphoribosyltransferase. *Biochemistry* 40:8043–8054.

Wang, L. and Smith, D. L. (2005). Capsid structure and dynamics of a human rhinovirus probed by hydrogen exchange mass spectrometry. *Protein Sci* 14:1661–1672.

Wells, J. A. (1996). Binding in the growth hormone receptor complex. *Proc Natl Acad Sci USA* 93:1–6.

Xiao, H., Kaltashov, I. A., and Eyles, S. J. (2003). Indirect assessment of small hydrophobic ligand binding to a model protein using a combination of ESI MS and HDX/ESI MS. *J Am Soc Mass Spectrom* 14:506–515.

Zhang, Z. and Smith, D. L. (1993). Determination of amide hydrogen exchange by mass spectrometry: a new tool for protein structure elucidation. *Protein Sci* 2:522–531.

Zhang, Z. and Smith, D. L. (1996). Thermal-induced unfolding domains in aldolase identified by amide hydrogen exchange and mass spectrometry. *Protein Sci* 5:1282–1289.

7

Elucidation of Protein–Protein and Protein–Ligand Interactions by NMR Spectroscopy

Hans Robert Kalbitzer*, Werner Kremer, Frank Schumann, Michael Spörner, and Wolfram Gronwald

7.1. INTRODUCTION

Following the first detection of nuclear resonance signals by Felix Bloch and independently by Edward Purcell in the year 1945, the impact and the applications of nuclear magnetic resonance (NMR) grew rapidly. In a few years, NMR in solution became an essential analytical method in chemistry. For a widespread application in biology and medicine, a technical development, the construction of superconducting magnets and a methodological development, the introduction of the Fourier spectroscopy was essential. The importance of magnetic resonance methods is also reflected by the fact that four Nobel Prizes were awarded for crucial achievements in the past. The first in 1952 to Felix Bloch and Edward Purcell (invention of the method), the second in 1999 to Richard Ernst (FT-NMR and multidimensional NMR spectroscopy), the third in 2002 to Kurt Wüthrich (NMR-structure determination of proteins), and the fourth in 2003 to Paul Lauterbur and Peter Mansfield (magnetic resonance imaging). Today in biochemistry, the characterization and structure determination of biological macromolecules such as

H. R. KALBITZER, W. KREMER, F. SCHUMANN, M. SPÖRNER, and W. GRONWALD • Institute of Biophysics and Physical Biochemistry, University of Regensburg, D-93053 Regensburg, Germany and *Corresponding author: Universität Regensburg, Institut für Biophysik und Physikalische Biochemie, Universitätsstraße 31, D-93053 Regensburg, Germany; e-mail: hans-robert.kalbitzer@biologie.uni-regensburg.de

proteins and nucleic acids in solution is probably the most prominent application of NMR and a considerable number of excellent monographies and reviews on this topic have appeared in the past (see e.g., Wüthrich, 1986; Hausser and Kalbitzer, 1991; Schwalbe, 2003; Bonvin *et al.*, 2005; Kay, 2005). In future, the detection and quantification of small molecules in biofluids and living organisms (*in vivo* NMR spectroscopy) may play an important role in system biology and metabonomics (for a review see e.g., Bollard *et al.*, 2005; Lindon *et al.*, 2005; Robertson *et al.*, 2005).

7.2. PRINCIPLES OF NMR STRUCTURE DETERMINATION

The theory of NMR spectroscopy is well developed and can be described sufficiently well by nonrelativistic quantum mechanics (see e.g., Abragam, 1961; Ernst *et al.*, 1987; Cavanagh *et al.*, 1996). For many applications, semi-classical descriptions are sufficient. Although NMR can be performed in any aggregate state, in biology solution NMR dominates, as this is for most cases the physiological environment (Table 7.1).

Solution state NMR is also suited for high-throughput protein structure determination as it is done in structural proteomics. The bottleneck here certainly is the protein production and the automation of data evaluation and structure calculation (Gronwald and Kalbitzer, 2004). The solid-state NMR spectroscopy (for a review see e.g., Luca *et al.*, 2003; McDermott, 2004) differs from the solution NMR spectroscopy essentially by the fact that anisotropic contributions of the interactions have to be considered that are averaged out in solution by the rapid Brownian molecular motions. However, using magic angle spinning (MAS) these contributions can be strongly suppressed.

Table 7.1. Structure determination by solution NMR spectroscopy and X-ray crystallography

NMR spectroscopy	X-ray crystallography
Advantages	*Disadvantages*
Structure in solution	Structure in a crystal lattice
– Quasi physiological conditions	– Nonphysiological conditions
– No crystallization required	– Crystallization required
Dynamics observable in the picosecond to kilosecond time scale	Motion in the crystal lattice, motion, and disorder difficult to distinguish
Fast detection of interactions sites	–
Disadvantages	*Advantages*
Size limitation	No size limitation
Locally differing accuracy	High precision of the coordinates (if high-quality crystals are available)

7.2.1. Fundamentals of NMR Spectroscopy

7.2.1.1. Spins and Magnetic Moments

Almost all atomic nuclei have a spin **J**, which is always connected with a magnetic moment $\boldsymbol{\mu}_I$ by

$$\boldsymbol{\mu}_I = \gamma_I \mathbf{J} = \hbar\gamma_I \mathbf{I}. \tag{7.1}$$

The constant of proportionality γ_I is usually called gyromagnetic ratio (although it is the magnetogyric ratio) and is a fundamental natural constant that has a specific value for each type of atomic nucleus. **I** is normally used when the nuclear spin is expressed in units of Planck's constant $\hbar(\hbar = h/2\pi)$. The energy E_m of a magnetic dipole in a static external magnetic field $\mathbf{B}_0$ is

$$E_m = \langle \mathbf{H}_Z \rangle = -\langle \boldsymbol{\mu}_I \rangle \mathbf{B}_0 = -\hbar\gamma_I \langle I_Z \rangle \mathbf{B}_0 = -\hbar\gamma_I m_I \mathbf{B}_0. \tag{7.2}$$

With $\mathbf{H}_Z$ the Hamilton operator of the nuclear Zeeman interaction and m_I the magnetic quantum number. Usually, the direction of the external magnetic field $\mathbf{B}_0$ defines the z-axis of the system. The energy of the electromagnetic quanta $\hbar\omega$ that can induce transitions between the different energy levels is for a spin-1/2 particle

$$\hbar\omega_I = -\hbar\gamma_I \mathbf{B}_0(-1/2 - 1/2) = \hbar\gamma_I \mathbf{B}_0. \tag{7.3}$$

And the corresponding frequency, $\omega = 2\pi\nu$, is

$$\omega = \omega_I = \gamma_I \mathbf{B}_0. \tag{7.4}$$

Equation (7.4) is called the resonance condition, the frequency ω is the resonance frequency. In Table 7.2, the properties of some nuclei important for biological NMR are summarized.

Table 7.2. Properties of some biologically important atomic nuclei[a]

Isotope	Nuclear spin	$\gamma_I[\mu T^{-1} s^{-1}]$	ν_I at 14.092 T (MHz)	Natural abundance (%)	Relative sensitivity (%)
^{1}H	½	267.52	600.0	99.985	100.00
^{2}H	1	41.065	92.1	0.015	0.96
^{3}H	½	285.53	640.0	–	121.36
^{12}C	0	–	–	99.89	–
^{13}C	½	67.266	150.9	0.11	1.59
^{14}N	1	19.325	43.3	99.63	0.10
^{15}N	½	–27.108	60.8	0.37	0.10
^{16}O	0	–	–	99.76	–
^{17}O	5/2	–36.267	81.4	0.04	2.91
^{18}O	0	–	–	0.20	–
^{19}F	½	251.67	564.5	100.00	83.34
^{31}P	½	108.29	242.9	100.00	6.63

[a] At a given magnetic field, the NMR sensitivity is in first approximation proportional $\gamma_I^3 I(I+1)$. The sensitivity of ^{1}H-nucleus was set to 100%

For the NMR detection, the time-dependent change of the macroscopic magnetization **M** is important, as it is responsible for the induction of the signal voltage. For many cases, the equation of motion can be described by Bloch's equation

$$\frac{\mathrm{d}M_x}{\mathrm{d}t} = \gamma_\mathrm{I}(\mathbf{M} \times \mathbf{B})_x - \frac{M_x}{T_2}, \tag{7.5a}$$

$$\frac{\mathrm{d}M_y}{\mathrm{d}t} = \gamma_\mathrm{I}(\mathbf{M} \times \mathbf{B})_y - \frac{M_y}{T_2}, \tag{7.5b}$$

$$\frac{\mathrm{d}M_z}{\mathrm{d}t} = \gamma_\mathrm{I}(\mathbf{M} \times \mathbf{B})_z - \frac{M_0 - M_z}{T_1}. \tag{7.5c}$$

The first term in Eq. (7.5) describes the precession of the magnetization around the magnetic field, that leads after a perturbation the equilibrium state to the induction signal. The second term describes the disappearance of the transverse components of the magnetization M_x and M_y after a perturbation of the equilibrium with the time constant T_2 (transverse relaxation time) and the return of M_z to the equilibrium value M_0 with the time constant T_1 (longitudinal relaxation time).

The longitudinal relaxation is also called spin-lattice relaxation, as energy has to be exchanged with the environment (lattice). The transverse relaxation essentially consists of a loss of the phase coherence between the individual spins and is therefore also called spin–spin relaxation.

Longitudinal relaxation requires a time-dependent modulation of the responsible interactions that is described by the spectral density $J(\omega)$. In liquids, a main source is the Brownian motion of the molecules, especially the rotational reorientation of the molecule and the internal motions of the atoms in the macromolecule. An important spectral density is defined by the so-called model-free approach (Lipari and Szabo, 1982a,b), where $J(\omega)$ is defined as

$$J(\omega) = \frac{2}{5}\left(\frac{S^2\tau_\mathrm{M}}{1 + (\tau_\mathrm{M}\omega)^2} + \frac{(1 - S^2)\tau}{1 + (\tau\omega)^2}\right),$$
$$\tau^{-1} = \tau_\mathrm{M}^{-1} + \tau_\mathrm{e}^{-1}, \tag{7.6}$$

where S is the generalized order parameter representing the amplitude of motion, τ_e is the effective correlation time, which is a measure of the rate of the motion, and τ_M is the correlation time for the special case when the overall motion can be described by one single correlation time.

7.2.1.2. The NMR Spectrum

The interaction of the nuclear spins with the external field only provides the possibility for an elementary analysis as the integral of the resonance line is proportional to the number of spins in the sample. For applications in biology and chemistry, we need additional information from other physical interactions.

Here, five interactions are of importance (1) the chemical shift interaction, (2) the indirect spin–spin coupling interaction, (3) the dipolar interaction, (4) the quadrupolar interaction, and (5) the nuclear–electron interaction.

In solution, the chemical shift leads to changes in the resonance frequencies of the individual spins that are dependent on the chemical and spatial structure of the molecule. The external magnetic field $\mathbf{B}_0$ at the location of the nuclear spin is modified by the electron shell to the effective magnetic field $\mathbf{B}'$,

$$\mathbf{B}' = \mathbf{B}_0(1 - \sigma), \tag{7.7a}$$

with the shielding constant σ. The resonance condition is then (compare Eq. 7.4):

$$\omega = \gamma_\mathrm{I}\mathbf{B}' = \gamma_\mathrm{I}\mathbf{B}_0(1 - \sigma). \tag{7.7b}$$

As the shielding effects are usually very small, the resonance frequency is usually given relative to the resonance frequency $\omega_\mathrm{R} = 2\pi\omega_\mathrm{R}$ of a reference substance R and expressed in parts per million (ppm)

$$\delta = 10^6\frac{\omega - \omega_\mathrm{R}}{\omega_\mathrm{R}} = 10^6\frac{\nu - \nu_\mathrm{R}}{\nu_\mathrm{R}}. \tag{7.8}$$

δ is usually taken as symbol for the relative frequency expressed in ppm. The indirect spin–spin coupling is transferred *via* the polarization of the coupled electrons, its magnitude decreases rapidly with the number of intervening bonds. In the ^{1}H-NMR, usually the coupling is too small to be observable when the nuclear spins are separated by more than four bonds. The spin–spin coupling leads to a line splitting of the resonance lines (multiplets), the corresponding J-coupling constants can be determined from the multiplet patterns. The observation of multiplet patterns allows the conclusion that the contributing spins are close (usually less than three to four bonds) in the covalent structure.

The dipolar and the quadrupolar interactions cause line broadening in one-dimensional spectra. The quadrupolar interaction is only existing for nuclei $I \geq 1$, here it leads to such a strong line broadening that the signals can barely be observed. Therefore, nuclei such as ^{14}N do not play a role in biological NMR. The nuclear–electron interaction (hyperfine interaction) is only observable when unpaired electrons are present in the system. This is not the case in most biologically relevant samples.

7.2.1.3. Detection of NMR Signals

Three principally different detection methods for NMR signals were developed in the past, the cw-NMR spectroscopy, the stochastic spectroscopy, and the Fourier spectroscopy. In the cw-NMR spectroscopy, the magnetic field or the frequency of the electromagnetic HF-field is shifted slowly and the HF-absorption in resonance is detected. In the stochastic NMR spectroscopy, randomly spaced weak pulses are used for the excitation of the spin system and the response is analyzed by a correlation analysis. In the Fourier spectroscopy, well-defined strong pulses are used, the time-domain signal called free induction decay (FID) is detected and is then transformed to a frequency domain spectrum with the aid of the Fourier transformation. Only the last method has a practical importance in biological NMR spectroscopy. The time-domain signal $S(t)$ of a spin with the resonance frequency ω_0 can be described by

$$S(t) = A\cos(\omega_0 t)e^{-t/T_2} \tag{7.9}$$

The frequency domain signal is a Lorentzian

$$S(\omega) = \frac{A}{\pi}\frac{1/T_2}{(1/T_2)^2 + (\omega - \omega_0)^2} \tag{7.10}$$

with the half line width at half height $\Delta\omega_{1/2} = 1/T_2$ (in frequency units ν, $\Delta\nu_{1/2} = 1/2\pi T_2$).

7.2.2. Multidimensional NMR spectroscopy of Proteins and Assignment of Resonance Lines

As proteins contains a very large number of atoms, it is hopeless to expect that the resonance lines of the individual atoms can be observed separately in one-dimensional spectra. Only the multidimensional NMR spectroscopy allows to observe the individual resonance lines in the multidimensional frequency space separately and to assign them to specific atoms. This is the precondition for the structure determination of macromolecules by NMR. In addition, the multidimensional NMR permits the selection of specific physical interactions between the nuclear spins and to observe them in the spectrum.

The second dimension is created by the introduction of a second time-variable in the pulse program. Two-dimensional Fourier transformation of the matrix of time-domain data leads to a two-dimensional frequency domain spectrum. This scheme can be applied to n dimensions by the introduction of additional time-variables (Figure 7.1).

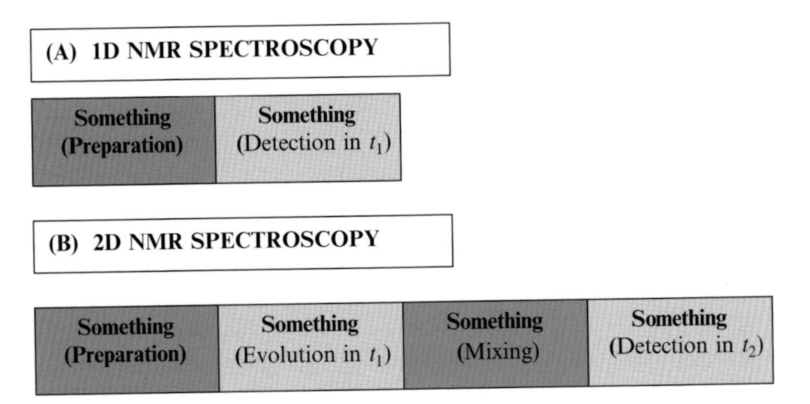

Figure 7.1. Schematic representation of one- and two-dimensional NMR spectra.

The spin system can be manipulated by high-frequency and gradient pulses (''something'') during the different periods of the measurement. In the preparation phase, the spin system is manipulated by one or several HF-pulses, in the detection phase, the time-domain signal $S(t_1)$ evolves and is recorded (Figure 7.1A). After Fourier transformation of $S(t_1)$, the one-dimensional frequency domain data are obtained. The preparation phase is followed by an evolution time t_1 that is incremented from spectrum to spectrum. In the mixing time, the spin system is again manipulated by suitable HF-pulses and the signal $S(t_1, t_2)$ is recorded (Figure 7.1B). The N spectra are Fourier transformed to obtain a mixed time-domain frequency domain signal $S(t_1, \omega_2)$. A second Fourier transformation according to parameter t_1 leads then to the 2D spectrum $S(\omega_1, \omega_2)$.

A signal $S(\omega_1, \omega_2)$ contained in a standard 2D-spectrum demonstrates that the interaction selected by the pulse sequence exists between a nuclear spin with a resonance frequency ω_1 and a spin with a resonance frequency ω_2. The volume of the peak is a measure for the strength of the interaction. For COSY-type spectra the observation of a 2D-signal (cross peak) means that the nuclei are J-coupled and therefore are close in the covalent structure. In the NOESY-experiment the observation of a cross peak implies that a considerable dipolar coupling exists, that is, the two nuclear spins are not separated too much in space (in practice <0.5–0.6 nm). The volume of the cross peak again represents the strength of the nuclear Overhauser effect (NOE).

Multidimensional experiments are not limited to a single type of atoms such as protons. By multidimensional experiments on ^{15}N und ^{13}C enriched proteins, additional frequency domains are accessible and the spectral resolution as well as the information content is increased considerably. For the structure determination of larger proteins, isotope enrichment is indispensable. Important is the heteronuclear single quantum coherence (HSQC) experiment that correlates the resonance frequency of a proton with that of the directly bonded hetero nucleus, e.g., the amide proton resonance with amide nitrogen resonance. Important experiments are 3D-experiments that are used to obtain the sequential assignment. They are usually named according to the protein resonances they correlate. A typical example is the HNCA that correlates the resonance of an amide proton with the amide nitrogen of amino acid *i* and the C^α resonances of the amino acids *i* and $i - 1$, and thus allows to identify resonances of amino acids which are connected *via* a peptide bond. Table 7.3 summarizes a selection of important NMR experiments.

7.2.3. Structure Calculation of Proteins from NMR Data

The NMR structure determination is still mainly based on two pieces of information extracted from the NMR data, the determination of interatomic distances d_{ij}, and dihedral angles. This information usually has to be completed by independent information from other sources describing the physical model including the covalent structure of the protein (Table 7.4). In addition, a number of other NMR-derived restraints can be measured that help to obtain a more refined structure.

Table 7.3. Important *n*D experiments

Experiment	Correlated resonances	References
COSY	Directly J-coupled spins	Jeener (1971), Aue *et al.* (1976)
TOCSY	All spins of a J-coupled spin system	Braunschweiler and Ernst (1983)
NOESY	All spins in close neighbourhood (<0.5 nm)	Jeener *et al.* (1979)
HSQC	Protons with directly bound ^{15}N or ^{13}C. Fingerprint	Müller (1979), Bodenhausen and Ruben (1980), Bax *et al.* (1990a), Norwood *et al.* (1990)
HNCA	H_i^N with N_i^H and C_i^α and C_{i-1}^α	Kay *et al.* (1990), Grzesiek and Bax (1992a), Farmer *et al.* (1992)
HNCO	H_i^N with N_i^H and C'_{i-1}	Kay *et al.* (1990), Grzesiek and Bax (1992a), Muhandiram and Kay (1994)
HN(CO)CA	H_i^N with N_i^H and C_{i-1}^α	Grzesiek and Bax (1992a), Bax and Ikura (1991)
HCCH-TOCSY	H^α with *via* ^{13}C coupled protons of the same amino acid	Bax *et al.* (1990b), Olejniczak *et al.* (1992)
CBCA(CO)NH	C_{i-1}^α and C_{i-1}^β. with N_i^H and H_i^N	Grzesiek and Bax (1992c, 1993),
CBCANH	C_1^α, C_i^β, C_{i-1}^α and C_{i-1}^β with N_i^H and H_i^N	Grzesiek and Bax (1992b)
HN(CO)CACB	H_i^N with N_i^H and C_{i-1}^α and C_{i-1}^β	Yamazaki *et al.* (1994)
HNCACB	H_i^N with N_i^H and C_i^α, C_i^β, C_{i-1}^α and C_{i-1}^β	Wittekind and Mueller (1993), Muhandiram and Kay (1994)

7.2.3.1. The Nuclear Overhauser Effect

Interatomic distances d_{ij} between two atoms i and j can be obtained from the dipolar coupling that is proportional to d_{ij}^{-3}. Usually it is measured by the NOE by two- or three-dimensional ($n > 1$) NOESY spectroscopy. The magnitude of the NOE in dependence on the mixing time τ_m in the NOESY experiment is approximated for large molecules with large rotational correlation times τ_c by

$$\frac{dNOE_{ij}}{d\tau_m}(\tau_m \to 0) = -\left(\frac{\mu_0}{4\pi}\right)^2 \frac{\hbar^2}{10} \gamma_i^2 \gamma_j^2 \tau_c \left\langle d_{ij}^{-6} \right\rangle. \tag{7.11}$$

Since in a macromolecule all spins are coupled, an exact description requires the simultaneous solution of a large set of differential equations (complete relaxation matrix formalism). Because of the importance of these simulation, different solutions to this problem were published in the past (see e.g., Keepers and James, 1984; Boelens *et al.*, 1989; van de Ven *et al.*, 1991; Zhu and Reid, 1995; Görler *et al.*, 1999) and are contained in the software packages for the automated structure determination such as AUREMOL (Gronwald and Kalbitzer, 2004).

7.2.3.2. Indirect Spin–Spin Coupling

Dihedral angles can be determined from the J-coupling constants that can be obtained from two-dimensional spectra. In good approximation the Karplus equation (Karplus, 1959),

Table 7.4. Restraints used for the NMR structure calculation

Data	Information
External restraints describing the physical model	
Covalent structure of amino acids and ligands (bond lengths, bond angles, dihedral angles, van der Waals radii, connectivity of atoms)	Possible positions of atoms in the conformational space
Amino acid sequence and possible ligands	Definition of the whole molecular complex
Disulfide bridges and other side-chain connections	Further restriction of the conformational space
Interatomic potentials (e.g., electrostatic potentials, Lennard–Jones potentials)	Further restriction of the conformational space
Homology models	Faster optimization
NMR derived data	
NOE	Interatomic distances
J-coupling	Dihedral angles
Chemical shift	Dihedral angles, interatomic distances, relative orientations of groups
Residual dipolar couplings	Interatomic distances and orientations
HNCO-type polarization transfer	Pairwise hydrogen bonding
Temperature dependence of amide chemical shifts	Hydrogen bonds
Hydrogen exchange rates	Hydrogen bonds
Nonaveraged chemical shift anisotropy	Orientations of bond vectors, distance information

$$J = A\cos^2\theta - B\cos\theta + C \tag{7.12a}$$

holds, with the empirical constants A, B, C, and the torsion angle θ. In proteins, the definition of θ has to be adapted to the definition of the dihedral angles ϕ, ψ, and χ. For the angle ϕ

$$\theta = \phi - 60 \tag{7.12b}$$

holds and an often used parameter set A, B, and C for the coupling constant $^3J_{\text{HN-H}\alpha}$ between the H^N and the H^α-spin is 6.4 Hz, 1.4 Hz, and 1.9 Hz, respectively (Pardi *et al.*, 1984).

7.2.3.3. Residual Dipolar Coupling

When a biomolecule tumbles isotropically in solution, the dipolar coupling between the nuclei in a molecule fluctuates and is averaged out. It is only visible as a relaxation mechanism that gives rise to the NOE and to homogeneous line broadening. In solids, it leads to a distance- and orientation- dependent line splitting and thus to a very strong inhomogeneous line broadening. By a weak alignment of the molecules in anisotropic solvents, the dipolar interactions are not completely averaged out and the information contained in the dipolar couplings can be retained without losing the narrow lines of liquid state spectra.

The resulting residual dipolar coupling (RDC) of the nuclei I and S leads to a line splitting of the I and S resonance lines that can be observed in normal 2D spectra. The residual dipolar coupling between a nitrogen or a carbon and its directly bonded proton(s) can be observed through ^{1}H, ^{15}N–HSQC and ^{1}H, ^{13}C–HSQC spectra, respectively. RDCs between protons result in an additional line splitting in NOESY or COSY spectra. The line splitting D_{IS} is given by

$$D_{\mathrm{IS}}(\theta, \phi) = A_a^{\mathrm{IS}}\left[(3\cos^2\theta - 1) + \frac{3}{2}R(\sin^2\theta\cos 2\phi)\right], \tag{7.13}$$

$$A_a^{\mathrm{IS}} = -\left(\frac{\mu_0 h}{16\pi^3}\right)S\gamma_I\gamma_S\langle d_{\mathrm{IS}}^{-3}\rangle A_a \tag{7.14}$$

with the permeability in a vacuum, μ_0, Planck's constant, h, and, the gyromagnetic ratios of the two nuclei involved γ_I, γ_S. D_{IS} depends on the internuclear distance d_{IS}, and the angles θ and ϕ between the internuclear bond vector and the axes of the alignment tensor. The molecular alignment tensor describes the partial alignment of the protein. A_a^{IS} and R are the axial and rhombic components of the molecular alignment tensor A. The order parameter S describes the internal motion of the internuclear vector (Lipsitz and Tjandra, 2004). For spins of atoms that are directly bonded as in the amide groups, the interatomic distance is known and the information on the orientation of the bond vector in the internal reference system can be derived from the RDCs. In NOESY or COSY-type spectra, the distance between not directly bonded protons is in general also unknown and can be obtained from the RDCs.

For obtaining a small alignment in the mean time different systems were proposed, which have all their advantages and disadvantages (see Table 7.5).

7.2.3.4. Restrained Simulated Annealing and Energy Minimization

Beginning with a large number of randomly selected initial structures, structures are searched for those that fulfill all experimental restraints. Mainly two different methods are used in combination, the minimization of an error function (usually the potential energy V) by classical minimization algorithms or by a variant of the classical molecular dynamics (MD). The experimental restraints are usually coded as pseudopotentials V_{p} (see e.g., Altieri and Byrd, 2004; Baran *et al.*, 2004; Gronwald and Kalbitzer, 2004). The pseudopotentials, V_{p}, are defined in such a way that they have global minimum when the experimental restraints are fulfilled.

In molecular dynamics, the Newtonian equations of motion are solved simultaneously for all N atoms by numerical integration. The acceleration $\mathrm{d}^2\mathbf{r}_i/\mathrm{d}t^2$ of the ith atoms with the mass m_i is calculated from the force $\mathbf{F}_i$ that is the spatial derivative of the V as

$$\mathbf{F}_i = m_i\frac{\mathrm{d}^2\mathbf{r}_i}{\mathrm{d}t^2} = -\mathrm{grad}_i\, V(\mathbf{r}_i, \ldots, \mathbf{r}_N). \tag{7.15}$$

Table 7.5. Weak orientation of proteins for measuring residual dipolar couplings (see Prestegard *et al.*, 2004)

Orientational system	Remarks	References
Lipid bicelles	+ Well-defined system – Limited temperature stability	Tjandra and Bax (1997)
Viruses	+ Easy preparation, sample recovery – Only for negatively charged biomolecules	Hansen *et al.* (1998)
Stressed polyacrylamide gel	+ Easy preparation, sample recovery, larger proteins – Difficult to align homogeneously, broad lines	Chou *et al.* (2002)
Diamagnetic orientation in the external magnetic field	– Very small couplings, very high magnetic fields	Brunner (2001)
Paramagnetic orientation in the external magnetic field	+ No compatibility problems – Very small degree of alignment	Brunner (2001)

When the initial coordinates $\mathbf{r}_i$ of all atoms i and their velocity $\mathbf{v}_i$ at time t_0 are known, the positions at the time $t + \Delta t$ and velocities can be approximated for small time intervals Δt (in praxis a few femtoseconds) by

$$\mathbf{r}_i(t + \Delta t) = \mathbf{r}_i(t) + \mathbf{v}_i\left(t + \frac{\Delta t}{2}\right)\Delta t, \tag{7.16a}$$

$$\mathbf{v}_i\left(t + \frac{\Delta t}{2}\right) = \mathbf{v}_i\left(t - \frac{\Delta t}{2}\right) + \frac{\mathrm{d}^2\mathbf{r}_i}{\mathrm{d}t^2}(t)\Delta t. \tag{7.16b}$$

The calculations can also be performed in the internal coordinate system of the torsion angles that decreases the degrees of freedom and thus reduces the computational time.

For increasing the speed of the convergence, usually a simulated annealing protocol is used where the system is heated to high temperatures (e.g., 1000 K) and cooled down slowly. Simultaneously, the van der Waals radii are set to small values and slowly increased during the cooling down process. A protocol that combines simulated annealing with a refinement in explicit water has been proposed recently that gives structures with superior quality (Linge *et al.*, 2003a; Nabuurs *et al.*, 2004).

7.3. NMR-BASED METHODS FOR THE STUDY OF PROTEIN–PROTEIN AND PROTEIN–LIGAND INTERACTIONS

The most informative but also the most difficult method for studying the protein–protein or protein–ligand interaction is the complete structure determination of the complex by NMR methods. However, for many applications, it is sufficient to map the interaction site on the surface of the protein. This information can then be used to dock the ligand on the protein structure. In principle, any NMR parameter that changes when a molecule interacts with the target protein can be used to describe an intermolecular interaction. This means that a large variety of methods exists that are useful for describing the interaction of a protein with a small ligand or another biomacromolecule.

7.3.1. Fast NMR Screening and Detection of Interactions with Small Molecules

The detection and investigation of the interaction between proteins and small ligands becomes more and more important as screening method in drug discovery and drug design. For fast screening purposes it is important to obtain meaningful binding data within a few minutes with a minimum quantity of protein. The highest sensitivity is obtained when changes of the ligand signals are observed, which is added in high relative amounts compared with the protein partner. In addition, the relatively sharp lines of the small compound grant a high NMR sensitivity. Most NMR screening methods are well suited to detect weak and intermediately strong interactions. To identify potential lead compounds with significant affinity to target protein, the following methods are commonly used: chemical shift and intensity perturbation, NOE pumping, the saturation transfer difference (STD) technique with a T_{1p} filter, and the water ligand observed *via* gradient spectroscopy (WaterLOGSY) experiment. In the following, we give a very short overview of the commonly used methodologies, for more detailed reviews see, e.g., Meyer and Peters (2003) and Carlomagno (2005).

7.3.1.1. Binding-Induced Changes of Chemical Shifts and Intensities

Using simple 1D NMR experiments, the binding of small molecules to a protein can directly be detected by chemical shift changes, line widths changes, or the decrease of intensities of compound signals. The spectral changes observed after interaction are dependent on the NMR timescale that is determined by the frequency difference $\Delta\omega$ between the signals of the free and the complexed form and the exchange correlation time τ_e. Under fast exchange conditions (weak binding), $|\Delta\omega\tau_e| \ll 1$ holds and the binding usually leads to a change in the chemical shift and an increase in line widths. Slow exchange ($|\Delta\omega\tau_e| \gg 1$) leads to the decrease of the signals of the free ligands, whereas new resonances appear at other positions in the spectrum but are often too weak to be detected. The protein background signal can be suppressed when the ligands are labeled with nuclei that do not occur in the protein. Here, fluorine is a well-suited nucleus (Tengel *et al.*, 2004). When the signals of the small compounds are known (as it is usually the case), a number of compounds can be screened simultaneously and the binding compound can be identified from the specific spectral pattern. Such a simultaneous screening is usually performed also with the other methods described subsequently.

7.3.1.2. Polarization Transfer

Polarization can be transferred from the protein occurring in low concentration to the ligand occurring in high concentration or vice versa. STD experiments are based on the NOE driven polarization transfer from the receptor spins to the ligand spins in the transient complex. In the STD experiment, the resonances of the target protein are saturated selectively by an irradiation frequency of <-1 or >8 ppm, where normally no proton resonances of the small ligand are found. The polarization is distributed *via* spin diffusion to all spins of the protein and then to the bound ligand. In large complexes, a negative NOE is observed, that is, the signal intensity is decreased. When the bound ligand exchanges rapidly with the free ligand the intensity change is transferred to the free ligand and can be observed as a reduction of its signal. The STD scheme can also be incorporated in two-dimensional experiments as it was shown for 2D-STD-TOCSY (Mayer and Meyer, 2001).

The *NOE pumping* experiments (Chen and Shapiro, 2000) are one-dimensional experiments, in which a loss of intensity of the ligand signals can be detected that is caused by the NOE-mediated magnetization transfer from the ligand to the target. This technique was originally applied to detect the complex formation between several fatty acids and human serum albumin. The NOE pumping methodology can be used as an alternative to STD methods especially when the selective irradiation of the target resonances is not possible.

7.3.1.3. Binding-Induced Changes of the Water Shell

The water molecules located in the interaction sphere between the target receptor and the bound ligand can be used to identify the ligand–protein interaction by performing a saturation transfer from the solvent. In the *NOE-ePHOGSY* experiment, the bulk water magnetization is either selectively inverted or saturated and transferred to the complex bound water by the exchange process. This results in a negative NOE enhancement of the protons of the bound ligand. In the *Water-LOGSY* experiment, the setup is modified using pulse-field gradients (Dalvit *et al.*, 2001). The excess in terms of sensitivity allows measurements with protein concentrations of less than 100 nM.

In Figure 7.2, an example of a Water-LOGSY experiment is shown. Adding of Ras protein to a solution of the small compound cyclen does not lead to a change of the spectrum (spectra A and B). When repeating the experiment with Zn^{2+}-cyclen, the binding can be detected, clearly, by the change of the sign of the signals belonging to the metal-cyclen derivative (spectra C and D). The advantage of this method for fast screening is that in a single experiment many different compounds can be tested and binding compounds can be directly identified if the corresponding resonances are known.

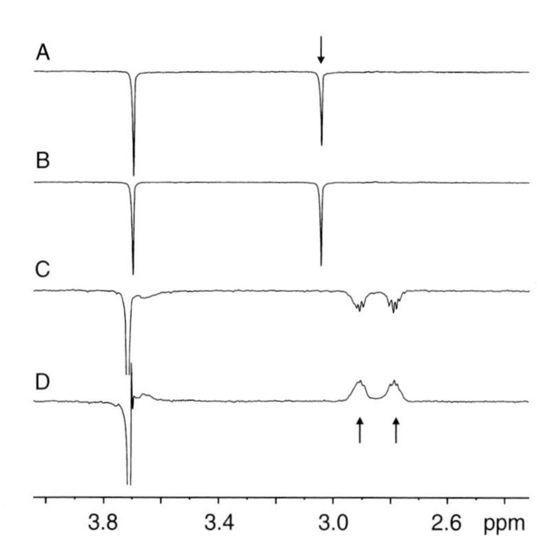

Figure 7.2. Water-LOGSY experiments of 5 HM cyclen in absence (A), and presence of 50 HM Ras protein (B). The same experiment was performed on Zn^{2+}-cyclen in the absence (C) and the presence of Ras protein (D). In contrast to the spectra A and B, the binding of Zn^{2+}-cyclen can be detected in spectrum D by the positive signals marked by arrows. The ^1H signal at 3.7 corresponds to Tris buffer.

7.3.2. Identification of Interaction Sites

7.3.2.1. Chemical Shift Perturbation

The observation of protein chemical shift changes induced after addition of an interaction partner proves that an interaction exists and gives additional information on the interaction itself. The chemical shift (resonance frequency) of a given spin in the protein reflects the chemical environment and responds very sensitively even to small structural changes. The interaction of a ligand with a protein causes chemical shift changes by different mechanisms, namely by (1) a direct dia- or paramagnetic effect of the magnetic field of the ligand on the field sensed by a nucleus of the protein, (2) by polarization of the electronic shell close to the nucleus, and (3) most importantly by a localized structural change close to the nucleus.

A simple example is shown in Figure 7.3. Wild type Ras $\cdot$ Mg^{2+} $\cdot$ GppNHp occurs in two different structural states that can be detected by two sets of phosphorus resonance lines of the bound nucleotide. When the Ras-binding domain (RBD) of the effector Raf-kinase or Byr2-kinase binds to the protein, the set of resonance lines assigned to state 2 increases in intensity, whereas the other resonance lines disappear. This shows that Raf-RBD interacts selectively with state 2 of Ras.

The interaction surface can be mapped in more detail by two-dimensional NMR spectroscopy when the isotope-enriched protein is titrated with an unlabeled interaction partner and the perturbation of the chemical shifts is recorded. The unlabeled molecule can be either a small molecule or a biological macromolecule

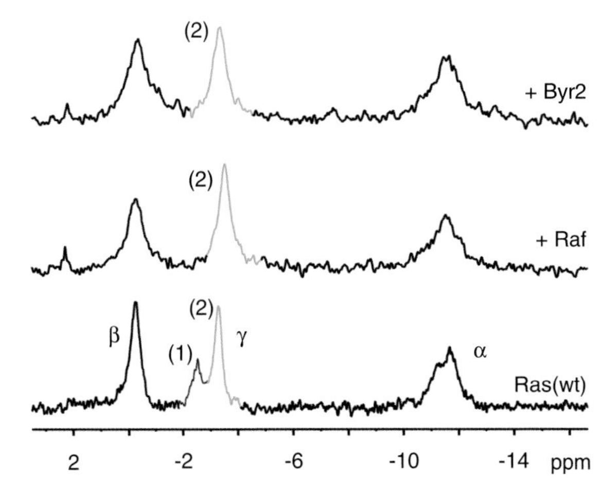

Figure 7.3. ^{31}P NMR spectra of Ras in the absence and presence of effectors. ^{31}P NMR spectra of Ras·Mg^{2+}·GppNHp (*bottom*), ^{31}P NMR spectra of Ras·Mg^{2+}·GppNHp in the presence of Raf-RBD (*middle*) and of Byr2-RBD (*top*).

such as a protein or RNA. The basic experiment used for chemical shift mapping is the two-dimensional heteronuclear $^{15}N-^{1}H$ HSQC experiment, which correlates the chemical shifts of amide nitrogens and amide protons of the protein backbone. The resulting spectrum, which is also called the "NMR fingerprint", shows for every backbone amide group one cross peak at a typical position in the 2D map. As an example, the 800-MHz spectrum of the ^{15}N-enriched RBD of the Byr2-kinase of *Schizosaccharomyces pombe* is shown (Figure 7.4). After titration with Ras $\cdot$ Mg^{2+} $\cdot$ GppNHp from human or from *S. pombe*, several resonances change their position by interaction with the Ras-protein. Surprisingly small differences are observable for the two Ras proteins, although the amino acids in the interaction site of the two Ras are identical.

Often the chemical shift changes in the observed nuclei (^{1}H, ^{13}C, ^{15}N) induced by ligand binding are considered simultaneously and a combined chemical shift change is calculated. Since the sensitivity to ligand binding and the size of the associated chemical shift changes $\Delta\delta_{ji}$ differ for different atom types i (e.g., the amide proton and the amide nitrogen of amino acid j), the chemical shift changes have to be weighted properly. Often the combined chemical shift change S_j is expressed as

$$S_j = \sqrt{\frac{1}{N_a}\sum_{i=1}^{N_a}(w_i\Delta\delta_{ji})^2} \tag{7.17}$$

with w_i, a weighting factor. The division by N_a is often omitted and does not play a role when N_a is the same for all residues j of the protein under consideration.

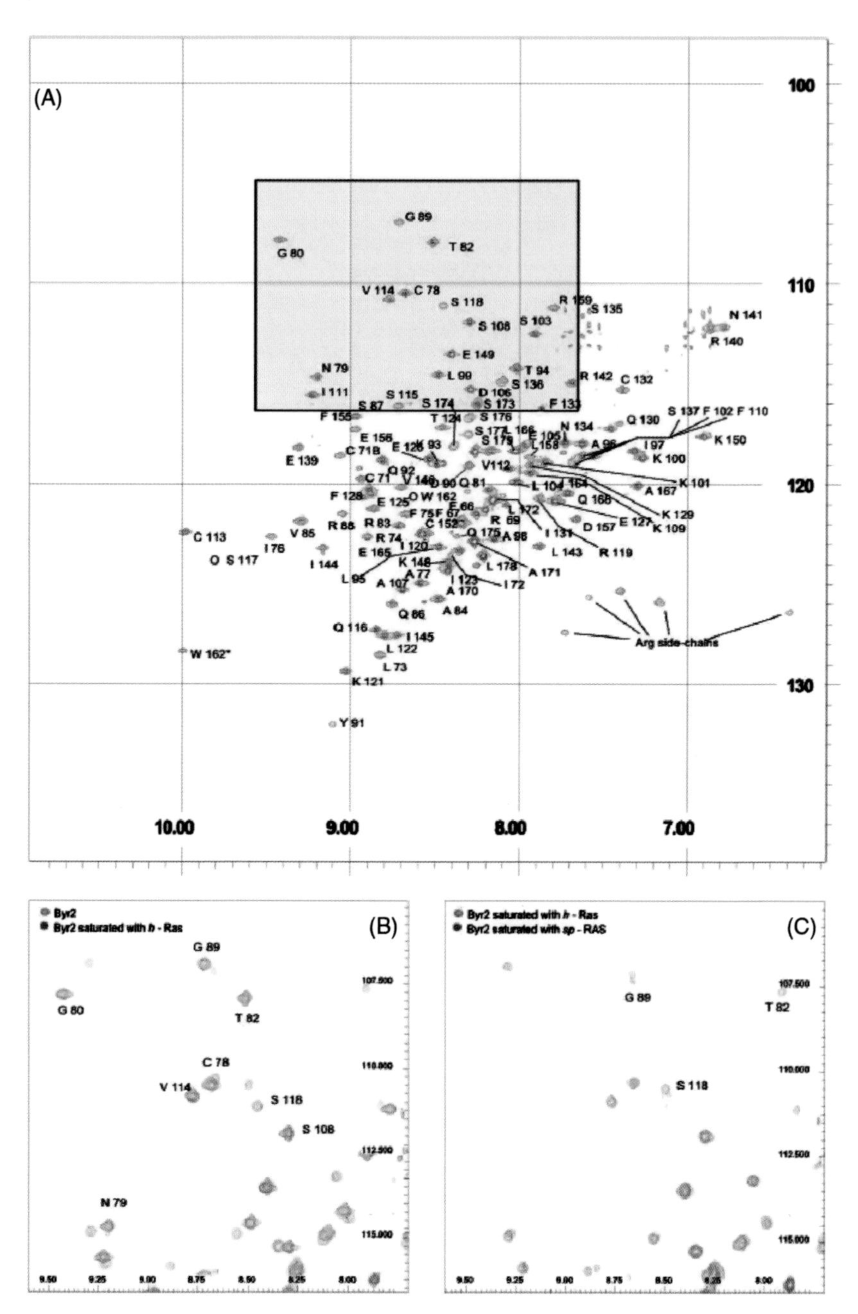

Figure 7.4. Complex of Byr2-RBD with *human* H-Ras and Ras from *S. pombe*. (A) ^{1}H, ^{15}N-TROSY-HSQC spectrum of 0.73 mM Byr2-RBD recorded at 800-MHz proton resonance frequency. (B) Spectral region marked in (A) after addition of human Ras · Mg^{2+} · GppNHp. (C) Comparison of spectral changes induced by human Ras and *S. pombe* Ras (Gronwald *et al.*, 2001, with permission).

When chemical shifts are expressed in ppm (what we assume in this chapter) then a suitable estimate for the weighting factors is given by (Geyer *et al.*, 1997)

$$w_i = \frac{\gamma_i}{\gamma_1} \tag{7.18}$$

with γ_i and γ_1, the gyromagnetic ratio of nucleus i and the proton 1, respectively. Various other schemes to calculate the combined chemical shifts can be found in literature, however, the differences in the prediction of interaction sites are rather small, especially since the definition of significance levels is also not trivial.

Resonances that experience significant shift perturbation can be mapped on to a known structure. This is shown in Figure 7.5 again for Byr2. As discussed already for small ligands, the chemical shift difference $\Delta\omega$ and the exchange correlation time τ_e determine the spectral changes. In principle, in an HSQC-spectrum of a

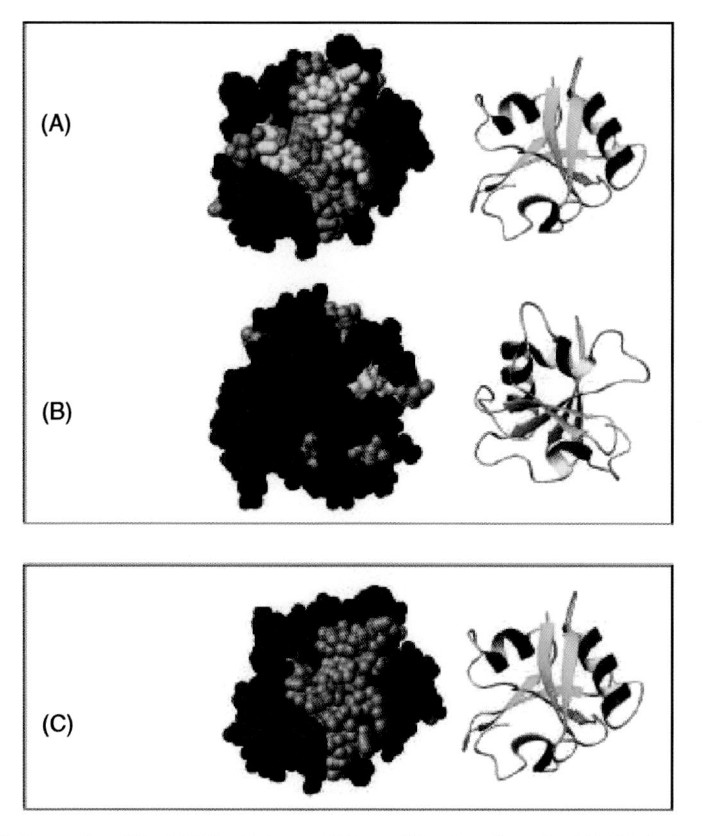

Figure 7.5. Interaction of Byr2-RBD with *human* H-Ras and Ras from *S. pombe*. (A), (B) Amino acids that shift with addition of human or *S. pombe* Ras · Mg^{2+} · GppNHp are labeled in orange and yellow. (C) Comparison of spectral changes induced by human Ras and *S. pombe* Ras, residues labeled in red show a larger chemical shift change in Ras from *S. pombe* (Gronwald *et al.*, 2001, with permission).

protein simultaneously fast, intermediate, and slow exchange can be observed since $\Delta\omega$ varies from one amide group to the other. When small ligands bind to a protein, a severe line broadening can be caused in the intermediate time regime where $\Delta\omega\tau_e$ is approximately 1. When a protein interacts with another protein, an additional line broadening is caused by the increase in molecular mass (see later). In the fast exchange regime, the observed line width $\Delta\nu_{1/2}$ is the population weighted average of the line widths $\Delta\nu_{1/2}^{(1)}$ and $\Delta\nu_{1/2}^{(2)}$ in the free and complexed form.

$$\Delta\nu_{1/2} = p_1\Delta\nu_{1/2}^{(1)} + p_2\Delta\nu_{1/2}^{(2)} \tag{7.19}$$

In the slow exchange regime, the line width in the complexed state is much higher since the mass of the complex is much higher (see later). As an additional effect, the intensity (volume) of the cross peaks decreases due to the increased transverse relaxation the polarization transfer.

Modern chemical shift mapping is used since the early 1990s and well reviewed (Pellecchia *et al.*, 2000; Stevens *et al.*, 2001; Zuiderweg, 2002; Clarkson and Campbell, 2003; Gao *et al.*, 2004). First studies of protein–protein interfaces were published in 1993, when the surface of the phosphocarrier protein HPr with the A-domain of enzyme II was mapped (Chen *et al.*, 1993). Only a couple of years later, the same HPr protein was titrated against the approximately 30 kDa N-terminal domain of enzyme I (EIN), which already reaches the limits of the above-mentioned HSQC technique (Garrett *et al.*, 1997), due to massive line broadening proportional to the molecular mass of the protein complex. Because of that many groups were mapping the binding site of proteins with peptides representing a subunit of the whole protein (Grzesiek *et al.*, 1996; McKay *et al.*, 1997, 1998). Chemical shift mapping is a very efficient method to study binding of small peptides and other ligand molecules like drug candidates (Shuker *et al.*, 1996; Clarkson and Campbell, 2003).

A major step to overcome the size problem was the development of the transverse relaxation-optimized spectroscopy technique (TROSY) (see later) (Riek *et al.*, 1999). Using HSQCs, it is possible to study protein interactions up to a molecular size of approximately 30 kDa. The TROSY experiment made it feasible, for example, to measure the kinetics of the complex formation of the lectin chaperone calreticulin (CRT) and ERp57, a thiol-disulfide oxidoreductase, a system of 66.5 kDa (Frickel *et al.*, 2002). Those technical advances have allowed its application to protein complexes as large as the 900 kDa GroEL–GroES complex (Fiaux *et al.*, 2002).

Generally, one cannot use chemical shift changes to know what exactly happens at the interface as only a small part of the chemical shift changes are due to direct effects. Usually binding also induces local and global conformational changes that in turn lead to chemical shift changes. Especially global conformational changes may lead to a wrong interpretation of the data. The chemical shift perturbation measurements just yield the connectivity of the putative interaction site. When in the extreme case at two opposite sides of the proteins chemical shift changes occur, it is often likely that the largest predicted interaction site represents the true solution, whereas the other chemical shift changes are caused by long-range conformational changes.

7.3.2.2. Amide Exchange

Another possibility for the mapping of protein interfaces uses the protection of amide-hydrogen exchange. Secondary structures in proteins are stabilized by characteristic hydrogen bonds. Amide hydrogen exchange is usually catalyzed by the water hydroxyl ions. Amide hydrogens, involved in hydrogen bonds or located in the core of a protein, are usually shielded from the attack by OH^- ions. If proteins interact with ligands, the chemical environment of the residues in the interaction surface is altered. Formerly solvent exposed amide hydrogens are now shielded from the solvent molecules. If one adds deuterons they normally exchange with hydrogens and the NMR signals vanish in the HSQC map and of course also in the ^1H spectrum. How fast that happens gives a hint whether an amide group is solvent exposed or not. Good solvent accessibility allows a fast exchange of the amide protons for other protons or deuterons. The exchange can take place on timescales ranging from minutes to months. It is to be expected that a protein–protein interaction should shield the interface area from exchange processes and the corresponding NMR signals of amino acids located in that interface disappear more slowly (ratio calculated from the intensities/volumes of the cross peaks in the 2D spectrum) than solvent exposed hydrogens (Walsh *et al.*, 2003).

It was, for example, possible to monitor the amyloid fibril formation of the bovine PI(3)-K SH3 domain by measuring the H/D exchange after 4 and 15 days under quite drastic conditions (pH 1.6). The extent of exchange rises only by 8% after 15 days, suggesting that the protein core is stable during fibril formation even under conditions where normally, in the monomer case, unfolding occurs. Most residues are protected from solvent exchange within the fibrillar structure, consistent with the resistance against proteolytic cleavage (Carulla *et al.*, 2005). This example shows that H/D exchange by NMR is also an advantageous experiment to monitor protein folding/unfolding processes.

7.3.2.3. Residual Dipolar Couplings

It is common to measure RDCs for NH vectors from differences in coupled $^1H, ^{15}N$ HSQC spectra in the absence or presence of alignment medium. This method is used to obtain relative domain orientations in multidomain proteins (Fischer *et al.*, 1999; Mueller *et al.*, 2000; Meier *et al.*, 2004) or relative protein orientations in multiprotein complexes (Clore, 2000). When the structures of the individual proteins/domains are known, the relative orientations of all dipolar vectors within the individual proteins/domains are given. The measured dipolar couplings, in turn, reflect the orientation of the dipoles with respect to the magnetic field. It is then possible to carry out a computer procedure rotating these individual proteins/domains to fit the measured dipolar couplings.

If A_a^{IS} and R are found to be equal for the different proteins/domains, it strongly indicates that the individual proteins/domains move together as a single entity (see 7.2.3.3). Then, the orientations of the ordering tensors (the axes) for the different molecules must be collinear, which defines the orientation of the individual molecules with respect to each other.

The method was for the first time demonstrated for a true two-protein complex with the 40-kDa complex of enzyme I and the histidine phosphocarrier protein (Dosset *et al.*, 2001). While this complex had been previously determined from thousands of NOEs (see earlier), it was much more efficient to dock the molecules together using only a few intermolecular NOEs and a few hundred dipolar coupling restraints for each of the proteins. The methods have been implemented in computer programs (Brüschweiler *et al.*, 1995).

Orientational constraints can also be obtained from anisotropic ^{15}N relaxation in proteins. The measurements to obtain the relative orientation of domains are much more involved but measure similar parameters. The relative orientation of loosely connected subdomains in individual proteins, such as the relative orientation of SH2 and SH3 domains in a combined SH2–SH3 construct, have been determined with this method (Fushman *et al.*, 1999a). Recently, workers were using RDCs extracted from unassigned spectra to identify protein interfaces (Mettu *et al.*, 2005).

7.3.2.4. Paramagnetic Perturbation

Paramagnetic mapping appears to be an interesting tool for the determination of protein–ligand interactions as it is the strong electron magnetic moment that provides relatively intense spectral effects. Paramagnetic perturbations can be classified in two categories (1) shift changes induced by the magnetic anisotropy of the center and (2) line broadening induced by the hyperfine coupling. The hyperfine coupling contains two parts, the Fermi contact part that is transmitted *via* the connected electronic system and the through-space dipolar contribution. The shift changes and the line broadening caused by the dipolar contribution are dependent on the distance $d_{\mathrm{p}i}$ between the paramagnetic center p and the nuclear spin i. In first approximation, the paramagnetically induced shift changes are dependent on the distance as $d_{\mathrm{p}i}^{-3}$ and the line broadening as $d_{\mathrm{p}i}^{-6}$. The relative contribution of the two effects depends on the intrinsic properties of the paramagnetic center used.

The method can only be used when a paramagnetic center is present in the sample, it may either occur naturally in the protein or its ligand or can be introduced artificially. Typically, naturally occurring paramagnetic centers are iron ions in heme groups or manganese ions in catalytic centers. As artificially introduced paramagnetic centers often serve stable organic radicals covalently linked to the protein, paramagnetic ions that replace a naturally occurring metal ion, or are bound to an artificially created metal-binding site. For these applications, often the diamagnetic Mg^{2+} ion bound physiologically to the protein is replaced by the paramagnetic Mn^{2+} ion or the diamagnetic Ca^{2+} ion is replaced by trivalent lanthanides as Gd^{3+} ions.

The line broadening effect is experienced by protons up to 1–1.5 nm away from the paramagnetic probe. The technique of attaching spin-labels to cysteine residues in an otherwise cysteine-free protein, affording only one spin-label per monomer can be used to study domain organization and interacting parts in a complex. The free stable radical nitroxide (spin-label) TEMPO can interact specifically with cysteine groups engineered on the protein surface. TEMPO is extensively used in EPR site-specific spin-labeling studies (Hubbell *et al.*, 1996). Paramagnetically labeled peptides are used to identify binding sites (Wang *et al.*, 1998; Jain *et al.*, 2001).

7.3.2.5. Modulation of Nuclear Relaxation

Recently a Japanese group (Takahashi *et al.*, 2000; Nakanishi *et al.*, 2002; Takeuchi *et al.*, 2003; Shimada, 2005) has developed a new method termed cross-saturation and transferred cross-saturation experiment to precisely identify the interface of larger protein–protein complexes. The principle of the cross-saturation method is pretty simple: protein I contains the residues, which we would like to identify as interface residues, and is labeled uniformly with ^{2}H and ^{15}N. This protein forms a complex with a nonlabeled target protein II. The complex of protein I and protein II contains one molecule with low proton density (highly deuterated protein I) and one molecule with high proton density (unlabeled protein II). As in the STD method (Mayer and Meyer, 2001), the aliphatic proton resonances are irradiated nonselectively using an RF field. All protons in protein II should saturate relatively quickly due to spin diffusion. The highly deuterated protein I (^{2}H and ^{15}N labeled) should not be influenced by the RF field, but the saturation of protein II should be transferred to the doubly labeled protein I *via* cross-relaxation through the complex interface. The saturation is limited to the interface, if the proton density and thus the propensity of spin diffusion is relatively slow in protein I and in the solvent. The residues at the interface of protein I are identified *via* observing a reduction in the peak intensities in TROSY-type spectra. The cross-saturation and transferred cross-saturation experiments are extremely sensitive in 10% $^{1}H_2O$–90% $^{2}H_2O$. In case of the complex between ^{2}H, ^{15}N-labeled Ig-binding domain B of the Staphylococcal protein A (termed FB) with a nonlabeled Fc fragment of the human myeloma protein IgG(κ), Shimada and co-workers (Shimada, 2005) could observe small intensity ratios for residues in the helices I and II with a pattern that every third and fourth residue showed smaller values suggesting that one side of each of the helices I and II is interacting with the Fc fragment. This fine pattern was lost in 90% $^{1}H_2O$–10% $^{2}H_2O$ (Shimada, 2005) due to spin diffusion.

Oscillations of the protein structure and random movement of the molecule relative to the magnetic field cause the relaxation of the magnetic nuclei inside the protein. Hence, differences in the mobility of parts of the protein backbone and side-chains and anisotropy of rotational diffusion of the protein result in a different relaxation behavior of the involved nuclei (^{1}H, ^{13}C, ^{15}N). Dynamics by NMR may provide a link between structure, function, and thermodynamics (Stone, 2001). Measuring the relaxation rates R_1, R_2 of ^{15}N, and the steady state ^{1}H, ^{15}N NOE at different static magnetic fields has become a standard method to characterize the dynamics of the protein backbone (Fushman *et al.*, 1999b; Palmer, 2001). Analyzing this data using the model-free approach (Lipari and Szabo, 1982a,b) delivers easy-to-interpret parameters of molecular motion. Motions faster than the overall tumbling time (τ_c) of the molecule (ps–ns timescale) and the generalized order parameter (S^2) are characteristic for the amplitude of motion of these internal fluctuations. Slower motions on the μs–ms timescale are represented in the conformational exchange term R_{ex} (Wang and Palmer, 2003; Palmer, 2004).

Binding of an interaction partner leads at first to a change in the overall rotational correlation time (τ_c) of the new protein complex in comparison to the ''free'' protein, causing changes in the measured relaxation parameters. On the other hand, one can imagine that the interaction changes the mobility of the involved side-chains and the protein backbone at the interface. Compensatingly, binding may even induce dynamic changes in regions distant from the binding site.

NMR dynamics experiments performed on a variety of systems have yielded information about the effect of ligand binding on motional properties of the protein backbone as well as side-chain methyl groups (Nicholson *et al.*, 1992; Farrow *et al.*, 1994; Stivers *et al.*, 1996; Yu *et al.*, 1996; Gagne *et al.*, 1998; Kay *et al.*, 1998; Zhang *et al.*, 1998; Zidek *et al.*, 1999; Lee *et al.*, 2000; Loh *et al.*, 2001; Finerty *et al.*, 2002). For some proteins, ligand binding leads to a reduction in fast timescale dynamics (Zhang *et al.*, 1998; Zidek *et al.*, 1999; Finerty et al., 2002), whereas in other instances, the overall flexibility of the backbone can remain unchanged or even increase (Stivers *et al.*, 1996; Yu *et al.*, 1996; Kay *et al.*, 1998; Zidek *et al.*, 1999). In many cases, these studies provide insights into the thermodynamics of binding. Nevertheless, it appears that protein dynamics is too much an intrinsic part of the binding process to make it a reliable tool for interface mapping.

7.3.2.6. Observation of Dynamic Equilibria by High-Pressure NMR Spectroscopy

There are generally two responses of proteins toward the application of pressure. They result from the property that pressure favors states with a smaller specific volume. Thus, any equilibrium in a population of conformers of proteins, including low-lying excited states and partially or completely unfolded conformations, connected with a nonzero volume change is shifted toward the more compact state by the application of hydrostatic pressure. The volume decrease in proteins observed upon partial or complete unfolding of the native structure depends on three kinds of interactions: (1) Ionic pairs in aqueous solution are strongly destabilized by hydrostatic pressure due to the electrostrictive effect of the separated charges, thus the overall volume change favors the dissociation and disruption of ionic interactions under pressure. (2) In a similar fashion, the exposure of hydrophobic groups to water disturbs the loosely packed structure of pure water and leads to a hydrophobic solvation layer, which is assumed to be more densely packed. This exposure of hydrophobic residues occurring during the unfolding of proteins is favored at elevated pressure. (3) Evolutionary selected areas with nonoptimized packing density through van der Waals' forces are key points where volume fluctuations can be probed by pressure, this is displaying the high anisotropy of compressibility found in proteins.

Applying these principles to protein complexes leads to the prediction that the increase of the surface by dissociation is favored by high pressure. As an example, the protein transthyretin (TTR) can be viewed (Silva *et al.*, 2001). Under normal conditions native TTR forms a stable tetramer. Compression of TTR leads to a monomeric molten-globule-like intermediate. Upon decompression TTR forms a tetrameric preaggregate with a conformation different from the native form leading to fibrillar aggregation. A repeated compression changes the conformation from fibrillar structures back to a monomeric molten-globule-like intermediate state. The structure of such an intermediate state with preamyloidogenic properties can be characterized by keeping it at a temperature of 4°C and avoiding the aggregation that way. In this sense, an elegant alternative to investigating multiple misfolding pathways of the proteins involved in neurodegenerative diseases with denaturing agents is applying hydrostatic pressure that allows studying intermediate forms in the folding pathway of such proteins different from those obtained by denaturing agents. As seen in case of the prion protein, pressure induces scrapie-like prion protein misfolding and amyloid fibril formation (Torrent *et al.*, 2003, 2004, 2005; Cordeiro *et al.*, 2004). Irreversible aggregation of *sha*PrP(90-231) was observed above 450 MPa, and incubation of *sha*PrP(90-231) at 600 MPa overnight led to the formation of amyloid fibrils, whereas pressures up to 200 MPa lead to reversible effects and recovery of the original structure after pressure release (Torrent *et al.*, 2003, 2004, 2005; Cordeiro *et al.*, 2004).

Fine-tuning of temperature and pressure would lead to a stabilization of the intermediate preamyloidogenic conformations from proteins involved in neurodegenerative diseases and subsequently to a structural characterization of such an intermediate. The combination with NMR spectroscopy would allow a structure determination of preamyloidogenic intermediates at the atomic level. In addition, high-pressure NMR spectroscopy can yield local information about mechanical and dynamical properties of proteins and can be used to stabilize folding and unfolding intermediates. Especially, at pressures of 200 MPa, the phase behavior of water allows even the observation of protein denaturation in aqueous solution at temperatures down to 255 K. Recent advances in high-pressure NMR spectroscopy on folding intermediates and denatured states of proteins as well as in the sampling of the conformational space of proteins by high-pressure NMR spectroscopy were pioneered by group of Akasaka (Akasaka and the Li, 2001; Akasaka 2003a,b 2006; Kamatari *et al.*, 2004) often in cooperation with the Regensburg group (Kalbitzer *et al.*, 2000; Inoue *et al.*, 2000; Kachel *et al.*, 2006).

Although in general monomerization is preferred by high pressure, under special conditions polymerization is preferred when the decrease of the partial volume by the increase of the surface in the monomer is compensated by other mechanisms. Such a case was observed for the viral protein YFID (Bombke *et al.*, to be published).

7.3.3. Three-Dimensional Structures of Protein Complexes

7.3.3.1. NMR Spectroscopy of Very Large Proteins and Protein Complexes

NMR has been shown to provide detailed structural information for biological macromolecules in solution with molecular masses up to 30 kDa. Protein–protein complexes tend to have very large molecular masses. A prerequisite for a successful application of NMR to high-molecular-mass proteins above 100 kDa is high-quality spectra with good sensitivity and spectral resolution. For studies on high-molecular-mass proteins and macromolecular complexes, a number of limiting factors arise which are (1) high complexity of the spectra due to extensive signal overlap, (2) the sensitivity goes down due to low solubility of the solute, and (3) the rapid transverse relaxation leads to even lower sensitivity and line broadening of the signals. In first approximation for larger molecules, the relaxation rate $1/T_2$ and thus the linewidth $\Delta\nu_{1/2} = 1/(\pi T_2)$ is proportional to the rotational correlation time τ_c. For a spherical rigid molecule, the Stokes–Einstein equation holds

$$\tau_c = \frac{V\eta}{kT}, \tag{7.20}$$

with V the volume of the protein, η the viscosity, and T the absolute temperature.

To avoid the overlap of spectral signals, many groups introduced new techniques like fractional deuteration (Gardner and Kay, 1998; Lian and Middleton, 2001), amino acid selective labeling (Kainosho, 1997), specific protonation of methyl groups in otherwise perdeuterated proteins (Tugarinov and Kay, 2003), and segmental labeling through protein splicing (Yamazaki *et al.*, 1998; Xu *et al.*, 1999; Yu, 1999). All these methods intend to reduce the complexity of the spectra. Low sensitivity, either as a result of low solubility or increased transverse relaxation, has been addressed by the manufacturers through the development of NMR cryogenic probes and in addition through novel experimental approaches which reduce the influence of transverse relaxation. TROSY uses the effect that in J-coupled systems, the dipolar relaxation and the relaxation by chemical shift anisotropy cancel for some of the multiplet components and lead in first order to vanishing line widths (Pervushin, 1997). In addition, polarization transfer from protons to heteronuclei for large molecules is performed by cross-correlated relaxation-enhanced polarization transfer (CRINEPT) (Riek *et al.*, 1999) and cross-correlated relaxation-induced polarization transfer (CRIPT) (Brüschweiler and Ernst, 1992; Dalvit, 1992; Riek *et al.*, 1999) more effectively. These novel techniques made studies on macromolecular complexes with molecular masses up to

900 kDa possible (Wider and Wüthrich, 1999; Fiaux *et al.*, 2002; Fernandez and Wider, 2003; Wider, 2005). Recently, the direct NMR observation of the substrate protein human dihydrofolate reductase (hDHFR) bound to the chaperonin GroEL (800 kDa) was shown by the Wüthrich group (Horst *et al.*, 2005). They measured the buildup of the NMR signals of hDHFR by different magnetization transfer mechanisms (INEPT and CRIPT buildup curves) to characterize the internal dynamics of the bound substrate (Horst *et al.*, 2005).

Any detailed study of protein structure, protein–ligand interaction, and protein dynamics by NMR spectroscopy involves the assignment either of the backbone and/or the side-chain signals. The novel experimental approaches described previously are best suited for highly deuterated protein samples and thus the preparation of them is one of the key steps in NMR studies of large proteins. There are many protocols for *in vitro* refolding of large proteins overexpressed in D_2O-based media (Creighton, 1997; Gardner and Kay, 1998; Tugarinov *et al.*, 2004). If the protein refolding is complete, TROSY-type spectra for $^1H - ^{15}N$ correlation can be used to establish the structural integrity of the protein in comparison to unfolded references. In the past years, the complete backbone assignments of a number of high-molecular-mass systems have been achieved which include the 110-kDa homo-octameric protein 7,8-dihydroneopterin aldolase (DHNA) (Salzmann *et al.*, 2000), the 8-stranded β-barrel membrane proteins OmpX (Fernandez *et al.*, 2001a,b, 2002, 2004), OmpA (Arora *et al.*, 2001), PagP (Hwang *et al.*, 2002, 2004) dissolved in lipid detergents (50–60 kDa range), malate synthase G (82 kDa, Tugarinov *et al.*, 2005), maltose binding protein (Wemmer and Williams, 1994; Kay, 2001; Millet *et al.*, 2003; Wemmer, 2003), and the 300-kDa cylindrical protease ClpP (Sprangers *et al.*, 2005). In this molecular mass range (up to 100 kDa), a 1 mM solution of a protein (concentration generally used for NMR studies) would result in up to 10% by mass protein, which is not possible for many cases. Consequently, many NMR experiments have to be performed with concentrations ranging well below 1 mM from 0.3 to 0.5 mM.

Especially, large monomeric proteins exhibit significant peak overlap which was addressed by the Kay group through the development of 4D TROSY experiments such as 4D TROSY-HNCACO [correlating $C^{\alpha}(i)$, $C'(i)$, $N^H(i)$, and $H^N(i)$, and in some cases $C^{\alpha}(i-1)$, $CO(i-1)$, $N^H(i)$, and $H^N(i)$], the 4D TROSY-HNCOCA [correlating $C^{\alpha}(i-1)$, $C'(i-1)$, $N^H(i)$, $H^N(i)$] (Yang and Kay, 1999), the 4D TROSY-HNCO$_{i-1}$Ca$_i$ [correlating $C^{\alpha}(i)$, $C'(i-1)$, $N^H(i)$, $H^N(i)$ and in some cases $C^{\alpha}(i-1)$, $C'(i-1)$, $N^H(i)$, $H^N(i)$] (Konrat *et al.*, 1999), and the 4D ^{15}N, ^{15}N-edited NOESY (Grzesiek *et al.*, 1995; Venters *et al.*, 1995). In perdeuterated proteins, the side-chain assignment is difficult due to the absence of protons bound to carbons. Either the introduction of specific protonated methyl groups (Gardner and Kay, 1997; Goto *et al.*, 1999) or the introduction of cryogenic probes with a sensitivity for ^{13}C of 800:1 comparably to room temperate probes with a sensitivity for ^1H of 800:1 allows the recording of protonless NMR spectra (Bermel *et al.*, 2005). This has in addition the advantage of exploiting the longer T_2 times for the ^{13}C nucleus for NMR studies compared with the rapid transverse relaxation and thus very short T_2 times for the ^1H nucleus in high-molecular-mass proteins. Here ^{13}C correlation and ^{13}C TOCSY spectra can be used to get a complete assignment (Bertini *et al.*, 2004).

The structure determination by NMR spectroscopy is based mainly on the NOEs between ^1H nuclei in the proteins. As pointed out earlier, large macromolecular complexes and high-molecular-mass proteins are very often completely deuterated thus allowing only to collect NOE data for protons of the exchanging amides. The incorporation of specific protonated methyl groups of Ile, Leu, and Val residues (Gardner and Kay, 1997; Goto *et al.*, 1999) allows to gain additional distance information by ^1H–^1H NOE connectivities. The measurement of RDC (Tjandra and Bax, 1997; Brunner, 2001; Prestegard *et al.*, 2004) is a valuable source of structural restraints for accurate structure determinations. The development of cryogenic probes for the ^{13}C nucleus allows again to measure ^{13}C–^{13}C NOESY spectra with the advantages mentioned previously (Bertini *et al.*, 2004).

Solid-state NMR is also a promising method to determine structures of large protein complexes when MAS is used to reduce the line width (for a review, see e.g., Luca *et al.*, 2003; McDermott, 2004).

7.3.3.2. Computational Methods for Calculation of Very Large Structures and Docking of Ligands

For large protein complexes, it is often very difficult or even impossible to obtain enough NMR data to calculate high-quality structures from them. But even when, a complete NMR structure determination is feasible, it may be mandatory to obtain structures using a limited data set only. This is a typical situation in proteomics and in high-throughput screening. Here, the combination of experimental NMR data with knowledge from other sources by suitable computational techniques is especially important. As it is easier to obtain 3D structures of the single components of a complex, one can obtain a picture of the corresponding biomolecular complex in a relatively short amount of time by using the 3D structures of the isolated molecules to calculate the complex structure by docking techniques. During the last few years, substantial progress has been made in the field of *ab initio* docking as reviewed by Halperin *et al.* (2002), Brooijmans and Kuntz (2003), Wodak and Janin (2003), and Vajda and Camacho (2004), where no other input than the 3D structures of the single molecules are used. But to date these techniques are still unreliable in obtaining the correct complex structures especially in cases where larger structural changes occur upon complex formation. The performance of several of the currently available docking algorithms is evaluated by the *c*ritical *a*ssessment of *pr*edicted *i*nteractions (CAPRI) competition (Janin, 2005).

7.3.3.2.1. Calculation of Structures from a Limited Set of Experimental Data Before one can apply docking techniques to obtain protein–protein or protein–ligand complexes one has to know the 3D structures of the individual components. One possible way to speed up the protein structure determination process is to reduce the required number of restraints and/or to use only restraints that are relatively easy available, e.g., backbone dihedral angles, chemical shifts, residual dipolar couplings, hydrogen bonds, or H^N–H^N NOEs. These methods should be applicable in particular to cases where one is more interested in the global fold of the molecule than in a highly detailed structure. In the easiest application, one can check if the trial protein adopts a previously known fold (Annila *et al.*, 1999). Other methods rely on the combination of modeling and NMR techniques. In one approach, the ROSETTA *ab initio* protein structure prediction method is supplemented with sparse NMR data such as chemical shift and NOE data (Bowers *et al.*, 2000). In the ROSETTA method, protein structures are assembled from fragments of known structures with sequences similar to the target protein (Simons *et al.*, 1997, 1999). Additionally, residual dipolar couplings can be applied for fragment selection (Delagio *et al.*, 2000; Andrec *et al.*, 2002). Fragments whose predicted residual dipolar couplings best fit the set of measured values are selected. In addition, unassigned residual dipolar couplings and unassigned backbone chemical shifts can be combined with the ROSETTA method to obtain an iterative approach automated sequential (ITAS) resonance line assignments as well as medium resolved three-dimensional structures (Jung *et al.*, 2004). Instead of using fragments for model building, it is possible to use reduced protein representations where each residue is modeled by the C^α, C^β, and the side-chain center of mass together with a knowledge-based force field that imposes a strong bias toward predicted secondary structure to obtain structures of sufficient quality using only sparse additional distance information from NOE data (Li *et al.*, 2003). It should be noted that for a successful application of this approach, a high-quality secondary structure prediction is required. In a different approach that combines modeling techniques with experimental data, it is demonstrated that fold prediction by protein threading can be shown to be improved by including experimental distance constraints obtained from, e.g., mass spectroscopy or NOE measurements (Albrecht *et al.*, 2002).

Other methods rely solely on the use of NMR data in combination with standard structure calculation procedures such as CNS or X-PLOR. Common to these methods is that relatively few distance information from NOE data are used. For small proteins such as Ubiquitin, it was shown that three different residual dipolar coupling measurements together with sparse long-range H^N–H^N NOE contacts are sufficient to define the global fold (Giesen *et al.*, 2003). Moreover, it was demonstrated that residual dipolar couplings alone are sufficient to check if a protein adopts a previously known fold (Annila *et al.*, 1999). In fully deuterated proteins, it is possible due to decreased spin diffusion to obtain long range NOEs between amide protons corresponding to distances up to 0.8 nm. Together with secondary structure restraints obtained from a chemical shift analysis, this information is sufficient to calculate structures of medium precision (Koharudin *et al.*, 2003). If a protein with selectively protonated side-chain methyl groups is available, additional distance information is obtained for the structure determination process. The new *I*ntelligent *S*tructural *I*nformation *C*ombination (ISIC) algorithm developed by us (Brunner *et al.*, 2006) aims at the high-resolution protein NMR structure determination using only a limited set of experimental data together with additional information from other structural sources such as homologous X-ray structures. The key point here is the combination of the available information from the other sources with the experimental data of the current project ensuring that only relevant information is used and no wrong bias is produced. ISIC is a part of the larger AUREMOL software package aiming at the automated three-dimensional structure determination in solution (Gronwald and Kalbitzer, 2004).

7.3.3.2.2. Docking of Small Ligands and Proteins on the Basis of Limited Experimental Data One avenue to increase the success rate of docking algorithms is to supplement them with additional experimental data such as the localization of the interfaces. An excellent review on this subject has recently been published (Van Dijk *et al.*, 2004). As has been described in more detail in previous parts of this chapter, NMR is especially well suited to map residues at the complex interfaces employing, for example, chemical shift perturbation experiments (Zuiderweg, 2002). A problem arising when using chemical shift perturbations is that chemical shift changes are not only observed for the residues directly participating in binding but also for residues where structural changes take place upon binding. The latter residues can in some cases be far removed from the binding site. Usually the user has to manually discriminate between these two types of residues using the assumption that residues that define the binding interface should cluster in one place. Residues participating in binding can also be defined from H/D exchange measurements or can, for example, be defined from cross-saturation experiments (Takahashi *et al.*, 2000). Residual dipolar couplings can be used to define the orientation of the individual complex molecules with respect to each other (Bax, 2003). This information can also be extracted from ^{15}N relaxation measurements in case of diffusion anisotropy (Fushman *et al.*, 2004). In case that one of the molecules contains a paramagnetic ion, for example, from a paramagnetic tag, one can observe the so-called pseudocontact shifts that provide very useful long-range information for the docking process (Zuiderweg, 2002).

Two components are needed in computerized docking algorithms: first, a sampling method that usually generates a large number of possible conformations and second, a scoring scheme that selects which of the resulting complex structures is most probably closest to the true structure. One important point in docking procedures is if the individual molecules are kept rigid during the docking allowing no overlap between the molecules (hard rigid body docking) or allowing some overlap (soft rigid body docking) or if side-chain and/or backbone flexibility is allowed. The molecules can be either explicitly represented using an atomic model or a grid representation can be used where the individual molecules are projected onto a three-dimensional grid (Katchalski-Katzir *et al.*, 1992). In the latter case, rigid body docking can be performed by computing correlations such as surface complementarity using fast Fourier transform methods. The main advantage here is computational speed compared with methods where the molecule is explicitly treated and, for example, Monte Carlo and molecular dynamics methods or genetic algorithms can be used for sampling. For scoring purposes, mostly force-field-based approaches are used to calculate a pseudoenergy of a given configuration. For this purpose, the energy is calculated for each atom–atom pair or residue–residue pair at the interface and subsequently the single energy terms are added together. The force fields can be based on physical data, e.g., van der Waals radii and electrostatic potentials or force fields can be knowledge-based. In the latter case, potentials are generated from a database of experimentally solved complexes by counting, for example, how often a specific residue–residue pair is found at the interface. In addition, terms such as buried surface area and desolvation energy are often added. Experimental data derived from NMR measurements can now be introduced only in the scoring stage to select the most meaningful configurations or they can additionally be used during the sampling stage to produce fewer configurations that are far removed from the true complex structure.

In the following, we first discuss in some more details methods where experimental data have only been used for scoring. For small systems, systematic grid searches can be performed for all possible orientations and the resulting low-energy structures are then checked for compatibility with experimental data (Adams *et al.*, 1996). Another grid-based method is used in the program BiGGER where fast heuristically optimized Boolean type operations (OR, AND, etc.) are employed to generate possible solutions (Palma *et al.*, 2000) that can be further filtered using, for example, chemical shift perturbation data. Fast Fourier transform methods such as the programs DOT and FTDOCK have been used together with NMR data. In one application, the top solutions generated with DOT were further filtered using experimental H/D exchange data (Anand *et al.*, 2003). Residual dipolar couplings and chemical shift perturbations were used to select a correct complex structure from a very large array of possible docked structures generated using the program FTDOCK (Dobrodumov and Gronenborn, 2003). Other docking approaches, such as DOCK (Meng *et al.*, 1993) and AUTODOCK which was used together with chemical shift perturbation data for scoring (Sachchidanand *et al.*, 2002), use an explicit search in conformational space. Additionally, chemical shifts back-calculated from the possible solutions using, for example, SHIFTX (Neal *et al.*, 2003) and SHIFTS (Xu and Case, 2001) can be compared with experimentally determined shifts of the complex to determine the quality of the docking. This procedure was applied for a manually docked complex that was then further refined by incorporating chemical shifts as restraints in a simulated annealing protocol (Stamos *et al.*, 2004).

Experimental data can also be used in the initial stage to drive the docking process having the advantage that the percentage of correct or almost correct solutions should be increased compared to the above approaches. This is of special importance in cases where the number of possible conformations prevents a proper sampling, as it is commonly the case when main-chain and/or side-chain flexibility is introduced. Therefore, in one application one can use the additional information solely to limit the search space and within this limited search space the docking is performed as without any experimental data (Schneidman-Duhovny *et al.*, 2003). The TREEDOCK approach uses experimentally determined anchor points (Fahmy and Wagner, 2002), where anchor points are defined as pairs of atoms one from each molecule that are always in contact. In fast Fourier transform methods, one can employ increased weights for grid points representing certain residues that were experimentally found to be involved in binding (Ben-Zeev and Eisenstein, 2003). In experimental NMR structure determination of complexes, intermolecular distance information from NOE data is widely used as restraints in structure calculation. Similarly, various docking approaches also use distance restraints to drive the docking process. Distance restraints are defined between residues for which it was experimentally shown that they are in close contact. A variety of experimental information such as mutagenesis data, chemical shift perturbations, pseudocontact shifts, and so on, can be used to define these residues. For example, using distance restraints obtained from double mutations a model of the interferon–receptor complex has been created (Roisman *et al.*, 2001). However, some of the experimental data such as chemical shift perturbations are highly ambiguous as they provide information only about residues participating in the interface but not about specific contacts. A possible solution to this problem comes from NMR spectroscopy where one often has to deal with ambiguous NOE assignments, and a variety of programs such as ARIA (Linge *et al.*, 2003b), CYANA (Güntert, 2004), and KNOWNOE (Gronwald *et al.*, 2002) have been developed to treat such cases. Extending the ARIA approach to ambiguous interaction restraints, the program HADDOCK (Dominguez *et al.*, 2003) was developed. An ambiguous interaction restraint $d_{i\mathrm{AB}}$ with a maximum value of 0.2–0.3 nm is defined here between any atom m of an interface residue i of molecule A and any atom n of all interface and neighboring residues k (N_{resB} in total) of molecule B (and vice versa) (Figure 7.6).

For amino acids supposed to be involved in a protein–protein interaction, ambiguous interaction restraints are defined with a maximum value of 0.2–0.3 nm between any atom of an active residue i (shown in red) of protein A and any atom of all active (shown in red) and passive (shown in green) residues x, y, z of protein B. Active residues are defined as residues that clearly participate in binding, whereas passive residues are located in the surrounding of active residues and might participate in binding.

During the docking process, the effective distance $d^{\mathrm{eff}}_{i\mathrm{AB}}$ that corresponds to one ambiguous interaction restraint is calculated from the intermediary complex structures according to

$$d^{\mathrm{eff}}_{i\mathrm{AB}} = \left(\sum_{m_{i\mathrm{A}}=1}^{N_{\mathrm{atoms}}} \sum_{k=1}^{N_{\mathrm{resB}}} \sum_{n_{k\mathrm{B}}=1}^{N_{\mathrm{atoms}}} \frac{1}{d^{6}_{m_{i\mathrm{A}} n_{k\mathrm{B}}}} \right)^{-1/6}, \tag{7.21}$$

where N_{atoms} indicates all atoms of a selected residue. An ambiguous interaction restraint is fulfilled if the corresponding effective distance is smaller than its

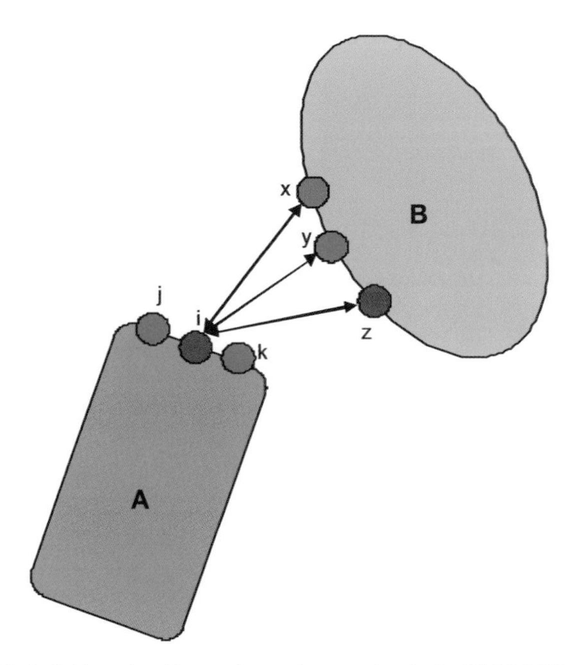

Figure 7.6. Definition of ambiguous interaction restraints in the HADDOCK procedure.

maximum value of 0.2–0.3 nm. Ambiguous interaction restraints are incorporated as an additional energy term of the global energy that is minimized during sampling. HADDOCK uses a three-stage docking procedure, in which increasing amounts of flexibility are introduced. In the most extreme application, fully flexible models can be defined, for example, for the docking of flexible peptides.

Figure 7.7 shows one example of the application of HADDOCK to calculate a protein–protein complex from NMR chemical shift perturbation data. As input for the docking process, the NMR structure of AF6 and the crystal structure of Rap1A were used. The interacting residues were defined from the residues which shift in the spectrum of ^{15}N-enriched AF6 after addition of Rap1A · Mg^{2+} · GppNHp used.

To date, mostly direct experimental data such as chemical shift perturbations are used to define interface residues. However, recently new computational methods have been developed based on the use of sequence conservation. For example, phylogenetic trees can be employed for this purpose (Armon *et al.*, 2001).

Future developments are necessary to treat complexes that are themselves dynamic, this might be especially important for weak and transient interactions. So far, relatively few methods such as CombDock (Inbar *et al.*, 2003) exist that treat the docking of multisubunit complexes, therefore, some future developments such as the extension of already existing programs can be expected in this area.

In conclusion, one can say that docking methods provide reliable structural information when combined with a sufficient amount of error free experimental data.

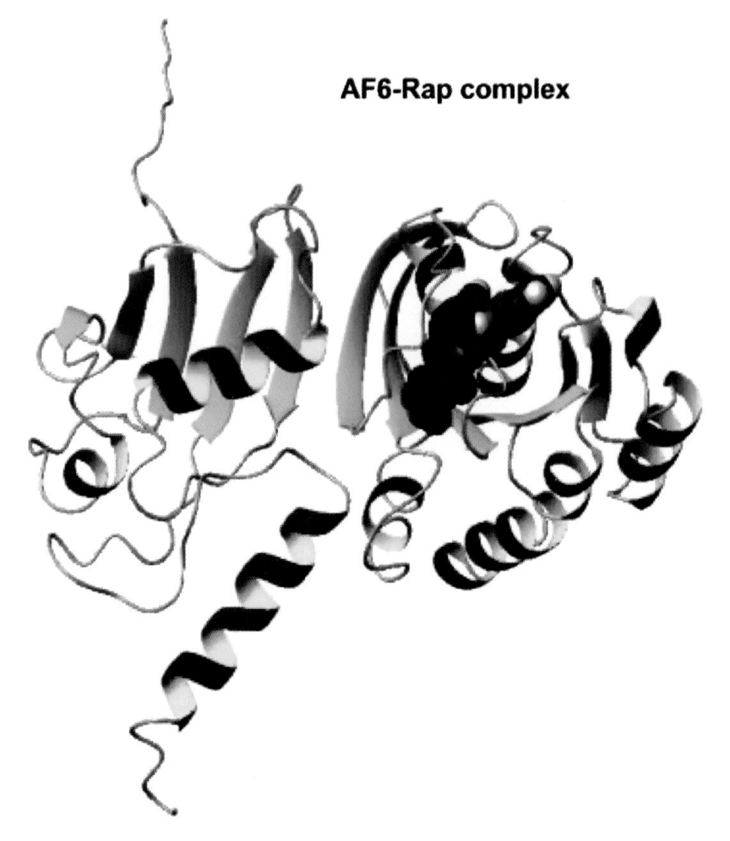

Figure 7.7. AF6 (*left*) in complex with Rap1A (*right*) obtained by using HADDOCK.

The main advantage is the drastically reduced amount of time required compared to full experimental complex structure determination. In addition, experimental data such as chemical shift perturbations can also be obtained for weak and transient complexes that are difficult to solve by standard NMR or X-ray techniques. Possible applications of models obtained by docking include the selection of residues targeted for mutagenesis or when flexibility is allowed expected structural changes upon complex formation can be modeled.

REFERENCES

Abragam, A. (1961). *Principles of Nuclear Magnetism.* Oxford University Press, London.

Adams, P. D., Engelman, D. M., and Brünger A. T. (1996). Improved prediction for the structure of the dimeric transmembrane domain of glycophorin A obtained through global searching. *Proteins* 26:257–261.

Akasaka, K. (2003a). Exploring the entire conformational space of proteins by high-pressure NMR. *Pure Appl Chem* 75:927–936.

Akasaka, K. (2003b). Highly fluctuating protein structures revealed by variable-pressure nuclear magnetic resonance. *Biochemistry* 42:10875–10885.

Akasaka, K. (2006). Probing conformational fluctuation of proteins by pressure perturbation. *Chem Rev* 106:1814–1835.

Akasaka, K. and Li, H. (2001). Low-lying excited states of proteins revealed from nonlinear pressure shifts in ^{1}H and ^{15}N NMR. *Biochemistry* 40:8665–8671.

Albrecht, M., Hanisch, D., Zimmer, R., and Lengauer, T. (2002). Improving fold recognition of protein threading by experimental distance constraints. *In Silico Biol* 2:1–12.

Altieri, A. S. and Byrd, R. A. (2004). Automation of NMR structures determination of proteins. *Curr Opin Struct Biol* 14:547–553.

Anand, G. S., Law, D., Mandell, J. G., Snead, A. N., Tsigelny, I., Taylor, S. S., Ten Eyck, L. F., and Komives, E. A. (2003). Identification of the protein kinase A regulatory $R'\alpha$-catalytic subunit interface by amide $H/^{2}H$ exchange and protein docking. *Proc Natl Acad Sci USA* 100:13264–13269.

Andrec, M., Harano, Y., Jacobson, M. P., Friesner, R. A., and Levy, R. M. (2002). Complete protein structure determination using backbone residual dipolar couplings and sidechain rotamer prediction. *J Struct Funct Genomics* 2:103–111.

Annila, A., Aito, H., Thulin, E., and Drakenberg, T. (1999). Recognition of protein folds via dipolar couplings. *J Biomol NMR* 14:223–230.

Armon, A., Graur, D., and Ben-Tal, N. (2001). ConSurf: an algorithmic tool for the identification of functional regions in proteins by surface mapping of phylogenetic information. *J Mol Biol* 307: 447–463.

Arora, A., Abildgaard, F., Bushweller, J. H., and Tamm, L. K. (2001). Structure of the outer membrane protein A transmembrane domain by NMR spectroscopy. *Nat Struct Biol* 8:334–338.

Aue, W. P., Bartholdi, E., and Ernst, R. R. (1976). Two-dimensional spectroscopy application to nuclear magnetic resonance. *J Chem Phys* 64:2229–2246.

Baran, M. C., Huang, Y. J., Moseley, H. N. B., and Montelione, G. T. (2004). Automated analysis of protein NMR assignments and structures. *Chem Rev* 104:3542–3555.

Bax, A. (2003). Weak alignment offers new NMR opportunities to study protein structure and dynamics. *Prot Sci* 12:1–16.

Bax, A., Clore, G. M., and Gronenborn, A. M. (1990b). ^{1}H–^{1}H correlation via isotropic mixing of ^{13}C magnetization, a new three-dimensional approach for assigning ^{1}H and ^{13}C spectra of ^{13}C-enriched proteins. *J Magn Reson* 88:425–431.

Bax, A. and Ikura, M. (1991). An efficient 3D NMR technique for correlating the proton and 15N backbone amide resonances with the alpha-carbon of the preceding residue in uniformly 15N/13C enriched proteins. *J Biomol NMR* 1:99–104.

Bax, A., Ikura, M., Kay, L. E., Torchia, D. A., and Tschudin, R. (1990a). Comparison of different modes of two-dimensional reverse-correlation NMR for the study of proteins. *J Magn Reson* 86:304–318.

Ben-Zeev, E. and Eisenstein, M. (2003). Weighted geometric docking: incorporating external information in the rotation–translation scan. *Proteins* 52:24–27.

Bermel, W., Bertini, I., Duma, L., Felli, I. C., Emsley, L., Pierattelli, R., and Vasos, P. R. (2005). Complete assignment of heteronuclear protein resonances by protonless NMR spectroscopy. *Angew Chem Int Ed* 117:3149–3152.

Bertini, I., Felli, I. C., Kümmerle, R., Moskau, D., and Pierattelli, R. (2004). ^{13}C–^{13}C NOESY: an attractive alternative for studying large macromolecules. *J Am Chem Soc* 126:464–465.

Bodenhausen, G. and Ruben, D. J. (1980). Natural abundance nitrogen-15 NMR by enhanced heteronuclear spectroscopy. *Chem Phys Lett* 69:185–189.

Boelens, R., Koning, T. M. G., van der Marel, G. A., van Boom, J. H., and Kaptein, R. (1989). Iterative procedure for structure determination from proton–proton NOEs using a full relaxation matrix approach. Application to a DNA octamer. *J Magn Reson* 82:290–308.

Bollard, M. E., Stanley, E. G., Lindon, J. C., Nicholson, J. K., and Holmes E. (2005). NMR-based metabonomic approaches for evaluating physiological influences on biofluid composition. *NMR Biomed* 18:143–162.

Bonvin, A. M. J. J., Boelens, R., and Kaptein, R. (2005). NMR analysis of protein interactions. *Curr Opin Chem Biol* 9:501–508.
Bowers, P. M., Strauss, C. E. M., and Baker, D. (2000). De novo protein structure determination using sparse NMR data. *J Biomol NMR* 18:311–318.
Braunschweiler, L. and Ernst R. R. (1983). Coherence transfer by isotropic mixing: application to proton correlation spectroscopy. *J Magn Reson* 53:521–528.
Brooijmans, N. and Kuntz, I. D. (2003). Molecular recognition and docking algorithms. *Annu Rev Biophys Biomol Struct* 32:335–373.
Brunner, E. (2001). Residual dipolar couplings in protein NMR. *Concepts Magn Reson* 13:238–259.
Brunner, K., Gronwald, W., Trenner, J. M., Neidig, K.-P., and Kalbitzer, H. R. (2006). A general method for the properly biased improvement of solution NMR structures by the use of related X-ray data, the AUREMOL-ISIC algorithm. *BMC-Struct Biol* 6:14.
Brüschweiler, R. and Ernst, R. R. (1992). Molecular dynamics monitored by cross-correlated cross relaxation of spins quantized along orthogonal axes. *J Chem Phys* 96:1758–1766.
Brüschweiler, R., Liao, X., and Wright, P. E. (1995). Long-range motional restrictions in a multidomain zinc-finger protein from anisotropic tumbling. *Science* 268:886–889.
Carlomagno, T. (2005). Ligand–target interactions: what can we learn from NMR? *Annu Rev Biophys Biomol Struct* 34:245–266.
Carulla, N., Caddy, G. L., Hall, D. R., Zurdo, J., Gairi, M., Feliz, M., Giralt, E., Robinson, C. V., and Dobson, C. M. (2005). Molecular recycling within amyloid fibrils. *Nature* 436:554–558.
Cavanagh, J., Fairbrother, W. J., Palmer, A. G., and Skelton, N. J. (1996). *Protein NMR Spectroscopy*. Academic press, San Diego.
Chen, A. D. and Shapiro, M. J. (2000). NOE pumping. 2. A high-throughput method to determine compounds with binding affinity to macromolecules by NMR. *J Am Chem Soc* 122:414–415.
Chen, Y., Reizer, J., Saier, M. H., Jr., Fairbrother, W. J., and Wright, P. E. (1993). Mapping of the binding interfaces of the proteins of the bacterial phosphotransferase system, HPr and IIAglc. *Biochemistry* 32:32–39.
Chou, J. J., Kaufman, J. D., Stahl, S. J., Wingfield, P. T., and Bax, A. (2002). Micelle-induced curvature in a water-insoluble HIV-1 Env peptide revealed by NMR dipolar coupling measurement in stretched polyacrylamide gel. *J Am Chem Soc* 124:2450–2451.
Clarkson, J. and Campbell, I. D. (2003). Studies of protein–ligand interactions by NMR. *Biochem Soc Trans* 31:1006–1009.
Clore, G. M. (2000). Accurate and rapid docking of protein–protein complexes on the basis of intermolecular nuclear Overhauser enhancement data and dipolar couplings by rigid body minimization. *Proc Natl Acad Sci USA* 97:9021–9025.
Cordeiro, Y., Kraineva, J., Ravindra, R., Lima, L. M., Gomes, M. P. B., Foguel, D., Winter, R., and Silva, J. L. (2004). Hydration and packing effects on prion folding and {beta}-sheet conversion: high pressure spectroscopy and pressure perturbation calorimetry studies. *J Biol Chem* 279:32354–32359.
Creighton, T. E. (1997). *Protein Structure, A practical Approach*, second edition; *Protein Function A practical Approach*, second edition; Two volume set ISBN 01996362006 (Pbk).
Dalvit, C. (1992). ^{1}H to ^{15}N polarization transfer via ^{1}H CSA-^{1}H-^{15}N dipole dipole cross correlation. *J Magn Reson* 97:645–650.
Dalvit, C., Fogliatto, G., Stewart, A., Veronesi, M., and Stockman, B. (2001). WaterLOGSY as a method for primary NMR screening: practical aspects and range of applicability. *J Biomol NMR* 21:349–359.
Delagio, F., Kontaxis, G., and Bax, A. (2000). Protein structure determination using molecular fragment replacement and NMR dipolar couplings. *J Am Chem Soc* 122:2142–2143.
Dobrodumov, A. and Gronenborn, A. (2003). Filtering and selection of structural models: combining docking and NMR. *Proteins* 53:18–32.
Dominguez, C., Boelens, R., and Bonvin, A. M. J. J. (2003). HADDOCK: a protein–protein docking approach based on biochemical or biophysical information. *J Am Chem Soc* 125:1731–1737.
Dosset, P., Hus, J. C., Marion, D., and Blackledge, M. (2001). A novel interactive tool for rigid-body modeling of multi-domain macromolecules using residual dipolar couplings. *J Biomol NMR* 20:223–231.

Ernst, R. R., Bodenhausen, G., and Wokaun, A. (1987). *Principles of Nuclear Magnetic Resonance in One and Two Dimensions*. Oxford University Press, London.

Fahmy, A. and Wagner, G. (2002). TreeDock: a tool for protein docking based on minimizing van der Waals energies. *J Am Chem Soc* 124:1241–1250.

Farmer, B. T., II, Venters L. D., Spicer L. D., Wittekind M. G., and Müller, L. (1992). A refocused and optimised HNCA: increased sensitivity and resolution in large macromolecules. *J Biomol NMR* 2:195–202.

Farrow, N. A., Muhandiram, R., Singer, A. U., Pascal, S. M., Kay, C. M, Gish, G., Shoelson, S. E., Pawson, T., Forman-Kay, J. D., and Kay, L. E. (1994). Backbone dynamics of a free and phosphopeptide-complexed Src homology 2 domain studied by 15N NMR relaxation. *Biochemistry* 33:5984–6003.

Fernández, C., Adeishvili, K., and Wüthrich, K. (2001b). Transverse relaxation-optimized NMR spectroscopy with the outer membrane protein OmpX in dihexanoyl phosphatidylcholine micelles. *Proc Natl Acad Sci USA* 98:2358–2563.

Fernández, C., Hilty, C., Bonjour, S., Adeishvili, K., Pervushin, K., and Wüthrich, K. (2001a). Solution NMR studies of the integral membrane proteins OmpX and OmpA from *Escherichia coli*. *FEBS Lett* 504:173–178.

Fernández, C., Hilty, C., Wider, G., Güntert, P., and Wüthrich, K. (2004). NMR structure of the integral membrane protein OmpX. *J Mol Biol* 336:1211–1221.

Fernández, C., Hilty, C., Wider, G., and Wüthrich, K. (2002). Lipid–protein interactions in DHPC micelles containing the integral membrane protein OmpX investigated by NMR. *Proc Natl Acad Sci USA* 99:13533–13537.

Fernández, C. and Wider, G. (2003). TROSY in NMR studies of the structure and function of large biological macromolecules. *Curr Opin Struct Biol* 13:570–580.

Fiaux, J., Bertelsen, E. B., Horwich, A. L., and Wüthrich, K. (2002). NMR analysis of a 900K GroEL GroES complex. *Nature* 418:207–211.

Finerty, P. J., Jr., Muhandiram, D. R., and Forman-Kay, J. D. (2002). Side chain dynamics of the SAP SH2 domain correlate with a binding hot spot and a region with conformational plasticity. *J Mol Biol* 322:605–620.

Fischer, M. W. F., Losonczi, J. A., Weaver, J. L., and Prestegard, J. H. (1999). Domain orientation and dynamics in multidomain proteins from residual dipolar couplings. *Biochemistry* 38:9013–9022.

Frickel, E. M., Riek, R., Jelesarov, I., Helenius, A., Wüthrich, K., and Ellgaard, L. (2002). TROSY-NMR reveals interaction between ERp57 and the tip of the calreticulin P-domain. *Proc Natl Acad Sci USA* 99:1954–1959.

Fushman, D., Tjandra, N., and Cowburn, D. (1999b). An approach to direct determination of protein dynamics from 15N NMR relaxation at multiple fields, independent of variable 15N chemical shift anisotropy and chemical exchange contributions. *J Am Chem Soc* 121:8577–8582.

Fushman, D., Varadan, R., Assfalg, M., and Walker, O. (2004). Determining domain orientation in macromolecules by using spin-relaxation and residual dipolar coupling measurements. *Prog NMR Spectrosc* 44:189–214.

Fushman, D., Xu, R., and Cowburn, D. (1999a). Direct determination of changes of interdomain orientation on ligation: use of the orientational dependence of 15N NMR relaxation in Abl SH(32). *Biochemistry* 38:10225–10230.

Gagne, S. M., Tsuda, S., Spyracopoulos, L., Kay, L. E., and Sykes, B. D. (1998). Backbone and methyl dynamics of the regulatory domain of troponin C: anisotropic rotational diffusion and contribution of conformational entropy to calcium affinity. *J Mol Biol* 278:667–686.

Gao, G., Williams, J. G., and Campell, S. L. (2004). Protein–protein interaction analysis by nuclear magnetic resonance spectroscopy. *Methods Mol Biol* 261:79–92.

Gardner, K. H. and Kay, L. E. (1997). Production and incorporation of ^{15}N, ^{13}C, ^{2}H (^{1}H-1 methyl) isoleucine into proteins for multidimensional NMR studies. *J Am Chem Soc* 119:7599–7560.

Gardner, K. H. and Kay, L. E. (1998). The use of ^{2}H, ^{13}C, ^{15}N multidimensional NMR to study the structure and dynamics of proteins. *Ann Rev Biophys Biomol Struct* 27:357–406.

Garrett, D. S., Seok, Y. J., Peterkofsky, A., Clore, G. M., and Gronenborn, A. M. (1997). Identification by NMR of the binding surface for the histidine-containing phosphocarrier protein HPr on the N-terminal domain of enzyme I of the *Escherichia coli* phosphotransferase system. *Biochemistry* 36:4393–4398.

Geyer, M., Herrmann, C., Wohlgemuth, S., Wittinghofer, A., and Kalbitzer, H. R. (1997). Structure of the Ras-binding domain of RalGEF and implications for Ras binding and signalling. *Nat Struct Biol* 4:694–699.

Giesen, A. W., Homans, S. W., and Brown, J. M. (2003). Determination of protein global folds using backbone residual dipolar coupling and long-range NOE restraints. *J Biomol NMR* 25:63–71.

Görler, A., Gronwald, W., Neidig, K. P., and Kalbitzer, H. R. (1999). Computer assisted assignment of 13C or 15N edited 3D-NOESY-HSQC spectra using back calculated and experimental spectra. *J Magn Reson* 137:39–45.

Goto, N. K., Gardner, K. H., Mueller, G. A., Willis, R. C., and Kay, L. E. (1999). A robust and cost-effective method for the production of Val, Leu, Ile (δ1)methyl-protonated ^{15}N-, ^{13}C-, ^{2}H-labeled proteins. *J Biomol NMR* 13:369–374.

Gronwald, W., Huber, F., Grünewald, P., Spörner, M., Wohlgemuth, S., Herrmann, C., Wittinghofer, A., and Kalbitzer, H. R. (2001). Solution structure of the Ras-binding domain of the protein kinase Byr2 from *Schizosaccharomyces pombe*. *Structure* 9:1029–1041.

Gronwald, W., Moussa, S., Elsner, R., Jung, A., Ganslmeier, B., Trenner, J., Kremer, W., Neidig, K. P., and Kalbitzer, H. R. (2002). Automated assignment of NOESY NMR spectra using a knowledge based method (KNOWNOE). *J Biomol NMR* 23:271–287.

Gronwald, W. and Kalbitzer, H. R. (2004). Automated structure determination of proteins by NMR spectroscopy. *Prog NMR Spectrosc* 44:33–96.

Grzesiek, S. and Bax, A. (1992a). Improved 3D triple-resonance NMR techniques applied to a 31 kDa protein. *J Magn Reson* 96:432–440.

Grzesiek, S. and Bax, A. (1992b). An efficient experiment for sequential backbone assignment of medium-sized isotopically enriched proteins. *J Magn Reson* 99:201–207.

Grzesiek, S. and Bax, A. (1992c). Correlating backbone amide and side chain resonances in larger proteins by multiple relayed triple resonance NMR. *J Am Chem Soc* 114:6291–6293.

Grzesiek, S. and Bax, A. (1993). Amino acid type determination in the sequential assignment procedure of uniformly ^{13}C/^{15}N-enriched proteins. *J Biomol NMR* 3:185–204.

Grzesiek, S., Stahl, S. J., Wingfield, P. T., and Bax, A. (1996). The CD4 determinant for downregulation by HIV-1 Nef directly binds to Nef. Mapping of the Nef binding surface by NMR. *Biochemistry* 35:10256–10261.

Grzesiek, S., Wingfield, P., Stahl, S., Kaufman, J., and Bax, A. (1995). Four-dimensional 15N-separated NOESY of slowly tumbling perdeuterated 15N-enriched proteins. Application to HIV-1 Nef. *J Am Chem Soc* 117:9594–9595.

Güntert, P. (2004). Automated NMR structure calculation with CYANA. *Methods Mol Biol* 278:353–378.

Halperin, I., Ma, B., Wolfson, H., and Nussinov, R. (2002). Principles of docking: an overview of search algorithms and a guide to scoring functions. *Proteins* 47:409–443.

Hansen, M. R., Mueller, L., and Pardi, A. (1998). Tunable alignment of macromolecules by filamentous phage yields dipolar coupling interactions. *Nat Struct Biol* 5:1065–1074.

Hausser, K. H. and Kalbitzer H. R. (1991). *NMR in Medicine and Biology*. Springer-Verlag, New York, ISBN 0-387-53195-5.

Horst, R., Bertelsen, E. B., Fiaux, J., Wider, G., Horwich, A. L., and Wüthrich, K. (2005). Direct NMR observation of a substrate protein bound to the chaperonin GroEL. *Proc Natl Acad Sci USA* 102:12748–12753.

Hubbell, W. L., Mchaourab, H. S., Altenbach, C., and Lietzow, M. A. (1996). Watching proteins move using site-directed spin labeling. *Structure* 4:779–783.

Hwang, P. M., Bishop, R. E., and Kay, L. E. (2004). The integral membrane enzyme PagP alternates between two dynamically distinct states. *Proc Natl Acad Sci USA* 101:9618–9623.

Hwang, P. M., Choy, W., Lo, E. I., Chen, L., Forman-Kay, J. D., Raetz, C. R. H., Prive, G. G., Bishop, R. E., and Kay, L. E. (2002). Solution structure and dynamics of the outer membrane enzyme PagP by NMR. *Proc Natl Acad Sci USA* 99:13560–13565.

Inbar, Y., Benyamini, H., Nussinov, R., and Wolfson, H. (2003). Protein structure prediction via combinatorial assembly of sub-structural units. *Bioinformatics* 19:i158–i168.

Inoue, K., Yamada, H., Akasaka, K., Herrmann, C., Kremer, W., Maurer, T., Döker, R., and Kalbitzer, H. R. (2000). Pressure-induced local unfolding of the Ras binding domain of RalGDS. *Nat Struct Biol* 7:547–550.

Jain, N. U., Venot, A., Umemoto, K., Leffler, H., and Prestegard, J. H. (2001). Distance mapping of protein-binding sites using spin-labeled oligosaccharide ligands. *Protein Sci* 10:2393–2400.

Janin, J. (2005). The targets of CAPRI rounds 3–5. *Proteins* 60:170–175.

Jeener, J. (1971). *Lecture Ampère Summer School.* Baske Polje, Yugoslavia.

Jeener, J., Meier, B. H., Bachmann, P., and Ernst R. R. (1979). Investigation of exchange processes by two-dimensional NMR spectroscopy. *J Chem Phys* 71:4546–4553.

Jung, Y.-S., Sharma, M., and Zweckstetter, M. (2004). Simultaneous assignment and structure determination of protein backbones by using NMR dipolar couplings. *Angew Chem Int Ed* 43:3479–3481.

Kachel, N., Kremer, W., Zahn, R., and Kalbitzer, H. R. (2006). Observation of intermediate states of the human prion protein by high pressure NMR spectroscopy. *BMC Struct Biol* 16:6.

Kainosho, M. (1997). Isotope labelling of macromolecules for structural determinations. *Nat Struct Biol* 4:858–861.

Kalbitzer, H. R., Görler, A., Li, H., Dubouskii, P., Hengstenberg, W., Kowolik, C., Yamada, H., and Akasaka, K. (2000). ^{15}N and ^{1}H NMR study of histidine containing, protein (Hpr) from staphylococcus carnosus at high pressure. *Prot Sci* 9:673–703.

Kamatari, Y. O., Kitahara, R., Yamada, H., Yokoyama, S., and Akasaka, K. (2004). High-pressure NMR spectroscopy for characterizing folding intermediates and denatured states of proteins. *Methods* 34:133–143.

Katchalski-Katzir, E., Shariv, I., Eisenstein, M., Frisem, A. A., Flalo, C., and Vasker, I. A. (1992). Molecular surface recognition: determination of geometric fit between proteins and their ligands by correlation techniques. *Proc Natl Acad Sci USA* 89:2195–2199.

Karplus, M. (1959). Contact electron spin coupling of nuclear moments. *J Chem Phys* 30:11.

Kay, L. E. (2001). Nuclear magnetic resonance methods for high molecular weight proteins: a study involving a complex of maltose binding protein and beta-cyclodextrin. *Meth Enzymol* 339:174–203.

Kay, L. E. (2005). NMR studies of protein structure and dynamics. *J Magn Reson* 173:193–207.

Kay, L. E., Ikura, M., Tschudin, R., and Bax, A. (1990). Three-dimensional triple-resonance NMR spectroscopy of isotopically enriched proteins. *J Magn Reson* 89:496–514.

Kay, L. E., Muhandiram, D. R., Wolf, G., Shoelson, S. E., and Forman-Kay, J. D. (1998). Correlation between binding and dynamics at SH2 domain interfaces. *Nat Struct Biol* 5:156–163.

Keepers, J. W. and James, T. L. (1984). A theoretical study of distance determination from NMR. Two-dimensional nuclear Overhauser effect spectra. *J Magn Reson* 57:404–426.

Koharudin, L. M, Bonvin, A. M. J. J., Kaptein, R., and Boelens, R. (2003). Use of very long-distance NOEs in a fully deuterated protein: an approach for rapid protein fold determination. *J Magn Reson* 163:228–235.

Konrat, R., Yang, D., and Kay, L. E. (1999). A 4D TROSY-based pulse scheme for correlating $^{1}HN_{i}$, $^{15}N_{i}$, $^{13}C_{\alpha i}$, $^{13}C'_{i-1}$ chemical shifts in high molecular weight, ^{15}N, ^{13}C, ^{2}H labeled proteins. *J Biomol NMR* 15:309–313.

Lee, A. L., Kinnear, S. A., and Wand, A. J. (2000). Redistribution and loss of side chain entropy upon formation of a calmodulinpeptide complex. *Nat Struct Biol* 7:72–77.

Li, W., Zhang, Y., Kihara, D., Huang, Y. J., Zheng, D., Montelione, G., Kolinski, A., and Skolnick, J. (2003). TOUCHSTONEX: protein structure prediction with sparse NMR data. *Proteins* 53: 290–306.

Lian, L. Y. and Middleton, D. A. (2001). Labelling approaches for protein structural studies by solution-state and solid-state NMR. *Prog NMR Spec* 39:171–190.

Lindon, J. C., Keun, H. C., Ebbels, T. M. D., Pearce, J. M. T., Holmes, E., and Nicholson, J. K. (2005). The consortium for metabonomic toxicology (COMET): aims, activities and achievements. *Pharmacogenomics* 6:691–699.

Linge, J. P., Habeck, M., Rieping, W., and Nilges, M. (2003b). ARIA: automated NOE assignment and NMR structure calculation. *Bioinformatics* 19:315–316.

Linge, J. P., Williams, M. A., Spronk, C. A. E. M., Bonvin, A. M. J. J., and Nilges, M. (2003a). Refinement of protein structures in explicit solvent. *Proteins* 50:496–506.

Lipari, G. and Szabo, A. (1982a). Model-free approach to the interpretation of nuclear magnetic resonance relaxation in macromolecules. 1. Theory and range of validity. *J Am Chem Soc* 104:4546–4559.

Lipari, G. and Szabo, A. (1982b). Model-free approach to the interpretation of nuclear magnetic resonance relaxation in macromolecules. 2. Analysis of experimental results. *J Am Chem Soc* 104:4559–4570.

Lipsitz, R. S. and Tjandra, N. (2004). Residual dipolar couplings in NMR structure analysis. *Annu Rev Biophys Biomol Struct* 33:387–413.

Loh, A. P., Pawley, N., Nicholson, L. K., and Oswald, R. E. (2001). An increase in side chain entropy facilitates effector binding: NMR characterization of the side chain methyl group dynamics in Cdc42Hs. *Biochemistry* 40:4590–4600.

Luca, S., Heise, H., and Baldus, M. (2003). High-resolution solid-state NMR applied to polypeptides and membrane proteins. *Accounts Chem Res* 36:858–865.

Mayer, M. and Meyer, B. (2001). Group epitope mapping by saturation difference NMR to identify segments of a ligand in direct contact with protein receptor. *J Am Chem Soc* 123:6108–6117.

McDermott, A. E. (2004). Structural and dynamic studies of proteins by solid-state NMR spectroscopy: rapid movement forward. *Curr Opin Struct Biol* 14:554–561.

McKay, R. T., Pearlstone, J. R., Corson, D. C., Gagne, S. M., Smillie, L. B., and Sykes, B. D. (1998). Structure and interaction site of the regulatory domain of troponin-C when complexed with the 96–148 region of troponin-I. *Biochemistry* 37:12419–12430.

McKay, R. T., Tripet, B. P., Hodges, R. S., and Sykes, B. D. (1997). Interaction of the second binding region of troponin I with the regulatory domain of skeletal muscle troponin C as determined by NMR spectroscopy. *J Biol Chem* 272:28494–28500.

Meier, S., Guthe, S., Kiefhaber, T., and Grzesiek, S. (2004). Foldon, the natural trimerization domain of T4 fibritin, dissociates into a monomeric A-state form containing a stable beta-hairpin: atomic details of trimer dissociation and local beta-hairpin stability from residual dipolar couplings. *J Mol Biol* 344:1051–1069.

Meng, E. C., Gschwend, D. A., Blaney, J. M., and Kuntz, I. D. (1993). Orientational sampling and rigid-body minimization in molecular docking. *Proteins* 17:266–278.

Mettu, R. R., Lilien, R. H., and Donald, B. R. (2005). High-throughput inference of protein–protein interfaces from unassigned NMR data. *Bioinformatics* 21(Suppl 1):i292–i301.

Meyer, B. and Peters, T. (2003). NMR spectroscopy techniques for screening and identifying ligand binding to protein receptors. *Angew Chem Int Ed* 42:864–890.

Millet, O., Hudson, R. P., and Kay, L. E. (2003). The energetic cost of domain orientation in maltose-binding protein as studied by NMR and fluorescence spectroscopy. *Proc Natl Acad Sci USA* 100:12700–12705.

Mueller, G. A., Choy, W. Y., Yang, D., Forman-Kay, J. D., Venters, R. A., and Kay, L. E. (2000). Global folds of proteins with low densities of NOEs using residual dipolar couplings: application to the 370-residue maltodextrin-binding protein. *J Mol Biol* 300:197–212.

Muhandiram, D. R. and Kay, L. E. (1994). Gradient-enhanced triple-resonance three-dimensional NMR experiments with improved sensitivity. *J Magn Reson* Series B 103:203–216.

Müller, L. (1979). Sensitivity enhanced detection of weak nuclei using heteronuclear multiple quantum coherence. *J Am Chem Soc* 101:4481–4484.

Nabuurs, S. B., Nederveen, A. J., Vranken, W., Doreleijers, J. F., Bonvin, A. M. J. J., Vuister, G. W., Vriend, G., and Spronk, C. A. E. M. (2004). DRESS: a Database of REfined Solution NMR Structures. *Proteins* 55:483–486.

Nakanishi, T., Miyazawa, M., Sakakura, M., Terasawa, H., Takahashi, H., and Shimada, I. (2002). Determination of the interface of a large protein complex by transferred cross-saturation measurements. *J Mol Biol* 318:245–249.

Neal, S., Nip, A. M., Haiyan, Z., and Wishart, D. S. (2003). Rapid and accurate calculation of protein ^{1}H, ^{13}C and ^{15}N chemical shifts. *J Biomol NMR* 26:215–240.

Nicholson, L. K., Kay, L. E., Baldisseri, D. M., Arango, J., Young, P. E., Bax, A., and Torchia, D. A. (1992). Dynamics of methyl groups in proteins as studied by proton-detected ^{13}C NMR spectroscopy. Application to the leucine residues of staphylococcal nuclease. *Biochemistry* 31:5253–5263.

Norwood, T. J., Boyd, J., Heritage, J. E., Soffe, N., and Campbell, I. D. (1990). Comparison of techniques for ^{1}H-detected heteronuclear ^{1}H–^{15}N spectroscopy. *J Magn Reson* 87:488–501.

Olejniczak, E. T., Xu, R. X., and Fesik, S. W. (1992). A 4D HCCH-TOCSY experiment for assigning the side chain ^{1}H and ^{13}C resonances of proteins. *J Biomol NMR* 2:655–659.

Palma, P. N., Krippahl, L., Wampler, J. E., and Moura, J. J. G. (2000). BiGGER: a new (soft) docking algorithm for predicting protein interactions. *Proteins* 39:372–384.

Palmer, A. G., III. (2001). NMR probes of molecular dynamics: overview and comparison with other techniques. *Annu Rev Biophys Biomol Struct* 30:129–155.

Palmer, A. G., III. (2004). NMR characterization of the dynamics of biomacromolecules. *Chem Rev* 104:3623–3640.

Pardi, A., Billeter, M., and Wüthrich, K. (1984). Calibration of the angular-dependence of the amide proton-C-alpha proton coupling constants, 3JHN-H-alpha, in a globular protein—use of 3JHN-H-alpha for identification of helical secondary structure. *J Mol Biol* 180:741–751.

Pellecchia, M., Stevens, S. Y., Vander Kooi, C. W., Montgomery, D. H., Feng, E. H., Gierasch, L. M., and Zuiderweg, E. R. P. (2000). Structural insights into substrate binding by the molecular chaperone DnaK. *Nat Struct Biol* 7:298–303.

Pervushin, K., Riek, R., Wider, G., and Wüthrich, K. (1997). Attenuated T2 relaxation by mutual cancellation of dipole–dipole coupling and chemical shift anisotropy indicates an avenue to NMR structures of very large biological macromolecules in solution. *Proc Natl Acad Sci USA* 94:12366–12371.

Prestegard, J. H., Bougault, C. M., and Kishore, A. I. (2004). Residual dipolar couplings in structure determination of biomolecules. *Chem Rev* 104:3519–3540.

Riek, R., Wider, G., Pervushin, K., and Wüthrich, K. (1999). Polarization transfer by cross-correlated relaxation in solution NMR with very large molecules. *Proc Natl Acad Sci USA* 96:4918–4923.

Robertson, D. G., Lindon, J. C., and Nicholson, J. K. (2005). *Metabonomics in Toxicity Assessment.* Marcel Dekker Inc., ISBN 0824726650.

Roisman, L. C., Pieler, J, Trosset, J. Y., Scheraga, H. A., and Schreiber, G. (2001). Structure of the interferon–receptor complex determined by distance constraints from double-mutant cycles and flexible docking. *Proc Natl Acad Sci USA* 98:13231–13236.

Sachchidanand, L. O., Staunton D., Mulloy B., Forster M. J., Yoshida K., and Campbell I. D. (2002). Mapping the heparin-binding site on the $^{13-14}$F3 fragment of fibronectin. *J Biol Chem* 277:50629–50635.

Salzmann, M., Pervushin, K., Wider, G., Senn, H., and Wüthrich, K. (2000). NMR assignment and secondary structure determination of an octameric 110 kDa protein using TROSY in triple resonance experiments. *J Am Chem Soc* 122:7543–7548.

Schneidman-Duhovny, D., Inbar, Y., Polak, V., Shatsky, M., Halperin, I., Benyamini, H., Barzilai A., Dror, O., Haspel, N., Nussinov, R., and Wolfson, H. (2003). Taking geometry to its edge: fast unbound rigid (and Hinge-Bent) docking. *Proteins* 52:107–112.

Schwalbe, H. (2003). Kurt Wüthrich, the ETH Zürich, and the development of NMR spectroscopy for the investigation of structure, dynamics, and folding of proteins. *ChemBioChem* 4:135–142.

Shimada, I. (2005). NMR techniques for identifying the interface of a larger protein–protein complex: cross-saturation and transferred cross-saturation experiments. *Meth Enzym* 394:483–506.

Shuker, S. B., Hajduk, P. J., Meadows, R. P., and Fesik, S. W. (1996). Discovering high-affinity ligands for proteins: SAR by NMR. *Science* 274:1531–1534.

Silva, J. L., Foguel, D., and Royer, C. A. (2001). Pressure provides new insights into protein folding, dynamics and structure. *Trends Biochem Sci* 26:612–618.

Simons, K. T., Kooperberg, C., Huang, E., and Baker, D. (1997). Assembly of protein tertiary structures from fragments with similar local sequences using simulated annealing and bayesian scoring functions. *J Mol Biol* 268:209–225.

Simons, K. T., Ruczinski, I., Kooperberg, C., Fox, B. A., Bystroff, C., and Baker, D. (1999). Improved recognition of native-like protein structures using a combination of sequence-dependent and sequence-independent features of proteins. *Proteins* 34:82–95.

Sprangers, R., Gibrun, A., Hwang, P. M., Houry, W. A., and Kay, L. E. (2005). Quantitative NMR spectroscopy of supramolecular complexes: dynamic side pores in ClpP are important for product release. *Proc Natl Acad Sci USA* 102:16678–16683.

Stamos, J., Eigenbrot, C., Nakamura, G. R., Reynolds, M. E., Yin, J., Lowman, H. B., Fairbrother, W. J., and Starovasnik, M. A. (2004). Convergent recognition of the IgE binding site on the high affinity IgE receptor. *Structure* 12:1289–1301.

Stevens, S. Y., Sanker, S., Kent, C., and Zuiderweg, E. R. P. (2001). Delineation of the allosteric mechanism of a cytidylyltransferase exhibiting negative cooperativity. *Nat Struct Biol* 8:947–952.

Stivers, J. T., Abeygunawardana, C., and Mildvan, A. S. (1996). ^{15}N NMR relaxation studies of free and inhibitor-bound 4-oxalocrotonate tautomerase: backbone dynamics and entropy changes of an enzyme upon inhibitor binding. *Biochemistry* 35:16036–16047.

Stone, M. J. (2001). NMR relaxation studies of the role of conformational entropy in protein stability and ligand binding. *Acc Chem Res* 34:379–388.

Takahashi, H., Nakanishi, T., Kami, K., Arata, Y., and Shimada, I. (2000). A novel NMR method for determining the interfaces of large protein–protein complexes. *Nat Struct Biol* 7:220–223.

Takeuchi, K., Yokogawa, M., Matsuda, T., Sugai, M., Kawano, S., Kohno, T., Nakamura, H., Takahashi, H., and Shimada, I. (2003). Structural basis of the KcsA K^+ channel and agitoxin2 pore-blocking toxin interaction by using the transferred cross-saturation method. *Structure (Camb)* 11:1381–1392.

Tengel, T., Fex, T., Emtenas, H., Almqvist, F., Sethson, I., and Kihlberg, J. (2004). Use of 19F NMR spectroscopy to screen chemical libraries for ligands that bind to proteins. *Org Biomol Chem* 2:725–731.

Tjandra, N. and Bax, A. (1997). Direct measurement of distances and angles in biomolecules by NMR in a dilute liquid crystalline medium. *Science* 278:1111–1114.

Torrent, J., Alvarez-Martinez, M. T., Heitz, F., Liautard, J. P., Balny, C., and Lange, R. (2003). Alternative prion structural changes revealed by high pressure. *Biochemistry* 42:1318–1325.

Torrent, J., Alvarez-Martinez, M. T., Harricane, M. C., Heitz, F., Liautard, J. P., Balny, C., and Lange, R. (2004). High pressure induces scrapie-like prion protein misfolding and amyloid fibril formation. *Biochemistry* 43:7162–7170.

Torrent, J., Alvarez-Martinez, M. T., Liautard, J. P., Balny, C., and Lange, R. (2005). The role of the 132–160 region in prion protein conformational transitions. *Protein Sci* 14:956–967.

Tugarinov, V., Choy, W. Y., Orekhov, V. Y., and Kay, L. E. (2005). Solution NMR-derived global fold of a monomeric 82-kDa enzyme. *Proc Natl Acad Sci USA* 102:622–627.

Tugarinov, V., Hwang, P. M., and Kay, L. E. (2004). Nuclear magnetic resonance spectroscopy of high-molecular weight proteins. *Ann Rev Biochem* 73:107–146.

Tugarinov, V. and Kay, L. E. (2003). Ile, Leu, and Val methyl assignments of the 723-residue malate synthase G using a new labelling strategy and novel NMR methods. *J Am Chem Soc* 125:13868–13878.

Vajda, S. and Camacho, C. J. (2004). Protein–protein docking: is the glass half-full or half-empty. *TRENDS Biotechnol* 22:110–116.

Van de Ven, F. J. M., Blommers, M. J. J., Schouten, R. E., and Hilbers, C. W. (1991). Calculation of interproton distances from NOE intensities. A relaxation matrix approach without requirement of a molecular model. *J Magn Reson* 94:140–151.

Van Dijk, A. D. J., Boelens, R., and Bonvin, A. M. J. J. (2004). Data-driven docking for the study of biomolecular complexes. *FEBS J* 272:293–312.

Venters, R. A., Metzler, W. J., Spicer, L. D., Mueller, L., and Farmer, B. T. (1995). Use of 1HN-1HN NOEs to determine protein global folds in perdeuterated proteins. *J Am Chem Soc* 117:9592–9593.

Walsh, S. T., Cheng, R. P., Wright, W. W., Alonso, D. O., Daggett, V., Vanderkooi, J. M., and DeGrado, W. F. (2003). The hydration of amides in helices; a comprehensive picture from molecular dynamics, IR, and NMR. *Protein Sci* 12:520–531.

Wang, H., Kurochkin, A. V., Pang, Y., Hu, W., Flynn, G. C., and Zuiderweg, E. R. P. (1998). NMR solution structure of the 21 kDa chaperone protein DnaK substrate binding domain: a preview of chaperone–protein interaction. *Biochemistry* 3:7929–7940.

Wang, C. and Palmer, A. G., III. (2003). Solution NMR methods for quantitative identification of chemical exchange in ^{15}N-labeled proteins. *Magn Reson Chem* 41:866–876.

Wemmer, D. E. (2003). The energetics of structural change in maltose-binding protein. *Proc Natl Acad Sci USA* 100:12529–12530.

Wemmer, D. E. and Williams, P. G. (1994). Use of nuclear magnetic resonance in probing ligand–macromolecule interactions. *Meth Enzymol* 239:739–767.

Wider, G. (2005). NMR techniques used with very large biological macromolecules in solution. *Meth Enzymol* 394:382–398.

Wider, G. and Wüthrich, K. (1999). NMR spectroscopy of large molecules and multimolecular assemblies in solution. *Curr Opin Struct Biol* 9:594–601.

Wittekind M. and Mueller L. (1993). HNCACB, a high-sensitivity 3D NMR experiment to correlate amide-proton and nitrogen resonances with the alpha- and beta-carbon resonances in proteins. *J Magn Reson* Series B 101:201–205.

Wodak, S. J. and Janin, J. (2003). Structural basis of macromolecular recognition. *Adv Protein Chem* 61:9–73.

Wüthrich K. (1986). *NMR of Proteins and Nucleic Acids*. Wiley Interscience, ISBN 0-471-82893-9.

Xu, R., Ayers, B., Cowburn, D., and Muir, T. W. (1999). Chemical ligation of folded recombinant proteins: segmental isotopic labeling of domains for NMR studies. *Proc Natl Acad Sci USA* 96:b388–393.

Xu, X. P. and Case, D. A. (2001). Automated prediction of 15N, 13Calpha, 13Cbeta and 13C' chemical shifts in proteins using a density functional database. *J Biomol NMR* 21:321–333.

Yamazaki, T., Lee, W., Arrowsmith, C. H., Muhandiram, D. R., and Kay, L. E. (1994). A suite of triple resonance NMR experiments for the backbone assignment of 15N, 13C, 2H labeled proteins with high sensitivity. *J Am Chem Soc* 116:11655–11666.

Yamazaki, T., Otomo, T., Oda, N., Kyogoku, Y., Uegaki, K., Ito, N., Ishino, Y., and Nakamura, H. (1998). Segmental isotope labelling for protein NMR using peptide splicing. *J Am Chem Soc* 120:5591–5592.

Yang, D. and Kay, L. E. (1999). TROSY triple-resonance four-dimensional NMR spectroscopy of a 46 ns tumbling protein. *J Am Chem Soc* 121:2571–2575.

Yu, H. T. (1999). Extending the size limit of protein nuclear magnetic resonance. *Proc Natl Acad Sci USA* 96:332–334.

Yu, L., Zhu, C. X., Tse-Dinh, Y. C., and Fesik, S. W. (1996). Backbone dynamics of the C-terminal domain of *Escherichia coli* topoisomerase I in the absence and presence of single-stranded DNA. *Biochemistry* 35:9661–9666.

Zhang, W., Smithgall, T. E., and Gmeiner, W. H. (1998). Selfassociation and backbone dynamics of the hck SH2 domain in the free and phosphopeptide-complexed forms. *Biochemistry* 37:7119–7126.

Zhu, L. and Reid, B. (1995). An improved NOESY simulation program for partially relaxed spectra: BIRDER. *J Magn Reson* Series B 106:227–235.

Zidek, L., Novotny, M. V., and Stone, M. J. (1999). Increased protein backbone conformational entropy upon hydrophobic ligand binding. *Nat Struct Biol* 6:1118–1121.

Zuiderweg, E. R. P. (2002). Mapping protein–protein interactions in solution by NMR spectroscopy. *Biochemistry* 41:1–7.

8

Application of Isothermal Titration Calorimetry in Exploring the Extended Interface

John E. Ladbury* and Mark A. Williams

8.1. INTRODUCTION

Biomolecular interactions can now be routinely characterized on an atomic level. The use of high-resolution structural information from X-ray crystallographic and NMR spectroscopic studies permits an understanding of the changes in the positioning of atoms as a complex is formed. Furthermore, knowing where individual atoms are likely to be in three-dimensional space allows an educated estimation of the positioning of short-range noncovalent bonds that form between the interacting partners. These structural details can be combined with the measurement of thermodynamic parameters of biomolecular complex formation derived from isothermal titration calorimetry (ITC). This method enables the highly accurate determination of the change in enthalpy (ΔH) as well as the strength of interaction via the change in free energy (ΔG). This combination of structural detail and thermodynamic data can allow a complete understanding of the biomolecular interaction and the forces that drive it. However, there is an increasing number of reports of binding events that are accompanied by structural and dynamic changes of the interacting partners or perturbation of the surrounding solvent. Conformational changes associated with binding range from differences in the local fluctuations

J. E. LADBURY • Department of Biochemistry & Molecular Biology, University College London, Gower Street, London WC1E 6BT, UK M. A. WILLIAMS • Department of Biochemistry & Molecular Biology, University College London, Gower Street, London WC1E 6BT, UK and *Corresponding author: Tel: +44 207 679 7012; fax: +44 207 679 7193; e-mail: j.ladbury@biochem.ucl.ac.uk

of structure between the free components and the complex, to the folding of domains or entire proteins that are intrinsically unfolded in the unbound state. Equally profound may be rearrangements of components of the solvent, for example, changes in the hydration, or interaction with solvent ions, of the interacting macromolecules. The contribution of these changes in structure and solvation to the overall thermodynamics of binding can be considerable, despite the fact that in many cases these occur at sites distal from the actual biomolecular binding interface. Consequently, the characterization and thermodynamic measurement of the effect of this ''extended interface'' is essential if a full understanding of the biomolecular interaction is to be obtained.

Data derived on the formation of biomolecular complexes have to be carefully interpreted in the light of the events occurring outside the contact surfaces. A number of methods lend themselves to the measurement of the effects of these extended interactions (Ladbury and Williams, 2004; Dyson and Wright, 2005). In this chapter, we focus on ITC and demonstrate how this experimental method can be used to extract data on the extended interface.

8.2. ISOTHERMAL TITRATION CALORIMETRY: GENERAL PRINCIPLES

Measurements of biomolecular interactions using ITC provide a direct determination of the change in heat energy between the unbound and the bound states (the change in enthalpy, ΔH). As two components are mixed in a calorimeter cell, the heat output, or intake, is accurately measured. If a single mixing step were to be used with one interacting component to completely bind to the other, then the heat obtained will be the total change in enthalpy (ΔH) for the interaction. If the concentration of one of the components is known, then this total heat can be given in molar terms, i.e., the molar enthalpy change. However, as the name of the instrument suggests, the interaction of one component with another is usually completed over the course of a titration. The serial injection of aliquots of one component into the calorimeter cell containing the other, in a concentration regime where the binding sites of the component in the cell are gradually saturated, produces a titration curve of heats per injection, or per mole of injected material (Figure 8.1). From this, in the knowledge of the respective concentrations of the interacting molecules (and in some cases, the stoichiometry of the interaction), the fraction of bound material can be determined at any point in the titration. As a result, the equilibrium binding constant ($K_B = 1/\text{equilibrium dissociation constant} = 1/K_D$) can be calculated. From the ΔH and the K_B, a full thermodynamic characterization of an interaction can be determined at a given temperature based on the following relationships. The change in free energy of binding (ΔG) can be determined from Eq. (8.1).

$$\Delta G = -RT \ln K_B, \quad (8.1)$$

where R is the gas constant ($8.3178\,\mathrm{J\,mol^{-1}\,K^{-1}}$) and T is the experimental temperature. This value can then be used to determine the change in entropy (ΔS) from Eq. (8.2).

$$\Delta S = \frac{\Delta H - \Delta G}{T}. \tag{8.2}$$

Repetition of the experiment over a range of temperatures (and at constant pressure) yields the change in constant pressure heat capacity (ΔC_p) that accompanies binding.

$$\Delta C_p = \frac{d\Delta H}{dT}. \tag{8.3}$$

The value of the ΔC_p term in respect of interpretation of the extended interface will be clarified later; however, the reported correlation of this term with the burial of

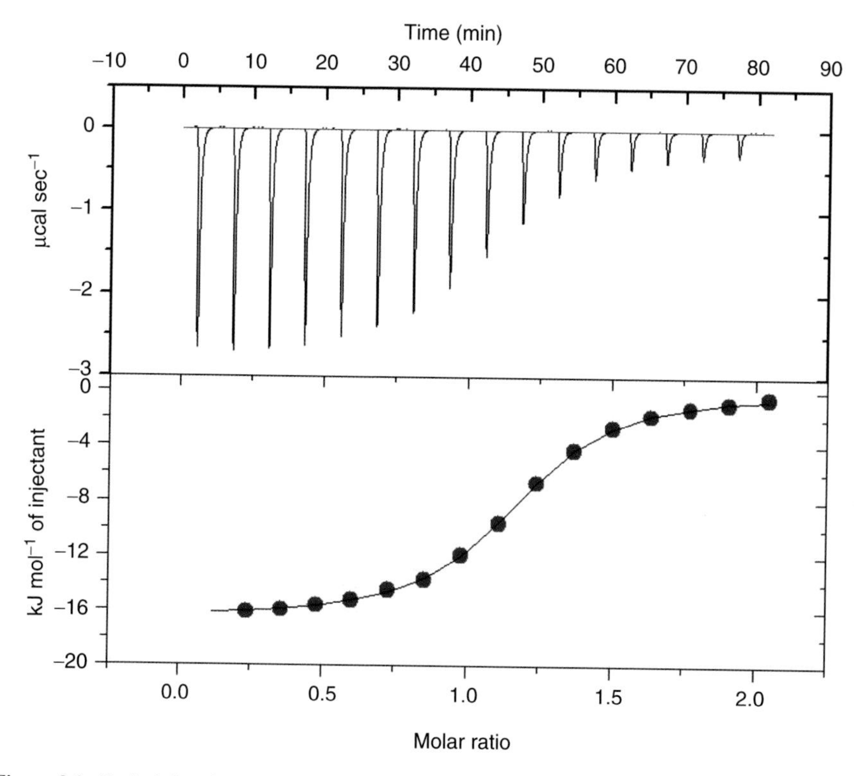

Figure 8.1. Typical data from an isothermal titration calorimetry (ITC) experiment. Upper panel: Raw data from the ITC experiment using a MicroCal (Northampton, Mass, USA) VP-ITC instrument. Lower panel: The peaks from the upper panel raw data are integrated plotted with respect to the concentrations of the interacting components as molar heats (y-axis) and molar ratio (x-axis). Fitting of this curve gives the parameters derived in the text.

biomolecular surface area can provide a tool for identifying additional effects which may not be apparent in the structure of the binding site.

Several reviews have been published which deal extensively with the instrument design, various applications of the instrument, and the basic experimental procedure (Ladbury, 1995; Ladbury and Chowdhry, 1996; Jelasarov and Bosshard, 1999; Leavitt and Freire, 2001; Ward and Holdgate, 2001; Thomson and Ladbury, 2004).

Data from an ITC experiment include the contributions, not only from the interaction between the two binding components, but also from any other event that might have an associated heat. Thus the measured ΔH includes the effects of any additional structural perturbation, changes in numbers of waters interacting with the biomolecules, the heats of binding or solvation of any ions, the heat of ionization of any protonation event, and so on. The thermodynamic parameters determined from an ITC experiment are therefore described as the ''observed'' or ''apparent'' values and given a subscript ''obs'' or ''app'' accordingly (obs is adopted in this chapter).

The thermodynamic parameters derived from ITC can provide information on the extended interface in several ways; most commonly, these involve varying the conditions (e.g., pH, ionic strength, temperature) under which the titrations are carried out to reveal the effect of a specific process on the binding interaction. In this chapter, we focus on ITC experiments that provide insight into the contribution of the ionic environment, protein conformational change, and the contribution of water molecules to the thermodynamic parameters of the interaction.

8.3. BINDING AND RELEASE OF SOLVENT IONS

The interactions of components of the solvent are of fundamental importance in dictating the outcome of biomolecular complex formation. In some cases, the positions of molecules or atoms derived from the solvent can be observed in high-resolution structures. However, their quantitative contribution to the energetics of binding clearly cannot be established merely from a pictorial representation of the contact interface. ITC provides a method to determine the overall energetics of an interaction; however, the parsing of this into the independent effects of the biomolecules forming the complex and the binding or release of various components of the solvent is more of a challenge. The interaction of anions and cations from the solvent can have a dramatic effect on the formation of complexes, particularly in cases where charged chemical groups play a role in dictating the specificity of the interacting biomolecules. Exploring this aspect of the extended interface can be facilitated by ITC experimentation. In particular, ITC can be used to explore the relationship between the affinity and the salt concentration, thus isolating the contribution of anion and cation rearrangement to binding. The underlying idea is that if, for example, the solvent ions have a lower energy in the presence of a complex than in the presence of the separate interacting biomolecules, then increasing the salt concentration favors the formation of the complex.

In the case of protein–DNA interactions, the correlation between the affinity of an interaction and the concentration of salt has been established by Record and colleagues (Record *et al.*, 1978; Ha *et al.*, 1989, 1992). A modified version of the derived empirical relationship has been adopted (Eq. (8.4); O'Brien *et al.*, 1998).

$$\log K_{\mathrm{B}} = \log K_{\mathrm{B,ref}} - A \log[\mathrm{Salt}] + 0.016B[\mathrm{Salt}], \tag{8.4}$$

where the values of A and B are the number of ions and the number of water molecules, respectively, released to bulk solvent (negative number) or bound to the complex (positive number). The effect of ion release is manifested in a negative slope and ion binding as a positive slope in a plot of log $K_{\mathrm{B,obs}}$ against log [salt]. In most systems investigated, the value of A is negative, highlighting the fact that the formation of the protein–DNA complex interface requires the removal of ions that were previously bound to the interacting biomolecules.

Intracellular salt concentrations vary significantly from approximately 200 mM in human tissue to in excess of 2 M in some halophilic archaea. These different salt environments require modification of the recognition mechanism, particularly in interactions of biomolecules in which charge plays an important role. The halophilic and thermophilic archaea *Pyrococcus woesei* (*Pw*) prevails in an environment where the salt concentration is in excess of 0.5 M and the temperature is above 100°C. Investigation of the interaction of the TATA-binding protein (TBP) from this organism with a cognate sequence of DNA using ITC revealed the effects of high ionic strength on complex formation. Titration experiments were performed in increasing concentrations of NaCl (0.1–1.9 mM; Bergqvist *et al.*, 2001, 2002). The slope of the plot of log $K_{\mathrm{B,obs}}$ against log [salt] was found to be positive, i.e., the binding becomes tighter with increasing concentrations of salt (see Figure 8.2). This was the opposite effect to that observed for the homologous protein from yeast (Petri *et al.*, 1995). This suggested the presence of a mode of recognition which had not been previously observed. In the case of the *Pw*TBP, fitting the data to Eq. (8.4) revealed that approximately two additional ions were bound to the complex (cf. the free protein and DNA), whereas for the yeast protein three ions were released upon complex formation. The behavior of the yeast protein is more typical of protein–DNA interactions and is explained by the exclusion of ions from the region around the electronegative DNA upon binding the protein. In the case of the *Pw*TBP, the tighter binding at higher salt concentration implies that ions interact more favorably with the complex and hence at higher concentrations the binding is enhanced. The *Pw*TBP differs from the yeast form in that it has a number of electronegative residues near to the protein–DNA contact surface. They, together with the electrostatic focusing effect of the low dielectric of the protein, appear to enhance the electronegativity of the complex, producing an extended region centered on the DNA minor groove which is favorable to cation (Na^+) binding (Figure 8.2). Recent studies of DNA itself suggest that Na^+ ions are localized to the most electronegative parts of DNA, particularly the minor groove. However, they are not immobilized in these sites and any particular site is only partially occupied (Denisov and Halle, 2000; Hud and Polack, 2001).

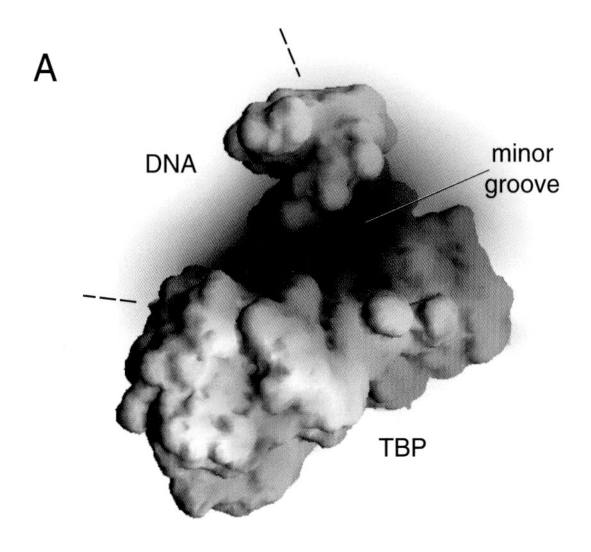

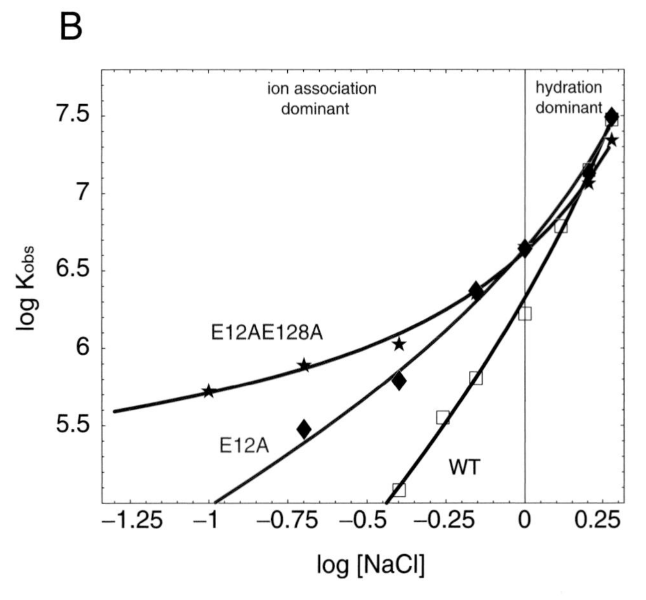

Figure 8.2. Salt effects on the binding thermodynamics of the *P. woesei* TATA-binding protein (TBP) binding to cognate DNA. (A) Space-filling model of the wild-type protein–DNA complex showing its highly negative local electrostatic field, which results in cation uptake into the complex (dotted lines show direction of the axis of the highly curved DNA). (B) Salt dependence of the affinity, showing the unusual increase in binding affinity with increasing salt for the wild-type protein. Mutations that reduce the electronegativity of the complex progressively diminish the cation-binding capability of the complex and its salt dependence.

Modeling studies of the TBP–DNA complex have extended this view of monovalent ion behavior, and envisage ions of restricted mobility partially occupying many sites on the complex (Bergqvist *et al.*, 2002, 2003). Support for this interpretation comes from mutational studies in which negatively charged residues, hypothesized to enhance the interaction with cations, were identified on the protein from X-ray crystallographic data and mutated to neutral amino acids. The substitution of a glutamate residue (E12) of the protein for an alanine (E12A) reduced the slope of the log $K_{B,obs}$ against log [salt] plot (see Figure 8.2). An additional mutation, E128A, reduced the positive slope of the log $K_{B,obs}$ against log [salt] plot further to the point whereby at low salt concentrations (<0.5 M) the line is almost horizontal (see Figure 8.2), indicating that this double mutant protein (E12AE128A) does not sequester cations to the complex (extended) interface. Thus, the relationship between the affinity and the salt concentration provides a direct route to quantifying the sequestration of ions in complex formation, a phenomenon that is clearly related to the changes in electronegativity upon complexation.

8.4. CHANGES IN HEAT CAPACITY REVEAL THE EXTENDED INTERFACE

As described earlier, the measurement of the ΔH upon formation of a complex over a range of temperatures provides the means to determine the ΔC_p. Early investigations revealed a correlation between the change in solvation of apolar or hydrophobic surface area and the change in heat capacity in protein folding or unfolding in common with the heat capacity effects of transfer of small hydrocarbon molecules into aqueous solution (Ha *et al.*, 1989; Livingstone *et al.*, 1991). The correlation was subsequently refined to account for both apolar and polar surfaces (Spolar *et al.*, 1992).

$$\Delta C_p = (1.34)\Delta A_{np} - (0.58)\Delta A_p \,(\text{J mol}^{-1}\text{K}^{-1}), \tag{8.5}$$

where ΔA_{np} and ΔA_p are the changes in nonpolar and polar solvent accessible surface areas (Å^2) on going from the free to the bound state, respectively. The correlation was seen to be a potential tool for predicting the changes in surface area upon biomolecular complex formation (Spolar and Record, 1994). Thus, determination of a ΔC_p would enable a prediction of the amount of surface removed from exposure to solvent. In many cases, this correlation has been found to accurately predict the size of the binding interface; however, in some cases, a discrepancy is found between the measured value of ΔC_p and that predicted from the structure of the complex. A major source of such discrepancies is the association of a conformational change, which may occur outside the binding interface, with the binding event. Such a conformational change is very likely to give rise to additional changes in solvent exposure, which may not be apparent from the structure of the complex. Indeed, such a discrepancy is often taken as evidence for conformational change coupled to the binding. However, care needs to be taken when making such an attribution, because there are a number of other effects that occur on binding, which can give rise to a change in heat capacity. Other mechanisms that provide potential explanations for the discrepancies include, linked protonation and ion-binding events (Lohman *et al.*, 1996; Kozlov and Lohman, 1998) and ''stiffening'' of ''soft'' vibrational modes at the interface particularly from entrapped water molecules (Sturtevant, 1977; Ladbury *et al.*, 1994; Morton and Ladbury, 1996). Of course, all of these could also be derived from interactions outside the defined binding site, and consequently, discrepancy between the correlation of the burial of biomolecular surface and ΔC_p is an important tool in identifying possible effects of the extended interface.

8.4.1. Conformational Changes in Protein–DNA Interactions

The octamer element DNA sequence is recognized by the POU DNA-binding domain of the Oct-1 transcription factor. Oct-1 is ubiquitously expressed in mammalian cells and regulates the transcription of small nuclear RNA genes and the histone H2B (Herr, 1992). In addition to its role as a transcription factor, Oct-1 is able to regulate the expression of both general and cell type-specific genes by recruiting other transcription factors or coactivators. For example, Oct-1 interacts with the B cell-specific protein Bob-1 to regulate transcription of immunoglobulin genes (see Lundbäck, 2000). The POU domain consists of two conserved and structurally independent subdomains that contact the DNA on opposite faces of the octamer motif. The domains are connected by a flexible linker, which appears to ensure an implicit cooperativity in binding (Herr and Cleary, 1995). The X-ray crystal structure of the Oct-1 or POU protein in complex with a promoter DNA sequence (H2B) shows that both the subdomains use a helix-turn-helix motif to recognize the functional groups in the DNA major groove. ITC was used to investigate the binding of the POU domain from Oct-1 to the octamer motif within the Igκ promoter (Lundbäck *et al.*, 2000). ΔC_p was determined by measuring ΔH_{obs} over a temperature range of 12–35°C (the protein and the DNA were shown to be stable above this temperature range by UV spectroscopic and differential scanning calorimetric methods). The ΔH_{obs} was endothermic at all temperatures measured and strongly temperature dependent. The $\Delta C_{p,obs}$ was $-3.47\,kJ\,mol^{-1}\,K^{-1}$. Several studies by ITC have demonstrated that the formation of highly complementary protein–DNA complexes is generally accompanied by large negative changes in heat capacity (Takeda *et al.*, 1992; Ladbury *et al.*, 1994; Hyre and Spicer, 1995; Merabet and Ackers, 1995; Lundbäck and Härd, 1996; O'Brien *et al.*, 1998). In most reported interactions of this type, large amounts of surface area are secluded from bulk solvent. The structure of Oct-1 bound to the H2B promoter was used to assess the burial of surface area. The oligonucleotide in this complex differs from the Igκ promoter region only in the region flanking the octamer element, and thus a high level of structural homology could be expected. Using the published structure and assuming a rigid body interaction, the change in polar (2,236 Å^2) and nonpolar (1,018 Å^2) surface area was calculated and related to the empirical correlation (Eq. (8.5); Spolar *et al.*, 1992; Spolar and Record, 1994) to give a ΔC_p of $-50.4\,\mathrm{J\ mol^{-1}\ K^{-1}}$. This small value is due to the highly polar nature of the interaction. The large discrepancy between this calculated value and the experimentally determined value (discussed earlier) suggests that some other event is occurring that greatly reduces the ΔC_p of the complex. In this case, it was proposed that concomitant folding of the protein on binding is responsible, at least in part, for the additional contribution to ΔC_p. Published solution structural data on both POU subdomains show the POU homeodomain of Oct-1 to be partially disordered. However, on forming the complex with DNA, part of the recognition helix appears to become more ordered. This ordering involves the burial of surface area from about 10 amino acids. On the basis of previously reported data, the observed discrepancy in ΔC_p would require a folding and burial of approximately 26 amino acids. The additional burial of surface area was believed to be derived from conformational changes within the flexible linker and possibly interactions between the domains. Proteolytic cleavage experiments showed that one of the POU subdomains became significantly more protected from digestion on binding with DNA. This is consistent with the reduction of exposure of amino acids concomitant with a local folding or a tightening of structure on complex formation.

The evaluation of ΔC_p via surface area burial calculations and the lack of correlation with empirical data can provide some indication of or insight into a coupled local folding event outside the defined binding site. Interestingly, the proposed folding event provides a ΔC_p term that is temperature independent, i.e., there is no deviation from linearity in the plot of ΔH_{obs} versus temperature. This suggests that the folding equilibrium is not temperature dependent, i.e., the same amount of protein goes from the unfolded to the folded state at all temperatures investigated. This is not the case when whole proteins or domains of proteins undergo a coupled folding event (see later).

8.4.2. The Role of Water Molecules in the Complex Interface

The relationship between surface area burial and ΔC_p has been shown not to hold if water molecules are present in the biomolecular interface. This was originally observed in experiments performed on the *trp* repressor–*trp* operator protein–DNA interaction (Ladbury *et al.*, 1994). Using the correlation previously observed (Spolar and Record, 1994), together with high-resolution structures of the complex and of the free protein and DNA, the predicted $\Delta C_p(-1.8\,\mathrm{kJ\ mol^{-1}\ K^{-1}})$ was significantly lower than that obtained by ITC measurements of the variation in ΔH_{obs} over a temperature range of 10–40°C ($-3.98\,\mathrm{kJ\ mol^{-1}\ K^{-1}}$). The structure of the protein–DNA complex (solved to a resolution of better than 2.0 Å) showed that the interfacial interactions were almost entirely mediated by water molecules. In most cases, these water molecules were significantly restricted in the interface making discrete hydrogen bonds between the two macromolecules. The orientated hydrogen bonds, which these water molecules make between interacting macromolecules, restrict their motion in comparison to bulk water and in particular reduce the available ''soft'' vibrational modes. This reduces the C_p for each water molecule and enhances the negative ΔC_p for the formation of the complex (Sturtevant, 1977; Williams and Ladbury, 2003). The large number of water molecules found in the *trp* repressor–*trp* operator protein–DNA interaction provides a very large discrepancy between the ΔC_p derived from the empirical correlation and the experimentally observed ΔC_p. Other protein–DNA interactions also show a discrepancy in the ΔC_p values when water molecules are present (Morton and Ladbury, 1996).

The interaction of *Pw*TBP with a cognate oligonucleotide shows a large negative temperature dependence of the ΔH_{obs}. Previous ITC experimentation on site-directed mutants of this protein had shown the role of the extended interface based on cation association in stabilizing the complex (see section 8.3). The interactions of the wild-type protein and a series of mutants with amino acids altered, neighboring the protein–DNA interface, were investigated over a range of temperatures (35–55°C) and the respective ΔC_p values determined (Bergqvist *et al.*, 2004). The mutations in the protein were designed to perturb the positioning of cations, as described earlier. The E12A and E128A mutations removed approximately one cation each from the binding interface. The values of ΔC_p determined for these mutants were different. However, in both cases the number of cations removed from the protein–DNA interface was identical (see Table 8.1 and Figure 8.3). Furthermore, the E128A mutant had a similar ΔC_p value to the wild-type interaction. This implied that the change in heat capacity differences were not attributable to the effects of surrounding cations. On the basis of this well-understood (but limited) system, it has been hypothesized that the sequestration of cations generally has little effect on the value of ΔC_p.

The reason for the large change in the ΔC_p value resulting from the E12A mutation was sought through analysis of the known structure. The high-resolution X-ray crystal structure of the complex shows that E12 is not only important in the interaction with cations (see earlier), but is also involved in sustaining a network of

Table 8.1. The change in constant pressure heat capacity for the binding of *P. woesei* TATA-binding protein (TBP), and mutants thereof, to cognate DNA

Protein	$\Delta C_{p,obs}$(kJ mol^{-1} K^{-1})
Wild type	-3.2 ± 0.2
E128A	-2.9 ± 0.1
E12A	-1.9 ± 0.2
E12AE128A	-2.1 ± 0.1
Q103A	-2.3 ± 0.1

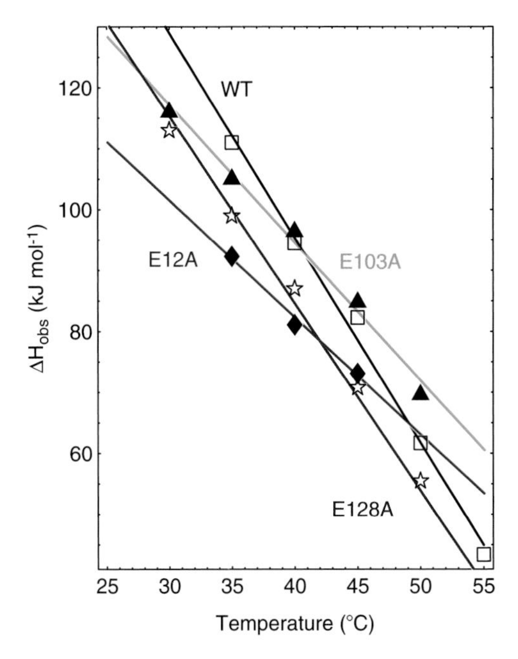

Figure 8.3. Temperature dependence of the change in enthalpy of *P. woesei* TATA-binding protein (TBP) (and mutants thereof) binding to cognate DNA. Large changes in the slope (change in heat capacity, ΔC_p) arise from point mutations at pseudosymmetrical sites in the structures of E12 and E103.

water molecules (DeDecker *et al.*, 1996; Kosa *et al.*, 1997), as shown in Figure 8.4. Interestingly, the binding site of the oligonulceotide has an axis of ''pseudosymmetry'', which means that a glutamine residue, Q103, occupies a position that is structurally similar to that occupied by E12 with respect to interactions with the DNA (see Figure 8.4). Mutation of this Q103 residue to alanine has an identical effect on the ΔC_p to that of the E12A mutation (Figure 8.3). Further investigation of the structure reveals that this residue also sustains a water network similar to that around E12 (Figure 8.4). Thus, because the common feature of both residues is that they are involved in supporting water networks, it seems plausible that the difference in the total observed ΔC_p is derived from the ΔC_p of the bound water. The networks of water molecules centered on E12 and Q103 are, unlike in the *trp*–repressor complex, on the surface of the protein forming a continuum with the bulk solvent water. They are in a ''hyperpolar'' environment (Bergqvist *et al.*, 2004), bridging between the phosphate backbone and the polar side chains of the protein. The network of strong hydrogen bonds in this environment is expected to reduce their heat capacity dramatically; however, the number of water molecules are too few to fully account for the large ΔC_p change arising from these point mutations in comparison with wild type. Consequently, it is hypothesized that the restricted surface layer water molecules also affect the hydrogen bonding of

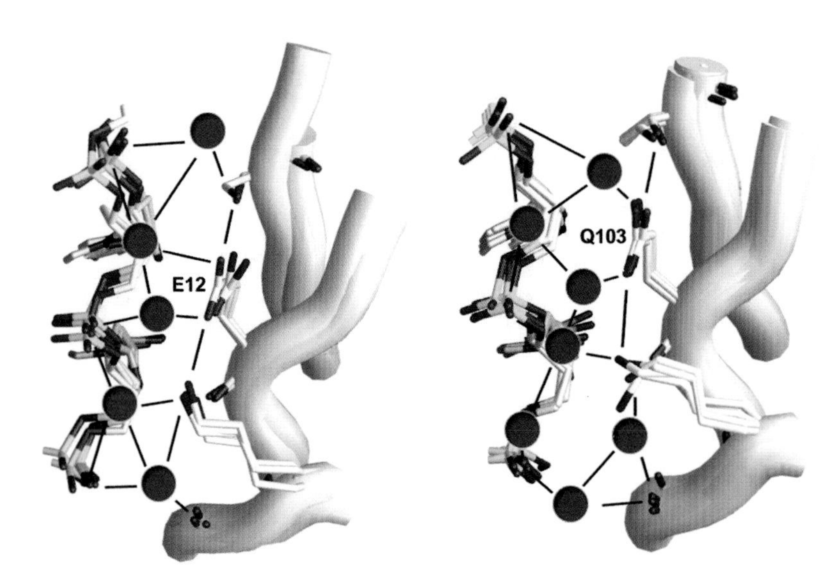

Figure 8.4. Hydration networks between TATA-binding protein (TBP) and the phosphate backbone of DNA. A superposition of the structure of TBP from *P. woesei, H. sapiens, A. thaliana,* and *S. cerevisiae* reveals conserved networks of strongly hydrogen-bonded water molecules on the surface of the complex centered on E12 and Q103. Mutation of either of these residues causes a large rise in the heat capacity of the complex, which is hypothesized to be due to disruption of the cooperative interactions of this network and neighboring bulk water molecules.

the layer of bulk water in contact with them. This explanation remains to be proven, but is supported in a general way by neutron scattering and infrared studies that imply that the effects of the protein surface extend for 2–3 solvent layers into the bulk solution. This explanation and other studies showing large heat capacity effects from point mutations (Jing *et al.*, 2002) undermine the idea that thermodynamic effects can be generally accounted for by the relationship based on surface area buried in an interface and reinforce the notion that the whole system including the surrounding solvent must be investigated in detail.

8.5. THERMODYNAMIC EFFECTS OF PROTEIN STRUCTURAL PERTURBATION ON BINDING

8.5.1. Conformational Changes in Protein–Protein Interactions

The idea that discrepancies between measured thermodynamic data and predictions based on structural features of the biomolecular recognition event can be used to identify and discriminate between different models of that process is applicable to not only ΔC_{p}, but also to other thermodynamic quantities, notably ΔG. Again, care must be taken to consider, and account for in the model, all of the effects of events that may occur simultaneously with the binding. Some examples from signaling complexes serve to illustrate this strategy.

Specificity in the interactions between proteins and their targets in intracellular signal transduction is fundamental to biological function. In the absence of a high level of specificity in the linear processing of signals, the potential for cross-interaction increases and could ultimately produce an inappropriate cellular response. The issue of specificity in tyrosine kinase-mediated signaling is particularly fascinating since many of the proteins involved in transducing signals interact through similar domains. Several of these individual domains are found in large numbers of proteins. They are structurally very similar and seem to bind promiscuously (Ladbury and Arold, 2000). For example, in a typical human cell, Src homology 2 (SH2) domains are found in around 100 different proteins implicated in tyrosine kinase-mediated signaling. SH2 domains consist of approximately 100 amino acids and have high levels of primary, secondary, and tertiary structure homology throughout the family. Their cognate ligands require a phosphorylated tyrosine (pY) residue for ''high-affinity'' binding. The pY binds in a deep electropositive ''pocket'', which is well-conserved throughout the SH2 domain family. Specificity in the SH2 domain interaction has been reported as invoked by ligand residues proximal to the pY. However, binding studies using ITC have revealed limited differences in the affinity of binding of these domains to a variety of tyrosyl-phosphorylated ligands (Bradshaw *et al.*, 1998, 1999; Chung *et al.*, 1998; Arold and Ladbury, 2000; Henriques and Ladbury, 2001; O'Rourke and Ladbury, 2003). Due to the apparent promiscuity in the interaction of these independent domains, it is clear that specificity in signaling pathways cannot be achieved by a simple linear chain of domain–ligand interactions.

One way of enhancing specificity and selectivity of the interactions of proteins involved in tyrosine kinase-mediated signaling could be by their binding with effector proteins through more than one domain. This multiple domain binding has been shown to occur in a number of SH2 domain-containing proteins. The hypothesized enhancement in specificity of the binding of two domains to the cognate ligand should be derived from the combined ΔG of the interactions of the individual SH2 domains as well as a favorable contribution from linkage of the domains. This additional contribution to ΔG derived from the expectation that once one domain is bound the entropic cost of restricting the other domain is already paid for by virtue of the linker. The thermodynamic parameters associated with tandem domain interactions have been explored to assess whether the juxtaposition of domains provides the hypothesized improvement in specificity. Any conformational change in the linker region associated with the positioning of the domains is a property of the extended interface since it is outside the direct binding site of the protein–ligand complex.

The Syk family of cytosolic kinases is comprised of two nonreceptor protein tyrosine kinases, Syk and ZAP70, which play important roles in signaling by antigen and immunoglobulin receptors. Deletion of either of these proteins results in arrested development of immune cells and functional defects for a variety of immune receptors. The ζ-chain-activating protein, ZAP70, plays a role in T-cell activation. It possesses two SH2 domains separated by 65 residues that bind to the doubly phosphorylated immunoreceptor tyrosine-based activation motif (ITAM) on the ζ-subunit of the CD3 T-cell receptor. The activated ITAM has a consensus amino acid sequence pYXX(L/I)-$X_{7/8}$-pYXX9(L/I). Each pY moiety binds to one SH2 domain. ITC-binding studies using monophosphorylated peptides showed weak binding to the intact protein (O'Brien *et al.*, 2000). The bisphosphorylated peptide bound with a stoichiometry of unity and an affinity that was nearly four orders of magnitude tighter than the monophosphorylated form. However, when this binding is reconciled in terms of the thermodynamic parameters, the ΔG shows only an additive effect. There is no apparent cooperative effect from the linkage of the tandem domains. The interactions of both mono- and bisphosphorylated peptides are accompanied by an unfavorable change in entropy. As expected, the interaction of the bisphosphopeptide with ZAP70 is significantly more entropically favorable than that with the monophosphopeptides. The lack of cooperativity and the additive value of ΔG_{obs} for the bis- versus the monophosphorylated peptide interaction seem to be due to the lower-than-expected ΔH_{obs} term. The structural effects of complex formation have to be considered to reconcile these thermodynamic data. The crystal structure reveals of the ZAP70–ITAM complex that ZAP70 is unique among the tyrosine kinase proteins in that the N-terminal pY pocket involves extensive interactions between the two SH2 domains (Hatada *et al.*, 1995; Graef *et al.*, 1997; Ottinger *et al.*, 1998). The binding of the ITAM appears to result in the formation of this pocket. In addition, it is known that for the tandem domain interaction to occur, the linker region has to undergo a significant conformational change to allow the formation of the N-terminal SH2 domain. This linker region is thought to fold into a coiled-coil structure in the presence of the bisphosphorylated peptide. Consequently, the formation of the peptide-bound complex involves not only the direct interactions between the domains and the peptide, but the formation of additional structures at other locations diminishing. This failure of the formation tandem domain complex to provide the expected cooperative effect due to the lower-than-expected ΔH_{obs} term of the bisphosphopeptide, compared with the monophosphopeptide interactions was hypothesized to be due to the reduction in the net contribution from noncovalent bonds formed in the structure in adjusting to the appropriate conformation to accommodate the bisphosphopeptide. Although the detail of the complex suggests that structure has to be formed in the assembly of the complex, it is also likely that a significant amount of other interactions outside the defined binding site are lost in forming a suitable steric match for the peptide.

The recruitment of the tandem SH2 domains of Syk to the immune cell receptor via its interaction with ITAMs has been studied (Fütterer *et al.*, 1998; Gruzca *et al.*, 1999). Unlike ZAP70, which is T-cell-specific, Syk is ubiquitously expressed in hemopoietic cells and can be activated by a large number of immune response receptors. The crystal structure of Syk protein, in the absence of the kinase domain, bound to a cognate ITAM, contains six copies of the protein in the asymmetric unit.

These reveal an inherent flexibility of the orientation of the two SH2 domains. It was also shown that, unlike ZAP70, the SH2 domains do not form a common N-terminal SH2 domain pY-binding site, suggesting that the SH2 domains function as independent units (Kumaran *et al.*, 2003). ITC experiments were performed over a range of temperatures. The ΔH_{obs} for the interaction revealed a nonlinear temperature dependence, suggesting the possibility of a temperature-dependent conformational change coupled to binding. CD experiments show no apparent change in the secondary structure of the protein and hence the conformational change is not due to wholesale, or partial, unfolding. The authors hypothesized that two conformers of the unligated Syk protein existed; with one conformer predominating at the lowest temperatures and the other at higher temperatures. Binding of the ITAM favored one of these conformations. The SH2 domains themselves move as rigid bodies, but a large variation in the relative orientation of these domains is permitted with little energetic difference between the forms. The authors suggested that the conformational equilibrium may facilitate Syk in conveying inherent flexibility to the protein, allowing the two SH2 domains to sample a continuum of relative orientations between the two conformers, and thus adapt to the various lengths between possible pY residues in ITAMs of different protein ligands involved in a variety of signal transduction pathways. This model was confirmed by inserting a disulphide bond between the SH2 domains, thus restricting their relative motion. This resulted in a linear temperature dependence of the ΔH_{obs}.

8.5.2. Coupled Folding and Binding Equilibria

Previously in this chapter, we highlighted systems in which the extended interface comprises conformational changes in localized sites in the protein, or involves solvent effects proximal to the complex interface. However, there are a growing number of examples in which the extended interface encompasses whole proteins, or domains thereof. In these cases, the protein, or the domain, exists in a state of equilibrium between an unfolded (or partially unfolded) state and a folded state, the structure of which corresponds to that in complex with the ligand. Thus, the formation of the biomolecular interface involves the coupled equilibria of folding and binding.

Analysis of several complete genomes indicates that intrinsically disordered proteins are highly prevalent, and that the proportion of proteins containing disordered regions increases with the complexity of the organism (Dunker *et al.*, 2002; Ward *et al.*, 2004). Interestingly, such analysis reveals that proteins involved in eukaryotic signal transduction, or that are linked to oncogenesis, have an increased propensity for intrinsic disorder (Iakoucheva *et al.*, 2002). The evolution of an increased proportion of disordered structure in higher organisms implies that such regions, including those for which binding reactions are linked to polypeptide folding or unfolding, confer functional advantages. At present, these advantages are not clearly understood (Dyson and Wright, 2005), although the possibility that inherent conformational flexibility allows adaption to a variety of ligands and that mutual recognition of multidomain proteins is enabled or enhanced by their connection via flexible segments (Zhou, 2003), seems likely to be important in some cases. The formation of structure could also act as a ''switch'' in a signaling system in the same way that other, less-extensive, conformational changes have been found to function. It has also been suggested that in some cases the unfolded state of the unliganded protein helps the cell to reprocess these proteins via proteolysis (Tompa, 2003). This could be particularly important in respect of signaling proteins, whereby rapid removal of these proteins would reduce or stop a signaling pathway once the signal had been transduced.

Several thermodynamic and kinetic advantages have been hypothesized for the coupling of folding to binding for single domains. The ''fly-casting'' mechanism (Shoemaker *et al.*, 2000) imagines a situation in which an encounter between a ligand and part of an unfolded protein creates a folding nucleus onto which the rest of the protein folds. This mechanism envisages an on-rate enhancement arising from the greater physical extent of the unfolded protein, increasing the encounter rate with the ligand. The mechanism has been used to provide an explanation of the decrease in rate of formation of the $p27^{Kip1}$/cyclin A/cyclin-dependent kinase-2 tertiary complex (Verkhivker *et al.*, 2003) in response to structurally stabilizing mutations of the natively unfolded p27. It seems likely that such effects are important in other signaling or regulatory complexes. However, the rate enhancement relies on the possible (largely unfolded) encounter complexes having significant binding affinity; e.g., in the case of p27, a few short stretches of sequence, which are spatially separated on the large contact surface of folded form, seem to be responsible for most of the binding affinity. Very commonly, however, substantial binding affinity only arises when a localized binding site is created by a particular local spatial arrangement of residues that are distant in sequence. For proteins that form such binding sites, the fly-casting mechanism does not provide a mechanism for coupling folding to binding. Instead, it is necessary to think about the effect of the ligand on the folding equilibrium of a protein.

Under any given condition, a polypeptide in equilibrium between a folded and an unfolded state will have populations of both forms, albeit the folded population may be very small. In the case where the cognate ligand binds to a site that exists only in folded form, upon addition of the ligand, the law of mass action dictates that as the ligand becomes bound the equilibrium will be restored by folding of unfolded polypeptide. This shifting of the folding equilibrium does not require any particular kinetic mechanism or thermodynamic characteristics for the interaction, except that the binding free energy is sufficiently large to stabilize the folded form. Considering this equilibrium, it is obvious that complex formed as a result of any coupled folding and binding process will always have weaker affinity than that if the components had been folded, due to the energetic penalty of folding (which is associated with the unfavorable reduction in entropy arising from the formation of the protein's structure).

This generalized thermodynamic view gives additional insights into how coupling binding to folding may modulate the formation of multiprotein complexes.

Firstly, the lowered affinity will mean that in particular concentration ranges for the free species the amount of complex can be significantly reduced compared with that of folded species. Secondly, because the observed affinity is equal to the ratio of on and off rates, the enhancement of the on rate for a protein that uses the fly-cast mechanism will be necessarily accompanied by a greater increase in the off rate (cf. hypothetical interaction between folded species). This makes the formation and dissociation of complexes faster than that had the equivalent folded protein been involved. Thirdly, because the structure of the final complex is the same, whatever the kinetic mechanism, the specificity of the interaction is unaltered by the coupling of the folding event. This combination of faster association and dissociation but unaltered specificity may well be important in cases where the signaling pathway has to be highly specific to avoid cross talk with other related pathways but also has to be able to dissociate rapidly when the signal is complete. It should be noted, however, that the increase in association and dissociation rates is a feature of a particular mechanism and not a general feature of coupled folding and binding (contrary to some assumptions in the literature). In the case in which a ligand binds only to the folded form of an otherwise unfolded protein, there are fewer diffusional encounters between ligands and binding sites than that had all of the protein been folded, and consequently the on rate is reduced. Thus, we see that coupling folding to binding can modulate the kinetics of formation of complexes in either direction, depending on the mechanism of the coupling. Presently, very few coupled folding and binding events have been investigated in thermodynamic and kinetic detail. The best understood from the thermodynamic viewpoint is the binding of the tetratricopeptide repeat (TPR) domain of protein phosphatase 5 (Ppp5) to its cognate peptide ligand from Hsp90.

TPR domain-containing proteins are involved in cell-cycle regulation, transcription control, protein import, and signal transduction. TPR domains contain between 1 and 16 tandem copies of the basic 34 amino acid repeat. Each repeat forms a helix-turn-helix structure, and the repeats stack one upon the other to form an elongated superhelix. Ppp5 has a TPR domain that has been implicated in its recruitment to the chaperone protein Hsp90 and in the autoregulation of its phosphatase activity (Chen *et al.*, 1994, 1996). NMR spectroscopic structural studies discovered that the Ppp5–TPR domain in isolation was only partially folded in the ambient to physiological temperature range (Cliff *et al.*, 2005). Binding of a pentapeptide (MEEVD) comprising the cognate sequence from Hsp90 stabilized the folded form of the domain (Cliff *et al.*, 2005). ITC was used to determine the thermodynamic parameters of binding. Over a range of temperatures, a striking temperature dependence of the ΔC_p of binding was observed (Figure 8.5). This curvature in the plot of ΔH_{obs} against temperature is characteristic of the presence of the coupled equilibria of folding and binding (Thomson *et al.*, 1994). The ΔH_{obs} is the sum of the contributions from both the folding and the binding processes. ΔH_{obs} at any given temperature (T) can thus be represented by Eq. (8.6).

$$\Delta H_{obs}(T) = \Delta H_{bind}(T) + f_U(T) \times \Delta H_{fold}(T), \tag{8.6}$$

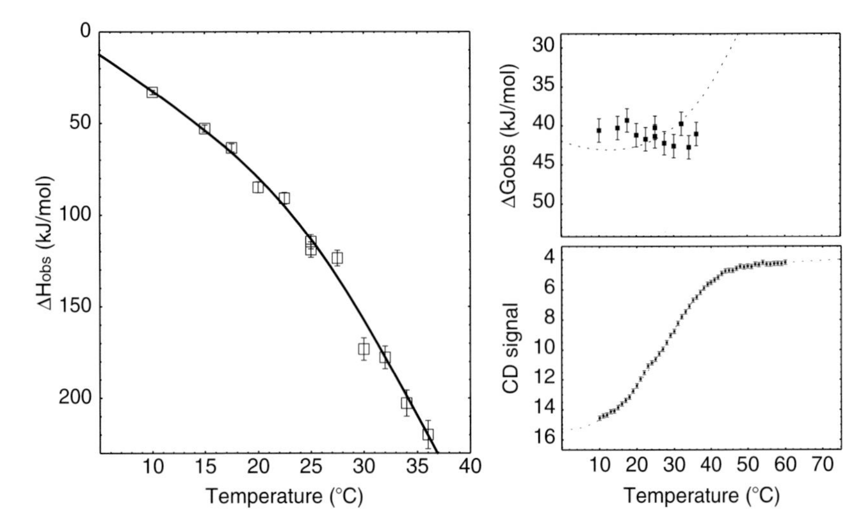

Figure 8.5. A Gibbs–Helmholtz model for the folding and the binding of the protein phosphatase 5–tetratricopeptide repeat (Ppp5–TPR) domain to its cognate peptide. The temperature dependence of the thermodynamic parameters of the coupled folding–binding event and the CD data showing the melting curve of the apodomain can be simultaneously fit to a single thermodynamic model. The best fit model is most strongly constrained by the enthalpy and CD, which are the more accurately experimentally determined quantities. The constraints from the CD data allow the thermodynamics to be accurately extrapolated to higher temperatures, despite the fact that isothermal titration calorimetry (ITC) data are not attainable much above the T_M, e.g., ΔG_{obs} varies little across a wide temperature range, but is expected to increase rapidly as the proportion of unfolded apoprotein approaches 100%.

where the fraction of protein unfolded, f_U, is dependent on the equilibrium constant at that temperature, $K_{fold}(T)$, or equivalently the change in free energy $\Delta G_{fold}(T)$ of folding in a manner given by the relationship:

$$f_U(T) = \frac{[U]_T}{[U]_T + [F]_T} = \frac{1}{1 + K_{fold}(T)} = \frac{1}{1 + \exp\left(-\Delta G_{fold}(T)/RT\right)}. \quad (8.7)$$

The nonlinear change in the fraction unfolded with temperature embodied in this equation means that curvature of the enthalpy change versus temperature plot is found even though the two contributing processes have constant heat capacity.

By definition, the folded and the unfolded states have the same energy at the melting temperature (T_M), i.e., $\Delta G_{fold}(T_M) = 0$ and therefore $\Delta H_{fold}(T_M) = T_M \Delta S_{fold}(T_M)$. Consequently, taking the T_M as the reference temperature gives

$$f_U(T) = \frac{1}{1 + \exp\left(\frac{-1}{R}\left(\Delta H_{fold}(T_M)\left(\frac{1}{T} - \frac{1}{T_M}\right) + \Delta C_{p,fold}\left(1 - \frac{T_M}{T} - \ln\left(\frac{T}{T_M}\right)\right)\right)\right)}. \quad (8.8)$$

To render the temperature dependence explicit, the standard expression for the Gibbs free energy, $\Delta G = \Delta H - T\Delta S$ can be used, together with the equations for the enthalpy and the entropy under conditions of constant heat capacity to derive expressions for $\Delta H_{\text{obs}}(T)$ and $\Delta G_{\text{obs}}(T)$ (Cliff *et al.*, 2005).

A reference temperature, T_{ref}, for the binding process can be arbitrarily fixed to, for example, 25°C and thus an expression for $\Delta H_{\text{obs}}(T)$ in terms of only five unknown thermodynamic parameters, $\Delta H_{\text{fold}}(T_{\text{M}})$, $\Delta C_{\text{p,fold}}$, $\Delta H_{\text{bind}}(T_{\text{ref}})$, $\Delta C_{\text{p,bind}}$, and T_{M}, can be derived.

$$\begin{aligned}\Delta H_{\text{obs}}(T) = \Delta H_{\text{bind}}(T_{\text{ref}}) + \Delta C_{\text{p,bind}}(T - T_{\text{ref}}) + f_{\text{U}}(T)\,(\Delta H_{\text{fold}}(T_{\text{M}})\\ + \Delta C_{\text{p,fold}}(T - T_{\text{M}})).\end{aligned} \tag{8.9}$$

Noting that the apparent binding constant K_{obs} can be written as a product of the folding and the binding constants and the mole fraction unfolded, $K_{\text{obs}} = K_{\text{bind}} \times K_{\text{fold}}/(1 + K_{\text{fold}})$, we can use the standard definition $K = \exp(-\Delta G/RT)$ to derive $\Delta G_{\text{obs}}(T)$ in terms of the fundamental thermodynamic parameters

$$\begin{aligned}\Delta G_{\text{obs}}(T) &= \Delta G_{\text{bind}}(T) + RT\ln\left(1 + \exp\left(\frac{G_{\text{bind}} - (T)}{RT}\right)\right)\\ &= \Delta H_{\text{bind}}(T_{\text{ref}}) - T\left(\Delta S_{\text{bind}}(T_{\text{ref}}) + \Delta C_{\text{p,bind}}\ln\left(\frac{T}{T_{\text{ref}}}\right)\right)\\ &\quad + RT\ln\left(1 + \exp\left(\frac{1}{R}\left(\Delta H_{\text{fold}}(T_{\text{M}})\left(\frac{1}{T} - \frac{1}{T_{\text{M}}} + \Delta C_{\text{p,fold}}\left(1 - \frac{T_{\text{M}}}{T} - \ln\frac{T}{T_{\text{M}}}\right)\right)\right)\right)\right).\end{aligned} \tag{8.10}$$

As there are more data points than unknown parameters, such equations can be fit using a nonlinear regression method to the experimentally observed enthalpy and free energy changes, and the parameters describing the separate folding and binding events estimated. However, there is a high degree of covariance among these parameters and consequently the derived estimates would have large associated errors. Thus, the overall molecular recognition event cannot be accurately deconvolved using the ITC data alone.

In their study of ribonuclease S protein binding an S-peptide, Thomson *et al.* (1994) fixed the value of $\Delta C_{\text{p,fold}}$ to that obtained by differential scanning calorimetry (DSC) studies of unfolding of the apoprotein, and then fit for the remaining parameters, to improve their estimates. In principle, it is also possible to use DSC data on the folding of an apoprotein more extensively to obtain estimates of T_{M}, $\Delta H_{\text{fold}}(T_{\text{M}})$, and $\Delta C_{\text{p,fold}}$, and subtract the folding contribution from the observed enthalpy at each temperature and fit the resulting data for the binding parameters using Eq. (8.9). However, in the case of the Ppp5–TPR domain, accurate DSC data cannot be obtained as the apoprotein aggregates significantly at the high protein concentrations required for the DSC experiments. Another possible strategy is to assume that in the lower part of the temperature range the apoprotein is fully folded and uses a linear fit of the data to Eq. (8.9) to obtain the binding parameters $\Delta H_{\text{bind}}(T_{\text{ref}})$ and $\Delta C_{\text{p,bind}}$, which can then be fixed in the fit (Thomson *et al.*, 1994). However, for the Ppp5–TPR domain, there is considerable curvature in the ΔH versus T plot even at low temperatures, so this approach is also not useful in this case.

To solve these problems, a method, based on combining the ITC data with circular dichroism studies of the folding of the apoprotein, was developed to allow a global fit for the thermodynamic parameters (Cliff *et al.*, 2005; this method is available at www.biochem.ucl.ac.uk/bsm/biophysics). This method has the considerable advantage that investigations can be carried out at low concentrations, similar to those used in the ITC experiments. Additionally, the precision of the estimates of the thermodynamic parameters of the folding and the binding processes resulting from this global fit is significantly greater than those of the methods discussed earlier.

The temperature dependence of the CD signal of a protein at a particular wavelength can be used to monitor its extent of folding. The temperature dependence of a CD signal (in this case, the characteristic α-helical signal at 222 nm) for a two-state folding equilibrium is related to the fraction of unfolded protein as follows:

$$\mathrm{CD_{obs}}(T) = (1-f_\mathrm{U}(T))(\mathrm{CD_F}(T_\mathrm{M}) + m_\mathrm{F}(T-T_\mathrm{M})) + f_\mathrm{U}(T)(\mathrm{CD}(T_\mathrm{M}) + m_\mathrm{U}(T-T_\mathrm{M})), \tag{8.11}$$

where $\mathrm{CD_F}$ and $\mathrm{CD_U}$ refer to the CD signals for the folded and the unfolded states, respectively. In this general case, the CD signals of the two states are assumed to vary linearly with temperature according to $\mathrm{CD} = \mathrm{CD}(T_\mathrm{M}) + m(T-T_\mathrm{M})$, where m is a temperature coefficient. A necessary condition for the two-state analysis of these melting data to be valid is that the temperature-dependent denaturation is reversible. We see that all three sets of observations—temperature-dependent enthalpy change, free energy change, and CD signal (described by Eqs (8.9), Eq. (8.10), and Eq. (8.11), respectively)—are dependent on ΔH_fold, $\Delta C_\mathrm{p,fold}$, and T_M. In addition, both the observed free energy and the enthalpy changes have ΔH_bind and $\Delta C_\mathrm{p,bind}$ as common parameters. Thus, a global least-squares fit to all of the temperature-dependent experimental observations can be used. A consequence of such a global fitting is that parameters common to two or more measurements are more precisely defined than by separate fits, with the observations that are most strongly dependent on particular parameters contributing most to their constraint. In this case, the unknown parameters are $\Delta H_\mathrm{bind}(T_\mathrm{ref})$, $\Delta H_\mathrm{fold}(T_\mathrm{M})$, $\Delta C_\mathrm{p,bind}$, T_M, $\Delta S_\mathrm{bind}(T_\mathrm{ref})$, and the CD signals for the folded and the unfolded states of the apoprotein. The reference temperature for the binding data (T_ref) was fixed as 25°C. This method provided the thermodynamic parameters listed in Table 8.2.

Interestingly, the data show that the contribution to the ΔG of the interaction from the folding event is very small. In this particular case, the large unfavorable $T\Delta S$ associated with the formation of the ordered protein structure is largely compensated for by favorable enthalphic contributions from new intramolecular contacts. Thus we see that, although the structural effects of coupling folding to binding for this TPR domain are dramatic, the thermodynamic (and probably also the kinetic) effects are rather mild (less than a factor of 10 variation from the equivalent folded protein). For this particular domain, any benefits of intrinsic disorder seem most likely to arise from phenomena associated with structure

Table 8.2. Estimates of the thermodynamic quantities if the separate folding and binding processes of the protein phosphatase 5-tetratricopeptide repeat (Ppp5–TPR) domain at 25°C

Parameter (units)	Value
ΔG_{bind}(kJ mol^{-1})	-43.2 ± 6.4
ΔH_{bind}(kJ mol^{-1})	-74.4 ± 4.5
$-T\Delta S_{\text{bind}}$(kJ mol^{-1})	31.2 ± 4.7
$\Delta C_{\text{p,bind}}$(kJ mol^{-1} K^{-1})	-3.05 ± 0.28
ΔG_{fold}(kJ mol^{-1})	-0.64 ± 0.53
ΔH_{fold}(kJ mol^{-1})	-90.0 ± 16.1
$-T\Delta S_{\text{fold}}$(kJ mol^{-1})	88.3 ± 16.3
$\Delta C_{\text{p,fold}}$(kJ mol^{-1} K^{-1})	-4.68 ± 1.10
f_{U}	0.436 ± 0.052

(switching, reprocessing) rather than modulation of the rates of formation or stability of complexes. The mildness of the effect also raises the possibility that some disorder–order transitions may be functionally and hence evolutionarily neutral.

In principle, the protocol described earlier can be used to establish the thermodynamic parameters for coupled equilibria for any system in which the ΔH_{obs} of the coupled folding and binding can be measured by ITC and for which there is a separate method for measurement of this temperature dependence of folding of the apoprotein.

8.6. SUMMARY

In this chapter, the use of ITC has been highlighted in terms of its use in investigating the extended interface. The ITC instrument is able to measure the total change in heat energy associated with complex formation. Any effects that occur outside the direct macromolecular interface are included in the total heat value. The value of ITC measurement in exploring the extended interface is only apparent when the individual contributions of these effects can be teased away from those occurring in forming the direct interfacial contacts. In some cases, these are qualitatively apparent in the form of unusual changes in thermodynamic parameter (e.g., the changes experienced by interSH2 domain polypeptides on binding). In other cases, the data can be attributed quantitatively to the particular phenomenon (e.g., the effects of ion and water sequestration of release or the folding of domains on binding a cognate ligand). The use of ITC coupled with other biophysical and structural methods provide an enormously powerful resource in our attempts to characterize the ever-growing list of recognition processes involving the extended interface.

REFERENCES

Bergqvist, S., O'Brien, R., and Ladbury, J. E. (2001). Site-specific cation binding mediates the TATA binding protein–DNA interaction from a hyperthermophilic archaeon. *Biochemistry* 40:2419–2425.

Bergqvist, S., Williams, M. A., O'Brien, R., and Ladbury, J. E. (2002). Reversal of halophilicity in a protein–DNA interaction by limited mutation strategy. *Struct Fold Des* 10:629–637.

Bergqvist, S., Williams, M. A., O'Brien, R., and Ladbury, J. E. (2003). Halophilic adaptation of protein–DNA interactions. *Biochem Soc Trans* 31:677–680.

Bergqvist, S., Williams, M. A., O'Brien, R., and Ladbury, J. E. (2004). Heat capacity effects of water molecules and ions at a protein–DNA interface. *J Mol Biol* 336:829–842.

Bradshaw, J. M., Gruzca, R. A., Ladbury, J. E., and Waksman, G. (1998). Probing the ''two-pronged plug two-holed socket'' model for the mechanism of binding of the Src SH2 domain to phosphotyrosyl peptides: a thermodynamic study. *Biochemistry* 37:9083–9090.

Bradshaw, J. M., Mitaxov, V., and Waksman, G. (1999). Investigation of phosphotyrosine recognition by the SH2 domain of the Src kinase. *J Mol Biol* 293:971–985.

Chen, M. X., Mcpartlin, A. E., Brown, L., Chen, Y. H., Barker, H. M., and Cohen, P. T. W. (1994). A novel human protein serine/threonine phosphatase, which possesses 4 tetratricopeptide repeat motifs and localizes to the nucleus. *EMBO J* 13:4278–4290.

Chen, M. S., Silverstein, A. M., Pratt, W. B., and Chinkers, M. (1996). The tetratricopeptide repeat domain of protein phosphatase 5 mediates binding to glucocorticoid receptor heterocomplexes and acts as a dominant negative mutant. *J Biol Chem* 271:32315–32320.

Chung, E., Henriques, D., Renzoni, D., Zvelebil, M., Bradshaw, J. M., Waksman, G., Robinson, C. V., and Ladbury, J. E. (1998). Mass spectral and thermodynamic studies reveal the role of water molecules in complexes formed between SH2 domains and tyrosyl phosphopeptides. *Structure* 6:1141–1151.

Cliff, M. J., Williams, M. A., Brooke-Smith, J., Barford, D., and Ladbury, J. E. (2005). Molecular recognition via coupled folding and binding in a TPR domain. *J Mol Biol* 346:717–732.

DeDecker, B. S., O'Brien, R. Fleming, P. J., Geiger, J., Jackson, S. P., and Sigler, P. B. (1996). The crystal structure of a hyperthermophilic archael TATA-box binding protein. *J Mol Biol* 264:1072–1084.

Denisov, V. P. and Halle, B. (2000). Sequence-specific binding of counterions to B-DNA. *Proc Natl Acad Sci USA* 97:629–633.

Dunker A. K., Brown, C. J., Lawson, J. D., Iakoucheva, L. M., and Obradovic, Z. (2002). Intrinsic disorder and protein function. *Biochemistry* 41:6573–6582.

Dyson H. J. and Wright, P. E. (2005). Intrinsically unstructured proteins and their functions. *Nat Rev Mol Cell Biol* 6:197–208.

Fütterer, K., Wong, J., Grucza, R. A., Chan, A. C., and Waksman, G. (1998). Structural basis for Syk tyrosine kinase ubiquity in signal transduction pathways revealed by the crystal structure of its regulatory SH2 domains bound to a dually phosphorylated ITAM peptide. *J Mol Biol* 281:523–537.

Graef, I. A., Holsinger, L. J., Diver, S., Schreiber, S. L., and Crabtree, G. R. (1997). Proximity and orientation underlie signalling by the non-receptor tyrosine kinase ZAP70. *EMBO J* 16:5618–5628.

Grucza, R. A., Futterer, K., Chan, A. C., and Waksman, G. (1999). Thermodynamic study of the binding of the tandem SH2 domain of the Syk kinase to a dually phosphorylated ITAM peptide: evidence for two conformers. *Biochemistry* 38:5024–5033.

Ha, J.-H., Capp, M. W., Hohenwalter, M. D., Baskerville, M., and Record, M. T. Jr. (1992). Thermodynamic stoichiometries of participation of water, cations and anions in specific and non-specific binding of lac repressor to DNA. Possible thermodynamic origins of the ''glutamate effect'' on protein–DNA interactions. *J Mol Biol* 228:252–264.

Ha, J.–H., Spolar, R. S., and Record, M. T. (1989). The role of the hydrophobic effect in stability of site-specific protein–DNA complexes. *J Mol Biol* 209:801–816.

Hatada, M. H., Lu, X., Laird, E. R., Green, J., Morgenstern, J. P., Lou, M., Marr, C. S., Phillips, T. B., Ram, M. K., Theriault, K., Zoller, M. J., and Karas, J. L. (1995). Molecular basis for the interaction of the protein tyrosine kinase ZAP-70 with the T-cell receptor. *Nature* 377:32–38.

Henriques, D. A., and Ladbury, J. E. (2001). Inhibitors to the Src SH2 domain: a lesson in structure-thermodynamic correlation in drug design. *Arch Biochem Biophys* 390:158–168.

Herr, W. (1992). In: McKnight, S. and Yamamoto, K. (eds), *Transcriptional Regulation.* Cold Spring Harbor Laboratory Press, Cold Spring Harbor, New York, pp. 243–265.

Herr, W. and Cleary, M. A. (1995). The POU domain: versatility in transcriptional regulation by a flexible two-in-one DNA-binding domain. *Genes Dev* 9:1679–1693.

Hud, N. V. and Polak, M. (2001). DNA–cation interactions: the major and minor grooves are flexible ionophores. *Curr Opin Struct Biol* 11:293–301.

Hyre, D. E. and Spicer, L. D. (1995). Thermodynamic evaluation of binding interactions in the methionine repressor system of *Escherichia coli* using isothermal titration calorimetry. *Biochemistry* 34:3212–3221.

Iakoucheva, L. M., Brown, C. J., Lawson, J. D., Obradovic, Z., and Dunker A. K. (2002). Intrinsic disorder in cell-signaling and cancer-associated proteins. *J Mol Biol* 323:573–584.

Jelesarov, I. and Bosshard, H. R. (1999). Isothermal titration calorimetry and differential scanning calorimetry as complementary tools to investigate the energetics of biomolecular recognition. *J Mol Recognit* 12:3–18.

Jing, H. I., Cooper, A., and Perham, R. N. (2002). Thermodynamic analysis of the binding of component enzymes in the assembly of the pyruvate dehydrogenase complex of *Bacillus stearothermophilus. Protein Sci* 11:1091–1100.

Kosa, P. F., Ghosh, G., DeDecker, B. S., and Sigler, P. B. (1997). The 2.1 angstrom crystal structure of an archaeal pre-initiation complex: TATA-box-binding protein/transcription factor (II) B core/TATA-box. *Proc Natl Acad Sci USA* 94:6042–6047.

Kozlov, A. G. and Lohman, T. M. (1998). Calorimetric studies of *E. coli* SSB protein-single-stranded DNA interactions. Effects of monovalent salts on binding enthalpy. *J Mol Biol* 278:999–1014.

Kumaran, S., Grucza, R. A., and Waksman, G. (2003). The tandem Src homology 2 domain of the Syk kinase: a molecular device that adapts to interphosphotyrosine distances. *Proc Natl Acad Sci USA* 100:14828–14833.

Ladbury, J. E. (1995). Counting the calories to stay in the groove. *Structure* 3:635–639.

Ladbury, J. E. and Arold, S. (2000). Searching for specificity in SH domains. *Chem Biol* 7:R3–R8.

Ladbury, J. E. and Chowdhry, B. Z. (1996). Sensing the heat: the application of isothermal titration calorimetry to thermodynamic studies of biomolecular interactions. *Chem Biol* 3:791–801.

Ladbury, J. E. and Williams, M. A. (2004). The extended interface: measuring non-local effects in biomolecular interactions. *Curr Opin Struct Biol* 14:562–569.

Ladbury, J. E., Wright, J. G., Sturtevent, J. M., and Sigler, P. B. (1994). A thermodynamic study of the *trp* repressor–operator interaction. *J Mol Biol* 238:669–681.

Leavitt, S. and Freire, E. (2001). Direct measurement of protein binding energetics by isothermal titration calorimetry. *Curr Opin Struct Biol* 11:560–566.

Livingstone, J. R., Spolar, R. S., and Record, M. T. Jr. (1991). Contribution to the thermodynamics of protein folding from reduction in the water-accessible non-polar surface area. *Biochemistry* 30:4237–4244.

Lohman, T. M., Overman, L. B., Ferrari, M. E., and Kozlov, A. G. (1996). A highly salt-dependent enthalpy change for *Escherichia coli* SSB protein–nucleic acid binding due to ion–protein interactions. *Biochemistry* 35:5272–5279.

Lundbäck, T., Chang, J.-F., Phillips, K., Luisi, B., and Ladbury, J. E. (2000). Characterisation of sequence-specific DNA-binding by the transcription factor Oct-1. *Biochemistry* 39:7570–7579.

Lundbäck, T. and Härd, T. (1996). Salt dependence of the free energy, enthalpy, and entropy of non-sequence specific DNA binding. *J Phys Chem* 100:17690–17695.

Merabet, E. and Ackers, G. K. (1995). Calorimetric analysis of λ cI repressor binding to DNA operator sites. *Biochemistry* 34:3212–3221.

Morton, C. J. and Ladbury, J. E. (1996). Water mediated protein–DNA interactions: the relationship of thermodynamics to structural detail. *Protein Sci* 5:2115–2118.

O'Brien, R., DeDecker, B., Fleming, K. G., Sigler, P. B., and Ladbury, J. E. (1998). The effect of salt on the TATA binding protein–DNA interaction from a hyperthermophilic archeon. *J Mol Biol* 279:117–125.

O'Brien, R., Rugman, P., Renzoni, D., Layton, M., Handa, R., Hilyard, K., Waterfield, M. D., Driscoll, P. C., and Ladbury, J. E. (2000). Alternative modes of binding of proteins with tandem SH2 domains. *Protein Sci* 9:570–579.

O'Rourke, L. and Ladbury, J. E. (2003). Specificity is complex and time consuming: Mutual exclusivity in tyrosine kinase-mediated signalling. *Accounts of Chemical Research* 36:410–416.

Ottinger E. A., Botfield, M. C., and Shoelson, S. E. (1998). Tandem SH2 domains confer high specificity in tyrosine kinase signaling. *J Biol Chem* 273:729–735.

Petri, V., Hsieh, M., and Brenowitz, M. (1995). Thermodynamic and kinetic characterization of the TATA binding protein to the adenovirus E4 promoter. *Biochemistry* 34:9977–9984.

Record, M. T. Jr., Anderson, C. F., and Lohman, T. M. (1978). Thermodynamic analysis of ion effects on the binding and conformational equilibria of proteins and nucleic acids: roles of ion association or release, screening and ion effects on water. *Q Rev Biophys* 11:103–178.

Shoemaker, B. A., Portman, J. J., and Wolynes, P. G. (2000). Speeding molecular recognition by using the folding funnel: the fly-casting mechanism. *Proc Natl Acad Sci USA* 97:8868–8873.

Spolar, R. S., Livingstone, J. R., and Record, M. T. Jr. (1992). Use of hydrocarbon and amide transfer data to estimate contributions to thermodynamic functions of protein folding from the removal of non-polar and polar surface from water. *Biochemistry* 31:3947–3955.

Spolar, R. S. and Record, M. T. Jr. (1994). Coupling of local folding to site-specific binding of proteins to DNA. *Science* 263:777–784.

Sturtevant, J. M. (1977). Heat capacity and entropy changes in processes involving proteins. *Proc Natl Acad Sci USA* 74:2236–2240.

Takeda, Y., Ross, P. D., and Mudd, C. P. (1992). Thermodynamics of Cro protein–DNA interactions. *Proc Natl Acad Sci USA* 89:8180–8184.

Thomson, J. A. and Ladbury, J. E. (2004). Isothermal titration calorimetry: a tutorial. In: Ladbury, J. E. and Doyle M. L. (eds), *Biocalorimetry 2: Applications of Calorimetry in the Biological Sciences*. Wiley, Chichester, Sussex, UK, pp. 37–58.

Thomson, J., Ratnaparkhi, G. S., Varadarajan, R., Sturtevant, J. M., and Richards, F. M. (1994). Thermodynamic and structural consequences of changing a sulfur atom to a methylene group in the M13N1e mutation in ribonuclease-S. *Biochemistry* 33:8587–8593.

Tompa, P. (2003). *J Mol Struct* 666:361–371.

Verkhivker, G. M., Bouzida, D., Gehlaar, D. K., Rejto, P. A., Freer, S. T., and Rose, P. W. (2003). Simulating disorder–order transitions in molecular recognition of unstructured proteins: where folding meets binding. *Proc Natl Acad Sci USA* 100:5148–5153.

Ward, W. H. J. and Holdgate, G. (2001). Isothermal titration calorimetry in drug discovery. In: King, F. D. and Oxford, A. W. (eds), *Progress in Medicinal Chemistry*, Vol. 38. Elsevier Science, Amsterdam, pp. 309–376.

Ward, J. J., Sodhi, J. S., McGuffin, L. J., Buxton, B. F., and Jones, D. T. (2004). Prediction and functional analysis of native disorder in proteins from the three kingdoms of life. *J Mol Biol* 337:635–645.

Williams, M. A. and Ladbury J. E. (2003). Hydrogen bonds in protein–ligand interactions. In: Böhm, H-J. and Schneider, G. (eds), *Protein–Ligand Interactions: From Molecular Recognition to Drug Design. Methods and Principles in Medicinal Chemistry Series*. Wiley-VCH, Weinheim, Germany, pp. 137–161.

Zhou, H-X. (2003). Quantitative account of the enhanced affinity of two linked scFvs specific for different epitopes on the same antigen. *J Mol Biol* 329:1–8.

9

Solvent Mediated Protein–Protein Interactions

Christine Ebel

9.1. INTRODUCTION

A solvent is an obligatory companion of biochemical and biophysical solution studies. Solutes such as salts, sugars, water-soluble small organic molecules, or polymers are often used in rather large concentrations to change macromolecular interactions. They can facilitate the macromolecule manipulation for a better description of its function. For example, solubility, folding, association, aggregation, crystallization, and so on, can be modulated by changes in the solvent composition. *In vivo*, changes in the concentrations of the macromolecules, ligands, and also small solutes and water control the cellular processes. Osmolytes can be accumulated in molar amount in organisms in response to various cellular stresses.

Basic reviews exist in the field (Eisenberg, 1976, 1994; Timasheff, 1993, 1998; Parsegian *et al.*, 1995; Record *et al.*, 1998c; Parsegian, 2002; Schellman, 2003; Cayley and Record, 2004; Rand, 2004); other fundamental and/or recent works are cited within the text. Our objective here is not to be exhaustive but to provide a basic understanding and some tools concerning solvent-mediated protein–protein interactions for nonspecialist biochemists and biophysicists. The aim is to help readers to relate some of the approaches that are described in the literature (however, e.g., analysis in terms of osmotic or depletion forces is not described).

Approaches are various because of the weakness of the considered interactions. Thermodynamics in general considers the energetics related to solution composition, determined by molecular interactions and in an equivalent way spatial distribution of the molecules within the solution. This allows the various measurements that are made on solutions containing macromolecules in solution to

C. EBEL • Institut de Biologie Structurale UMR 5075 CEA-CNRS-UJF, 41 rue Jules Horowitz, 38027 Grenoble Cedex 1 France. Tel: +33 4 38 78 96 38; fax: +33 4 38 78 54 94; E-mail: christine.ebel@ibs.fr

be explained in the framework of thermodynamics. Depending on the systems studied, different languages are chosen, either using particles and complexes—a monomer, a dimer, a folded, and an unfolded protein binding a number of water and ligand molecules—or statistical thermodynamics describing particle distributions. The best description rationalizes the experimental data and gives sense to the experimental features.

After this introduction, Section 9.2 describes the thermodynamic background, based on the general Gibbs–Duhem equation. The thermodynamic equations are precisely derived, using parameters that are precisely defined: components, concentrations, volumes, and chemical potentials.

Section 9.3 presents the protein interacting with the solvent. First, a thermodynamic description is given using preferential solvent interaction parameters, which are rigorously defined and are also experimentally accessible parameters. We present the meaning of the different preferential solvent interaction parameters used in the literature and how they relate. Second, the solvated protein is described in terms of water and cosolvent distributions, with mainly two approaches: one in which solvent components are localized in a domain closed to the macromolecule; the other where the contribution of solvent fluctuations in the whole sample are considered.

Section 9.4 addresses intermacromolecular interactions, first in terms of discrete associated species—described with association constants—then in terms of weak interactions—described with second virial coefficients—and how the weak interactions can be related to the spatial distribution of macromolecule and effective intermacromolecular potentials.

Section 9.5 is the key section because it describes the thermodynamic linkage between macromolecule–solvent and macromolecule–macromolecule interactions.

Section 9.6 outlines some models used for the description of solvation—those used to interpret the preferential binding parameters in terms water and cosolvent binding and/or exchange—and for the origin of solvation—affinity of cosolvent for the macromolecule. This also includes a discussion of properties of the solvent, solvophobic/osmophobic effects, and the effect of the size of the cosolvent, with a mention of polyelectrolyte and salt effects.

Section 9.7 gives some examples of perturbation of macromolecular equilibrium by cosolvents: the general effects of salts and osmolytes on macromolecular stability and interactions; how the actions of osmolytes *in vivo* and the adaptation of halophilic proteins to high salts are understood; It also gives one example of linkage of detergent binding and protein oligomerization.

Section 9.8 describes the general principles of different techniques: density; osmometry; sedimentation equilibrium and velocity; scattering of light, X-rays, and neutrons. The formalism that allows us to obtain separately the parameters related to macromolecule–solvent interactions and to the macromolecule itself, and the macromolecule molar mass and macromolecule–macromolecule interactions will be given.

9.2. THERMODYNAMICS BACKGROUND

9.2.1. Components

We consider a three-component system: the solvent composed of water (component 1), one type of macromolecule (component 2), and one additional small solute: the cosolvent (component 3). When components 2 and 3 are electrolytes, they have to be defined as electroneutral combination of species (Eisenberg, 1976).

9.2.2. The Gibbs–Duhem Equation

For an open system that exchanges work with its environment (reservoir) only in terms of expansion/compression and entrance/removal of material, in a reversible way and for infinitesimal processes, the fundamental equation of thermodynamics leads to the Gibbs–Duhem equation

$$S\ \mathrm{d}T - V\ \mathrm{d}P + n_\mathrm{J}\ \mathrm{d}\mu_\mathrm{J} = 0. \tag{9.1}$$

The entropy S, volume V, and number of moles n_J are extensive variables, which depend on the size of the system. The temperature T, pressure P, and chemical potential μ_J are intensive variables, independent of the size of the system. Expressed in molal units (m_J and V_m are defined in Section 9.2.3 and 9.2.4) and for the case of the three-component system considered here, at constant temperature, the Gibbs–Duhem equation is written as

$$V_\mathrm{m}\,\mathrm{d}P + m_1^*\ \mathrm{d}\mu_1 + m_2\ \mathrm{d}\mu_2 + m_3\ \mathrm{d}\mu_3 = 0. \tag{9.2}$$

9.2.3. Concentrations

Depending on the experimental and theoretical approaches used, concentrations are expressed in various scales. We consider a solution of total volume V comprising components J with mole numbers n_J and molar mass M_J. The concentrations C_J (molarity expressed in mole/L or M) or $c_\mathrm{J} = 1,000\ C_\mathrm{J} \times M_\mathrm{J}$ (c_J expressed in g/mL) are volume based. The concentrations m_J (molality is expressed as mole of J per kg of water; units: m), w_J (weight molalities: gram of J per gram of water), and mole ratio n_J/n_1 (mole of J per mole of water) associate relative numbers of mole or grams in the solution and are insensitive to volume variation. They are of particular interest in thermodynamic derivations of the Gibbs–Duhem equation at constant pressure P and temperature T (see later). Note that m_1 is equal to 55.5 mol/kg (1,000/18 is the number of grams in 1 kg of water divided by its molar mass). We label it as m_1^*, to indicate that it is a constant (Anderson *et al.*, 2002). In the following, the subscript $^\circ$ is given on the condition of infinite dilution of macromolecule [except for μ_J°, Eq. (9.5a, b)].

9.2.4. Volume

The partial molal volume and specific volume of the component J are noted $\overline{V}_{\rm J}$ (mL/mole) and $\overline{v}_{\rm J} = \overline{V}_{\rm J}/M_{\rm J}$ (mL/g), respectively. If V_m (mL) is the volume of the solution containing 1 kg of water, $\overline{V}_2 = (\partial V_{\rm m}/\partial m_2)_{PTm_3}$ and $\overline{V}_3 = (\partial V_{\rm m}/\partial m_3)_{PTm_2}$. The indices indicate the constancy of the parameters P, T, and m_2 or m_3. The partial volume of a component describes not only the changes in the volume of the solution related to the molecular occupation of the component, but also the changes in the volume occupied by other components in the solution caused by its presence. For example, the partial specific volumes of proline, salts, nucleic acids, and acidic proteins increase with the solvent salt concentration (Eisenberg, 1990; Kernel *et al.*, 1999; Courtenay *et al.*, 2000; Ebel *et al.*, 2002). In the case of electrolytes, this arises from water electrostriction being larger in dilute solution, which can be understood considering that in high salt most water molecules are already under the influence of a salt ion. The conversion between the molality and molarity scales depends on knowledge of $V_{\rm m}$

$$C_{\rm J} = 1000 m_{\rm J}/V_{\rm m}. \tag{9.3a}$$

$V_{\rm m}$ can be calculated using the solution density, ρ (g/mL) or $\overline{v}_{\rm J}$ values

$$V_{\rm m} = (1000 + M_2 m_2 + M_3 m_3)/\rho = 1000 \times (\overline{v}_1 w_1 + \overline{v}_2 w_2 + \overline{v}_3 w_3). \tag{9.3b}$$

9.2.5. Chemical Potential

The chemical potential $\mu_{\rm J}$ of the component J describes the change in the Gibbs free energy of the solution with respect to the change in concentration of the component J

$$\mu_{\rm J} = (\partial G/\partial n_{\rm J})_{PTn_{{\rm J}' \neq {\rm J}}}. \tag{9.4}$$

Expressed in terms of molality of J, it can be written using R, the gas constant; $\mu_{\rm J}^\circ$, the standard chemical potential; and the activity $a_{\rm J} = m_{\rm J}\gamma_{\rm J}$, with $\gamma_{\rm J}$ the activity coefficient (the values of $\mu_{\rm J}^\circ$, $a_{\rm J}$, and $\gamma_{\rm J}$ are related to the choice of the concentration scale)

$$\mu_{\rm J} = \mu_{\rm J}^\circ + RT \ln a_{\rm J} = \mu_{\rm J}^\circ + RT \ln m_{\rm J} + RT \ln \gamma_{\rm J} \tag{9.5a}$$

If the component J is defined as a combination of species (in the case of an electrolyte), and if $\nu_{\rm kJ}$ is the number of mole of species k included per mole of component J, the chemical potential is

$$\mu_{\rm J} = \mu_{\rm J}^\circ + RT \sum_{\rm k} \nu_{\rm kJ} \ln m_{\rm k} + RT \ln \gamma_{\rm J}. \tag{9.5b}$$

The $\gamma_{\rm J}$ value depends on the definition of the component in terms of species, which in general can be made in different ways, provided the electroneutrality of the components is preserved. $\mu_{\rm J}$ is related, *via* RT ln $m_{\rm J}$ or $RT \sum_{\rm k} \nu_{\rm kJ} \ln m_{\rm k}$, to the number of species in the solution. This contribution to μ_J is expected from ideal mixing entropy. The species can be common to different components of the solution, for example in the case of an ion common to a polyelectrolyte macromolecule and solvent salt: m_k then is expressed as a function of the concentrations of both components in solution (Eisenberg, 1976; Costenaro *et al.*, 2002). The excess potential $RT \ln \gamma_J$ is in general a function of pressure, temperature, and solution composition. This is related to all the intermolecular interactions, including interactions of the component with itself (Anderson *et al.*, 2002).

9.3. DESCRIPTION OF THE SOLVATION OF MACROMOLECULES IN TWO-COMPONENT SOLVENT

9.3.1. Preferential Cosolvent and Water-Binding Parameters

Cosolvent-binding parameters—or the related preferential hydration parameters—are key thermodynamic parameters that can be modeled in a structural description of water and cosolvent binding. They can be obtained from the experiment, as will be described in Chapter 8. They express the change in the concentration of cosolvent related to the presence of the macromolecule $(\partial m_3/\partial m_2)$. Depending on the experimental conditions, they can be defined in different ways, such as $(\partial m_3/\partial m_2)_{TP\mu_3}$, $(\partial m_3/\partial m_2)_{TP\mu_1}$, $(\partial m_3/\partial m_2)_{TP\mu_2}$, or $(\partial m_3/\partial m_2)_{T\mu_1\mu_3}$. The indices express the conditions of constancy of temperature, pressure, and/or chemical potential of the mentioned component.

9.3.1.1. $(\partial m_3/\partial m_2)_{TP\mu_3}$

In the condition of constant temperature, pressure, and chemical potential of a cosolvent, the preferential binding parameter is expressed as

$$(\partial m_3/\partial m_2)_{TP\mu_3} = -\frac{(\partial \mu_2/\partial m_3)_{TPm_2}}{(\partial \mu_3/\partial m_3)_{TPm_2}} = -(\partial \mu_2/\partial \mu_3)_{TPm_2}. \tag{9.6}$$

Because this relation is fundamental for our purpose (see Section 9.5), we demonstrate it here. Since μ_3 is a function of T, P, m_2 and m_3

$$\begin{aligned} d\mu_3 &= (\partial \mu_3/\partial T)_{Pm_2m_3}\, dT + (\partial \mu_3/\partial P)_{Tm_2m_3}\, dP \\ &\quad + (\partial \mu_3/\partial m_2)_{TPm_3} dm_2 + (\partial \mu_3/\partial m_3)_{TPm_2} dm_3. \end{aligned} \tag{9.6a}$$

When T and P are constant, this reduces to

$$d\mu_3 = (\partial \mu_3/\partial m_2)_{TPm_3}\, dm_2 + (\partial \mu_3/\partial m_3)_{TPm_2}\, dm_3. \tag{9.6b}$$

Then, we consider the variation of μ_3 with m_2 at constant T and P

$$(\partial \mu_3/\partial m_2)_{TP} = (\partial \mu_3/\partial m_2)_{TPm_3} + (\partial \mu_3/\partial m_3)_{TPm_2}\, (\partial m_3/\partial m_2)_{TP}. \tag{9.6c}$$

When μ_3 is constant, $(\partial\mu_3/\partial m_2)_{TP} = 0$ and $(\partial m_3/\partial m_2)_{TP}$ is written as $(\partial m_3/\partial m_2)_{TP\mu_3}$

$$0 = (\partial\mu_3/\partial m_2)_{TPm_3} + (\partial\mu_3/\partial m_3)_{TPm_2}(\partial m_3/\partial m_2)_{TP\mu_3}. \qquad (9.6d)$$

Because $\mu_2 = (\partial G/\partial m_2)_{PTm_3}$ and $\mu_3 = (\partial G/\partial m_3)_{PTm_2}$, the cross-differentiation relationship applies (Cantor and Schimmel, 1980) and gives $(\partial\mu_3/\partial m_2)_{TPm_3} = (\partial\mu_2/\partial m_3)_{TPm_2}$, leading, after rearrangement to Eq. (9.6). This tells us that $(\partial m_3/\partial m_2)_{TP\mu_3}$ is strictly related to the relative changes in the chemical potentials of the macromolecule and the cosolvent with cosolvent concentration. However, there is no experimental device that simultaneously maintains both constant pressure and chemical potential of a cosolvent. Now we examine how the experimentally accessible parameters $(\partial m_3/\partial m_2)_{T\mu_1\mu_3}$ and $(\partial m_3/\partial m_2)_{TP\mu_1}$ relate to $(\partial m_3/\partial m_2)_{TP\mu_3}$.

9.3.1.2. $(\partial m_3/\partial m_2)_{T\mu_1\mu_3}$

Dialysis equilibrium with an infinite reservoir is the experimental device that corresponds to conditions of constant chemical potentials of cosolvent and water (Figure 9.1A). In that case, the presence of the macromolecule induces osmotic pressure Π in the dialysis bag. $(\partial m_3/\partial m_2)_{T\mu_1\mu_3}$ is related to $(\partial m_3/\partial m_2)_{TP\mu_3}$ (Eisenberg, 1976)

$$(\partial m_3/\partial m_2)_{T\mu_1\mu_3} = (\partial m_3/\partial m_2)_{TP\mu_3} - (\overline{V}_3/(\partial\mu_3/\partial m_3)_{TPm_2}).(\mathrm{d}\Pi/\mathrm{d}m_2). \qquad (9.7)$$

To have an estimate of the difference between $(\partial m_3/\partial m_2)_{T\mu_1\mu_3}$ and $(\partial m_3/\partial m_2)_{TP\mu_3}$, we omit the nonideality terms and consider a case where component 3 is a salt that dissociates: $(\partial\mu_3/\partial m_3)_{TPm_2}$ reduces to $RT(\sum_k \nu_{3k})/m_3$. The limit of $(\mathrm{d}\Pi/\mathrm{d}m_2)$ at low m_2 is RT/V_m°. Thus, at low protein concentration, the second term of the equation can be approximated as $\overline{V}_3^\circ m_3/(V_m^\circ \sum_k \nu_{3k})$. We consider a salt such as NaCl that dissociates into $\sum_k \nu_{3k} = 2$ species, with $\overline{V}_3 = 25\,\mathrm{mL}$ and $m_3 = 1\,\mathrm{m}$ (1 mole/kg water, i.e., a rather large concentration), $V_m^\circ = 1{,}000\,\mathrm{mL}$. It provides a value for the second term of Eq. (9.7) of 0.01 mole of cosolvent per mole of macromolecule, which is below the usual experimental limit (Casassa and Eisenberg, 1964). To give an order of magnitude, experimental values of $(\partial m_3/\partial m_2)_{T\mu_1\mu_3})$ typically range between some units to some tenths of units (Timasheff, 1993; Courtenay *et al.*, 2000; Ebel *et al.*, 2002). Increasing the molar mass of the cosolvent, especially if it is a neutral compound, would increase this number, since $\overline{V}_3 = \bar{v}_3^\circ M_3$, but, even for the protein bovine serum albumin (BSA) at about 3×10^{-3} m - (more than 100 mg/mL !), in the presence of various osmolytes in the multimolal range, $(\partial m_3/\partial m_2)_{T\mu_1\mu_3}$ and $(\partial m_3/\partial m_2)_{TP\mu_3}$ could not be distinguished (Courtenay *et al.*, 2000).

9.3.1.3. $(\partial m_3/\partial m_2)_{TP\mu_1}$

The condition of constant chemical potential of water and pressure is that experimentally obtained in isopiestic distillation or vapor pressure osmometry

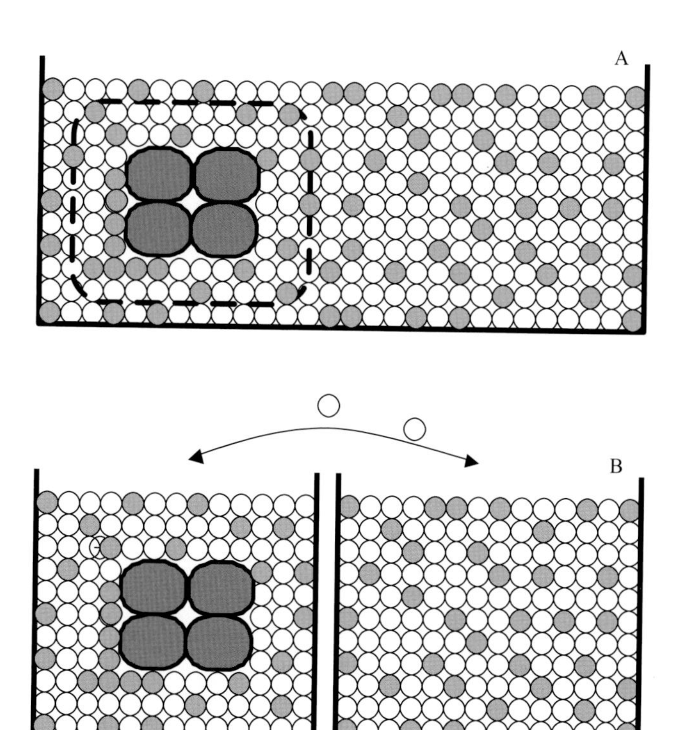

Figure 9.1. Schematic representation of dialysis and isopiestic distillation. Water and cosolvent are represented by white and grey balls, respectively. (A) Dialysis: The dialysis bag is permeable to water and cosolvent, their chemical potentials are fixed by their values in the reservoir. The bag is not permeable to macromolecule and the pressure in the compartment containing the macromolecule differs from that outside by the osmotic pressure Π. (B) Isopiestic distillation: The chemical potential of water is determined by its value in a reservoir; the pressure is the same for the two compartments.

(Figure 9.1B). For BSA at about 3×10^{-3} m (more than 100 mg/mL), in the presence of various osmolytes in the multimolal range, $(\partial m_3/\partial m_2)_{TP\mu_1}$ differs by about 10 mole/mole from $(\partial m_3/\partial m_2)_{TP\mu_3} \approx (\partial m_3/\partial m_2)_{T\mu_1\mu_3}$ (Courtenay *et al.*, 2000). This is not negligible. Anderson *et al.* (2002) revisited the thermodynamic relations between $(\partial m_3/\partial m_2)_{TP\mu_1}$ and $(\partial m_3/\partial m_2)_{TP\mu_3}$. At high macromolecular dilution, the equation reduces to

$$(\partial m_3/\partial m_2)_{TP\mu_3} - (\partial m_3/\partial m_2)_{TP\mu_1} \approx \left[\frac{m_2(\partial \mu_2/\partial m_2)_{TPm_3}}{m_3(\partial \mu_3/\partial m_3)_{TPm_2}}\right]. \tag{9.8}$$

$(\partial \mu_2/\partial m_2)_{Pm_3}$ and $(\partial \mu_3/\partial m_3)_{Pm_2}$ depend on the number of species released by the cosolvent and macromolecule in the solution (see Section 9.2.5). Thus, even at high dilution and independent of the interactions of the macromolecule with the cosolvent, the difference can be significant, particularly if the macromolecule is charged and does not have a common ion with the cosolvent. Equation (9.8) has been developed for different practical cases and numerical applications are given in Anderson *et al.* (2002).

9.3.1.4. Preferential Hydration Parameter

The preferential hydration parameter when written as $\partial m_1/\partial m_2$ does not have an obvious meaning, since m_1 by definition is constant. It would have to be written $(\partial(n_1/n_3)/\partial(n_2/n_3))$, n_1, n_2, and n_3 being numbers of molecules, i.e., with reference to the concentration of a component other than water. Often used when the phenomenon of hydration is emphasized, it is related to $\partial m_3/\partial m_2$ by the solvent composition through the ratio m_1/m_3 at constant μ_1 and μ_3 (Eisenberg, 1976)

$$(\partial m_1/\partial m_2)_{T\mu_1\mu_3} = -(m_1/m_3)(\partial m_3/\partial m_2)_{T\mu_1\mu_3}. \tag{9.9}$$

9.3.1.5. Units

$\partial m_3/\partial m_2$ and $\partial m_1/\partial m_2$ (mole/mole), often named Γ_3 and Γ_1, can be converted to their analogs $\partial w_3/\partial w_2$ and $\partial w_1/\partial w_2$ in g/g units, often named ξ_3 and ξ_1, respectively

$$\partial w_3/\partial w_2 = (M_3/M_2)(\partial m_3/\partial m_2) \tag{9.10a}$$

$$\partial w_1/\partial w_2 = (M_1/M_2)(\partial m_1/\partial m_2) \tag{9.10b}$$

with

$$(\partial w_1/\partial w_2)_{T\mu_1\mu_3} = -(1/w_3)(\partial w_3/\partial w_2)_{T\mu_1\mu_3}. \tag{9.10c}$$

9.3.2. Description of the Solvated Protein in Terms of Water and Cosolvent Distribution

Equation (9.6) tells us that because of the interactions between the molecules in solution, the solvent composition changes as a consequence of the presence of the macromolecule. Since the cosolvent-binding parameters in the conditions of constant μ_3 and P on one hand and of constant μ_1 and μ_3 on the other are experimentally indistinguishable (Section 9.3.1.2), we focus now on the dialysis experiment. The preferential binding is a measure of the difference in the cosolvent molality in the dialysis bag, m_3^{local}, and the bulk, m_3^{bulk}

$$(\partial m_3/\partial m_2)_{T\mu_1\mu_3} = \frac{\left(m_3^{\text{local}} - m_3^{\text{bulk}}\right)}{m_2}. \tag{9.11}$$

It corresponds to an equivalent number of moles of solute that would have to be added (or removed) with respect to 1 mole of the macromolecule to maintain the constancy of the chemical potentials of water and cosolvent. The preferential hydration parameter corresponds to an equivalent number of moles of water that would have to be added (or removed) with respect to 1 mole of the macromolecule in the (same) dialysis experiment.

9.3.2.1. The Two-Domain Approach

In a structural approach in which we consider a domain close to the protein containing N_1 moles of water and N_3 moles of solute per mole of macromolecule, the preferential solvent-binding parameters are expressed as

$$(\partial m_3/\partial m_2)_{T\mu_1\mu_3} = N_3 - N_1 \times (m_3^{\text{bulk}}/m_1^*). \quad (9.12a)$$

$$(\partial m_1/\partial m_2)_{T\mu_1\mu_3} = N_1 - N_3 \times (m_1^*/m_3^{\text{bulk}}). \quad (9.12b)$$

The size of the domain is not important. A combination of solvent molecules in the domain producing the same composition as the bulk does not contribute to $(\partial m_3/\partial m_2)_{T\mu_1\mu_3}$ (Figure 9.2). The same equation can be written in terms of weight concentrations, using water and cosolvent numbers B_1 and B_3, in a gram of water and cosolvent, respectively, per gram of macromolecule

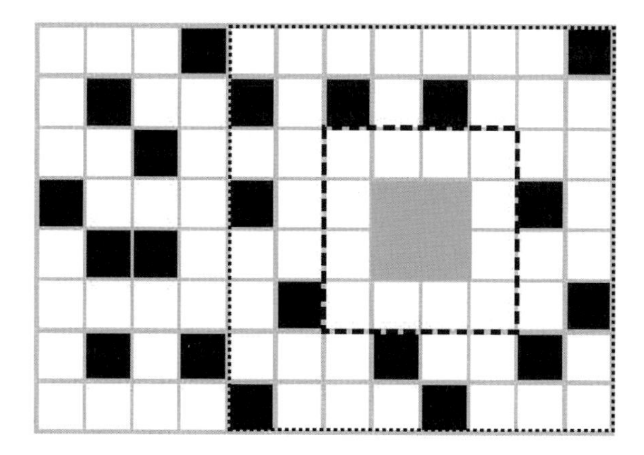

Figure 9.2. Understanding preferential solvent-binding parameters. Small white square: water (1) small black square: small solute (3) large gray square: macromolecule (2) In this example, the ratio of water on solute m_1/m_3 is 3 in the bulk solvent and 12 molecules of water are bound to the macromolecule. They have to be added with the (nude) macromolecule to maintain the chemical potential of the solvent constant: $(\partial m_3/\partial m_2)_{T\mu_1\mu_3} = 12$. On the other hand, we can consider that four molecules of small solute, previously associated with the 12 molecules of water in the solvent, were removed on the introduction of the macromolecule, so that $(\partial m_3/\partial m_2)_{T\mu_1\mu_3} = -4$. The preferential solute-binding parameter and preferential hydration parameter are related by the solvent composition. Note that the hypothesis of strong binding is not needed, nor the definition of the limit of the domain which can contain unperturbed solvent: the same value for the preferential binding parameters are obtained if counting the molecules of water in the three domains defined by the dashed line, dotted line, or whole figure. From Costenaro and Ebel (2002), with IUCr's copyright permission (http://journals.iucr.org/).

$$(\partial w_3/\partial w_2)_{T\mu_1\mu_3} = B_3 - B_1 \times (w_3^{\text{bulk}}/w_1^{\text{bulk}}). \tag{9.12c}$$

$$(\partial w_1/\partial w_2)_{T\mu_1\mu_3} = B_1 - B_3 \times (w_1^{\text{bulk}}/w_3^{\text{bulk}}). \tag{9.12d}$$

Some models frequently found in the literature are given in Section 9.6.

9.3.2.2. Excess Solvation Numbers and Cosolvent in Terms of Radial Distribution Functions

The statistical thermodynamics approach theory provides a direct link between microscopic (structural) and macroscopic (thermodynamic) pictures of solvation (Chitra and Smith, 2001; Parsegian, 2002; Shimizu, 2004). The radial distribution function $g_{2J}(r)$ counts—in a normalized way—the number of molecules of component J localized at a distance r from component 2 (Figure 9.3A). $g_{2J}(r)$ is above/below 1 if this number exceeds/is less than that expected for an ideal distribution. The Kirkwood–Buff integral $G_{2J} = \int_0^\infty [g_{2J}(r) - 1]\, 4\pi r^2 \mathrm{d}r$ is positive if there is an excess of J around the component 2, when compared with the distribution expected statistically, and negative in the opposite case. G_{2J} relates to the excess solvation numbers N_{21} and N_{23}, defined as $N_{21} = C_1 G_{21}$ and $N_{23} = C_3 G_{23}$, and to the preferential solvent-binding parameter

$$(\partial m_3/\partial m_2)_{PT\mu_3} = C_3(G_{23} - G_{21}) = N_{23} - (C_3/C_1)N_{21}. \tag{9.13}$$

The excess solvation numbers N_{23} and N_{21} differ from the above-mentioned N_3 and N_1 in particular by the fact that each of them contains a negative contribution arising from the volume occupied by the macromolecule. However, their contributions to $(\partial m_3/\partial m_2)_{PT\mu_3}$ cancel out.

9.4. MACROMOLECULE–MACROMOLECULE INTERACTIONS

9.4.1. In Terms of Discrete Species

We consider here an equilibrium in solution between reactants R_1, $R_2 \ldots$ and products P_1, $P_2 \ldots$, neglecting all interactions with the solvent components. At constant pressure and temperature, for a chemical equilibrium

$$\nu_{R_1} R_1 + \nu_{R_2} R_2 + \cdots <=> \nu_{P_1} P_1 + \nu_{P_2} P_2 + \cdots. \tag{9.14}$$

Restricting to a system with R_1, R_2, P_1, and P_2, the equilibrium condition $\Delta G = 0$ is expressed as

$$\nu_{P_1}\mu_{P_1} + \nu_{P_2}\mu_{P_2} - \nu_{R_1}\mu_{R_1} - \nu_{R_2}\mu_{R_2} = 0. \tag{9.15a}$$

Developing the chemical potential according to Eq. (9.5a) gives

$$\nu_{P_1}\mu^\circ_{P_1} + \nu_{P_2}\mu^\circ_{P_2} - \nu_{R_1}\mu^\circ_{R_1} - \nu_{R_2}\mu^\circ_{R_2} = \mathrm{RT}[\nu_{P_1}\ln a_{P_1} + \nu_{P_2}\ln a_{P_2} - \nu_{R_1}\ln a_{R_1} - \nu_{R_2}\ln a_{R_2}], \tag{9.15b}$$

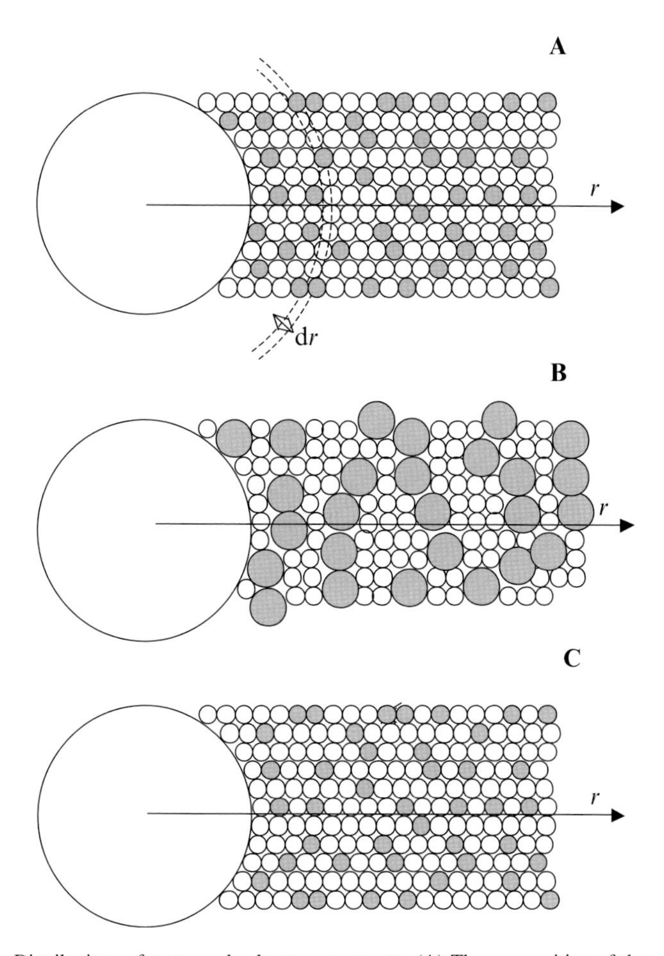

Figure 9.3. Distributions of water and solvent components. (A) The composition of the solvent is not affected by the macromolecule and the size of cosolvent is similar to that of water. (B) The composition of the solvent is not affected by the macromolecule but the size of cosolvent is large compared with that of water. (C) There is accumulation of water on the macromolecule surface, which can have various origins (see the text). The radial distribution functions $g_{21}(r)$, $g_{23}(r)$, respectively, count the numbers of water, of cosolvent molecules, respectively, at a distance between r to $r + dr$ and compare these numbers to that expected for a statistical distribution of the particles. For distances from the macromolecule surface below the ''radius'' of the cosolvent in Figure B, cosolvent is excluded by the effect of its size. In Figures B and C, $(\partial m_3/\partial m_2)_{T\mu_1\mu_3}$ is negative.

which can be rewritten as

$$\Delta G^\circ = -RT \ln K, \tag{9.15c}$$

relating two constant values. ΔG° is linked to chemical potentials at infinite dilution

$$\Delta G^\circ = \nu_{P_1}\mu^\circ_{P_1} + \nu_{P_2}\mu^\circ_{P_2} - \nu_{R_1}\mu^\circ_{R_1} - \nu_{R_2}\mu^\circ_{R_2}. \tag{9.15d}$$

K, the association constant, is expressed as a function of the activity coefficients

$$K = \frac{\left[a_{\mathrm{P}_1}^{\nu_{\mathrm{P}_1}} \times a_{\mathrm{P}_2}^{\nu_{\mathrm{P}_2}}\right]}{\left[a_{\mathrm{R}_1}^{\nu_{\mathrm{R}_1}} \times a_{\mathrm{R}_2}^{\nu_{\mathrm{R}_2}}\right]} = \left(\frac{\left[m_{\mathrm{P}_1}^{\nu_{\mathrm{P}_1}} \times m_{\mathrm{P}_2}^{\nu_{\mathrm{P}_2}}\right]}{\left[m_{\mathrm{R}_1}^{\nu_{\mathrm{R}_1}} \times m_{\mathrm{R}_2}^{\nu_{\mathrm{R}_2}}\right]}\right) \times \left(\frac{\left[\gamma_{\mathrm{P}_1} \times \gamma_{\mathrm{P}_2}\right]}{\left[\gamma_{\mathrm{R}_1} \times \gamma_{\mathrm{R}_2}\right]}\right). \tag{9.15e}$$

Restraining to ideal conditions, $\gamma_{\mathrm{J}} = 1$

$$K = \frac{\left[m_{\mathrm{P}_1}^{\nu_{\mathrm{P}_1}} \times m_{\mathrm{P}_2}^{\nu_{\mathrm{P}_2}}\right]}{\left[m_{\mathrm{R}_1}^{\nu_{\mathrm{R}_1}} \times m_{\mathrm{R}_2}^{\nu_{\mathrm{R}_2}}\right]}. \tag{9.16}$$

For an autoassociation process: ν Monomer <=> Multimer, it would reduce to $K = m_{\mathrm{multimer}}/m^{\nu}_{\mathrm{monomer}}$. The units of K are $m^{(\nu_{\mathrm{P}_1}+\nu_{\mathrm{P}_2}-\nu_{\mathrm{R}_1}-\nu_{\mathrm{R}_2})}$ ($m^{\nu-1}$ in the case of the monomer–multimer equilibrium). K can be converted to other units of scale, the most familiar and appropriate to biochemists being the molarity scale (the value of μ_{J}° depends also on the scale chosen). In addition, dissociation constants $= 1/K$ are generally considered. The units that are used do not change the value of $\partial \ln K$, which gives the change in free energy at equilibrium (Eq. (9.15c)), when changing, for example, the solvent composition (Section 9.5).

9.4.2. Weak Interactions

When association constants are weak, it could be more appropriate to consider interparticle interactions in terms of nonideality. Nonideality is quantified through the measurement of second virial coefficients. These are related to the osmotic pressure, Π, which is a change in the pressure related to the presence of the macromolecule, when maintaining the chemical component of the solvent components constant, as in a dialysis experiment. In these conditions, i.e., with $\mathrm{d}T = 0$, $\mathrm{d}P = \mathrm{d}\Pi$, $\mathrm{d}\mu_1 = \mathrm{d}\mu_3 = 0$, the Gibbs–Duhem equation reduces to $V\mathrm{d}P = n_2\mathrm{d}\mu_2$, from which

$$V_{\mathrm{m}}(\mathrm{d}\Pi/\mathrm{d}m_2) = m_2(\partial\mu_2/\partial m_2)_{T\mu_1\mu_3}. \tag{9.17}$$

Π can also be expressed in terms of changes in water activity (at constant P), with the only reasonable assumption that water partial specific volume is invariant when increasing the pressure from P to $P + \Pi$ (Eisenberg, 1976; Costenaro and Ebel, 2002). This is because the partial molal volume $\overline{V_{\mathrm{J}}}$ is by definition $(\partial\mu_{\mathrm{J}}/\partial P)_{Tn\mathrm{J}}$. This conversion is experimentally made as soon as the sample containing the macromolecule is separated from the solvent bath, after a dialysis experiment or after gel filtration.

For an ideal (dilute) solution of macromolecule, osmotic pressure Π is directly related to the number of macromolecules in the solution, as is pressure P for an ideal (dilute) gas: $PV = nRT$. In the molality scale, in the limit of infinite macromolecule dilution: $\Pi\, V_{\mathrm{m}}^{\circ} = m_2RT$. In Table 9.1, we report expressions of osmotic pressure and osmotic pressure derivative when using different concentration units. For concentrations expressed by weight, osmotic pressure is related to macromolecule molar mass. We present in Section 9.8 that $(\partial\Pi/\partial c_2)_{T\mu_1\mu_3}$ is obtained through sedimentation and scattering experiments.

For a real solution, the deviation from ideal behavior gives information about the weak interparticle interactions. The osmotic pressure may be expanded in integral powers of concentration of the macromolecule by using virial coefficients. Depending on the concentration units used, different second virial coefficients are expressed (Table 9.1). We refer now to A_2, which is obtained experimentally from sedimentation and scattering experiments (Section 9.8). If A_2 is positive, weak interactions are globally repulsive. Conversely, if A_2 is negative, they are globally attractive. A given macromolecule concentration corresponds to a larger osmotic pressure when there are repulsive interactions between the macromolecules and to a smaller one if there are attractive intermacromolecular interactions.

9.4.3. Weak Interactions in Terms of the Radial Distribution Function

Osmotic pressure is related to the equilibrium spatial distribution of the macromolecules. Restricting the consideration to quasispherical particles, A_2 is related to the pair correlation function $g_{22}(r)$, which expresses the probability of finding macromolecules at a distance r

$$A_2 = -(N_A/2M_2^2)\int_0^\infty [g_{22}(r) - 1]4\pi r^2 \mathrm{d}r. \tag{9.18}$$

A negative value of A_2 is related to an excess of macromolecule when compared with a statistical repartition, reflecting attraction between the macromolecule. For weakly interacting spheres, the pair correlation function $g_{22}(r)$ can be related, in terms of MacMillan–Mayer theory (1945), to the pair interaction potential of mean force $W(r)$

$$g_{22}(r) = \exp\left[\frac{-W(r)}{kT}\right]. \tag{9.19}$$

$W(r)$ is the difference in free energy between the solution with two macromolecules at separation r and the same at infinite dilution. It is an effective pair potential, which includes the interactions of the solvent components. It also implicitly contains many-body correlations, related to the presence of all but the two considered macromolecules: only at infinite dilution it coincides with the pair potential that can be defined between two infinitely diluted macromolecules (Belloni, 1991). $g_{22}(r)$ determines the structure factor measured in scattering techniques (see Section 9.8). $g_{22}(r)$ and $W(r)$ can be computed from assumptions on all the direct potentials between all the solution molecules: macromolecules and solvent components (Belloni, 1991; Malfois *et al.*, 1996; Tardieu *et al.*, 1999; Bostrom *et al.*, 2005). Because two macromolecules cannot occupy the same place, $W(r)$ contains always a repulsive contribution of excluded volume, corresponding to a positive contribution in A_2.

Table 9.1. Second Virial Coefficients

Concentration unit	Osmotic pressure	Osmotic pressure derivative	Second virial coefficient
C_2 mole/L	$\Pi 1{,}000/(RT) = C_2 + B^{c} C_2^2 + \cdots$	$(\mathrm{d}\Pi/\mathrm{d}C_2)1{,}000/(RT) = 1 + 2B^{c} C_2 + \cdots$	B^{c}
m_2 mole/kg of water	$\Pi V_{m}^{o}/(RT) = m_2 + B^{w} m_2^2 + \cdots$	$(\mathrm{d}\Pi/\mathrm{d}m_2)V_{m}^{o}/(RT) = 1 + 2B^{w} m_2 + \cdots$	$B^{w} = 1{,}000B^{c}/V_{m}^{o} - (\partial V_{m}/\partial m_2)^{o}/V_{m}^{o}$
c_2 g/mL	$\Pi/(RT) = c_2/M_2 + A^{c} c_2^2 + \cdots$	$(\mathrm{d}\Pi/\mathrm{d}c_2)/(RT) = 1/M2 + 2A^{c} c_2 + \cdots$	$A^{c} = A_2 = 1{,}000B^{c}/M_2^2$
w_2 g/g of water	$\Pi V_{m}^{o}/(1{,}000RT) = w_2/M_2 + A^{w} w_2^2 + \cdots$	$(\mathrm{d}\Pi/\mathrm{d}w_2)V_{m}^{o}/(1{,}000RT) = 1/M_2 + 2A^{w} w_2 + \cdots$	$A^{w} = 1{,}000B^{w}/M_2^2$

The derivations are made at constant temperature and chemical potential of the solvent components. (Cassassa and Eisenberg, 1964).

9.5. LINKAGE BETWEEN PROTEIN–SOLVENT AND PROTEIN–PROTEIN INTERACTIONS

In the presence of cosolvent, the macromolecular equilibrium and equilibrium conditions given in Eq. (9.15) describes properly the system only if the chemical potential of the reactant and products are affected in the same extent by a change in the solvent composition. Said in another way, it means that the macromolecular reaction does not affect the chemical potential of the solvent components. However, in general, the chemical potentials of the solvent components and macromolecule evolve with the solvent composition. The preferential binding parameter $(\partial m_3/\partial m_2)_{TP\mu_3}$ relates these mutual changes (Eq. (9.6)).

The behavior of macromolecule in solution is characterized experimentally by the apparent association constant, K_{obs}, which describes the concentrations of the macromolecular species in solution. K_{obs} may change with the solvent composition

$$K_{\text{obs}} = \left[\frac{m_{\text{P}_1}^{\nu_{\text{P}_1}} \times m_{\text{P}_2}^{\nu_{\text{P}_2}}}{m_{\text{R}_1}^{\nu_{\text{R}_1}} \times m_{\text{R}_2}^{\nu_{\text{R}_2}}}\right]. \tag{9.20}$$

From the Wyman linkage relationships, the change in K_{obs} when increasing the cosolvent concentration (i.e., a_3) can be related to the preferential binding parameters of the different macromolecules in solution

$$\begin{aligned}\partial \ln K_{\text{obs}}/\partial \ln a_3 = {} & \nu_{\text{P}_1}(\partial m_3/\partial m_{\text{P}_1})_{TP\mu_3} + \nu_{\text{P}_2}(\partial m_3/\partial m_{\text{P}_2})_{TP\mu_3} \\ & - \nu_{\text{R}_1}(\partial m_3/\partial m_{\text{R}_1})_{TP\mu_3} - \nu_{\text{R}_2}(\partial m_3/\partial m_{\text{R}_2})_{TP\mu_3}.\end{aligned} \tag{9.21}$$

In a condensed form

$$\partial \ln K_{\text{obs}}/\partial \ln a_3 = \Sigma\Delta(\partial m_3/\partial m_{\text{P}})_{TP\mu_3}, \tag{9.22}$$

where $\Sigma\Delta(\partial m_3/\partial m_{\text{P}})_{TP\mu_3}$ gives the change in preferential binding parameter on the reaction. In a structural approach, if the reaction is accompanied by changes ΔN_1 and ΔN_3 in the number of water and cosolvent molecules ''associated'' to the macromolecules (two-domain model), following Eq. (9.12):

$$\partial K_{\text{obs}}/\partial \ln a_3 = \Delta N_3 - \Delta N_1(m_3^{\text{bulk}}/m_1^*)_{TP\mu_3}. \tag{9.23}$$

Among the species involved in an equilibrium or an interconversion system, those interacting in the more efficient way with the cosolvent and/or less efficiently with water, will be stabilized by increasing the cosolvent content in the solvent.

9.6. MODELS FOR DESCRIBING SOLVATION AND ITS ORIGINS

We now report some models of solvation—mainly within the two-domain approach—which are used to interpret or predict the effects of cosolvent on macromolecular equilibrium. Any region of space that is statistically occupied with water and solute in the composition of the bulk leads to a null contribution in $(\partial m_3/\partial m_2)_{T\mu_1\mu_3}$. Restricting ourselves to the region of solvent that is perturbed by the macromolecule provides a solvation model, in terms of water and cosolvent numbers (N_1, N_3), often referred to as ''binding sites'' (Eq. (9.12)). Because one value of $(\partial m_3/\partial m_2)_{T\mu_1\mu_3}$ (or the related $(\partial m_1/\partial m_2)_{T\mu_1\mu_3}$) is related to an infinity of mathematical solutions (N_1, N_3), it is not possible to obtain a unique structural model in terms of water and cosolvent numbers from one measurement—or even one set of measurements with same technique.

9.6.1. Models Describing Solvation

Preferential binding parameters are interpreted in the framework of models.

9.6.1.1. The ''Only Hydration'' Model

In this model, a region of space is considered, which is occupied only by water, and corresponding to the total exclusion of cosolvent (see Figure 9.3B). Considering N_1 water-binding sites, Eq. (9.12a and b) reduce to $N_1 = (\partial m_1/\partial m_2)_{T\mu_1\mu_3}$; $N_1 = -(\partial m_3/\partial m_2)_{T\mu_1\mu_3}/m_3^{\text{bulk}}$. In an equilibrium A<=>B, the macromolecular form A or B corresponding to the less hydrated form is stabilized to a larger extent at high cosolvent concentration: $\partial \ln K_{\text{obs}}/\partial \ln a_3 = -\Delta N_1 m_3^{\text{bulk}}$. The changes in preferential solvent binding can be interpreted in terms of displaced water or of change in accessible surface area (ASA). N_1 can be understood as apparent hydration number (Parsegian *et al.*, 2000; Timasheff, 2002).

9.6.1.2. The Fixed Hydration Model

A typical value for hydration is 0.3–0.4 g of water by grams of macromolecule. Hydration can also be defined from the exposed surface area by considering one molecular layer of hydration. If hydration is considered as constant, changes in preferential solvent binding are interpreted in terms of cosolvent binding. It is particularly appropriate for ''strong'' ligand (cosolvent) binding, which modulates the macromolecular equilibrium properties at low cosolvent concentration (when, in Eq. (9.12a), $N_1 m_3^{\text{bulk}}/m_1^*$ is negligible compared with N_3) an example is given in Section 9.7.5 for the case of a detergent-solubilized protein, or for practical purposes such as estimating molar mass from buoyant molar mass (see Section 9.8).

9.6.1.3. The Invariant Particle Model

If a set of experimental $(\partial m_3/\partial m_2)_{T\mu_1\mu_3}$ values obtained at different solute concentrations are linear with m_3/m_1^*, then the particle can be described as an invariant particle, i.e., with constant N_3 and N_1 values (infinitively strongly bound salt and water molecules) in Eq. (9.12a and b) (Eisenberg, 2000).

9.6.1.4. Exchangeable-Solvent-Binding Sites Partition

Solvent-binding sites that exchange water and cosolvent with a common exchange equilibrium constant were defined (Schellman, 2003). In addition to these sites, there could be a number of hydration sites and sites that are "neutral," i.e., statistically occupied by water and cosolvent molecules in the same proportion as in the bulk solvent. The local-bulk solute partitioning model considers a partition coefficient of the cosolvent between the local domain (whose size is proportional to the exposed surface of the macromolecule) and the bulk solvent. These models can integrate nonideality of the cosolvent, which can have important effects because the concentrations of cosolvent are often in the molar range. A detailed analysis is of the exchangeable solvent-binding sites and partition models was given recently in Felitsky and Record (2004).

9.6.2. Models for the Origin of the Solvation

Why do the solvent-binding parameters of macromolecules depend on the type of cosolvent?

9.6.2.1. Affinity of the Cosolvent for the Macromolecule

Some solutes can have a weak or strong affinity for the macromolecule as is described for some examples given subsequently. In the case of strong affinity, interactions are well described in terms of ligand-binding and can be modeled from measurements at various cosolvent concentrations; in the case of weak interactions, they are unresolved from hydration: a weak cosolvent affinity can be described by a change in apparent hydration number N_1 or in partition coefficient (Record *et al.*, 1998c; Timasheff, 1998; Ebel *et al.*, 2000).

9.6.2.2. Properties of the Solvent Itself

As an example, the effect of salts (described subsequently, Section 9.7) can be understood in large part in view of the properties of the solvent itself in the complex interplay between solvent–solvent (Collins, 1997; Dill *et al.*, 2005) and solvent–macromolecule (Moelbert *et al.*, 2004) interactions. Stabilizing ions—those with the highest charge density—are kosmotropic, i.e., they order water in comparison with the state of pure water, while destabilizing ions are choatropic. The kosmotropic effect of salting-out salts was related to the fact that they increase the surface tension of water, which leads to an energy cost for forming a cavity in the solvent to accommodate the macromolecule: this disfavors macromolecular forms of larger surface. The surface tension argument can be expressed in terms of preferential binding parameters. An increase in the surface tension corresponds to a relative increase in the water concentration or cosolvent depletion at the interface (Timasheff, 1993).

9.6.2.3. Solvophobic/Osmophobic Effects

These terms describe unfavorable interactions of chemical groups with solute or solute-containing solvents, which result in preferential hydration. The salting out (precipitating) effects of methyl pentane-1-4 diol are correlated to its repulsion by charged groups (Pittz and Timasheff, 1978; Costenaro *et al.*, 2001); the stabilizing effects of osmolytes on proteins can be related to their unfavorable interactions with the peptide groups (see Section 9.7.2) (Liu and Bolen, 1995; Bolen and Baskakov, 2001).

9.6.2.4. Effects of Size of Large Cosolvent Molecules

Large cosolvent molecules are expected from geometry considerations to be excluded from the surface of macromolecules. This is illustrated in Figure 9.3, where components are considered as impenetrable spheres. The same statistical repartition of water and cosolvent in the solvent is drawn in Figure 9.3B (large cosolvent) and Figure 9.3A (water and cosolvent of similar size). Because of the forbidden overlap of macromolecule and cosolvent molecules, a zone expanding from the surface of the macromolecule to a distance corresponding to the half-diameter of the cosolvent only contains water (consider the ''centers'' of the molecules in Figure 9.3B). Here, the figure illustrates the fact that water exclusion is only a consequence of geometrical consideration, and not of a preferential affinity of the macromolecule for water when compared with the large cosolvent. The thermodynamic consequence of this steric exclusion—the crowding effect—is the same as if there was water-binding (Figure 9.3C): it favors macromolecular forms with smallest surface. Increasing the concentration of the large cosolvent or of the macromolecule itself stabilizes folding and macromolecular assemblies, *in vitro* (polymers are frequently used as precipitating agents for crystallization of biological macromolecules) or *in cellulo* (Ellis and Minton, 2003; Hall and Minton, 2003; Minton, 2005).

9.6.2.5. Polyelectrolyte and Salt Effect

The specific effects related to polyelectrolyte behavior are not detailed here; however, examples concerning nucleic acids and their interactions (Record *et al.*, 1978, 1998c; Eisenberg, 1990) are given in Section 9.7.4.

9.7. EXAMPLES OF PERTUBATION OF MACROMOLECULAR EQUILIBRIUM BY COSOLVENTS

9.7.1. The Salting Out Effects in the Hofmeister Series

The Hofmeister series describes the effect of neutral salts on the structure and conformational stability of macromolecules in solution (for a review, see Von Hippel and Schleich, 1969). Salting-out salts are precipitating and stabilize the folded and auto-associated states of macromolecules. Salting-in salts are solubilizing and favor their dissociation and unfolding. The effects of anions and cations are in the molar range and are additive. They are approximately proportional to the cosolvent concentration, and overcome specific effects due to the composition or conformation of the macromolecules. Salts, but also individual anions and cations can be classified from salting-out to salting-in. As emphasized by Timashef and coworkers, salting out to salting-in characteristics are well correlated with the values of solvent-binding parameters: for example, the values of $(\partial m_3/\partial m_2)_{T\mu_1\mu_3}$ are 3, -17, -22, and -35 mole/mole for BSA in the presence of 1M NaSCN, NaCl, $NaCH_3CO_2$, and Na_2SO_4, respectively, which range from salting-in to salting-out salts (Timasheff, 1993). With the assumption that these parameters do not depend on the nature of the protein surface, the stabilization of the folded, associated, and then precipitated forms of macromolecules by salting-out agents is related to their decreased exposed surface when compared with the unfolded, dissociated, or solubilized forms. The solvent-binding parameters of macromolecules depend on the type of salt. On the one hand, certain ions have some affinity for groups of the macromolecule: for example, cations for carboxylates and phosphates, explaining also some pH-dependent features. On the other hand, salting-out salts increase the surface tension of water, which disfavors macromolecular forms of larger surface (Timasheff, 1993, 1998; Benas *et al.*, 2002; Ebel *et al.*, 2002; Retailleau *et al.*, 2002; Apetri and Surewicz, 2003; Bostrom *et al.*, 2005).

9.7.2. Osmolytes and Macromolecule Stability

Osmolytes are small organic molecules or ions, which are accumulated in the cytoplasm of cells or micro or macroorganisms subjected to water-stress (osmotic shock, dehydration). They protect them against various physical stresses such as pressure, high and low temperatures, chemical stresses such as oxygen radicals or to cellular stresses such as the accumulation of proteins in cells defective in regulatory proteolytic function. Osmolytes are highly water soluble. Organic osmolytes can be polyols, amino acids and derivatives, methylamine, methylsulfonium, urea and so on. K^+ is the main osmolyte ion, which occurs in addition to organic solutes, but some halophilic microorganisms accumulate molar concentration of salts in their cytoplasm (da Costa *et al.*, 1998; Lee and Goldberg, 1998; Record *et al.*, 1998a,b; Wood, 1999; Oren, 2002; Cayley and Record, 2004; Yancey, 2004).

The terms ''osmolyte'' or ''osmoprotectant'' have been used with different meanings in the literature and often have been restricted to organic stabilizing osmolytes. ''Compatible solute'' indicates the fact that macromolecules or biochemical reactions are apparently not perturbed by its presence in large amounts. ''Compensatory,'' ''counteracting solute,'' or ''chemical chaperones'' emphasize a stabilizing efficiency that counterbalances the effects of another destabilizing cosolvent. While some of the osmolytes are destabilizing, such as urea, or known to inhibit nucleic acid–protein interactions, such as K^+, a large number of osmolytes are stabilizing for folded forms. Examples are: N, N, N-trimethyl glycine, named glycine betaine or betaine; trimethyl amine N-oxide: TMAO, 1,4,5, 6-Tetrahydro-2-methyl-4-pyrimidinecarboxylic acid: Ectoine. Stabilizing osmolytes can be used *in vitro* to force unfolded proteins to fold to native-like species (Baskakov and Bolen, 1998)—they can perturb folding (Chilson and Chilson, 2003)—or stabilize a given macromolecular state (Bennion *et al.*, 2004). As in the case of salts, the effects of osmolytes appear to be additive *in vitro* and *in vivo* (Baskakov *et al.*, 1998). The preferential cosolvent-binding parameters of proteins is proportional to osmolyte concentration—$(\partial m_3/\partial m_2)_\mu$ proportional to m_3^{bulk}—which can be described in terms of a concentration-independent apparent number of bound water molecules or a partition coefficient for the osmolytes (Courtenay *et al.*, 2000; Ebel *et al.*, 2000). Stabilizing osmolytes are described to be preferentially excluded (betaine, proline, TMAO, trehalose, glutamate...), whereas urea—as guanidinium salts—accumulate near the surfaces of folded proteins (Timasheff, 1998; Courtenay *et al.*, 2000, 2001). The stabilizing effect of sorbitol or trehalose was related to the larger value of the preferential hydration for the unfolded form when compared with the folded one (Xie and Timasheff, 1997). We have compared the preferential hydration of a protein in the presence of sugars of increasing size and observed the preferential exclusion increasing with the size of the sugars, but less than expected from pure geometrical considerations (Ebel *et al.*, 2000).

From solubility measurement of model compounds in water and solution containing the cosolvent, Bolen found that it is the unfavorable transfer of the peptide bond to the solution containing the stabilizing cosolvent (osmophobic effect) that explains mainly their ability to stabilize globular proteins (Liu and Bolen, 1995; Bolen and Baskakov, 2001). Solubility measurements on model compounds, analyzed in terms of transfer free energy, gave an explanation for the stabilizing effects of osmolytes: they were shown to act mainly through their unfavorable interaction with the peptide backbone (osmophobic effect), which raises the free energy of the denaturated states. Structurally, this feature is correlated to a reduced dimension of the unfolded states in the presence of an osmolyte (Bolen and Baskakov, 2001). The degree of betaine exclusion was related to the amounts of polar amide and of anionic ASAs, with common characteristic values $(\partial m_3/\partial m_2)_\mu/(m_3^{\text{bulk}}\text{ASA})$ of -1.1 and -4.0 $/M/\mathring{A}^2$, respectively (Felitsky and Record, 2004; Felitsky *et al.*, 2004).

9.7.3. Osmolytes *In vivo*

Any modification of macromolecule stability and macromolecule–macromolecule interactions must affect intracellular processes. There must be compensatory effects when a stabilizing or destabilizing osmolyte is accumulated in the cells. Such compensation has been found experimentally in the case of TMAO and urea, accumulated by marine cartilaginous fishes (Baskakov *et al.*, 1998). Record and coworkers, who measured the concentration of osmolytes and evaluated the amount of ''free'' water for *Escherichia coli* subjected to osmotic stresses, found that cell growth—which decreases with external osmolality but is highly stimulated by betaine or proline—is primarily determined by the amount of free cytoplasmic water ($\mu L/mg$). The correlation between large changes in cytoplasmic concentrations of K^+ and biopolymer suggests that the destabilization of the protein–nucleic acid interactions by K^+ is compensated by crowding effect: reducing available volume favors macromolecular interactions. The efficiency of betaine as an osmoprotectant (its capacity to increase internal osmolality at a given concentration) can be related to its strong exclusion from the anionic surfaces of the cell biopolymers; It would perturb *in vivo* biopolymer processes only moderately, because only a few of them involve changes in exposed anionic surfaces (Record *et al.*, 1998a; Cayley *et al.*, 2000; Cayley and Record, 2003, 2004; Felitsky *et al.*, 2004).

9.7.4. Adaptation of Halophilic Proteins

Halophilic proteins—from the extreme halophilic organisms that accumulate molar concentrations of salts, mainly K^+, Na^+, and Cl^-- are not stable at "low" salt, below a concentration that depends on the salt type in the solvent: for example, below 2M NaCl. Their amino acid composition is characterized by a high content of acidic residues (Ng *et al.*, 2000; Baliga *et al.*, 2004). The solvent interactions of halophilic malate dehydrogenase were measured in the presence of various salts for the native folded form. They show a linear dependence of $(\partial m_3/\partial m_2)_\mu$ with m_3^{bulk}, which differs from the usual behavior described for osmolytes (Section 9.7.2) since there is in general a nonnull extrapolation at infinite dilution of the salt (component 3). In NaCl or $MgCl_2$, $(\partial m_3/\partial m_2)_\mu$ is positive at low salt—indicating more salt close to the protein than in the bulk—and negative at high salt, indicating more water close to the protein than in the bulk. In different solvent with a given salt at various concentrations, the data can be interpreted in terms of salt- and water-binding sites (saturated when the salt concentration is enough for the protein to be folded), the numbers depending on the type of salt in the solvent. For NaCl it corresponds to a constant local concentration of 3.4M while the bulk concentration is varied from 2M to 5M. This behavior can be related to the very acidic character of halophilic proteins, which allows ion affinity for the protein surface. The results suggest that, in NaCl and KCl, the folded form of the protein is stabilized at high salt by ion binding in addition to nonspecific hydration effects (Ebel *et al.*, 1999, 2002). We measured second virial coefficients A_2 in a variety of solvent conditions. We found that A_2 is lowered in conditions where the composition of the solvent in the local domain differs from the bulk solvent, which was rationalized as follows: it is known that $(\partial m_3/\partial m_2)_\mu \neq 0$ contributes as a negative entropic terms in A_2. The two situations of cosolvent depletion and water depletion (cosolvent accumulation) at the macromolecular interface have the same consequence: an effective macromolecular attraction. We think that the macromolecular surfaces of halophilic proteins have evolved in order to avoid water enrichment at their surfaces and thus preserve their solubility in the crowded and salted cytoplasm of halophilic cells (Costenaro and Ebel, 2002; Costenaro *et al.*, 2002; Ebel and Zaccai, 2004).

9.7.5. Membrane Proteins

The association state of membrane proteins is often the subject of controversy. This is mainly due to the fact that their extraction and purification require large amount of detergent or other amphiphilic compounds (Moller and le Maire, 1993; le Maire *et al.*, 2000; Popot *et al.*, 2003). Detergent can be denaturating, i.e., they can dissociate membrane protein assemblies, consisting of different proteins and cofactors or lipids, relevant to the function. Different protocols of extractions (e.g., using different types and concentrations of detergent) can therefore lead to the stabilization of different membrane protein assemblies. Detergent and lipids could be associated to the proteins in amounts that can be very large and depend on the protein and type of detergent used. The protein EmrE is a recently described example (Butler *et al.*, 2004; Winstone *et al.*, 2005). In addition, detergent micelles coexist always with protein complexes. Solutions of complexes of membrane proteins and detergent thus consist of at least three-component systems. They have to be characterized using appropriate rigorous methods. The concentrations of detergent and lipids are expected to modulate macromolecular association if they do not interact to the same extent with the different macromolecular forms. Conversely, a change in macromolecular association constants with the concentration of detergent or lipid necessarily reflects different interactions of the small molecules for the different forms of the protein.

For a detergent solubilized lipoprotein (Josse *et al.*, 2002), the apparent dimer dissociation constant $K_d = C^2_{monomer}/C_{dimer}$ was found to change with detergent concentration in the solution. We analyzed the detergent concentration dependence of K_d considering the model of monomer- and dimer-binding detergent in different amounts: $N_{det,\,monomer}$ and $N_{det,\,dimer}$ are the numbers of bound detergent to the monomer and dimer. Here compound 3 is labeled "det." Because m^{bulk}_{det} is of the order of 10^{-3} m and m^*_1 is 55.55 m (there is 55.55 mole water/kg water), $N_1 m^{bulk}_{det}/m^*_1$ is negligible compared with N_{det}: a change in hydration would not affect K_d. Equation (9.23), which relates K_d to the activity $a_{det,\,monomer}$ of detergent in the monomer form, simplifies

$$\partial \ln K_d / \partial \ln a_{det,\,monomer} = 2N_{det,\,monomer} - N_{det,\,dimer}. \tag{9.24a}$$

Why is $a_{det,\,monomer}$ related to the detergent monomer? Because the values of N_{det} address moles of monomer of detergent, activity also refers to that of the "free" monomer detergent. It is approximately the monomer concentration and does not reflect the total number of detergent molecules in solution, because detergent molecules auto-associate to form micelles: the concentration of detergent monomer is increasing only very slightly above the critical micellar concentration (CMC). Considering for the detergent micelle an aggregation number N_{agg} and a micelle formation association constant $K_{agg} = a_{det,\,micelle}/a_{det,\,monomer}{}^{N_{agg}}$ (Eq. (9.15e)), with $a_{det,\,micelle}$ the activity of the detergent expressed in terms of micelles, at vanishing concentration of macromolecule, and omitting nonideality terms, Eq. (9.24a) gives

$$\begin{aligned}\partial \ln K_d / \partial \ln a_{det,\,micelle} &= \partial \ln K_d / \partial \ln C_{det,\,micelle} \\ &= (2N_{det,\,monomer} - N_{det,\,dimer})/N_{agg}.\end{aligned} \tag{9.24b}$$

$C_{\text{det, micelle}}$ (mole/L) can be estimated at vanishing concentration of macromolecule from the total concentration of detergent $C_{\text{det, total}}$—expressed in terms of mole/L of monomer—and *CMC*: $C_{\text{det, micelle}} = (C_{\text{det, total}} - \text{CMC})/N_{\text{agg}}$. We plotted ln K_d as a function of ln $C_{\text{det, micelle}}$ and found a linear dependence, which was used to determine the number of molecules of detergent released on auto-association of the protein. This number corresponded to about 100 molecules of detergent, i.e., one micelle, a feature that is the object of interesting discussions (Fisher *et al.*, 2003; Fleming *et al.*, 2004).

9.8. METHODS THAT ALLOW THE CHARACTERIZATION OF THE PROTEIN AS A SOLVATED PARTICLE

This last section describes different techniques that allow the extraction of information about solvent–protein interactions. Some of these techniques are also usually used to measure molar masses and characterize interactions between macromolecules.

9.8.1. Density

Precise density meters are commercially available with an accuracy of 2×10^{-6} mL/g (density meter DMA5000 from PAAR, Austria). The measurement uses less than 1 mL of solution. The sample is filled into an oscillating tube, whose resonant frequency is proportional to the square of its mass. Given that the volume of the tube is known, the density of the sample is determined. Most of the uncertainties of the density measurements are related to sample preparation: density measurements need carefully prepared samples and reference solvents in the case of multicomponent solvents. The protocol of preparation of the samples has to be rigorously defined (e.g., evaporation of water from a dialysis bag can affect significantly the measurement). Density measurements can be conducted in different ways.

9.8.1.1. Measurements at Constant Molalities

They provide values for the partial specific volume $\bar{v}_2$, by dissolving the pure dried macromolecule in a solvent of given composition, or using a stock solution containing the macromolecule dissolved in pure water (Cohen and Eisenberg, 1968; Kernel *et al.*, 1999). The densities ρ and ρ° of two solutions with the same solvent composition but with and without, respectively, the macromolecule at concentration c_2, are compared. At low c_2, with the assumption that $(\partial\rho/\partial c_2)_{PTm_3} = (\rho - \rho^\circ)/c_2$

$$(\partial\rho/\partial c_2)_{PTm_3} = 1 - \rho^\circ\bar{v}_2. \tag{9.25a}$$

When using macromolecule molar concentration C_2, partial molal volume $\bar{V}_2$ is obtained

$$(\partial\rho/\partial C_2)_{PTm_3} = 1 - \rho^\circ\bar{V}_2. \tag{9.25b}$$

Unless the diluted macromolecule is studied in buffer without salt, its partial specific volume is generally independent of its concentration. However, the value of $\overline{v}_2$ is not only related to the presence of the macromolecule (see Section 9.2.4).

9.8.1.2. Measurements at Constant Chemical Potential of the Solvent Components

These conditions are realized by extensive dialysis (the reference solvent being the solvent out side the bag) or size exclusion chromatography. With the reasonable assumptions that $(\partial\rho/\partial c_2)_{T\mu_1\mu_3} = (\rho - \rho^\circ)/c_2$ (similar to earlier ones) and also that $(\partial\rho/\partial P)_{Tc_2c_3}(\partial P/\partial c_2)_{T\mu_1\mu_3}$ is negligible compared with $(\partial\rho/\partial c_2)_{T\mu_1\mu_3}$

$$(\partial\rho/\partial C_2)_{T\mu_1\mu_3} = (M_2 - \rho^\circ\overline{V}_2) + (\partial m_3/\partial m_2)_{T\mu_1\mu_3} \times (M_3 - \rho^\circ\overline{V}_3) \tag{9.26a}$$

$$(\partial\rho/\partial c_2)_{T\mu_1\mu_3} = (1 - \rho^\circ\overline{v}_2) + (\partial w_3/\partial w_2)_{T\mu_1\mu_3} \times (1 - \rho^\circ\overline{v}_3). \tag{9.26b}$$

Or, using preferential hydration parameters

$$(\partial\rho/\partial C_2)_{T\mu_1\mu_3} = (M_2 - \rho^\circ\overline{V}_2) + (\partial m_1/\partial m_2)_{T\mu_1\mu_3} \times (M_1 - \rho^\circ\overline{V}_1), \tag{9.26c}$$

$$(\partial\rho/\partial c_2)_{T\mu_1\mu_3} = (1 - \rho^\circ\overline{v}_2) + (\partial w_1/\partial w_2)_{T\mu_1\mu_3} \times (1 - \rho^\circ\overline{v}_1). \tag{9.26d}$$

The preferential binding parameter is therefore an experimental parameter that measures the changes in the buoyant properties of the macromolecule. When written in terms of water (N_1 in mole/mole or B_1 in g/g) and cosolvent (N_3 or B_3) in the perturbed solvent

$$(\partial\rho/\partial C_2)_{T\mu_1\mu_3} = (M_2 - \rho^\circ\overline{V}_2) + N_1(M_1 - \rho^\circ\overline{V}_1) + N_3(M_3 - \rho^\circ\overline{V}_3), \tag{9.27a}$$

$$(\partial\rho/\partial C_2)_{T\mu_1\mu_3} = (M_2 + N_1M_1 + N_3M_3) - \rho^\circ(\overline{V}_2 + N_1\overline{V}_1 + N_3\overline{V}_3), \tag{9.27b}$$

$$(\partial\rho/\partial c_2)_{T\mu_1\mu_3} = (1 - \rho^\circ\overline{v}_2) + B_1(1 - \rho^\circ\overline{v}_1) + B_3(1 - \rho^\circ\overline{v}_3), \tag{9.27c}$$

$$(\partial\rho/\partial c_2)_{T\mu_1\mu_3} = (1 + B_1 + B_3) - \rho^\circ(\overline{v}_2 + B_1\overline{v}_1 + N_3\overline{v}_3). \tag{9.27d}$$

For this ‘‘solvated particle’’ or ‘‘complex,’’ a molar mass M_c and a partial specific volume $\overline{v}_c$ can be defined

$$M_c = M_2 + N_1M_1 + N_3M_3, \tag{9.28a}$$

$$\overline{v}_c = \left(\frac{\overline{v}_2 + B_1\overline{v}_1 + N_3\overline{v}_3}{1 + B_1 + B_3}\right). \tag{9.28b}$$

9.8.2. Vapor Pressure Osmometry

One protocol of isopiestic distillation (Figure 9.1B) consists of the preparation of two samples containing identical amounts of water and cosolvent and, for one of them, in addition, the macromolecule. The difference in weights before and after isopiestic equilibration is directly interpreted in terms of $(\Delta m_1/\Delta m_2)_{TP\mu_1} = -(m_1/m_3)(\partial m_3/\partial m_2)_{TP\mu_1}$ (Eisenberg, 1976). The principle of vapor pressure osmometry, the preparation of the sample and interpretation of the data are clearly described in Zhang *et al.* (1996) and Courtenay *et al.* (2000). The activity a_1 of water in the sample solution is measured from the measurement of a precise difference in temperature between the sample solution and a small volume of pure water condensing on a thermocouple. This difference is related to the ratio of the water vapor pressures of over pure water and over the solution. It is expressed in osmolality (Osm): $\text{Osm} = -55.5 \ln a_1$.

Measurements of solution osmolality as a function of the solute concentration in the presence and absence of macromolecules at concentrations of typically 100 mg/mL provide values of $(\Delta m_3/\Delta m_2)_{TP\mu_1} = (\partial m_3/\partial m_2)_{TP\mu_1}$ at a given solute concentration (corresponding to the absence of macromolecule). The experimental values of $(\partial \text{Osm}/\partial m_3)_{TPm_2} = 0$ and $(\partial \text{Osm}/\partial m_3)_{TPm_2} = 0$ performed (or interpolated) at the same solvent composition is used to obtain $(\partial m_3/\partial m_2)_{TP\mu_3}$ and $(\partial m_3/\partial m_2)_{TP\mu_3}$.

9.8.3. Sedimentation Equilibrium

In analytical ultracentrifugation, the transport of macromolecule subjected to a centrifugal field is followed (Eq. (9.1) is not valid!). In sedimentation equilibrium, the solution containing the macromolecule is centrifuged at a given angular velocity ω for a sufficient time (typically 24 h in standard protocols) to obtain the equilibrium condition, in which the concentration profile does not evolve with time (Figure 9.4) (see Chapter 10). The macromolecule concentration c_2 is measured as a function of radial position r in the ultracentrifuge. The equilibrium condition corresponds to an invariance of the total potential—composed of chemical gravitational and electrostatic terms—throughout the system for each species. Considering that electroneutrality condition is satisfied locally, which is obtained in practice, and in view of the definition as electroneutral species of the components, the electrostatic contribution is eliminated. The equilibrium condition is written: $d\mu_2 + M_2 d\phi = 0$, with $\phi = -(1/2)\omega^2 r^2$ the gravitational field potential, from which (Eisenberg, 1976)

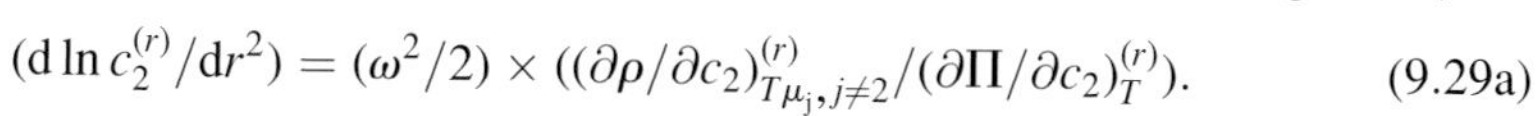

$$(d \ln c_2^{(r)}/dr^2) = (\omega^2/2) \times ((\partial \rho/\partial c_2)^{(r)}_{T\mu_j, j\neq 2}/(\partial \Pi/\partial c_2)^{(r)}_T). \qquad (9.29a)$$

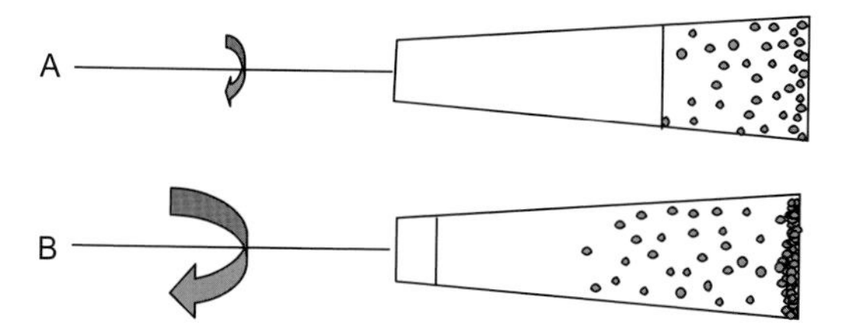

Figure 9.4. Schematic representation of a sedimentation experiment. (A) Sedimentation equilibrium. (B) Sedimentation velocity.

The superscript (r) is a reminder that the variable depends on the position r in the ultracentrifuge, and is omitted below. This equation is general for any component among any number of components of any molecular weight, without restrictions to thermodynamic nonideality or to incompressibility (pressure effects for example are manifested in the dependence of the variables). This equation reduces, in the case of dilute noninteracting macromolecule in a three-component system (see Table 9.1), to

$$(\mathrm{d}\ln c_2^{(r)}/\mathrm{d}r^2) = (\omega^2/2)\times(\partial\rho/\partial c_2)^{(r)}_{T\mu_1\mu_3}\times M_2. \tag{9.29b}$$

9.8.4. Svedberg Equation in Sedimentation Velocity

In sedimentation velocity experiments, measurements are performed as a function of time t, whereas the macromolecule is transported by the action of ultracentrifugation (Figure 9.4) (see Chapter 16). The Lamm equation describes the transport in the ultracentrifuge for a two-component system

$$(\partial c_2/\partial t) = -(1/r)\times\partial/\partial r[r(c_2 s\omega^2 r - D(\partial c_2/\partial r))]. \tag{9.30}$$

The sedimentation coefficient s measures the velocity of the macromolecule per unit centrifugal field $\omega^2 r$. The diffusion coefficient of the macromolecule D measures the relaxation of a gradient of macromolecule concentration. The Svedberg equation relates s° and D°, obtained or extrapolated at vanishing macromolecule concentration, to the molar mass M_2 of the macromolecule and its buoyant factor $(1-\rho^\circ\overline{v}_2)$

$$s^\circ/D^\circ = M_2(1-\rho^\circ\overline{v}_2)^\circ/RT. \tag{9.31a}$$

For a three-component system, there could be interactions between the flows of the different solute components—via hydrodynamic and thermodynamic coupling. In general, the expression for s is rather complex. It simplifies considerably under the assumption that cosolvent 3 is at sedimentation equilibrium. In that case

$$s/D = (\partial\rho/\partial m^2)_{TP\mu_3}/(\partial\Pi/\partial m_2), \tag{9.31b}$$

which can be approximated, neglecting terms contributing to less than 1% to

$$s/D = (\partial\rho/\partial c^2)_{TP\mu_1\mu_3}/(\partial\Pi/\partial c_2). \tag{9.31c}$$

D contains contributions from D_{22} and D_{23} (which relate the macromolecular flow to $(\mathrm{d}c_2/\mathrm{d}r)$ and $(\mathrm{d}c_3/\mathrm{d}r)$, respectively) (Eisenberg, 1976). The concentration dependences of s, D, and $(\partial\Pi/\partial c_2)$ were examined in Solovyova *et al.* (2001).

9.8.5. Small Angle Scattering of Light (LS), X-Rays (SAXS), or Neutrons (SANS)

The geometry of scattering experiments is schematized on Figure 9.5. The sample—a solution of macromolecule—is illuminated by a beam of plane-wave monochromatic light. A small part of the light is scattered elastically (at the same

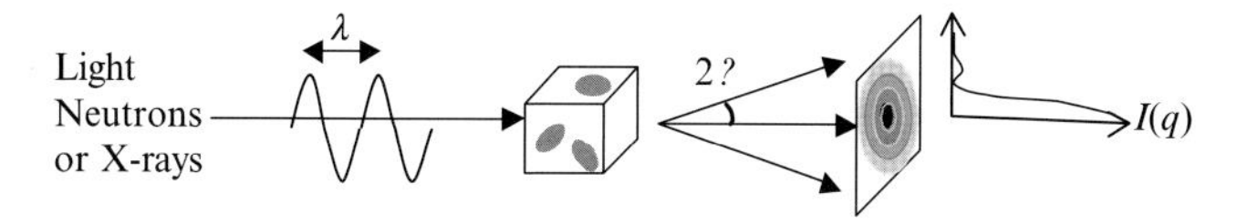

Figure 9.5. Schematic representation of a small angle scattering experiment.

wavelength) and isotropically (because of the disorder of the orientation of the macromolecules in the sample). The scattering curve $I(q)$ is obtained after radial averaging and subtraction of background scattering. The modulus of the scattering vector q is related to the scattering angle θ and wavelength λ. $q = 4\pi/\lambda \times \sin\theta/2(q = 4\pi n_0/\lambda \times \sin\theta/2$ for LS, with n_0 the refractive index of the solvent). The wavelength λ is typically 600 nm for light, ~0.1 nm for X-rays, or ~1 nm for neutrons. In consequence, for a given sample, only $q \sim 0$ part of $I(q)$ is investigated by light scattering as opposed to X-rays or neutrons scattering.

The shape and intensity of the scattering curve is related to the real space fluctuations in refractive index, electron density, or neutron scattering length density, respectively, for the three related radiations: the scattered intensity is the Fourier transform of the spatial correlation function of these ''densities'' (Koch *et al.*, 2003). In general, these ''densities'' depend on the local chemical composition, and are different for the solvent and the particle (and even within the particle, or in the perturbed solvent). The spatial correlation function counts the distances between infinitesimal volumes given their ''density'' differs from the ''bulk'' solvent. For a homogeneous solution of globular (quasispherical) particles, the scattering curve is a function of the form factor—which is dependent of the macromolecule solvent interactions and measures the spatial correlations within the particle—and of the structure factor, related to the radial distribution of interparticle distances—the pair correlation function $g_{22}(r)$, that is, to intermacromolecular interactions. New data treatment procedures allow now to obtain, from the form factor, *ab initio* topologies of macromolecules at ~1 nm resolution from X-rays and neutron data (Svergun and Koch, 2002). Statistical mechanics allows the calculation, from defined interparticle potentials, of particle distribution and thus structure factors that can be compared with experimental ones (see Section 9.4.3). More details on SAXS and SANS can be found in Chapter 11.

9.8.6. Forward Intensity in LS, SAXS, and SANS

Now we focus on the forward scattering at $q = 0$, which has to be extrapolated from other data, since in practice it is superimposed to the nonscattered incident beam and masked. In the case of noninteracting particles, the Guinier approximation (Eq. (9.32)) allows the determination of the radius of gyration of scattering density contrast (Rg) and the forward intensity $I(0)$ from the linear plot at small angle ($Rgq<1$)

$$\ln I(q) = \ln I(0) - (1/3)Rg^2q^2 \tag{9.32}$$

The forward intensity $I(0)$ can be obtained in the same way in the case of real solution at finite concentration c_2, while an apparent radius of gyration Rg_{app} is derived from the slope of the Guinier plot. For light scattering data, angular dependence of $I(q)$ occurs only for very large particle ($Rg>20$ nm) and Zimm plots are generally used for representing the angular and concentration dependence of $I(q)$.

To a very good approximation, the forward intensities, normalized in the appropriate way and obtained from LS ($\Delta R(0)$), SANS ($I_N(0)$), and SAXS ($I_{el}(0)$), can be presented in a similar fashion (Eisenberg, 1981)

$$\Delta R(0)/c_2 = (1/kT)(\partial n/\partial c_2)^2_{TP\mu_3}/(\partial \Pi/\partial c_2) \tag{9.33a}$$

$$I_N(0)/c_2 = (1/kT)(\partial \rho_N/\partial c_2)^2_{TP\mu_3}/(\partial \Pi/\partial c_2) \tag{9.33b}$$

$$I_{el}(0)/c_2 = (1/kT)(\partial \rho_{el}/\partial c_2)^2_{TP\mu_3}/(\partial \Pi/\partial c_2). \tag{9.33c}$$

At vanishing macromolecule concentration, the following relationships hold

$$\Delta R(0)/c_2 = (1/N_A)(\partial n/\partial c_2)^2_{TP\mu_3} M_2 \tag{9.34a}$$

$$(I_N(0)/c_2) = (1/N_A)(\partial \rho_N/\partial c_2)^2_{TP\mu_3} M_2, \tag{9.34b}$$

$$(I_{el}(0)/c_2) = (1/N_A)(\partial \rho_{el}/\partial c_2)^2_{TP\mu_3} M_2. \tag{9.34c}$$

The terms of contrast are, depending on the property studied: the refractive index increment $(\partial n/\partial c_2)_{TP\mu_3}$, the neutron scattering length density increment $(\partial \rho_N/\partial c_2)_{TP\mu_3}$, the electron density increment $(\partial \rho_{el}/\partial c_2)_{TP\mu_3}$. They can be expressed as functions of the preferential cosolvent-binding parameters $(\partial m_3/\partial m_2)_{T\mu_1\mu_3}$ as shown below.

9.8.7. Refractive Index Increments

The refractive index increment $(\partial n/\partial m_2)_{TP\mu_3}$ can be expressed as

$$(\partial n/\partial m_2)_{TP\mu_3} = (\partial n/\partial m_2)_{TPm_3} + (\partial m_3/\partial m_2)_{TP\mu_3}(\partial n/\partial m_3)_{TPm_2}. \tag{9.34}$$

It is in practice indistinguishable from $(\partial n/\partial c_2)_{T\mu_1\mu_3}$ that could be measured by dialysis. In the limit of low concentration c_2 (Eisenberg, 1976), it can be expressed as

$$(\partial n/\partial c_2)^\circ_{TP\mu_3} \approx (\partial n/\partial c_2)^\circ_{TPm_3} + (\partial n/\partial c_3)_{TPm_2=0}\,(1 - c_3\bar{v}_3)\cdot(\partial w_3/\partial w_2)_{T\mu_1\mu_3}. \tag{9.35}$$

It is important to realize that $(\partial n/\partial c_2)^\circ_{TPm_3}$ is strongly dependent on the solvent composition (Wen and Arakawa, 2000). An estimation of $(\partial n/\partial c_2)^\circ_{TPm_3}$, in a solvent with the cosolvent at concentration c_3 and of refractive index n_{c_3} can be obtained from $(\partial n/\partial c_2)^\circ_{TPc_3=0}$, corresponding to a solvent condition without cosolvent and of refractive index n_{m_0} (in the approximation of the invariance of the partial specific volume with solvent composition), as

$$(\partial n/\partial c_2)^{\circ}_{TPm_3} = (\partial n/\partial c_2)^{\circ}_{TPc_3=0} - c_3\overline{\nu}_2(\partial n/\partial c_3), \tag{9.36a}$$

or

$$(\partial n/\partial c_2)^{\circ}_{TPm_3} = (\partial n/\partial c_2)^{\circ}_{TPc_3=0} - \overline{\nu}_2(n_{m_3} - n_{m_0}). \tag{9.36b}$$

9.8.8. Neutron Scattering Length Density Increment and Electron Density Increment

$(\partial \rho_N/\partial c_2)_{TP\mu_3}$ and $(\partial \rho_{el}/\partial c_2)_{TP\mu_3}$ are in practice indistinguishable from $(\partial \rho_N/\partial c_2)_{P\mu_1\mu_3}$ and $(\partial \rho_{el}/\partial c_2)_{P\mu_1\mu_3}$. They can be expressed in forms analogous to those given for the increment of mass density (Eqs. 9.26a–9.26d or 9.27a–9.27d). This is done by using the neutron scattering length per gram of solute J b_J (cm/g), the scattering length density of the solvent ρ°_N (cm/mL), the number of electron per gram of solute l_J (electron/g), and electron density of the solvent ρ°_{el} (electron/mL). All these parameters can be obtained from chemical composition and/or concentrations. Using $(\partial w_3/\partial w_2)_{P\mu_1\mu_3}$ for example

$$(\partial \rho_N/\partial c_2)_{P\mu_1\mu_3} = (b_2 - \rho^{\circ}_N\overline{\nu}_2) + (\partial w_3/\partial w_2)_{P\mu_1\mu_3}(b_3 - \rho^{\circ}_N\overline{\nu}_3), \tag{9.37}$$

$$(\partial \rho_{el}/\partial c_2)_{P\mu_1\mu_3} = (l_2 - \rho^{\circ}_{el}\overline{\nu}_2) + (\partial w_3/\partial w_2)_{P\mu_1\mu_3}(l_3 - \rho^{\circ}_{el}\overline{\nu}_3). \tag{9.38}$$

ACKNOWLEDGMENT

I am extremely grateful to G. Zaccai for help in the preparation of the manuscript and helpful discussions.

REFERENCES

Anderson, C. F., Courtenay, E. S., and Record, M. T., Jr. (2002). Thermodynamic expressions relating different types of preferential interaction coefficients in solutions containing two solute components. *J Phys Chem B* 106:418–433.

Apetri, A. C. and Surewicz, W. K. (2003). Atypical effect of salts on the thermodynamic stability of human prion protein. *J Biol Chem* 278:22187–22192.

Baliga, N. S., Bonneau, R., Facciotti, M. T., Pan, M., Glusman, G., Deutsch, E. W., Shannon, P., Chiu, Y., Weng, R. S., Gan, R. R., *et al.* (2004). Genome sequence of Haloarcula marismortui: a halophilic archaeon from the Dead Sea. *Genome Res* 14:2221–2234.

Baskakov, I. and Bolen, D. W. (1998). Forcing thermodynamically unfolded proteins to fold. *J Biol Chem* 273:4831–4834.

Baskakov, I., Wang, A., and Bolen, D. W. (1998). Trimethylamine-N-oxide counteracts urea effects on rabbit muscle lactate dehydrogenase function: a test of the counteraction hypothesis. *Biophys J* 74:2666–2673.

Belloni, L. (1991). Interacting monodisperse and polydisperse spheres. In: Lindner, P. and Zemb, T. (eds), *Neutron, X-ray and Light Scattering: Introduction to An Investigative Tool for Colloidal and Polymeric Systems*. Elsevier Science Publishers B.V., New York, pp. 135–155.

Benas, P., Legrand, L., and Ries-Kautt, M. (2002). Strong and specific effects of cations on lysozyme chloride solubility. *Acta Crystallogr D Biol Crystallogr* 58:1582–1587.

Bennion, B. J., DeMarco, M. L., and Daggett, V. (2004). Preventing misfolding of the prion protein by trimethylamine N-oxide. *Biochemistry* 43:12955–12963.

Bolen, D. W. and Baskakov, I. V. (2001). The osmophobic effect: natural selection of a thermodynamic force in protein folding. *J Mol Biol* 310:955–963.

Bostrom, M., Tavares, F. W., Finet, S., Skouri-Panet, F., Tardieu, A., and Ninham, B. W. (2005). Why forces between proteins follow different Hofmeister series for pH above and below pI. *Biophys Chem* 117:217–224.

Butler, P. J., Ubarretxena-Belandia, I., Warne, T., and Tate, C. G. (2004). The *Escherichia coli* multidrug transporter EmrE is a dimer in the detergent-solubilised state. *J Mol Biol* 340:797–808.

Cantor, C. R. and Schimmel, P. R. (1980). *Biophysical chemistry*. Freeman, San Francisco.

Casassa, E. F. and Eisenberg, H. (1964). Thermodynamic analysis of multicomponent solutions. *Adv Protein Chem* 19:287–395.

Cayley, D. S., Guttman, H. J., and Record, M. T., Jr. (2000). Biophysical characterization of changes in amounts and activity of *Escherichia coli* cell and compartment water and turgor pressure in response to osmotic stress. *Biophys J* 78:1748–1764.

Cayley, S. and Record, M. T., Jr. (2003). Roles of cytoplasmic osmolytes, water, and crowding in the response of *Escherichia coli* to osmotic stress: biophysical basis of osmoprotection by glycine betaine. *Biochemistry* 42:12596–12609.

Cayley, S. and Record, M. T., Jr. (2004). Large changes in cytoplasmic biopolymer concentration with osmolality indicate that macromolecular crowding may regulate protein-DNA interactions and growth rate in osmotically stressed *Escherichia coli* K-12. *J Mol Recognit* 17:488–496.

Chilson, O. P. and Chilson, A. E. (2003). Perturbation of folding and reassociation of lactate dehydrogenase by proline and trimethylamine oxide. *Eur J Biochem* 270:4823–4834.

Chitra, R. and Smith, P. E. (2001). Preferential interactions of cosolvents with hydrophobic solutes. *J Phys Chem B* 105:11513–11522.

Cohen, G. and Eisenberg, H. (1968). Deoxyribonucleate solutions: sedimentation in a density gradient, partial specific volumes, density and refractive index increments, and preferential interactions. *Biopolymers* 6:1077–1100.

Collins, K. D. (1997). Charge density-dependent strength of hydration and biological structure. *Biophys J* 72:65–76.

Costenaro, L. and Ebel, C. (2002). Thermodynamic relationships between protein-solvent and protein-protein interactions. *Acta Crystallogr D Biol Crystallogr* 58:1554–1559.

Costenaro, L., Zaccai, G., and Ebel, C. (2001). Understanding protein crystallisation by dilution: the ternary NaCl-MPD-H2O system. *J Cryst Growth* 232:102–113.

Costenaro, L., Zaccai, G., and Ebel, C. (2002). Link between protein-solvent and weak protein-protein interactions gives insight into halophilic adaptation. *Biochemistry* 41:13245–13252.

Courtenay, E. S., Capp, M. W., Anderson, C. F., and Record, M. T., Jr. (2000). Vapor pressure osmometry studies of osmolyte-protein interactions: implications for the action of osmoprotectants in vivo and for the interpretation of ''osmotic stress'' experiments in vitro. *Biochemistry* 39: 4455–4471.

Courtenay, E. S., Capp, M. W., and Record, M. T., Jr. (2001). Thermodynamics of interactions of urea and guanidinium salts with protein surface: relationship between solute effects on protein processes and changes in water-accessible surface area. *Protein Sci* 10:2485–2497.

da Costa, M. S., Santos, H., and Galinski, E. A. (1998). An overview of the role and diversity of compatible solutes in bacteria and archaea. *Adv Biochem Eng Biotechnol* 61:117–153.

Dill, K. A., Truskett, T. M., Vlachy, V., and Hribar-Lee, B. (2005). Modeling water, the hydrophobic effect, and ion solvation. *Annu Rev Biophys Biomol Struct* 34:173–199.

Ebel, C., Costenaro, L., Pascu, M., Faou, P., Kernel, B., Proust-De Martin, F., and Zaccai, G. (2002). Solvent interactions of halophilic malate dehydrogenase. *Biochemistry* 41:13234–13244.

Ebel, C., Eisenberg, H., and Ghirlando, R. (2000). Probing protein-sugar interactions. *Biophys J* 78:385–393.

Ebel, C., Faou, P., Kernel, B., and Zaccai, G. (1999). Relative role of anions and cations in the stabilisation of halophilic malate dehydrogenase. *Biochemistry* 38:9039–9047.

Ebel, C. and Zaccai, G. (2004). Crowding in extremophiles: linkage between solvation and weak protein-protein interactions, stability and dynamics, provides insight into molecular adaptation. *J Mol Recognit* 17:382–389.

Eisenberg, H. (1976). *Biological Macromolecules and Polyelectrolytes in Solution.* Clarendon Press, Oxford.

Eisenberg, H. (1981). Forward scattering of light, X-rays and neutrons. *Q Rev Biophys* 14:141–172.

Eisenberg, H. (1990). Solution properties of DNA: sedimentation, scattering of light, X-rays and neutrons, and viscometry. In: Saenger, W. (ed.), *Landolt-Börnstein: Numerical Data and Functional Relationships in Science and Technology, New Series Biophysics-nucleic Acids.* Springer, Berlin, pp. 257–276.

Eisenberg, H. (1994). Protein and nucleic acid hydration and cosolvent interactions: establishment of reliable baseline values at high cosolvent concentrations. *Biophys Chem* 53:57–68.

Eisenberg, H. (2000). Analytical ultracentrifugation in a Gibbsian perspective. *Biophys Chem* 88:1–9.

Ellis, R. J. and Minton, A. P. (2003). Cell biology: join the crowd. *Nature* 425:27–28.

Felitsky, D. J., Cannon, J. G., Capp, M. W., Hong, J., Van Wynsberghe, A. W., Anderson, C. F., and Record, M. T., Jr. (2004). The exclusion of glycine betaine from anionic biopolymer surface: why glycine betaine is an effective osmoprotectant but also a compatible solute. *Biochemistry* 43:14732–14743.

Felitsky, D. J. and Record, M. T., Jr. (2004). Application of the local-bulk partitioning and competitive binding models to interpret preferential interactions of glycine betaine and urea with protein surface. *Biochemistry* 43:9276–9288.

Fisher, L. E., Engelman, D. M., and Sturgis, J. N. (2003). Effect of detergents on the association of the glycophorin a transmembrane helix. *Biophys J* 85:3097–3105.

Fleming, K. G., Ren, C. C., Doura, A. K., Eisley, M. E., Kobus, F. J., and Stanley, A. M. (2004). Thermodynamics of glycophorin A transmembrane helix dimerization in C14 betaine micelles. *Biophys Chem* 108:43–49.

Hall, D. and Minton, A. P. (2003). Macromolecular crowding: qualitative and semiquantitative successes, quantitative challenges. *Biochim Biophys Acta* 1649:127–139.

Josse, D., Ebel, C., Stroebel, D., Fontaine, A., Borges, F., Echalier, A., Baud, D., Renault, F., Le Maire, M., Chabrieres, E., and Masson, P. (2002). Oligomeric states of the detergent-solubilized human serum paraoxonase (PON1). *J Biol Chem* 277:33386–33397.

Kernel, B., Zaccai, G., and Ebel, C. (1999). Determination of partial molal volumes, and salt and water binding of highly charged biological macromolecules (tRNA, halophilic protein) in multimolar salt solutions. *Prog Colloid Polym Sci* 113:168–175.

Koch, M. H., Vachette, P., and Svergun, D. I. (2003). Small-angle scattering: a view on the properties, structures and structural changes of biological macromolecules in solution. *Q Rev Biophys* 36:147–227.

le Maire, M., Champeil, P., and Moller, J. V. (2000). Interaction of membrane proteins and lipids with solubilizing detergents. *Biochim Biophys Acta* 1508:86–111.

Lee, D. H. and Goldberg, A. L. (1998). Proteasome inhibitors cause induction of heat shock proteins and trehalose, which together confer thermotolerance in Saccharomyces cerevisiae. *Mol Cell Biol* 18:30–38.

Liu, Y. and Bolen, D. W. (1995). The peptide backbone plays a dominant role in protein stabilization by naturally occurring osmolytes. *Biochemistry* 34:12884–12891.

Malfois, M., Bonneté, F., Belloni, L., and Tardieu, A. (1996). A model of attractive interactions to account for fluid-fluid phase separation of protein solutions. *J Chem Phys* 105:3290–3300.

McMillan, W. G. and Mayer, J. E. (1945). The statistical thermodynamics of multicomponent systems. *J Chem Phys* 13:276–305.

Minton, A. P. (2005). Influence of macromolecular crowding upon the stability and state of association of proteins: Predictions and observations. *J Pharm Sci* 94:1668–1675.

Moelbert, S., Normand, B., and De Los Rios, P. (2004). Kosmotropes and chaotropes: modelling preferential exclusion, binding and aggregate stability. *Biophys Chem* 112:45–57.

Moller, J. V. and le Maire, M. (1993). Detergent binding as a measure of hydrophobic surface area of integral membrane proteins. *J Biol Chem* 268:18659–18672.

Ng, W. V., Kennedy, S. P., Mahairas, G. G., Berquist, B., Pan, M., Shukla, H. D., Lasky, S. R., Baliga, N. S., Thorsson, V., Sbrogna, J., *et al.* (2000). Genome sequence of Halobacterium species NRC-1. *Proc Natl Acad Sci USA* 97:12176–12181.

Oren, A. (2002). *Halophilic Microorganisms and their Environments*. Kluwer Academic, Dordrecht; Boston, MA.

Parsegian, V. A. (2002). Protein-water interactions. *Int Rev Cytol* 215:1–31.

Parsegian, V. A., Rand, R. P., and Rau, D. C. (1995). Macromolecules and water: probing with osmotic stress. *Methods Enzymol* 259:43–94.

Parsegian, V. A., Rand, R. P., and Rau, D. C. (2000). Osmotic stress, crowding, preferential hydration, and binding: A comparison of perspectives. *Proc Natl Acad Sci USA* 97:3987–3992.

Pittz, E. P. and Timasheff, S. N. (1978). Interaction of ribonuclease A with aqueous 2-methyl-2, 4-pentanediol at pH 5.8. *Biochemistry* 17:615–623.

Popot, J. L., Berry, E. A., Charvolin, D., Creuzenet, C., Ebel, C., Engelman, D. M., Flotenmeyer, M., Giusti, F., Gohon, Y., Hong, Q., *et al.* (2003). Amphipols: polymeric surfactants for membrane biology research. *Cell Mol Life Sci* 60:1559–1574.

Rand, R. P. (2004). Probing the role of water in protein conformation and function. *Philos Trans R Soc Lond B Biol Sci* 359:1277–1284; discussion 1284–1275.

Record, M. T., Jr., Anderson, C. F., and Lohman, T. M. (1978). Thermodynamic analysis of ion effects on the binding and conformational equilibria of proteins and nucleic acids: the roles of ion association or release, screening, and ion effects on water activity. *Q Rev Biophys* 11:103–178.

Record, M. T., Jr., Zhang, W., and Anderson, C. F. (1998c). Analysis of effects of salts and uncharged solutes on protein and nucleic acid equilibria and processes: a practical guide to recognizing and interpreting polyelectrolyte effects, Hofmeister effects, and osmotic effects of salts. *Adv Protein Chem* 51:281–353.

Record, M. T., Jr., Courtenay, E. S., Cayley, D. S., and Guttman, H. J. (1998a). Responses of *E. coli* to osmotic stress: large changes in amounts of cytoplasmic solutes and water. *Trends Biochem Sci* 23:143–148.

Record, M. T., Jr., Courtenay, E. S., Cayley, S., and Guttman, H. J. (1998b). Biophysical compensation mechanisms buffering *E. coli* protein-nucleic acid interactions against changing environments. *Trends Biochem Sci* 23:190–194.

Retailleau, P., Ducruix, A., and Ries-Kautt, M. (2002). Importance of the nature of anions in lysozyme crystallisation correlated with protein net charge variation. *Acta Crystallogr D Biol Crystallogr* 58:1576–1581.

Schellman, J. A. (2003). Protein stability in mixed solvents: a balance of contact interaction and excluded volume. *Biophys J* 85:108–125.

Shimizu, S. (2004). Estimating hydration changes upon biomolecular reactions from osmotic stress, high pressure, and preferential hydration experiments. *Proc Natl Acad Sci USA* 101:1195–1199.

Solovyova, A., Schuck, P., Costenaro, L., and Ebel, C. (2001). Non-ideality by sedimentation velocity of halophilic malate dehydrogenase in complex solvents. *Biophys J* 81:1868–1880.

Svergun, D. I. and Koch, M. H. (2002). Advances in structure analysis using small-angle scattering in solution. *Curr Opin Struct Biol* 12:654–660.

Tardieu, A., Le Verge, A., Malfois, M., Bonneté, F., Finet, S., Riès-Kautt, M., and Belloni, L. (1999). Proteins in solution: from X-ray scattering intensities to interaction potentials. *J Crystal Growth* 196:193–203.

Timasheff, S. N. (1993). The control of protein stability and association by weak interactions with water: how do solvents affect these processes? *Annu Rev Biophys Biomol Struct* 22:67–97.

Timasheff, S. N. (1998). Control of protein stability and reactions by weakly interacting cosolvents: the simplicity of the complicated. *Adv Protein Chem* 51:355–432.

Timasheff, S. N. (2002). Protein hydration, thermodynamic binding, and preferential hydration. *Biochemistry* 41:13473–13482.

Von Hippel, P. H. and Schleich, T. (1969). The effects of neutral salts on the structure and conformational stability of macromolecules in solution. In: Timasheff, S. N. and Fasman, G.D. (eds), *Structure of Biological Macromolecules.* Marcel Dekker, NY, pp. 417–575.

Wen, J. and Arakawa, T. (2000). Refractive index of proteins in aqueous sodium chloride. *Anal Biochem* 280:327–329.

Winstone, T. L., Jidenko, M., le Maire, M., Ebel, C., Duncalf, K. A., and Turner, R. J. (2005). Organic solvent extracted EmrE solubilized in dodecyl maltoside is monomeric and binds drug ligand. *Biochem Biophys Res Commun* 327:437–445.

Wood, J. M. (1999). Osmosensing by bacteria: signals and membrane-based sensors. *Microbiol Mol Biol Rev* 63:230–262.

Xie, G. and Timasheff, S. N. (1997). Mechanism of the stabilization of ribonuclease A by sorbitol: preferential hydration is greater for the denatured then for the native protein. *Protein Sci* 6:211–221.

Yancey, P. H. (2004). Compatible and counteracting solutes: protecting cells from the Dead Sea to the Deep sea. *Sci Prog* 87:1–24.

Zhang, W., Capp, M. W., Bond, J. P., Anderson, C. F., and Record, M. T., Jr. (1996). Thermodynamic characterization of interactions of native bovine serum albumin with highly excluded (glycine betaine) and moderately accumulated (urea) solutes by a novel application of vapor pressure osmometry. *Biochemistry* 35:10506–10516.

10

Sedimentation Equilibrium Analytical Ultracentrifugation for Multicomponent Protein Interactions

Peter Schuck

10.1. INTRODUCTION

Sedimentation equilibrium (SE) analytical ultracentrifugation is one of the oldest approaches to measure the molar mass of macromolecules in solution (Svedberg and Fahraeus, 1926; Svedberg and Pedersen, 1940; Schachman, 1959). It may also be considered as one of the most direct methods, as it consists solely in the application of a gravitational force and the observation of the concentration gradients established in solution. In contrast to the complementary approach of sedimentation velocity (SV) analytical ultracentrifugation (reviewed in Chapter 16), the time-course of sedimentation is not of interest in SE. Of concern in SE is only the final distribution attained after a long time, which is independent of all kinetic or hydrodynamic factors, but measures directly the macromolecular mass.

Historically, the measurement of protein mass was an important application of SE (Svedberg and Fahraeus, 1926; Schachman, 1992), even though in time it was replaced in most cases by more rapid approaches, such as routine SDS-PAGE electrophoresis or mass spectroscopy, the last exhibiting far superior precision and resolution (see Chapters 5 and 15). Nevertheless, in some cases (e.g., very large proteins, glycoproteins, and protein–detergent complexes) SE is still applied as a powerful tool for measuring the molar mass. What is of much more current interest is the aspect of SE that the measurement establishes strict thermodynamic

P. SCHUCK • National Institute for Biomedical Imaging and Bioengineering, National Institutes of Health, Building 13, Room 3N17, 13 South Drive, Bethesda, Maryland, MD 20892, USA. Tel: 301 4351950; Fax: 301 4801242; E-mail: pschuck@ helix.nih.gov

equilibrium, which makes it uniquely suited for the characterization of reversible protein complexes. It can be used to study protein self-association, as well as heterogeneous associations, and elucidating the reaction schemes and energetics of protein interactions is the main application of modern SE. During the last decades, significant improvement has taken place in the data analysis due to the availability of modern computational techniques, which make the increasingly complex systems accessible for study.

SE takes place in free solution, in the absence of any matrix, and the interpretation is directly based on first principles. Generally, it does not require protein modifications, and can be applied to a wide range of solution conditions. An experiment provides data over a complete isotherm at once (due to the gradient established), and the range of detectable equilibrium dissociation constants is from low nanomoles to millimoles. The molar masses of the macromolecules under study and their complexes can range from kilodalton to greater than megadalton. Although the experiment spans usually one or more days, the extended experimental time can sometimes be an advantage when studying equilibria with slow interconversions of species (reactions that are associated with large activation energies).

The present chapter introduces the main ideas of SE analytical ultracentrifugation for the study of protein interactions. In particular, it highlights the aspects that represent limitations or provide opportunities for studying complex systems. These include problems resulting from exponential curve fitting, and the approaches to overcome these limitations by imposing mass conservation constraints or exploiting multiple signals in the study of heterogeneous associations. Techniques for studying detergent-solubilized membrane proteins are also described briefly. The use of SE is illustrated with a brief review of selected applications in the literature, and with example applications of a receptor–ligand system exhibiting multiple site binding, and a mixed self- and ternary heteroassociating system.

Additional introductory texts and more general and representative literature reviews relating to SE can be found (Arisaka, 1998; Cole and Hansen, 1999; Laue and Stafford, 1999; Minton, 2000; Lebowitz *et al.*, 2002; Behlke and Ristau, 2003; Cole, 2004; Ucci and Cole, 2004; Howlett *et al.*, 2006). For practical information, the reader is referred to the protocols (Laue, 1999; Balbo and Schuck, 2005), the commentary (Schuck and Braswell, 2000), and internet resources (www.analyticalultracentrifugation.com/sedphat/sedphat.htm). For historic accounts, see Elzen (1988), Schachman (1992), and van Holde (2004).

10.2. BASIC PRINCIPLES

Sedimentation equilibrium analytical ultracentrifugation uses the same apparatus as sedimentation velocity described in detail in Chapter 16. It can be summarized as an ultracentrifuge with integrated ultraviolet (UV) absorbance optical and Rayleigh interferometric detection system to visualize the concentration distributions of protein during sedimentation (for details, see Figure 16.1). In contrast to SV, relatively lower rotor speeds are used, such that diffusional transport balances the migration arising from the gravitational force.

In SE, the time-course of sedimentation is not of interest except to the extent of verifying that the sedimentation equilibrium has been attained, i.e., whether the sedimentation, diffusion, and chemical reaction fluxes are in steady state. This can be a slow process, and usually shorter solution columns are used than in SV to reduce the required time. Typically, the time-scale to attain SE in 5 mm columns is on the order of days. For this reason, frequently, SE ultracentrifugation runs are conducted at low temperature. SE is tolerant to initial convection and, therefore, any shape of the solution column can be employed.

In principle, SE could be viewed as a special case of sedimentation velocity analytical ultracentrifugation conducted at lower centrifugal forces. In fact, it can be conducted in this fashion and the time-course to equilibrium may be interpreted with the same methods outlined later in the SV chapter (Chapter 16). As compared with a sedimentation velocity setup, the lower rotor speeds provide relatively small sedimentation fluxes and the larger experimental time allows for significantly more diffusion, such that hydrodynamic separation is strongly diminished and thermodynamic information is dominant. A low-resolution sedimentation coefficient distribution may be derived to extract limited information on hydrodynamic shapes of the sedimenting proteins. Archibald has shown that the slope of the concentration gradients close to the meniscus and the bottom of the solution column report about the molar mass of the solute *at all times* (Archibald, 1947), which can be exploited for analysis, for example, for rapidly degrading proteins or protein complexes that cannot be maintained for sufficiently long time to study by conventional SE (Schachman, 1959; Schuck and Millar, 1998). [An alternative approach is the rapid attainment of SE in very short solution columns, which permit the estimation of a single weight-average molecular weight using the data from the entire solution column (Correia and Yphantis, 1992).]

However, the view of SE as an extrapolation of SV to infinite time, although theoretically correct, would be a misperception of the capabilities of SE, in particular for the study of multicomponent mixtures. Several different features emerge in analytical ultracentrifugation under conditions of SE:

1. The equilibrium concentration gradients can be interpreted entirely by thermodynamic principles alone. Many processes can be described for which the precise analysis of the kinetics of sedimentation would be very complicated or beyond current knowledge. The ability to rely on thermodynamic principles becomes very important when dealing with factors causing nonideal sedimentation, for example, thermodynamic contributions from steric repulsion of proteins at very high concentrations (Zorrilla *et al.*, 2004) or electrostatic interactions (Roark and Yphantis, 1971; Wills and Winzor, 1992), and systems with multicomponent solvents, such as denaturants or detergents (Eisenberg, 1976; Lee *et al.*, 1979) (which usually

form negligible density gradients at the low rotor speeds typically employed in SE). SE distributions of ideally sedimenting interacting multi-component mixtures are analytically tractable and experimentally accessible, because they are free of parameters of chemical reaction kinetics and coupled migration. Interestingly, for ideally sedimenting interacting systems, even though the evolution of concentration distributions while attaining SE is dependent on the interaction and is different from a mixture of noninteracting species, the final SE distributions are indistinguishable from a superposition of a noninteracting mixture comprising all occurring species and complexes at equivalent concentrations (compare Figure 10.1 and see section 10.4).

2. The composition-dependent weight- (or signal-) average molar mass $M(c)$ is measured directly in SE, as opposed to the more indirect assessment as a ratio of sedimentation and diffusion coefficient in SV. SE provides a whole

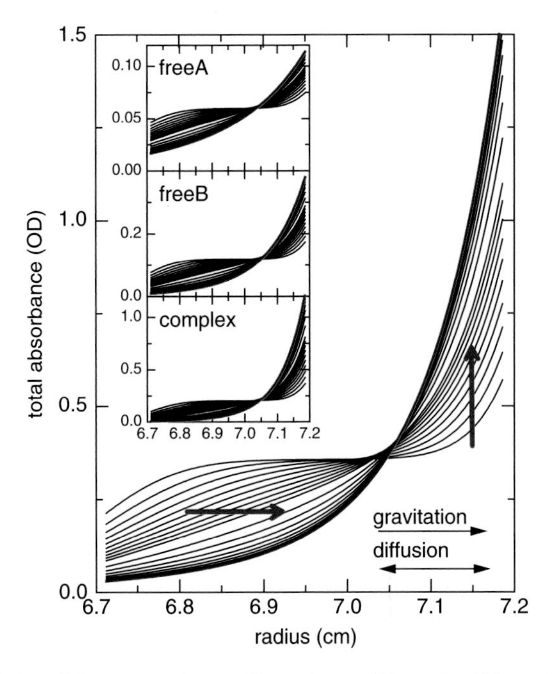

Figure 10.1. Evolution of the concentration profiles under conditions of an SE experiment, for a system of protein A (50 kDa, 3.5S) interacting with another protein B (100 kDa, 5S), forming a 6.5 S-complex. Curves shown are in 1 h intervals for the first 10 h, and then 3 h intervals. The red arrows indicate the movement with time. The final steady-state distribution of SE, when the effect of the gravitational force is balanced by diffusion in the concentration gradient, is indicated in red. Only these red traces are the subject of SE analysis. The distributions are calculated for A, B, and AB to be in chemical equilibrium following mass action law at all times. The *inset* shows the partial concentration of free A, free B, and the complex. Protein loading concentrations are 2 μM each, and K_D of the interactions is 1 μM.

isotherm of $M(c)$ in each experiment (within each solution column), ranging from very low concentrations to concentrations far exceeding the loading concentration.

3. Experiments can be conducted such that SE is established sequentially at different rotor speeds. The redistribution of sedimenting material at different gravitational forces provides an additional data dimension, which can be extremely helpful in unraveling interacting systems.
4. The problems of numerical data analysis are completely different from SV. Conceptually, instead of relatively well-conditioned decomposition into functions resembling step functions that are displaced with time, SE analysis relies on the decomposition of data into a sum of exponential contributions. Therefore, SE has much poorer size resolution for different species. For these reasons, SE can be regarded as an ultracentrifugation technique that is much different and in many ways is orthogonal to SV.

Another analytical ultracentrifugation approach that also relies on a steady state between sedimentation and diffusion attained after long time is buoyant density centrifugation. Here, a dynamic density gradient is produced by the redistribution of a solvent component (e.g., cesium chloride at high concentration) in the centrifugal field (Meselson *et al.*, 1957). In equilibrium, macromolecules have migrated into a band, which is located in the gradient isopycnically at the position of neutral buoyancy, and is approximately Gaussian in shape with a half-width determined by the macromolecular diffusion coefficient (Fujita, 1962; Schuck, 2004). This technique also has a long history (Meselson and Stahl, 1958; Schachman, 1992), in particular, in the study of nucleic acids, and is currently still an important tool for the study of genome composition (Clay *et al.*, 2001, 2003). Similarly, it can be expected that density gradient sedimentation is still a very powerful approach to assess the composition of proteins in complex with other macromolecules of dissimilar density. However, because the analysis of such experiments is much different from the standard SE analysis, the absence of a density gradient is assumed in the discussion that follows.

10.3. EXPERIMENTAL

In practice, the amount of material needed for a SE experiment is typically on the order of 0.1 mg, dependent on the protein size, extinction coefficient, and optical detection method used. Most commonly, absorption optical detection of aromatic amino acids or peptide bond in the UV, or of intrinsic or extrinsic chromophores in the visible spectrum is used, along with refractive index sensitive laser interferometry. These two optical detection methods provide great flexibility regarding buffer choice. It should be noted, however, that SE requires some salt (usually $>$10 mM) to neutralize long-range repulsive electrostatic forces that could otherwise dominate the sedimentation equilibrium distribution. For protocols describing and considerations for the choice of optical system and other aspects of the experimental design, as well as the practical steps involved in setting up an SE experiment, see Balbo and Schuck (2005).

Clearly, for binding constants to be determined, conditions need to be established where it is possible to detect both free and complex species. Otherwise, if the binding is too tight, only the stoichiometry of the complex may be determined. The detection limit for proteins is typically $<10\ \mu g/mL$, and this determines the upper limit of affinities dependent on the extinction coefficient of the proteins under study.

Greater than 95% purity of the protein is usually required. It is highly desirable to perform size-exclusion chromatography as the last preparative step; this removes (among other contaminants) small molar mass impurities that may go undetected by SDS-PAGE, but distribute throughout the solution column and may establish an additional signal gradient. If unrecognized, this can lead to erroneous parameter estimates in the data analysis. Very large molar mass impurities are sometimes less of a concern, as they may sediment to the bottom of the solution column and remain outside the radial window of detection. In many laboratories, SV analytical ultracentrifugation is routinely applied before the SE experiment, to assess the purity of the material and to obtain additional information on the protein interactions, such as the number and approximate size of species involved (for details, see Chapter 16).

It is noteworthy that SE involves the measurement of protein concentrations, and therefore, it does not depend on the precise knowledge of the loading concentration. (Also, if a combination of refractive index and absorbance detection is used, the molar extinction coefficients can be measured as a part of the SE experiment). In particular, when quantitatively reconciling the results of different biophysical techniques, differences in the propagation of errors from the protein concentration can be a major concern. This is a more difficult problem if fractions of the protein preparation may be partially misfolded and incapable of participating in self-association or heteroassociation. Fractions of incompetent proteins can be taken into account in the quantitative analysis of SE, and, in favorable cases, can be estimated from the experimental data (Vistica *et al.*, 2004).

10.4. THEORY

10.4.1. Single Species in the Gravitational Field

The chemical potential of an ideally sedimenting particle in the centrifugal field may be written as

$$\mu(r) = \mu_0 + RT \ln c - \frac{1}{2} M\omega^2 r^2, \tag{10.1}$$

(van Holde *et al.*, 1998) where M is the molar mass, ω is the angular velocity, and r is the distance from the center of rotation. In sedimentation equilibrium, the gradient $(\mathrm{d}\mu/\mathrm{d}r)_\mathrm{T}$ vanishes, such that the concentration profile assumes a Boltzmann distribution:

$$c(r) = c(r_0)\exp\left(M_b\omega^2(r^2 - r_0^2)/2RT\right) \qquad (10.2)$$

with M_b denoting the so-called buoyant molar mass, $M\mathrm{d}\rho/\mathrm{d}c$, (with $\mathrm{d}\rho/\mathrm{d}c$ denoting the density increment of the particle), and with r_0 an arbitrarily chosen reference radius. Thus, from an experimentally measured concentration profile of a single, ideally sedimenting protein, the buoyant molar mass M_b can be determined. Thermodynamically deeper and more detailed derivations of Eq. (10.2) can be found, for example, in Fujita (1962) and Eisenberg (1976). In particular, it is instructive to consider the relationship to the concentration dependence of the osmotic pressure in SE, $\mathrm{d}\Pi/\mathrm{d}c_2 = (\omega^2/2)(\mathrm{d}\rho/\mathrm{d}c_2)(\mathrm{dln}(c)/\mathrm{d}r^2)^{-1}$ (see Chapter 9; Eisenberg, 1976), which highlights the relationship of SE to osmotic pressure in dialysis equilibrium, and the relationship to particles numbers (Eisenberg, 2000).

It is usually convenient to replace $M\mathrm{d}\rho/\mathrm{d}c$ with the product $M(1-\bar{\nu}\rho)$, where $\bar{\nu}$ is the partial-specific volume (the inverse density) of the protein. Following Archimedes principle, this term represents the difference between the mass of the protein and the buoyancy arising from the mass of the displaced solvent, $M\bar{\nu}\rho$. However, it should be noted that the partial-specific volume $\bar{\nu}$ is not a well-defined thermodynamic quantity to the extent that it can be dependent on solvent conditions (Eisenberg, 1976, 2003; Lee *et al.*, 1979), and corrections for preferential solvent interactions may have to be applied (see Ebel *et al.*, 2000). Nevertheless, for most commonly used aqueous buffer systems with moderate ionic strength and density close to unity, the contributions of preferential interactions to the buoyant molar mass of the protein are negligible (Lebowitz *et al.*, 2003). The role of solvation as a driving force in protein interactions is reviewed in Chapter 9.

In practice, it is usually possible to predict $\bar{\nu}$ of a protein with good precision from the amino acid composition (Durchschlag, 1986), such that Eq. (10.2) reveals the molar mass M. Approaches to account for the contributions of prosthetic groups and carbohydrates or other protein modifications are well developed (Durchschlag, 1986; Shire, 1992; Fairman *et al.*, 1999; Lewis and Junghans, 2000). The value of $\bar{\nu}$ (or $\mathrm{d}\rho/\mathrm{d}c$) can be determined experimentally by densitometry (Kratky *et al.*, 1973; Durchschlag, 1986), although the sufficiently precise measurement frequently takes impractical amounts of material (Lebowitz *et al.*, 2003). In many cases, M is known with great precision from the amino acid sequence or from mass spectroscopy, and the experimentally determined quantity M_b can be used, conversely, to determine the value of $\bar{\nu}$ of the protein in the buffer conditions applied. This may not be of interest, for example, for heterogeneous protein associations, where the buoyant molar masses of both proteins can be determined in separate experiments: It should be noted that the buoyant molar mass is the only quantity required for the analysis of protein interactions, and not $\bar{\nu}$. Cases where the prediction of $\bar{\nu}$ is more critical are, for example, studies of weak reversible self-associations or studies with large oligomers with unknown stoichiometry.

10.4.2. Density Variation

Similar to the contrast variation in the scattering techniques, which are reviewed in Chapter 11, it is possible to adjust the solution density and gain information simultaneously on the partial-specific volume and the molar mass (Edelstein and Schachman, 1967) from the dependence of the buoyant molar mass on solution density:

$$M_b(\rho) = M^*(1-\bar{v}\rho). \tag{10.3}$$

This is possible, for example, with mixtures of H_2O and D_2O or D_2O^{18}, but requires the estimation of the increase in the protein mass M^* due to H/D exchange. A completely different density variation approach is isopycnic sedimentation, as described earlier. However, under typical conditions of SE of proteins, density gradients are negligible due to the relatively low centrifugal fields.

Tanford and Reynolds have extended the Edelstein–Schachman technique in a significant way for the study of protein–detergent complexes (Reynolds and Tanford, 1976; Reynolds and McCaslin, 1985). Many nonionic detergents are transparent in the UV, which makes SE a powerful tool to study detergent-solubilized membrane proteins (Schubert and Schuck, 1991; Howlett, 1992). The buoyant molar mass of a protein–detergent complex may be divided into contributions from the protein and the detergent

$$M_b = M^{*(P)}\left[\left(1-\bar{v}^{(P)}\rho\right) + \delta\left(1-\bar{v}^{(D)}\rho\right) + \beta\left(1-\bar{v}^{(H)}\rho\right)\right] \tag{10.4}$$

with $\bar{v}^{(D)}$ as the partial-specific volume of the detergent and δ the amount of detergent bound to the protein (g/g). The second term in Eq. (10.4) describes the buoyant molar mass of the detergent in the protein–detergent complex, and the third term formally describes the contribution from preferential hydration. In aqueous solvents, the last term typically vanishes as the density of the solvation shell is close to the density of the solvent. Many detergents have densities that are in the range accessible for density matching with H_2O–D_2O mixtures, and when ρ is chosen at $1/\bar{v}^{(D)}$, the detergent becomes neutrally buoyant and invisible in SE. In this case, M_b is determined entirely by the protein buoyant molar mass, which can reveal, for example, the oligomeric state of a detergent-solubilized membrane protein. For detergents that have a density greater than D_2O, and for cases where the addition of D_2O is impractical, the solvent density may be raised by adding an additional solvent component, such as sucrose (Mayer *et al.*, 1999; Lustig *et al.*, 2000; Center *et al.*, 2001). In this case, preferential hydration can occur, and care must be taken to estimate its contribution to the detergent and to the protein.

The density variation approach can be used to study both the degrees of detergent binding and the self-association properties of detergent-solubilized membrane proteins (Schubert *et al.*, 1994; Hellstern *et al.*, 2001; Noy *et al.*, 2003; Gohon *et al.*, 2004; Lear *et al.*, 2004; Stanley and Fleming, 2005). For heterogeneous associations, in principle, the experimentally determined buoyant molar mass of the protein–detergent complex can be taken as a basis for the analysis of the interaction.

The following sections outline the theory for reversible self-association and hetero-association equilibria by SE.

10.4.3. Self-Associations

One of the unique features of SE analytical ultracentrifugation is that the experiment establishes a concentration gradient, and protein–protein interactions are observed over a very large concentration range at once in a single experiment. For interacting proteins, the SE condition requires the sedimenting species to be in chemical equilibrium, as well as centrifugal equilibrium. For a reversible self-associating monomer–dimer system, this requires that the radial distribution of the monomer, $c_1(r)$, and dimer, $c_2(r)$, follows the mass action law

$$c_2(r) = K_{12}(c_1(r))^2 \tag{10.5}$$

with the association equilibrium constant K_{12}. As the rotor speeds in SE are relatively low, the maximum pressures are typically <1 MPa and the pressure dependence of the association constants can be neglected. If we assume that the radial distribution of the monomer $c_1(r)$ follows Eq. (10.2), we obtain

$$c_2(r) = K_{12}(c_1(r_0))^2 \exp(2M_b\xi(r)), \tag{10.6}$$

(with the abbreviation $\xi(r) = \omega^2(r^2 - r_0^2)/2RT$), which can be recognized as the radial distribution of an ideally sedimenting particle of twice the buoyant molar mass of the monomer. Because the dimer molar mass is double the monomer molar mass, and as changes in the buoyancy contributions of the hydration on dimerization are usually absent or negligible (they would lead to pressure-dependent equilibrium constants), it can be seen that the Boltzmann distributions in the form of Eq. (10.2) allow the centrifugal and chemical equilibrium to be fulfilled simultaneously. Accordingly, the total concentration distribution of a monomer–dimer system is

$$c(r) = c_1(r_0)\exp(M_b\xi(r)) + K_{12}(c_1(r_0))^2 \exp(2M_b\xi(r)). \tag{10.7}$$

These radial distributions are illustrated in Figure 10.2 (blue lines in Panel B). Deviations from these distributions can occur at high protein concentrations, where additional repulsive interactions have to be taken into account. Theoretical and experimental sedimentation equilibrium techniques for studying proteins in highly concentrated solutions have been developed (Rivas *et al.*, 1999, 2001; Zorrilla *et al.*, 2004). Analogous considerations can be used to derive the radial concentration distribution for multistep self-associations, which result in similar multiexponential expressions where each n-mer follows a Boltzmann distribution (Eq. 10.2) with the n-fold monomer buoyant molar mass, and with an amplitude determined by mass action law. This takes the general form

$$c(r) = \sum_n K_{1n}(c_1(r_0))^n \exp(nM_b\xi(r)) \tag{10.8}$$

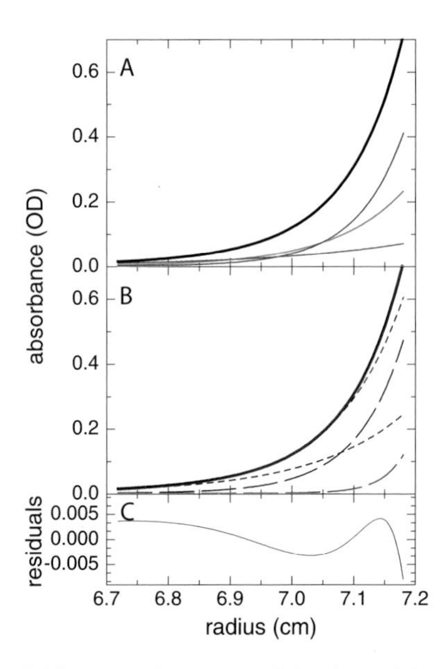

Figure 10.2. Contributions of different species to the total signal. Panel A: Heterogeneous association of two molecules A (50 kDa, blue) and B (100 kDa, green), forming a 1:1 complex AB (red). The loading concentrations of A and B are equimolar at $K_D = 1\ \mu$M. The total signal is shown as a black line. Panel B: Monomer–dimer self-association of 75 kDa protein with $\log_{10}(K_{12}) = 5.7$, at a loading concentration of 2.3 μM. The total is shown as a blue line, and the monomer and dimer contributions are shown as blue dotted and dashed lines, respectively. This same panel also shows the monomer–trimer self-association of a molecule of 103 kDa, with $\log_{10}(K_{13}) = 9.4$, at a loading concentration of 1.5 μM. The total is shown as solid red line, the monomer and trimer contributions as red dotted and dashed lines, respectively. The solid blue and red lines virtually superimpose (residuals are shown in Panel C), illustrating the difficulty of exponential decomposition of traces from single SE experiments without prior knowledge. These can be overcome by using knowledge of the molar mass, global modeling experiments conducted at a large range of concentrations and rotor speeds, mass conservation constraints, and stringent criteria for the goodness-of-fit.

(with K_{1n} the equilibrium association constant from monomer to n-mer, and $K_{11} = 1$). For indefinite isodesmic self-associating systems where the addition of a monomer to any oligomer is governed by the same equilibrium constant K, the sedimentation equilibrium distribution can be written in the closed form

$$c(r) = \frac{c_1(r_0)\exp(M_b\xi(r))}{(1-Kc_1(r_0)\exp(M_b\xi(r)))^2}. \tag{10.9}$$

Many systems exhibit such self-association behavior, including, for example beta-lactoglobulin A, the bacterial cell division protein FtsZ, and HIV Rev (Adams and Lewis, 1968; Rivas *et al.*, 2000; Surendran *et al.*, 2004). Extensions have been derived and applied for decaying indefinite self-associations (Chatelier, 1987; Rivas *et al.*, 2000) and for isodesmic associations with a different binding constant for the initial formation of the first oligomer (Howlett *et al.*, 1973; Morris and Ralston, 1989).

10.4.4. Heterogeneous Associations

While studying heterogeneous interactions, it is necessary to consider the different signal contributions of different protein components. For example, for mixtures of two protein components A and B in reversible equilibrium with an association constant K_{AB}, the signal obtained in SE from the total concentration gradient is

$$a_\lambda(r) = c_{\mathrm{A}}(r_0)\varepsilon_\lambda^{(\mathrm{A})} d \exp\left(M_{\mathrm{b}}^{(\mathrm{A})}\xi(r)\right) + c_{\mathrm{B}}(r_0)\varepsilon_\lambda^{(\mathrm{B})} d \exp\left(M_{\mathrm{b}}^{(\mathrm{B})}\xi(r)\right) + K_{\mathrm{AB}} c_{\mathrm{A}}(r_0) c_{\mathrm{B}}(r_0)\left(\varepsilon_\lambda^{(\mathrm{A})} + \varepsilon_\lambda^{(\mathrm{B})}\right) d \exp\left(\left(M_{\mathrm{b}}^{(\mathrm{A})} + M_{\mathrm{b}}^{(\mathrm{B})}\right)\xi(r)\right), \quad (10.10)$$

where $a_\lambda(r)$ is the radial dependence of the measured signal at wavelength λ, $\varepsilon_\lambda^{(\mathrm{A})}$ and $\varepsilon_\lambda^{(\mathrm{B})}$ are the extinction coefficients at that wavelength, d is the optical pathlength, $M_{\mathrm{b}}^{(\mathrm{A})}$ and $M_{\mathrm{b}}^{(\mathrm{B})}$ are the buoyant molar masses of A and B, respectively, (Figure 10.2).

Similarly, the concentration distributions of multicomponent mixtures exhibiting heterogeneous interactions or exhibiting both self-association and heteroassociation can be derived, with one Boltzmann exponential term for each free species and for each homooligomer or heterooligomer

$$a_\lambda(r) = \sum_{\mathrm{i,j}} K_{\mathrm{ij}}(c_{\mathrm{A}}(r_0))^{\mathrm{i}}(c_{\mathrm{B}}(r_0))^{\mathrm{j}}\left(\mathrm{i}\varepsilon_\lambda^{(\mathrm{A})} + \mathrm{j}\varepsilon_\lambda^{(\mathrm{B})}\right)\exp\left\{\left(\mathrm{i}M_{\mathrm{b}}^{(\mathrm{A})} + \mathrm{j}M_{\mathrm{b}}^{(\mathrm{B})}\right)\xi(r)\right\} \quad (10.11)$$

with K_{ij} the equilibrium constant for the formation of species (i, j) from the free monomers of A and B ($K_{10} = K_{01} = 1$). The complexity of the system under study is limited essentially only by the capability of resolving the different exponentials in the data analysis, which is described later.

10.4.5. Nonideal Sedimentation

The above description of SE assumes the absence of additional protein interactions arising from electrostatic or steric repulsive forces. This holds true typically for dilute solutions of relatively globular proteins (<1 mg/mL) at moderate ionic strengths. For weak interactions, it may be necessary to increase the protein concentration above this limit, and to account for repulsive interactions through the application of a first-order term of a virial expansion

$$c(r) = c(r_0)\exp\left(-2BM[c(r) - c(m)] + M_b\omega^2(r^2 - r_0^2)/2RT\right) \quad (10.12)$$

(with B the second virial coefficient, and m the meniscus position of the solution column) (Haschemeyer and Bowers, 1970; Holladay and Sohpianolpoulos, 1972). Phenomenologically, these interactions produce distinct deviations from the exponential distributions by reducing the gradient at high concentrations. Nevertheless, the mass action law is valid, and from this principle, equations for interacting systems can be derived. For globular proteins, the repulsive contributions can be estimated as $BM = 4\bar{v}$ ($\sim$3–4 mL/g), but more elongated proteins exhibit significantly larger values (Tanford, 1961). The characteristic pattern of a reduced slope at higher concentration may be masked, partially, by attractive interactions causing formation of complexes that increase the gradients, making the analysis of association under nonideal conditions significantly more difficult. An example is the study of the monomer–dimer self-association equilibrium of the human cytomegalovirus protease, where the dimer constitutes the active species ($K_D = 60\,\mu M$), and the strong thermodynamic nonideal sedimentation was observed (Cole, 1996).

Another origin of thermodynamic nonideal sedimentation is the presence of high concentrations of unrelated macromolecules or molecular crowding. Interest in such conditions has been stimulated by the high intracellular concentrations. In the solutions of high total macromolecular concentration, optical detection is limited, for example, due to the large refractive index gradients. Fractionation of SE gradients established in preparative centrifuges followed by postcentrifugal determination of protein concentration in each fraction was developed as an experimental tool to study the energetics of protein–protein interactions in crowded media (Rivas *et al.*, 1999). A theoretical description can be found in Zorrilla *et al.* (2004). This has been applied, for example, to the study of the bacterial cell division protein FtsZ, where crowding could be shown to significantly enhance self-assembly (Rivas *et al.*, 2001).

10.5. DATA ANALYSIS

Fitting of experimental data with exponentials is a problem encountered in many biophysical disciplines (see Chapters 4, 7, and 13). The basic problem of fitting exponentials is illustrated in Panel B of Figure 10.2 with the superposition of the solid red and blue lines: over a limited signal range, the sum of two exponentials is very similar to a combination of two different exponentials, or sometimes even to a single exponential with intermediate exponent. Reliably distinguishing multiple exponential terms can be extraordinarily difficult (Varah, 1985). Fortunately, for several reasons the decomposition of the measured data into exponential contributions appears more advantageous in SE than in other techniques, and the simplicity of the physical setup of SE experiments and the first-principle based theory allow the quality of fit to be carefully scrutinized. Many of the advanced techniques of SE in the study of complex interacting systems revolve around the strategies to arrive at a well-conditioned data analysis, and discerning between a correct and impostor models.

Importantly, the exponents are frequently known in SE, as they are determined directly by the buoyant molar mass, which can either be experimentally measured separately or predicted based on amino acid composition and mass spectroscopy (see above). This allows up to four species to be determined in SE from a single concentration profile with ordinary signal–noise level, assuming that the molar masses of species are well separated. In comparison with dynamic light scattering, fluorescence correlation spectroscopy or other diffusion-based approaches, the size resolution is higher in SE because M is observed directly, as opposed to the Stokes radius $R_S \sim M^{-1/3}$. Therefore, fractional contributions from monomers and dimers, for example, can be readily distinguished from a single scan. Without prior knowledge or additional data sets, however, it can be very difficult to distinguish a monomer–trimer from monomer–dimer–tetramer self-association. It is highly useful to conduct sedimentation velocity studies before SE, as the analysis of the sedimentation velocity profiles can in many cases reveal the mode of association, and/or the number of species present in solution.

The principle of global analysis of many concentration profiles obtained for a range of conditions is an important component of SE analysis. Typically, several samples covering a range of loading concentrations occupy different positions in the rotor and are studied the same centrifugation run. Even though each single SE profile provides information over a whole concentration range, the global analysis of multiple gradients from different loading concentrations can greatly improve the ability to distinguish different association models. In a preliminary state of the analysis, it offers the possibility of examining the isotherm of weight-average or cell-average, molar mass $M_w(c)$ as a function of loading concentration, which can be helpful in gaining information on the association scheme, such as the minimum oligomer size participating and an estimate of K_D. Usually, however, to extract only a weight-average molar mass sacrifices more detailed information from the concentration profiles, in particular from SE in longer solution columns, and the direct global analysis based on known or hypothesized interaction modes is advantageous.

The information on interacting systems can be further increased by acquiring equilibrium profiles sequentially at different rotor speeds. This forces redistribution of the proteins and chemical equilibrium to be established in different concentration gradients. Generally, data with high curvature and ''meniscus depletion'' conditions exhibit a relatively high precision in the molar mass, define the baseline absorbance well, and enable the characterization of the smallest species (Yphantis, 1964), but they require data acquisition at steep gradients and close to the bottom of the solution column. Conversely, low-speed SE (Richards *et al.*, 1968) reports on the largest species present and permits the observation of the majority of the material. Inclusion of profiles at multiple rotor speeds into a global analysis takes advantage of both configurations and improves the ability to distinguish between different models. Importantly, the information on the redistribution at different rotor speeds can also allow refining the analysis by unraveling radial-dependent baseline signals (Vistica *et al.*, 2004).

When studying heterogeneous protein interactions, it is important to exploit differences in the absorbance spectra of the different components. In combination with the refractive index sensitive detection system, absolute differences in the extinction coefficients can be exploited, as well as relative differences, for example, at 250 and 280 nm arising from different ratios of tryptophan and tyrosine residues. Even better discrimination can frequently be achieved if extrinsic chromophoric labels can be attached to one or more protein components. Technically, multiple radial absorbance distributions at different wavelengths can be collected (Osborne *et al.*, 1980; Lewis *et al.*, 1993; Howlett, 1998; Vistica *et al.*, 2004), as well as multiple absorbance spectra at different radii within the gradient (Schuck, 1994; Noy *et al.*, 2005). Both approaches can allow following the concentration gradient of each component separately. Therefore, a global multiwavelength or multisignal analysis greatly enriches the information content of the SE experiment.

An important approach for the analysis of interacting systems is mass balance considerations. It is possible to calculate the total amount of material in the centrifuge cell by volume integration of the fitted concentration gradient from meniscus to bottom, and to calculate the equivalent uniformly distributed loading concentration $\bar{c}$ that corresponds to the same total amount. For a single species, integration of Eq. (10.2) over the sector-shaped solution volume gives

$$\bar{c} = c(r_0)\frac{\exp(M_b\xi(b)) - \exp(M_b\xi(m))}{M_b(\xi(b) - \xi(m))}, \quad (10.13)$$

where m and b are the meniscus and the bottom of the solution column, respectively. Numerical integration can be used to determine $\bar{c}$ for different solution column shapes. For interacting systems, the summation of the $\bar{c}$ terms for the free and each oligomeric species (considering their stoichiometry) should lead to the actual loading concentration of each component (Fujita, 1975).

A difficulty in the mass conservation constraints is that the actual loading concentrations, as well as the physical limits of the solution column, may not be known with sufficient precision. The first problem can be circumvented by requiring the total amount of material to remain constant only during redistribution among the different rotor speeds (Roark, 1976). This imposes constraints only on the basis of the experimentally measured concentration of soluble material in SE, and frees the mass conservation analysis from errors in the *a priori* knowledge of the protein concentration. The second problem can be addressed by treatment the geometric limits of the solution column, in particular, at the bottom of the solution column where a large fraction of the material is located in SE, as a fitting parameter (Vistica *et al.*, 2004). It is usually well determined if gradients at three different rotor speeds are available.

This approach can provide a set of powerful constraints and significantly improve the data analysis. For example, this eliminates or alleviates the problem of poorly distinguishable concentration distributions from species with similar molar mass but different composition. For example, consider the determination of the binding constant of two similar-sized proteins A and B forming a 1:1 complex AB. Unless the free species of both components can be distinguished spectrally, from the analysis of SE gradients without mass conservation constraints only a lower limit for the association constant can be determined. (This is due to the similarity of the exponential distribution of the free proteins.) However, mass conservation constraints using the knowledge of the loaded material allows to quantitatively assign the relative populations of free A and free B, and thus to determine the binding constant. Similarly, when A and B have very different buoyant molar mass, the mass conservation constraints can overcome limitations arising from too small relative differences between the larger free species and the complex.

Finally, for heterogeneous interactions it is possible to design the experiment such that the mixtures in the different rotor positions of the ultracentrifuge run are either serial dilutions or titrations, an approach developed by Vistica *et al.* (2004). In this case, the molar ratio of the equivalent loading concentrations of the components, or the equivalent loading concentration of the component that was held constant, respectively, can be constrained in the global analysis of the data from the different samples. This can further reduce the number of parameters to be fitted, reduce the parameter correlations, and improve the discrimination among different sedimentation models. Here, it is important to strike a balance between constraints that may be affected by preparative imperfections (such as dilution errors affecting the absolute protein concentrations), and constraints that are more robust (such as the invariant molar ratio of two proteins after dilution of a stock mixture).

In summary, global analyses are performed typically including 10–50 individual scans from the exponential concentration profiles obtained at different loading concentrations, rotor speeds, and detection systems. The fitting parameters include those of the interaction model, concentration parameters (which can include the relative activity of the loaded protein components), and parameters local to each scan (such as baseline signals) and local to each sample column (such as the meniscus and the bottom). Generally, Marquardt-Levenberg and simulated annealing optimization have proven to be particularly useful algorithms. Obviously, the error surface to be minimized can be very complicated. However, using the constraints outlined earlier, the number of floating parameters can be reduced, and interacting multicomponent system can be studied with many different species and protein components. Even if some constraints do not strictly hold, they may be useful in the optimization process. For example, Philo has introduced the use of mass conservation principles as scalable constraints to simplify the error surface during the initial stages of the optimization (Philo, 2000). Statistical weights for the analysis of absorbance data are discussed by Lewis and Reily (2004).

10.6. APPLICATIONS

To illustrate the potential of SE, two applications with different focus and different variations of SE analysis are reviewed in some detail. This is followed by a selective representation of different types of SE applications in the literature.

10.6.1. Self- and Heterogeneous Associations of Band 3 Protein

The band 3 protein (AE1) is an ubiquitous anion exchanger, which is found in large numbers on the human erythrocyte membrane (Passow, 1992). It could be regarded as a prototype of an allosteric, multifunctional membrane protein. Human erythrocyte band 3 (∼100 kDa) has been studied by SE analytical ultracentrifugation in detergent solution, and was found to exist in a self-association equilibrium between monomers, dimers, and tetramers (Schubert *et al.*, 1992). In SE density matching experiments with band 3 protein reconstituted functionally into lipid vesicles, it was found that the monomer is the functional unit for anion exchange (Lindenthal and Schubert, 1991). The intracellular domain of the band 3 protein exists in different conformations, and represents a binding site for a number of intracellular proteins, including hemoglobin (with a different affinity for oxy- and deoxyhemoglobin), band 4.1, ankyrin, aldolase, D-glyceraldehyde-3-phosphate dehydrogenase (GAPDH), and phosphofructokinase (PFK) (Low, 1992), with binding constants of GAPDH and aldolase (binding to erythrocyte membranes) in the micromolar range (Kelley and Winzor, 1984). The band 3 protein was proposed to be the locus of a multienzyme complex of glycolytic enzymes, and *in vivo* evidence for the existence of a regulated complex has been found by immunofluorescence staining (Campanella *et al.*, 2005).

Interestingly, the self-association of the membrane protein and the heterogeneous association with cytoplasmic proteins, as well as the binding of transport inhibitors are linked (Schubert *et al.*, 1992; Schuck *et al.*, 1995; von Ruckmann and Schubert, 2002). This was established in a series of SE studies using mixtures of band 3 protein in detergent solution and extrinsically labeled intracellular ligands (no label was required for hemoglobin). In these experiments, the absorption distribution in the visible spectrum exclusively describes the distribution of the species containing dye-labeled ligand, and in a double exponential fit, with the first species describing the known molar mass of the free ligand, the second exponential component determines the average molar mass of the band 3–ligand complex. Due to the large molar mass difference of the band 3 oligomers relative to the ligand molar mass, it could be established that the band 3 tetramer is the only binding site for these ligands. Furthermore, with the exception of ankyrin, the molar mass of the complex was dependent on the band 3–ligand molar ratio, indicating that up to four ligands could bind per tetramer at very low band 3–ligand ratios (Schubert *et al.*, 1992; von Ruckmann and Schubert, 2002). At very high band 3–ligand ratios, there was no indication of ligand occupancy of oligomers other than the tetramer (von Ruckmann and Schubert, 2002). Due to the high complexity of this linked association system, the main question addressed in these studies is that of the functional unit of the membrane protein and the number of ligands capable of binding, but not a quantitative analysis of the binding energetics.

Further, this approach was applied to study ternary protein complexes, in mixtures of band 3 simultaneously with both the glycolytic enzymes aldolase and (fluorescently labeled) GAPDH (von Ruckmann and Schubert, 2002). Aldolase exhibited a higher affinity to band 3 and was able to displace GAPDH from band 3 tetramers when aldolase was in excess, consistent with a partial overlap of binding sites or negative cooperativity mediated through conformational changes of band 3. However, at substoichiometric concentrations of GAPDH, the dependence of the

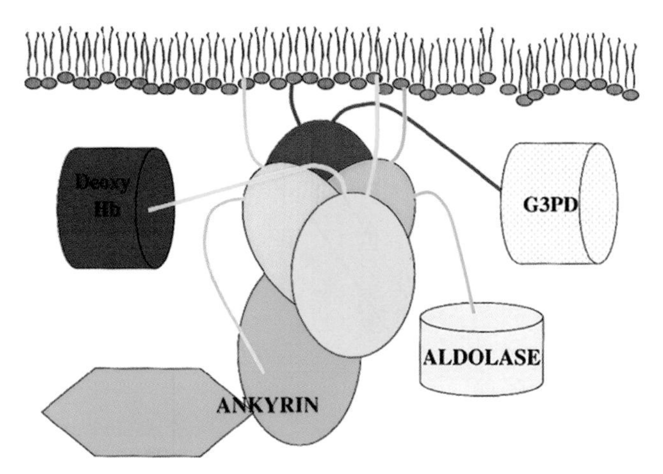

Figure 10.3. Cartoon of the cytoplasmic domains of a band 3 tetramer, and the possible organization of a peripheral protein complex. G3PD: glyceraldehyde-3-phosphate dehydrogenase; Deoxy Hb: deoxyhemoglobin [Reprinted from Zhang *et al.* (2000), with permission].

average molecular weight of the GAPDH-containing complexes on aldolase concentration showed a significant increase at substoichiometric aldolase concentrations, indicating the formation of ternary complexes of band 3 tetramers occupied simultaneously by both ligands (Figure 10.3). This has implications for the function of band 3 tetramer cytoplasmic domains as a nucleus of an active multienzyme complex (Campanella *et al.*, 2005).

With this system, under some conditions SV analytical ultracentrifugation (Chapter 16) exhibited the difficulty of reaction kinetics on the time-scale of the SV experiment, in which case the observed *s*-values of the reaction boundaries do not directly reflect only the size of the complexes formed (von Ruckmann and Schubert, 2002). Interestingly, this is in contrast to the behavior of some signal-transduction and the cell-surface receptor systems illustrated in Chapter 16, where the complexes exhibited longer lifetimes (lower dissociation rate constants), making SV more suitable and permitting resolving different complexes.

10.6.2. High- and Low-Affinity Sites of the Fc Receptor FcRn

FcRn is a Fc receptor, which was originally thought to be primarily responsible for the transplacental transfer of gammaglobulins (IgG) from the mother to the young, but has been found to be involved more generally in transporting IgG within and across cells of diverse origin, and to regulate IgG levels throughout the body (Rodewald and Kraehenbuhl, 1984; Ober *et al.*, 2004; Vaccaro *et al.*, 2005). From symmetry considerations of the IgG molecule, it was hypothesized that two FcRn molecules may bind to one Fc (or IgG) molecule, and cocrystal structures of two

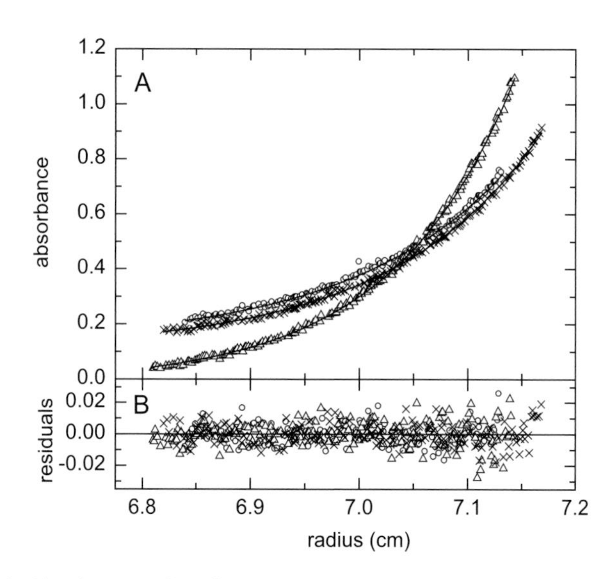

Figure 10.4. (A) Absorbance profiles of FcRn (circles), IgG (triangles), and Fc (crosses) separately in sedimentation equilibrium at a rotor speed of 12,000 rpm, and at a detection wavelength of 250 nm. Best-fit curves (solid lines) are obtained with buoyant molar mass values of 13,500 Da for FcRn, 40,660 Da for IgG, and 14,500 Da for Fc. (B) Residuals of the fits [Reprinted from Schuck *et al.* (1999), with permission].

different 2:1 complexes in different geometry were found (Burmeister *et al.*, 1994). Solution studies, however, yielded conflicting results, and an asymmetric binding model was proposed with negative cooperativity of the two sites mediated through conformational changes in the Fc (Burmeister *et al.*, 1994; Ghetie and Ward, 1997; Weng *et al.*, 1998). Sedimentation equilibrium was used to determine the binding stoichiometry and the energetics of the interaction (Schuck *et al.*, 1999).

Figure 10.4 shows the measured SE distributions of FcRn, IgG, and the Fc fragments when studied separately. These data were used to determine the buoyant molar mass of the glycosylated proteins. It also demonstrates the quality of fit that can be routinely achieved when modeling SE data. The SE absorbance distributions of mixtures of FcRn with Fc show the formation of both 1:1 and 2:1 complexes, which can be discerned from the curvature of the measured distributions (Figure 10.5). The dashed lines in Figure 10.5 illustrate the calculated contributions of the 2:1 complex. For comparison, the residuals of the best-fit excluding the formation of 2:1 complexes are shown in Panel C, a model that must be rejected because of the systematic deviations. The macroscopic binding constants were determined from the populations of the different species, which produced values of <130 nM and 6 μM for sites 1 and 2, respectively. This corresponds to a more than tenfold difference of microscopic affinity constants (assuming equivalent sites), or at least 1.3 kcal/mol lower binding energy for binding the second FcRn to the preformed 1:1 complex, when compared with the first binding event.

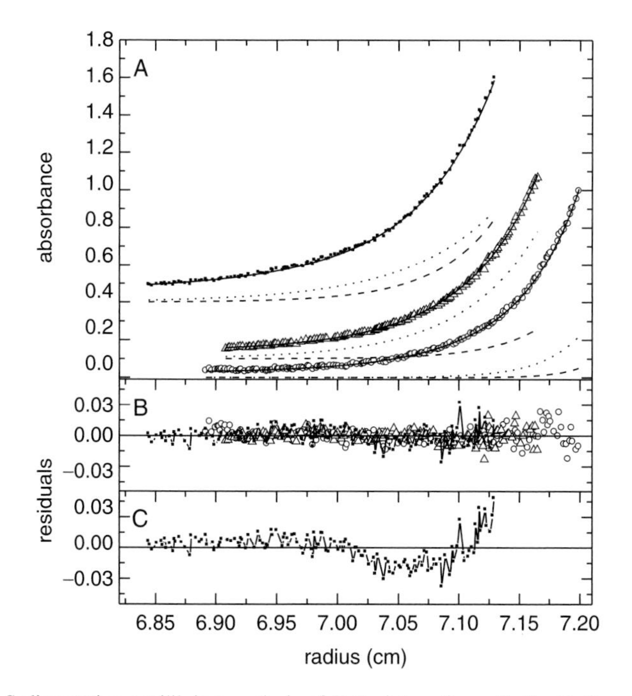

Figure 10.5. Sedimentation equilibrium analysis of FcRn interaction with the Fc fragment. Panel A: Experimental absorbance profiles of FcRn–Fc mixtures at total concentrations of 0.56 μM FcRn and 0.47 μM Fc at 18,000 rpm, scanned at 230 nm (circles), 1.95 μM FcRn and 1.35 μM Fc at 13,000 rpm, scanned at 280 nm (triangles) (for clarity, data are shown with an offset of 0.1 OD), and 5.5 μM FcRn and 2.8 μM Fc at 12,000 rpm, scanned at 250 nm (squares) (offset by 0.4 OD). Global analysis led to best-fit distributions as indicated by the solid lines, with $K_A^{(1:1)}$ for 1:1 complex formation of $6.5\times10^8\ M^{-1}$, and an affinity $K_A^{(2:1)}$ for the addition of a second FcRn molecule (formation of 2:1 complex) of $1.64\times10^5\ M^{-1}$ (dashed lines). The theoretical distributions are shown with the same offset as the fitted data. To illustrate the contribution of the mixed complexes to the observed absorption profiles at the different loading concentrations, the calculated distribution of the 1:1 complex was indicated by the dotted line; and the 2:1 complex by the dashed line. Panel B: Residuals of the fit from Panel A. Panel C: Residuals of the best-fit to the highest concentration data are shown in A assuming the absence of 2:1 complex formation [Reprinted from Schuck *et al.* (1999), with permission].

The statistical accuracy of the binding constants can be studied using projections of the error surface [constraining K_d to different values while floating all other unknown parameters (Bevington and Robinson, 1992)] (Figure 10.6). In general, this is more advantageous than Monte-Carlo error analysis, due to the requirement for automated nonlinear optimization in the latter approach, which can frequently fail with highly ill-behaved error surfaces of global sedimentation equilibrium fits. In the present case of the Fc–FcRn interaction, binding of the first molecule is too tight and the buoyant molar masses of Fc and FcRn are too similar to allow the discrimination of unbound molecules. For this reason, only an upper limit for K_d of site 1 could be determined. In contrast, the affinity of site 2 was well determined by

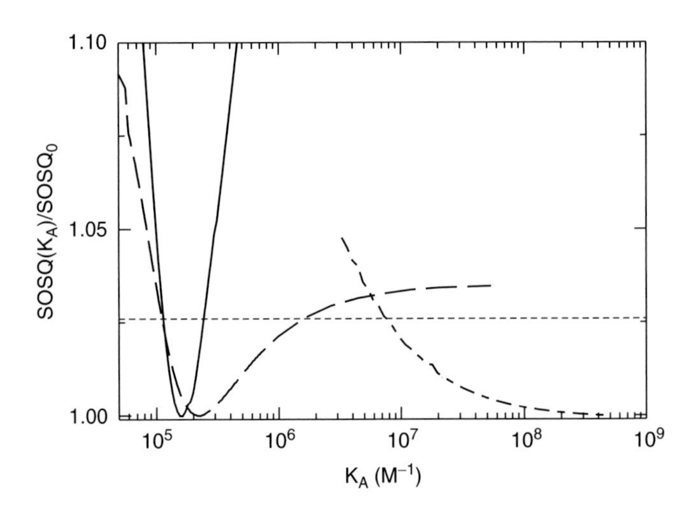

Figure 10.6. Projections of the error surface of the global sedimentation equilibrium analyses for the association constant of the 1:1 FcRn–Fc complex formation (dash-dotted line), the 2:1 FcRn-Fc complex formation (solid line), and the 2:1 FcRn–IgG complex formation (dashed line). These curves show the relative increase in the sum of the squared residuals (SOSQ) of the global fit obtained when constraining the parameter value of K_A to a nonoptimal value, while optimizing at each point of the curve all other parameters. Their minimum value shows the best-fit solution for the binding constants, whereas their shapes visualize the information content of the data on the binding constants. The horizontal dotted line indicates the level of increase in the SOSQ that corresponds to the probability of one standard deviation according to F-statistics ($n>1,200$). Its intersection with the SOSQ curves indicates the uncertainty of the derived optimal parameter values [Reprinted from Schuck *et al.* (1999), with permission].

the SE data. A different limitation of SE analysis is encountered with the FcRn–IgG interaction, where due to the larger molar mass of the IgG, the relative mass difference of the 1:1 and 2:1 complex (25%) is only half that in the FcRn–Fc interaction (52%). As a consequence, the error interval for the affinity of the second site is larger (Figure 10.6; Schuck *et al.*, 1999).

10.6.3. More Applications of SE

It is far outside the scope of the current chapter to provide an exhaustive review of SE applications in the literature, as there has been a wealth of important SE studies throughout many decades. Therefore, only small selections of references are highlighted as examples for the application of specific SE approaches.

Receptor–ligand interactions are among the applications of SE studies with mass conservation constraints. For example, Philo and coworkers have used SE with mass conservation constraints to determine that erythropoietin receptor can form complexes with erythropoietin with both 1:1 and 2:1 stoichiometry (Philo *et al.*, 1996). The binding constant of the second site was determined to be ~1000-fold weaker than that of the first site. This could reconcile apparent discrepancies with previous results from size-exclusion chromatography, where only the high-affinity site remained occupied. To further examine the thermodynamic parameters of the two sites, isothermal titration calorimetry was applied. Using the binding constants predetermined from the SE analysis, significant differences in the enthalpy of binding for the two sites were found. Philo and coworkers discussed in detail the relationship of their results with those obtained by other methods, as well as the implications for the role of receptor dimerization in signal transduction in the context of membrane-embedded receptors. (This topic is addressed further in Chapter 3 on Solid Phase Detection Methods.)

Another SE study on cell-surface receptor interactions that is of technical interest is the interaction of LY49C natural killer cell receptor with its MHC ligand (Dam *et al.*, 2003). It was found by crystallography and in solution by sedimentation velocity that two MHC molecules can bind to the LY49C dimer. Equilibrium constants were determined by SE and found to be in the low micromolar range, without evidence of cooperativity. In this study, a global multiwavelength analysis with soft mass conservation constraints was applied, permitting the bottom position of the solution column to be determined as a floating parameter from the data obtained at different rotor speeds (Vistica *et al.*, 2004). In addition, relationships of the loading concentrations between different cells in the same ultracentrifuge run were exploited, which in conjunction with the multiwavelength detection, helped to overcome the similarity in molar mass of the receptor fragment (~32 kDa) and its ligand (~45 kDa). Figure 10.7 shows a subset of the sedimentation equilibrium data acquired at different wavelengths, and the fit representing part of the global model and the calculated contributions of the 2:1 (MHC per Ly49C dimer) complex. The comparison of panels A–B and C–D highlights the different radial range that can be analyzed with a multisignal approach, and the shift of the relative populations of the 1:1 and 2:1 complexes across the solution column in SE.

Binding studies in SE can be conducted over a range of temperatures, and thermodynamic parameters can be derived (Ghirlando *et al.*, 1995; Yoo and Lewis, 2000). This requires considering the temperature dependence of the partial-specific volume of the binding partners, which can be measured experimentally by SE. An example is the study of the stoichiometry and thermodynamics of the interaction of the Fc fragment of human IgG_1 and the low-affinity receptor FcγRIII (Ghirlando *et al.*, 1995), where the temperature dependence of the binding constants indicate enthalpy–entropy compensation, and the presence of favorable enthalpy at physiological temperatures.

SE has been used extensively to study nucleic acids and their interactions. The application of SE to the study of sequence nonspecific protein–nucleic acid interactions has been reviewed by Ucci and Cole (2004). Protein–nucleic acid interactions frequently lend themselves to multiwavelength or multisignal analyses. The spectral properties of the protein can be enhanced with 5-hydroxytryptophan as an intrinsic probe, reducing the overlap of the nucleotide extinction with the protein extinction (Laue *et al.*, 1993). Generally, the presence of hyper- or hypochromicity can be assessed independently; if neglected, this would lead to errors in the estimated

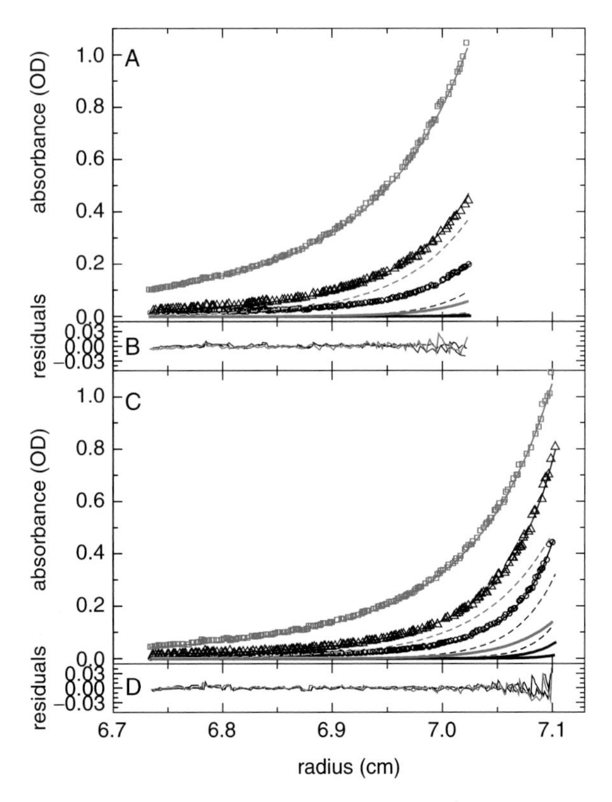

Figure 10.7. Sedimentation equilibrium profiles of Ly49C with H-2K^b-OVA. The binding constant was 3 μM, as determined from a global fit of SE data acquired in a titration series of loading concentrations, using a model for two equivalent sites. For each sample, SE data were acquired at rotor speeds of 15,000 rpm (green), 20,000 rpm (blue), and 25,000 rpm (black), and at different wavelengths. Mass conservation constraints were applied for the redistribution within a sample at the different rotor speeds, and for the relationships of effective loading concentrations across different samples. Shown are SE data (symbols) from one cell, acquired at 280 nm (A and B) and 250 nm (C and D), and the best-fit distributions (solid lines). Also shown are the calculated contributions from the 1:1 complex (dotted lines), and the 2:1 complex (bold solid lines).

complex population and the corresponding binding constants. An example of a multiwavelength study of a protein–DNA interaction is the *Escherichia coli* regulatory protein TyrR, which undergoes a ligand-induced dimer–hexamer self-association (Wilson *et al.*, 1994). The binding of a 42 bp oligonucleotide containing *a TyrR box to TyrR* was examined using chromophorically labeled oligonucleotide. A single nucleotide was found to bind to both the TyrR dimer, and the TyrR hexamer. The latter result was obtained using a competition SE assay, in which the mixtures of nonlabeled oligonucleotide were used to compete the labeled oligonucleotide bound to TyrR hexamers. An analogous competition assay was conducted in parallel using steady-state fluorescence anisotropy (see Chapter 13).

10.7. CONCLUSIONS

SE is a powerful tool for the rigorous thermodynamic characterization of interacting protein systems in free solution. Both protein self-association and hetero-association can be studied. SE allows discriminating the populations of different protein species and their complexes in a concentration gradient, while maintaining strict thermodynamic equilibrium. Binding constants and binding energies can be evaluated through the application of mass action law. In many ways, SE can be viewed as orthogonal to sedimentation velocity (described in Chapter 16), but SE measures molar masses directly, and is insensitive to the kinetics of interaction, and to the presence of species in different conformational states. The main current limitation of SE in the application to complex interacting systems is the unambiguous mathematical modeling of the exponential signal distributions. Nevertheless, it is possible to characterize multistep binary and ternary protein interactions, in particular when using techniques such as multi-wavelength and multisignal detection, as well as modern mass conservation and global analysis approaches. As documented by the large literature of SE in the study of protein interactions, it is a very flexible approach that can be applied to a great variety of protein systems, including detergent-solubilized membrane proteins, and more general protein–ligand, protein–nucleic acid interactions.

ACKNOWLEDGMENT

I thank Drs. Marc Lewis, Allen Minton, and Teresa Magone for a critical reading of the manuscript.

REFERENCES

Adams, E. T. Jr. and Lewis, M. S. (1968). Sedimentation equilibrium in reacting systems. VI. Some applications to indefinite self-associations. Studies with beta-lactoglobulin A. *Biochemistry* 7:1044–1053.

Archibald, W. J. (1947). A demonstration of some new methods of determining molecular weights from the data of the ultracentrifuge. *J Phys Colloid Chem* 51:1204–1214.

Arisaka, F. (1998). [Sedimentation equilibrium]. *Tanpakushitsu Kakusan Koso* 43:2238–2244.

Balbo, A. and Schuck, P. (2005). Analytical ultracentrifugation in the study of protein self-association and heterogeneous protein–protein interactions. In: Golemis, E. and Adams, P. D. *Protein–Protein Interactions*. Cold Spring Harbor Laboratory Press, Cold Spring Harbor, New York, pp. 253–277.

Behlke, J. and Ristau, O. (2003). Sedimentation equilibrium: a valuable tool to study homologous and heterogeneous interactions of proteins or proteins and nucleic acids. *Eur Biophys J* 32:427–431.

Bevington, P. R. and Robinson, D. K. (1992). *Data Reduction and Error Analysis for the Physical Sciences*. McGraw-Hill, New York.

Burmeister, W. P., Huber, A. H., and Bjorkman, P. J. (1994). Crystal structure of the complex of rat neonatal Fc receptor with Fc. *Nature* 372:379–383.

Campanella, M. E., Chu, H., and Low, P. S. (2005). Assembly and regulation of a glycolytic enzyme complex on the human erythrocyte membrane. *Proc Natl Acad Sci U S A* 102:2402–2407.

Center, R. J., Schuck, P., Leapman, R. D., Arthur, L. O., Earl, P. L., Moss, B., and Lebowitz, J. (2001). Oligomeric structure of virion-associated and soluble forms of the simian immunodeficiency virus envelope protein in the pre-fusion activated conformation. *Proc Natl Acad Sci U S A* 98:14877–14882.

Chatelier, R. C. (1987). Indefinite isoenthalpic self-association of solute molecules. *Biophys Chem* 28:121–128.

Clay, O., Carels, N., Douady, C., Macaya, G., and Bernardi, G. (2001). Compositional heterogeneity within and among isochores in mammalian genomes. I. CsCl and sequence analyses. *Gene* 276: 15–24.

Clay, O., Douady, C. J., Carels, N., Hughes, S., Bucciarelli, G., and Bernardi, G. (2003). Using analytical ultracentrifugation to study compositional variation in vertebrate genomes. *Eur Biophys J* 32: 418–426.

Cole, J. L. (1996). Characterization of human cytomegalovirus protease dimerization by analytical ultracentrifugation. *Biochemistry* 35:15601–15610.

Cole, J. L. (2004). Analysis of heterogeneous interactions. *Methods Enzymol* 384:212–232.

Cole, J. L. and Hansen, J. C. (1999). Analytical ultracentrifugation as a contemporary biomolecular research tool. *J Biomol Tech* 10:163–176.

Correia, J. J. and Yphantis, D. A. (1992). Equilibrium sedimentation in short solution columns. In: Harding, S. E., Rowe, A. J., and Horton, J. C. *Analytical Ultracentrifugation in Biochemistry and Polymer Science*. The Royal Society of Chemistry, Cambridge, UK, pp. 231–252.

Dam, J., Guan, R., Natarajan, K., Dimasi, N., Chlewicki, L. K., Kranz, D. M., Schuck, P., Margulies, D. H., and Mariuzza, R. A. (2003). Variable MHC class I engagement by Ly49 NK cell receptors revealed by the crystal structure of Ly49C bound to H-2Kb. *Nat Immunol* 4:1213–1222.

Durchschlag, H. (1986). Specific volumes of biological macromolecules and some other molecules of biological interest. In: Hinz, H.-J. *Thermodynamic Data for Biochemistry and Biotechnology*. Springer, Berlin, pp. 45–128.

Ebel, C., Eisenberg, H., and Ghirlando, R. (2000). Probing protein–sugar interactions. *Biophys J* 78: 385–393.

Edelstein, J. and Schachman, H. (1967). Simultaneous determination of partial specific volumes and molecular weights with microgram quantities. *J Biol Chem* 242:306–311.

Eisenberg, H. (1976). *Biological Macromolecules and Polyelectrolytes in Solution*. Clarendon Press, Oxford.

Eisenberg, H. (2000). Analytical ultracentrifugation in a Gibbsian perspective. *Biophys Chem* 88:1–9.

Eisenberg, H. (2003). Modern analytical ultracentrifugation in protein science: look forward, not back. *Protein Sci* 12:2647–2649; discussion 2649–2650.

Elzen, B. (1988). *Scientists and Rotors. The Development of Biochemical Ultracentrifuges*. Enschede, Dissertation, University Twente.

Fairman, R., Fenderson, W., Hail, M. E., Wu, Y., and Shaw, S.-Y. (1999). Molecular weights of CTLA-4 and CD80 by sedimentation equilibrium ultracentrifugation. *Anal Biochem* 270:286–295.

Fujita, H. (1962). *Mathematical Theory of Sedimentation Analysis*. Academic Press, New York.

Fujita, H. (1975). *Foundations of Ultracentrifugal Analysis*. John Wiley & Sons, New York.

Ghetie, V. and Ward, E. S. (1997). FcRn: the MHC class I-related receptor that is more than an IgG transporter. *Immunol Today* 18:592–598.

Ghirlando, R., Keown, M. B., Mackay, G. A., Lewis, M. S., Unkeless, J. C., and Gould, H. J. (1995). Stoichiometry and thermodynamics of the interaction between the Fc fragment of human IgG1 and its low-affinity receptor FcγRIII. *Biochemistry* 34:13320–13327.

Gohon, Y., Pavlov, G., Timmins, P., Tribet, C., Popot, J. L., and Ebel, C. (2004). Partial specific volume and solvent interactions of amphipol A8–35. *Anal Biochem* 334:318–334.

Haschemeyer, R. H. and Bowers, W. F. (1970). Exponential analysis of concentration or concentration difference data for discrete molecular weight distributions in sedimentation equilibrium. *Biochemistry* 9:435–445.

Hellstern, S., Pegoraro, S., Karim, C. B., Lustig, A., Thomas, D. D., Moroder, L., and Engel, J. (2001). Sarcolipin, the shorter homologue of phospholamban, forms oligomeric structures in detergent micelles and in liposomes. *J Biol Chem* 276:30845–30852.

Holladay, L. A. and Sohpianolpoulos, A. J. (1972). Nonideal associating systems. I. Documentation of a new method for determining the parameters from sedimentation equilibrium data. *J Biol Chem* 247:427–439.

Howlett, G. J. (1992). Sedimentation analysis of membrane proteins. In: Harding, S. E., Rowe, A. J., and Horton, J. C. *Analytical Ultracentrifugation in Biochemistry and Polymer Science*. The Royal Society of Chemistry, Cambridge, UK, pp. 470–483.

Howlett, G. J. (1998). Analysis of hetero-associating systems by sedimentation equilibrium. *Chemtracts Biochem Mol Biol* 11:950–959.

Howlett, G. J., Nichol, L. W., and Andrews, P. R. (1973). Sedimentation equilibrium studies on indefinitely self-associating systems. *N*-methylacetamide in carbon tetrachloride. *J Phys Chem* 77:2907–2912.

Howlett, G. J., Minton, A. P., and Rivas, G. (2006). Analytical ultracentrifugation for the study of protein association and assembly. *Curr Opin Chem Biol* 10:430–436.

Kelley, G. E. and Winzor, D. J. (1984). Quantitative characterization of the interactions of aldolase and glyceraldehyde-3-phosphate dehydrogenase with erythrocyte membranes. *Biochim Biophys Acta* 778:67–73.

Kratky, O., Leopold, H., and Stabinger, H. (1973). The determination of the partial-specific volume of proteins by the mechanical oscillator technique. *Methods Enzymol* 17:98–110.

Laue, T. M. (1999). Analytical centrifugation: equilibrium approach. *Curr Protocols Protein Sci* 20.3.1–20.3.13.

Laue, T. M., Senear, D. F., Eaton, S., and Ross, J. B. (1993). 5-hydroxytryptophan as a new intrinsic probe for investigating protein–DNA interactions by analytical ultracentrifugation. Study of the effect of DNA on self-assembly of the bacteriophage lambda cI repressor. *Biochemistry* 32:2469–2472.

Laue, T. M. and Stafford, W. F. I. (1999). Modern applications of analytical ultracentrifugation. *Annu Rev Biophys Biomol Struct* 28:75–100.

Lear, J. D., Stouffer, A. L., Gratkowski, H., Nanda, V., and Degrado, W. F. (2004). Association of a model transmembrane peptide containing gly in a heptad sequence motif. *Biophys J* 87:3421–3429.

Lebowitz, J., Lewis, M. S., and Schuck, P. (2002). Modern analytical ultracentrifugation in protein science: a tutorial review. *Protein Sci* 11:2067–2079.

Lebowitz, J., Lewis, M. S., and Schuck, P. (2003). Back to the future: a rebuttal to Henryk Eisenberg. *Protein Sci* 12:2649–2650.

Lee, J. C., Gekko, K., and Timasheff, S. N. (1979). Measurements of preferential solvent interactions by densimetric techniques. *Methods Enzymol* 61:26–49.

Lewis, M. S. and Junghans, R. P. (2000). Ultracentrifugal analysis of the molecular mass of glycoproteins of unknown or ill-defined carbohydrate composition. *Methods Enzymol* 321:136–149.

Lewis, M. S. and Reily, M. M. (2004). Estimation of weights for various methods of the fitting of equilibrium data from the analytical ultracentrifuge. *Methods Enzymol* 384:232–242.

Lewis, M. S., Shrager, R. I., and Kim, S.-J. (1993). Analysis of protein–nucleic acid and protein–protein interactions using multi-wavelength scans from the XL-A analytical ultracentrifuge. In: Schuster, T. M. and Laue, T. M. *Modern Analytical Ultracentrifugation*. Birkhäuser, Boston, pp. 94–115.

Lindenthal, S. and Schubert, D. (1991). Monomeric erythrocyte band 3 protein transports anions. *Proc Natl Acad Sci U S A* 88:6540–6544.

Low, P. S. (1992). Band 3: calorimetry, cytoskeletal associations, roli in metabolic regulation, and role in aging. In: Bamberg, E. and Passow, H. *The Band 3 Proteins: Anion Transporters, Binding Proteins and Senescent Antigens*. 2. Elsevier, Amsterdam, pp. 219–225.

Lustig, A., Engel, A., Tsiotis, G., Landau, E. M., and Baschong, W. (2000). Molecular weight determination of membrane proteins by sedimentation equilibrium at the sucrose or nycodenz-adjusted density of the hydrated detergent micelle. *Biochim Biophys Acta* 1464:199–206.

Mayer, G., Ludwig, B., Muller, H. W., van den Broek, J. A., Friesen, R. H. E., and Schubert, D. (1999). Studying membrane proteins in detergent solution by analytical ultracentrifugation: different methods for density matching. *Prog Colloid Polymer Sci* 113:176–181.

Meselson, M. and Stahl, F. W. (1958). The replication of DNA in *Escherichia coli. Proc Natl Acad Sci U S A* 44:671.

Meselson, M., Stahl, F., and Vinograd, J. (1957). Equilibrium sedimentation of macromolecules in density gradients. *Proc Natl Acad Sci U S A* 43:581–588.

Minton, A. P. (2000). Quantitative characterization of reversible macromolecular associations via sedimentation equilibrium: an introduction. *Exp Mol Med* 32:1–5.

Morris, M. and Ralston, G. B. (1989). A thermodynamic model for the self-association of human spectrin. *Biochemistry* 28:8561–8567.

Noy, D., Calhoun, J. R., and Lear, J. D. (2003). Direct analysis of protein sedimentation equilibrium in detergent solutions without density matching. *Anal Biochem* 320:185–192.

Noy, D., Discher, B. M., Rubtsov, I. V., Hochstrasser, R. M., and Dutton, P. L. (2005). Design of amphiphilic protein maquettes: enhancing maquette functionality through binding of extremely hydrophobic cofactors to lipophilic domains. *Biochemistry* 44:12344–12354.

Ober, R. J., Martinez, C., Vaccaro, C., Zhou, J., and Ward, E. S. (2004). Visualizing the site and dynamics of IgG salvage by the MHC class I-related receptor, FcRn. *J Immunol* 172:2021–2029.

Osborne, J. C., Powell, G. M., and Brewer, H. B. (1980). Analysis of the mixed association between human apolipoproteins A-I and A-II in aqueous solution. *Biochim Biophys Acta* 619:559–571.

Passow, H. (1992). The band 3 proteins. An introduction. In: Bamberg, E. and Passow, H. *The Band 3 Proteins: Anion Transporters, Binding Proteins and Senescent Antigens*. 2. Elsevier, Amsterdam, pp. 1–6.

Philo, J. S. (2000). Sedimentation equilibrium analysis of mixed associations using numerical constraints to impose mass or signal conservation. *Methods Enzymol* 321:100–120.

Philo, J. S., Aoki, K. H., Arakawa, T., Narhi, L. O., and Wen, J. (1996). Dimerization of the extracellular domain of the erythropoietin (EPO) receptor by EPO: One high-affinity and one low-affinity interaction. *Biochemistry* 35:1681–1691.

Reynolds, J. A. and McCaslin, D. R. (1985). Determination of protein molecular weight in complexes with detergent without knowledge of binding. *Methods Enzymol* 117:41–53.

Reynolds, J. A. and Tanford, C. (1976). Determination of molecular weight of protein moiety in protein–detergent complexes without prior knowledge of detergent binding. *Proc Natl Acad Sci U S A* 73:4467–4470.

Richards, E. G., Teller, D. C., and Schachman, H. K. (1968). Ultracentrifuge studies with Rayleigh interference optics. II. Low-speed sedimentation equilibrium of homogeneous systems. *Biochemistry* 7:1054–1076.

Rivas, G., Fernandez, J. A., and Minton, A. P. (1999). Direct observation of the self-association of dilute proteins in the presence of inert macromolecules at high concentration via tracer sedimentation equilibrium: theory, experiment, and biological significance. *Biochemistry* 38:9379–9388.

Rivas, G., Fernandez, J. A., and Minton, A. P. (2001). Direct observation of the enhancement of noncooperative protein self-assembly by macromolecular crowding: indefinite linear self-association of bacterial cell division protein FtsZ. *Proc Natl Acad Sci U S A* 98:3150–3155.

Rivas, G., Lopez, A., Mingorance, J., Ferrandiz, M. J., Zorrilla, S., Minton, A. P., Vicente, M., and Andreu, J. M. (2000). Magnesium-induced linear self-association of the FtsZ bacterial cell division protein monomer. The primary steps for FtsZ assembly. *J Biol Chem* 275:11740–11749.

Roark, D. E. (1976). Sedimentation equilibrium techniques: multiple speed analyses and an overspeed procedure. *Biophys Chem* 5:185–196.

Roark, D. E. and Yphantis, D. A. (1971). Equilibrium centrifugation of nonideal systems. The Donnan effect in self-associating systems. *Biochemistry* 10:3241–3249.

Rodewald, R. and Kraehenbuhl, J. P. (1984). Receptor-mediated transport of IgG. *J Cell Biol* 99: 159s–164s.

Schachman, H. K. (1959). *Ultracentrifugation in Biochemistry*. Academic Press, New York.

Schachman, H. K. (1992). Is there a future for the ultracentrifuge? In: Harding, S. E., Rowe, A. J., and Horton, J. C. *Analytical Ultracentrifugation in Biochemistry and Polymer Science*. Royal Society of Chemistry, Cambridge, pp. 3–15.

Schubert, D. and Schuck, P. (1991). Analytical ultracentrifugation as a tool for studying membrane proteins. *Prog Colloid Polym Sci* 86:12–22.

Schubert, D., Huber, E., Lindenthal, S., Mulzer, K.-H., and Schuck, P. (1992). The relationships between the oligomeric structure and the functions of human erythrocyte band 3 protein: the functional unit for the binding of ankyrin, hemoglobin and aldolase and for anion transport. In: Bamberg, E. and Passow, H. *The Band 3 Proteins: Anion Transporters, Binding Proteins and Senescent Antigens*. 2. Elsevier, Amsterdam, pp. 209–217.

Schubert, D., Tziatzios, C., Broeck, J. A. v. d., Schuck, P., Germeroth, L., and Michel, H. (1994). Determination of the molar mass of pigment-containing complexes of intrinsic membrane proteins: problems, solutions and application to the light-harvesting complex B800/820 of *Rhodospirillum molischianum*. *Prog Colloid Polym Sci* 94:14–19.

Schuck, P. (1994). Simultaneous radial and wavelength analysis with the Optima XL-A analytical ultracentrifuge. *Prog Colloid Polym Sci* 94:1–13.

Schuck, P. (2004). A model for sedimentation in inhomogeneous media. I. Dynamic density gradients from sedimenting co-solutes. *Biophys Chem* 108:187–200.

Schuck, P. and Braswell, E. H. (2000). Measurement of protein interactions by equilibrium ultracentrifugation. In: Coligan, J. E., Kruisbeek, A. M., Margulies, D. H., Shevach, E. M., and Strober, W., (eds), *Current Protocols in Immunology*. John Wiley & Sons, New York, pp. 18.8.1–18.8.22.

Schuck, P. and Millar, D. B. (1998). Rapid determination of molar mass in modified Archibald experiments using direct fitting of the Lamm equation. *Anal Biochem* 259:48–53.

Schuck, P., Legrum, B., Passow, H., and Schubert, D. (1995). The influence of two anion-transport inhibitors, 4,4′-diisothiocyanatodihydrostilbene-2,2′-disulfonate and 4,4′-dibenzoylstilbene-2, 2′-disulfonate, on the self-association of erythrocyte band 3 protein. *Eur J Biochem* 230:806–812.

Schuck, P., Radu, C. G., and Ward, E. S. (1999). Sedimentation equilibrium analysis of recombinant mouse FcRn with murine IgG1 and Fc fragment. *Mol Immunol* 36:1117–1125.

Shire, S. (1992). *Determination of Molecular Weight of Glycoproteins by Analytical Ultracentrifugation*. Palo Alto, CA, Beckman Instruments.

Stanley, A. M. and Fleming, K. G. (2005). The transmembrane domains of ErbB receptors do not dimerize strongly in micelles. *J Mol Biol* 347:759–772.

Surendran, R., Herman, P., Cheng, Z., Daly, T. J., and Ching Lee, J. (2004). HIV Rev self-assembly is linked to a molten-globule to compact structural transition. *Biophys Chem* 108:101–119.

Svedberg, T. and Fahraeus, R. (1926). A new method for the determination of the molecular weight of the proteins. *J Am Chem Soc* 48:320–438.

Svedberg, T. and Pedersen, K. O. (1940). *The ultracentrifuge*. Oxford University Press, London.

Tanford, C. (1961). *Physical Chemistry of Macromolecules*. Wiley, New York.

Ucci, J. W. and Cole, J. L. (2004). Global analysis of non-specific protein–nucleic interactions by sedimentation equilibrium. *Biophys Chem* 108:127–140.

Vaccaro, C., Zhou, J., Ober, R. J., and Ward, E. S. (2005). Engineering the Fc region of immunoglobulin G to modulate in vivo antibody levels. *Nat Biotechnol* 23:1283–1288.

van Holde, K. E. (2004). Sedimentation equilibrium and the foundations of protein chemistry. *Biophys Chem* 108:5–8.

van Holde, K. E., Johnson, W. C., and Ho, P. S. (1998). *Principles of Physical Biochemistry*. Upper Saddle River, Prentice Hall.

Varah, J. M. (1985). On fitting exponentials by nonlinear least squares. *SIAM J Sci Stat Comput* 6:30–44.

Vistica, J., Dam, J., Balbo, A., Yikilmaz, E., Mariuzza, R. A., Rouault, T. A., and Schuck, P. (2004). Sedimentation equilibrium analysis of protein interactions with global implicit mass conservation constraints and systematic noise decomposition. *Anal Biochem* 326:234–256.

von Ruckmann, B. and Schubert, D. (2002). The complex of band 3 protein of the human erythrocyte membrane and glyceraldehyde-3-phosphate dehydrogenase: stoichiometry and competition by aldolase. *Biochim Biophys Acta* 1559:43–55.

Weng, Z., Gulukota, K., Vaughn, D. E., Bjorkman, P. J., and DeLisi, C. (1998). Computational determination of the structure of rat Fc bound to the neonatal Fc receptor. *J Mol Biol* 282: 217–225.

Wills, P. R. and Winzor, D. J. (1992). Thermodynamic nonideality and sedimentation equilibrium. In: Harding, S. E., Rowe, A. J., and Horton, J. C. *Analytical Ultracentrifugation in Biochemistry and Polymer Science*. Royal Society of Chemistry, Cambridge, pp. 311–330.

Wilson, T. J., Maroudas, P., Howlett, G. J., and Davidson, B. E. (1994). Ligand-induced self-association of the *Escherichia coli* regulatory protein TyrR. *J Mol Biol* 238:309–318.

Yoo, S. H. and Lewis, M. S. (2000). Interaction of chromogranin B and the near N-terminal region of chromogranin B with an intraluminal loop peptide of the inositol 1,4,5-trisphosphate receptor. *J Biol Chem* 275:30293–30300.

Yphantis, D. A. (1964). Equilibrium ultracentrifugation of dilute solutions. *Biochemistry* 3:297–317.

Zhang, D., Kiyatkin, A., Bolin, J. T., and Low, P. S. (2000). Crystallographic structure and functional interpretation of the cytoplasmic domain of erythrocyte membrane band 3. *Blood* 96:2925–2933.

Zorrilla, S., Jimenez, M., Lillo, P., Rivas, G., and Minton, A. P. (2004). Sedimentation equilibrium in a solution containing an arbitrary number of solute species at arbitrary concentrations: theory and application to concentrated solutions of ribonuclease. *Biophys Chem* 108:89–100.

11

Structure Analysis of Macromolecular Complexes by Solution Small-Angle Scattering

D. I. Svergun and P. Vachette

11.1. INTRODUCTION

The challenge of the "postgenomic" era, when vast numbers of genome sequences are available, led to structural genomics initiatives aiming at large-scale expression and purification for subsequent structure determination using X-ray crystallography and NMR spectroscopy (Edwards *et al.*, 2000; Gerstein *et al.*, 2003). These initiatives have already yielded unprecedented number of high-resolution models for isolated proteins and/or their domains, and these numbers are expected to grow rapidly in the coming years. More difficulties have been encountered in the study of macromolecular complexes, which are usually too large for the structural NMR studies, and they often possess inherent structural flexibility, which make them difficult to crystallize.

As most important cellular functions are accomplished by macromolecular complexes, it would be beneficial to determine the structures of entire assemblies whenever possible, and the focus on modern structural genomics is currently being shifted toward the study of complex structures. This includes new crystallographic initiatives complemented by the use of methods yielding structural information in solution at lower resolution. In particular, cryoelectron microscopy (cryo-EM)-based shape analysis provides a framework for docking the high-resolution models of individual proteins or domains (Sali *et al.*, 2003). This approach allows one to obtain excellent results in many cases, but application of cryo-EM is usually limited to relatively large macromolecular aggregates (starting from 200 or 300 kDa).

D. I. SVERGUN Hamburg Outstation, European Molecular Biology Laboratory, Notkestrasse 85, 22603, Hamburg, Germany, and Institute of Crystallography, Russian Academy of Sciences, Leninsky pr. 59, 117333 Moscow, Russia • P. VACHETTE. Centre national de la recherche scientifique, IBBMC, UMR 8619, Université Paris-sud, Bâtiment 430, 91405 Orsay Cedex, France.

This chapter describes one of the most efficient methods for structural characterization of macromolecular complexes in solution, a diffraction technique called small-angle scattering (SAS). Both X-rays and neutrons are employed, and the two methods are referred to as SAXS and SANS, respectively. SAS allows one to study the low-resolution structure of native particles in nearly physiological solutions and to analyze structural changes in response to variation in external conditions. For establishing the three-dimensional structural models, this technique needs monodisperse solutions of purified macromolecules but normally does not require special sample preparation. Similar to cryo-EM method, SAS yields information about the overall shape of the macromolecule, and each of the two techniques has advantages and shortcomings. On the one hand, three-dimensional cryo-EM images may provide more detailed particle shapes than the *ab initio* SAS reconstructions. On the other hand, scattering experiments and data processing are much faster than cryo-EM analysis, and SAS is applicable to a broader range of conditions and sizes (from a few kilodaltons to hundreds of megadaltons). Rigid body modeling driven by the fit to the experimental data, a SAS counterpart of the EM-based docking, allows one to construct models of the quaternary structure of complex particles. Further, the contrast variation method, especially powerful in SANS, thanks to the use of hydrogen–deuterium exchange, yields unique information about the internal structure of multicomponent complexes. Unlike most other structural methods, SAS is able to quantitatively characterize equilibrium and nonequilibrium mixtures and monitor kinetic processes such as (dis)assembly and (un)folding.

Recently, the power of SAS has been boosted by the significant improvements in instrumentation (most notably, by the high brilliance synchrotron radiation sources) accompanied by the development of novel data analysis methods. These developments made it possible to significantly improve resolution and reliability of the models constructed from the SAXS and the SANS data. In this chapter, after a short reminder of the main theoretical and experimental aspects of X-ray and neutron scattering (briefly mentioning also some concepts of static light scattering [SLS]), basic data processing procedures are presented. The major section is devoted to the advanced data analysis approaches and their implementation into software packages, and is complemented by a series of examples illustrating the applications to different biological systems.

11.2. MAIN THEORETICAL AND EXPERIMENTAL ASPECTS

This section briefly describes the basic theoretical and experimental aspects of SAS required for the normal user to understand the principles of the technique as applied to solutions of biological macromolecules. It is beyond the scope of this chapter to give a complete presentation of the physical basis underlying the application of X-ray and neutron scattering to macromolecular solutions and other disperse systems. The reader is referred to textbooks (Feigin and Svergun, 1987) or to recent reviews (Koch *et al.*, 2003; Svergun and Koch, 2003) for a full description of the theoretical principles, mathematical apparatus, and instrumentation issues.

11.2.1. Basics of a SAS Experiment

Conceptually, a SAS experiment is rather simple, as illustrated in Figure 11.1. Solution of macromolecules placed in a sample container (e.g., quartz capillary or cuvette) is illuminated by a collimated monochromatic beam of X-rays or neutrons, and the intensity of the scattered beam is measured as a function of the scattering angle. For solutions, the scattering is usually isotropic due to random orientation of particles, and it is in principle sufficient to measure the scattering intensity by a linear detector along a line perpendicular to the beam direction, but two-dimensional detectors are more often used, which yield better accuracy after radial average. A similar measurement is performed in the solvent-filled container, and solvent scattering is subtracted from that of the macromolecular solution. The difference scattering pattern is due to the dissolved particles and contains information about their structure and, for concentrated solutions, also about interparticle interactions.

The SAXS and the SANS experiments are usually performed at large-scale facilities. Experimental stations completely or partially dedicated to biological SAXS are available at all major synchrotrons (beamline ID02 at ESRF, Grenoble, BioCAT at ANL, Argonne, X33 at DESY, Hamburg, BL45A at Spring-8, Himeji, and many others). Laboratory SAXS cameras, although yielding much lower flux, are also available from various firms (e.g., the NanoSTAR camera from the Bruker Group, Ultima III instrument from Rigaku, and Kratky camera from Hecus M. Braun and A. Paar). Biological SANS can be done exclusively on research reactors or pulsed sources, e.g., at the stations D11 and D22 (ILL, Grenoble), NG3 SANS instrument at NIST, Washington, and LOQ station at ISIS, Oxford. The requirements for the purity of the sample experiments are similar for both

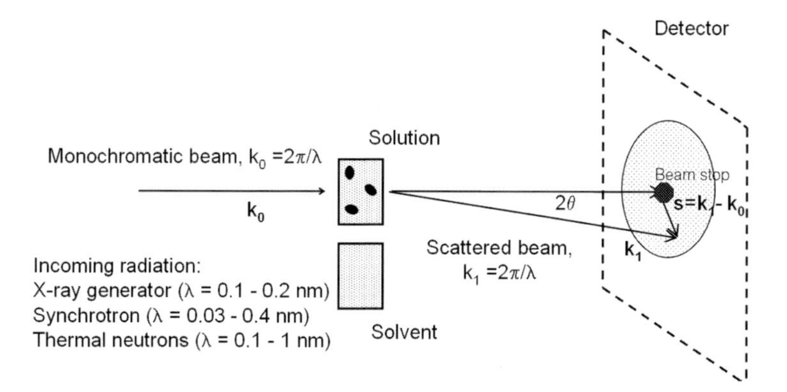

Figure 11.1. General scheme of a small-angle scattering (SAS) experiment.

radiations. For structure analysis (shape, quaternary structure), monodisperse samples, i.e., solutions containing only one molecular species with no higher-order aggregates, are required, with monodispersity greater than 90%. Sample monodispersity must be verified by other methods (native gel filtration, dynamic light scattering, and analytical ultracentrifugation) before the SAS experiment. Sample concentration should be determined as precisely as possible (accuracy better than 10% is desirable, as the concentrations are required to estimate the molecular mass [MM] of the solute). Typical concentrations required are in the range of ~0.5 to 10–20 mg/mL, and a concentration series is usually measured to get rid of interparticle interference effects. The sample volume per measurement is ~50–100 μL for SAXS and ~100–300 μL for SANS, so that ~2–10 mg of purified material is usually required for a complete structural study.

11.2.2. Basics of X-Ray, Neutron, and Light Scattering

The scattering phenomenon emerges because of interactions of incoming X-rays or neutrons with atoms in the sample. For structural SAS, only elastic scattering effects are relevant, where there is no energy exchange between the radiation and the atoms, i.e., the radiation wavelength λ of the scattered beam is equal to that of the incoming beam. Although the physical mechanisms of elastic X-ray and neutron scattering by matter are different, they can be conveniently described by the same mathematical formalism. When the sample is illuminated by a monochromatic plane wave with wavevector $k_0 = |\mathbf{k_0}| = 2\pi/\lambda$, the atoms within the object interacting with the incident radiation become sources of spherical waves. For elastic scattering, the modulus of the scattered wave $k_1 = |\mathbf{k_1}| = k_0$. The amplitude of the wave scattered by each atom is described by its scattering length, f (the name reflects the fact that this characteristic has the units of length, i.e., centimeter). X-rays interact with electrons, and the atomic scattering length f_x is proportional to the number of electrons. The interaction of neutrons with the nuclear potential is described by the nuclear scattering length f_p, which does not display systematic dependence on the atomic number, being rather sensitive to the isotopic content instead. Table 11.1 presents the X-ray and neutron scattering lengths of some biologically relevant atoms and isotopes, where the case of hydrogen and deuterium deserves special attention. Whereas for X-rays the two isotopes are indistinguishable (both have one electron), a large difference is observed between H and D atoms in the case of neutron scattering. This difference provides the basis for the use of selective deuteration and measurements in water and heavy water mixtures for the analysis of the internal structure of complex particles.

The earlier considerations give a simplified picture of interactions of X-ray and neutrons with matter, leaving aside the phenomena of anomalous (wavelength-dependent) X-ray scattering and spin-dependent neutron scattering. The former effect is linked with the wavelength-dependent X-ray scattering effects of atoms in the vicinity of their absorption edges and allows one to gain information about

Table 11.1. X-ray and neutron scattering lengths of some atoms or isotopes

Atom	H	D	C	N	O	P	S	K	Au
Atomic mass	1	2	12	14	16	30	32	38	197
N electrons	1	1	6	7	8	15	16	19	79
f_X,10^{-12} cm	0.282	0.282	1.69	1.97	2.16	3.23	4.51	5.30	22.3
f_N,10^{-12} cm	−0.374	0.667	0.665	0.940	0.580	0.510	0.280	0.370	0.760

the distribution of these atoms. The latter effect emerges because of the interactions between atomic spins in the sample and the spins of the incoming neutrons, and yields structure-related signal only for polarized neutron beam and polarized target (otherwise, the spin-dependent signal is just a flat incoherent background). The reader is referred to original papers and reviews describing application of the advanced techniques making use of these effects in SAXS (Stuhrmann and Notbohm, 1981; Stuhrmann *et al.*, 1991) and SANS (Stuhrmann *et al.*, 1986; Stuhrmann, 1991).

The scattering process involves a transformation from the ''real'' space (coordinates $\mathbf{r}$ of the object) to the ''reciprocal'' space, i.e., coordinates of scattering vectors $\mathbf{s} = (s, \Omega) = \mathbf{k}_1 - \mathbf{k}_0$. Here, the momentum transfer $s = 4\pi\lambda^{-1}\sin(\theta)$, where 2θ is the scattering angle and Ω is the direction of the scattering vector. The momentum transfer is often denoted by other letters, e.g., q or, in neutron scattering, Q; note that in some publications, s denotes the quantity $2\lambda^{-1}\sin(\theta)$. The transformation is mathematically described by the Fourier operator

$$A(\mathbf{s}) = \Im[\rho(\mathbf{r})] = \int \rho(\mathbf{r}) \exp(i\mathbf{sr})d\mathbf{r}, \tag{11.1}$$

where the function $\rho(\mathbf{r})$ is the scattering length density distribution equal to the total scattering length of the atoms per unit volume and $A(\mathbf{s})$ is the scattering amplitude. Experimentally, one measures the scattering intensity, i.e., the number of photons or neutrons scattered in the given direction $\mathbf{s}$, which is proportional to the squared amplitude $I(\mathbf{s}) = [A(\mathbf{s})]^2$.

In biological SAS studies, relatively hard X-ray photons with the wavelength λ of ~0.10–0.15 nm or thermal neutrons ($\lambda \sim 0.20$–1.0 nm) are typically employed. These wavelengths are significantly smaller than the sizes of macromolecules. Following the reciprocal properties of Fourier transformation in Eq. (11.1), where a product of $\mathbf{sr}$ is entered, the larger the size in real space, the shorter the corresponding reciprocal space vector. The real space resolution of the scattering data can be estimated as $d = 2\pi/s$, which is a counterpart of the well-known Bragg equation in crystallography. The scattering at small angles, i.e., at small s, therefore provides information about large distances (much larger than the wavelength), i.e., about overall particle structure.

The situation is different for another scattering technique, SLS, employed for macromolecular solutions, where visible light is used and the scattering emerges because of the difference in refractive index between macromolecule and solvent (Bohren and Huffman, 1983). The laser wavelength λ is typically about a few hundred nanometers, and the value of the momentum transfer, which is inversely proportional to λ, remains small even for large angles. As the macromolecular particle sizes are also much smaller than λ, an angle-independent, so-called Rayleigh scattering, is usually observed, which can be used to obtain molecular mass of the solute, similar to the way it is done with X-rays or neutrons (see Section 11.2.4). In fact, light was already used to determine the molecular mass of proteins back in 1930s, (Putzeys and Brosteaux, 1935) and prompted to some extent the subsequent use of X-rays. For larger particles, angle-dependent (Mie) scattering also yields information about the overall particle size and the second virial coefficient describing the interparticle interactions (see Section 11.2.6).

11.2.3. Types of Macromolecular Systems

In the studies of disordered objects such as solutions of biomacromolecules, the scattering intensity is isotropic and depends only on the momentum transfer s. To obtain the net scattering from the macromolecules, the solvent scattering is subtracted from that of the solution, so that the particle structure is described by the excess scattering density distribution $\Delta\rho(\mathbf{r}) = \rho(\mathbf{r}) - \rho_s$, where ρ_s is the scattering density of the solvent. There are three important cases:

1. A dilute solution of noninteracting identical particles, which are randomly distributed, and their positions and orientations are not correlated. The intensity from the entire ensemble is proportional to the scattering from a single particle averaged over all orientations: $I(s) = <I(\mathbf{s})>_\Omega = [\Im[\Delta\rho(\mathbf{r})]^2]_\Omega$. These "particle scattering" data bear information about the macromolecular shape and internal structure. Typical objects of this kind are monodisperse dilute solutions of purified proteins, nucleic acids, or macromolecular complexes and the major task of SAS is the structural determination of such systems at low resolution (1–2 nm).
2. For nonideal solutions of identical particles, which are randomly oriented but interact with each other, local correlations between the neighboring particles must be taken into account. The scattering intensity can be written as $I_S(s) = I(s) \times S(s)$, where $S(s)$ is the "structure factor" related to the interaction potential between the particles. Typical task of SAS for semi-dilute macromolecular solutions is the analysis of interparticle interactions, e.g., at the onset of crystallization.
3. Mixtures of particles of different types. In this case, the scattering intensity will be a linear combination of the contributions from the components weighted by their volume fractions in the system. Typical objects of this kind are equilibrium systems like oligomeric mixtures of proteins or nonequilibrium systems in the process of (dis)assembly or (un)folding. Usual tasks of SAS are to determine the volume fractions of the components

in the system, if their scattering patterns are available, or to find the number of components based on the measurements in different conditions.

Main equations are briefly presented in the following section, which allow one to compute the structural parameters of particles from the experimental data. The recent advances in the three-dimensional modeling methods are also considered in the next section.

11.2.4. Overall Parameters

Some parameters (so-called invariants) can be computed from the SAS data without model assumptions, and for monodisperse systems they are directly related to the weight and the geometrical characteristics of the particle. The best-known parameter is the radius of gyration R_g derived from the so-called Guinier plot (Guinier, 1939),

$$I(s) \cong I(0)\exp\left(-\frac{1}{3}R_g^2 s^2\right). \tag{11.2}$$

This approximation is valid for very small angles (in the range of $s < 1.3/R_g$) and the Guinier plot is very useful at the first stages of data analysis. For ideal monodisperse systems, the plot ($\ln(I(s))$ versus s^2) should be a linear function, whose intercept gives the forward scattering *I(0)* and the slope yields the R_g. Deviations from linearity may point to interparticle interactions or polydispersity of the sample (Figure 11.2). Note that for SLS, even in the case of Mie scattering, the entire spectrum is well within the Guinier range, so that only two parameters can be extracted from the data (in SLS, another representation called Debye

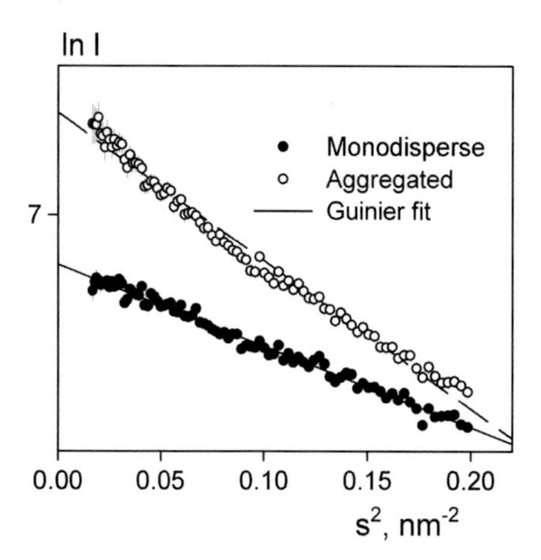

Figure 11.2. A linear Guinier plot from a monodisperse solution and a curved plot from a polydisperse solution.

plot is employed [Bohren and Huffman, 1983], which has similar meaning to the Guinier plot).

Introducing the notion of particle contrast, i.e., average difference between the particle and the solvent scattering length density, $\Delta\rho = \langle\Delta\rho(\mathbf{r})\rangle = \langle\rho(\mathbf{r})\rangle - \rho_s$, the forward scattering is $I(0) = (\Delta\rho)^2 V$, where V is the particle volume. As the contrast is defined by the chemical composition of the particle and the solvent, the $I(0)$ value permits one to evaluate the molecular mass of the solute from a comparison with the scattering of a reference sample or a secondary standard in SAXS (Kratky and Pilz, 1972) or incoherent scattering from water in SANS (Zaccai and Jacrot, 1983).

The value of R_g characterizes the overall particle size and anisometry. Similar equations hold for very elongated and for very flat particles

$$sI(s) \cong I_C(0)\exp\left(-\frac{1}{2}R_c^2 s^2\right);\ s^2 I(s) \cong I_T(0)\exp\left(-R_t^2 s^2\right), \tag{11.3}$$

to compute the radii of gyration of the cross-section R_c and of the thickness R_t.

Another important overall parameter is the so-called Porod invariant (Porod, 1982)

$$Q = \int_0^\infty s^2 I(s)\mathrm{d}s = 2\pi^2 \int_V (\Delta\rho(\mathbf{r}))^2 \mathrm{d}\mathbf{r}. \tag{11.4}$$

For homogeneous particles, $Q = 2\pi^2(\Delta\rho)^2 V$, and, taking into account that $I(0) = (\Delta\rho)^2 V^2$, the particle volume is $V = 2\pi^2 I(0) Q^{-1}$ (note that this ''Porod'' volume corresponds to a hydrated particle, i.e., that seen with its solvation shell). Further, for particles with sharp boundaries, $I(s)$ should decay as $s^{-4}(\pi S/QV)$, where S is the particle surface, which allows one to estimate the specific surface of the particle S/V. These equations hold for homogeneous particles but can also be used with some caveats, e.g., to analyze X-ray scattering from proteins with MM >30 kDa. To reduce the influence of the scattering from inhomogeneities, an appropriate constant is subtracted from the experimental data using the approximation $s^4 I(s) \approx Bs^4 + A$.

Fourier transformation of the scattering intensity yields the distance distribution function of the particle

$$p(r) = \frac{r^2}{2\pi^2}\int_0^\infty s^2 I(s)\frac{\sin sr}{sr}\mathrm{d}r, \tag{11.5}$$

which is a spherically averaged autocorrelation function of the excess scattering density $\Delta\rho(\mathbf{r})$. This function is equal to zero for the distances r exceeding the maximum particle diameter $D_{\max}$, which allows one to estimate this parameter. Further, the appearance of $p(r)$ provides visual information about the particle shape, as illustrated in Figure 11.3, presenting typical scattering patterns and distance distribution functions of geometrical bodies with the same maximum size. Globular

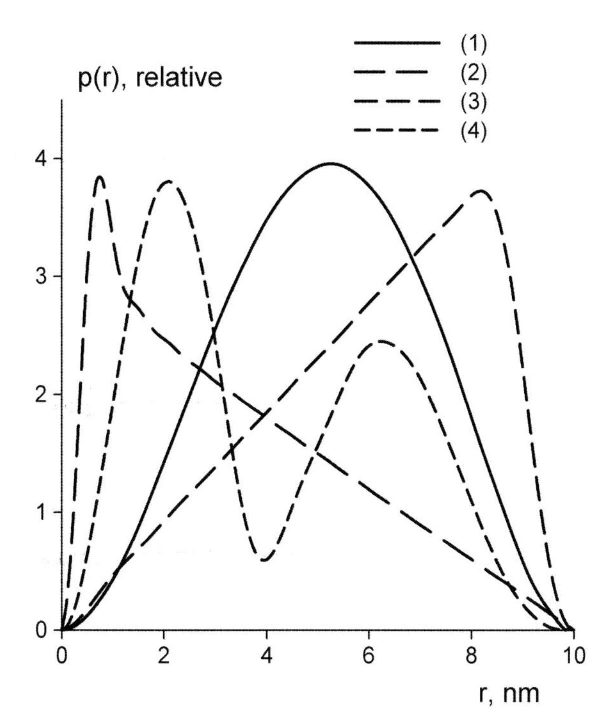

Figure 11.3. Distance distribution functions from different geometrical bodies with the same maximum size: (1) solid sphere, (2) a long cylinder, (3) a hollow sphere, and (4) a dumbbell structure (two separated solid spheres).

particles (curve 1) display bell-shaped $p(r)$ functions with a maximum at about $D_{max}/2$. For elongated particles, the maximum is shifted toward smaller distances corresponding to the radius of the cross-section (curve 2). A maximum shifted toward distances larger than $D_{max}/2$ is usually indicative of a hollow particle (curve 3). Particles consisting of two separated subunits may display two maxima, the first corresponding to the intrasubunit distances and the second yields the separation between the subunits (curve 4).

Reliable computation of $p(r)$ is a necessary step in the analysis of SAS data but direct implementation of Eq. (11.5) is usually difficult as the experimental data $I(s)$ are only measured in a limited angular interval $[s_{min}, s_{max}]$ rather than $[0, \infty]$. Moreover, the data may contain smearing effects due to the finite beam size, divergence, and/or polychromaticity. To evaluate the distribution function, an indirect Fourier transformation technique first proposed by Glatter (1977) is usually used. The idea of the method is that the $p(r)$ function is parameterized on the interval $[0, D_{max}]$ by a set of orthogonal functions, and the coefficients of the parameterization are determined by fitting the experimental data to a linear combination of Fourier-transformed functions. Smoothness of $p(r)$ must be appropriately taken into account to obtain stable solutions but this approach does not require extrapolation of the data beyond the experimentally measured interval and allows one to account for instrumental smearing. Different implementations of the indirect transformation method in computer programs are available (Glatter, 1977; Moore, 1980; Svergun *et al.*, 1988; Svergun, 1992).

11.2.5. Contrast Variation

The use of contrast variation in SAS was first proposed by Stuhrmann and Kirste (1965), where the excess particle density was represented as $\rho(\mathbf{r}) = \Delta\rho\ \rho_C(\mathbf{r}) + \rho_F(\mathbf{r})$. Here $\rho_C(\mathbf{r})$ is the shape function equal to 1 inside the particle and 0 outside, and $\rho_F(\mathbf{r}) = \rho(\mathbf{r}) - <\rho(\mathbf{r})>$ represents the density fluctuations around the average value (Figure 11.4). The scattering intensity is then written as a sum of three basic scattering functions

$$I(s) = (\Delta\rho)^2 I_C(s) + 2\Delta\rho I_{CF}(s) + I_F(s), \tag{11.6}$$

where $I_C(s)$, $I_F(s)$, and $I_{CF}(s)$ are the scattering from the particle shape, the fluctuations, and the cross-term, respectively. The contributions from the overall shape and the internal structure of particles can be separated using measurements in solutions with different solvent densities (i.e., for different $\Delta\rho$). The so-called Stuhrmann plot of $\pm[I(0)]^{1/2}$ versus ρ_s should yield a straight line (Stuhrmann and Kirste, 1965) intercepting zero at the matching point of the particle, where the solvent density equals the average particle density (the sign of the square root coincides with the sign of the contrast). For the radius of gyration, one can write (Ibel and Stuhrmann, 1975)

$$R_g^2 = R_c^2 + \frac{\alpha}{\Delta\rho} - \frac{\beta}{(\Delta\rho)^2}, \tag{11.7}$$

where, R_c is the radius of gyration of the particle shape, α is the second moment of the internal structure (positive or negative depending on whether higher scattering density regions are on the periphery or in the particle center, respectively), and β describes the displacement of the center of mass with the contrast. For homogeneous particles, both α and β are equal to zero. These parameters are evaluated from

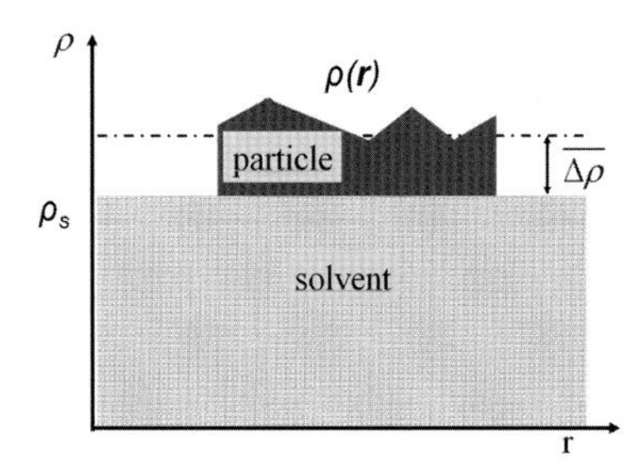

Figure 11.4. Schematic representation of the concept of contrast.

the plot of R_g versus $(\Delta\rho)^{-1}$ and provide overall information about the density distribution within the particle.

For a complex particle consisting of two components with distinctly different contrasts $\Delta\rho_1$ and $\Delta\rho_2$, the scattering intensity can be written in an alternative form

$$I(s) = (\Delta\rho_1)^2 I_1(s) + 2\Delta\rho_1\Delta\rho_2 I_{12}(s) + (\Delta\rho_2)^2 I_2(s), \quad (11.8)$$

where $I_1(s)$ and $I_2(s)$ are the scattering intensities from the two components, and $I_{12}(s)$ is the cross-term. Measurements at the matching point of one component would therefore exclusively yield information about the other one. The experimental radius of gyration of such a two-component system is

$$R_g^2 = w_1 R_{g1}^2 + (1-w_1)R_{g2}^2 + w_1(1-w_1)L^2, \quad (11.9)$$

where $w_1 = \Delta\rho_1 V_1(\Delta\rho_1 V_1 + \Delta\rho_2 V_2)$, V_1, V_2 and R_{g1}, R_{g2} are the volumes and radii of gyration of the components, respectively, and L denotes the separation between the centers of the scattering length distributions.

Contrast variation is most often used in SANS exploiting the difference in the scattering between hydrogen and deuterium. Neutron contrast variation in H_2O–D_2O mixtures allows match out of all major components of biological macromolecules, as illustrated in Table 11.2. Further, the scattering length density of per-deuterated material grown in the fully or partially deuterated cell culture significantly differs from that of the protonated macromolecules (Table 11.2). Contrast variation alone and its combination with specific deuteration is, therefore, a very effective method for the study of biomolecular complexes. A classical example is the selective labeling of protein pairs in the 30S ribosomal subunit of *Escherichia coli*, which yielded a three-dimensional map of the positions of 21 ribosomal proteins by triangulation (Capel *et al*., 1987).

Table 11.2. X-ray and neutron scattering densities of components of biological complexes

	X-rays		Neutrons		
Component	ρ, 10^{10} cm^{-2}	Matching solvent	ρr in H_2O, 10^{10} cm^{-2}	ρ in D_2O, 10^{10} cm^{-2}	Matching % D_2O
H_2O	9.42	–	−0.6	–	–
D_2O	9.42	–	6.4	–	–
Lipids	8.46	–	0.3	−6.0	≈15%
Proteins	11.8	65% sucrose	1.8	3.1	≈40%
D-Proteins	11.8	65% sucrose	6.6	8.0	–
Nucleic acids	15.5	–	3.7	4.8	≈70%
D-Nucleic acids	15.5	–	6.6	7.7	–

Note: For X-rays, the scattering length density is often expressed in terms of electron density (the number of electrons per Å^3; 1 electron/Å^3 = 2.82×10^{11} cm^{-2}). The increase of the scattering length density of the same component in D_2O compared with H_2O is due to the H/D substitution of exchangeable protons.

11.2.6. Interacting Systems

Unless explicitly specified, what precedes only dealt with ideal solutions of identical particles. Under the twofold assumption of ideality and monodispersity, the intensity scattered by the solute is proportional to the intensity scattered by a single particle, thereby justifying the retrieval of structural information regarding the particle directly from experimental SAS data. In real solutions, solute molecules are mutually exerting and experiencing various interaction forces.[1] In this case, as mentioned in Section 11.2.3, the experimental intensity can be expressed as the product of two terms

$$I_{\exp}(c,s) = I(s) \times \mathrm{SF}(c,s). \tag{11.10}$$

This separation holds for globular particles, which can be considered roughly spherical on the scale of their average separation. The first term corresponds to the shape of the particles (form factor) whereas the second one reflects their spatial distribution (structure factor). The effect of the structure factor in the case of strong interactions, attractive and repulsive, is illustrated in Figure 11.5.

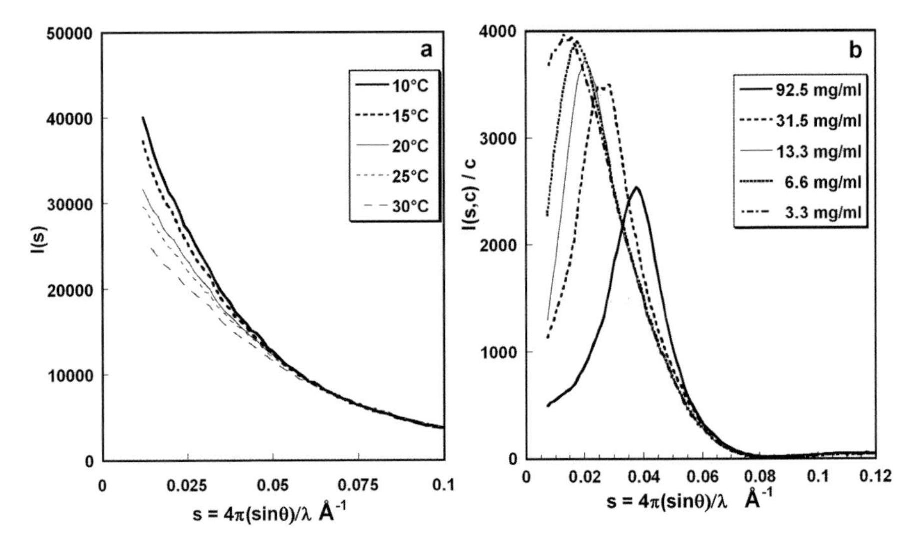

Figure 11.5. Effect of interparticle interactions on the scattering pattern. (A) 160 mg/mL solution of γ-crystallins in 50 mM phosphate buffer, pH 7.0, as a function of temperature (courtesy of A. Tardieu); and (B) ATCase solutions in 10 mM borate buffer pH 8.3 at different concentrations. Here, repulsive interactions are strong enough to give rise to a sharp lateral maximum (so-called correlation peak) and to cause a marked decrease in intensity at very small s-values.

[1] A general presentation of protein–protein interactions in solution can be found in Chapter 9 by C. Ebel to which the reader is referred. This section is restricted to the ways in which interactions are dealt with using SAS.

The X-ray structure factor at the origin is related to the osmotic pressure of the solution by the following relationship:

$$S(c,0) = \left(\frac{RT}{M}\right)\left(\frac{\partial \Pi}{\partial c}\right)^{-1}, \tag{11.11}$$

where $R = 8.31\,\mathrm{J\,mol^{-1}\,K^{-1}}$ is the gas constant and M is the molecular mass in dalton. The osmotic pressure can be expanded as a power series of the concentration c. For structural studies in solution, physicochemical parameters such as pH and ionic strength can be adjusted to produce slight repulsive interactions. The repulsive regime ensures monodispersity by preventing protein association. When interactions are weak and the protein concentration is low, a first-order (linear) expansion is used

$$\frac{1}{S(c,0)} = 1 + 2MA_2c. \tag{11.12}$$

In the case of net repulsive interactions, the particles tend to be evenly distributed whereas the attractive interactions lead to large fluctuations in their distribution with particle-rich regions surrounded by depleted regions. Accordingly, the osmotic pressure is higher (repulsion) or lower (attraction) than in the ideal case, the structure factor at the origin is lower or higher than 1, and A_2, the second virial coefficient, is positive or negative. Measurements are performed at several concentrations, and extrapolation of the scattering curve to zero concentration using Eq. (11.12) yields the "ideal" value of the intensity at the origin $I_{\mathrm{ideal}}(0)$. The same procedure is applied to the first points at small angles, which are most sensitive to interactions and exhibit concentration dependence. This provides a simple and effective way of correcting for residual interparticle interactions.

The structure factor is not only a nuisance in the determination of the form factor; it also contains information about the forces that determine the spatial distribution of macromolecules in solution. A systematic study of the influence of physicochemical parameters on the structure factor can therefore yield new insight into these forces, provided that the form factor be known beforehand, or be experimentally determined in preliminary measurements. This can contribute to a better understanding of physiological properties of interacting systems, such as eye-lens transparency [for a recent review see Bloemendal *et al.* (2004)], and also of protein crystallization, certainly a major driving force behind the study of interactions between proteins, if one considers that crystal production remains the bottleneck in many structural studies (Finet *et al.*, 2003).

Based on SLS results, it had been proposed that proteins crystallize under conditions where the second virial coefficient is slightly negative, corresponding to weak attractive forces between molecules (George and Wilson, 1994). The success of the so-called crystallization slot contributed to turn SLS into a routine tool for rapid screening of crystallization conditions. However, SAXS, using much shorter wavelength, is not restricted to the origin of the scattering pattern but yields the entire structure factor curve. There is no direct way, though, to retrieve unambiguously the information it contains. Numerical simulations of the structure factor, initially developed in liquid state physics (Hansen and McDonald, 1986; Belloni, 1991), must be applied to solutions of macromolecules. In these simulations, only interactions between pairs of macromolecules are taken into account, although all interactions are mediated via the surrounding solvent and ions (see Chapter 9). The different interactions are each represented using a Yukawa potential (Cummings and Monson, 1985) of the form

$$\frac{u(r)}{k_B T} = J\left(\frac{\sigma}{r}\right) \exp\left(-\frac{(r-\sigma)}{d}\right), \tag{11.13}$$

which depends on the hard-sphere diameter σ, the potential depth (J in units of kT), and its range d (k_B is the Boltzmann constant). These parameters are determined by a trial-and-error procedure in which the structure factor is calculated for various combinations of values (Malfois *et al.*, 1996).

In brief, statistical mechanics models based on the Ornstein–Zernicke (OZ) and the hypernetted chain (HNC) integral equations are used to calculate $S(c,s)$, which is related by Fourier transform to the particle distribution $g(r)$

$$S(c, s) = 1 + \rho \int 4\pi r^2 (g(r) - 1)(\sin rs/rs) dr, \tag{11.14}$$

where ρ is the number density of particles in the solution. The OZ relationship between the Fourier transforms of the total and the direct correlation functions $h(r) = g(r) - 1$ and $c(r)$ can be written as

$$(1 + \mathrm{FT}(h(r)))\,(1 - \mathrm{FT}(c(r))) = 1, \tag{11.15}$$

while the HNC equation is

$$g(r) = \exp\left[-u(r)/k_B T + h(r) - c(r)\right], \tag{11.16}$$

where $u(r)$ is the interaction potential between macromolecules. The OZ equation is solved by iteratively using the closure relationship in Eq. (11.16) followed by the calculation of the structure factor using Eq. (11.14) (Belloni, 1993).

If one tries to confront this phenomenological potential to the actual physical forces between molecules in solution, three contributions must be considered, as originally proposed by Derjaguin, Landau, Verwey, and Overbeek in their study of colloids (the so-called DLVO potential) (Verwey and Overbeek, 1948); a hard-sphere term that accounts for the fact that proteins are mutually impenetrable, coulombic repulsion, and van der Waals interactions. Beyond screening coulombic repulsions, salts also contribute an extra attraction that increases with decreasing temperature and depends on the nature of the anion, following the Hofmeister series (Ducruix *et al.*, 1996). However, salts cannot always make the net interactions attractive, even slightly as required for crystallization, especially when dealing with large multisubunit assemblies. In such cases, crystallographers often resort to polymer addition, mostly polyethylene glycols (PEGs) with molecular masses between 0.4 and 20 kDa. The resulting additional attraction, actually well-documented in the field of colloids and usually called ''depletion force,'' is of entropic origin (Lekkerkerker, 1997). The range of this additional potential depends on the size of the added polymer. Numerical simulations were performed using a ''two-component'' approach to account for the coexistence in the solution of two types of macromolecules, the polymer and the protein, and of the corresponding three pair potentials (Vivares *et al.*, 2002).

The approach has intrinsic limits as neither the interaction potentials nor the calculation of the structure factors take into account departures from a spherical shape of the protein shape, as pointed out by Lomakin *et al.* (1999), although the spherical averaging of interactions in solution strongly reduces the impact of the geometrical factor. The charge distribution on the protein surface is also neglected in the expression of the potentials whereas ions are represented by point charges. This might explain the failure to obtain a fully quantitative description of the structure factors (see Section 11.4.4).

11.2.7. Scattering from Mixtures

The earlier considerations were referring to monodisperse systems, where the experimental intensity is the spherically averaged scattering from a single particle (accounting, for interacting systems, also for the structure factor $S(s)$). In this section, we consider the case of noninteracting mixtures. If a system contains K distinct noninteracting components, the scattering pattern is a linear combination

$$I(s) = \sum_{k=1}^{K} \nu_k I_k(s), \tag{11.17}$$

where ν_k and $I_k(s)$ are the volume fraction and the scattering intensity of the kth component, respectively. Computation of the invariants is still possible for mixtures, although they have different meaning than for the monodisperse systems. The effective values of R_g, MM, and V are averages over the ensemble weighted by the volume fractions of the species, and the $p(r)$ functions of mixtures are weighted superpositions of the distance distributions of individual components, whereas D_{max} corresponds to the maximum distance of the largest particle in the mixture.

When neither number nor intensities of the components are known *a priori*, but multiple data sets are recorded from the system with varying volume fractions of the components, useful information can be extracted by model-independent analysis using singular value decomposition (SVD) (Golub and Reinsh, 1970). This approach yields the so-called singular vectors (fictitious scattering curves providing an orthogonal basis for linear representation of the entire set of data) and their associated singular values (weights in this basis). The number of nonrandom singular vectors with significant singular values provide the minimum number of independent curves required to represent the entire data set, i.e., the number of significant components in the mixture. Applications of SVD to analyze, e.g., the number of intermediates during folding and assembly processes is presented in Section 11.4.

If the number of components and their scattering intensities are known, the volume fractions can be readily found by a linear least-squares fit to the experimental data minimizing discrepancy

$$\chi^2 = \frac{1}{N-1} \sum_{j=1}^{N} \left[\frac{\mu I(s_j) - I_{\text{exp}}(s_j)}{\sigma(s_j)} \right]^2, \tag{11.18}$$

where N is the number of experimental points, $I_{\text{exp}}(s)$ and $\sigma(s)$ are the experimental intensity and the standard deviation, respectively, and μ is the scaling factor. This approach is useful, e.g., to quantitatively characterize systems like oligomeric equilibrium mixtures of proteins with known high-resolution structure. SVD analysis and linear decomposition programs are publicly available, e.g., as programs SVDPLOT and OLIGOMER, respectively (Konarev *et al.*, 2003).

11.3. RECENT DEVELOPMENTS IN DATA ANALYSIS METHODS

The parameters of the macromolecules directly determined from the experimental data, as described in the previous section, provide limited information about the overall particle structure. In the past, analysis in terms of three-dimensional models was limited to simple geometrical bodies (e.g., ellipsoids, cylinders, etc.) or was performed by a trial-and-error modeling (Glatter and Kratky, 1982; Feigin and Svergun, 1987) using information from other methods such as electron microscopy as a constrain (Pilz *et al.*, 1972; Tardieu and Vachette, 1982). The recent decade brought a breakthrough in SAS data analysis methods, which tremendously improved the resolution and the reliability of the models constructed from the experimental data. In particular, it is now possible to reconstruct three-dimensional low-resolution shape and domain structure of macromolecules *ab initio*. Advanced methods have been developed to build models of complexes from the high-resolution structures of individual components by interactive and/or automated rigid body fitting of SAS data, and these methods have been successfully applied in practice. This progress matches the advance in instrumentation allowing one to collect precise SAXS and SANS curves over a wide angular range and to cut the time resolution down to submillisecond on third-generation synchrotron sources in the studies of protein and nucleic acid folding. The novel data analysis methods for the interpretation of the scattering patterns from macromolecular solutions in terms of three-dimensional structures are presented in this section. Possible limitations of the SAS-based modeling of complex particles are discussed and the advantages of synergistic use of SAS with other experimental and computational techniques are also demonstrated. The new methods presented here are illustrated by practical applications to various types of macromolecular complexes in the next section.

11.3.1. *Ab Initio* Methods

Reconstruction of three-dimensional structure from isotropic scattering data *ab initio* is possible only for monodisperse solutions without interactions, and only at a low resolution (1–2 nm). At this resolution, the search is usually limited to homogeneous models (shape determination). Given the loss of information due to the spherical average of the scattering data, unambiguous shape reconstruction is fundamentally impossible and care must be taken not to overinterpret the data. In the past, shape modeling was done on a trial-and-error basis by computing scattering patterns from different shapes and comparing them with the experimental data. The first and very elegant *ab initio* approach was proposed by Stuhrmann (1970a), where the shape was described by an angular envelope function expressed using spherical harmonics. This approach was further developed by Svergun and Stuhrmann (1991) and the first publicly available program SASHA was written by Svergun *et al.* (1997). It was demonstrated (Svergun *et al.*, 1996) that under certain circumstances a unique three-dimensional shape can be extracted from the SAS data (up to an enantiomorphic shape, which always gives the same scattering curve; this ambiguity holds also for all *ab initio* methods described later).

The exponentially growing number crunching capacity of modern computers made it possible to use Monte-Carlo-type search for *ab initio* analysis, which permits one to construct yet more detailed models. The idea of this search was first proposed by Chacon *et al.* (1998, 2000) and implemented in a program DALAI_GA. Here, a sphere with diameter D_{max} is filled with a large number $M \gg 1$ of densely packed beads. Each bead belongs either to the particle (index $= 1$) or to the solvent (index $= 0$), and the shape is thus described by a binary string of length M. Starting from a random string, a genetic algorithm searches for a model that fits the data. We briefly describe here a more general approach (Svergun, 1999), which is also suitable for the analysis of multicomponent complexes (*ab initio* shape determination is a particular case of this procedure).

Let us consider a particle consisting of K components with distinctly different scattering length densities (e.g., for a nucleoprotein complex $K = 2$, and these components would be the protein and the nucleic acid moieties). A volume sufficiently large to enclose this particle (e.g., a sphere of radius $R = D_{max}/2$) is filled with densely packed small spheres of radius $r_0 \ll R$ (called beads or "dummy atoms"). Each bead is assigned an index X_j indicating the component ("phase") to which it belongs [X_j ranges from 0 (solvent) to K]. As the bead positions are fixed, the structure of such a model is completely described by a phase assignment (configuration) vector X with $M \approx (R/r_0)^3$ components. If the beads of the kth phase have contrast $\Delta\rho_k$, the scattering intensity is

$$I(s) = \left\langle \left[\sum_{k=1}^{K} \Delta\rho_k A_k(s) \right]^2 \right\rangle_\Omega = \sum_{k=1}^{K} \Delta\rho_k^2 I_k(s) + 2 \sum_{n>k} \Delta\rho_k \Delta\rho_n I_{kn}(s), \qquad (11.19)$$

where $A_k(s)$ and $I_k(s)$ are the scattering amplitude and intensity, respectively, from the volume occupied by the kth phase, and $I_{kn}(s)$ are the cross-terms [cf. Eq. (11.8)].

To rapidly evaluate the scattering from such a model, spherical harmonics are employed. Each three-dimensional scattering amplitude from the individual phase is represented as a series

$$A(s) = \sum_{l=0}^{\infty} \sum_{m=-l}^{l} A_{lm}(s) Y_{lm}(\Omega). \tag{11.20}$$

Here, the spherical harmonics are angle-dependent functions $Y_{lm}(\Omega)$ defined on the surface of the unit sphere, and $A_{lm}(s)$ are radially dependent functions. This representation, first introduced in SAS by Harrison (1969) and Stuhrmann (1970b), is extremely useful to describe the isotropic scattering and is also employed in many advanced algorithms described later. Most importantly, the use of spherical harmonics allows one to perform the orientation average analytically. For the multiphase model, a closed expression for the spherically averaged intensity is written as

$$I(s) = 2\pi^2 \sum_{l=0}^{\infty} \sum_{m=-1}^{l} \left\{ \sum_{k=1}^{K} \left[\Delta\rho_k A_{lm}^{(k)}(s)\right]^2 + 2 \sum_{n>k} \Delta\rho_k A_{lm}^{(k)}(s) \Delta\rho_n \left[A_{lm}^{(n)}(s)\right]^* \right\}, \tag{11.21}$$

where the partial amplitudes are

$$A_{lm}^{(k)}(s) = i^l \sqrt{2/\pi}\ \nu_a \sum_j j_l(sr_j) Y_{lm}^*(\omega_j). \tag{11.22}$$

Here, the sum runs over the beads of the kth phase, $(r_j\,\omega_j) = \mathbf{r}_j$ are their polar coordinates, and $\nu_a = (4\pi r_0^3/3)/0.74$ is the displaced volume per bead.

If a set of $N_C \geq 1$ contrast variation curves $I_{\exp}^{(j)}(s)$, $j = 1, \ldots, N_C$ is available, one can search for a configuration X fitting the multiple curves simultaneously, i.e., minimizing the overall discrepancy

$$\chi_{ov}^2(X) = \sum_{j=1}^{N_C} \chi_j^2, \tag{11.23}$$

where the individual discrepancies are between the experimental and the calculated curves for the given contrast, as defined in Eq. (11.18).

The bead models contain thousands of beads, and many configurations may be found compatible with the experimental data. To constrain the solution, a penalty term $P(X)$ is introduced ensuring compactness and connectivity of the individual components in the resulting model. The goal function to be minimized takes the form $f(X) = \chi^2 + \alpha P(X)$, where $\alpha > 0$ is the penalty weight ensuring proper account of the constrain.

Given the large number of variables, the minimization can only be done using a Monte-Carlo-type search. One of the most suitable methods is simulated annealing (SA) (Kirkpatrick *et al.*, 1983). The idea in this method is, having started from a random vector X, to perform random modifications of this vector X, always moving to configurations that decrease $f(X)$ but sometimes also to those increasing $f(X)$. The probability of accepting this last type of move decreases in the course of the minimization (the system is ''cooled''). Initially, the temperature is high and the changes are almost random whereas at the end a configuration corresponding (nearly) to the minimum of the goal function is reached.

In the multiphase *ab initio* analysis program MONSA (Svergun and Nierhaus, 2000), assignment of a single bead is changed at each move and the amplitudes in Eqs (11.21) and (11.22) are only updated but not recalculated. This, together with the use of spherical harmonics, accelerates the computations significantly and permits to run SA procedures requiring millions of function evaluations in reasonable time (depending on the task and on the computer, a few hours to a few days). A full-scale application of this method developed to analyze the contrast variation data from ribosomes is given in Section 11.4.2. In the particular case of a single-component particle ($K = 1$), this approach reduces to the *ab initio* shape determination procedure implemented in the program DAMMIN (Svergun, 1999). This program allows to use *a priori* information about the particle (e.g., anisometry, symmetry) into account (Petoukhov and Svergun, 2003). It has been actively used for practical shape determination by different groups (Funari *et al.*, 2000; Egea *et al.*, 2001; Fujisawa *et al.*, 2001; Sokolova *et al.*, 2001; Aparicio *et al.*, 2002; Scott *et al.*, 2002; Arndt *et al.*, 2003; Bugs *et al.*, 2004; Hammel *et al.*, 2004b; Dainese *et al.*, 2005). Other Monte-Carlo-based *ab initio* approaches have also been proposed, including the original genetic algorithm program DALAI_GA (Chacon *et al.*, 1998, 2000), and the two programs, which do not restrict the search space, spheres modeling program GA_STRUCT (Heller *et al.*, 2002) and a ''give-n-take'' procedure SAXS3D (Bada *et al.*, 2000).

An intrinsic limitation of the earlier *ab initio* methods is the assumption of uniform scattering length density within the given component (or uniform particle density for the shape determination of single-phase component). This not only limits the resolution but also the reliability of the models, as only a restricted portion of the data can be fitted. A more versatile approach to build protein models from SAXS data has recently been proposed (Svergun *et al.*, 2001), where the protein is represented by an assembly of M dummy residues (DRs). The *ab initio* structure analysis method, implemented in the program GASBOR, starts from a randomly distributed gas of DRs in a spherical search volume of diameter D_{max}. Contrary to the earlier bead modeling, the positions of the residues are not fixed but their number is assumed to be known *a priori* from the sequence (which is usually the case). The task is to find the coordinates of M DRs fitting the experimental data and building a protein-like structure. During the simulated annealing, the DRs are randomly relocated within the search volume, and the compactness criterion used in shape determination is replaced by a requirement for the model to have a ''chain-compatible'' spatial arrangement of DRs. In particular, given that the C_α atoms of neighboring residues in the primary sequence are separated by ~0.38 nm, it is required that each DR would have two neighbors at a distance of 0.38 nm. The use of DRs to represent the protein structure is valid up to a resolution of ~0.5 nm, and the program permits one to construct models fitting the experimental data up to this resolution. These models are, especially for smaller (<100 kDa) proteins, more detailed than the models constructed out of beads, and the DR modeling method has also found numerous applications for proteins and protein complexes (Grossmann *et al.*, 2002; Witty *et al.*, 2002; Solovyova *et al.*, 2004; Davies *et al.*, 2005; Shi *et al.*, 2005).

Running a Monte-Carlo shape determination program several times from random starts produces a manifold of models corresponding to nearly identical scattering patterns. The models obtained in independent runs can be superimposed and averaged to obtain a most probable and an averaged model. This analysis is done automatically in the program package DAMAVER (Volkov and Svergun, 2003) employing the program SUPCOMB (Kozin and Svergun, 2001) to align and superimpose two arbitrary low- or high-resolution models represented by ensembles of points. This procedure also allows one to assess the uniqueness of the solution (Volkov and Svergun, 2003).

11.3.2. Computation of Scattering Patterns from Atomic Models

Comparisons between experimental SAS curves and those evaluated from atomic models have been often used to validate theoretically predicted models and verify the structural similarity between macromolecules in crystals and in solution (Ninio *et al.*, 1972; Pavlov *et al.*, 1986; Mueller *et al.*, 1990). Moreover, if high-resolution models of individual fragments or subunits in a complex are available from crystallography or NMR, rigid body refinement can be employed to model the quaternary structure of the complex. Accurate evaluation of the scattering from atomic models is, however, not a trivial task as the solvent contribution must be adequately taken into account. In a general form, the scattering from a particle in solution can be written as

$$I(s) = \left\langle \left| A_{\rm a}(s) - \rho_{\rm s} A_{\rm s}(s) + \delta\rho_{\rm b} A_{\rm b}(s) \right|^2 \right\rangle_\Omega, \tag{11.24}$$

where $A_a(s)$ is the scattering amplitude from the particle in vacuum, $A_s(s)$ and $A_b(s)$ are, respectively, the scattering amplitudes from the particle volume inaccessible to the solvent and from the hydration shell, both with unit density. Here, the density of the bound solvent ρ_b may differ from that of the bulk ρ_s, so that the contrast of the hydration shell $\delta\rho_b = \rho_b - \rho_s$ may differ from zero. The programs CRYSOL (Svergun *et al.*, 1995) for X-rays and CRYSON (Svergun *et al.*, 1998) for neutrons represent the hydration shell by a 0.3-nm thick layer surrounding the macromolecule with an adjustable density ρ_b. Similar to Eqs (11.20) and (11.22), spherical harmonics are used to compute the scattering so that the spherical averaging is done analytically. These partial amplitudes can be further employed to speed up computations in rigid body modeling (see Section 11.3.3). Given the atomic coordinates, e.g., from the Protein Data Bank (Berman *et al.*, 2000), the two programs either fit the experimental scattering curve using two free parameters, the particle volume and the contrast of the hydration layer $\delta\rho_b$, or predict the scattering pattern using the default values of these parameters. Joint X-ray and neutron studies on proteins with known structure demonstrated that the hydration layer has a density of 1.05–1.20 times than that of the bulk (Svergun *et al.*, 1998), and this experimental finding has recently been corroborated by molecular dynamics calculations (Merzel and Smith, 2002).

11.3.3. Rigid Body Modeling

Very often, atomic models of individual subunits comprising a macromolecular complex are available whereas the high-resolution structure of the entire complex cannot be determined. In this situation, rigid body modeling against SAS data can be very useful. This method is similar to docking in cryo-EM studies in as much as the model is constructed from individual subunits with known structure; however, there are also major differences. In SAS, resolution of the *ab initio* models is usually not sufficient for accurate positioning of the subunits, so that determining the low-resolution shape of the complex usually does not allow to unambiguously solve the docking problem. Instead, rigid body modeling in SAS is directly driven by the fit to the experimental data. We illustrate the idea of rigid body modeling by a complex of two subunits, A and B (Figure 11.6). If one fixes subunit A and translates and rotates subunit B, the scattering intensity of the particle is (Svergun, 1991)

$$I(s, \alpha, \beta, \gamma, \mathbf{u}) = I_a(s) + I_b(s) + 4\pi^2 \sum_{L=0}^{\infty} \sum_{m=-l}^{l} \mathrm{Re}\left[A_{lm}(s) C^*_{lm}(s)\right], \qquad (11.25)$$

where $I_a(s)$ and $I_b(s)$ are the scattering intensities from the subunits A and B, respectively. $A_{lm}(s)$ is the partial amplitudes of the fixed subunit A, and $C_{lm}(s)$ is those of the shifted and the rotated subunit B. The structure and the scattering

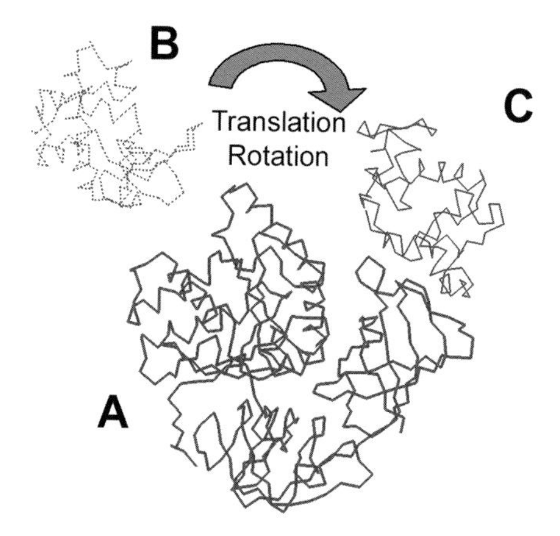

Figure 11.6. Rigid body modeling for a complex of two proteins. Protein A (bold trace) is kept fixed, protein B is rotated and moved from its reference position (dashed trace) to position C. This operation depends on six parameters and changes only the cross-term between the two proteins.

intensity from such a complex depend on the six parameters, the Euler rotation angles α, β, and γ, and shift vector $\boldsymbol{u}$. These parameters can be refined such that the computed intensity, Eq. (11.25), fits best the experimental scattering data from the complex. Given the amplitudes from both subunits computed using CRYSOL or CRYSON, rapid algorithms have been developed (Svergun, 1991, 1994) to evaluate the complex scattering for arbitrary rotations and displacements of the second subunit. This principle can be easily generalized for complexes consisting of several subunits.

This fast rigid body modeling algorithm was coupled with three-dimensional graphic programs ASSA for major UNIX platforms (Kozin *et al.*, 1997; Kozin and Svergun, 2000) and MASSHA for Wintel-based machines (Konarev *et al.*, 2001), where the subunits can be interactively translated and rotated as rigid bodies whereas observing corresponding changes in the fit to the experimental data. A limited automated refinement mode is available for performing an exhaustive search in the vicinity of the current configuration.

Several approaches were used for automated rigid body modeling. An ''automated constrained fit'' procedure (Boehm *et al.*, 1999; Furtado *et al.*, 2004; Sun *et al.*, 2004) generates thousands of possible bead models from high-resolution models in the exhaustive search for the best fit. In the ellipsoidal modeling (Krueger *et al.*, 1998; Wall *et al.*, 2000), the domains are first positioned as triaxial ellipsoids into which atomic models are docked using information from other methods, molecular dynamics and energy minimization (Tung *et al.*, 2002; Priddy *et al.*, 2005). Automated and semiautomated procedures based on screening randomly or systematically generated models, also against multiple contrast variation data sets, were employed by different authors (Heller *et al.*, 2003; Nollmann *et al.*, 2004; King *et al.*, 2005).

A suite of global refinement programs based on spherical harmonics calculations has recently been developed (Petoukhov and Svergun, 2005), where the choice of the search algorithm depends on the complexity of the object. An exhaustive grid search is used for hetero- and homodimeric particles and for symmetric oligomers formed by identical subunits (program GLOBSYMM). For complexes containing more than one subunit per asymmetric unit, heuristic Monte-Carlo-type algorithms are employed. In particular, the program SASREF uses simulated annealing to find the positions and the orientations of the subunits forming interconnected assembly without steric clashes. To this end, the domains are randomly moved and rotated to minimize the sum of the discrepancy between the calculated and the experimental scattering profiles and a penalty for disconnectivity and for the overlaps between the subunits. If several scattering data sets (e.g., from partial constructs or from contrast variation studies) are available, the program performs simultaneous fitting of the multiple data sets, which significantly increases the information content and thus the reliability of the resulting model. SASREF further allows one to account for known interfaces (e.g., binding sites) between subunits by restraining the correspondent interresidue distances. Information about symmetry and/or anisometry of the complex, if available, can also be taken into account.

It must be noted that although the complexes built by rigid body analysis are composed of atomic structures, they remain as low-resolution models, and obtaining ambiguous solutions is still possible. The use of information from other methods and the analysis of biochemical feasibility of the obtained models (e.g., by assessing the quality of the interfaces) are required to rank possible multiple models providing similar fits to the experimental data.

11.3.4. Reconstruction of Missing Fragments

Inherent flexibility and conformational heterogeneity of macromolecules or complexes can often result in the absence of loops and even entire domains in structures determined using X-ray crystallographic or NMR methods. Further, in the studies of multidomain or multisubunit macromolecules consisting of globular domains or subunits linked by flexible loops, high-resolution structures or homology models may be available for the individual domains or subunits, but usually not for the linkers. In this case, a combined rigid body and *ab initio* modeling approach can be employed to determine the overall structure of the entire assembly. The idea of this approach is to simultaneously find optimal positions and orientations of the domains or subunits moved as rigid bodies and probable conformations of the flexible linkers or missing domains. The unknown portions of the structure can be represented as beads or DRs. The scattering from the entire complex contains two types of terms

$$I(s) = 2\pi^2 \sum_{l=0}^{\infty} \sum_{m=-l}^{l} \left| \sum_k A_{lm}^{(k)}(s) + \sum_i D_{lm}^{(i)}(s) \right|^2, \qquad (11.26)$$

where $A_{lm}^{(k)}(s)$ and $D_{lm}^{(i)}(s)$ are the partial amplitudes of the domains with known structure and from the missing portions, respectively.

Using this idea, methods were developed (Petoukhov *et al.*, 2002) to add missing loops or domains by fixing the known structure and building the unknown regions to fit the experimental scattering data obtained from the entire particle (program suite CREDO). Simulated annealing was employed to minimize a scoring function containing the discrepancy between the experimental and calculated patterns and the relevant penalty terms. If interface location between known and unknown parts is unavailable, the missing domain is represented by a gas of DRs. If the interface is known, loops or domains are represented as interconnected chains (or ensembles of residues with spring forces between the C_α atoms), attached to known position(s) in the available structure. Residue-specific constraints were employed to ensure native-like folds of the missing fragments.

In a more recent program BUNCH (Petoukhov and Svergun, 2005), missing linkers or protein domains with unknown structure are substituted by a flexible chain of interconnected DRs. As in the earlier modeling of multisubunit complexes, SA is employed for global minimization. A single modification of the system is performed by a random rotation of the part of the structure between two randomly selected DRs about the axis connecting these DRs, or alternatively, a single DR is selected dividing the entire chain into two parts and the smaller part is rotated by a random angle about a random axis drawn through this DR. Appropriate penalties are added to ensure native-like folds of the missing fragments. If the scattering data from partial constructs or deletion mutants are available, the method is able to fit multiple scattering data sets by the scattering computed from the relevant portions of the model. Symmetry of the particle can be taken into account so that only the symmetrically independent part will be modified and the rest will be generated automatically. BUNCH is primarily oriented toward single-chain proteins or symmetric assemblies containing one polypeptide chain per asymmetric part, but can also be used for macromolecular complexes consisting of several subunits, when not all the structures of the subunits are known. In this case, not only missing loops within one single subunit can be reconstructed but also the shape(s) of the missing subunit(s) can be restored. Note that all methods using DRs use average X-ray form factor of a DR and are currently not usable for neutrons.

11.4. EXAMPLES OF PRACTICAL APPLICATIONS

There are many excellent applications of SAXS and SANS to study the structure of macromolecular complexes, and a complete review of these would be beyond the scope of this chapter. The examples given later illustrate some recent applications to different types of complexes where modern analysis methods were employed. In these selected examples, SAS was often used together with other methods to construct consensus models and to address important functional questions.

11.4.1. Protein–Protein Complexes

11.4.1.1. Z1Z2–Telethonin Complex

The giant muscle protein titin is the largest gene product found in the human genome, with a MM in the range of 3–4 MDa (Labeit and Kolmerer, 1995; Bang *et al.*, 2001). Titin spans across one-half of a muscle sarcomere; its N-terminus is located within the Z-disk defining the border between adjacent sarcomeres, whereas its C-terminus is located in the M-line at the center of the sarcomere. Titin acts as a molecular ruler within the sarcomeric units by providing many spatially defined binding sites for other sarcomeric proteins over the distance of an entire half-sarcomere (Sanger and Sanger, 2001). Telethonin, a 167-residue protein with unknown structure, interacts specifically with the 2 Z-disk Ig-like domains (Z1Z2) at the N-terminus of titin. The molecular basis of the interaction between the N-terminus of titin and telethonin is a key for understanding the anchoring mechanism of titin within the Z-disk. Synchrotron radiation X-ray scattering was employed to study the solution structures of Z1Z2 and its complexes with telethonin (the scattering patterns from different constructs are given in Figure 11.7, top panel).

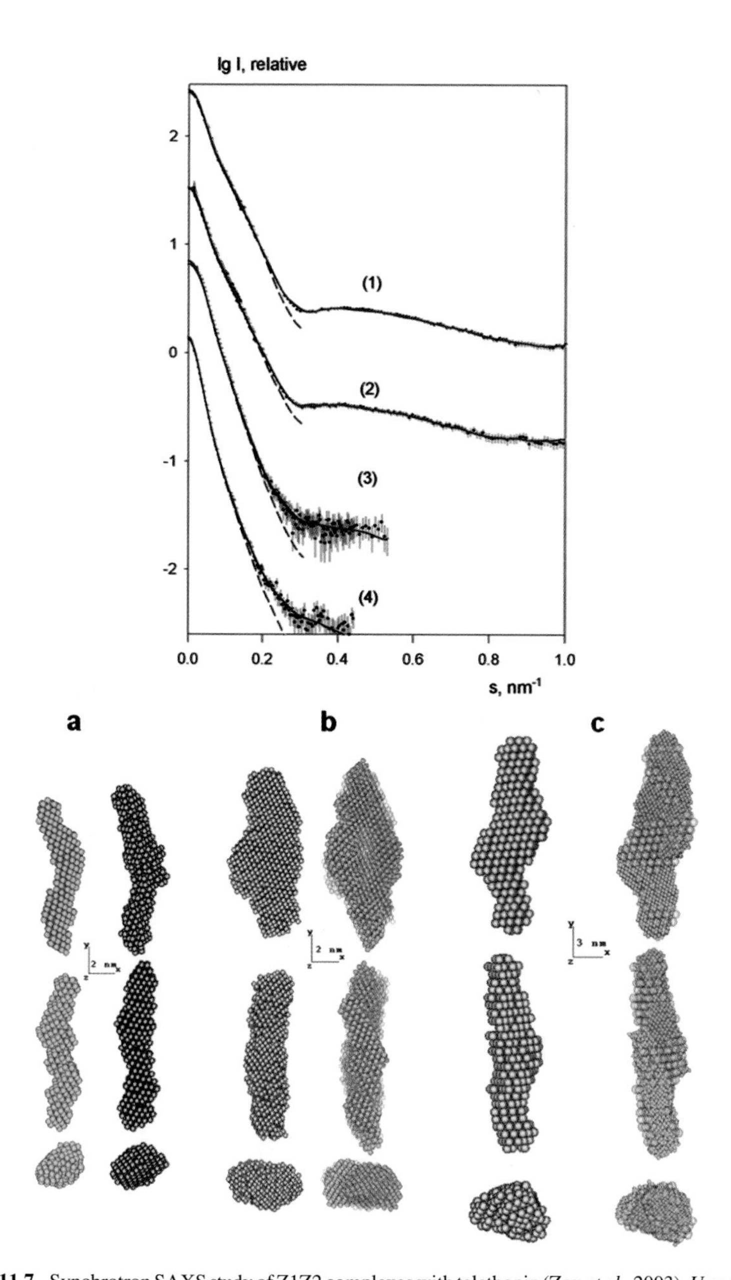

Figure 11.7. Synchrotron SAXS study of Z1Z2 complexes with telethonin (Zou *et al.*, 2003). *Upper panel*: X-ray scattering patterns from Z1Z2 (1) His-Z1Z2 (2), TE(90)–Z1Z2 (3), and TE(167)–Z1Z2 (4). The experimental data are given as dots with error bars, dummy atom fits as red-dashed lines, dummy residue fits as blue full lines. The scattering patterns are displaced by one logarithmic unit for better visualization. *Lower panel*: *ab initio* low-resolution models: (a) Models of Z1Z2 (cyan beads, left) and His-Z1Z2 (brown beads, *right*; the extra seven residues due to the His-tag correspond to the extra volume on the top of the molecule in upper and middle rows). (b) Shape of TE(90)–Z1Z2 (yellow beads, *left*) and this model as semitransparent beads superimposed with two antiparallel models of Z1Z2 (cyan and green beads, *right*). (c) The model of TE(167)–Z1Z2 (yellow beads, *left*) and two models of TE(90)–Z1Z2 (cyan and green beads) tentatively positioned inside the TE(167)–Z1Z2 model (*right*). In all panels, the middle and the bottom rows are rotated counter clockwise by 90° around the *Y* and the *X*-axes, respectively.

Z1Z2 and Histidine (His)-Z1Z2 (with a seven-residue long His-tag attached to the N-terminal end of Z1) appear to be elongated as judged from the R_g and D_{max} values. A slight increase in both values is associated with the His-tag presence. Low-resolution models were constructed *ab initio* from the scattering data using DAMMIN and GASBOR (Figure 11.7a). The His-tag was localized at the tip of the Z1 domain by the comparison of models of native Z1Z2 (cyan) and His-tagged Z1Z2 (brown). Although its location at the end of an elongated molecule facilitates the detection by SAXS, this is illustrative of the possibility to detect even small peptide binding to a 200-residue-long protein. Next, the complex between a truncated telethonin construct lacking the C-terminus and Z1Z2 [TE(90)–Z1Z2] was studied by SAXS. Both the molecular mass estimate derived from the intensity at the origin and the Porod volume estimate suggest a 1:2 stoichiometry for the TE(90)–Z1Z2 complex. Both R_g and D_{max} are close to those of Z1Z2. The bead and DR-type *ab initio* models are very similar and can accommodate two Z1Z2 molecules in an antiparallel association with telethonin acting as a central linker (Figure 11.7b). The complex of full-length telethonin with Z1Z2 [TE(167)–Z1Z2] also appears to have a 1:2 stoichiometry at concentrations <1 mg/mL, but dimerizes at higher concentrations, as indicated by the values of the molecular mass and the Porod volume nearly twice as large as that of the TE(90)–Z1Z2 complex, and confirmed by the larger values of both the radius of gyration and the maximal diameter. Two models of TE(90)–Z1Z2 can be positioned as in Figure 11.7c inside the model of TE(167)–Z1Z2 complex, and the additional volume in the central part of TE(167)–Z1Z2 may accommodate the 154 residues of the 2 telethonin molecules missing in the truncated TE(90)–Z1Z2 complex. These results suggest a cross-linking function for telethonin, connecting two titin molecules at their N-termini leading to a telethonin-mediated autoanchoring of titin dimers in the Z-disk. This example therefore vividly illustrates how functional questions can be addressed by a low-resolution technique such as SAXS. Interestingly, a later high-resolution structure of the complex of TE(90) with Z1Z2 in the crystal fully confirmed both the complex stoichiometry and the cross-linking of Z1Z2 via telethonin (Zou *et al.*, 2006).

11.4.1.2. A Cellusomal Enzyme and its Complex with Cohesin

Cellulosomes are large (0.7–2 MDa) multiprotein complexes from anaerobic bacteria such as *Clostridium cellulolyticum* that degrade cellulose. They are composed of a number of cellulases bound to a cellulose-binding module and to several cohesin modules through a tight ($K_a \geq 10^9\ M^{-1}$) interaction with dockerin, a complementary module of the catalytically active moiety. A striking feature of these complexes is their enhanced catalytic efficiency as compared with that of individual cellulases. The high-resolution structures of isolated modules have been determined including cohesin modules, cellulase, and a cohesin–dockerin complex. However, no entire complex comprising a cellulase with its dockerin module complexed to its cognate cohesin has been crystallized, most likely due to conformational heterogeneity of the particle. Therefore, the system was approached using SAXS (Hammel *et al.*, 2004a). The scattering pattern of the isolated catalytic module Cel48F is identical to that calculated from the atomic coordinates using CRYSOL, showing that the conformation in solution does not differ from that in the crystal. The entire cellulase with its dockerin module and the intervening linker was then studied in solution. The large R_g and D_{max} values show that the protein adopts an extended conformation (Figure 11.8a). The overall shape of the particle was determined *ab initio* using bead and DR modeling. The resulting models are very similar, displaying a larger and a smaller globular domain connected by a narrow region. The atomic models of the catalytic and the dockerin module nicely fit into each globule.

In a second modeling stage, the atomic structure of the catalytic module was used as a starting point, which was completed using the program CREDO. All models exhibited a stretched region of the same length ending into a small globular region and only differed by the orientation of the stretched region and the attached small globular unit (Figure 11.8b). Interestingly, the number of residues in the stretched region was close to that predicted for the linker. After modeling the linker residues starting from the resulting C_α positions using the molecule display and manipulation program TURBO (Roussel and Cambillau, 1989), the scattering curve of the complete atomic model yields a good fit to the experimental data. Finally, as the variable position of the linker suggests that the molecule exhibits a certain degree of flexibility, an attempt was made to fit the experimental pattern by a linear combination of the various models. The improved fit suggests that the protein adopts various conformations.

The study of the entire free cellulase was complemented in the same work (Hammel *et al.*, 2004a) by a study of its complex with a cohesin. The radius of gyration does not increase on cohesin binding whereas the maximum diameter appears to be significantly smaller than its value for the free enzyme. This means that the complexed cellulase is more compact than the free state. Models for the overall shape of the molecule exhibit two globular units in close proximity, in which the atomic structures of individual domains nicely fit. As performed for the free cellulase, the missing parts of the structures and the linker were modeled *ab initio*. All modeled fragments, although slightly different, were localized very close to the average shape. A complete atomic model in solution was then built using TURBO, the scattering curve of which was in good agreement with the experimental pattern. Complex modeling therefore shows that its more compact shape is due to the condensed conformation adopted by the linker joining the catalytic module and the dockerin–cohesin unit. Finally, as carried out for the free cellulase, the known crystal structure of the catalytic module was taken as a starting point, and the rest of the molecule (linker with the dockerin–cohesin complex) was modeled using the program CREDO, to check that no major rearrangement occurred as compared with the crystal structure. The resulting models differed only in the orientation of the modeled part, whereas the shape of the globular unit was consistent with the crystal structure of the isolated dockerin–cohesin complex. This suggests some flexibility of the linker allowing rigid body movements of the dockerin–cohesin unit, but of much reduced amplitude as compared with the free

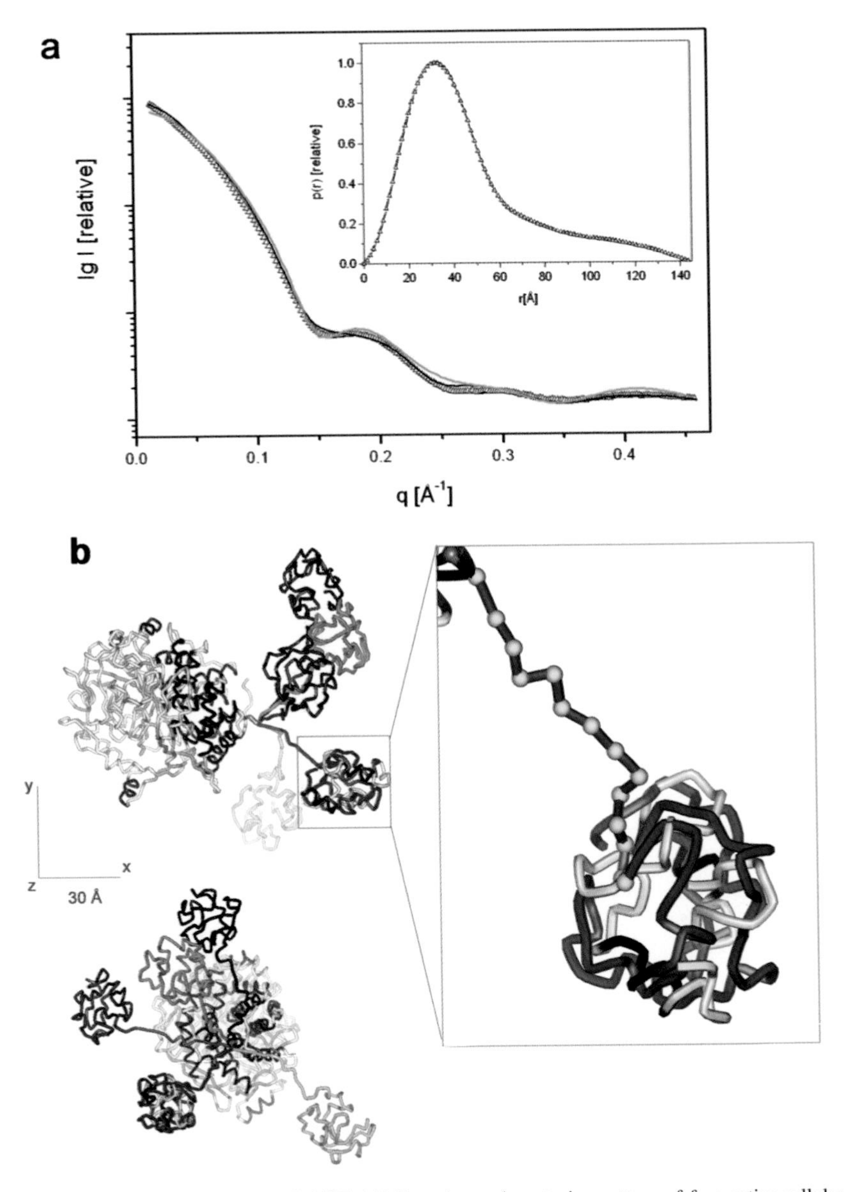

Figure 11.8. Free entire cellulase Cel48F. (A) Experimental scattering pattern of free entire cellulase (black triangles) together with the averaged calculated scattering curve of the CREDO models given by OLIGOMER (red line) and with the scattering curve calculated from the final atomic model (green line; see text for details). *Inset*: Pair distribution function $p(r)$ derived from the experimental scattering pattern. (B) Two orthogonal views of five CREDO models of the linker-dockerin moiety displayed in different colors together with the crystal structure of the catalytic module colored according to secondary structure elements (Cα tube representation). The inset shows the crystal structure of the dockerin domain (white Cα tube with red helices) superimposed over one CREDO model in green. The Cα atoms of the linker region are represented as yellow spheres. Courtesy of V. Receveur-Bréchot.

protein. The flexibility was also confirmed by the improved fit obtained using a mixture of protein conformations. Complementary information regarding rigid body movements of the dockerin and dockerin–cohesin complex with respect to the catalytic module was derived from a normal mode analysis of the two final models. The mean square displacements corresponding to the 100 lowest frequency normal modes were calculated. They were significantly higher in the linker and the dockerin–cohesin regions than in the catalytic module. Furthermore, the linker and the dockerin regions display much larger mean square displacements values in the free state than in the complex with cohesin, in agreement the results from the SAXS study.

The case of flexible molecules such as the cellulase just presented calls for some comments. Indeed, the flexibility of the linker and thereby of the whole molecule, strongly suggested by the SAXS analysis, means that the molecule explores an ensemble of conformations (Heller, 2005). As mentioned by the authors, the experimental pattern is the average of the scattering patterns of all accessible conformations weighted by their respective fractional concentration. The scattering patterns of the various models obtained using the modeling approach are all nearly identical to the experimental pattern. However, there is no reason to assume that all accessible conformations do fit the experimental scattering pattern. The conformational space explored by the molecule is therefore likely to be significantly greater than that indicated by the procedure used. For instance, in the present case, some more compact or more extended configurations are probably explored at times by the free cellulase. At that point, SAXS should be complemented by other techniques in solution such as NMR or fluorescence approaches, which might provide invaluable information.

11.4.2. Nucleoprotein Complexes

11.4.2.1. L20c–RNA Binding Site Complex

In bacteria, the expression of ribosomal proteins is often feedback regulated at the translation level by the binding of the protein to its own mRNA. This is true for several ''core proteins'' from the 50S subunit, which bind directly to the 23S rRNA, independently of other proteins. L20, one of those core proteins, binds to two distinct sites of its mRNA that both resemble its binding site on 23S rRNA, as predicted in the classical Nomura model; in the presence of excess unbound rRNA, the repressor r-protein would be displaced from its mRNA, thereby allowing translation to proceed. Interestingly, both sites are apparently independently bound by L20, and both are required for repression of expression of L20. This peculiar ''dual-site'' organization of the operator raises a number of questions: (i) How many L20 molecules are required to bind to each site? (ii) Why is simultaneous binding required for repression? (iii) What is the mechanism allowing communication between the two sites? The three-dimensional structures of free and ribosome-bound forms of L20 have been reported. They reveal a segmented organization, with a globular C-terminal domain (L20C) that interacts with helices

40 and 41 of the 23S rRNA and can repress translation by itself, and a highly cationic N-terminal domain, which is disordered in the free state and folds as a long helical shaft within the ribosome.

The interaction of L20C with either its target site on the 23S rRNA or one of its two binding sites within its operon (oRNA) has been studied using a combination of heteronuclear NMR and SAXS (Raibaud *et al.*, 2003). Titration of ^{15}N-labeled L20C was followed with [^{15}N,^{1}H] correlation spectroscopy. The spectral changes observed in the rRNA–L20C and oRNA–L20C complexes appear strikingly similar, suggesting that oRNA–L20C and rRNA–L20C form similar, stable, and specific complexes. A three-dimensional NOESY-TROSY experiment on the complex shows that the last 30 residues of L20C (helices α3 and α4) are unaffected by RNA addition, in keeping with observations in the ribosome context, where the surface of interaction between L20C and the 23S rRNA corresponds to the first two helices and loops of the domain. Preliminary ^{15}N-^{1}H HSQC experiments performed on the L20C–rRNA and L20C–oRNA complexes revealed that all the amide linewidths were very broad for the expected size of the complex ($\sim$20 kDa), even after size-exclusion chromatography, ruling out significant aggregation or heterogeneity of the formed complex and suggesting the formation of a higher-order molecular structure. To characterize its composition and conformation, solutions of free RNA and gel-filtration purified L20C–rRNA complex were studied by SAXS. The molecular mass estimates were consistent with a monomeric free RNA and a dimer of L20C–rRNA complexes (Figure 11.9a). From the $p(r)$ function R_g and D_{max} estimates of 26.0 $\pm$ 0.1 Å and 85 $\pm$ 3 Å were derived, much larger than the corresponding values of $\sim$18.5 Å and 60 Å, respectively, calculated for a (1:1) complex of L20C and RNA, further supporting the view that the complex dimerizes under the conditions used for measurements.

Modeling of the L20c–rRNA dimer was performed assuming that L20C recognized the rRNA fragment as observed in the full ribosome context. A complete

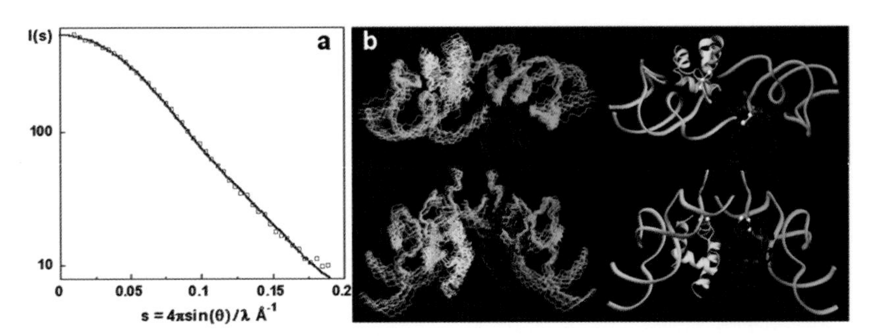

Figure 11.9. (A) Experimental scattering pattern from the L20C–rRNA complex (squares) with the calculated scattering pattern for the best-fitting model of the complex (continuous line). (B) The superimposed backbones of the eight resulting RNA–L20C dimers are shown on the left. On the right is shown a ribbon drawing of the best-fitting structure. Bottom view and top view are rotated by 90° about a horizontal axis. One L20C–rRNA complex is shown in yellow/cyan and the other in red/green. Potential bridging residues, located in loop 2, are shown as white spheres [from Raibaud *et al.* (2003)].

atomic model of the L20C–rRNA complex was therefore constructed using the 50S crystal structure and the NMR model of free L20C. The actual modeling of the dimer made use of a set of constraints directly derived from the NMR and SAXS results:

(i) *Symmetry constraint:* In the NMR spectra of the L20C–RNA complex, only one single set of peaks is observed for each residue in the protein. This strongly suggests that the dimer is symmetrical.
(ii) *Bridging constraint:* L20C or the RNA do not dimerize by themselves. This indicates that the dimer assembly must involve RNA–protein contacts.
(iii) *Exposed C-terminus constraint:* In the TROSY spectrum of the complex, residues within the C-terminal helix of L20C (residues 102–117) appear globally unaffected, in contrast with other parts of the protein, for which large chemical shift displacements are observed. Thus, the C-terminal helix is unlikely to be in close contact with the other complex in the dimer, but, rather, remains exposed to the solvent.
(iv) *Radius of gyration constraint:* The radius of gyration of the dimer (26.0 Å) is known from the SAXS experiment.

Forming a dimer with two identical monomers related by a twofold symmetry axis is a problem with 3 degrees of freedom. To systematically sample the conformational space, a Monte-Carlo approach was used to randomly generate symmetric dimers. The resulting dimers were then selected according to the earlier geometrical constraints. More specifically, we enforced that the calculated radius of gyration should fall within 24.5 and 27.5 Å, that the bridging distance should be 3 Å or less and that the C-terminus distance should be at least 5 Å. Overall, ~1,300 random dimers were generated, 100 of which met the earlier geometrical criteria. These 100 dimers could be grouped into only six topological families, based on pairwise rmsd clustering. Each family contained several dimers, suggesting that the sampling process was rather exhaustive. Finally, the ability of the dimers to account for the SAXS data was evaluated. The theoretical scattering profile was calculated for each of the 100 selected dimers and fitted to the experimental data using CRYSOL. Only 8 of the 100 dimers gave high-quality fits to the experimental data, with χ values under 2. All models belonged to the same topological family and are shown in Figure 11.9b. It thus appears that the earlier constraints define only one allowable global arrangement for the assembly of L20C–rRNA dimers. The resulting models indeed share a similar topology, in which the two RNA stems form a V-shape, with helices 40 of the rRNA at the tip, with an angle of $85° \pm 15°$ (Fig. 4.7). The axes of the two helices 40 are separated by 18 ± 6 Å. The two L20C proteins are located inside the V-shaped assembly. Although these models only provide crude structural information on the possible arrangement within the dimer of complexes, some consistent features are conserved in all of them. For instance, the "bridging" contacts involving one L20C and the rRNA from the other complex always involve residues within loop 2 of L20C (around R90-K91).

Turning back to the repression of its operon by L20, the translation repressor would thus be the molecular assembly of two L20C bound to the two sites on the messenger RNA, thereby explaining the simultaneous requirement for two distinct sites. This assembly would be necessary to efficiently hinder the binding of the ribosome to the initiation codon and/or to enhance the stability of the L20–mRNA complex.

11.4.2.2. 70S Ribosome *E. coli*

Ribosomes, supramolecular complexes (MM ~2.5 MDa) responsible for protein synthesis in all organisms, have been classical SAS objects for ~30 years. Each of the two unequal ribosomal subunits is a complex assembly of proteins and nucleic acids, and the ribosome may be dis- and reassembled into individual components. This makes the ribosome a very attractive object not only for contrast variation studies using H_2O–D_2O mixtures but also for selective deuteration. One of the most striking examples was the protein triangulation (Capel *et al.*, 1987), already mentioned in Section 11.2.5. A comprehensive study of the hybrid 70S *E. coli* ribosomes was the first practical application of the multiphase-simulated annealing bead modeling (Svergun, 1999) described in Section 11.3.1. A total of 42 SAXS and SANS data sets (Figure 11.10, left panel) were collected from H_2O–D_2O mixtures from reconstituted ribosomes, where the proteins and the ribosomal RNA moieties in the subunits were either protonated or deuterated in all possible combinations. The search volume for the multiphase model was defined by a cryo-EM model of the ribosome (Frank *et al.*, 1995) and filled by densely packed beads of radius 0.5 nm. Each bead is assigned either to solvent, to protein, or to ribosomal RNA moieties to simultaneously fit all the available scattering curves. The resulting model (Figure 11.10, *top row*) represents the volumes occupied by the rRNA and the protein moieties in the entire ribosome at a resolution of 3 nm (Svergun and Nierhaus, 2000). The predicted protein–RNA map is in remarkably good agreement with the later high-resolution crystallographic ribosomal models from other species (Ban *et al.*, 2000; Schluenzen *et al.*, 2000; Yusupov *et al.*, 2001; Brodersen *et al.*, 2002). Figure 11.10 presents a comparison of this map with the recent crystallographic structure of the *E. coli* 70S ribosome at 0.95 nm resolution (Vila-Sanjurjo *et al.*, 2003). The results obtained for ribosome confirm that such a multiphase modeling approach to the analysis of contrast variation data should be useful in future studies of macromolecular complexes.

11.4.3. Equilibrium Mixtures of Complexes and Components

11.4.3.1. Fd–Fd–GltS Complex

Glutamate synthase (GltS) is a member of the Ntn-type glutamine-dependent amidotransferase class, which, together with glutamine synthetase, forms the main ammonia assimilation pathway in bacteria and plants. Ferredoxin-dependent GltS

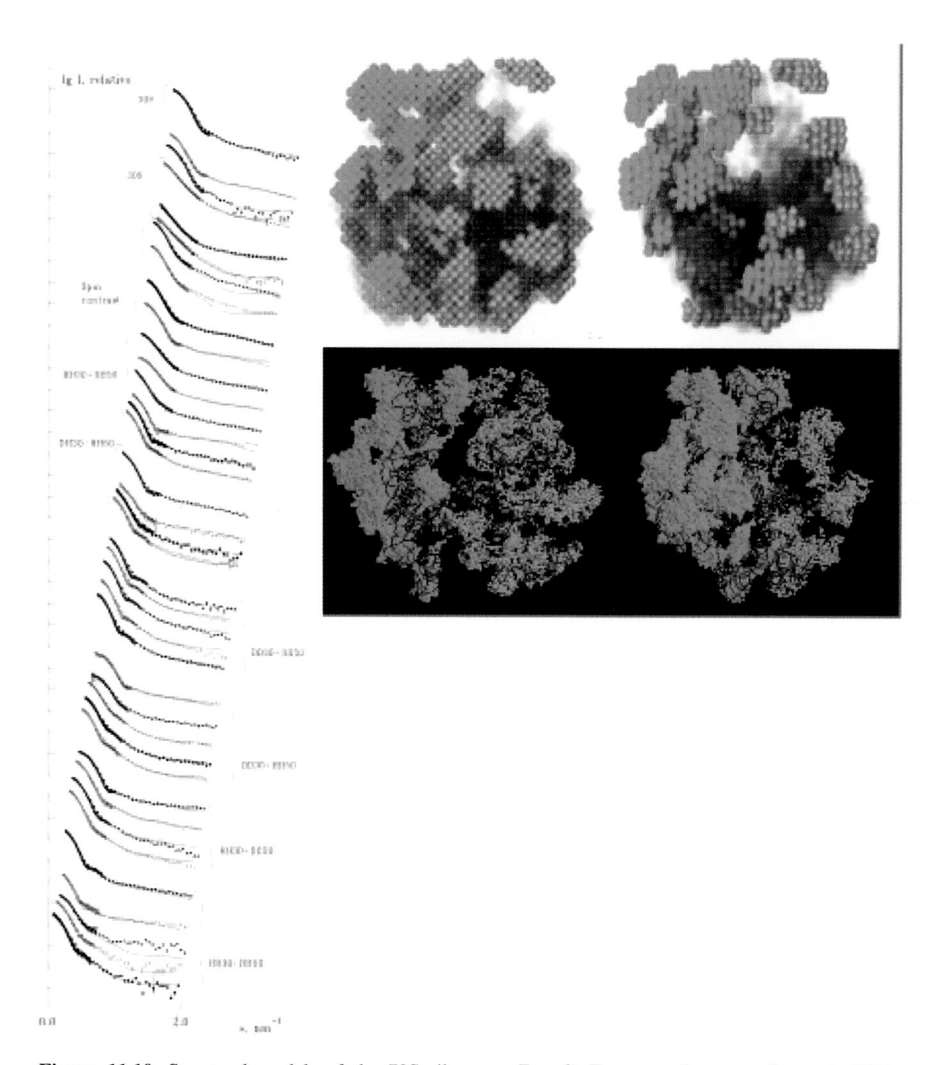

Figure 11.10. Structural models of the 70S ribosome *E. coli*. *Top row*: the map of protein–RNA distribution at 3 nm resolution derived from SAXS and SANS data within the cryoelectron microscopy (cryo-EM) defined search volume. The beads representing proteins are displayed as solids, those representing RNA as semitransparent bodies (bead radius, $r_0 = 0.5$ nm). Magenta and red, ribosomal RNA moiety in the 30S and 50S, respectively; green and cyan, protein moiety in the 30S and 50S, respectively. The model fits 39 neutron scattering data sets from hybrid 70S ribosomes and free subunits (displayed in the *left* panel; experimental data are presented as dots with error bars, the fits as solid lines) and three SAXS curves (not shown). *Bottom row*: the model of the 70S *E. coli* ribosome in the crystal at a resolution of 0.95 nm (Vila-Sanjurjo *et al.*, 2003), PDB codes 1pnx and 1pny. The RNA molecules are displayed as nucleotide backbones, ribosomal proteins in a space-filling mode (the color codes are the same as for the SANS model). The right view is rotated counterclockwise by 45° around the vertical axis.

(Fd–GltS) from *Synechocystis* sp. is active as a monomer in which the noncovalent binding of two reduced ferredoxin (Fd) molecules provides the electrons for the formation of L-glutamate. Crystal structures of both GltS and Fd have been determined at 2.0 Å resolution.

The stoichiometry and the quaternary structures of the Fd–GltS complex were studied using SAXS (van den Heuvel *et al.*, 2003). The scattering patterns of free Fd and Fd–GltS were recorded together with scattering from mixtures of Fd and Fd–GltS at molar ratios of 1:1 and 2:1 (see Figure 11.11). The estimated molecular mass of free Fd is 12 kDa, close to that of the monomeric protein (10 kDa). Furthermore, the scattering pattern calculated from the crystallographic model of Fd is in fair agreement with the experimental curve, suggesting that Fd is essentially monomeric in solution and that its global shape is similar to that in the crystal. The scattering curve calculated from the crystallographic model of Fd–GltS displays minor systematic deviations from the data.

Addition of Fd to Fd–GltS causes small but consistently detected changes in the overall structural parameters. Equimolar mixtures (1:1) yielded higher molecular masses and nearly the same R_g value as Fd–GltS alone whereas the 2:1 Fd:Fd–GltS

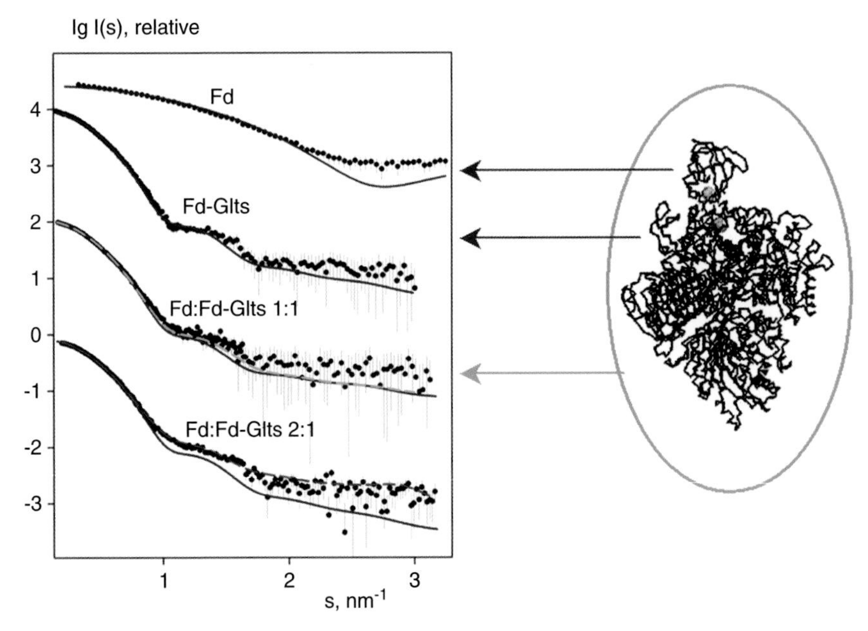

Figure 11.11. X-ray scattering patterns from ferredoxin (Fd) (1), Fd–GltS (glutamate synthase) (2) and their mixtures with ratios 1:1 (3) and 2:1 (4) (dots with error bars); scattering of the crystallographic models of Fd (1) and Fd–GltS (2–4) (continuous lines). Broken lines represent the scattering from a tentative model of Fd:Fd–GltS 1:1 complex with Fd positioned in the vicinity of the Fd–GltS 3Fe–4S cluster (3) and the best fit to the Fd:Fd–GltS 2:1 scattering data yielded by the mixture of Fd and Fd–GltS 1:1 (4). The scattering patterns are appropriately displaced in the logarithmic scale for a better visualization.

mixture systematically exhibited lower molecular masses and R_g values. These observations suggest that Fd and Fd–GltS form a 1:1 complex. Indeed, a 1:1 mixture yields a nearly monodisperse solution of a complex of increased molecular mass. In contrast, a 2:1 stoichiometry leads to a mixture of Fd–Fd–GltS complex with free Fd molecules, thereby lowering the effective molecular mass and the R_g value.

Once the stoichiometry has been determined, tentative models were built using the crystallographic models of Fd and Fd–GltS so as to fit the experimental scattering pattern of the 1:1 Fd:Fd–GltS complex. Sequence alignments and the relative position of the 3Fe–4S cluster and FMN cofactor suggest that Fd must bind near residues 907–933 that form the so-called Fd-binding loop. Positioning the Fd molecule close to that loop and to the 3Fe–4S cluster of Fd–GltS allowed fitting the scattering data with a χ-value of 1.0.

Finally, the assumption that the 2:1 Fd:Fd–GltS is a mixture of free Fd and Fd:Fd–GltS complex was tested using the program OLIGOMER. The corresponding scattering data were represented as a linear combination of free Fd and of the 1:1 Fd:Fd–GltS solution. The best fit ($\chi = 1.1$) was obtained for volume fractions of 0.066 for Fd and 0.934 for 1:1 Fd:Fd–GltS in good agreement with the expected mass fractions in an equimolar mixture of Fd and 1:1 Fd:Fd–GltS complex (0.054 and 0.946, respectively), thereby confirming the hypothesis.

In conclusion, this example illustrates the interest of coupling crystal studies of individual proteins and solution studies of their complex in the case where the latter does not crystallize. The resulting stoichiometry of the complex has fundamental implications for the function of the system because two Fd molecules are required to complete the catalytic steps. The 1:1 stoichiometry therefore implies that two Fd molecules bind consecutively to the catalytically active Fd–GltS.

11.4.3.2. BPTI Decamer Formation

A study of BPTI decamers observed at low pH is an interesting case of stable oligomer formation combining diffraction and SAXS in solution (Hamiaux *et al.*, 2000). Crystallization of bovine pancreatic trypsin inhibitor (BPTI) at acidic pH in the presence of thiocyanate, chloride, and sulfate ions leads to three different polymorphs in P21, P6422, and P6322 space groups. The same decamer with ten BPTI molecules organized through two perpendicular twofold and fivefold axes forming a well-defined compact object is found in the three polymorphs. This is at variance with the monomeric crystal forms observed at basic pH. Does this decamer exist in solution as a stable species or only transiently before crystal incorporation? Are smaller oligomers formed during decamer assembly? To answer these questions SAXS measurements were performed on both undersaturated and supersaturated BPTI solutions at pH 4.5 using all three anions. Data showed the oligomerization of BPTI molecules under all investigated conditions (Figure 11.12a). Accordingly, scattering patterns were analyzed in terms of the formation of discrete oligomers ($n = 1$–10). In addition to the monomer, a dimer, a pentamer, and a decamer were identified within the crystal structure and their scattering patterns were computed using CRYSOL (Figure 11.12b). The experimental curves were then analyzed as linear combinations of these theoretical patterns using a nonlinear curve fitting procedure. The results, confirmed by gel filtration experiments, unambiguously demonstrate the coexistence of only two types of BPTI particles in solution, monomers and decamers, without any evidence for other intermediates (Figure 11.12a). The fraction of decamers increases with salt concentration (i.e., when reaching and crossing the solubility curve). This suggests that crystallization of BPTI at acidic pH is a two-step process whereby decamers first form in under- and supersaturated solutions followed by the growth of what are best described as ''BPTI decamer'' crystals.

Some light is shed on decamer assembly from the crystal structures. In all three cases (thiocyanate, chloride, or sulfate), the same anion-binding site is found within a decamer particle, located at a dimer–dimer interface in the decamer. Furthermore, a major interaction is found between two BPTI molecules involved in an inter-pentamer dimer. It is therefore likely that such dimers may form in solution as protein–protein attractions increase with anion concentration. The next step would be the association of two dimers through anion binding, and cooperatively

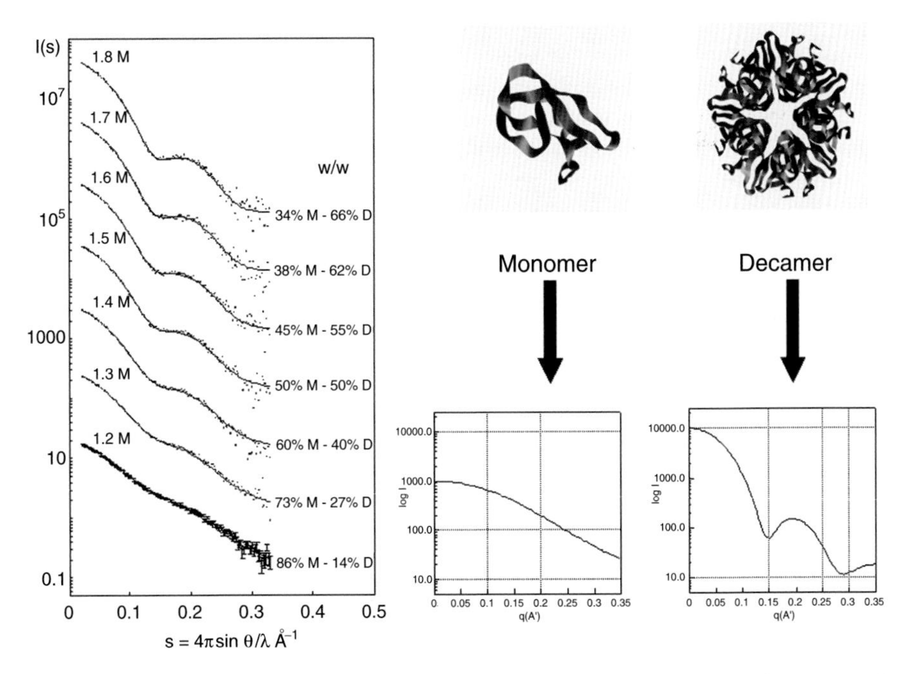

Figure 11.12. Bovine pancreatic trypsin inhibitor (BPTI) decamer formation. (A) Scattering patterns of 30 mg/mL BPTI solutions containing NaCl at concentrations from 1.2 to 1.8 M. Dots represent experimental intensities; the continuous line represents the best fit using a linear combination of the theoretical scattering patterns of monomer and decamer calculated using CRYSOL (see text for details). (B) Ribbon representation of BPTI monomer and decamer with their calculated scattering pattern. Courtesy of C. Hamiaux.

the whole decamer particle is built by association of five dimers around a fivefold symmetry axis. The absence of dimers, never detected in our SAXS experiments, merely indicates that decamer assembly is a very cooperative process; as soon as dimers form in solution, they immediately interact with anions and form a decamer.

11.4.4. Analysis of Weak Interactions in Solution Using Structure Factors

Under conditions where Coulombic repulsion is negligible, small proteins like lysozyme at high ionic strength or γ-crystallins close to the isoelectric point exhibit a liquid–liquid phase separation when temperature is lowered below a critical value. A SAXS study led to the determination of the structure factor as a function of temperature (Figure 11.13a). An attractive Yukawa potential was shown to account for the structure factor and the phase separation at low temperature. In both cases, a qualitative agreement between calculated and experimental structure factors was obtained (see Figure 11.13 for the case of γ-crystallins) for a value of σ yielding a particle volume close to or equal to the protein dry volume, a mean value of 3 Å for the potential range, and a potential depth of around −2.7 kT (Malfois *et al.*, 1996), in agreement with Monte-Carlo simulations (Lomakin *et al.*, 1996). Calculations showed that the reported values of the Yukawa potential parameters correspond to the contribution from the van der Waals interactions. This confirms the validity of the approach, which met with similar successes in other studies of the

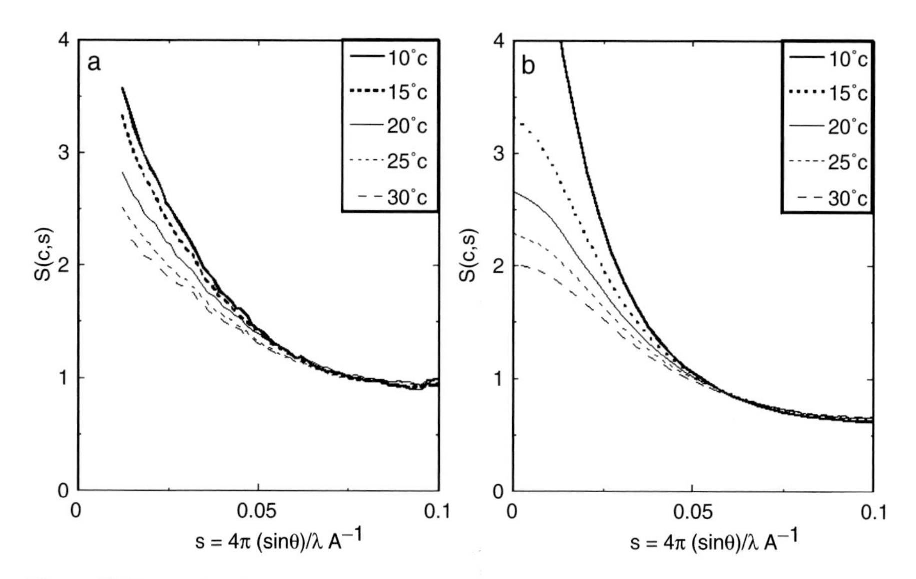

Figure 11.13. (A) Experimental structure factors of γ-crystallins derived from the curves in Figure 11.5a. (B) Theoretical structure factors calculated using a Yukawa potential with a diameter of 36 Å, a potential range of 3 Å, and a potential depth of −2.63 kT at 10°C. The temperature series is obtained by modifying only the temperature. Courtesy of A. Tardieu.

dependence on pH, temperature, ionic strength, and protein concentration of interactions between small proteins in solution (Tardieu *et al.*, 1999). In the cases where coulombic repulsions could not be neglected, they were also described by a second Yukawa potential, with a depth J related to the charge of the protein Z_{p}. The additional ''depletion potential'' associated with the presence of PEG was studied in detail in the case of urate oxydase, a 128-kDa tetrameric enzyme (Vivares *et al.*, 2002). Although temperature-induced and salt-induced attractions are always short range (a few Angstroms), the depth and the range of the depletion potential vary with the size and the concentration of the polymer. Interestingly, the range of the potential can be comparable with the macromolecular dimensions, as observed in the case of urate oxydase in the presence of 8-kDa PEG.

A recent SAXS and SANS study investigated the content of concentrated lysozyme solutions at low ionic strength, where the long-range repulsive electrostatic potential is only weakly screened (Stradner *et al.*, 2004). The structure factor of the solution exhibits a correlation peak because of strong repulsive interactions, the position of which is essentially independent of the protein concentration but shifts to lower s-values with decreasing temperature. A second peak at larger angles could be seen in SANS data, whose position and height do not change either with concentration or with temperature. These observations have been interpreted by proposing that proteins self-assemble into small clusters with an aggregation number dependent on concentration. The driving force for assembly into clusters is short-range attraction. The low ionic strength ensures that the Debye length is larger or comparable with the cluster size, thereby providing a stabilizing mechanism against gelation and determining a finite aggregation number. In this model, the first (lower s) peak corresponds to cluster–cluster correlations whereas the second peak reflects the positional correlations of monomers within each cluster. The temperature dependence of the position of the cluster–cluster correlation peak is interpreted in terms of the formation of fewer but larger aggregates with decreasing temperature, which shifts the balance in favor of attractive interactions. Without questioning the main features of the model, a direct correspondence between peak positions and average distance cannot be taken for granted, and it might be interesting to perform a more elaborate analysis in terms of interaction potentials.

11.4.5. Time-Resolved Studies

Kinetic studies have been undertaken from the early days of synchrotron radiation use for SAXS in view of the high flux available on these installations as compared with laboratory equipment. Processes amenable to time-resolved SAXS studies (TR-SAXS) can be classified into three broad categories: conformational transitions of (generally) multisubunit proteins or supramolecular assemblies such as viruses, assembly phenomena, and (un)folding processes of proteins or RNA. The latter class represents today the overwhelming majority of time-resolved studies that have recently been reviewed (Doniach, 2001; Svergun and Koch, 2003). They are the driving forces behind most instrumental developments around time-resolved experiments at third-generation sources. Like all time-resolved measurements, they imply recording and analysis of mixtures of different particles, the complexity of which can only be meaningfully addressed by a combination of different approaches, either structural or spectroscopic, the latter reporting on the kinetics of different aspects of the process (circular dichroism, fluorescence, etc.). The combined study by TR-CD and TR-SAXS of the folding of acid-denatured cytochrome *c* illustrates the requirement for coupled methods (Akiyama *et al.*, 2000, 2002). Using microfabricated mixing devices of the continuous flow type, time resolution of the order of a fraction of a millisecond can be achieved (Pollack *et al.*, 2001; Akiyama *et al.*, 2002). For slower processes, stopped-flow mixers are generally used as they require a smaller sample volume (Tsuruta *et al.*, 1989; Casselyn *et al.*, 2002).

Conformational transitions of allosteric enzymes such as aspartate transcarbamylase from *E. coli* (Tsuruta *et al.*, 1994, 2005) or GroEL (Inobe *et al.*, 2003) yielded new insights into the transition mechanisms. For instance, it has recently been shown that though substrate binding to ATCase active site triggers a concerted transition between two conformations (T and R), a basic feature of the Monod, Wyman, and Changeux (MWC) model for allostery (Monod *et al.*, 1965), the nucleotide activator ATP, does not alter the equilibrium between T and R as predicted by the same model, but changes the conformation of the substrate-bound enzyme (Tsuruta *et al.*, 2005). During critical stages of their life cycle such as maturation or decapsidation, viruses undergo large-scale quaternary structure changes. The kinetics of these phenomena have been studied using a combination of SAS and electron microscopy (Lata *et al.*, 2000; Perez *et al.*, 2000; Canady *et al.*, 2001; Lee *et al.*, 2004).

Assembly or dissociation phenomena studied by TR-SAXS involved virus capsids such as Brome mosaic virus (BMV), actin filaments and microtubule formation, and dynamic instabilities. The latter system has been extensively studied by a combination of cryo-EM and SAXS (Spann *et al.*, 1987; Mandelkow *et al.*, 1991). The assembly of BPTI decamers at acidic pH has been presented earlier. In a sequel of this study, the kinetics of the decamer dissociation process was studied using TR-SAXS (Hamiaux and Pérez, personal communication; Hamiaux, 2000). BPTI dissociation was triggered by changing salt concentration on a microdesalting column immediately before collecting SAXS patterns every minute. All scattering patterns share a common ''isoscattering'' point (see Figure 11.14, *left panel*), strongly suggesting that no third species beyond monomer and decamer contributed to the scattering by the solution. All experimental data could be approximated within experimental errors by linear combinations of the scattering pattern of the decamer found in crystals and that of the monomer. This analysis yields the fraction of BPTI molecules present as monomers in solution whose time evolution has been used to monitor dissociation kinetics. It appears to be perfectly monoexponential with a characteristic time of 8 min (Figure 11.14, *right panel*). Decamer dissociation therefore appears to be a highly cooperative process, complementing the conclusions of the equilibrium studies.

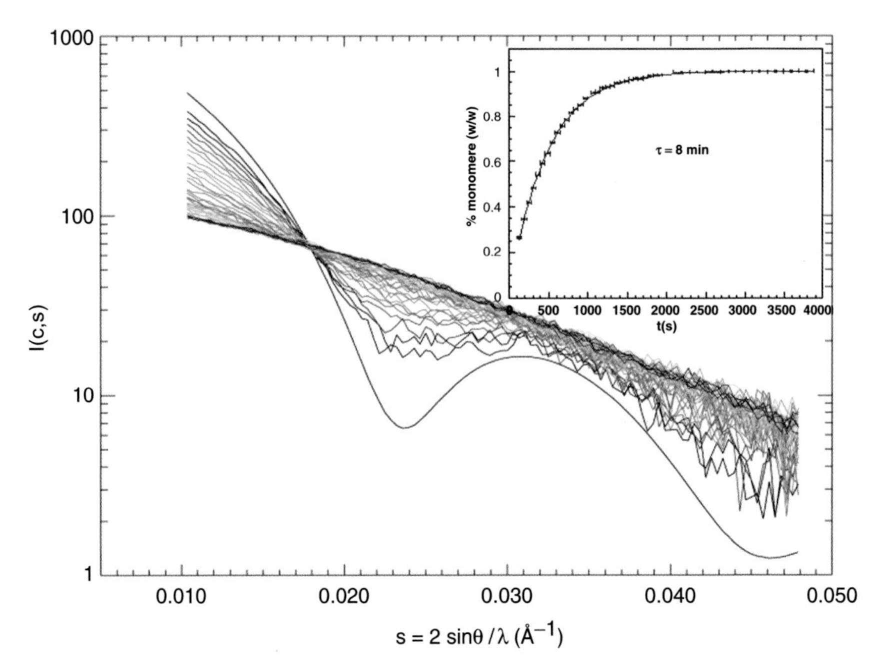

Figure 11.14. Scattering pattern of a 25 mg/mL solution of bovine pancreatic trypsin inhibitor (BPTI) recorded every minute after reduction of NaCl concentration down to 0.5 M in which the decamer dissociates. The noise-free solid line is the calculated scattering pattern of the decamer found in the crystal. *Inset*: time evolution of the fraction of monomer calculated by least-square fitting of each scattering pattern (see text for details). Courtesy of C. Hamiaux.

The kinetics of changes in phase transitions induced by intermolecular interactions can also be studied by SAXS: two recent examples are a liquid–liquid phase transition in α-crystallin solution induced by PEG addition (Finet *et al.*, personal communication) and the liquid–solid phase separation of BMV, an icosahedral plant virus, following PEG addition (Casselyn *et al.*, 2002, 2004). In the latter study, experimental conditions were chosen so as to produce microcrystals in great numbers (overnucleation), thereby allowing the authors to monitor the kinetics of early crystal formation and growth as a function of PEG size and concentration. Scattering patterns of 50-ms-long exposure were collected between 180 ms and 10–25 min at increasing intervals following the fast mixing of the viral solution with the precipitating agent in a stopped-flow device. Although the first pattern is practically identical to the form factor of the virus (Figure 11.15a), Bragg peaks due to the first microcrystals produced appear on the two-dimensional image after ~1 s. Within 1 min, the spherically average pattern appears to be strongly modulated by peaks typical of a powder diagram (Figure 11.15a). After division by the virus form factor, increasingly intense and sharp diffraction peaks are seen, which can be indexed in a face-centered cubic crystal form (Figure 11.15b). Assuming a Gaussian shape for each reflection, an estimate of the average radius of the

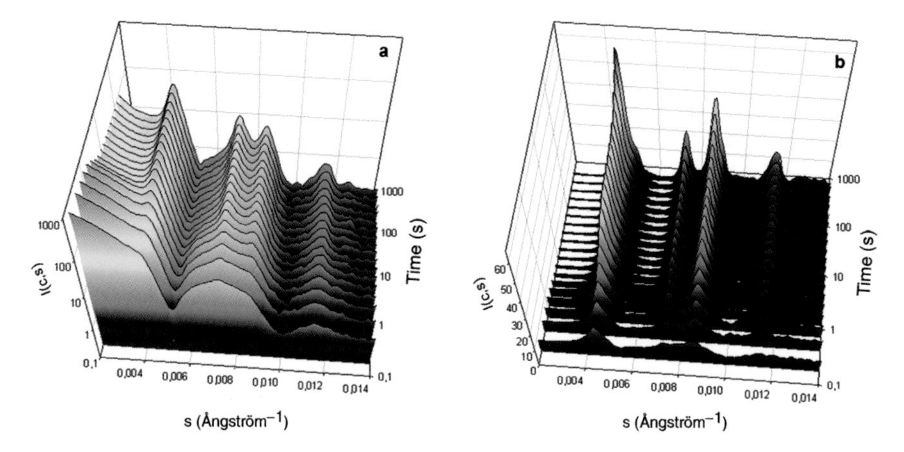

Figure 11.15. Brome mosaic virus (BMV) crystal growth. (A) Time evolution of the scattering patterns of an initially 20 mg/mL BMV solution containing 10% (w/v) polyethylene glycol (PEG) 3 K. (B) Corresponding structure factors derived by dividing the scattered intensities by the virus form factor. Note that in this figure $s = 2(\sin\theta)/\lambda$. Courtesy of H. Delacroix.

individual microcrystals could be derived, ranging between 500 and 1,500 Å as a function of virus concentration and PEG molecular mass (from 3 to 20 K). The lower value gives an upper limit to the size of the critical nucleus corresponding to 30–40 virus particles. Data analysis also shows that viruses are present in only two states, in solution or in crystals, revealing that the formation of periodic order proceeds without any detectable intermediate.

11.5. CONCLUSIONS

The potential of SAS as a structural technique has been recently boosted both from the point of view of experimental facilities and analysis methods. In SAXS, most notable is the advent of high brilliance third-generation synchrotron radiation sources; for SANS, further development of the detectors, neutron guides, and specific deuteration facilities are major factors. This experimental progress is accompanied by the development of novel data analysis algorithms, which made it possible to improve the resolution and the reliability of the models constructed from the SAS data. The technique allows *ab initio* low-resolution shape determination and rigid body modeling of the quaternary structure, the latter being more important for the study of macromolecular complexes. Advanced methods have been developed to build models of complexes from the high-resolution structures of individual components by interactive and/or automated fitting of SAXS data, and these methods have been successfully applied in practice, as illustrated in this chapter. Many useful data processing and modeling programs are publicly accessible on the Web (see Table 11.3). Further selected readings and useful links for SAS

Table 11.3. Publicly available programs for SAXS/SANS data analysis

Analysis of mixtures	Interacting systems[a]	*Ab initio* shape analysis
Oligomer	Fish[b]	DAMMIN
SVDPLOT	Mixture	GASBOR
		DALAI_GA[c]
		SAXS3D[d]
Data reduction and processing	Computation of characteristic functions	Addition of missing fragments
PRIMUS	GNOM	CREDO
		BUNCH
Computation of scattering from atomic models	Rigid body modeling	Models superposition and averaging
	MASSHA	
CRYSOL	ASSA	SUPCOMB
CRYSON	GLOBSYMM	DAMAVER
	SASREF	

Note: If no superscript is given, the program belongs to ATSAS 2.1 program suite, www.embl-hamburg.de/ExternalInfo/Research/Sax/

[a] We are not aware of general public programs available for computation of structure factors of protein solutions. The programs indicated below allow us to use structure factors for simple geometrical bodies

[b] http://www.isis.rl.ac.uk/largescale/LOQ/FISH/FISH_intro.htm

[c] akilonia.cib.csic.es/DALAI_GA2/

[d] www.cmpharm.ucsf.edu/~walther/saxs/

can be found, e.g., on the Web page of the Commission on Small-Angle Scattering of the International Union of Crystallography (http://www.iucr.org/iucr-top/comm/csas/index.html).

The major bottleneck for all rigid body refinement approaches based on data from low-resolution methods (whether using cryo-EM maps or SAXS data) lies in the potential ambiguity of the generated models. It is thus extremely important to synergistically use SAS with other experimental and computational techniques. In particular, SAS data can be usefully combined with the results of the NMR measurements of residual dipolar couplings that provide rotational orientation constraints for structural domains (Mattinen *et al.*, 2002). Computational rigid or semiflexible docking methods based on combination of energetics, shape complementarity, and biochemical and biophysical interaction data can provide effective tools to assess and rank the rigid body models built by SAS (Dominguez *et al.*, 2003). Contrast variation in SANS is expected to continue playing one of the major roles in the study of complexes, and efforts in providing users of large-scale facilities with the means for specific deuteration are already on the way (e.g., user-oriented joint EMBL-ILL deuteration laboratory in Grenoble, France). Overall, recent progress in SAXS and SANS, yielding information about structure and structural transitions in solution under nearly physiological conditions, makes them extremely useful for the study of macromolecular complexes in the postgenomic era.

REFERENCES

Akiyama, S., Takahashi, S., Ishimori, K., and Morishima, I. (2000). Stepwise formation of alpha-helices during cytochrome c folding. *Nat Struct Biol* 7:514–520.

Akiyama, S., Takahashi, S., Kimura, T., Ishimori, K., Morishima, I., Nishikawa, Y., and Fujisawa, T. (2002). Conformational landscape of cytochrome c folding studied by microsecond-resolved small-angle x-ray scattering. *Proc Natl Acad Sci USA* 99:1329–1334.

Aparicio, R., Fischer, H., Scott, D. J., Verschueren, K. H., Kulminskaya, A. A., Eneiskaya, E. V., Neustroev, K. N., Craievich, A. F., Golubev, A. M., and Polikarpov, I. (2002). Structural insights into the beta-mannosidase from *T. reesei* obtained by synchrotron small-angle X-ray solution scattering enhanced by X-ray crystallography. *Biochemistry* 41:9370–9375.

Arndt, M. H., de Oliveira, C. L., Regis, W. C., Torriani, I. L., and Santoro, M. M. (2003). Small angle x-ray scattering of the hemoglobin from *Biomphalaria glabrata*. *Biopolymers* 69:470–479.

Bada, M., Walther, D., Arcangioli, B., Doniach, S., and Delarue, M. (2000). Solution structural studies and low-resolution model of the *Schizosaccharomyces pombe* sap1 protein. *J Mol Biol* 300:563–574.

Ban, N., Nissen, P., Hansen, J., Moore, P. B., and Steitz, T. A. (2000). The complete atomic structure of the large ribosomal subunit at 2.4 A resolution. *Science* 289:905–920.

Bang, M. L., Centner, T., Fornoff, F., Geach, A. J., Gotthardt, M., McNabb, M., Witt, C. C., Labeit, D., Gregorio, C. C., Granzier, H., and Labeit, S. (2001). The complete gene sequence of titin, expression of an unusual approximately 700-kDa titin isoform, and its interaction with obscurin identify a novel Z-line to I-band linking system. *Circ Res* 89:1065–1072.

Belloni, L. (1991) *Interacting Monodisperse and Polydisperse Spheres*. Elsevier Science Publishers B.V., Amsterdam.

Belloni, L. (1993). Inability of the hypernetted chain integral equation to exhibit a spinodal line. *J Chem Phys* 98:8080–8095.

Berman, H. M., Westbrook, J., Feng, Z., Gilliland, G., Bhat, T. N., Weissig, H., Shindyalov, I. N., and Bourne, P. E. (2000). The protein data bank. *Nucleic Acids Res* 28:235–242.

Bloemendal, H., de Jong, W., Jaenicke, R., Lubsen, N. H., Slingsby, C., and Tardieu, A. (2004). Ageing and vision: structure, stability and function of lens crystallins. *Prog Biophys Mol Biol* 86:407–485.

Boehm, M. K., Woof, J. M., Kerr, M. A., and Perkins, S. J. (1999). The Fab and Fc fragments of IgA1 exhibit a different arrangement from that in IgG: a study by X-ray and neutron solution scattering and homology modelling. *J Mol Biol* 286:1421–1447.

Bohren, C. F. and Huffman, D. R. (1983) *Absorption and Scattering of Light by Small Particles*. Wiley, New York.

Brodersen, D. E., Clemons, W. M. Jr., Carter, A. P., Wimberly, B. T., and Ramakrishnan, V. (2002). Crystal structure of the 30 S ribosomal subunit from *Thermus thermophilus*: structure of the proteins and their interactions with 16 S RNA. *J Mol Biol* 316:725–768.

Bugs, M. R., Forato, L. A., Bortoleto-Bugs, R. K., Fischer, H., Mascarenhas, Y. P., Ward, R. J., and Colnago, L. A. (2004). Spectroscopic characterization and structural modeling of prolamin from maize and pearl millet. *Eur Biophys J* 33:335–343.

Canady, M. A., Tsuruta, H., and Johnson, J. E. (2001). Analysis of rapid, large-scale protein quaternary structural changes: time-resolved X-ray solution scattering of *Nudaurelia capensis* omega virus (NomegaV) maturation. *J Mol Biol* 311:803–814.

Capel, M. S., Engelman, D. M., Freeborn, B. R., Kjeldgaard, M., Langer, J. A., Ramakrishnan, V., Schindler, D. G., Schneider, D. K., Schoenborn, B. P., Sillers, I. Y., *et al.* (1987). A complete mapping of the proteins in the small ribosomal subunit of *Escherichia coli*. *Science* 238:1403–1406.

Casselyn, M., Finet, S., Tardieu, A., and Delacroix, H. (2002). Time-resolved scattering investigations of brome mosaic virus microcrystals appearance. *Acta Crystallogr* D 58:1568–1570.

Casselyn, M., Tardieu, A., Delacroix, H., and Finet, S. (2004). Birth and growth kinetics of brome mosaic virus microcrystals. *Biophys J* 87:2737–2748.

Chacon, P., Diaz, J. F., Moran, F., and Andreu, J. M. (2000). Reconstruction of protein form with X-ray solution scattering and a genetic algorithm. *J Mol Biol* 299:1289–1302.

Chacon, P., Moran, F., Diaz, J. F., Pantos, E., and Andreu, J. M. (1998). Low-resolution structures of proteins in solution retrieved from X-ray scattering with a genetic algorithm. *Biophys J* 74: 2760–2775.

Cummings, P. T. and Monson, P. A. (1985). Solution of the Ornstein–Zernike equation in the vicinity of the critical point of a simple fluid. *J Chem Phys* 82:4303–4311.

Dainese, E., Sabatucci, A., van Zadelhoff, G., Angelucci, C. B., Vachette, P., Veldink, G. A., Agro, A. F., and Maccarrone, M. (2005). Structural stability of soybean lipoxygenase-1 in solution as probed by small angle X-ray scattering. *J Mol Biol* 349:143–152.

Davies, J. M., Tsuruta, H., May, A. P., and Weis, W. I. (2005). Conformational changes of p97 during nucleotide hydrolysis determined by small-angle X-Ray scattering. *Structure* (Camb) 13:183–195.

Dominguez, C., Boelens, R., and Bonvin, A. M. (2003). HADDOCK: a protein–protein docking approach based on biochemical or biophysical information. *J Am Chem Soc* 125:1731–1737.

Doniach, S. (2001). Changes in biomolecular conformation seen by small angle X-ray scattering. *Chem Rev*, 101:1763–1778.

Ducruix, A., Guilloteau, J. P., Riиs-Kautt, M., and Tardieu, A. (1996). Protein interactions as seen by solution X-ray scattering prior to crystallogenesis. *J Cryst Growth* 168:28–39.

Edwards, A. M., Arrowsmith, C. H., Christendat, D., Dharamsi, A., Friesen, J. D., Greenblatt, J. F., and Vedadi, M. (2000). Protein production: feeding the crystallographers and NMR spectroscopists. *Nat Struct Biol* Suppl 7:970–972.

Egea, P. F., Rochel, N., Birck, C., Vachette, P., Timmins, P. A., and Moras, D. (2001). Effects of ligand binding on the association properties and conformation in solution of retinoic acid receptors RXR and RAR. *J Mol Biol* 307:557–576.

Feigin, L. A. and Svergun, D. I. (1987) *Structure Analysis by Small-Angle X-Ray and Neutron Scattering*. Plenum Press, New York, Vol xiii, 335 pp.

Finet, S., Vivares, D., Bonnete, F., and Tardieu, A. (2003). Controlling biomolecular crystallization by understanding the distinct effects of PEGs and salts on solubility. *Methods Enzymol* 368: 105–129.

Finet, S., Narayanan, T. and Tardieu, A. (personal communication).

Frank, J., Zhu, J., Penczek, P., Li, Y., Srivastava, S., Verschoor, A., Radermacher, M., Grassucci, R., Lata, R. K., and Agrawal, R. K. (1995). A model of protein synthesis based on a new cryo-electron microscopy reconstruction of the *E. coli* ribosome. *Nature* 376:441–444.

Fujisawa, T., Kostyukova, A., and Maeda, Y. (2001). The shapes and sizes of two domains of tropomodulin, the P-end-capping protein of actin-tropomyosin. *FEBS Lett* 498:67–71.

Funari, S. S., Rapp, G., Perbandt, M., Dierks, K., Vallazza, M., Betzel, C., Erdmann, V. A., and Svergun, D. I. (2000). Structure of free thermus flavus 5 S rRNA at 1.3 nm resolution from synchrotron X-ray solution scattering. *J Biol Chem* 275:31283–31288.

Furtado, P. B., Whitty, P. W., Robertson, A., Eaton, J. T., Almogren, A., Kerr, M. A., Woof, J. M., and Perkins, S. J. (2004). Solution structure determination of monomeric human IgA2 by X-ray and neutron scattering, analytical ultracentrifugation and constrained modelling: a comparison with monomeric human IgA1. *J Mol Biol* 338:921–941.

George, A. and Wilson, W. W. (1994). Predicting protein crystallization from a dilute solution property. *Acta Crystallogr* D 50:361–365.

Gerstein, M., Edwards, A., Arrowsmith, C. H., and Montelione, G. T. (2003). Structural genomics: current progress. *Science* 299:1663.

Glatter, O. (1977). A new method for the evaluation of small-angle scattering data. *J Appl Cryst* 10: 415–421.

Glatter, O. and Kratky, O. (1982) *Small Angle X-ray Scattering*. Academic Press, London, p. 515.

Golub, G. H. and Reinsh, C. (1970). Singular value decomposition and least squares solution. *Numer Math* 14:403–420.

Grossmann, J. G., Hall, J. F., Kanbi, L. D., and Hasnain, S. S. (2002). The N-terminal extension of rusticyanin is not responsible for its acid stability. *Biochemistry* 41:3613–3619.

Guinier, A. (1939). La diffraction des rayons X aux tres petits angles; application a l'etude de phenomenes ultramicroscopiques. *Ann Phys* (Paris) 12:161–237.

Hamiaux, C., 2000. Cristallogenese du BPTI a pH acide: etude des relations entre l'etat d'association des molecules en solution et a l'etat cristallin par diffusion et diffraction des rayons-X. PhD Thesis, Paris XI, Orsay, 173 pp.

Hamiaux, C., Perez, J., Prange, T., Veesler, S., Ries-Kautt, M., and Vachette, P. (2000). The BPTI decamer observed in acidic pH crystal forms pre-exists as a stable species in solution. *J Mol Biol* 297:697–712.

Hamiaux, C. and Pérez, J. (personal communication).

Hammel, M., Fierobe, H. P., Czjzek, M., Finet, S., and Receveur-Brechot, V. (2004a). Structural insights into the mechanism of formation of cellulosomes probed by small angle X-ray scattering. *J Biol Chem* 279:55985–55994.

Hammel, M., Walther, M., Prassl, R., and Kuhn, H. (2004b). Structural flexibility of the N-terminal beta-barrel domain of 15-lipoxygenase-1 probed by small angle X-ray scattering. Functional consequences for activity regulation and membrane binding. *J Mol Biol* 343:917–929.

Hansen, J. P. and McDonald, I. R. (1986) *Theory of Simple Liquids*. Academic Press, New York.

Harrison, S. C. (1969). Structure of tomato bushy stunt virus. I. The spherically averaged electron density. *J Mol Biol* 42:457–483.

Heller, W. T. (2005). Influence of multiple well defined conformations on small-angle scattering of proteins in solution. *Acta Crystallogr D Biol Crystallogr* 61:33–44.

Heller, W. T., Abusamhadneh, E., Finley, N., Rosevear, P. R., and Trewhella, J. (2002). The solution structure of a cardiac troponin C-troponin I-troponin T complex shows a somewhat compact troponin C interacting with an extended troponin I-troponin T component. *Biochemistry* 41: 15654–15663.

Heller, W. T., Finley, N. L., Dong, W. J., Timmins, P., Cheung, H. C., Rosevear, P. R., and Trewhella, J. (2003). Small-angle neutron scattering with contrast variation reveals spatial relationships between the three subunits in the ternary cardiac troponin complex and the effects of troponin I phosphorylation. *Biochemistry* 42:7790–7800.

Ibel, K. and Stuhrmann, H. B. (1975). Comparison of neutron and X-ray scattering of dilute myoglobin solutions. *J Mol Biol* 93:255–265.

Inobe, T., Arai, M., Nakao, M., Ito, K., Kamagata, K., Makio, T., Amemiya, Y., Kihara, H., and Kuwajima, K. (2003). Equilibrium and kinetics of the allosteric transition of GroEL studied by solution x-ray scattering and fluorescence spectroscopy. *J Mol Biol* 327:183–191.

King, W. A., Stone, D. B., Timmins, P. A., Narayanan, T., von Brasch, A. A., Mendelson, R. A., and Curmi, P. M. (2005). Solution structure of the chicken skeletal muscle troponin complex via small-angle neutron and X-ray scattering. *J Mol Biol* 345(4):797–815.

Kirkpatrick, S., Gelatt, C. D. Jr., and Vecci, M. P. (1983). Optimization by simulated annealing. *Science* 220:671–680.

Koch, M. H., Vachette, P., and Svergun, D. I. (2003). Small-angle scattering: a view on the properties, structures and structural changes of biological macromolecules in solution. *Q Rev Biophys* 36: 147–227.

Konarev, P. V., Petoukhov, M. V., and Svergun, D. I. (2001). MASSHA—a graphic system for rigid body modelling of macromolecular complexes against solution scattering data. *J Appl Crystallogr* 34:527–532.

Konarev, P. V., Volkov, V. V., Sokolova, A. V., Koch, M. H. J., and Svergun, D. I. (2003). PRIMUS—a Windows-PC based system for small-angle scattering data analysis. *J Appl Crystallogr* 36: 1277–1282.

Kozin, M. B. and Svergun, D. I. (2000). A software system for automated and interactive rigid body modeling of solution scattering data. *J Appl Crystallogr* 33:775–777.

Kozin, M. B. and Svergun, D. I. (2001). Automated matching of high- and low-resolution structural models. *J Appl Crystallogr* 34:33–41.

Kozin, M. B., Volkov, V. V., and Svergun, D. I. (1997). ASSA—a program for three-dimensional rendering in solution scattering from biopolymers. *J Appl Crystallogr* 30:811–815.

Kratky, O. and Pilz, I. (1972). Recent advances and applications of diffuse X-ray small-angle scattering on biopolymers in dilute solutions. *Q Rev Biophys* 5:481–537.

Krueger, J. K., Zhi, G., Stull, J. T., and Trewhella, J. (1998). Neutron-scattering studies reveal further details of the Ca^{2+}/calmodulin-dependent activation mechanism of myosin light chain kinase. *Biochemistry* 37:13997–4004.

Labeit, S. and Kolmerer, B. (1995). Titins: giant proteins in charge of muscle ultrastructure and elasticity [see comments]. *Science* 270:293–296.

Lata, R., Conway, J. F., Cheng, N., Duda, R. L., Hendrix, R. W., Wikoff, W. R., Johnson, J. E., Tsuruta, H., and Steven, A. C. (2000). Maturation dynamics of a viral capsid: visualization of transitional intermediate states. *Cell* 100:253–263.

Lee, K. K., Gan, L., Tsuruta, H., Hendrix, R. W., Duda, R. L., and Johnson, J. E. (2004). Evidence that a local refolding event triggers maturation of HK97 bacteriophage capsid. *J Mol Biol* 340: 419–433.

Lekkerkerker, H. N. W. (1997). Strong, weak and metastable liquids. *Physica A* 244:227–237.

Lomakin, A., Asherie, N., and Benedek, G. B. (1996). Monte-Carlo study of phase separation in aqueous protein solutions. *J Chem Phys* 104:1646–1656.

Lomakin, A., Asherie, N., and Benedek, G. B. (1999). Aleotopic interactions of globular proteins. *Proc Natl Acad Sci USA* 96:9465–9468.

Malfois, M., Bonnete, F., Belloni, L., and Tardieu, A. (1996). A model of attractive interactions to account for liquid–liquid phase separation of protein solutions. *J Chem Phys*, 105:3290–3300.

Mandelkow, E. M., Mandelkow, E., and Milligan, R. A. (1991). Microtubule dynamics and microtubule caps: a time-resolved cryo-electron microscopy study. *J Cell Biol* 114:977–991.

Mattinen, M. L., Paakkonen, K., Ikonen, T., Craven, J., Drakenberg, T., Serimaa, R., Waltho, J., and Annila, A. (2002). Quaternary structure built from subunits combining NMR and small-angle x-ray scattering data. *Biophys J* 83:1177–1183.

Merzel, F. and Smith, J. C. (2002). Is the first hydration shell of lysozyme of higher density than bulk water? *Proc Natl Acad Sci USA*, 99:5378–5383.

Monod, J., Wyman, J., and Changeux, J. P. (1965). On the nature of allosteric transitions. A plausible model. *J Mol Biol* 12:88–118.

Moore, P. B. (1980). Small-angle scattering: Information content and error analysis. *J Appl Cryst* 13: 168–175.

Mueller, J. J., Damaschun, G., and Schrauber, H. (1990). The highly resolved excess electron distance distribution of biopolymers in solution—calculation from intermediate-angle X-ray scattering and interpretation. *J Appl Cryst* 23:26–34.

Ninio, J., Luzzati, V., and Yaniv, M. (1972). Comparative small-angle x-ray scattering studies on unacylated, acylated and cross-linked *Escherichia coli* transfer RNA I Val. *J Mol Biol* 71:217–229.

Nollmann, M., He, J., Byron, O., and Stark, W. M. (2004). Solution structure of the Tn3 resolvase-crossover site synaptic complex. *Mol Cell* 16:127–137.

Pavlov, M., Sinev, M. A., Timchenko, A. A., and Ptitsyn, O. B. (1986). A study of apo- and holo-forms of horse liver alcohol dehydrogenase in solution by diffuse x-ray scattering. *Biopolymers* 25: 1385–1397.

Perez, J., Defrenne, S., Witz, J., and Vachette, P. (2000). Detection and characterization of an intermediate conformation during the divalent ion-dependent swelling of tomato bushy stunt virus. *Cell Mol Biol* (Noisy-le-grand) 46:937–948.

Petoukhov, M. V., Eady, N. A., Brown, K. A., and Svergun, D. I. (2002). Addition of missing loops and domains to protein models by x-ray solution scattering. *Biophys J* 83:3113–3125.

Petoukhov, M. V. and Svergun, D. I. (2003). New methods for domain structure determination of proteins from solution scattering data. *J Appl Crystallogr*, 36:540–544.

Petoukhov, M. V. and Svergun, D. I. (2005). Global rigid body modelling of macromolecular complexes against small-angle scattering data. *Biophys J* 89:1237–1250.

Pilz, I., Glatter, O., and Kratky, O. (1972). Small-angle x-ray-scattering studies on the substructure of Helix pomatia hemocyanin. *Z Naturforsch* B 27:518–524.

Pollack, L., Tate, M. W., Finnefrock, A. C., Kalidas, C., Trotter, S., Darnton, N. C., Lurio, L., Austin, R. H., Batt, C. A., Gruner, S. M., and Mochrie, S. G. J. (2001). Time resolved collapse of a folding protein observed with small angle x-ray scattering. *Phys Rev Lett* 86:4962–4965.

Porod, G., 1982. General theory. In: O. Glatter and O. Kratky (eds), *Small-Angle X-ray Scattering*. Academic Press, London, pp. 17–51.

Priddy, T. S., MacDonald, B. A., Heller, W. T., Nadeau, O. W., Trewhella, J., and Carlson, G. M. (2005). Ca^{2+}-induced structural changes in phosphorylase kinase detected by small-angle X-ray scattering. *Protein Sci* 14:1039–1048.

Putzeys, P. and Brosteaux, J. (1935). The scattering of light in protein solutions. *Trans Faraday Soc* 31:1314–1325.

Raibaud, S., Vachette, P., Guillier, M., Allemand, F., Chiaruttini, C., and Dardel, F. (2003). How bacterial ribosomal protein L20 assembles with 23S ribosomal RNA and its own messenger RNA. *J Biol Chem* 278:36522–36530.

Roussel, A. and Cambillau, C., 1989. TURBO-FRODO. In: S. Graphics (ed), *Silicon Graphics Geometry Partners Directory*. Silicon Graphics, Mountain View, CA.

Sali, A., Glaeser, R., Earnest, T., and Baumeister, W. (2003). From words to literature in structural proteomics. *Nature* 422:216–225.

Sanger, J. W. and Sanger, J. M. (2001). Fishing out proteins that bind to titin. *J Cell Biol* 154: 21–24.

Schluenzen, F., Tocilj, A., Zarivach, R., Harms, J., Gluehmann, M., Janell, D., Bashan, A., Bartels, H., Agmon, I., Franceschi, F., and Yonath, A. (2000). Structure of functionally activated small ribosomal subunit at 3.3 angstroms resolution. *Cell* 102:615–623.

Scott, D. J., Grossmann, J. G., Tame, J. R., Byron, O., Wilson, K. S., and Otto, B. R. (2002). Low resolution solution structure of the Apo form of *Escherichia coli* haemoglobin protease Hbp. *J Mol Biol* 315:1179–1187.

Shi, Y. Y., Hong, X. G., and Wang, C. C. (2005). The C-terminal (331–376) Sequence of *Escherichia coli* DNAJ is essential for dimerization and chaperone activity: a small angle x-ray scattering study in solution. *J Biol Chem* 280:22761–22768.

Sokolova, A., Malfois, M., Caldentey, J., Svergun, D. I., Koch, M. H., Bamford, D. H., and Tuma, R. (2001). Solution structure of bacteriophage PRD1 vertex complex. *J Biol Chem* 27:27.

Solovyova, A. S., Nollmann, M., Mitchell, T. J., and Byron, O. (2004). The solution structure and oligomerization behavior of two bacterial toxins: pneumolysin and perfringolysin O. *Biophys J* 87:540–552.

Spann, U., Renner, W., Mandelkow, E. M., Bordas, J., and Mandelkow, E. (1987). Tubulin oligomers and microtubule assembly studied by time-resolved X-ray scattering: separation of prenucleation and nucleation events. *Biochemistry* 26:1123–1132.

Stradner, A., Sedgwick, H., Cardinaux, F., Poon, W. C., Egelhaaf, S. U., and Schurtenberger, P. (2004). Equilibrium cluster formation in concentrated protein solutions and colloids. *Nature* 432:492–495.

Stuhrmann, H. B. (1970a). Ein neues Verfahren zur Bestimmung der Oberflaechenform und der inneren Struktur von geloesten globularen Proteinen aus Roentgenkleinwinkelmessungen. *Z Phys Chem Neue Folge* 72:177–198.

Stuhrmann, H. B. (1970b). Interpretation of small-angle scattering of dilute solutions and gases. A representation of the structures related to a one-particle scattering functions. *Acta Cryst* A 26:297–306.

Stuhrmann, H. B. (1991). Frozen spin targets in ribosomal structure research. *Biochimie* 73:899–910.

Stuhrmann, H. B., Goerigk, G., and Munk, B., 1991. Anomalous X-ray scattering. In: S. Ebashi, M. Koch, and E. Rubenstein (eds), *Handbook on Synchrotron Radiation*. Elsevier Science Publishers B.V., Amsterdam.

Stuhrmann, H. B. and Kirste, R. G. (1965). Elimination der intrapartikulaeren Untergrundstreuung bei der Roentgenkleinwinkelstreuung am kompakten Teilchen (Proteinen). *Z Phys Chem Neue Folge* 46:247–250.

Stuhrmann, H. B. and Notbohm, H. (1981). Configuration of the four iron atoms in dissolved human hemoglobin as studied by anomalous dispersion. *Proc Natl Acad Sci USA* 78:6216–6220.

Stuhrmann, H. B., Scharpf, O., Krumpolc, M., Niinikoski, T. O., Rieubland, M. and Rijllart, A. (1986). Dynamic nuclear polarisation of biological matter. *Eur Biophys J*, 14:1–6.

Sun, Z., Reid, K. B., and Perkins, S. J. (2004). The dimeric and trimeric solution structures of the multidomain complement protein properdin by X-ray scattering, analytical ultracentrifugation and constrained modelling. *J Mol Biol* 343:1327–1343.

Svergun, D. I. (1991). Mathematical methods in small-angle scattering data analysis. *J Appl Crystallogr* 24:485–492.

Svergun, D. I. (1992). Determination of the regularization parameter in indirect-transform methods using perceptual criteria. *J Appl Crystallogr* 25:495–503.

Svergun, D. I. (1994). Solution scattering from biopolymers: advanced contrast variation data analysis. *Acta Crystallogr* A 50:391–402.

Svergun, D. I. (1999). Restoring low resolution structure of biological macromolecules from solution scattering using simulated annealing. *Biophys J* 76:2879–86.

Svergun, D. I., Barberato, C., and Koch, M. H. J. (1995). CRYSOL—a program to evaluate X-ray solution scattering of biological macromolecules from atomic coordinates. *J Appl Crystallogr* 28:768–773.

Svergun, D. I. and Koch, M. H. J. (2003). Small angle scattering studies of biological macromolecules in solution. *Rep Progr Phys* 66:1735–1782.

Svergun, D. I. and Nierhaus, K. H. (2000). A map of protein–rRNA distribution in the 70 S *Escherichia coli* ribosome. *J Biol Chem* 275:14432–14439.

Svergun, D. I., Petoukhov, M. V., and Koch, M. H. J. (2001). Determination of domain structure of proteins from X-ray solution scattering. *Biophys J* 80:2946–2953.

Svergun, D. I., Richard, S., Koch, M. H., Sayers, Z., Kuprin, S., and Zaccai, G. (1998). Protein hydration in solution: experimental observation by x-ray and neutron scattering. *Proc Natl Acad Sci USA* 95:2267–2272.

Svergun, D. I., Semenyuk, A. V., and Feigin, L. A. (1988). Small-angle-scattering-data treatment by the regularization method. *Acta Crystallogr* A 44:244–250.

Svergun, D. I. and Stuhrmann, H. B. (1991). New developments in direct shape determination from small-angle scattering 1. Theory and model calculations. *Acta Crystallogr* A 47:736–744.

Svergun, D. I., Volkov, V. V., Kozin, M. B., and Stuhrmann, H. B. (1996). New developments in direct shape determination from small-angle scattering 2. Uniqueness. *Acta Crystallogr* A 52:419–426.

Svergun, D. I., Volkov, V. V., Kozin, M. B., Stuhrmann, H. B., Barberato, C., and Koch, M. H. J. (1997). Shape determination from solution scattering of biopolymers. *J Appl Crystallogr* 30:798–802.

Tardieu, A., Le Verge, A., Riиs-Kautt, M., Malfois, M., Bonnet, F., Finet, S., and Belloni, L. (1999). Proteins in solution: from X-ray scattering intensities to interaction potentials. *J Cryst Growth* 196:193–203.

Tardieu, A. and Vachette, P. (1982). Analysis of models of irregular shape by solution X-ray scattering: the case of the 50S ribosomal subunit from E. coli. *EMBO J* 1:35–40.

Tsuruta, H., Kihara, H., Sano, T., Amemiya, Y., and Vachette, P. (2005). Influence of nucleotide effectors on the kinetics of the quaternary structure transition of allosteric aspartate transcarbamylase. *J Mol Biol* 348:195–204.

Tsuruta, H., Nagamura, T., Kimura, K., Igarashi, Y., Kajita, A., Wang, Z. X., Wakabayashi, K., Amemiya, Y., and Kihara, H. (1989). Stopped-flow apparatus for x-ray scattering at subzero temperature. *Rev Sci Instrum* 60:2356–2358.

Tsuruta, H., Vachette, P., Sano, T., Moody, M. F., Amemiya, Y., Wakabayashi, K., and Kihara, H. (1994). Kinetics of the quaternary structure change of aspartate transcarbamylase triggered by succinate, a competitive inhibitor. *Biochemistry* 33:10007–10012.

Tung, C. S., Walsh, D. A., and Trewhella, J. (2002). A structural model of the catalytic subunit-regulatory subunit dimeric complex of the cAMP-dependent protein kinase. *J Biol Chem* 277:12423–12431.

van den Heuvel, R. H., Svergun, D. I., Petoukhov, M. V., Coda, A., Curti, B., Ravasio, S., Vanoni, M. A., and Mattevi, A. (2003). The active conformation of glutamate synthase and its binding to ferredoxin. *J Mol Biol* 330:113–128.

Verwey, E. J. W. and Overbeek, J. T. G. (1948) *Theory of the Stability of Lyophobic Colloids*. Elsevier, Amsterdam.

Vila-Sanjurjo, A., Ridgeway, W. K., Seymaner, V., Zhang, W., Santoso, S., Yu, K., and Cate, J. H. (2003). X-ray crystal structures of the WT and a hyper-accurate ribosome from *Escherichia coli*. *Proc Natl Acad Sci USA* 100:8682–8687.

Vivares, D., Belloni, L., Tardieu, A., and Bonnetи, F. (2002). Catching the PEG-induced attractive interaction between proteins. *Eur Phys J E Soft Matter*, 9:15–25.

Volkov, V. V. and Svergun, D. I. (2003). Uniqueness of ab initio shape determination in small angle scattering. *J Appl Crystallogr* 36:860–864.

Wall, M. E., Gallagher, S. C., and Trewhella, J. (2000). Large-scale shape changes in proteins and macromolecular complexes. *Annu Rev Phys Chem* 51:355–380.

Witty, M., Sanz, C., Shah, A., Grossmann, J. G., Mizuguchi, K., Perham, R. N., and Luisi, B. (2002). Structure of the periplasmic domain of *Pseudomonas aeruginosa* TolA: evidence for an evolutionary relationship with the TonB transporter protein. *EMBO J* 21:4207–4218.

Yusupov, M. M., Yusupova, G. Z., Baucom, A., Lieberman, K., Earnest, T. N., Cate, J. H., and Noller, H. F. (2001). Crystal structure of the ribosome at 5.5 A resolution. *Science* 292:883–96.

Zaccai, G. and Jacrot, B. (1983). Small angle neutron scattering. *Annu Rev Biophys Bioeng* 12:139–157.

Zou, P., Gautel, M., Geerlof, A., Wilmanns, M., Koch, M. H., and Svergun, D. I. (2003). Solution scattering suggests cross-linking function of telethonin in the complex with titin. *J Biol Chem* 278:2636–2644.

Zou, P., Pinotsis, N., Lange, S., Song, Y. H., Popov, A., Mavridis, I., Mayans, O. M., Gautel, M. and Wilmanns, M. (2006). Palindromic assembly of the giant muscle protein titin in the sacromeric Z-disk. Nature, 439:229–233.

12

Fluorescence Detection of Proximity

K. Wojtuszewski, J. J. Harvey, M. K. Han, and J. R. Knutson*

12.1. INTRODUCTION

Biochemical and biophysical analyses of reversible associations may be reduced to a few fundamental pieces of information. One is the degree to which the complexes are formed (vs. disassembled) in any particular setting (concentrations, temperature, salts, etc.) Another is the docking geometry and conformation of such complexes. The latter information can be gleaned from structural techniques such as X-ray crystallography solution, NMR spectroscopy, SANS, or SAXS (or in some cases, cryo-EM studies), but many of those detailed methods are performed in conditions that change the equilibria we mean to study in the first place; i.e., either concentrations or temperatures are outside of the physiologically interesting regime. The fact that interesting and relevant structures still emerge is a tribute to artistry in those methods and an indication that many complexes are hardy and long-lived. It would be advantageous, however, to also work with techniques that are capable of working on low concentration samples (needed to assess the stoichiometry of tight interactions) and that are sensitive to short-lived, flexible, fragile, or sparsely populated complexes.

A further benefit would accrue from techniques that translate into the kinetic regime and techniques that have the potential to be projected into living cells. Thus, this chapter briefly reviews two fluorescence phenomena (FRET and excimers) widely used in biochemistry to assess proximity, both quantitatively and on a binary (bound/free) basis. This chapter is by no means intended to serve as a comprehensive review; there are already several strong reviews (Clegg, 1992, 1995; Wu and

K. WOJTUSZEWSKI • Optical Spectroscopy Section, Laboratory of Molecular Biophysics, NHLBI, NIH, Building 10 Room 5D-14, Bethesda, MD 20892–1412. J. J. HARVEY and M. K. HAN • Excimus Biotech, Inc., Savage, MD 20763. J. R. KNUTSON • Optical Spectroscopy Section, Laboratory of Molecular Biophysics, NHLBI, NIH, Building 10 Room 5D-14, Bethesda, MD 20892–1412 and *Corresponding author: Tel: +301 496 2557; Fax: +301 480 4625; E-mail: jaysan@helix.nih.gov

Brand, 1994; Lehrer, 1997) we mention later. Instead, we focus on topics we find most relevant to the study of reversible macromolecular complexes and intriguing in general.

12.2. FRET

This most popular fluorescence technique for proximity assessment is useful in the 10–100 Å range, where it can give interprobe distance data accurate within a few Å. Thus, one focus of this chapter is to highlight recent advances in Förster resonance energy transfer (FRET) (sometimes called ''fluorescence'' resonance energy transfer, although R. E. Dale has pointed out that in each successful transfer, the donor does not fluoresce). We will give a brief introduction discussing the basic theory of FRET, then give examples of how FRET is routinely used, followed by some advances in the analysis of this excited-state complex. For more in-depth tutorials of FRET, we refer the reader to selected references listed at the end of this chapter.

Resonance energy transfer occurs when electronic energy is transferred from the excited state of a donor fluorophore (fluorescent molecule) to another molecule (*acceptor*) whose absorbance spectrum overlaps the donor emission. The transfer of energy is a nonradiative process because no real photons are emitted to pass between the two fluorescent probes. FRET is a result of long-range electronic transition dipole–dipole interactions in resonance (matched energy) between donor and acceptor molecules. Energy transfer stringently requires the energy gaps for transitions of the donor emission and acceptor excitation match. The efficiency of energy transfer declines with the distance (R) between the donor and the acceptor with $1/R^6$ dependence. The outcome of energy transfer is a decrease in the fluorescence intensity and lifetime of the donor ensemble and an increase in emission intensity (if any) of the acceptor. By monitoring either the energy of the emitted light, relative decrease in donor intensity or changed lifetimes, one can determine the extent of energy transfer.

12.2.1. Efficiency of Energy Transfer

The efficiency of energy transfer is the experimental observable used to determine the relative distance between the donor and the acceptor. Hence, one can use energy transfer to obtain valuable information on the structure of biological complexes. Energy is usually transferred over distances that range from 10–100 Å. With the design of new probes, or with multistep transfer, this value can increase beyond 200 Å (see later text). These independent solution measurements generate FRET distance constraints that may support current structural NMR and X-ray crystallography coordinates, or even more significantly, they may disagree if, for example, crystallographic packing forces have led to distorted conformations that are relaxed in dilute solution. For four decades now, FRET has been recognized as a ''spectroscopic ruler'' (Stryer and Haugland, 1967; Stryer, 1978). Prescient studies using flexible biopolymers instead of rigid molecular ladders appear in Brand and Witholt (1967), Edelhoch *et al.* (1967), and Conrad and Brand (1968).

The efficiency of energy transfer (E) is usually calculated from the yield remaining in an interacting donor population divided by the quantum yield of the donor in the absence of the acceptor. The following equation is used:

$$E = 1 - \frac{\phi_{\mathrm{DA}}}{\phi_{\mathrm{D}}}, \tag{12.1}$$

where the quantum yield in the presence of both donor and acceptor (ϕ_{DA}) divided by the quantum yield of donor only (ϕ_{D}) yields the efficiency of energy transfer (E). Efficiency is equal to one if all the energy from the donor is lost to the acceptor (i.e., the donor no longer emits).

Three key factors control the efficiency and rates of energy transfer for a given system. The first is the distance between the donor and acceptor probes. The second is the relative orientation for dipole–dipole interaction. This term (κ^2) contains the influence of mutual orientation of transition dipoles (and their separation vector) on the transfer rate (Dale and Eisinger, 1974). The third is the spectral overlap of the emission spectrum of the donor and the absorption spectrum of the acceptor. A summary of these factors is shown in Figure 12.1.

For the first term, Theodor Förster (1948) developed a theory that relates the rate of transfer (k_{T}) to a factor of $1/R^6$ and to $1/\tau_{\mathrm{D}}$, the lifetime of the donor molecules in the excited state:

$$k_{\mathrm{T}} = \left(\frac{1}{\tau_{\mathrm{D}}}\right)\left(\frac{R_0}{R}\right)^6. \tag{12.2}$$

In terms of the efficiency of energy transfer,

$$E = \frac{1}{\left(1 + \frac{R^6}{R_0^6}\right)}, \tag{12.3}$$

where R_0 is the Förster radius, the characteristic distance at which 50% of the energy is transferred to the acceptor (cf. Selvin, 1995). This value is usually nearly constant, except for environmentally sensitive probes akin to dansyl, and thus characteristic for a particular donor–acceptor pair. The factor of $1/R^6$ arises because the transfer probability is directly proportional to the square of the electric field produced by the donor and the electric dipole field decays as $1/R^3$ (Cantor and Schimmel, 1980).

12.2.2. Relative Orientation Factor

The orientation factor (κ^2) is dependent on the angles of the donor and the acceptor transition dipole vectors to each other and to the radius vector joining them. Depending on the orientation of these vectors, κ^2 can range from 0 to 4.

The FRET Process

***A.** Dependence on Mutual Orientation*

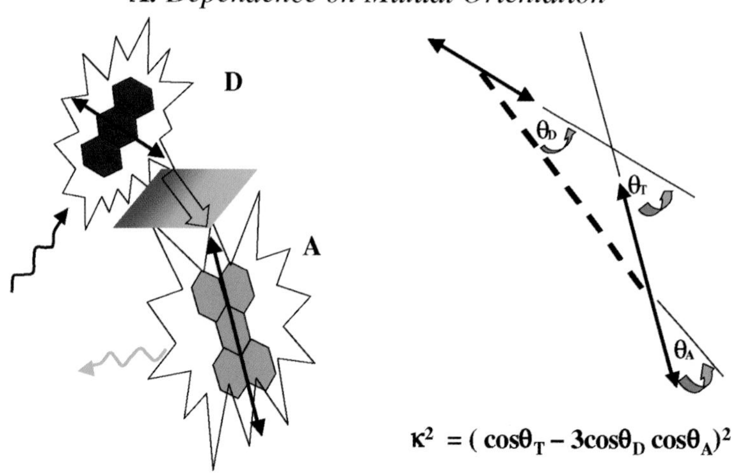

$$\kappa^2 = (\cos\theta_T - 3\cos\theta_D \cos\theta_A)^2$$

***B.** Efficiency and Overlap Integral*

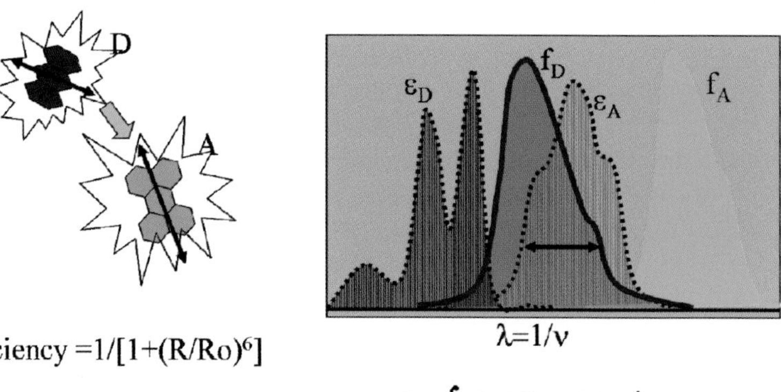

where:

$$J = \int f_D(\lambda)\varepsilon_A(\lambda)\lambda^4 d\lambda$$

Half-transfer (Forster) radius, Ro(A) $= 0.21\{\kappa^2 \eta^{-4} q_D J\}^{1/6}$

η is index of refraction and q_D is quantum yield of donor

Figure 12.1. Illustration of the overall FRET process: (A) Dependence on mutual orientation and (B) Efficiency and overlap integral.

The dependence on the angles between the donor and the acceptor transition dipoles is shown in Figure 12.1A.

$$\kappa^2 = (\cos\theta_T - 3\cos\theta_D\cos\theta_A)^2. \qquad (12.4)$$

θ_T is the angle between the emission transition dipole of the donor and the absorption transition dipole of the acceptor, and θ_D and θ_A are the angles between these dipoles and the vector joining the donor and the acceptor (Steinberg *et al.*, 1983; Clegg, 1992). Collinear dipole vectors have a κ^2 value of 4, whereas parallel vectors normal to r have a κ^2 value of 1. A κ^2 value of 0 results when vectors of the transition dipoles are perpendicular (Dale and Eisinger, 1974; van der Meer, 2002) and one must always be cognizant of this potential failure to transfer.

If energy transfer occurs between two molecules that are isotropic and/or rotating rapidly, the orientation factor tends toward 2/3. Exceptions to this widely used assumption can occur if one or both fluorophores are embedded in a rigid biopolymer. Experimentally, determining a relative orientation factor is difficult; generally donor and acceptor are arbitrarily assumed to attain ‘‘random dynamic’’ orientation. Various methodologies have been used to calculate relative orientation factors based on the donor and acceptor polarizations. A probability distribution can be calculated for κ^2 from the frequency for which the orientations of the donor and the acceptor occur over all possible directions in space. This, of course, assumes that all configurations have equal probability, and the assumption is difficult to verify. Still, this approach gives the investigator some sense of the risk of the assumption. For example, Haas and coworkers have shown that using a donor and acceptor pair that have limiting polarization, P, values of 0.07 and 0.4 (a low and relatively high polarization value), κ^2-induced error in the calculated end-to-end distribution in the labeled oligopeptide was negligible (i.e., true vs. apparent R) (Haas *et al.*, 1978). If acceptor polarization values are higher than 0.4, this approximation is less satisfactory. To relate P to the more familiar anisotropy (r),

$$P = \frac{3}{\left(\frac{2}{r} + 1\right)}. \qquad (12.5)$$

Fairly recently, detailed calculations and simulations were carried out with both depolarization factors close to zero, showing that the most probable distance is close to that expected from a κ^2 of 2/3 (van der Meer, 2002). van der Meer generated confidence limits by plotting ρ, the ratio of the actual donor–acceptor distance over the distance obtained from FRET, as a function of axial depolarization factors for the acceptor and the donor transition moments. A 67% confidence interval has been determined relative to the most probable distance (Figure 12.2).

For most protein labeling cases, a value for κ^2 of 2/3 is acceptable, and it is used in the following equation to calculate the Förster distance (R_0) between two fluorescent probes; cf. (Lakowicz, 1999):

$$R_0(\text{Å}) = 0.21\left[\kappa^2\eta^{-4}q_D J(\lambda)\right]^{1/6}, \qquad (12.6)$$

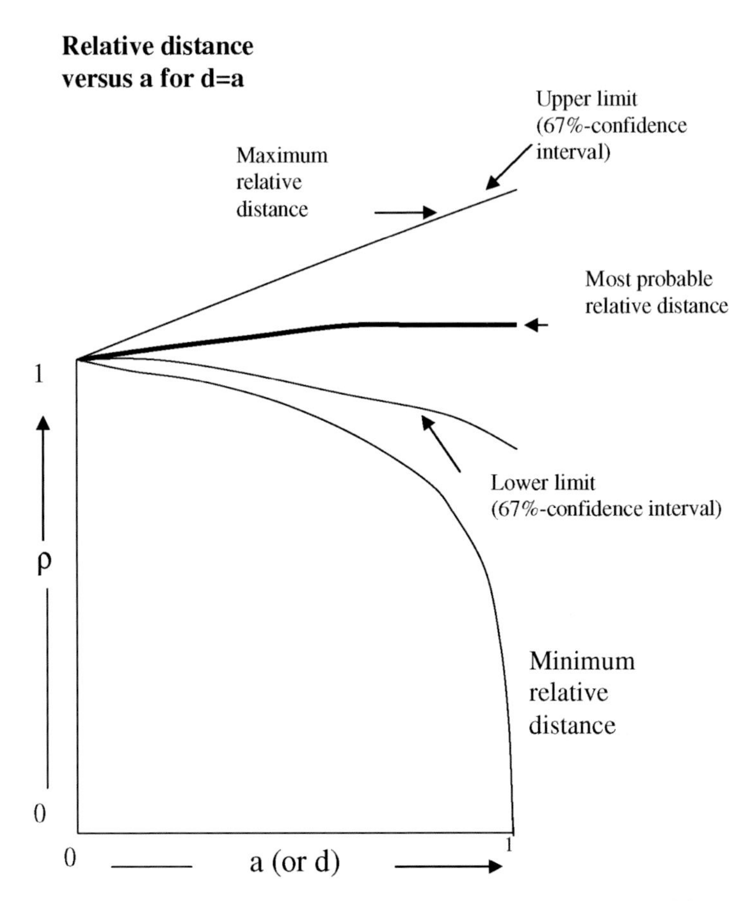

Figure 12.2. Assessing the orientation error. Relative distances ρ, the ratio of the actual donor–acceptor distance over the distance obtained from FRET data assuming $\kappa^2 = 2/3$ ($\rho = R/R'$) as a function of the axial depolarization factors. ρ_{MAX}, ρ_{MIN}, ρ_{hi}, ρ_{lo}, and ρ_{mp} are maximum, minimum, high, and low confidence limits and most probable relative distances. In this case, the two axial depolarization factors for the acceptor or donor transition moments (a and d) are equal to each other. Curves drawn as thicker lines represent values that are completely known, and curves drawn as thinner lines denote interpolated values. Redrawn from van der Meer (2002) Figure 9, with permission from Elsevier (Publisher).

where η is the refractive index of the medium, assumed to be 1.4 for biomolecules in aqueous solution. q_{D} is the quantum yield of the donor in the absence of the acceptor. The constant 0.21 is used when the wavelength is expressed in nanometers and the final value for R_0 can be expressed in Ångstroms. $J(\lambda)$ is the degree of spectral overlap between the donor emission and the acceptor absorption, expressed as $\mathrm{M}^{-1}\mathrm{cm}^{-1}(\mathrm{nm})^4$, described later.

12.2.3. Spectral Overlap

For energy transfer to take place, the donor emission spectrum must overlap with the acceptor absorbance spectrum. This spectral overlap $J(\lambda)$ is usually calculated from the fluorescence of the donor in the absence of the acceptor and the extinction coefficient ($\varepsilon_A(\lambda)$) of the acceptor as follows:

$$J(\lambda) = \int_0^{\infty} f_D(\lambda)\varepsilon_A(\lambda)\lambda^4 \, d\lambda = \frac{\int_0^{\infty} F_D(\lambda)\varepsilon_A(\lambda)\lambda^4 d\lambda}{\int_0^{\infty} F_D(\lambda) d\lambda}, \tag{12.7}$$

where $f_D(\lambda)$ is the fraction of the total donor fluorescence occurring at each wavelength from λ to $(\lambda + \Delta\lambda)$; and ($\varepsilon_A(\lambda)$) is the extinction coefficient spectrum of the acceptor ($M^{-1}cm^{-1}$). The calculated spectral overlap $J(\lambda)$ yields R_0 using Eq. (12.6).

12.2.4. General Excited-State Reaction Theory

FRET and excimer formation are both excited-state reactions. Proton transfer, tautomerization, and other excited-state reactions follow similar theoretical blueprints. Mathematically, one treats the system as a set of linear coupled ''compartments'' with known initial conditions, and each compartment is subject to kinetic processes that transfer population between excited-state species. Further discussion of compartmental analysis can be found in Godfrey (1983) and Szubiakowski *et al.* (2004). More generally, time-resolved measurement of each species is useful. Figure 12.3 shows several types of reactions.

12.2.4.1. One-Step FRET and Excimer Kinetics

The time dependence of each excited species depends on couplings to other compartments (species) and the losses of each to the ground state. These processes have been reviewed in Boens *et al.* (2000); basically, a set of excited populations ($\vec{P}$) is defined as a vector:

$$\vec{P}(t) = \{[A(t)],[B(t)],[C(t)],\ldots\}, \tag{12.8}$$

the components are subject to initial conditions and coupled ordinary differential equations

$$\dot{\vec{P}} = [T\vec{P}], \tag{12.9}$$

where T is the matrix that transfers populations between compartments, A, B, ..., which represent species in the excited state. Irreversible two-state transfer, the subcase appropriate to one-step FRET and excimer formation, is given by

$$\begin{aligned} \dot{A} &= -XA + 0(B) \\ \dot{B} &= k_{BA}A - YB \end{aligned} \qquad T = \begin{bmatrix} -X & 0 \\ k_{BA} & -Y \end{bmatrix}, \tag{12.10}$$

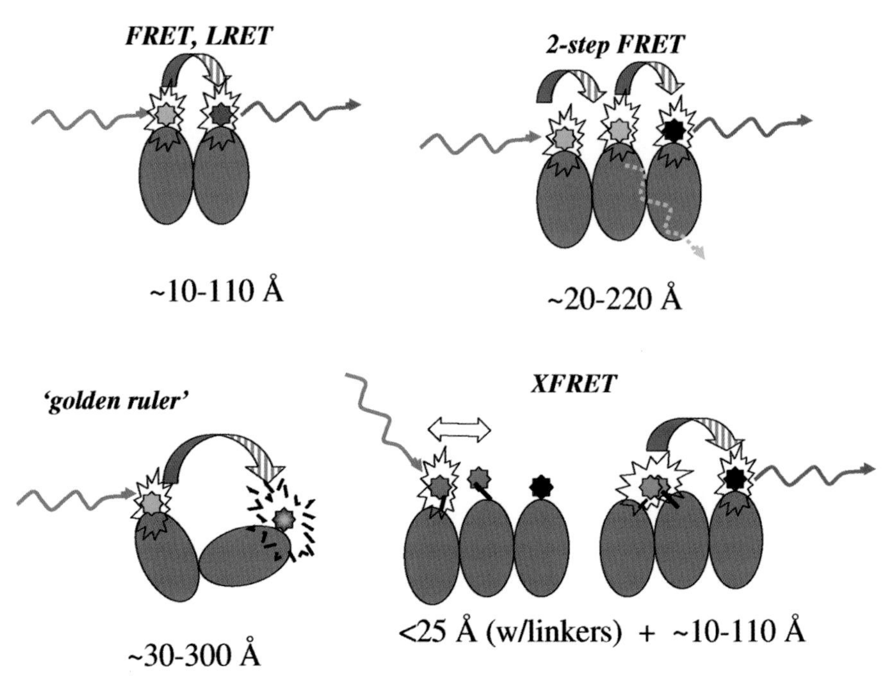

Figure 12.3. Schematic comparisons of excited-state fluorescence processes.

where X and Y are the total rates of depopulating A and B, respectively: X is the normal decay of A, k_A, and the FRET rate k_{BA}. Y is k_B in this case.

In the very simple excimer case, where A is a monomer and B is an excimer; k_{BA} is replaced by the formation rate $k_{BA}[A_0]$, where $[A_0]$ represents the concentration of the second (ground state) monomer that interacts with the excited monomer. An example of excimer formation is shown later under applications.

In both the cases, the solution is given by

$$\begin{aligned} A(t) &= a_1 e^{-Xt} + a_2 e^{-Yt} \\ B(t) &= b_1 e^{-Xt} + b_2 e^{-Yt} \end{aligned} \tag{12.11}$$

$$(X = k_A + k_{BA})$$
$$(Y = k_B).$$

The mathematics is explained in detail in the proton transfer work of Laws (Gafni *et al.*, 1976; Laws and Brand, 1979) and also in Davenport *et al.* (1986a). The coefficients a_i, b_i (fractional amplitudes) are derived from initial conditions; for example, no excimer is present at $t = 0$, so $b_1 = -b_2$.

The lack of feedback from B to A implies $a_2 = 0$. When examining decay surfaces and consequent decay-associated spectra (DAS), (Knutson *et al.*, 1982), the cases $X > Y$ versus $Y > X$ conveniently divide the problem; in the former case, the shortest lifetime (appearing as a rise in the B spectrum) contains the transfer rate. If $Y > X$, the decline, not the rise, of B yields X (Davenport *et al.*, 1986b). For FRET, the decline of the donor and enhancement of the acceptor imply proximity at or near R_0. For an excimer system, the emergence of a B spectrum implies that proximity between monomers permits excimer formation. For example, for pyrene iodoacetamide-labeled protein, the extended distances from the sulfur of Cys to the middle of the pyrene is about 9 Å (Lehrer, 1995). Thus, labeled regions in an intramolecular excimer-forming macromolecule are certain to be within 25 Å of each other. This is an important result, especially for the assembly of subunits. Decay analysis gives us more than the simple proximity; it allows us to detect multiple species with differing proximity and it helps us to quantify the rates in the system without assumptions inherent in the steady-state average. Hence, compartmental analysis of FRET and excimers lets us more reliably assess the real distance(s) present.

The approach using coupled ordinary differential equations (ODE) is fruitful for multiple compartments, but careful consideration of the model identifiably becomes paramount (Beechem *et al.*, 1991; Boens *et al.*, 2000).

12.2.4.2. XFRET Theory

Excimer FRET (XFRET) is a special case of energy transfer. XFRET is the sequential formation of an excimer that itself becomes the donor for FRET to a terminal acceptor, as indicated in Schematic 1. The key potential of XFRET is the selective detection of 'more-than-binary' interactions. The parameters of highest interest – the biochemical rate of excimer formation and the FRET rates – are related with the experimental observables in the following.

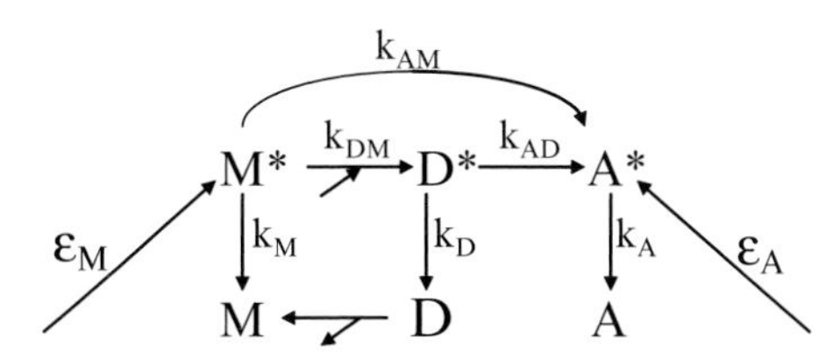

Schematic 1: M*, D*, and A* are excited monomer, excited dimer (excimer) and excited acceptor populations (* denotes the excited species). k_{AM} is the rate constant for FRET from excited monomer to acceptor, k_{DM} is the rate for excited monomer to combine with ground state monomer to form the excimer, and k_{AD} is the rate constant for FRET from the excimer to the acceptor (XFRET).

Let k_{MT} be the total depopulation rate constant of M*,

$$k_{MT} = k_M + k_{DM}[M] + k_{AM} \tag{12.12}$$

where [M] is the effective concentration of monomer in the ground state, and the rate constants are defined in Schematic 1. Similarly, the total rate constant of dimer depopulation is

$$k_{DT} = k_D + k_{AD} \tag{12.13}$$

and the total acceptor depopulation:

$$k_{AT} = k_A \tag{12.14}$$

(in these expressions we use the observation that FRET is essentially irreversible). For a three-state system, the time-course of populations of the excited states follows the differential equation system

$$\begin{bmatrix} M^*(t) \\ D^*(t) \\ A^*(t) \end{bmatrix} = \sum_{i=1}^{3} \begin{bmatrix} m_i \\ d_i \\ a_i \end{bmatrix} \left[e^{-\sigma_1 t} \quad e^{-\sigma_2 t} \quad e^{-\sigma_3 t} \right] \tag{12.15}$$

where σ_i are the eigenvalues of the system and are the inverses of the lifetimes seen. For our particular irreversible system shown in Schematic 1, $\vec{\sigma} = [k_{MT}, k_{DT}, k_{AT}]$. m_i, d_i, and a_i are the amplitudes for each lifetime term in each compartment. We solve for these amplitudes under the following boundary conditions: First, we use the initial populations that are in the excited state. The excimer is initially absent ($D^*(0) = 0$), and both monomer and acceptor are excited in proportion to their individual extinctions and concentrations: $M^*(0) = M_0^*$, $A^*(0) = A_0^*$, and $M_0^*/A_0^* = (\varepsilon_M[M]/\varepsilon_A[A])$. From this, we conclude that the amplitudes of each compartment sum to these values:

$$d_1 + d_2 + d_3 = 0, \quad m_1 + m_2 + m_3 = M_0^*, \quad a_1 + a_2 + a_3 = A_0^* \tag{12.16}$$

Next, we compute the initial rates (see Schematic 1):

$$\begin{aligned} \left.\frac{dM^*}{dt}\right|_{(t=0)} &= -k_{MT} M_0^* \\ \left.\frac{dD^*}{dt}\right|_{(t=0)} &= +k_{DM}[M] M_0^* \\ \left.\frac{dA^*}{dt}\right|_{(t=0)} &= +k_{AM} M_0^* - k_A A_0^* \end{aligned} \tag{12.17}$$

which, using Eq. (12.15), lead to

$$-\sigma_1 m_1 - \sigma_2 m_2 - \sigma_3 m_3 = \sigma_1 (m_1 + m_2 + m_3) \tag{12.18}$$

which is only true if $m_2 = m_3 = 0$, so

$$M^*(t) = M_0^* e^{-\sigma_1 t} = M_0^* e^{-k_{MT} t} \tag{12.19}$$

For the excimer, the same operations of combining Eqs. (12.17) and (12.15) yields

$$-\sigma_1 d_1 - \sigma_2 d_2 - \sigma_3 d_3 = +k_{DM}[M] M_0^* \tag{12.20}$$

D^* has no feedback term k_{DA} and is insensitive to k_A or A^*, so $d_3 = 0$ and $d_1 = -d_2$. Rewriting Eq. (12.20) with the convenient substitution $d_1 = -d_2 = \beta$, we find

$$D^*(t) = \beta e^{-k_{MT}t} + \beta e^{-k_{DT}t} \quad with \quad \beta = \frac{k_{DM}[M]M_0^*}{\sigma_2 - \sigma_1} \tag{12.21}$$

Only the acceptor A*(t) is influenced by all three compartments and is tri-exponential, since the terminal acceptor can be excited directly from the ground state, or FRET can come from monomer or excimer. (Spectral overlap can be engineered to favor either donor.) For the acceptor, the combination of Eqs. (12.17) and (12.15) yields

$$\sigma_1 a_1 + \sigma_2 a_2 + \sigma_3 a_3 = k_A A_0^* - k_{AM} M_0^* \tag{12.22}$$

Back-substitution of the previous compartment solutions of the monomer Eq. (12.19) and excimer Eq. (12.21), together with the initial condition (12.16), yields the final result

$$A^*(t) = \left\{\frac{k_{AM}M_0^* + k_{AD}\beta}{\sigma_3 - \sigma_1}\right\} e^{-k_{MT}t} + \left\langle \frac{k_{AD}\beta}{\sigma_3 - \sigma_2} \right\rangle e^{-k_{DT}t} + \left[A_0^* - \{\ldots\} - \langle\ldots\rangle\right] e^{-k_{AT}t} \tag{12.23}$$

The solutions of these equations provide us a mapping between the experimentally observed lifetimes and amplitudes (i.e., decay associated spectra – the sum of amplitudes multiplied by compartment spectra) and the biochemically useful unknowns: excimer formation rate and FRET (particularly XFRET) rates. The three-compartment analysis has been generalized to other systems with different connectivities (cf. 'two-step' FRET later). As mentioned above, the key potential of XFRET analysis is the selective detection of higher order interactions.

12.2.4.3. Global Analysis in FRET and Excimer Systems

Most of the FRET analysis done is based on the simplest formula, Eq. (12.1). Donor quenching is easy to observe in most systems and one can usually correct for the ''non-FRET quench'' (i.e., conformation change) with a suitable control, such as a molecule without acceptor dye (but otherwise as similar as possible). As long as the control really evokes the same emission as the system would have without FRET, this is sufficient to recover the average transfer efficiency.

In time-resolved analysis, however, more information is available than just a single average transfer rate. If a monoexponential donor is subject to quenching by acceptors at a variety of distances, the donor decay will become multiexponential with a transformed distribution of lifetimes, as first pointed out by Haas and Steinberg (1984). If distances are distributed, like $\rho(r)$, where $\int \rho(r)\mathrm{d}r = 1$ and $I_\mathrm{D}(t) = I_0 \mathrm{e}^{-t/\tau_\mathrm{D}}$ then,

$$I'(t) = I_0 \int \rho(R) \mathrm{e}^{-\left[t/\tau_\mathrm{D}\left(1+\left[\frac{R_0}{R}\right]\right)^6\right]} \mathrm{d}R.$$

The definition of ρ is the probability of finding donor and acceptor separated by r. In single-curve analysis, a presumed Gaussian (or other functional) distribution of distances, characterized by a few parameters, can be combined with the known τ_D to fit a donor decay. Considerably greater information accrues from global analysis of donor and acceptor decay, as first shown by Beechem and Brand (1985) and Beechem (1989). Global analysis of wavelength and viscosity even makes analysis of configuration diffusion (Haas and Steinberg, 1984) a possibility, that is, not only is a distribution of distance recoverable, but also a rate of distance fluctuation (Eis and Lakowicz, 1993). Of course, such complex models must always be compared statistically with simpler models invoking a few discrete distances, and Occam's razor must favor the simple view, unless either it fails to fit the data or is incompatible with other (e.g., NMR) knowledge about the system.

In excimer analysis, as for general excited-state reaction theory, the global analysis of decay versus wavelength is always a first step of merit. The characteristic shapes of DAS and their related species-associated spectra (SAS) are useful to diagnose X>Y versus Y>X, reversibility, and heterogeneity (Davenport *et al.*, 1986a).

More formal SAS analysis with compartmental models represents the best way to obtain confidence limits on the rates and amounts of excimer (or other excited-state reactions such as proton transfer, exciplex, or FRET) (Beechem *et al.*, 1985). These targeted analyses should also prove fruitful as they now begin to be applied to the more challenging extensions: two-step FRET and XFRET.

12.3. APPLICATIONS

12.3.1. Distance Measurements: FRET and DNA Bending

As described earlier, the most frequent goal of an FRET measurement is to quantify a distance. This can be useful in studying molecular conformations and in ascertaining the mode of assembly of complexes. For example, by fluorescently labeling two particular regions of a protein, one can gain structural insight from the distance between the two probes. Various studies have demonstrated determining a bend angle in DNA or RNA based on the end-to-end distance between the donor and the acceptor labels at oligonucleotide ends. This helps define structural aspects of the duplex and overall curvature that could either be sequence- or protein-induced.

One example is the degree of DNA bend induced by the prokaryotic protein, HU. The binding of the architectural DNA-binding protein HU leads to a stepwise change in the end-to-end distance of an oligonucleotide that was monitored by the efficiency of energy transfer. The method quantified the extent to which HU bends the duplex DNA (Wojtuszewski and Mukerji, 2003). Figure 12.4A shows the change in energy transfer efficiency for titration of the phased A-tract and non-A-tract duplex with HU protein. The plots of energy transfer efficiency as a function of HU concentration were also used to determine an equilibrium binding constant, which was in agreement with those determined by steady-state anisotropy and gel mobility shift assays (GMSA) (Wojtuszewski and Mukerji, 2003). The end-to-end distance relative to DNA contour length ($R/R(C)$) yields a bend angle for a 20-bp

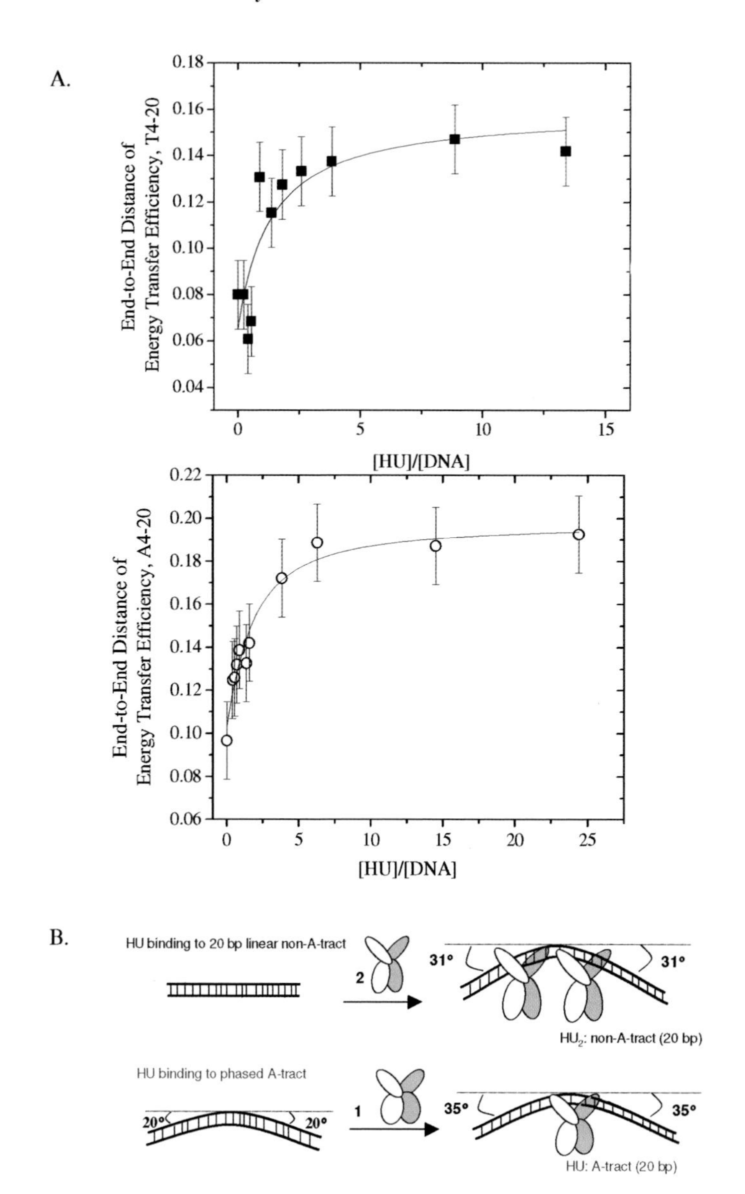

Figure 12.4. Energy transfer efficiencies (E) as a function of HU/DNA mole ratio. (A) A4–20 (○) and (B) T4–20 (■). A4–20 and T4–20 correspond to the phased A-tract (A_nT_n) and non-A-tract (T_nA_n), respectively. For precise sequence, please refer to the mentioned reference. Transfer efficiencies were calculated as described in the text. All samples were prepared in 10 mM TRIS–HCl, 0.1 mM EDTA, pH 7.6. Data shown are fit with binding isotherms as a visual aid; equilibrium-binding constants are 4.0 μM^{-1} for the A4–20 and 2.2 μM^{-1} for the T4–20 duplex and were determined as a function of HU concentration. Figure A is reprinted from Wojtuszewski and Mukerji (2003). (B) Diagram of the stoichiometry heterodimer HU protein (subunits are depicted as gray and white) binding to the 20-bp linear non-A-tract and prebent A-tract (unpublished drawing, courtesy of Ishita Mukerji).

phased A-tract duplex (A4–20) of $45 \pm 7°$ in the absence of HU and $70 \pm 3°$ in the presence of one HU dimer. In contrast, the bend angle calculated for a linear 20-bp T_4A_4 tract duplex was $62 \pm 4°$ after binding two HU dimers (Figure 12.4B). The use of FRET in this study was significant in two ways. First, it showed that HU avidly binds and further bends A-tract DNA, although it can also bend straight DNA (with higher occupancy and lower binding affinity). Second, these findings are consistent with the earlier reports that T_nA_n tracts are relatively straight, whereas A_nT_n tracts are curved (Hagerman, 1986; Haran and Crothers, 1989; Park and Breslauer, 1991). Measured bend angles of 22.5° per A-tract were in good agreement with cyclization and transient electric dichroism experiments, which measured a bend per A-tract of 28° and 18°, respectively, but were not consistent with electrophoretic studies in which an 11° bend per A-tract was observed (Calladine and Drew, 1986; Levene *et al.*, 1986; Ulanovsky *et al.*, 1986).

Along with measuring DNA bending, FRET has also become a valuable tool in characterizing complex DNA conformations such as four-way or Holliday junctions (Murchie *et al.*, 1989; Clegg *et al.*, 1992, 1993; Eis and Millar, 1993; McKinney *et al.*, 2004) and various RNA structures. Many of these experiments were achieved using traditional donor–acceptor pairs such as fluorescein and rhodamine. Parkhurst has also examined the three-dimensional (helical) distance versus sequence in protein–DNA complexes (Parkhurst *et al.*, 2001; Parkhurst, 2004).

12.3.2. Excimers

An excimer serves the experimenter as a transient fluorescent crosslink, minimally perturbing the ground state of macromolecules and forming opportunistically after one probe is excited. Excimer fluorescence, whereas less popular than FRET, provides definitive information about changes in proximity between many attachment sites in biological macromolecules.

To begin with the basics, certain fluorophores can form an excited-state dimer, or excimer, by a specific interaction between the excited-state monomer and a ground-state monomer. The ability of particular molecules to generate excimers was first reported by Kaspar and Förster (Förster and Kasper, 1955). Excimers are a subset of the more general process of forming an exciplex (excited-state complex), but this term is generally reserved for heterocomplexes and the phenomenon has not been as popular in biochemistry.While excimer fluorescence has been observed for benzene, toluene, and phenol (including para-substituted phenols), excimer fluorescence of pyrene-labeled proteins has been used most often to investigate intramolecular and intermolecular proximity and protein conformational change (for a keystone review, see (Lehrer, 1997)). Site-directed mutagenesis can be used to introduce amino acid residues (such as Cys) that selectively react with pyrene derivatives, such as pyrene maleimide or iodoacetamide.

Mechanistically, excimer emission originates from nearly planar stacked pyrenes oriented only in the correct configuration. Normally, pyrenes that are proximate will equilibrate between unstacked and nearly stacked configurations in the ground state. After monomer excitation, a time window is opened during which the stacked pyrenes must adopt the correct orientation. If this configuration cannot be achieved because of conformational constraints, then the interaction may result in quenching of the monomer fluorescence without formation of a productive exciplex.

12.3.2.1. Excimers and Protein–Protein Interactions

Classic studies employing pyrene fluorescence for the study of protein interactions were performed by Lehrer and colleagues. A series of papers investigated the interactions between tropomyosin (Betcher-Lange and Lehrer, 1978; Graceffa and Lehrer, 1980; Lehrer, 1994), troponin and tropomyosin (Ishii and Lehrer, 1991), and tropomyosin, actin, and myosin (Ishii and Lehrer, 1985, 1987, 1990). For an additional example, see the study on PM-tubulin, in which stoichiometry and association/dissociation were monitored relative to excimer formation, (Panda and Bhattacharyya, 1992). A more recent study of excimer formation is presented in Yang et al. (2005), using molecularly engineered ''light-switching'' excimer aptamer probes for the detection of platelet-derived growth factor (PDGF).

In the initial Lehrer papers, rabbit skeletal or cardiac tropomyosin was labeled at Cys190 with N-(1-pyrene)-maleimide (PM). Tropomyosin (Tm) is a rod-like coiled-coil that lies on the actin filament and functions as a component of the Ca^{2+}-dependent regulatory system. The two chains, α and β, can combine into $\alpha\alpha$ and $\alpha\beta$. The Cys190-SH groups on each chain are adjacent, and can form interchain disulfide bonds. Solutions of PM-labeled Tm-$\alpha\alpha$ form pyrene excimers. In the presence of 4 M Gn-HCl, Tm completely unfolds; the two chains separate and excimer emission is lost. A second series of experiments demonstrated that excimer formation is temperature-dependent. They deduced that pyrenes were already close to the required excimer configuration, similar to crystalline pyrene. In the resulting model, tropomyosin fluctuates between a more open state near Cys190 capable of forming excimer (the ''X'' state) and a more stable closed chain state unable to form an excimer (the ''N'' or native/monomer state). Increases in the temperature or salt concentration shifted the equilibrium toward ''X.''

Troponin is a complex of troponin T (TnT, the Tm-binding subunit), troponin I (the inhibitory subunit), and troponin C (the Ca^{2+}-binding subunit). The binding of troponin to pyrene maleimide-labeled Tm increased the monomer fluorescence 30–35% and decreased the excimer fluorescence 35–40%. Of particular interest was the fact that troponin bound eight times more strongly to labeled Tm than unlabeled Tm, showing that at times, the hydrophobic perturbation of pyrene may come into play. Multiple fragment studies led to an attachment model in which the TnT2 (C-terminal half) portion binds to the region including Cys190, while TnT1 (N-terminal half) binds to the C-terminal end of Tm.

The binding of F-actin to PM-labeled Tm decreased excimer fluorescence by about 40%, while increasing monomer fluorescence by about 20%. This observation indicated that the binding of F-actin shifted the equilibrium between the N and X states of Tm more toward the N state. When the S1 subfragment of myosin was added to a mixture of PM-labeled Tm complexed with F-actin, time-dependent changes of both monomer and excimer occurred. Within the first few seconds, monomer fluorescence increased approximately 18% without affecting the excimer, followed by a slow phase in which the monomer fluorescence increased somewhat further, along with a parallel decrease in the excimer fluorescence at about the same rate. Clearly, excimers are windows into complicated conformation change in large complexes.

Both proximity techniques have advantages in studies of macromolecular association; excimers require covalent attachment of only one type of label, but their effective range with common linkers ($<\sim 25$ Å) is shorter than FRET. An example of a dimeric protein conformation probed by *both* FRET and excimer techniques is EcoRI (Liu *et al.*, 1998; Watrob *et al.*, 2001). In this study, the characteristic excimer emission band of pyrene-labeled EcoRI indicated facile stacking of the two pyrene rings in the homodimer.

Other examples of how protein–protein interactions can be investigated using pyrene fluorescence include the assembly of translin monomers into a functional multimeric assembly (Han *et al.*, 2002) and the dimerization of enzyme I (Han *et al.*, 1990). Translin is a nucleic acid binding protein that recognizes chromosomal breakpoints and may be involved in chromosomal translocation. The protein is a 228-amino acid single-strand binding protein containing a putative transmembrane helix and a leucine zipper domain in the carboxy terminus. It was postulated that the leucine zipper participates in the protein–protein interactions of translin, which may be critical for its DNA-binding activity. The results of sedimentation velocity experiments indicated that translin existed as an octamer and bound single-stranded DNA with a 1:1 stoichiometry (Lee *et al.*, 2001). However, this data did not resolve the quaternary structure of the octamer. Two site-directed mutants of translin were constructed for pyrene labeling of either the amino-terminal (C225S, pyrene on Cys58), or carboxyl-terminal (C58S, pyrene on Cys225) domains using pyrene maleimide. Intermolecular excimer fluorescence, as evidenced by fluorescence emission at 480 nm, was observed only from pyrene labeling of the carboxyl-terminal, and not for pyrene labeled on the amino-terminal. Given the C-terminal location of the Cys225 residue, these results suggested that translin subunits interact in a ''tail-to-tail'' configuration to form an octamer, as depicted in Figure 12.7.

12.3.2.2. Example of an XFRET Target

To determine if XFRET from pyrene excimers to fluorescein could be observed, a solution of C58S pyrene maleimide (fully) labeled translin was titrated with increasing concentrations of fluorescein isothiocyanate (FITC) labeled single-stranded DNA. Figure 12.5A shows pyrene excimer emission at 480 nm in the native octamer. As the protein solution is titrated with increasing concentrations of FITC-labeled DNA, excimer emission diminishes by energy transfer to fluorescein. Figure 12.5B shows that at saturation, approximately 90% of pyrene excimer fluorescence was quenched by fluoresceinated ss-DNA.

A.

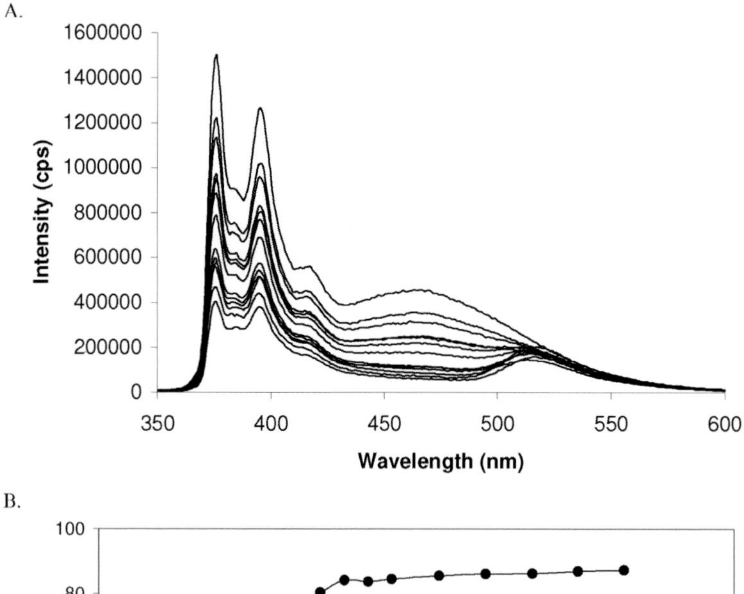

B.

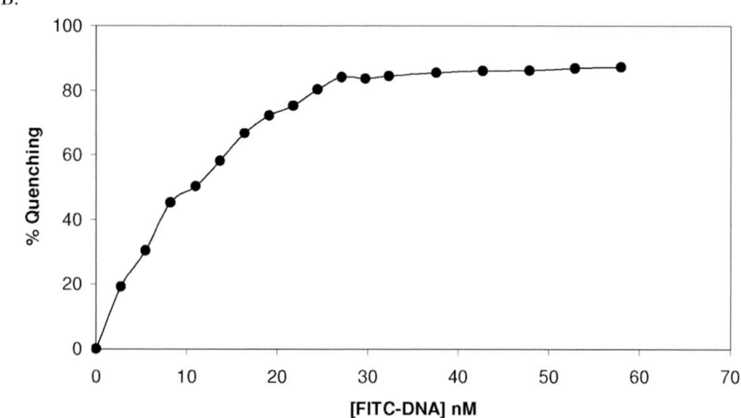

Figure 12.5. (A) Titration of pyrene-labeled Cys225 translin with FITC-labeled single-stranded DNA. The translin concentration was equivalent to 24 nM of octamer. At saturation the final concentration of FITC-labeled ss-DNA was 60 nM. The intensity of pyrene excimer at 480 nm decreases as FITC-ssDNA concentration increases. (B) Titration of pyrene-labeled Cys225 translin with FITC-labeled single-stranded DNA. In this figure the quenched fluorescence emission intensity at 480 nm is plotted versus the concentration of FITC-labeled ss-DNA.

Unfortunately, the rise in the sensitized emission of the acceptor did not exactly match this decline, indicating the presence of *other quenching mechanisms*. Static quenching and other acceptor complications are addressed for FRET in a recent paper (Majumdar *et al*., 2005). Nevertheless, the ability to selectively generate functional overlap in the excimer–acceptor pair makes it possible to selectively define ternary interactions; that is, only those proximate subunits creating excimers are being tested for their mutual proximity to the acceptor (Han, MK, Harvey, JJ, Knutson, JR, to be submitted). This selects for *functional* assembly, and offers advantages akin to those of two-step FRET in that regard.

Meanwhile, a series of traditional FRET experiments were performed to determine the average distance between Cys58 and Cys225 AEDANS (5-((((2-iodoacetyl)amino)ethyl)amino)naphthalene-1-sulfonic acid) labeled translin octamers and FITC-labeled DNA.

In Figure 12.6A, a solution of Cys58 AEDANS-labeled translin octamers was titrated with increasing concentrations of FITC-labeled DNA. In the absence of fluorescein, maximal AEDANS emission at 490 nm is seen. As labeled DNA is titrated into the solution the AEDANS emission decreases because of FRET to FITC, with a concomitant increase in fluorescein emission at 520 nm. At saturation, 60% of the AEDANS fluorescence is quenched by FRET to fluorescein. Using an R_0 of 49 Å for the AEDANS–FITC pair (Birmachu *et al.*, 1989; Jona *et al.*, 1990) the calculated distance between Cys58 and the DNA is 45.8 Å, although, again, the sensitized emission did not climb as much, because of static acceptor quenching in the complex (Birmachu *et al.*, 1989; Jona *et al.*, 1990) (data not shown).

Figure 12.6B shows the results of the same experiment as in Figure 12.6A, except that the translin octamers were labeled with AEDANS at Cys225. In this experiment 82% of the AEDANS fluorescence is quenched by FRET to fluorescein, resulting in a calculated distance between Cys225 and the DNA of 38.1 Å. Taken together, these data suggest that the translin octamer is arranged in a doughnut configuration (Figure 12.7), with the carboxyl termini of each monomer close enough to form a central hole through which the single-stranded DNA passes, while the amino-terminal portion of each monomer lies on the perimeter of the assembly. Of course, we have only measured two distances (and assumed cylindrical symmetry) in this simple model. More detailed (especially time-resolved) studies will be needed to further define the geometry of the octamer.

12.3.3. Proximity and FRET

12.3.3.1. FRET as a Sensor

In many biochemical problems, one does not really need to quantify distance; one merely needs to *ascertain* proximity. Hence, a positive FRET has great usefulness as a molecular sensor. If two probes have the spectral overlap described earlier, energy transfer happens if and only if they are in a few nanometer-proximity. In this application, the researcher need not carry out detailed calculations, worry about labeling efficiency, or correct for small environmental effects. A binary high/low FRET signal is enough for high-throughput screening of drugs, proteins, lipids, and nucleic acid molecules for their target site. This method has both *in vivo* and *in vitro* uses.

Heyduk and Heyduk use FRET as a sensor with ''molecular beacons'' to screen and quantify sequence-specific DNA-binding proteins. The term ''beacon'' originally referred to self-quenched (by FRET) oligonucleotides that lose FRET on

A.

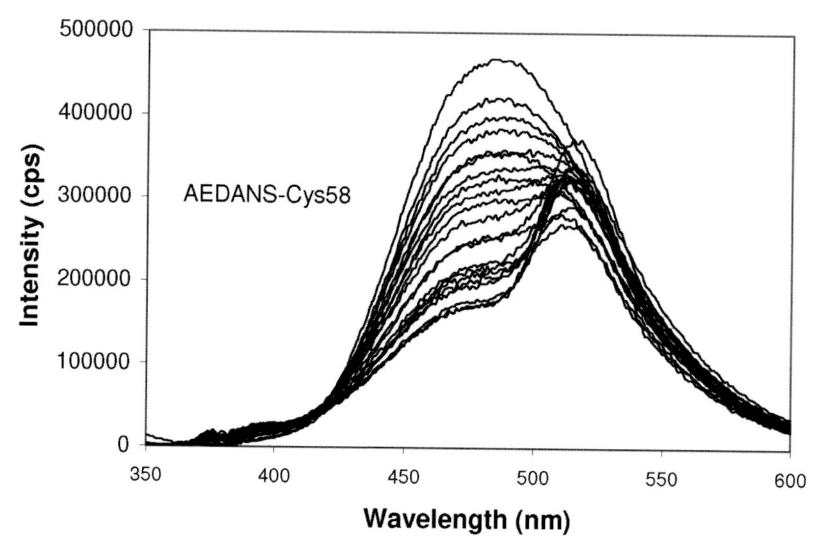

B.

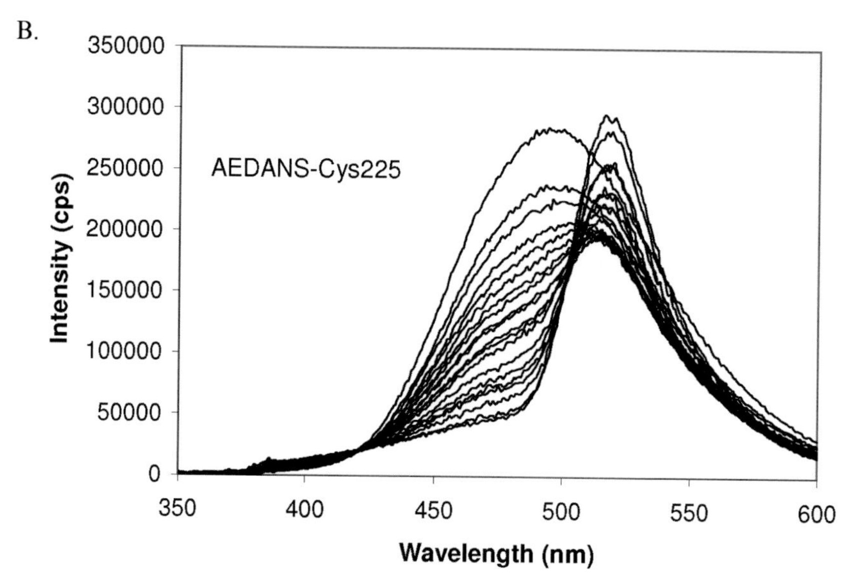

Figure 12.6. (A) Cys58 AEDANS-labeled translin octamer was titrated with increasing concentrations of FITC-labeled ss-DNA. Increasing DNA concentrations resulted in a decrease in AEDANS emission intensity (490 nm). At saturation 60% of the AEDANS fluorescence is quenched by FRET to FITC, resulting in a calculated distance between Cys58 and the DNA end of 45.8 Å. (B) Similar titration, but with Cys225 AEDANS-labeled translin octamer. At saturation 82% of the AEDANS fluorescence is quenched (490 nm) by FRET to FITC (520 nm), resulting in a calculated distance between Cys225 and the DNA end of 38.1 Å.

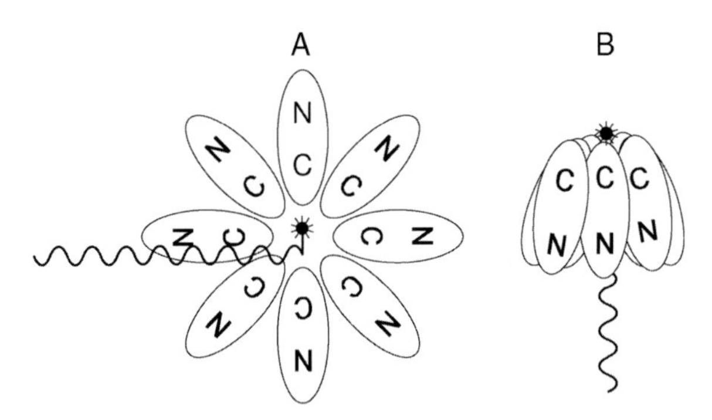

Figure 12.7. One potential model of the translin octamer assembly consistent with the FRET data. Figure A shows a ''top down'' version of the model in Figure B. The model illustrates that the carboxyl termini of the octamer are closer to the FITC-labeled ss-DNA than the amino terminal ends.

unfolding with hybridization. However, the term can mean any binding-dependent fluorescence change, whether employing FRET or direct quenching of a fluorophore. They designed their particular ''molecular beacons'' to detect a DNA-binding protein. Two beacons are made such that each contains half of a specific DNA-binding protein site. Once the protein of interest is added, the two sites come together and produce FRET. In the design of these beacons, one can incorporate the donor and the acceptor molecule external to or within the protein-binding site. In the presence of the protein of interest, its high affinity for the complete specific sequence drives the annealing of the two DNA fragments (Heyduk and Heyduk, 2002, 2005).

More generally, dequenching of fluorophores upon ''beacon'' elongation/hybridization is widely used (Tyagi and Kramer, 1996; Tyagi *et al.*, 1998; Bonnet *et al.*, 1999; Abravaya *et al.*, 2003). A common example of this behavior is the use of molecular beacons for the detection of specific nucleic acid sequences (Tyagi and Kramer, 1996). Beacons designed for this purpose contain a sequence complementary to the intended target embedded within the loop of a hairpin structure. In solution, the molecular beacon remains folded in a relatively nonfluorescent hairpin conformation, which juxtaposes a fluorophore and quencher linked to opposite ends of the linear probe sequence. Quenching of the fluorophore is normally not due to FRET, but to direct quenching by molecules such as 4-((4-(dimethylamino)phenyl)azo)benzoic acid (DABCYL) and the ''black hole quenchers.'' The use of excitonic coupling to provide quenching more efficient than FRET is discussed in Packard *et al.* (1997). The beacon will unfold and hybridize only to a single-stranded target, which includes a sequence fully complementary to that of the probe loop. On hybridization, the beacon unfolds to form a probe: target duplex wherein the fluorophore and quencher are separated. This application can be extended to identify the presence of single base changes [single nucleotide polymorphisms (SNPs)] within the target nucleic acid (Marras *et al.*,

2003). If an SNP is present, the complementary sequence within the loop of the beacon cannot fully hybridize to its intended target and some of the beacon will remain in a folded, quenched conformation. Design of the molecular beacon must take into account the GC content and length of the stem sequence as well as the temperature at which the assay is to be performed.

Since the indicating fluorophores are usually at low concentrations in these assays, the separation of donor–acceptor by cleavage/diffusion away into dilute solution is more effective than simple elongation. Hence, cleavage-dependent fluorescence release technologies such as Taqman and CataCleave (catalytically cleavable) have become available. The former is irrevocably tied to amplification (Heid *et al.*, 1996) whereas the latter is uncoupled from amplification and can be used with or without PCR (Lee *et al.*, 1994, 1995; Lee and Han, 1997; Harvey *et al.*, 2004).

CataCleave probes are catalytically cleavable fluorescence probes that can be used for real-time detection of specific DNA sequences (including targets harboring SNPs, insertions, or deletions). The probes consist of a chimeric DNA–RNA–DNA sequence where one DNA arm is labeled with a fluorescence donor and the other is labeled with a quencher. The RNA portion of the probe is normally four nucleotides long. Fluorescent donor intensity of the intact probe is heavily suppressed due to FRET. On hybridization of the probe to its complementary sequence, RNase H can cleave the RNA-containing portion of the probe, but does not cleave the target. After cleavage, the two halves dissociate from their target and diffuse away from each other, resulting in the enhancement of donor fluorescence. Figure 12.8A demonstrates the CataCleave process.

Target DNA detection with a CataCleave probe is a cyclic process, involving multiple rounds of probe-binding, RNase H-mediated cleavage, and probe fragment dissociation from the same target locus. Since the cleaved probes dissociate at the reaction temperature, the reaction can be performed isothermally or in conjunction with nucleic acid amplification procedures such as polymerase chain reaction (PCR), rolling circle amplification (RCA), and nucleic acid sequence based amplification (NASBA). Cleavage of multiple probes catalyzed by a single target generates an amplified signal that increases detection sensitivity. The initial velocity of probe cleavage is linear and varies in a concentration-dependent manner for both single- and double-stranded DNA targets. Experimental initial velocity data can be compared with a standard curve constructed using known concentrations of target to determine the quantity of target present in the sample. In addition, the probes can be used to detect mutations in the target DNA sequence because efficient cleavage of the probe requires a fully matched target sequence. Interestingly, mismatches between the RNA portion of the probe and its complementary sequence result in a significant reduction in the cleavage rate. This single base resolution is demonstrated by the results presented in Figure 12.8B. The figure shows that a CataCleave probe hybridized to a fully complementary target will be efficiently cleaved by RNase H, whereas the same probe hybridized to a target containing a single base mismatch shows minimal cleavage (Harvey *et al.*, 2004).

A.

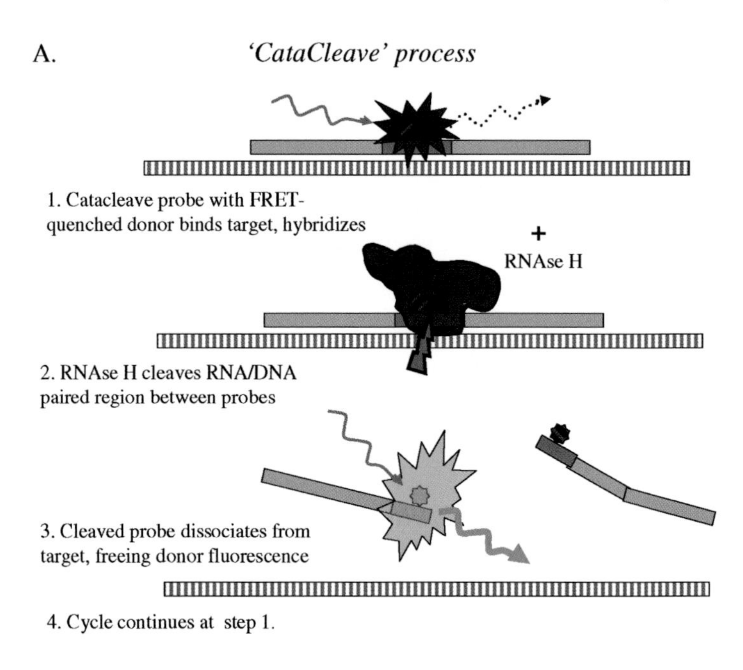

B.

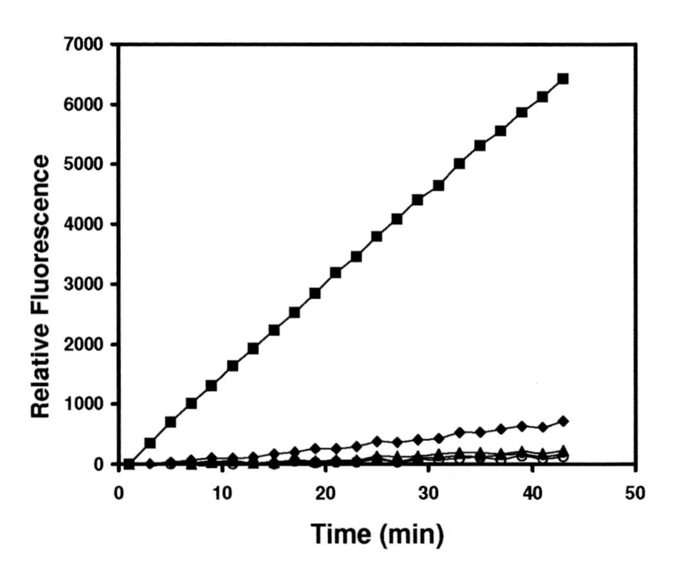

Figure 12.8. (A) The CataCleave (catalytically cleavable) process. (B) A 24-mer CataCleave probe with the RNA sequence 5′-GAGA-3′ was designed and tested with single-stranded 24-mer DNA templates including internal sequences that were either fully complementary (3′-CTCT-5′ ■), or contained single base mismatches (3′-CTTT-5′ ◆, 3′-CTCC-5′ ●, 3′-CGCT-5′ ▲, 3′-ATCT-5′ ○) to the probe RNA moiety.

Note: Figure 12.8B is reprinted from Harvey *et al.* (2004), Figure 3, with permission from Elsevier).

12.3.4. Some Newer Developments

12.3.4.1. LRET

A recent development in FRET application has been the resurgence of LRET (luminescence RET) where the donor is a long-lived luminophore that does not emit from its singlet. Usually, this donor is a transition/rare earth metal ion chelated by extrinsic organic groups and covalently bound to the macromolecule, although Tb^{3+} can occupy native Ca^{2+}-binding sites as well. The theoretical concepts are analogous; the R_0 distances are similar to ordinary FRET. The main advantages of LRET are:

1. Observation of donor quenching in a time-resolved manner can be done with relatively inexpensive instrumentation.
2. Background/autofluorescence is generally much weaker in the μs–ms regime where measurements are carried out, so LRET allows for simpler time-gated detection and the consequent suppression of nonluminophore (short-lived) backgrounds.
3. κ^2 is less problematic since the donor is symmetric.

Together these features make precise distance measurement possible; an example of its most recent use (for small distance changes) is found in Posson *et al.* (2005).

12.3.4.2. FRET to Metal (Au) Nanoparticles (''Golden Ruler'')

As mentioned earlier, the overlap integral is an important determinant of the half-transfer distance, R_0. Since organic (especially aromatic) fluorophores have $\varepsilon < 80{,}000$ ($cm^{-1}M^{-1}$) in the visible, R_0 is constrained (for reasonable Stokes' shift matching) to $\sim<60$ Å, and distances out to ~80 Å are within routine reach of the FRET calculation (Eq. (12.3) earlier).

In the mid 1990s, references to an article by E. Heilweil revealed that colloidal gold (ca. 30–50 Å diameter) had interesting nonlinear characteristics and very high molar extinction ($\varepsilon > 10^{6-7}$) (Heilweil, 1985). Since molar extinction directly multiplies the overlap integral, Eq. (12.6) tells us that R_0 (Å) will scale up (compared with excellent fluorophores) by $(10^6/8 \times 10^4)^{1/6} \sim 1.5$ or $(10^6/10^4)^{1/6} > 2$ when compared with typical probes. Thus, we anticipated that FRET to gold colloids would reach 50–100% further than contemporary FRET pairs. This was confirmed by attaching fluorescein-labeled dsDNA oligonucleotides (30 bp) with biotinylated complementary strands to streptavidin-coated gold, generating a model with an expected dye-gold distance of $>\sim100$ Å. The steady-state quenching we first observed was only about 1/3, a disappointment until this was found to be a trivial artifact: $\sim2/3$ of the streptavidin was free, generating a large unquenched fraction. Time-resolved [via time correlated single photon counting (TCSPC)] measurements of fluorescein lifetime in the same system demonstrated that the gold conjugates were, in fact, >98% quenched, in agreement with $R_0 > 150$ Å. This increased range led us to select the title ''golden ruler'' (Walczak *et al.*, 1997). We have confirmed the long reach of acceptor gold with thiolated-DNA (with more reliable affinity) more recently (Fernandez, C., Wojtuszewski, K., and Knutson, J. R., unpublished observations). Other groups have more fully taken note of (and exploited) colloidal gold as an FRET acceptor (Dubertret *et al.*, 2001; Yun *et al.*, 2005).

The former, in particular, explores reconciling the measured transfer with mechanisms beyond simple dipole–dipole. Since the surface plasmon resonance band of nanoparticles is now tunable in nanoshells (Averitt *et al.*, 1999a, b) FRET to metal should become a more popular tool in biochemistry—for many *different* fluorophores emitting at different wavelengths, even in the near infrared.

12.3.4.3. Two-Step FRET

Watrob *et al.* (2003) have also taken FRET to a new level by building a theoretical and practical framework for ternary systems: a terminal acceptor is added, so two acceptors are present and sequential FRET can occur. The terminal acceptor can be excited directly and through FRET from either the original donor or the intermediate acceptor. They derive both time-resolved and steady-state expressions for the efficiency of each of the transfer steps (and the consequent relay efficiency).

Since the Barkley lab study, two-step FRET has emerged in *in vivo* studies of protein–protein interactions. Galperin *et al.* (2004) have expressed proteins with monomeric red fluorescent protein 1 (mRFP) as an FRET acceptor paired with cyan fluorescent protein (CFP) or yellow fluorescent protein (YFP). These three donor–acceptor pairs ''3-FRET'' were used microscopically to observe the molecular machinery in single endosomal compartments.

Theoretically, the two-step FRET system is very similar to the XFRET equations earlier, except in XFRET, no direct excitation of excimer species occurs, whereas the ''intermediate acceptor'' can be directly excited. In both cases, overlap of donor with terminal acceptor and consequent transfer must be considered.

This technique, like XFRET, can be used to ascertain the presence of functional *ternary* complexes. It can, however, also extract geometric information about the triangle formed by the donor and both acceptors. Spectral characteristics of the three probes can be selected to ''turn off or on'' donor to final acceptor transfer; meanwhile, the relay steps can be maintained. This technique has implications for macromolecular (particularly DNA) bending. It also provides the means to extend the reach of FRET by demanding sequential steps—a ''FRET extension ladder.'' While they expanded the equations for only two steps, extension to more steps is straightforward, although practical spectral widths of organic fluorophores probably set an upper limit of four heterogeneous steps. Their method has already found use in other biological systems (Clamme and Deniz, 2005; Jager *et al.*, 2005; Majumdar *et al.*, 2005).

12.3.4.4. Alternating-Laser Excitation of Single Molecules

FRET has already gained popularity in single-molecule (small ensemble) distance measurement. When excitation spectra of both donor and acceptor are well known, the single-molecule fluorescence spectroscopy (SMFS) (Weiss, 1999; Michalet *et al.*, 2003) technique that uses alternating-laser excitation (ALEX) of single molecules is well suited to determine dynamic distance measurements and stoichiometry (Kapanidis *et al.*, 2005). Using ALEX helps eliminate the need for separate donor and acceptor control experiments. In their method, they alternate rapidly between two excitation wavelengths, the predominantly donor and acceptor excitation wavelengths, to obtain signals for each donor–acceptor pair. Electrooptical modulation was used to interleave pulses for alternating-laser excitation. Four emission rates result from their signature profile corresponding to f_{Dexc}^{Dem}, f_{Aexc}^{Dem}, f_{Dexc}^{Aem}, f_{Aexc}^{Aem}. Two fluorescence ratios obtained from the emission signatures are FRET efficiency E and distance-independent ratio S, which reports on donor–acceptor relative stoichiometry. It seems there are many advantages to using this ALEX for spFRET (single pair FRET) analysis; in particular, incompletely labeled FRET pairs are discriminated. For a detailed review refer to Kanpanidis *et al.* (2005).

Another example of spFRET is shown in the work of Deniz *et al.* (1999). As discussed in chapter 1, single-molecule fluorescence permits the detection and characterization of individual species within a distribution, thus recovering the information lost in the ensemble average used in conventional techniques (Johnson *et al.*, 2005). Deniz *et al.* were able to separate subpopulations of a restriction endonuclease cleavage reaction and hairpin unfolding by observing ratiometric spFRET. From these variations, they obtained distributions of energy transfer efficiency of freely diffusing single molecules without complications from surface interactions typical in single-molecule experiments. They point out this may be used to study nonequilibrium conditions and hence characterize reaction intermediates (Deniz *et al.*, 1999). Single-pair FRET is a rapidly expanding field; see references in Deniz *et al.* (1999), Hohng and Ha (2005), and Johnson *et al.* (2005).

Avoiding traditional FRET controls (those that require preparation of matched, purely labeled, and unlabeled systems) is a key advantage, since the FRET experiment can now be carried out on unique biological samples (difficult to match) such as cells or even tissues. Global analysis of ensemble FRET with incomplete labeling (Knutson *et al.*, 1992) addresses some of the same problems, but it requires external manipulation of the stoichiometry; in contrast, ALEX uses the *natural* variation of stoichiometry already present in the small ensemble to achieve separation of variables. ALEX and the related pulsed interleaved excitation (PIE) methods (Lamb *et al.*, 2005; Muller *et al.*, 2005) will likely become widely used.

12.4. CONCLUSION

As we warned at the beginning, this has been, of necessity, a truncated and biased review focused on some of our favorite (especially new) topics in fluorescence detection and proximity. We have neglected a wide variety of topics; as a few examples, in FRET, we slighted homotransfer and complex transfer of excitons (such as found in antennary photosynthetic complexes), multipole field effects, effective refractive index, mobile/diffusion enhanced FRET, aggregation within membranes, and so on. For excimers, we have neglected the study of domains in bilayers, both extrinsic and intrinsic excimers and exciplexes in nucleic acid structures, etc. We are, however, confident that the other reviews cited (and/or current internet searching) can guide the reader to remedy these deficits.

Disclaimer: The US Government makes no endorsement (implied or inferred) of any commercial products mentioned herein. The views expressed in the chapter do not necessarily represent the views of the NIH, DHHS, nor the United States.

ACKNOWLEDGMENTS

This research was supported (in part) by the Intramural Research Program of the NIH, NHLBI. We also thank (for advice) C. Royer, M. Hawkins, I. Mukerji, L. Brand, L. Davenport, M. Ameloot, M. Barkley, and R. Dale.

REFERENCES

Abravaya, K., Huff, J., Marshall, R., Merchant, B., Mullen, C., Schneider, G., and Robinson, J. (2003). Molecular beacons as diagnostic tools: technology and applications. *Clin Chem Lab Med* 41(4): 468–474.

Averitt, R., Westcott, S., and Halas, N. (1999a). Ultrafast optical properties of gold nanoshells. *J Opt Soc Am B* 16:1814–1823.

Averitt, R., Westcott, S., and Halas, N. (1999b). Linear optical properties of gold nanoshells. *J Opt Soc Am B* 16:1824–1832.

Beechem, J. M. (1989). Simultaneous determination of intramolecular distance distributions and conformational dynamics by global analysis of energy-transfer measurements. *Biophys J* 55(6): 1225–1236.

Beechem, J. M., Ameloot, M., and Brand, L. (1985). Global analysis of fluorescence decay surfaces-excited-state reactions. *Chem Phys Lett* 120(4–5):466–472.

Beechem, J. M. and Brand, L. (1985). Time-resolved fluorescence of proteins. *Annu Rev Biochem* 54(1):43–71.

Beechem, J. M., Gratton, E., Ameloot, M., Knutson, J. R., and Brand, L. 1991, The global analysis of fluorescence intensity and anisotropy decay data: second-generation theory and programs, in *Topics in Fluorescence Spectroscopy*, vol 2 J. R. Lakowicz, (ed.), Plenum Press, New York, pp. 241–305.

Betcher-Lange, S. L. and Lehrer, S. S. (1978). Pyrene excimer fluorescence in rabbit skeletal alphaalphatropomyosin labeled with N-(1-pyrene)maleimide. A probe of sulfhydryl proximity and local chain separation. *J Biol Chem* 253(11):3757–3760.

Birmachu, W., Nisswandt, F. L., and Thomas, D. D. (1989). Conformational transitions in the calcium adenosine-triphosphatase studied by time-resolved fluorescence resonance energy-transfer. *Biochemistry* 28(9):3940–3947.

Boens, N., Szubiakowski, J., Novikov, E., and Ameloot, M. (2000). Testing the identifiability of a model for reversible intermolecular two-state excited-state processes. *J Chem Phys* 112(19):8260–8266.

Bonnet, G., Tyagi, S., Libchaber, A., and Kramer, F. R. (1999). Thermodynamic basis of the enhanced specificity of structured DNA probes. *Proc Natl Acad Sci USA* 96(11):6171–6176.

Brand, L. and Witholt, B. (1967). Fluorescence measurements. *Methods Enzymol* XI (Enzyme structure, [Part A]):776–856.

Calladine, C. R. and Drew, H. R. (1986). Principles of sequence-dependent flexure of DNA. *J Mol Biol* 192(4):907–918.

Cantor, C. R. and Schimmel, P. R. (1980). Biophysical Chemistry: Part II: Techniques for the Study of Biological Structure and Function (Their Biophysical Chemistry; PT. 2), W H Freeman and Co.

Clamme, J. P. and Deniz, A. A. (2005). Three-color single-molecule fluorescence resonance energy transfer. *Chemphyschem* 6(1):74–77.

Clegg, R. M. (1992). Fluorescence resonance energy transfer and nucleic acids. *Methods Enzymol* 211:353–388.

Clegg, R. M. (1995). Fluorescence resonance energy transfer. *Curr Opin Biotechnol* 6(1):103–110.

Clegg, R. M., Murchie, A. I., and Lilley, D. M. (1993). The four-way DNA junction: a fluorescence resonance energy transfer study. *Braz J Med Biol Res* 26(4):405–416.

Clegg, R. M., Murchie, A. I., Zechel, A., Carlberg, C., Diekmann, S., and Lilley, D. M. (1992). Fluorescence resonance energy transfer analysis of the structure of the four-way DNA junction. *Biochemistry* 31(20):4846–4856.

Conrad, R. H. and Brand, L. (1968). Intramolecular transfer of excitation from tryptophan to 1-dimethylaminonaphthalene-5-sulfonamide in a series of model compounds. *Biochemistry* 7(2):777–787.

Dale, R. E. and Eisinger, J. (1974). Intramolecular distances determined by energy transfer. Dependence on orientational freedom of donor and acceptor. *Biopolymers* 13(8):1573–1605.

Davenport, L., Knutson, J. R., and Brand, L. (1986a). Excited-state proton transfer of equilenin and dihydroequilenin: interaction with bilayer vesicles. *Biochemistry* 25(5):1186–1195.

Davenport, L., Knutson, J. R., and Brand, L. (1986b). Studies of membrane heterogeneity using fluorescence associative techniques. *Faraday Discuss* (81):81–94.

Deniz, A. A., Dahan, M., Grunwell, J. R., Ha, T., Faulhaber, A. E., Chemla, D. S., Weiss, S., and Schultz, P. G. (1999). Single-pair fluorescence resonance energy transfer on freely diffusing molecules: observation of Förster distance dependence and subpopulations. *Proc Natl Acad Sci USA* 96(7):3670–3675.

Dubertret, B., Calame, M., and Libchaber, A. J. (2001). Single-mismatch detection using gold-quenched fluorescent oligonucleotides. *Nat Biotech* 19(4):365–370.

Edelhoch, H., Brand, L., and Wilchek, M. (1967). Fluorescence studies with tryptophyl peptides. *Biochemistry* 6(2):547–559.

Eis, P. S. and Lakowicz, J. R. (1993). Time-resolved energy transfer measurements of donor-acceptor distance distributions and intramolecular flexibility of a CCHH zinc finger peptide. *Biochemistry* 32(31):7981–7993.

Eis, P. S. and Millar, D. P. (1993). Conformational distributions of a four-way DNA junction revealed by time-resolved fluorescence resonance energy transfer. *Biochemistry* 32(50):13852–13860.

Förster, Th. (1948). Intermolecular energy migration and fluorescence. *Ann Phys* 2:55–75.

Förster, Th. and Kasper, K. (1955). Ein Konzentrationsumschlag der Fluoreszenz des Pyrens. *Zeitschrift fur Elektrochemie* 59(10):976–980.

Gafni, A., Modlin, R. L., and Brand, L. (1976). Nanosecond fluorescence decay studies of 2-hydroxy-1-naphthalene acetic acid-excited-state proton-transfer. *J Phys Chem* 80(8):898–904.

Galperin, E., Verkhusha, V. V., and Sorkin, A. (2004). Three-chromophore FRET microscopy to analyze multiprotein interactions in living cells. *Nat Methods* 1(3):209–217.

Godfrey, K. (1983). *Compartmental Models and Their Application*, Academic Press Inc., New York.

Graceffa, P. and Lehrer, S. S. (1980). The excimer fluorescence of pyrene-labeled tropomyosin. A probe of conformational dynamics. *J Biol Chem* 255(23):11296–11300.

Haas, E., Katchalski-Katzir, E., and Steinberg, I. Z. (1978). Effect of the orientation of donor and acceptor on the probability of energy transfer involving electronic transitions of mixed polarization. *Biochemistry* 17(23):5064–5070.

Haas, E. and Steinberg, I. Z. (1984). Intramolecular dynamics of chain molecules monitored by fluctuations in efficiency of excitation energy transfer. A theoretical study. *Biophys J* 46(4): 429–437.

Hagerman, P. J. (1986). Sequence-directed curvature of DNA. *Nature* 321(6068):449–450.

Han, M. K., Knutson, J. R., Roseman, S., and Brand, L. (1990). Sugar transport by the bacterial phosphotransferase system. Fluorescence studies of subunit interactions of enzyme I. *J Biol Chem* 265(4):1996–2003.

Han, M. K., Lin, P., Paek, D., Harvey, J. J., Fuior, E., and Knutson, J. R. (2002). Fluorescence studies of pyrene maleimide-labeled translin: excimer fluorescence indicates subunits associate in a tail-to-tail configuration to form octamer. *Biochemistry* 41(10):3468–3476.

Haran, T. E. and Crothers, D. M. (1989). Cooperativity in A-tract structure and bending properties of composite TnAn blocks. *Biochemistry* 28(7):2763–2767.

Harvey, J. J., Lee, S. P., Chan, E. K., Kim, J. H., Hwang, E. S., Cha, C. Y., Knutson, J. R., and Han, M. K. (2004). Characterization and applications of CataCleave probe in real-time detection assays. *Anal Biochem* 333(2):246–255.

Heid, C. A., Stevens, J., Livak, K. J., and Williams, P. M. (1996). Real time quantitative PCR. *Genome Res* 6(10):986–994.

Heilweil, E. J. (1985). Nonlinear spectroscopy and picosecond transient grating study of colloidal gold. *J Chem Phys* 82(11):4762–4770.

Heyduk, E. and Heyduk, T. (2005). Nucleic acid-based fluorescence sensors for detecting proteins. *Anal Chem* 77(4):1147–1156.

Heyduk, T. and Heyduk, E. (2002). Molecular beacons for detecting DNA binding proteins. *Nat Biotechnol* 20(2):171–176.

Hohng, S. and Ha, T. (2005). Single-molecule quantum-dot fluorescence resonance energy transfer. *Chemphyschem* 6(5):956–960.

Ishii, Y. and Lehrer, S. S. (1985). Fluorescence studies of the conformation of pyrene-labeled tropomyosin: effects of F-actin and myosin subfragment 1. *Biochemistry* 24(23):6631–6638.

Ishii, Y. and Lehrer, S. S. (1987). Fluorescence probe studies of the state of tropomyosin in reconstituted muscle thin filaments. *Biochemistry* 26(16):4922–4925.

Ishii, Y. and Lehrer, S. S. (1990). Excimer fluorescence of pyrenyliodoacetamide-labeled tropomyosin: a probe of the state of tropomyosin in reconstituted muscle thin filaments. *Biochemistry* 29(5): 1160–1166.

Ishii, Y. and Lehrer, S. S. (1991). Two-site attachment of troponin to pyrene-labeled tropomyosin. *J Biol Chem* 266(11):6894–6903.

Jager, M., Michalet, X., and Weiss, S. (2005). Protein-protein interactions as a tool for site-specific labeling of proteins. *Protein Sci* 14(8):2059–2068.

Johnson, C. K., Osborn, K. D., Allen, M. W., and Slaughter, B. D. (2005). Single-molecule fluorescence spectroscopy: new probes of protein function and dynamics. *Physiology* (*Bethesda*) 20:10–14.

Jona, I., Matko, J., and Martonosi, A. (1990). Structural dynamics of the $Ca2^{+}$-atpase of sarcoplasmic-reticulum—temperature profiles of fluorescence polarization and intramolecular energy-transfer. *Biochim Biophys Acta* 1028(2):183–199.

Kapanidis, A. N., Laurence, T. A., Lee, N. K., Margeat, E., Kong, X., and Weiss, S. (2005). Alternating-laser excitation of single molecules. *Acc Chem Res* 38(7):523–533.

Knutson, J. R., Chen, R. F., Porter, D. K., Hensley, P., Han, M. K., Kim, S. J., Wilson, S. H., Clague, M., and Williamson, C. K. (1992). Fluorescence quenching in proteins: some applications to protein-DNA and protein-lipid interactions. in *Proceedings of SPIE*, 1640: edn., J. Lakowicz, (ed.), pp. 102–117.

Knutson, J. R., Walbridge, D. G., and Brand, L. (1982). Decay-associated fluorescence spectra and the heterogeneous emission of alcohol dehydrogenase 37. *Biochemistry* 21(19):4671–4679.

Lakowicz, J. R. (1999). *Principles of Fluorescence Spectroscopy*, 2nd edn. Plenum Publishers, New York.

Lamb, D. C., Muller, B. K., and Brauchle, C. (2005). Enhancing the sensitivity of fluorescence correlation spectroscopy by using time-correlated single photon counting. *Curr Pharm Biotechnol* 6(5):405–414.

Laws, W. R. and Brand, L. (1979). Analysis of 2-State excited-state reactions—fluorescence decay of 2-Naphthol. *J Phys Chem* 83(7):795–802.

Lee, S. P., Censullo, M. L., Kim, H. G., Knutson, J. R., and Han, M. K. (1995). Characterization of endonucleolytic activity of HIV-1 integrase using a fluorogenic substrate. *Anal Biochem* 227(2):295–301.

Lee, S. P., Fuior, E., Lewis, M. S., and Han, M. K. (2001). Analytical ultracentrifugation studies of translin: analysis of protein–DNA interactions using a single-stranded fluorogenic oligonucleotide. *Biochemistry* 40 (46):14081–14088.

Lee, S. P. and Han, M. K. (1997). Fluorescence assays for DNA cleavage. *Methods Enzymol* 278: 343–363.

Lee, S. P., Porter, D., Chirikjian, J. G., Knutson, J. R., and Han, M. K. (1994). A fluorometric assay for DNA cleavage reactions characterized with BamHI restriction endonuclease. *Anal Biochem* 220(2):377–383.

Lehrer, S. S. (1994). The regulatory switch of the muscle thin filament: $Ca2^+$ or myosin heads? *J Muscle Res Cell Motil* 15(3):232–236.

Lehrer, S. S. (1995). Pyrene excimer fluorescence as a probe of protein conformational change. *Subcell Biochem* 24:115–132.

Lehrer, S. S. (1997). Intramolecular pyrene excimer fluorescence: a probe of proximity and protein conformational change. *Methods Enzymol* 278:286–295.

Levene, S. D., Wu, H. M., and Crothers, D. M. (1986). Bending and flexibility of kinetoplast DNA. *Biochemistry* 25(14):3988–3995.

Liu, W., Chen, Y., Watrob, H., Bartlett, S. G., Jen-Jacobson, L., and Barkley, M. D. (1998). N-termini of EcoRI restriction endonuclease dimer are in close proximity on the protein surface. *Biochemistry* 37(44):15457–15465.

Majumdar, Z. K., Hickerson, R., Noller, H. F., and Clegg, R. M. (2005). Measurements of internal distance changes of the 30S ribosome using FRET with multiple donor-acceptor pairs: quantitative spectroscopic methods. *J Mol Biol* 351(5):1123–1145.

Marras, S. A., Kramer, F. R., and Tyagi, S. (2003). Genotyping SNPs with molecular beacons. *Methods Mol Biol* 212:111–128.

McKinney, S. A., Tan, E., Wilson, T. J., Nahas, M. K., Declais, A. C., Clegg, R. M., Lilley, D. M., and Ha, T. (2004). Single-molecule studies of DNA and RNA four-way junctions. *Biochem Soc Trans* 32(Pt 1):41–45.

Michalet, X., Kapanidis, A. N., Laurence, T., Pinaud, F., Doose, S., Pflughoefft, M., and Weiss, S. (2003). The power and prospects of fluorescence microscopies and spectroscopies. *Annu Rev Biophys Biomol Struct* 32:161–182.

Muller, B. K., Zaychikov, E., Brauchle, C., and Lamb, D. C. (2005). Pulsed interleaved excitation. *Biophys J* 89(5):3508–3522.

Murchie, A. I., Clegg, R. M., von, K. E., Duckett, D. R., Diekmann, S., and Lilley, D. M. (1989). Fluorescence energy transfer shows that the four-way DNA junction is a right-handed cross of antiparallel molecules. *Nature* 341(6244):763–766.

Packard, B. Z., Toptygin, D. D., Komoriya, A., and Brand, L. (1997). Design of profluorescent protease substrates guided by exciton theory. *Methods Enzymol* 278:15–23.

Panda, D. and Bhattacharyya, B. (1992). Excimer fluorescence of pyrene-maleimide-labeled tubulin. *Eur J Biochem* 204(2):783–787.

Park, Y. W. and Breslauer, K. J. (1991). A spectroscopic and calorimetric study of the melting behaviors of a ''bent'' and a ''normal'' DNA duplex: [d(GA4T4C)]2 versus [d(GT4A4C)]2. *Proc Natl Acad Sci USA* 88(4):1551–1555.

Parkhurst, L. J. (2004). Distance parameters derived from time-resolved Förster resonance energy transfer measurements and their use in structural interpretations of thermodynamic quantities associated with protein-DNA interactions. *Methods Enzymol* 379:235–262.

Parkhurst, L. J., Parkhurst, K. M., Powell, R., Wu, J., and Williams, S. (2001). Time-resolved fluorescence resonance energy transfer studies of DNA bending in double-stranded oligonucleotides and in DNA-protein complexes. *Biopolymers* 61(3):180–200.

Posson, D. J., Ge, P., Miller, C., Bezanilla, F., and Selvin, P. R. (2005). Small vertical movement of a K+channel voltage sensor measured with luminescence energy transfer. *Nature* 436(7052): 848–851.

Selvin, P. R. (1995). Fluorescence resonance energy transfer. *Methods Enzymol* 246:300–334.

Steinberg, I. A., Hass, E., & Katchalski-Katzir, E. (1983). Long-range nonradiative transfer of electronic excitation energy, in *Time-Resolved Fluorescence Spectroscopy in Biochemistry and Biology*, R. B. Cundall and R. E. Dale, (eds), Plenum, New York, pp. 411–450.

Stryer, L. (1978). Fluorescence energy-transfer as a spectroscopic ruler. *Annu Rev Biochem* 47:819–846.

Stryer, L. and Haugland, R. P. (1967). Energy transfer: a spectroscopic ruler. *Proc Natl Acad Sci USA* 58:719–726.

Szubiakowski, J. P., Dale, R. E., Boens, N., and Ameloot, M. (2004). Identifiability analysis of models for reversible intermolecular two-state excited-state processes coupled with species-dependent rotational diffusion monitored by time-resolved fluorescence depolarization. *J Chem Phys* 121(16):7829–7839.

Tyagi, S., Bratu, D. P., and Kramer, F. R. (1998). Multicolor molecular beacons for allele discrimination. *Nat Biotechnol* 16(1):49–53.

Tyagi, S. and Kramer, F. R. (1996). Molecular beacons: probes that fluoresce upon hybridization. *Nat Biotech* 14(3):303–308.

Ulanovsky, L., Bodner, M., Trifonov, E. N., and Choder, M. (1986). Curved DNA: design, synthesis, and circularization. *Proc Natl Acad Sci USA* 83(4):862–866.

van der Meer, B. W. (2002). Kappa-squared: from nuisance to new sense. *Reviews in Mol Biotech* 82(3):181–196.

Walczak, W. J., Xiao, J. M., Kopetz, E. S., Lease, K., Grau, H., Lee, S. P., Han, M. K., and Knutson, J. R. (1997). ''Golden ruler'': very long-range resonance energy transfer to surface plasmon acceptors. *Biophys J* 72(2): Abstract TU367-TU367.

Watrob, H., Liu, W., Chen, Y., Bartlett, S. G., Jen-Jacobson, L., and Barkley, M. D. (2001). Solution conformation of EcoRI restriction endonuclease changes upon binding of cognate DNA and Mg^{2+} cofactor. *Biochemistry* 40(3):683–692.

Watrob, H. M., Pan, C. P., and Barkley, M. D. (2003). Two-step FRET as a structural tool. *J Am Chem Soc* 125(24):7336–7343.

Weiss, S. (1999). Fluorescence spectroscopy of single biomolecules. *Science* 283(5408):1676–1683.

Wojtuszewski, K. and Mukerji, I. (2003). HU binding to bent DNA: a fluorescence resonance energy transfer and anisotropy study. *Biochemistry* 42(10):3096–3104.

Wu, P. and Brand, L. (1994). Resonance energy transfer: methods and applications. *Anal Biochem* 218(1):1–13.

Yang, C. J., Jockusch, S., Vicens, M., Turro, N. J., and Tan, W. (2005). Light-switching excimer probes for rapid protein monitoring in complex biological fluids. *Proc Natl Acad Sci USA* 102(48): 17278–17283.

Yun, C. S., Javier, A., Jennings, T., Fisher, M., Hira, S., Peterson, S., Hopkins, B., Reich, N. O., and Strouse, G. F. (2005). Nanometal surface energy transfer in optical rulers, breaking the FRET barrier. *J Am Chem Soc* 127(9):3115–3119.

13

Steady-State and Time-Resolved Emission Anisotropy

K. Wojtuszewski and J. R. Knutson*

13.1. INTRODUCTION

Reversible associations of macromolecules, by definition, create fragile complexes whose shape and stoichiometry are important to their function; further, any real system is likely to contain multiple species. Although true structural techniques like X-ray crystallography and solution NMR spectroscopy can give atomic-level detail of the consensus forms of tight and rigid complexes, we also desire techniques applicable to dilute solution, dynamic equilibria, and flexible macromolecules. Fluorescence emission anisotropy has proven to be valuable in these areas. Fluorescence emission anisotropy allows us to directly measure Brownian rotations, and the hydrodynamics of those rotations depends on size, shape, and flexibility of the complex.

This chapter is meant to be complementary to existing reviews on fluorescence anisotropy. Its primary focus is to give our personal perspective on steady-state and time-resolved fluorescence anisotropy employed in characterizing macromolecular complexes. For a more detailed explanation on the theory behind these methods, one should refer to the references listed within the text.

13.2. THEORY OF TIME-RESOLVED EMISSION ANISOTROPY

The rotational diffusion of macromolecules can yield abundant information about size, shape, and/or local viscosity. Ideally, one would like a technique that accurately yields the entire duration of the angular correlation function for each of the principle axes of the molecule.

K. WOJTUSZEWSKI and J. R. KNUTSON • Optical Spectroscopy Section, Laboratory of Molecular Biophysics, NHLBI, NIH, Bethesda, MD 20892-1412. *Corresponding author: Tel: +301 496 2557; fax: +301 480 4625; e-mail: jaysan@helix.nih.gov

No currently available biophysical technique yields that much information, although emission anisotropy can provide most of the information required. First, a brief pulse of light creates a population of excited fluorophores whose transition moments are not distributed isotropically—a process called photoselection. In the usual (single photon) case, the exciting light is z-polarized, entering the sample along the x-axis. The photoselected ensemble will be chosen by the fact that the electric field of the excitation, E_z, projects on each absorber like $E = E_z \cos\theta$, where θ is the azimuthal angle. Since excitation will be proportional to E^2, completely horizontal oscillators have no chance of selection, and vertical oscillators are most favored. Figure 13.1 illustrates the photoselection process in the fluorimeter along with vertical and horizontal intensities as a function of time.

Before diffusive motion, electronic rearrangements can modify the initial angular pattern. In particular, if the molecules are randomly oriented and internal conversion (subpicosecond nonradiative transition of a molecule to its lowest

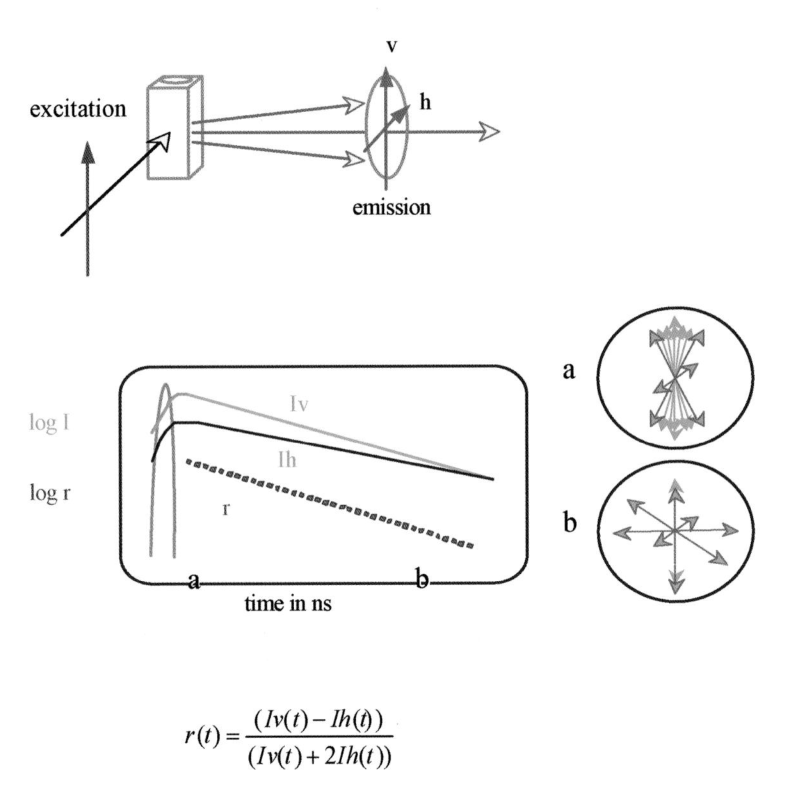

$$r(t) = \frac{(Iv(t) - Ih(t))}{(Iv(t) + 2Ih(t))}$$

Figure 13.1. Fluorescence emission anisotropy is a measure of how well a photoselected ensemble of excited states maintains alignment. An initial vertically polarized excitation pulse selects those probes whose oscillators have significant vertical orientation. (A) After a few nanoseconds of Brownian rotation, the ensemble is partly randomized and the emission process has also depleted the total number. (B) The emission intensities measured through vertical versus horizontal polarizers determine the anisotropy, $r(t)$.

excited state) (Sarkar *et al.*, 1999; Shen and Knutson, 2001a,b) leads to an internal rotation γ of the emitting oscillator in the molecule, then an ensemble with less polar nature will be created. In the case where γ approaches 90°, an equatorial rather than polar distribution will result.

Observation of the emitted light (usually traveling along the y-axis) leads to another electric field projection onto the vertical (z) and horizontal (x) polarizer axes used to analyze intensity. If one defines the anisotropy[1] as

$$r = \frac{(I_z - I_x)}{(I_z + 2I_x)}, \tag{13.1}$$

one can express the projection and internal rotation processes succinctly as

$$r_o = \frac{2}{5}\left(\frac{(3\cos^2\gamma - 1)}{2}\right), \tag{13.2}$$

where the subscript ($_o$) refers to anisotropy after internal conversion events, but before rotational diffusion.

Before the ultrafast laser became available, r_o was obtained by ensuring rotation was prevented by, e.g., measuring the anisotropy in rigid glasses (Mantulin and Weber, 1977).

13.2.1. Rotational Diffusion

The anisotropy of an initially photoselected ensemble is lost in solution through random angular diffusion. For example, a spherical macromolecule will lose its polarization according to

$$r(t) = r_o \langle P_2(\cos\psi)\rangle, \tag{13.3}$$

where $\langle\rangle$ represents ensemble averaging, $P_2(x)$ is the Legendre polynomial $(3x^2-1)/2$, and ψ is the angle between the initial and the final emission dipole configurations of the molecule. Following Stokes–Einstein relations for spheres,

$$r(t) = r_o e^{\frac{-t}{\phi}}, \tag{13.4}$$

where ϕ, the rotational correlation time, is

$$\phi = \frac{\eta V}{kT}, \tag{13.5}$$

where η is the viscosity and V is the molecular volume. The linearity of ϕ with volume makes a very convenient method to size globular proteins, which have similar density and hydration characteristics:

[1] The factor of 2 in the denominator comes from the need to divide by the total intensity. Since photons could be emitted with x, y, or z polarization and polarization is perpendicular to travel, I_y is invisible when we collect along the y-axis. Fortunately, the ensemble is symmetric with $I_y = I_x$.

$$\phi_{ns} \approx \frac{MW_{kD}}{2} \text{ (in room temperature buffers).} \tag{13.6}$$

This approximation comes from the work of Ph.Wahl (1983). Figure 13.2 illustrates the molecular weight effect on ϕ and $\log r$. Later, we will use the steady-state version of this equation; first, however we will deal with anisotropic rotations. (Emission anisotropy is defined in Eq. (13.1), and the term defines the anisotropy of the measured ensemble of polarized photons; anisotropic rotation, on the other hand, refers to anisotropy in the shape of the body and the kinetic consequences that result.)

13.2.2. Anisotropic Rotations

Nonspherical molecules will, of course, diffuse in a more complex fashion. The details of rotational motion can be obtained numerically for arbitrary shapes (Youngren and Acrivos, 1975; Allison, 1999). Proteins can (in some cases) be approximated by ensembles of "beads" for reduced computational load (Garcia de la Torre and Bloomfield, 1981).

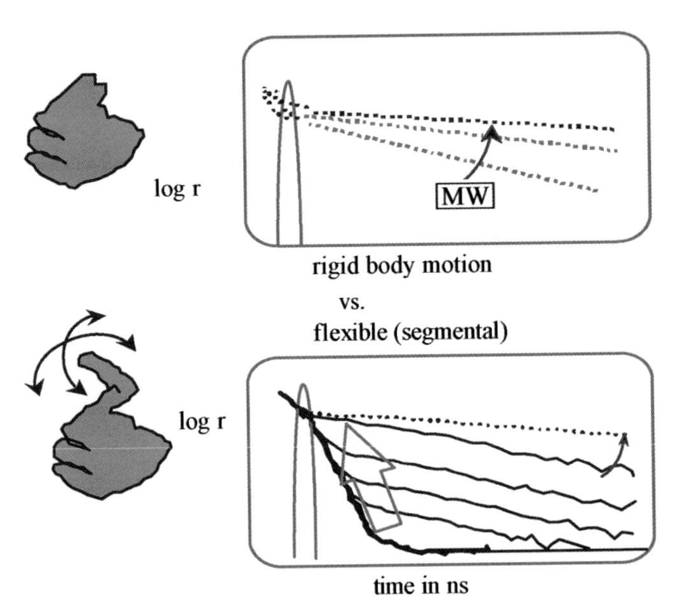

Figure 13.2. For a rigid, nearly spherical macromolecule, emission anisotropy decays (after excitation) as a monoexponential whose e-folding time, ϕ, is proportional to the volume with a rough correspondence in room temperature buffers: ϕ in nsec $\sim$ molecular weight in kD/2. If the fluorophore is on a flexible segment, it is the molecular weight of that segment that determines the first of (at least) two exponentials, the slowest being the bulk rotation. If a binding event leads to segment restriction, both ordering (green arrow) and mass increase (blue arrow) will increase anisotropy; the former by increasing the amplitude of slow (bulk) motion and the latter by slowing the bulk rate.

For emission anisotropy studies, it is often useful to further simplify the shape by approximating the macromolecules as ellipsoidal bodies with analytical formulae, as originally done by Chuang and Eisenthal (1972).

In fact, the even greater simplification of treating the body as an ellipsoid of revolution (Beechem *et al.*, 1986) with five exponential functions is still beyond all but the most exhaustive experimentation, so the collapse of similar (paired) terms into three exponentials is usually sought. The modeled macromolecule has an axial ratio and a unique axis. With the probe's emission and excitation oscillators defined by their azimuthal and mutual angles, the equation becomes

$$r(t) = \sum_{j=1}^{3} \beta_j e^{\frac{-t}{\phi_j}}, \tag{13.7}$$

where $r_o = \Sigma\beta_j$ (Barkley *et al.*, 1981) and $[\phi_1 = 1/(5D_\perp + D_\parallel)$, $\phi_2 = 1/(2D_\perp + 4D_\parallel)$, $\phi_3 = 1/(6D_\perp)]$.

The application of this theory to small molecules is straightforward compared with the use with macromolecules, largely owing to the symmetry inherent in the dye and the improved S/N one obtains when lifetime is comparable with ϕ (see later text). Nevertheless, axial ratios of certain prolate or oblate proteins have been recovered (Small and Isenberg, 1977b) from emission anisotropy decay.

The introduction of global nonlinear least-square (NLLS) analysis into time-resolved anisotropy (Beechem *et al.*, 1986; Knutson *et al.*, 1986) provided new avenues to recover the ϕ_j and thus, the $D_\parallel$ and $D_\perp$ values.

13.2.3. Global Multioscillator Approach

Basically, if one performs multiple experiments with dyes (intrinsic or covalently bonded or simply adsorbed) arranged at different angles with the body, a series of curves bearing the same ϕ_j but differing (generally unknown) β_j is obtained.

The global analysis requires the same ϕ_j parameters to be linked together in the analysis of these sets. More generally, the viscosity η can be included as a known scalar multiplier of ϕ_j.

The overdetermination of the ϕ_j set has led to an ability to recover axial ratios as low as ~3:1, where conventional analysis recovers crude estimates unless the ratio exceeds 6:1 (Beechem *et al.*, 1986).

The global approach is not a panacea, however, as local motion and other sources of heterogeneity can confound the analysis (Brunet *et al.*, 1994; Jameson and Sawyer, 1995). The models that compensate for some of those artifacts have been incorporated into recent code (Beechem *et al.*, 1991).

13.2.4. Other Sources of Multiexponentiality

The previous analysis was based on the assumption that, after internal conversion, the photophysics of the fluorophore was independent of motion and homogeneous throughout the ensemble. Further, the dye oscillator was presumed to be fixed in macromolecular coordinates. We address the latter assumption first.

13.2.5. Local Motion

Reporter groups are often attached to segments of a macromolecule that have independent (though restricted) mobility. Over the years, several models have been developed for the local flexibility. A summary of those models (applied to a protein) is given by Tcherkasskaya *et al.* (2000). In all cases, the approximate form is that of one β multiplying an exponential with ϕ representing the fast local motion and another term for the overall motion. Actually, the general form is

$$r = \exp\left(\frac{-t}{\phi_{\text{slow}}}\right)\left(\beta_2 + \beta_1 \exp\left(\frac{-t}{\phi_{\text{fast}}}\right)\right), \tag{13.8}$$

but the product $\exp(-t/\phi_{\text{slow}})\exp(-t/\phi_{\text{fast}})$ yields a composite $e^{-t/\phi_{sf}}$. Where $\phi_{sf} = \phi_s\phi_f/(\phi_s + \phi_f) \approx \phi_f$, when $\phi_s \gg \phi_f$. Fortunately, segmental mobility times for sidechains and linked dyes on proteins are often on the order of $\phi_f \sim 1$ ns, so the local rate is easily separated from overall motion when the molecular weight exceeds $\sim$6,000. Slower local motions, however, are certainly possible (especially in macromolecules with distinct domains linked by flexible regions) (Kim *et al.*, 1994). Analysis should include a local motion term in all unknown systems.

Local motion is not merely a nuisance, however; in some systems, local flexibility is an intrinsic feature of the independent macromolecule, and changes in that flexibility when bound indicate target-induced folding (Spolar and Record, Jr., 1994). We have used the consequent loss of fast anisotropy decay to observe "freezing" of an acidic activation domain (AAD) of the transcriptional activator VP16 (Shen and Knutson, 2001a,b). Flexibility changes are also observable in macromolecular motors during different steps in their cyclic association and processivity (Kim *et al.*, 1994).

13.2.6. "Associative" Decay Kinetics

Returning to the assumption of motion uncoupled from photophysics, let us consider an example of simple stark heterogeneity. A particular free fluorescent ligand rotates in less than a nanosecond and, owing to solvent quenching, has a short lifetime. When bound, the chromophore is protected and its lifetime (and yield) increases. Further, it rotates slowly, as if part of the macromolecule. Clearly, a mixture of fast and slow rotations will be seen, but the lifetime differences in the two environments add to the kinetic behavior as well.

Heterogeneity of steady-state anisotropy can sometimes be seen in a binding isotherm. Mathematically, the anisotropy function is properly normalized to total intensity (Kim *et al.*, 1994), so it is a linear sum of intensity-weighted contributions

$$r(t) = \frac{I_B(t)r_B(t) + I_F(t)r_F(t)}{I_B(t) + I_F(t)}, \tag{13.9}$$

where B and F refer to bound and free fluorescent ligand populations, respectively. In a similar problem (Ludescher *et al.*, 1987), the equivalent relations $f_B = I_B/(I_B+I_F)$ and $f_F = 1-f_B$ were used to model heterogeneous liquid crystalline phases

$$r(t) = f_B(t)r_B(t) + (1-f_B(t))r_F(t). \tag{13.10}$$

More generally, each component of the motion could have multiple lifetimes and each lifetime could link with multiple rates of motion. Hence, we developed an ‘‘associative decay model’’ (Knutson *et al.*, 1986; Beechem *et al.*, 1991) with an association matrix assigning ownership of lifetimes (τ_i) by correlation times (ϕ_i) and vice versa; each is tied to the ith subpopulation of fluorophores.

$$I_\Omega = \frac{1}{3}\sum_{i=1}^{N}\alpha_i e^{\frac{-t}{\tau_i}}\left(1 + (3\cos^2\Omega-1)\sum_{j=1}^{M}\beta_j e^{\frac{-t}{\phi_j}} M_{ij}\right), \tag{13.11}$$

where I_Ω is the intensity seen through a polarizer oriented at angle Ω from the vertical symmetry axis. Note that, at 54.7° (‘‘magic angle’’) the anisotropy term drops out; alternatively, $(I_{0^\circ} + 2I_{90^\circ})$ yields the same function. In normal (‘‘non-associative’’) modeling, $M_{ij} \equiv 1$ (scalar) and homogeneity are assumed. In associative systems like the binary dye environments described earlier,

$$M_{ij} = \delta_{ij}, \quad M_{ij} = \begin{bmatrix} 1\,0 \\ 0\,1 \end{bmatrix} \tag{13.12}$$

Associative curves have nonintuitive shapes, as shown in Figure 13.3.

The best way to detect associative decay processes, in cases where the curves are not obviously distorted, is manipulation of the system. For example, in the dye system one may alter stoichiometry with mass action, so consequent true changes in α_I would appear in the apparent β values of the analysis. Alternatively, a collisional quencher that does not modify the structure should alter τ_i and consequently the apparent ϕ: $(1/(1/\phi + 1/\tau)$ terms in Eq. (13.15) of an associative decay, whereas true ϕ are invariants. In some cases, the α–β correlation can be extracted from a maximum entropy map of these parameters (Brochon, 1994).

13.2.7. Sum, Difference, Vector, and Global Analyses

The traditional means to analyze I_{0° and I_{90° curves obtained in the laboratory was developed in the days of slow (and low-capacity) computers. The function $S(t) = I_{0^\circ} + 2I_{90^\circ}$, as mentioned before Eq. (13.11), depends only on α_i, τ_i. Hence that function (the ‘‘sum’’) was fit first, and the function $D(t) = I_{0^\circ} - I_{90^\circ}$ (the ‘‘difference’’) was subsequently fit. With α_i, τ_i held constant in the formula for S,

$$D(t) = [S(t)]r(t) = [S(t)]\sum \beta_j e^{\frac{-t}{\phi_j}}. \tag{13.13}$$

Of course, both sum and difference (S & D) are properly convolved with instrument response functions, and weights appropriate to the algebraic construction of S and D

Associative

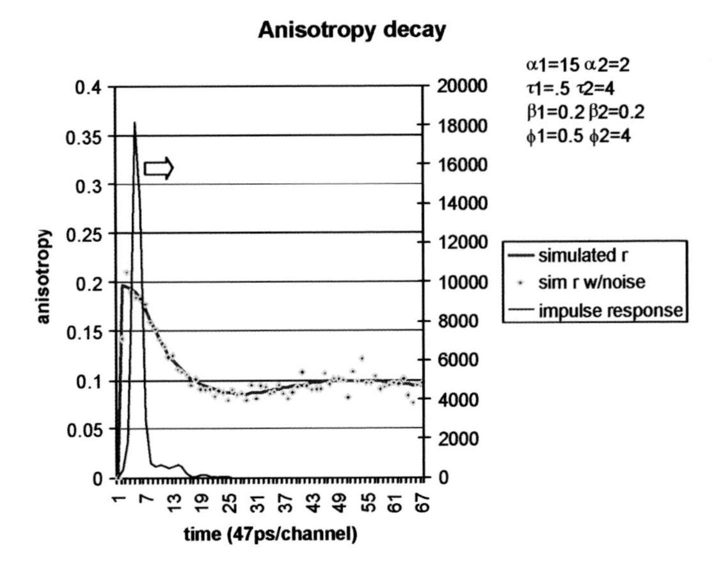

Non-Associative

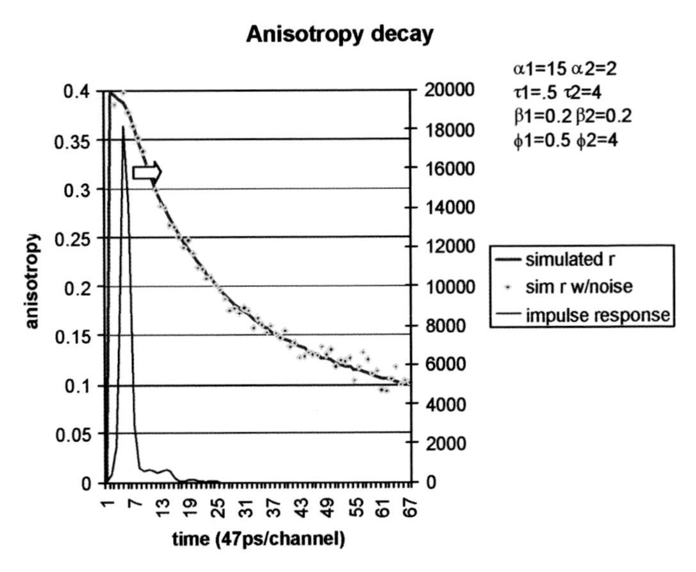

Figure 13.3. Comparison of associative and nonassociative curve shapes.

are used. S & D fitting is still useful and can be more intuitive than other methods, especially since the lifetime and the rotation parameters are segregated.

Vector analysis was first described by Dale and Gilbert (1983). Essentially, it demanded the simultaneous analysis of I_{0° and I_{90° for the $\alpha_i, \tau_i, \beta_j$, and ϕ_j common to both curves. Their insight into the transparency of weighting these data normally was a key element in the development of global NLLS analysis that followed.

Global analysis is simply the analysis of related experiments with parameters that can be held in common (either directly or via functional linkages) (Beechem *et al.*, 1991). When vector analysis was built into the early global anisotropy programs, the opportunity became clear to include the associative matrix M_{ij}. More recent global programs have opened the possibility to use $I_{0^\circ}, I_{55^\circ}$, and I_{90° together; further, a means to search the M_{ij} configuration for appropriate models (sometimes degenerate) has been developed (Feinstein *et al.*, 2003).

In all, analytical programs for anisotropy decay analysis are now widely available (e.g., our DecayFit or Globals Unlimited, LFD) that can address all forms of multiexponentiality and identify (in most cases) the physical origin of those terms. Of course, to achieve this mapping, it may be necessary to carry out multiple related experiments that vary either hydrodynamics (via temperature, viscosity) or lifetime kinetics (concentrations of ligands or quenchers).

13.2.8. Limits on Parameter Recovery

The recovery of hydrodynamic values that eventually yield shape and size depends strongly on experimental design. One of the strengths of time resolution via time correlated single photon counting (TCSPC) is the ability to easily simulate idealized experiments (O'Connor and Phillips, 1984). Photon statistics are often a key determinant of parameter recovery, and they can be simulated with Poisson noise. Hence, the experimenters can ''dry run'' analyses on relevant models. As an example, Ross and colleagues used multiple simulations to test the ability to recover protein sizes (single tumbling exponential with or without local motion) (Ross *et al.*, 1981). They found lifetime was a key limit on this ability, as ϕ/τ ratios much greater than ten were problematic in S & D analysis.

Lifetime certainly defines a time window for TCSPC; decay of ten lifetimes represents a 10^4 reduction in signal. One could imagine single experiment containing many billions of photons, but most collected curves contain $<10^7$. Most TCSPC instruments have intrinsic nonlinearity in counting that is small compared with photon noise until individual channels exceed $\sim 10^5$, and Selinger (Selinger and Hinde, 1983) has shown ideal time width in analysis is well under 10^3 channels. Keeping peak counts $<10^5$ has kept photon statistics valid (total counts $\ll 10^8$ per curve).

It is possible, of course, to refine TCSPC differential nonlinearity (DNL) beyond these limits, either electronically or via software correction (e.g., obtain and normalize the DNL pattern). Alternatively, other analysis methods such as method of moments (Small and Isenberg, 1977a; Solie *et al.*, 1980) can be made insensitive to DNL, but they lack confidence intervals (see later text). For our purposes in this section, we will confine discussions of recovery to ''conventional'' TCSPC experiments, e.g., peaks of $I_0(t) \sim 2 \times 10^4$ photons.

13.2.9. Confidence Limits

The recovery of nonlinear parameters from NLLS analysis is different from linear least-squares (LLS). The error hypersurface, which can be often simply viewed as parabolic in linear problems, takes on tortuous shapes for exponential fitting (Figure 5.2 in Beechem *et al.*, 1991). The ''error matrix'' that yields standard deviations in LLS is very optimistic in NLLS—misleadingly optimistic.

The proper confidence interval for nonlinear parameters can be obtained from ''support plane'' analysis; from the initial solution, purposefully fix each parameter at a series of values, whereas all others may adjust to minimize χ^2, the error metric. The χ^2 versus fixed value plot will exceed confidence setpoints we calculate in accordance with the F-test (Press *et al.*, 1990). More thorough searching along multiple vectors is useful in quantifying covariance (Straume and Johnson, 1992). The F-test depends not only on the number of parameters, but also on the number of data points and the confidence (e.g., 90%) required, as shown in Figure 13.4.

TCSPC analyses in past decades relied on a ''rule of thumb'' that these χ^2 setpoints were at 10% greater than χ^2_R minimum, a result from F-test tables for a handful of parameters and a few hundred data points. This 10% test does not

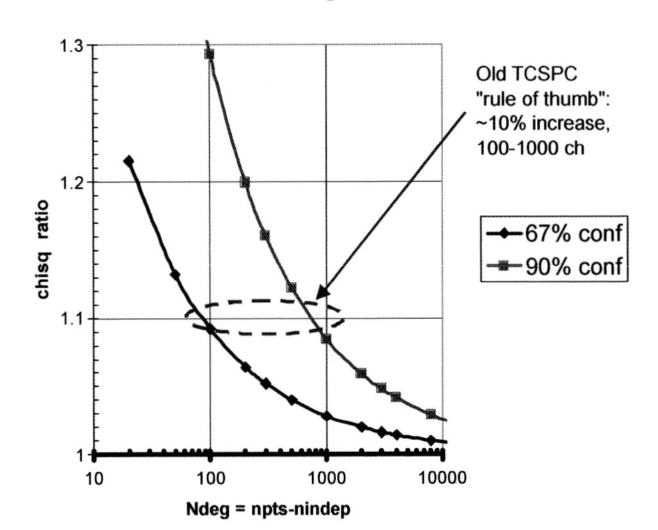

Figure 13.4. χ^2 ratio or F-ratios at 67 and 90% confidence levels as a function of the number of degrees of freedom (number of data – number of free parameters) (Press *et al.*, 1990). A 1.12 F-test ratio at 90% confidence exceeds the range of the χ^2 surface plot in Figure 13.6 later.

translate directly into phase fluorometry data (Knutson, 1992), but also needs to be adjusted for global analyses that may include thousands of data points.

In the simulations we present later for actin dimerization, we focus on difficult cases: $\phi \gg \tau$ and those ϕ are only separated by a factor of 2. The advantages that accrue from the reduced F-test threshold of simultaneous analysis will be more apparent there.

13.3. EXPERIMENTAL

13.3.1. Routine Measurements

13.3.1.1. Binding and Stoichiometry: Steady-State Measurements

Steady-state emission anisotropy is simply the time-averaged response of $r(t)$ over the entire fluorescence decay

$$r_{ss} = \frac{\int I(t)r(t)\mathrm{d}t}{\int I(t)\mathrm{d}t} \tag{13.14}$$

for a single lifetime / single correlation time case, this means

$$r_{ss} = \frac{r_o}{(1 + \tau/\phi)}. \tag{13.15}$$

Steady-state anisotropy measures the competition between rotation and decay; if "the lights go out" before much motion can happen, r is high (near r_o), whereas motion more rapid than emission decay brings r toward zero.

Steady-state anisotropy values have proven to be essential in characterizing macromolecule binding and shape of the molecule during monomer–dimer formation. An increase in anisotropy occurs as a function of ligand or macromolecule concentration, and the experimentalist can fit anisotropy changes to standard binding equations. One example of this is the titration of a protein relative to small duplex DNA labeled with a fluorescent probe. Without any protein bound, the fluorescent probe can freely rotate, and the anisotropy value is low. With increasing amounts of protein bound, the probe's rotational diffusion will decrease, and the anisotropy will increase. Relating size to anisotropy values may also provide insight into the stoichiometry and the number of complexes that are present in the solution (LeTilly and Royer, 1993; Hill and Royer, 1997).

13.3.1.2. Analysis of Binding Parameters Using Fluorescence Anisotropy

The fraction of bound species is determined from fluorescence anisotropy using the relationship

$$f_b = \frac{(r - r_o)}{(r_b - r_o)}, \tag{13.16}$$

where r, r_o, and r_b represents the anisotropy of the titrated sample, free ligand, and bound at saturation, respectively. When analyzing steady-state anisotropy results, it is important to realize that the final population is a mixture of bound and free states. In the bound state, the fluorophore may have an intensity change one should consider in the final calculation of the binding constant. Given below is the fraction of species bound f_b when we compensate for intensity changes of the label

$$f_b = \frac{(r - r_o)}{[(r_b - r_o)R + r - r_o]}, \tag{13.17}$$

where $R = \Phi_b/\Phi_o$, Φ_b is the relative quantum yield of bound ligand at saturation and Φ_o is the relative quantum yield of free ligand (Evett and Isenberg, 1969; Malencik and Anderson, 1988).

Mukerji and coworkers have used this method to determine the stoichiometry and the binding affinity of the nonspecific DNA-binding HU to DNA (Wojtuszewski *et al.*, 2001). HU, a histone-like DNA-binding protein found in bacteria, participates in a number of genomic events as an accessory protein and forms multiple complexes with DNA. The binding interaction with DNA was characterized by using a guanosine fluorescent analog, 3-methyl-8-(2-deoxy-β-D-ribofuranosyl)isoxanthopterin (3MI) (Hawkins *et al.*, 1997), directly incorporated into the DNA. Steady-state anisotropy measurements revealed three HU molecules bind to the 34 bp duplex, whereas two HU molecules bind to the 13 bp duplex (Figure 13.5A). An independent binding site model showed the first binding event for both duplexes are similar ($\sim 1 \times 10^6$ M^{-1}), indicating binding is independent of duplex length using Eq. (13.18) (Figure 13.5B). The equilibrium binding curves were fit assuming identical and noninteracting binding sites (Bujalowski and Lohman, 1987; Eftink, 1997).

$$r = r_o + \frac{nK[L](r_1 - r_o)}{1 + K[L]}. \tag{13.18}$$

This equation modified with Eq. (13.17) takes into account a change in fluorescence intensity for the ligand interacting with the macromolecule.

$$r = \frac{r_o(1 + K[L]) + nK[L](Rr_1 - r_o)}{1 + K[L] + nK[L](R - 1)}, \tag{13.19}$$

where r, r_0, and r_1 represent the anisotropy of the titrated sample, free ligand, and bound to macromolecule, respectively. K is the microscopic binding constant, n represents the number of ligands bound, and $[L]$ corresponds to the concentration of free ligand. Analytical ultracentrifugation studies of HU binding to the unlabeled duplexes confirmed the stoichiometry and binding affinities.

13.3.1.3. Rotational Correlation Time and Protein Molecular Weight

A valuable tool in determining the stoichiometry of heteromeric protein species is using the relationship that the molecular weight (in kD) of a fairly

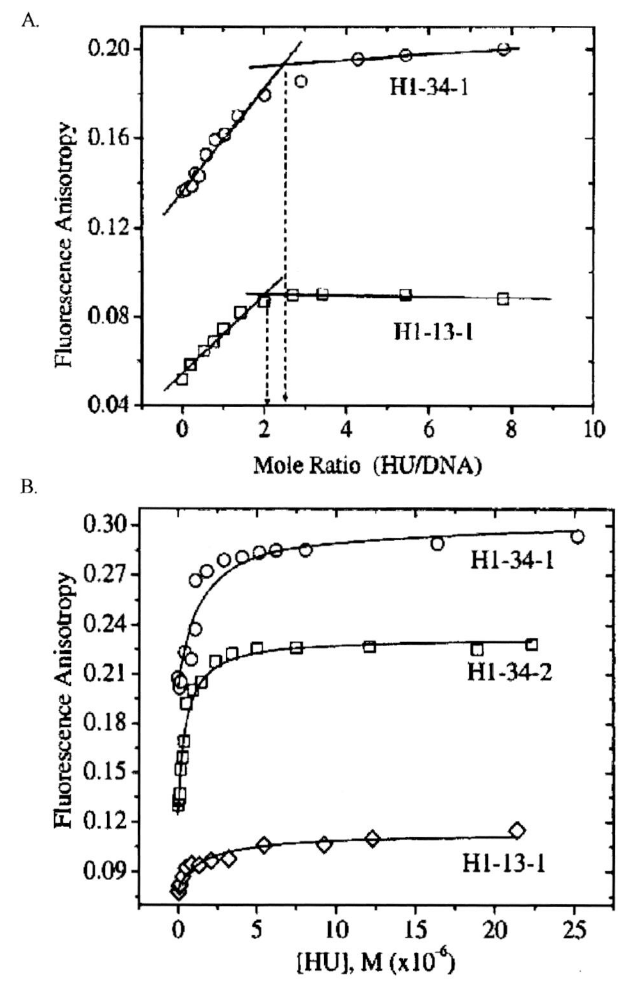

Figure 13.5. (A) Stoichiometric and (B) binding affinity of HU protein to 13-bp (H1-13-1) and 34-bp (H1-34-1 and 2) duplexes by anisotropy. Figures are reprinted from Wojtuszewski *et al.* (2001).

spherical protein is approximately twice the longest rotational correlation time (ϕ) in nanoseconds (Fig. 13.7, Ph.Wahl, 1983). Most ellipsoidal macromolecules with a chromophore rigidly attached generate three exponentials from Eq. (13.7). Shorter ϕ corresponds to internal local motions of the emission transition moments and the longest lifetime corresponds to the overall rotational diffusion of the molecule. We have used this relationship to confirm the molecular weight of a single-stranded DNA-binding protein, translin, forming an octamer. Translin's proposed function is recognizing chromosomal breakage points and may be involved in chromosomal translocation of lymphoid malignancies. Time-resolved anisotropy analysis of pyrene maleimide linked to Cys mutants of this protein at an emission wavelength of 400 nm revealed a long ϕ ($\geq$100 ns) corresponding to eight monomers, where a monomer is ~26 kD. Earlier studies had detected both octamer and dimer of this protein in gels; however, this anisotropy analysis did not require a term with a correlation time appropriate for a dimer at 30 ns. This study is also in agreement with earlier analytical ultracentrifugation studies (Wu *et al.*, 1998; Lee *et al.*, 2001).

13.3.2. Extensions

13.3.2.1. Protein–DNA Interactions and Their Effect on $r(t)$

Millar and coworkers have investigated the substrate distribution of DNA–polymerase interaction using time-resolved fluorescence anisotropy (Bailey *et al.*, 2001). Time-resolved anisotropy measurements of dansyl-labeled ligand were used to distinguish between a replicative and exonuclease complex for the polymerase. The advantage of using dansyl in this study is its sensitivity to local environment (it has a short lifetime in aqueous environment and long lifetime in a nonpolar environment). By observing signals linked to these different lifetimes, they isolated events in the exonuclease and the polymerase of Klenow complex. NLLS analysis of the complex anisotropy decays resolved the mole fractions bound to the polymerase and the exonuclease site.

Dansyl positioned in the exonuclease site is at a frayed end and translocation of the DNA draws the probe into a nonpolar environment, resulting in a long lifetime of 14 ns. When dansyl is located within processed DNA, there is an increase in local mobility in an aqueous environment resulting in a short lifetime of 4.7 ns. Over the course of decay, these two populations contribute to a ''dip and rise'' associative shape (cf. Figure 13.3) in the anisotropy decay curve (Bailey *et al.*, 2001). The population of ''fast moving/fast decaying'' fluorophores at early lifetimes results in a quick dip in the $r(t)$ curve, and the ''slow moving/long lived'' probes survive the early events and thus contribute to a rise in the anisotropy curve. In their analysis, assumptions made are that the buried and the exposed probes possess the same molar extinction coefficients and radiative rate constants, decay is significantly faster than site interconversion (ns vs. ms), and the ground-state populations of buried and exposed probes are in equilibrium. The equilibrium constant describing the partitioning between the two sites was calculated as fractions of buried and exposed probes (Bailey *et al.*, 2001). In a later study of this system, Millar *et al.* elegantly describe this partitioning with two distinct binding modes, and a simple 1:1 association model was shown insufficient to explain the primer or template modes defined in their structural studies. They were able to conduct a thermodynamic analysis on the steady-state and time-resolved anisotropy data incorporating the previous analysis with various single mutations (Bailey *et al.*, 2004).

13.3.2.2. Protein–Protein Interactions, Rotational Correlation Values and Confidence Intervals

Bubb *et al.* have used time-resolved anisotropy to investigate the characteristics of actobindin, a low molecular weight protein (9 kD) that binds to the G-actin monomer, induces the formation of actin dimers, and prevents nucleation and self-association in the process of forming polymers. Interestingly, these actin dimers are still able to join elongating filaments, suggesting actobindin is involved in the maintenance of unpolymerized actin that is available for filament elongation (Bubb *et al.*, 1994). The rotational correlation time for the actin monomer is $\phi_1 = 41$ ns. In the presence of increasing amounts of actobindin, a second rotational correlation term improved the fit ($\phi_2 = 93$ ns). This ϕ_2 falls within the theoretical range for a spherical actin dimer (2.5–4.8 times ϕ_1), and is significantly less than actin as a trimer (Bubb *et al.*, 1994).

The extraction of monomer–dimer population and hydrodynamics by Bubb *et al.* would not be possible without global analysis. Just as global analysis of varying dye angle experiments permits identification of moderate axial ratios (Beechem *et al.*, 1986), while single curve analysis of ellipsoidal shape is useful only for extremes, global analysis of reversible associations (varying populations rather than dye angles) is a necessity for resolving moderate mass increases.

We have simulated a system more difficult, but similar to that of Bubb *et al.* to illustrate this point; a lifetime of 33 ns limits the observation window, while experiments with varying amounts of monomer (80 kD $\sim$ 40 ns) and dimer (160 kD $\sim$ 80 ns) are present. In all cases, at least 25% of the signal comes from each (no ''pure'' monomer or dimer data). Data sets (vertical and horizontal curves) each contained $\sim 4 \times 10^6$ photons. In Figure 13.6, we plot the χ^2 for fixed ϕ, allowing all other parameters in the fit to vary without restraint (cf. Figure 4 in Tcherkasskaya *et al.*, 2000).

Two things are apparent from Figure 13.6

1. Global analysis confidence is tighter because combining related (but not identical) experiments narrow the χ^2_R basins.
2. Global analysis greatly increases the number of degrees of freedom in the fit; thus, the F-ratio ''confidence bar'' moves down considerably (e.g., 1.04 and 1.12 for single-experiment analysis at 67 and 90% confidence versus 1.01 and 1.03, respectively).

Clearly, single-experiment analysis could not resolve the monomer–dimer system; the χ^2_R ratio for degeneracy ($\phi_1 = \phi_2 = 66$ ns) is well below even a 67% confidence bar. The global analysis resolves the pair of ϕ_s with 90% confidence, since the degeneracy at $\sim$55 ns is resolved. Confidence limits from the support plane at 67% are 35.6–45.41 ns (average = 40.5 ns) for ϕ_1 and 69.2–88.5 ns (average = 78.8 ns) for ϕ_2. For comparison, we show also F-test-derived confidence limits (dashed bars) for a directed (rather than exhaustive) search of χ^2 space.

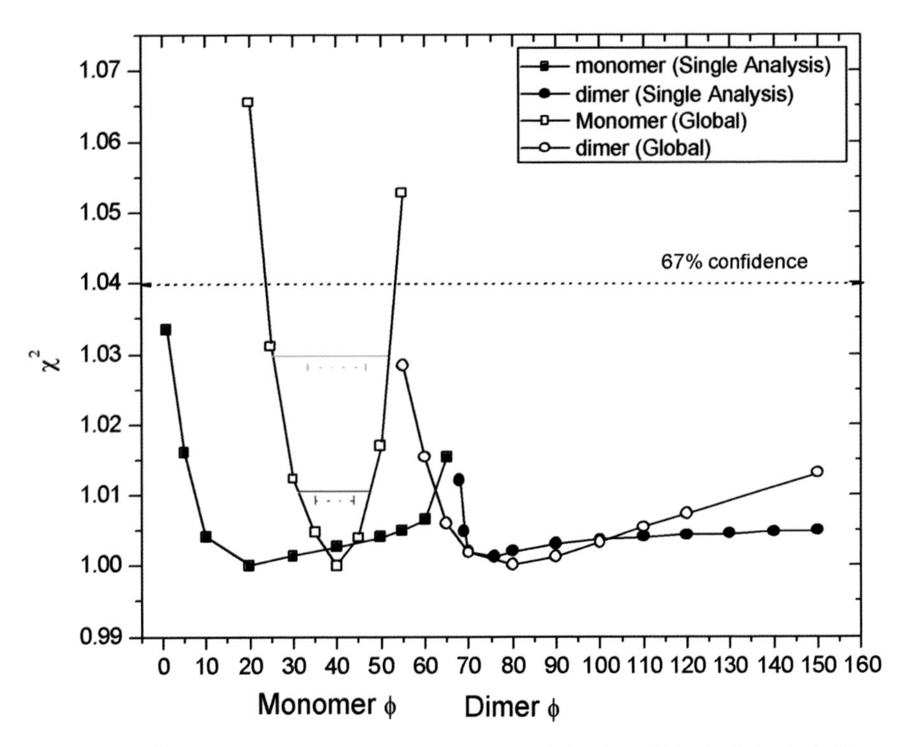

Figure 13.6. χ^2 surface for two relatively long rotational correlation times. This simulation is similar to the correlation times determined from actobindin induction of actin dimers (Bubb *et al.*, 1994). Single-curve analysis data for the monomer and dimer ϕ values are filled squares and circles, respectively. Confidence intervals are shown for the monomer only. In the single data analysis, a 67% confidence level from a F-test was based on 500 data points used in the simulation; confidence would be off the scale at a χ^2 value of 1.12. In the global analysis, 67 and 90% confidence levels are shown for the monomer only in red and green, respectively. The dotted colored bar represents limits from a directed search and the solid colored bars represent limits from an exhaustive search to the F-ratio target (Beechem *et al.*, 1991).

In Figure 13.4, we previously showed the threshold F-ratios for 67 and 90% confidence versus degrees of freedom (number of data – number of free parameters). The F-ratio $F(\nu, \nu-1, p)$ is used since one is comparing to a system with one less degree of freedom (the fixed parameter that is being manipulated). The advantage of obtaining several thousand data points is obvious there and in Figure 13.6 bar heights, but it is clear that increasing the number of data is less important than collection of data that offer a different view of the same phenomenon—i.e., the lower bar is not enough; the χ^2 surface must be constricted by overdetermination with proper experimental design.

Note in Figure 13.6 that one cannot resolve whether the larger partner is a dimer or tetramer at 90% confidence; it can only be done with 67% confidence since 160 ns is excluded at that level. We chose to simulate a difficult case; we used only 4×10^6 counts/experiment and only eight experiments. A longer-lived probe, better counting

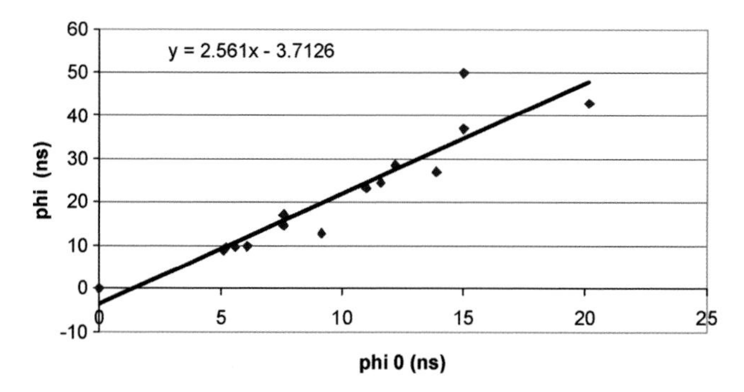

Figure 13.7. Rotational correlational time (ϕ) as a function of ϕ_0, the theoretical time for a dry sphere. (Data were extracted from Figure 3, Ph.Wahl, 1983.)

times, and more experiments (i.e., normal rigor) could certainly achieve 90% + confidence. This example was only presented as a guide to proper confidence analysis and choice of experimental design—e.g., adding counts in identical experiments is inferior to collections of multiple experiments, especially if one has an opportunity to thereby sample multiple manifestations of related kinetics.

13.4. PROSPECTS

Time-resolved and steady-state anisotropy can, with suitable probes (those capable of emitting multiple photons), be employed at ''single molecule'' (small ensemble) levels. Since the photoselection is of an oriented system (Dale, 1988), more potential information is available. An example of this new direction is shown by Osborn *et al.* (2005). The use of time-resolved anisotropy in the microscope also offers an avenue to quantify binding inside cells; Vishwasrao *et al.* (2005) have, in fact, used the fully associative anisotropy model to distinguish free and bound NADH within neurons. This area will likely be one to see significant growth.

In all, time-resolved anisotropy has not yet been pushed to its limits for reversible association studies, and we have not even discussed the possibilities created by varying extrinsic variables (temperature, viscosity, pressure, etc.). Perhaps the ultimate use in this arena will be global analysis of hydrodynamics from different experimental techniques; NLLS analysis of ultracentrifugation data and/or fluorescence correlation spectroscopy curves combined with time-resolved anisotropy data should open new doors into macromolecular hydrodynamics.

ACKNOWLEDGEMENTS

This research was supported (in part) by the Intramural Research Program of the NIH, NHLBI. We also thank for advice: R.E. Dale, L. Brand, L. Davenport, J.B.A. Ross, W.R. Laws, B. Selinger, J. Beechem, and M. Ameloot.

REFERENCES

Allison, S. A. (1999). Low Reynolds number transport properties of axisymmetric particles employing stick and slip boundary conditions. *Macromolecules* 32(16):5304–5312.

Bailey, M. F., Thompson, E. H., and Millar, D. P. (2001). Probing DNA polymerase fidelity mechanisms using time-resolved fluorescence anisotropy. *Methods* 25(1):62–77.

Bailey, M. F., van der Schans, E. J., and Millar, D. P. (2004). Thermodynamic dissection of the polymerizing and editing modes of a DNA polymerase. *J Mol Biol* 336(3):673–693.

Barkley, M., Kowalczyk, A., Brand, L., and Aoshima, M. (1981). Fluorescence decay studies of anisotropic rotations of small molecules. *J Chem Phys* 75:7, 3581–3593.

Beechem, J. M., Gratton, E., Ameloot, M., Knutson, J. R., and Brand, L. (1991). The global analysis of fluorescence intensity and anisotropy decay data: second-generation theory and programs. In: Lakowicz, J. R. (ed.), *Topics in Fluorescence Spectroscopy*, vol. 2. Plenum Press, New York, pp. 241–305.

Beechem, J. M., Knutson, J. R., and Brand, L. (1986). Global analysis of multiple dye fluorescence anisotropy experiments on proteins. *Biochem Soc Trans* 14(5):832–835.

Brochon, J. C. (1994). Maximum entropy method of data analysis in time-resolved spectroscopy. *Methods Enzymol* 240:262–311.

Brunet, J. E., Vargas, V., Gratton, E., and Jameson, D. M. (1994). Hydrodynamics of horseradish peroxidase revealed by global analysis of multiple fluorescence probes. *Biophys J* 66(2 Pt 1): 446–453.

Bubb, M. R., Knutson, J. R., Porter, D. K., and Korn, E. D. (1994). Actobindin induces the accumulation of actin dimers that neither nucleate polymerization nor self-associate. *J Biol Chem* 269(41): 25592–25597.

Bujalowski, W. and Lohman, T. M. (1987). A general method of analysis of ligand-macromolecule equilibria using a spectroscopic signal from the ligand to monitor binding. Application to *Escherichia coli* single-strand binding protein–nucleic acid interactions. *Biochemistry* 26(11): 3099–3106.

Chuang, T. J. and Eisenthal, K. B. (1972). Theory of fluorescence depolarization by anisotropic rotational diffusion. *J Chem Phys* 57(12):5094–5097.

Dale, R. E. (1988). Some aspects of excited-state probe emission spectroscopy for structure and dynamics of model and biological membranes. *Polarized Spectroscopy of Ordered Systems*, vol. C242. pp. 491–567. (NATO ASI Proceedings, Series C)

Dale, R. E. and Gilbert, C. W. (1983). Membrane structure and dynamics by fluorescence probe depolarization kinetics [Appendix: A vector method for the non-linear least squares reconvolution-and-fitting analysis of polarized fluorescence decay data]. In: Cundall, R. B. and Dale R. E. (eds), *Time-Resolved Fluorescence Spectroscopy in Biochemistry and Biology*. Plenum Press, New York, pp. 555–612.

Eftink, M. R. (1997). Fluorescence methods for studying equilibrium macromolecule–ligand interactions. *Methods Enzymol* 278:221–257.

Evett, J. and Isenberg, I. (1969). DNA–polylysine interaction as studied by polarization of fluorescence. *Ann N Y Acad Sci* 158(1):210–222.

Feinstein, E., Deikus, G., Rusinova, E., Rachofsky, E. L., Ross, J. B., and Laws, W. R. (2003). Constrained analysis of fluorescence anisotropy decay: application to experimental protein dynamics. *Biophys J* 84(1):599–611.

Garcia de la Torre, J. G. and Bloomfield, V. A. (1981). Hydrodynamic properties of complex, rigid, biological macromolecules: theory and applications. *Q Rev Biophys* 14(1):81–139.

Hawkins, M. E., Pfleiderer, W., Balis, F. M., Porter, D., and Knutson, J. R. (1997). Fluorescence properties of pteridine nucleoside analogs as monomers and incorporated into oligonucleotides. *Anal Biochem* 244(1):86–95.

Hill, J. J. and Royer, C. A. (1997). Fluorescence approaches to study of protein–nucleic acid complexation. *Methods Enzymol* 278:390–416.

Jameson, D. M. and Sawyer, W. H. (1995). Fluorescence anisotropy applied to biomolecular interactions. *Methods Enzymol* 246:283–300.

Kim, S. J., Lewis, M. S., Knutson, J. R., Porter, D. K., Kumar, A., and Wilson, S. H. (1994). Characterization of the tryptophan fluorescence and hydrodynamic properties of rat DNA polymerase beta. *J Mol Biol* 244(2):224–235.

Knutson, J. R. (1992). Alternatives to consider in fluorescence decay analysis. *Methods Enzymol* 210:357–374.

Knutson, J. R., Davenport, L., and Brand, L. (1986). Anisotropy decay associated fluorescence spectra and analysis of rotational heterogeneity. 1. Theory and applications. *Biochemistry* 25(7): 1805–1810.

Lee, S. P., Fuior, E., Lewis, M. S., and Han, M. K. (2001). Analytical ultracentrifugation studies of translin: analysis of protein-DNA interactions using a single-stranded fluorogenic oligonucleotide. *Biochemistry* 40(46):14081–14088.

LeTilly, V. and Royer, C. A. (1993). Fluorescence anisotropy assays implicate protein–protein interactions in regulating trp repressor DNA binding. *Biochemistry* 32(30):7753–7758.

Ludescher, R. D., Peting, L., Hudson, S., and Hudson, B. (1987). Time-resolved fluorescence anisotropy for systems with lifetime and dynamic heterogeneity. *Biophys Chem* 28(1):59–75.

Malencik, D. A. and Anderson, S. R. (1988). Association of melittin with the isolated myosin light chains. *Biochemistry* 27(6):1941–1949.

Mantulin, W. and Weber, G. (1977). Rotational anisotropy and solvent-fluorophore bonds: an investigation by differential polarized phase fluorometry. *J Chem Phys* 66:4092–4099.

O'Connor, D. V. and Phillips, D. (1984). *Time-Correlated Single Photon Counting*. Academic Press, Orlando.

Osborn, K. D., Zaidi, A., Urbauer, R. J., Michaelis, M. L., and Johnson, C. K. (2005). Single-molecule characterization of the dynamics of calmodulin bound to oxidatively modified plasma-membrane $Ca2^{+}$-ATPase. *Biochemistry* 44(33):11074–11081.

Ph.Wahl. (1983). Time-resolved fluorescence spectroscopy in biochemistry and biology. In: Cundall, R. B. and Dale, R. E. (eds), *Fluorescence Anisotropy Decay and Brownian Rotational Motion: Theory and Application in Biological Systems*. Plenum Press, New York, pp. 497–521.

Press, W. H., Flannery, B. P., Teukolsky, S. A., and Vetterling, W. T. (1990). Special functions. *Numerical Recipes*. Cambridge University Press, New York, pp. 155–190.

Ross, J. B., Schmidt, C. J., and Brand, L. (1981). Time-resolved fluorescence of the two tryptophans in horse liver alcohol dehydrogenase. *Biochemistry* 20(15):4369–4377.

Sarkar, N., Takeuchi, S., and Tahara, T. (1999). Vibronic relaxation of polyatomic molecule in nonpolar solvent: femtosecond anisotropy intensity measurements of the S-n and S-1 fluorescence of tetracene. *J Phys Chem A* 103(25):4808–4814.

Selinger, B. K. and Hinde, A. L. (1983). Least squares methods of analysis. I. Confidence limits. In: Cundall, R. B. and Dale R. E. (eds), *Time-Resolved Fluorescence Spectroscopy in Biochemistry and Biology*. Plenum Press, New York, pp. 129–141.

Shen, X. H. and Knutson, J. R. (2001a). Femtosecond emission anisotropy decay studies of perylene. *Biophys J* 80(1):360A.

Shen, X. H. and Knutson, J. R. (2001b). Femtosecond internal conversion and reorientation of 5-methoxyindole in hexadecane. *Chem Phys Lett* 339(3–4):191–196.

Small, E. W. and Isenberg, I. (1977a). Moment index displacement. *J Chem Phys* 66(8):3347–3351.

Small, E. W. and Isenberg, I. (1977b). Hydrodynamic properties of a rigid molecule: rotational and linear diffusion and fluorescence anisotropy. *Biopolymers* 16(9):1907–1928.

Solie, T. N., Small, E. W., and Isenberg, I. (1980). Analysis of nonexponential fluorescence decay data by a method of moments. *Biophys J* 29(3):367–378.

Spolar, R. S. and Record, M. T., Jr. (1994). Coupling of local folding to site-specific binding of proteins to DNA. *Science* 263(5148):777–784.

Straume, M. and Johnson, M. L. (1992). Analysis of residuals: criteria for determining goodness-of-fit. *Methods Enzymol* 210:87–105.

Tcherkasskaya, O., Ptitsyn, O. B., and Knutson, J. R. (2000). Nanosecond dynamics of tryptophans in different conformational states of apomyoglobin proteins. *Biochemistry* 39(7):1879–1889.

Vishwasrao, H. D., Heikal, A. A., Kasischke, K. A., and Webb, W. W. (2005). Conformational dependence of intracellular NADH on metabolic state revealed by associated fluorescence anisotropy. *J Biol Chem* 280(26):25119–25126.

Wojtuszewski, K., Hawkins, M. E., Cole, J. L., and Mukerji, I. (2001). HU binding to DNA: evidence for multiple complex formation and DNA bending. *Biochemistry* 40(8):2588–2598.

Wu, X. Q., Xu, L., and Hecht, N. B. (1998). Dimerization of the testis brain RNA-binding protein (translin) is mediated through its C-terminus and is required for DNA- and RNA-binding. *Nucleic Acids Res* 26(7):1675–1680.

Youngren, G. K. and Acrivos, A. (1975). Stokes flow past a particle of arbitrary shape: a numerical method of solution. *J Fluid Mech* 69:377–403.

14

Analysis of Protein–DNA Equilibria by Native Gel Electrophoresis

Claire A. Adams and Michael G. Fried*

14.1. INTRODUCTION

The electrophoretic mobility-shift assay (EMSA) is widely used to detect protein–nucleic acid interactions [for representative reviews, see Garner and Revzin (1986), Chodosh (1988), Fried (1989), Carey (1991), Lane et al. (1992), Fried and Garner (1998)]. A particular strength of this method is its ability to detect the simultaneous binding of several proteins to a single molecule of nucleic acid (Figure 14.1), or one (or more) protein(s) to several nucleic acids. Such interactions are of interest because nucleic acid complexes containing large number of proteins play central roles in important cellular transactions including (but not limited to) DNA replication, recombination and repair, and RNA transcription and processing. Although mobility-shift assays are often used for qualitative purposes, under appropriate conditions they can provide quantitative data for the determination of binding stoichiometries, affinities, and kinetics. Quantitative applications require the ability to resolve reaction components in high yield and a theory relating the observed electrophoretic band intensities to the formation constants of the molecular system under study. Here, we describe factors that contribute to the stability of protein–DNA complexes during native gel electrophoresis and one approach to the analysis of electrophoretic band intensities that is appropriate for quantitative analysis of complex formation. Factors that influence the relative electrophoretic

C. A. ADAMS • Department of Molecular and Cellular Biochemistry, University of Kentucky College of Medicine, Lexington, Kentucky 40536-0298 M. G. FRIED • Department of Molecular and Cellular Biochemistry, University of Kentucky College of Medicine, Lexington, Kentucky 40536-0298 and *Corresponding author: Tel: +859 323 1205; Fax: +859 323 1037; E-mail: michael.fried@uky.edu

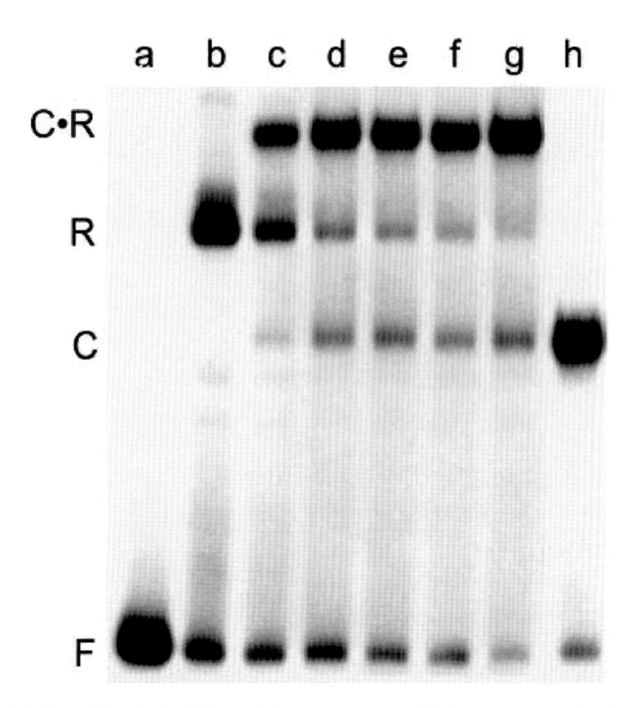

Figure 14.1. Binding of the *Escherichia coli lac* repressor and CAP proteins to a 214-bp lac promoter–operator DNA fragment. Addition of CAP to a solution containing the free DNA fragment (F), the 1:1 repressor–DNA complex (R) and samples c–g results in the formation of a ternary complex (C • R) and a small amount of 1:1 CAP–DNA complex (C). All reactions were carried out at $21 \pm 1°C$ in 10 mM Tris (pH 8.0), 1 mM EDTA, 20 μM cAMP, 5% glycerol, and 100 μg of bovine serum albumin/mL. All samples contained 2.7 nM DNA fragment. Samples b–g contained, in addition, 2.0 nM *lac* repressor, and samples c–g contained, in addition, 1.5, 3.1, 4.6, 6.2, and 7.7 nM CAP, respectively. Sample h contained 2.7 nM DNA and 6.2 nM CAP. Electrophoresis was performed as described (Hudson and Fried, 1990).

mobilities of complexes have been discussed elsewhere (Bading, 1988; Fried, 1989; Carey, 1991; Crothers *et al.*, 1991; Crothers and Drak, 1992; Lane *et al.*, 1992) and will not be detailed here.

14.2. CHOICE OF SUBSTRATE

The selection of a binding substrate requires assumptions about the DNA structures needed for the protein to bind in an appropriate manner. Often, the focus is on providing a particular sequence, without consideration of its structural context. If this context differs significantly from the one with which the protein evolved, the behavior of the system may not be representative of its function *in vivo*. Here, we briefly consider some of the compromises involved in selecting a DNA substrate of a particular size.

The availability of synthetic DNAs has exerted a dominant influence on the range of structures that are in routine use. The majority of mobility-shift assays are performed with short (~25 bp) DNAs. These are inexpensive to synthesize and can often be used with little purification. They offer three additional advantages. First, a large proportion of the residues in a short DNA can be part of a specific target sequence. The relative paucity of competing nonspecific sites can simplify the analysis of DNA-binding activities of moderate or low sequence specificity. Second, in systems containing several binding activities, it can be difficult to characterize the binding of a single component if the DNA substrate contains target sequences for other proteins. Third, as small DNAs have high electrophoretic mobilities, the mobility shift that accompanies protein binding is often large. Efficient resolution simplifies the detection and the quantitation of binding and makes short electrophoresis runs practical. As discussed later, short runs improve the recovery of complexes that would be lost to dissociation during electrophoresis.

Significant trade-offs are made to obtain these advantages. With short DNAs, the target sequence is never far from one or both molecular ends. Unprotected DNA ends are rare *in vivo* and are likely to provide atypical environments for the binding of most proteins. End structures are conformationally dynamic. Transient melting has been observed over a range of conditions (Gotoh *et al.*, 1979; Naritsin and Lyubchenko, 1991; Vallone and Benight, 2000) and duplex twist and base-pair tilt, pitch and roll are likely to deviate from typical B-form values. Together these factors may inhibit binding, or as seen with *Escherichia coli* RNA polymerase, create additional, artifactual binding sites (Melançon *et al.*, 1983). Finally, the electrostatic environment of DNA ends is unlikely to be representative of that found in the middle of a long duplex. Model calculations suggest that lower counterion densities are present in the solution near DNA ends than are present at comparable locations near the middle of a long duplex (Olmsted *et al.*, 1989). These calculations also suggest that the counterion gradient may extend as much as 10 bp inward from the molecular end. If this model is correct, only the central 5 bp of a 25-bp duplex would present a counterion environment that is typical of much longer DNAs. As electrostatic contributions to the binding free energy are often large (Record *et al.*, 1978, 1985), a target site located close to a DNA end may not be bound with affinity typical of the same sequence located near the center of a larger DNA.

In large DNAs, the proportion of base pairs that are close to an end is small but the number of competing binding sites is large. At equilibrium, a protein partitions over all available binding sites. Properties observed for this ensemble of interactions are averages, weighted by the number of competing sites and the relative affinity of the protein for each. This can be a problem if the goal of the experiment is to characterize the binding of a poorly specific protein with a target sequence that is located on a large DNA molecule. On the other hand, large DNAs allow interactions that are not always available with small templates. Important among these are the binding of several proteins to a single template (Fried and Crothers, 1981), the formation of looped complexes in which a single protein binds two sites on a DNA

(Krämer *et al.*, 1987, 1988; Oehler *et al.*, 1990; Lobell and Schleif, 1991), and binding to circular and superhelical molecules (Nordheim and Meese, 1988; Lobell and Schlief, 1990; Fried and Garner, 1998). Although the relative decrement in electrophoretic mobility resulting from protein binding typically decreases with increasing DNA length (Fried, 1989), the binding of single proteins to DNAs as large as 1000 bp is often observable. Larger DNAs approach the resolution limits of current polyacrylamide gel formulations, and stoichiometric discrimination is less good. Although agarose matrices can be used for mobility-shift assay (Berman *et al.*, 1987; Fried, 1989; Fried and Garner, 1998), their resolution is often too poor to allow the observation of single protein binding. Binding competition analyses[1] offer a partial solution to this problem.

14.3. DETECTION AND QUANTITATION OF COMPLEXES AND FREE DNA

Radioisotope-labeled nucleic acid. In the most widely used variants of the mobility-shift assay, binding is detected through the appearance of one or more protein–DNA complexes and a corresponding reduction of the intensity of the free DNA band (Figure 14.2). The ease with which DNA can be labeled with ^{32}P (Maxam and Gilbert, 1977), the great sensitivity with which this isotope can be detected (Fried, 1989; Fried and Garner, 1998), and the fact that ^{32}P is chemically indistinguishable from naturally occurring DNA phosphate have made this the label of choice for quantitative binding analyses. Other radioisotope labels, including

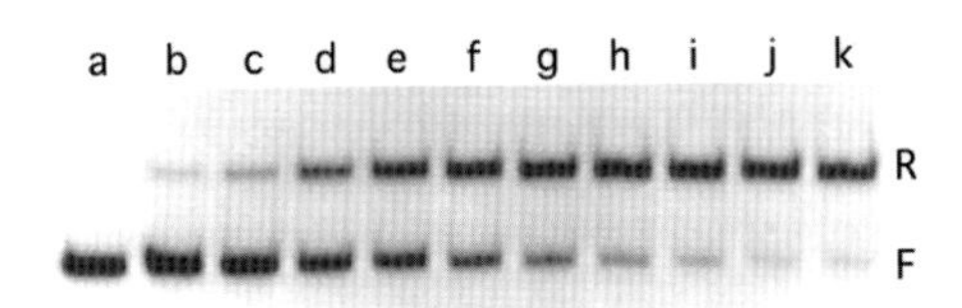

Figure 14.2. Binding of the *Escherichia coli lac* repressor to a 77 bp *lac* operator fragment. All samples contained 4.2×10^{-10} M *lac* operator DNA. Samples b–k also contained *lac* repressor at 1.1×10^{-11}M, 2.1×10^{-11}M, 1.1×10^{-10}M, 2.1×10^{-10}M, 5.3×10^{-10}M, 1.1×10^{-9}M, 2.1×10^{-9}M, 5.3×10^{-9}M, 1.1×10^{-8}M, 2.1×10^{-8}M, respectively. Reactions were carried out at 20 ± 1°C in 10 mM Tris (pH 8.0), 1 mM EDTA, 250 mM KCl, and 0.1 mg of bovine serum albumin/mL. Samples were resolved on a 10% gel. Band designations: F, free DNA; R, 1:1 repressor–DNA complex.

[1] Binding competition assays can be used to measure the relative affinities of proteins for DNAs that are too large to resolve in conventional gels (Fried and Crothers, 1981; Perri *et al.*, 1991). A complex formed with a short, ^{32}P-labeled reference DNA is titrated with unlabeled high molecular weight competitor. The transfer of protein to competitor is detected as a release of reference DNA. The ratio of binding affinities for reference and competitor DNAs can be calculated using Eq. (14.4). The use of a common reference DNA allows the relative affinities for several competitors to be compared (Fried *et al.*, 1996).

^{33}P and ^{35}S, have been used on occasion. Both offer longer half-lives than ^{32}P (25.4 days for ^{33}P and 87.1 days for ^{35}S compared with 14.7 days for ^{32}P), allowing a longer interval of use following the labeling reaction. In addition, the lower energy beta emissions of ^{33}P and ^{35}S result in the capture of somewhat higher-resolution images using film or phosphorimager technologies. However, the specific radioactivities of ^{33}P and ^{35}S in chemical forms (such as [γ-^{33}P]ATP) that are convenient for incorporation into DNA are less than that of ^{32}P, so these advantages are obtained at the cost of some reduction in assay sensitivity.

Radiosiotope-labeled protein. Although good methods are available for radioisotope labeling of proteins (including *in vivo*, enzymatic and cell-free and chemical techniques) (Lorand *et al.*, 1983; Means and Feeney, 1995; Maillet *et al.*, 1996; Shimba *et al.*, 2002; Torizawa *et al.*, 2004; Calero and Ghiso, 2005; de Boer *et al.*, 2005), the use of radioisotope-labeled proteins in EMSA is rare. Nucleic acid labeling procedures are more convenient and can produce higher specific activities (hence more sensitive detection) than comparable protein-labeling reactions. On the other hand, the use of proteins and nucleic acids with different radioisotopic labels provides a simple and rigorous approach to the determination of binding stoichiometry.

Film autoradiography. Autoradiography using X-ray film was the first method of image capture used in radioisotope-detected mobility-shift experiments (Fried and Crothers, 1981; Garner and Revzin, 1981), and this approach is still used occasionally. Its advantages are technical simplicity, wide availability of film and developer equipment, relatively low cost, high spatial resolution and great sensitivity.[2] A disadvantage is the relatively small dynamic range of film response to beta radiation.[3] In our experience, with commonly used film types, this range is rarely greater than two orders of magnitude. To detect gel bands differing in radioactivity by $\geq$ 100-fold calls for care in adjusting the duration of exposure and the use of calibration standards. A partial solution to this problem is to use the developed film as a guide for the excision of gel regions, which are then counted in a scintillation counter (modern scintillation counters have dynamic ranges on the order of 10^6). Molecules labeled with ^{32}P can be directly detected in gel slices without the use of scintillation cocktail, as a result of the Cerenkov effect. Counting efficiencies as high as 55% have been reported for ^{32}P-labeled molecules by this method (L'Annunziata, 2005). Scintillation cocktails are needed for efficient counting of radioisotopes that emit with lower energies. As counting efficiencies can be affected

[2] In our hands, 10^{-18} mole of a 5′-[^{32}P]-labeled DNA fragment can be detected as a unique gel band after extended (1 week) film exposure. Greater sensitivity is possible if the DNA is internally labeled, due to the larger number of ^{32}P atoms incorporated.

[3] This is due, in part, to the fact that the complete activation of a silver halide crystal in the photographic emulsion is a two-photon (or beta-particle) event. At low-radiation flux, an excited silver halide grain may decay to its ground state before a second ''hit'' occurs. At high flux, the maximum film response is obtained well before the end of the exposure time and some radiation reaching the emulsion goes unrecorded.

by colored species present in the sample, care is required to avoid interference from electrophoresis tracking dyes.

Phosphorimagers. Storage phosphor image-capture screens and detection hardware offer several advantages over film autoradiography. Although screens and readers from different manufacturers vary in efficiency, most of them used by us are fivefold to tenfold more sensitive to beta-radiation from ^{32}P-labeled DNA than typical autoradiography films. In many cases, this allows data collection to be accomplished in short exposures that eliminate the need to fix, freeze, or dry the gel to prevent diffusion of labeled species during data collection. The dynamic range of storage phosphor screens (10^5–10^6, Amersham Biosciences, 2002) is much greater than that of the film ($\sim 10^2$). Thus, electrophoretic bands differing by more than 100-fold in radioactivity are easily quantitated in one phosphorimager exposure. The same task would require two or more careful film exposures and the presence of bands of intermediate intensity to act as exposure controls. The spatial resolution of storage phosphor screens, typically in the range of 10–200 microns (Amersham Biosciences, 2002), can be as good as that of high-quality X-ray film. Further, the image dispersion due to the finite thickness of the gel (all parts of a gel band are not equally close to film or screen) means that resolution differences of film and screen are unlikely to be significant unless the screen is imaged at the lowest possible resolution.

Fluorescence. Radioisotopes require special procedures for safe handling and the disposal of waste. In addition, the short half-lives of ^{32}P,^{33}P, and ^{35}S limit the length of time that molecules labeled with these isotopes can be used. Nonradioactive methods of detection have been developed to avoid these disadvantages, although these approaches offer different compromises in terms of sensitivity and convenience. The use of noncovalent fluorescent dyes has long been a standard approach to the detection of nucleic acids in EMSA. Early assays used ethidium bromide (Garner and Revzin, 1981; Wu and Crothers, 1984; Garner and Revzin, 1986) and this continues to be a standard strategy today. More recently, covalent dimers of intercalating dyes have offered improved binding affinity for duplex DNA and thus higher assay sensitivity (Rye *et al.*, 1993). Other dyes, including the SYBR series (available from Molecular Probes), stain both single-stranded and duplex nucleic acids with high affinity. The recent development of protein-specific fluorophores, with emission characteristics that are readily distinguished from those of DNA-specific dyes, holds out the promise of simultaneous two-color quantitation of proteins and nucleic acids (Jing *et al.*, 2003, 2004). In addition, new instruments using confocal scanning have significantly improved the dynamic range and the sensitivity of fluorescence detection, compared with early methods involving photography (Amersham Biosciences, 2002). Currently, these advances allow in-gel quantitation of short DNA molecules in the nanomolar to picomolar concentration range (important variables include DNA size and the identity of fluorescent dye used). The slow dissociation of some dye complexes allows DNA to be stained in free solution before electrophoresis (Rye *et al.*, 1993). However, this convenience comes at the cost of changes in electrophoretic mobility and unknown and possibly large effects on protein binding. Staining following electrophoresis avoids these problems although it increases the background fluorescence of the gel. A drawback to both approaches is the unknown effect of protein binding on the quantum yield of bound dye. Quantitation error from this source can be expected to produce a systematic offset in measures of binding saturation.

The use of covalently attached fluorophores is a viable alternative to the staining methods described earlier. Short synthetic DNAs with 5′, 3′, or internal labels are readily available, and larger DNAs with fluorophores at or near the 5′ ends can be obtained by PCR using labeled primers. In addition, *in vitro* and *in vivo* strategies have been developed for the fluorescent labeling of proteins (Park and Raines, 1997, 2004). Because the fluorophores are covalently coupled to DNA and/or protein, these approaches are characterized by lower background fluorescence than those using postelectrophoresis staining. However, sensitivity is limited by the small number of fluorophores that can be incorporated into a DNA or protein molecule without significantly changing the properties of the system under study. Even with singly labeled molecules, we recommend control experiments with chemically unmodified species to ensure that the incorporation of labels does not change the binding in a meaningful way.

The use of fluorescence energy transfer (FRET) to measure the separation between the ends of short DNA molecules (Heyduk and Lee, 1990) or between a labeled protein and a DNA end (Mukhopadhyay *et al.*, 2003) in solution opens the possibility that such assays could be carried out within native gels. The combination of electrophoretic resolution of stoichiometric and/or conformational isomers and FRET characterization of resolved species opens new possibilities for the analysis of complex mixtures. Pioneering experiments using this approach have been described (Mukhopadhyay *et al.*, 2003).

Chromogenic and chemiluminescent detection methods. These approaches are closely related to western blotting (Burnette, 1981; Towbin *et al.*, 1989, 1992), and many of the reagents used are available commercially in kit form. In most cases, they require transfer of the nucleic acid or protein of interest to a membrane, which is subsequently incubated with an antibody against the molecule of interest,[4] which is then detected by a secondary antibody conjugated with an enzyme that catalyzes the chromogenic or chemiluminescent reaction. Sensitivities that rival those obtainable with radioisotope detection have been reported for variants of these methods (Berger *et al.*, 1993; Rodgers *et al.*, 2000). Two problems must be overcome before these approaches will be useful for quantitative binding analyses. The transfer of molecules from the gel to the membrane is not always accomplished in high yield, and the production of chromophoric or chemiluminescent signals depends on enzyme activity, initial substrate concentration, and reaction interval before the signal is read. Careful standardization of assay conditions and procedures may overcome these challenges.

[4] In another variant of this method, molecules labeled with biotin are detected by avidin or streptavidin conjugated with the enzyme responsible for the chromogenic or chemiluminescent reaction.

14.4. STABILITY OF COMPLEXES DURING ELECTROPHORESIS

In an ideal mobility-shift experiment, all macromolecular species should be resolved from one another and the quantity of each at the end of the gel run should be the same as that in the original sample. In reality, it is unusual for either condition to be fully met. It is not always possible to resolve conformational isomers[5] from one another, and under unfavorable conditions even stoichiometric isomers may not be fully resolved (Fried and Crothers, 1981; Fried, 1989). In addition, dissociation reactions may limit the recovery of species as unique bands at the end of the run. On the other hand, resolution and recovery are frequently good enough to be useful (Fried, 1989; Senear and Brenowitz, 1991; Fried and Daugherty, 1998), and it is often possible to optimize sample and electrophoresis conditions to enhance these characteristics (Fried, 1989; Fried and Garner, 1998). In addition, it is often possible to measure the rate at which individual complexes are lost during electrophoresis (Fried and Liu, 1994; Fried and Bromberg, 1997; Vossen and Fried, 1997). This allows assessment of any dissociation that may occur during electrophoresis, calculation of the concentrations that were present at the start of electrophoresis, and comparison of recovery between parallel experiments for assay optimization.

Electrophoresis running buffer composition. In principle, any buffer that allows adequate electrophoretic resolution and efficient recovery of complexes is likely to be suitable for quantitative mobility-shift studies. In many cases, standard electrophoresis buffers containing Tris-borate or Tris-acetate have performed well, in spite of the fact that these differ significantly from typical sample buffers. Of unknown and potentially significant effect is the mixing of sample and electrophoresis buffers during gel-loading and free electrophoresis. Control experiments in which samples are allowed to incubate for varying times in contact with running buffer have the potential to reveal the risk of protein redistribution at this stage (Fried, 1989). The use of running buffers that are similar in composition to the sample buffer may reduce undesirable effects of buffer exchange. As many buffers are compatible with high-resolution gel electrophoresis, the experiment need not be limited to the buffer compositions described in standard electrophoresis protocols.

The stabilities of many protein–DNA systems depend sensitively on environmental variables such as temperature, salt concentration, and water activity (Barkley, 1981; Fried and Crothers, 1981; Fried and Garner, 1998; Sidorova and Rau, 2004; Milev et al., 2005). These variables can be adjusted to minimize dissociation during electrophoresis. Examples of this strategy include the use of low [salt] running buffers (Fried and Crothers, 1981; Roder and Schweizer, 2001), operation at low (Carey, 1991) or even cryogenic temperatures (Bain and Ackers, 1998), and the inclusion of osmolytes such as glycerol or ethylene glycol in the gel

[5] Complexes of identical stoichiometry but differing in shape.

buffer (Fried and Garner, 1998). Where necessary, low molecular weight effectors (e.g., cAMP for the CAP protein) (Fried and Crothers, 1984a), di- and oligovalent cations, and reducing agents or nonionic detergents (Hendrickson and Schlief, 1984) can be included in gel and running buffers. When such reagents react with components of the gel polymerization reaction, they can be introduced after polymerization by diffusion[6] (for neutral reagents) or by electrophoresis (charged reagents) (Fried and Crothers, 1984a; Fried, 1989).

The period of free electrophoresis. In the simplest versions of the mobility-shift assay, samples at binding equilibrium are placed in the wells of a gel and the separation of components (and relaxation of the system toward a new equilibrium state) starts with the start of electrophoresis. Initially, there is a period of free electrophoresis until reaction components enter the gel. This interval is typically short and its duration is under experimental control. As shown in Figure 14.3, the duration of free electrophoresis depends on the molecular weight of the DNA and on the voltage gradient imposed. Other experiments have shown that it also depends on the conductivities of sample and electrophoresis buffers (Fried, unpublished results). Thus, one strategy for minimizing dead time involves the application of high voltages at the start of electrophoresis (Fried and Crothers, 1981; Carey, 1988; Fried, 1989). A second is the use of small sample volumes and narrow starting zones (Fried, 1989).

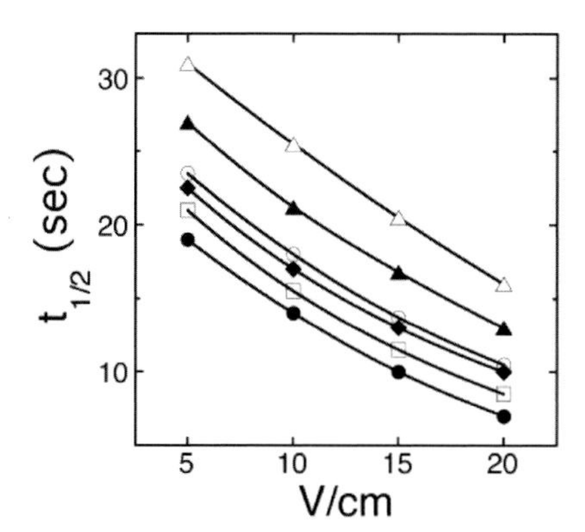

Figure 14.3. Relationship between loading voltage and $t_{1/2}$ for migration of DNA fragments into a polyacrylamide gel. Samples (10 μL) containing ^{32}P-labeled linear, duplex DNAs were applied to 5% polyacrylamide gels (acrylamide:bisacrylamide = 75), equilibrated with 45 mM Tris-borate (pH 8.4) and 2.5 mM EDTA buffer. The sample buffer was 45 mM Tris-borate (pH 8.4), 2.5 mM EDTA, and 5% glycerol. At intervals following the start of electrophoresis, the sample wells were rinsed with running buffer to remove unincorporated DNA. DNA bands were quantitated by scintillation counting of gel slices (Fried, 1989). Values of $t_{1/2}$ were determined from graphs of fraction incorporated as a function of time. DNA sizes: (●) 40 bp; (□) 82 bp; (◆) 203 bp; (○) 214 bp; (▲) 522 bp; and (△) 2,686 bp.

[6] Running gels in the horizontal ''submarine'' format can facilitate the introduction of neutral compounds by diffusion.

Matrix effects. The gel provides the sieving effect needed to resolve free nucleic acids and protein–nucleic acid complexes from one another. Migration through the gel accounts for a large fraction of the time required for electrophoresis. Fortunately, many complexes are significantly more stable within the gel than they are in free solution. An example of this effect is shown in Figure 14.4. The sources of this stabilization are poorly understood but results from a small number of protein–DNA systems suggest that it may be due in part to macromolecular crowding. Modest concentrations of linear polymers (polyacrylamide, polyethylene glycol, and dextran) stabilize CAP–DNA (Fried and Bromberg, 1997) and *lac* repressor–DNA complexes (Fried, unpublished result). Stabilization is greater with high molecular weight polymers than with equivalent weight concentrations of smaller polymers. In addition, complexes become increasingly stable as polymer concentrations increase. Both trends are expected for mechanisms that involve

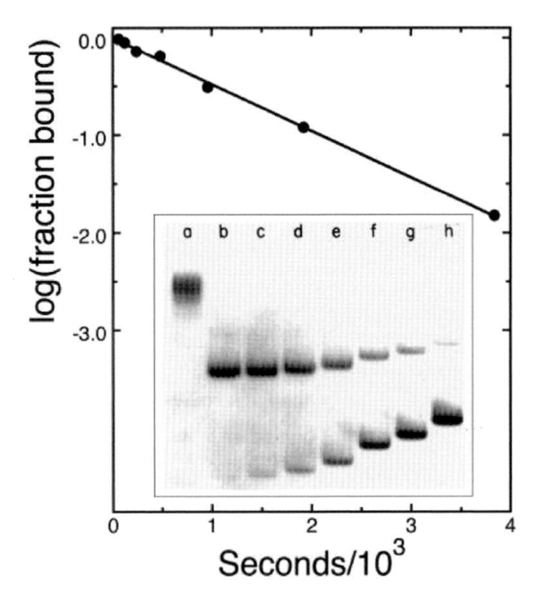

Figure 14.4. Lac repressor–operator complexes are more stable in a 5% polyacrylamide gel than in free solution. Graph: time course of dissociation of the 1:1 *lac* repressor–operator complex in free solution (data from the experiment are shown in the inset). *Inset*: samples a–h contained a ^{32}P-labeled 203-bp *lac* promoter–operator fragment (final concentration = 1.9×10^{-9}M) and wild-type *lac* repressor (final concentration = 1.1×10^{-8}M). Samples b–h contained, in addition, unlabeled 203-bp *lac* promoter–operator fragment (final concentration = 1.9×10^{-7}M) added at the start of the dissociation reaction. Reactions were carried out at 20 ± 1°C in a buffer consisting of 10 mM Tris (pH 8.0 at 20°C), 1 mM EDTA, 50 mM potassium chloride, and 5% glycerol. Incubation times for samples b–h were 1, 2, 4, 8, 16, 32, and 64 min, respectively. Samples were applied to a running 5% polyacrylamide gel cast and run in 45 mM Tris-borate (pH 8.4) and 2.5 mM EDTA. Total electrophoresis time was 150 min. The prominent band in the middle of the gel image is the 1:1 repressor–operator complex and that at the bottom is the free DNA. Although the half-life of the 1:1 complex is ~10 min in free solution, the corresponding gel band is well defined at the end of the gel run.

crowding (Zimmerman and Minton, 1993; Hall and Minton, 2003). The results with polymer solutions extend to the cross-linked polymers used as gel matrices. To date, all complexes that we have tested are significantly more stable in concentrated gels than in dilute gels[7] (Figure 14.5A). This fact can be exploited to optimize the recovery of complexes as resolved bands, at the end of electrophoresis.

The isolation of protein–DNA complexes within the gel matrix may make an additional contribution to their stability during gel electrophoresis. A wide variety of DNA-binding proteins have been found to dissociate from their initial DNA binding sites by DNA-catalyzed mechanisms.[8] The involvement of competing DNA in dissociation reactions is reflected in a kinetic reaction order (with respect to competing DNA) that is greater than zero. In addition, these complexes can also dissociate by the simpler competitor-independent pathway that is of order zero in DNA. Data for the free-solution dissociation of the *lac* repressor from *lac* operator DNA in the presence of competing operator DNA are shown in Figure 14.5B. Under this set of conditions, the apparent kinetic order in competitor is ~ 0.8, consistent with a mixed mechanism in which the dominant pathway is competitor dependent. Parallel reactions carried out within a 10% polyacrylamide gel have an apparent kinetic order of ~ 0.5, indicating that in the gel, a larger fraction of the dissociation reaction followed the slower competitor-independent pathway. Relatively concentrated gels appear to limit the DNA-dependent reaction more effectively than do less concentrated ones. Evidence of this kind, also obtained for the CAP–DNA system (Fried and Liu, 1994), suggests that gel matrices slow the encounter of complexes with competitor DNA, limiting the impact of DNA-catalyzed dissociation on the recovery of complexes at the end of electrophoresis. The observation that several proteins differing in structure and mechanism of DNA interaction have competitor-dependent dissociation kinetics suggests that this ‘‘sequestration’’ effect is likely to be relevant to many experimental systems.

These results lead to some practical conclusions. First, crowding and sequestration mechanisms predict that complexes are more stable in high-concentration gels than they are in more dilute matrices. Second, although dissociation reactions may be slower within a gel matrix (and thus easier to study), the relative contributions of competitor-dependent and competitor-independent pathways may not be the same as those prevailing in free solution. This may allow greater experimental access to competitor-independent pathways. Finally, the practice of including high

[7] It has been proposed that interactions with the gel matrix might contribute to the stability of migrating complexes (Fried and Crothers, 1981; Garner and Revzin, 1986). However, the polymer data cited earlier and the observation that complexes are stabilized by both polyacrylamide and agarose gels argue against mechanisms that require specific interactions between the gel matrix and the migrating complexes.

[8] These include *E. coli lac* repressor (Fried and Crothers, 1984b; Fickert and Muller-Hill, 1992; Hsieh and Brenowitz, 1997), CAP (Fried and Crothers, 1984b; Fried and Liu, 1994), integration host factor (Dhavan *et al*, 2002), single-strand binding protein (Kozlov and Lohman, 2002), glucocorticoid receptor (Zerby and Lieberman, 1977), and human O^6-alkylguanine DNA alkyltransferase (Fried, unpublished result).

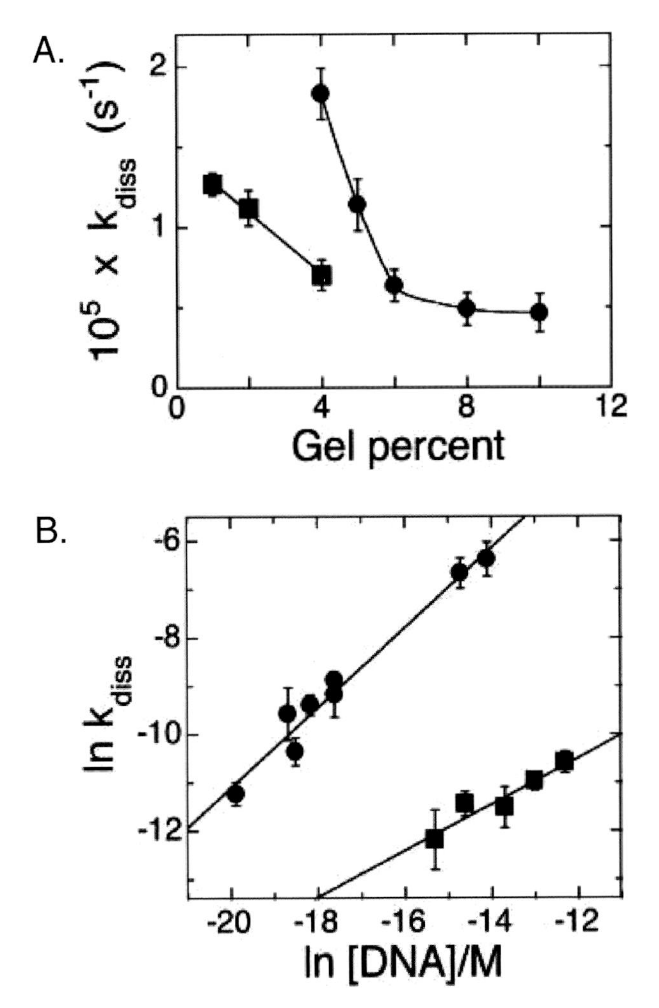

Figure 14.5. (A) Gel concentration influences the stability of 1:1 lac repressor–operator complexes during electrophoresis in polyacrylamide and agarose gels. Samples contained a ^{32}P-labeled 216-bp *lac* promoter–operator fragment (8.9×10^{-10}M) and wild-type *lac* repressor (7.8×10^{-10}M) equilibrated at $21 \pm 1°$C in 20 mM Tris-acetate (pH 8.0), 0.5 mM EDTA, 0.1 mg/mL bovine serum albumin, and 5% glycerol. Aliquots of the equilibrium mixture were applied at intervals to consecutive lanes of running polyacrylamide and agarose gels. Gel and electrophoresis buffers were 20 mM Tris-acetate (pH 8.0) and 0.5 mM EDTA. Gel sections containing the complex band, the free DNA band, and the region between bands (where DNA dissociated from complexes is expected to run) were excised and counted. Data were analyzed according to the rate equation $\ln([R \bullet O]/[R \bullet O]_0) = -kt$, in which $[R \bullet O]_0$ and $[R \bullet O]$ are the concentrations of repressor–operator complex at the start of the reaction and at time t, respectively. Symbols: data from agarose gels (■); data from polyacrylamide gels (●). The error bars are 95% confidence intervals. Redrawn from Vossen and Fried (1997), with permission. (B) Dependence of k_{diss} on DNA concentration for lac repressor–operator complexes dissociating in free solution and within a 10% polyacrylamide gel. Reactions carried out in solution (●) started with mixtures containing ^{32}P-labeled 216-bp *lac* promoter–operator fragment (8.9×10^{-10}M) and wild-type *lac* repressor

concentrations of nonspecific DNA in reaction mixtures, intended to reduce nonspecific binding to the target DNA sequence (Chodosh, 1988), may reduce the recovery of sequence-specific complexes that are present in the sample when it is applied to the gel. Use of a high molecular weight competitor that does not comigrate with complexes in the gel may avoid this unintended consequence.

14.5. MEASUREMENT OF STOICHIOMETRY

An accurate measure of stoichiometry is needed for the determination of binding activities and the development of realistic model(s) for quantitative analysis. These functions are so important that we routinely use two or more independent methods to measure stoichiometry. Several methods have been proposed for the determination of stoichiometries of complexes resolved in native gels. These include labeling of proteins and DNA with complementary radioisotopes, noncovalent staining with dyes or fluorophores, and Western blotting.[9] Where sample quantities are not limiting, rigorous free-solution methods such as analytical ultracentrifugation (Laue, 1995; Daugherty and Fried, 2003) and light scattering (Bloomfield and Lim, 1978; Burchard, 1992) should be considered in addition to native gel approaches. When the nucleic acid binding activity of the protein is independently known, the continuous variation method (Huang, 1982) can provide

Figure 14.5. *(Continued)* (7.8×10^{-10}M) equilibrated at 21 ± 1°C in 20 mM Tris-acetate (pH 8.0), 0.5 mM EDTA, 0.1 mg/mL bovine serum albumin, and 5% glycerol. Dissociation was initiated by addition of unlabeled 216-bp *lac* promoter–operator fragment. Aliquots were applied at intervals to running 10% polyacrylamide gels. Following the gel run, complex and free DNA bands were quantitated by scintillation counting. Data were analyzed according to the rate equation $\ln([R \bullet O]/[R \bullet O]_0) = -kt$, where $[R \bullet O]_0$ and $[R \bullet O]$ are the concentrations of the repressor–operator complex at the start of the reaction and at time t, respectively. The order of the reaction in competing DNA, as estimated from $\partial \ln k_{diss}/\partial \ln([\mathrm{DNA}]/M)$ is 0.83 ± 0.11. Reactions carried out in 10% polyacrylamide gels (■) started as equilibrium mixtures containing ^{32}P-labeled 216-bp *lac* promoter–operator fragment (8.9×10^{-10}M) and wild-type *lac* repressor (4.4×10^{-10}M) equilibrated at 21 ± 1°C in 10 mM Tris-acetate (pH 8.0), 1 mM EDTA, 0.1 mg/mL bovine serum albumin, and 5% glycerol. Aliquots were applied to a 10% polyacrylamide gel cast and run in 20 mM Tris-acetate (pH 8.0) and 2 mM EDTA. After electrophoresis for 90 min at 6.7 V/cm, aliquots of unlabeled *lac* 216 DNA were applied to the same sample wells and electrophoresis was continued. As the competitor has a higher mobility than the repressor–operator complex, the competitor band will pass through the migrating band of complex, allowing exchange to take place. Methods for calculating the exchange times for gels operating at different voltage gradients have also been described (Fried and Liu, 1994; Vossen and Fried, 1997). After the gel run, the concentrations of surviving $R \bullet O$ complex and released DNA were measured by band excision and scintillation counting. As in free solution, dissociation in the gel followed pseudo first-order kinetics. The order of the reaction in competing DNA, as estimated from $\partial \ln k_{diss}/\partial \ln([\mathrm{DNA}]/M)$ was 0.48 ± 0.09. Error bars represent 95% confidence intervals. Redrawn from Vossen and Fried (1997), with permission.

[9] Although the staining and Western-blotting methods are promising, they have not been widely tested to date and should be used with appropriate caution.

an estimate of stoichiometry. Serial dilution and Scatchard analyses, which provide information on both affinity and stoichiometry, is discussed in a later section.

Pitfalls of band counting. It is often assumed that the first complex to appear in a titration of nucleic acid with protein has a 1:1 protein:nucleic acid ratio, and subsequent complexes correspond to higher stoichiometries with each step representing the addition of a single equivalent of protein. Although this is the expected result when binding is noncooperative (cf., Figure 14.6A), highly cooperative binding can result in the appearance of a multiprotein complex in equilibrium with free nucleic acid, without detectable lower-stoichiometry intermediates.

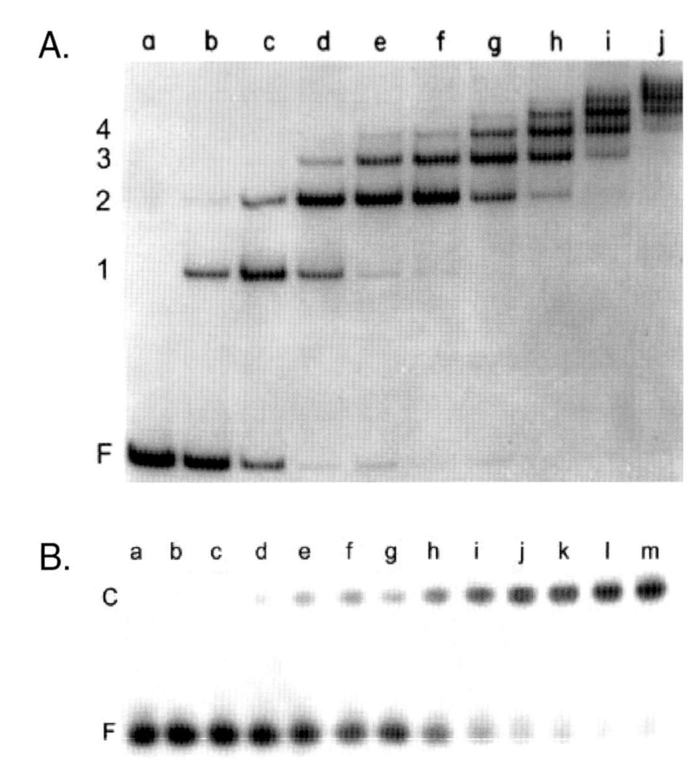

Figure 14.6. (A) Titration of a 203-bp lactose promoter–operator fragment with lac repressor. The DNA fragment concentration was 18.5 nM. Repressor concentrations were 0, 12.4, 24.8, 37.0, 49.4, 61.6, 74.0, 98.8, 123.4, and 148.0 nM for samples a–j, respectively. Samples were incubated for 30 min at room temperature in 10 mM Tris (pH 8.0 at 21°C), 1 mM EDTA, and 50 mM KCl and applied to a 5% polyacrylamide gel equilibrated with the same buffer. The repressor:fragment ratios of some complexes are given in the left margin. Band F denotes free DNA. Modified from Fried and Crothers (1981), with permission. (B) Titration of a 16-bp duplex DNA with alkylated human O^6 alkylguanine DNA alkyltransferase (AGT). Binding was carried out at 20°C in 10 mM Tris (pH 7.6), 1 mM DTT, and 10 μg/mL bovine serum albumin. Samples contained 8.75×10^{-7} M DNA duplex; samples b–l contained increasing concentrations of Cys^{145}-methyl derivative of human AGT protein, with 1.9×10^{-6}M $\leq$ [AGT] $\leq 5.1 \times 10^{-6}$M. Samples were equilibrated for 30 min before electrophoresis in a 10% polyacrylamide gel. Modified from Rasimas *et al.* (2003), with permission.

An example of this is the titration of a 16-bp duplex DNA with O^6 alkylguanine DNA alkyltransferase (AGT; Figure 14.6B). Here the only detectable complex has a 4:1 protein:DNA ratio (Rasimas *et al.*, 2003). A second assumption is that complexes differing in electrophoretic mobility will have different stoichiometries, with lower mobilities corresponding to higher protein–DNA ratios. Although correlations between electrophoretic mobility and stoichiometry have been observed for some molecular systems (Fried and Crothers, 1981; Bading, 1988), they are not an appropriate basis for assigning the stoichiometries of other protein–DNA interactions. Mobility correlations do not usually take into account the effects of DNA bending (Wu and Crothers, 1984; Crothers and Drak, 1992), looping (Krämer *et al.*, 1987; Schleif, 1992; Muller *et al.*, 1996), or protein charge (Carey, 1988), which can differ from system to system.

Quantitation of the protein–DNA ratio within the gel. If specific radioactivities are known, experiments in which protein and nucleic acid are labeled with different isotopes can permit the direct evaluation of the stoichiometric ratios of complexes (Fried and Crothers, 1984a). Arguably, this classical approach is still the most sensitive method available. Metabolic incorporation of radioisotopic amino acids has the advantage that the resultant protein can be chemically identical to the unlabeled wild-type species. Many methods of metabolic labeling with radioisotopes are available for proteins that can be expressed in *E. coli*. For other proteins, kits for cell-free translation with ^{35}S-methionine are commercially available. An alternative approach is the covalent labeling of protein with an extrinsic radioisotopic probe. These include iodination (Bolton and Hunter, 1973; Stern and Frieden, 1993; Smith and Peterson, 2003), reductive alkylation (Means, 1977; Means and Feeney, 1995), and phosphorylation (Clark *et al.*, 2002; Delucchi *et al.*, 2003). Although the use of extrinsic labels is convenient, labeling reactions have the potential to change DNA-binding activity and/or stoichiometry, so they should be used with appropriate caution. A variant of the extrinsic-labeling approach is the use of an isotopically labeled low molecular weight ligand that is bound tightly by the protein. If the ligand:protein stoichiometry is reliably known and if its dissociation during electrophoresis is sufficiently slow, the quantity of ligand present in electrophoretic bands at the end of the gel run should be proportional to the amount of protein present in the bands (Garner and Revzin, 1982).

Under favorable conditions, strongly absorbing dye–protein complexes, such as those obtained with Coomassie blue R-250 or fluorescent complexes like those obtained with SYPRO® dyes, can be quantitated following densitometric scanning (Fried and Crothers, 1981) or fluorescence imaging (Jing *et al.*, 2003, 2004). As with all staining techniques, samples containing known amounts of the protein(s) of interest should be run in the same gel to provide calibration data. In addition, it is necessary to verify that the response of the chosen stain is not modified by the presence of nucleic acid. For example, although silver staining is commonly used for detection of proteins following SDS-PAGE, nucleic acids react under some conditions (Lomholt and Frederiksen, 1987; Qu *et al.*, 2005) and might be expected to contribute to the staining of protein–DNA complexes.

Two-dimensional electrophoresis employing native mobility-shift conditions in the first dimension and denaturing SDS-PAGE in the second avoids this complication and can provide detailed information about the subunit composition of nucleic acid-bound proteins (Straney and Crothers, 1985).

Western blotting has also been used to detect proteins following electrophoretic resolution of complexes (Chen and Chang, 2001). Though this method is sensitive and flexible,[10] inefficiencies in the transfer of proteins to the membrane and in the binding of antibodies, as well as nonlinear amplification of signal at enzyme-catalyzed steps are the difficulties that remain to be overcome before this approach is useful for quantitative analysis.

Continuous variation (job) analysis. This widely used method is infrequently reviewed (Huang, 1982). It provides convenient measure of stoichiometry if the binding activity of the protein preparation is accurately known. In this approach, the molar ratio of active protein to DNA ($[P]_{tot}$: $[D]_{tot}$) is varied, whereas the sum of concentrations ($[P]_{tot} + [D]_{tot}$) is held constant. This is conveniently carried out by preparing solutions of protein and DNA with equal active concentrations and then mixing them in different volume ratios. The ratio that yields the greatest concentration of complex (the optimal combining ratio) is a measure of the binding stoichiometry. This can be estimated from a graph of [complex] as a function of the mole fraction of one component (Figure 14.7). An evident limitation of this approach is that large optimal combining ratios are not easy to resolve, as the corresponding mole fraction values are very closely spaced.

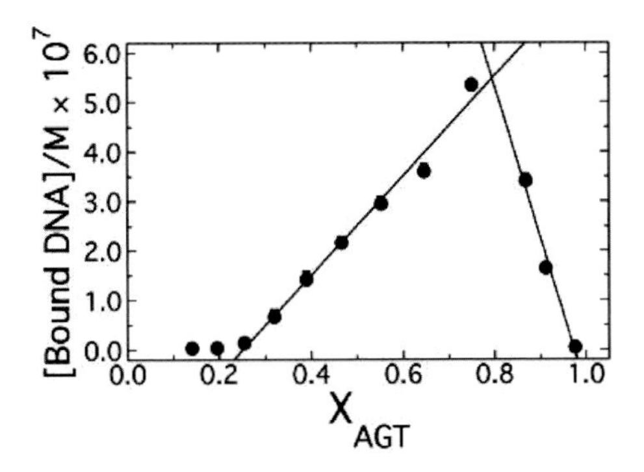

Figure 14.7. Continuous variation analysis of an O^6 alkylguanine DNA alkyltransferase (AGT)–DNA complex. The total macromolecular concentration was fixed ([AGT] + [DNA] = 6×10^{-6}M) and binding was carried out at 20 ± 1°C in 10 mM Tris (pH 7.6), 1 mM DTT, and 10 μg/mL bovine serum albumin. Error bars (almost obscured by data points) represent the maximum range of values obtained in three parallel measurements. The solid lines are least-squares fits to the rising and the falling subsets of the data. Their intersection yields an optimal combining ratio of 3.9. Modified from Rasimas *et al.* (2003), with permission.

[10] For instance, the expression of epitope-tagged proteins allows standard immunochemistry to be used for their purification and subsequent detection in complex with nucleic acids.

14.6. MEASUREMENT OF BINDING ACTIVITY

It is unusual for either the protein or the nucleic acid preparation to be fully competent for binding. One strategy for the determination of the fractional binding activity of the protein (β) is to titrate nucleic acid with protein at a nucleic acid concentration much larger than K_d. Under these conditions, protein binding should be very nearly stoichiometric. The initial slope of a graph of mole fraction of DNA bound as a function of the protein:DNA ratio should be equal to the product of $1/$(stoichiometry) and β (Fried and Crothers, 1981). A similar approach, useful if the binding stoichiometry is known, is to titrate a concentrated sample of protein ([protein] $\gg K_d$) with DNA. Under stoichiometric binding conditions, the molar ratio of protein to DNA at the point of protein saturation will be equal to m/β, where m is the stoichiometry and β is the fractional binding activity of the protein. The work of Melcher and Xu (2001) provides a good example of this approach.

It is sometimes found that a fraction of the nucleic acid remains unbound, even in the limit of high protein concentration. When synthetic oligonucleotides are used, such ''incompetent'' DNA may result from unintended side reactions that accompany synthesis or from incomplete deprotection during sample preparation. When DNA is obtained by PCR amplification, low polymerase fidelity may result in a subset of molecules with sequence modifications that make them incompetent. When DNAs are synthesized *in vivo*, heterogeneity may result from mutations that occur during cloning or vector propagation or from the action of cellular DNA modification systems. Finally, samples from all sources are subject to modification during isolation and storage. The discovery that a large fraction of a DNA sample is incompetent is cause for concern, as it opens to question whether the ''competent'' fraction has escaped modification that might alter its properties in a meaningful way. Even when a large fraction of DNA appears to be competent, it is useful to determine its mole fraction because errors in the active concentration of DNA are propagated throughout the analysis of binding.

14.7. MEASUREMENT OF ASSOCIATION CONSTANTS

Binding to a single site. Consider the simple case in which a DNA-binding protein (P) exists in only two states, bound to a specific DNA site (D) or free in solution. The protein distribution is governed by the equilibrium

$$P + D \overset{K}{\rightleftharpoons} PD, \tag{14.1}$$

in which $K = [PD]/[D][P]$. Addition of protein to a DNA solution results in binding as $1/[P]$ approaches K. At the mid-point of the reaction, $[PD]/[D] = 1$ and $1/[P] = K$. If $[P]$ is measurable at this point, K is easily determined. However, the high binding affinities of many protein–DNA interactions result in low values of $[P]$, which are difficult to measure or calculate with accuracy. Two strategies have been developed to overcome this problem. The first uses DNA at tracer concentrations ($[D] \ll 1/K$). Under these conditions, a negligible fraction of the added protein will be bound, even at DNA saturation, so that the total protein concentration $[P]_{\text{tot}}$ is a good approximation of the free protein concentration $[P]$. When it is not practical to use tracer concentrations, for example, if the method used to detect DNA is of low sensitivity or the value of K is very large, the best option is to use the solution of the conservation equations for the system, shown below for a 1:1 interaction.

$$Y = \frac{\left([P]_{\text{tot}} + \frac{1}{K} + [D]_{\text{tot}}\right) - \left(\left([P]_{\text{tot}} + \frac{1}{K} + [D]_{\text{tot}}\right)^2 - 4[P]_{\text{tot}}\,[D]_{\text{tot}}\right)^{1/2}}{2[D]_{\text{tot}}}. \tag{14.2}$$

Here, $Y = [PD]/([D] + [PD])$, and $[P]_{\text{tot}}$ and $[D]_{\text{tot}}$ are the total concentrations of protein and DNA, respectively. An example of this approach is shown in Figure 14.8.

Binding competition. Competition assays allow comparison of the stability of a protein–DNA complex with that of a similar complex containing a standard competitor. This analysis is usually carried out by titrating a preformed specific complex (PS) with a lower-affinity competitor[11] (C) and determining the fraction of the initial complex remaining after each step (Figure 14.9).This titration establishes the equilibrium

$$\text{PS} + \text{C} \rightleftharpoons \text{PC} + \text{S}. \tag{14.3}$$

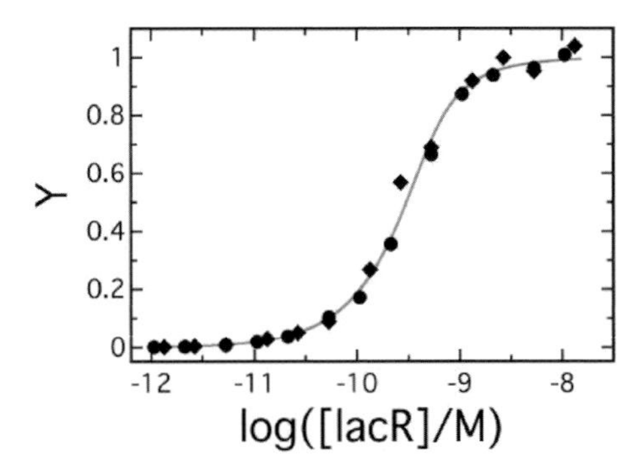

Figure 14.8. Evaluation of the association constant for a *lac* repressor–operator interaction. Data from experiments like the one shown in Figure 14.2. The smooth curve represents the fit of Eq. (14.2) to the data, with $K = 1.1 \pm 0.1 \times 10^{10}\ \text{M}^{-1}$. From Fried et al. (2002), with permission.

[11] The most commonly used competitors are repeating homopolymers such as poly d(A–T) or poly d(I–C), which offer the advantage of homogeneous binding (Fried, 1989). On the other hand, when the competitor is a genomic DNA from the same organism as the DNA-binding protein, the ratio of binding constants, K_s/K_c, reflects the relative strengths of interactions that determine the distribution of protein between specific and nonspecific genomic sites *in vivo*.

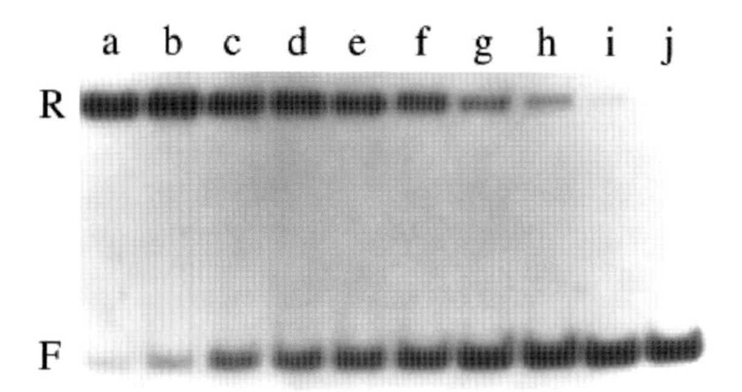

Figure 14.9. Binding competition assay. Titration of a specific *lac* repressor–operator complex (R) with sheared *Escherichia coli* DNA. Transfer of repressor to the competitor produces free *lac* operator DNA (F). Reactions were carried out in 10 mM Tris (pH 8.0), 1 mM EDTA, 250 mM KCl, 0.1 mg of bovine serum albumin/mL, and 5% ethylene glycol. All samples contained 2.2×10^{-11} M *lac* operator DNA and 3.4×10^{-11} M *lac* repressor. Samples b–j contained competing DNA at concentrations increasing from 3.9×10^{-7} M base pairs (sample b) to 9.7×10^{-3} M base pairs (sample j). Reproduced from Fried et al. (2002), with permission.

Here, S is the free specific DNA and PC is the protein complex formed with competitor. At each step of the titration, the ratio of equilibrium constants K_s/K_c is given by

$$\frac{K_S}{K_C} = \frac{[PS][C]}{[S][PC]}. \tag{14.4}$$

Here, K_s and K_c are the formation constants for the specific and the competitor complexes, respectively. Analysis of this equilibrium is simple as long as the binding to the competitor is weak compared with that of the specific DNA. This avoids problems of neighbor exclusion and competitor saturation (Fried and Crothers, 1981). When the specific DNA is labeled and competitor is unlabeled, $[PC]$ and $[C]$ can be calculated from the amount of protein transferred from specific DNA to competitor. After the ith titration step, $[PC] = [PS]_o - [PS]_i$, in which $[PS]_o$ is the concentration of specific complex at the start of the experiment and $[PS]_i$ is its concentration after the ith step. Similarly, the free competitor concentration can be calculated using $[C] = [C]_o - n[CP]$. Here, $[C]_o$ is the total concentration of competing binding sites[12] present at the ith step and n is the number of base pairs per nonspecific binding site.

Competition assays provide a means of comparing the affinities of a protein for DNAs that are too large or too heterogeneous to allow convenient analysis in polyacrylamide gels. A homogeneous DNA that is appropriate for gel resolution

[12] When small DNAs are used as competitor, the number of full-length sites per DNA molecule is $(N-n) + 1$, in which N is the number of base pairs per molecule and n is the binding site size of the protein (in base pairs; McGhee and von Hippel, 1974). For larger DNAs with $N \gg n$, the number of sites approaches the number of base pairs, N.

is used as a reference ligand, and the DNA to be tested is used as a competitor. If the protein's affinity for the reference DNA is accurately known, the measured value of K_s/K_c will allow calculation of a population-average association constant K_c for the ensemble of sites available in the competitor. When the goal is to compare average affinities for two or more large DNAs, it is convenient to perform parallel competition assays using the same reference ligand for each. For two DNAs A and B, the ratio of the relative affinities with respect to the reference yields the ratio of affinities for the two DNAs.

$$\frac{K_S/K_B}{K_S/K_A} = \frac{K_A}{K_B}. \tag{14.5}$$

Positively cooperative binding without resolution of stoichiometric isomers. When small DNAs are bound with strongly positive cooperativity, multiprotein complexes may be formed without significant accumulation of lower stoichiometry intermediates. An example of this is the 4:1 complex formed by alkylated human AGT protein with a 16-bp duplex (Figure 14.6B). For systems of this kind, we routinely use a serial dilution method for initial characterization of stoichiometry and the apparent association constant. For the concerted binding of m protein molecules to a single DNA ($m\mathrm{P} + \mathrm{D} \rightleftarrows \mathrm{P}_m\mathrm{D}$), the apparent association constant[13] is $K_{obs} = [P_mD]/\ [P]^m[D]$. Separating variables and taking logarithms gives the linear relationship

$$\ln\frac{[P_mD]}{[D]} = m\ln[P] + \ln K_{obs}. \tag{14.6}$$

Serial dilution allows variation of the ratio $[P_mD]/[D]$ by mass action,[14] without changing the input ratio of protein to DNA (Figure 14.10A). An initial estimate of $[P]$ can be calculated for each serial dilution step using the conservation relation $[P] = [P]_{tot} - m[P_mD]$, in which $[P]_{tot}$ is the total protein concentration and m is an initial estimate of the stoichiometry. An updated estimate of m is then obtained from the linear dependence of $\ln[P_mD]/[D]$ on $\ln[P]$ (Figure 14.10B). The new value of m is fed back into the conservation relation and the cycle is iterated until values of m and K_{obs} converge. This approach is particularly useful when $[P]$ must be calculated, because both $[P]_{tot}$ and $[P_mD]$ can be accurately known at each dilution step.

If the stoichiometry m of the cooperative complex is known, values of the association constant, the cooperativity parameter, and the binding site size can be evaluated using the finite-chain variant of the McGhee–von Hippel equation, as

[13] K_{obs} is the formation constant for the cooperative complex, containing contributions from both protein–DNA and protein–protein interactions.

[14] There are two advantages to this strategy. First, it is possible to start at concentrations of DNA and protein that are large enough to measure directly. Second, quantitation of DNA at each step of the dilution (e.g., from its radioactivity if ^{32}P label is used) allows experimental measure of the stepwise and cumulative dilution factors, from which the total protein concentration of the sample can be accurately calculated.

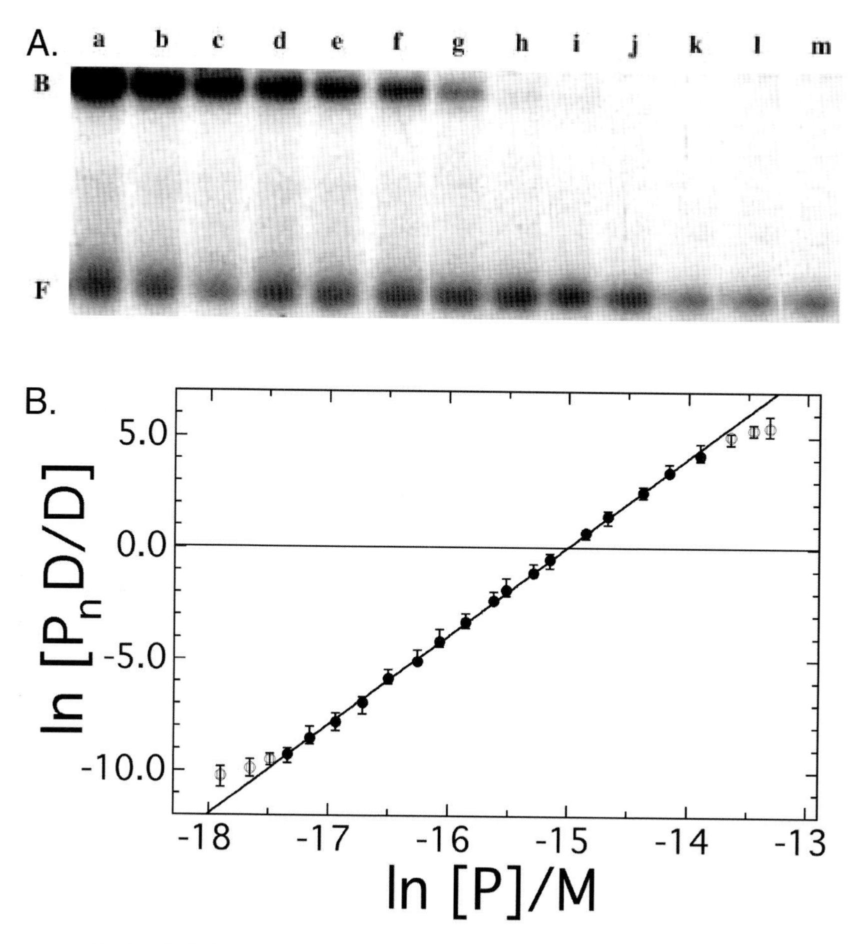

Figure 14.10. Serial dilution analysis of the interaction between human O^6 alkylguanine DNA alkyltransferase (AGT) and a 16-bp duplex. (A) Mobility-shift assay. The initial mixture contained $(His)_6$-tagged C145S mutant human AGT (3.5 μM) and double-stranded 16-mer (0.8 μM duplex) in 10 mM Tris (pH 7.6), 1 mM DTT, and 10 μg/mL bovine serum albumin (lane a). Serial dilutions were performed using the same buffer (dilution factor 0.85/step; lanes b–m), giving decreasing concentrations of both protein and DNA. Electrophoresis was carried out in a 10% polyacrylamide gel. Band designations: B, bound DNA; F, free DNA. From Rasimas *et al.* (2003), with permission. (B) Graph of the dependence of $\ln([P_nD]/[D])$ on $\ln[P]$/M. The error bars indicate the range of values obtained in three replicate experiments. The solid line represents a least-squares fit of Eq. (14.6) to the subset of data indicated by the filled symbols. This analysis returned $n = 3.9 \pm 0.1$ and $K_{obs} = 4.3(\pm 0.9) \times 10^{23}\ M^{-4}$.

described by Tsodikov *et al.* (2001). For a DNA containing N monomer units (bp or nt) this expression is

$$\frac{\nu}{[P]} = K(1 - n\nu)\left(\frac{(2\omega - 1)(1 - n\nu) + \nu - R}{2(\omega - 1)(1 - n\nu)}\right)^{n-1}\left(\frac{1 - (n+1)\nu + R}{2(1 - n\nu)}\right)^2\left(\frac{N - n + 1}{N}\right), \quad (14.7)$$

with $R = ([1 - (n+1)\nu]^2 + 4\omega\nu(1 - n\nu))^{1/2}$. Here, the binding density $\nu = m[P_mD]/N[D]_{tot}$, where $[P]$ is the free protein concentration, K is the equilibrium constant for the binding of a protein monomer for any available site on the DNA, n is the number of residues occupied by the protein to the exclusion of other protein molecules, and ω is the cooperativity parameter.[15] This expression has the same general form as the Scatchard equation (Scatchard, 1949). Binding data graphed as a Scatchard plot can be subjected to regression analysis using Eq. (14.7), to obtain fitted values of the parameters K, ω, and n. An example of this is shown in Figure 14.11.

The calculation of binding density (ν) requires knowledge of the binding stoichiometry, if DNA is the only molecule that is available for direct quantitation (as is the case with standard mobility-shift assays using ^{32}P-DNA and unlabeled protein). If the assay is performed under conditions in which the quantity of bound protein can be measured (as with radioisotope-labeled or fluorescent protein), binding density can be expressed in terms of the concentrations of bound protein and total DNA: $\nu = [P]_{bound}/N[D]_{tot}$. When this is the case, the ratio N/n may provide an estimate of the stoichiometry.

Cooperative binding with resolution of stoichiometric isomers. In systems containing a single protein, individual stoichometric steps may be detected when binding is weakly cooperative, noncooperative, or anticooperative (cf., Figure 14.7). In such cases, the relative stabilities of individual association states can be

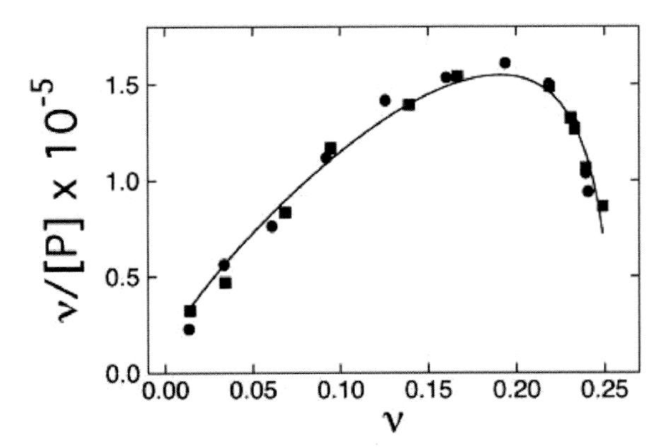

Figure 14.11. Scatchard plot: cooperative binding of human O^6 alkylguanine DNA alkyltransferase (AGT) to a 16-bp duplex. Reactions were carried out in 10 mM Tris (pH 7.6), 1 mM DTT, and 10 μg/mL bovine serum albumin. Data were obtained by serial dilution (•) and forward titration (■). The smooth curve is a least-squares fit of Eq. (14.7) to this data, returning $K = 1.2 \pm 0.2 \times 10^4$ M^{-1}, $\omega = 102 \pm 11$, and $n = 3.9 \pm 0.1$.

[15] The cooperativity parameter is the equilibrium constant for the process of moving a protein from an isolated binding site to the one immediately adjacent to another molecule of protein. When $\omega > 1$, proteins attract each other and binding is said to be positively cooperative, and when $\omega < 1$, proteins repel each other and binding is anticooperative, and when $\omega = 1$, binding is noncooperative.

assessed using the binding competition assay, even when several complexes are present in a solution at the same time (Fried and Crothers, 1981). In addition, average values of K, ω, and n can be obtained from forward titration or serial dilution data, using the finite-chain variant of the McGhee–von Hippel equation (Eq. (14.7); Tsodikov *et al.*, 2001). Because several stoichiometric isomers may be present at each step of the experiment, the average binding density ν must be calculated by summing the contributions of all resolved species

$$\nu = \frac{1}{N[D]_{\text{tot}}} \sum_{m} m[P_m D]. \tag{14.8}$$

Heterocooperativity. One of the strengths of the mobility-shift assay is its ability to resolve complexes containing different proteins bound to a single kind of DNA (Figure 14.1). Accurate measurement of the mole fractions of the corresponding species allows evaluation of heterogeneous cooperative interactions. Our approach follows standard thermodynamic reasoning. For an interacting protein–DNA system at equilibrium, the probability with which a given state (j) occurs is proportional to $\exp(-\Delta G_j/RT)$, where ΔG_j is the free energy change that accompanies the formation of state j from a reference state, R is the gas constant, and T is the temperature in degrees Kelvin (Hill, 1985; Wyman and Gill, 1990). If the reference state is that in which DNA is unbound, the term $\exp(-\Delta G_j/RT)$ for the association of a protein with a DNA site j is equal to $K_j[P]$, in which $[P]$ is the free protein concentration and K_j is the site-specific association constant. The sum of terms for all states of the system is the partition function, Q

$$Q = \sum_{j} \exp(-\Delta G_j/RT). \tag{14.9}$$

Thus, for the system shown in Figure 14.1 containing *lac* repressor protein (R), CAP protein (C) and a *lac* promoter DNA containing one high-affinity site for CAP (C1) and one for repressor (O1), the partition function is

$$Q = 1 + K_{\text{C1}}[C] + K_{\text{O1}}[R] + K_{\text{C1}}K_{\text{O1}}[C][R]\omega_{\text{C1O1}}. \tag{14.10}$$

Here the term 1 represents the free DNA state (when $\Delta G = 0$, $\exp(-\Delta G_j/RT) = 1$); $K_{\text{C1}}[C]$ represents the state in which the only interaction is the binding of CAP to site C1; $K_{\text{O1}}[R]$ represents the state in which the only interactions is that of repressor with site O1; and $K_{\text{C1}}K_{\text{O1}}[C][R]\omega_{\text{C1O1}}$ the state in which both CAP and repressor are bound to the same DNA. The cooperativity term $\omega_{\text{C1O1}} = \exp(-\Delta G_{\text{int}}/RT)$, where ΔG_{int} is the free energy of CAP–repressor interaction when each is bound to its corresponding DNA site.

The formation probability[16] for state j is equal to the term for state j divided by Q. Thus, the probability of the state with no proteins bound $P_0 = 1/Q$, that with

[16] As the samples employed in the mobility shift assay typically contain large number of molecules, the relative frequency with which a state is observed to occur will closely approximate its formation probability. Thus, under favorable conditions, normalized integrated band intensities provide useful measures of the formation probabilities of the free and protein-bound states of the DNA.

CAP protein only bound $P_C = K_{C1}[C]/Q$, that with *lac* repressor only bound $P_R = K_{O1}[R]/Q$, and that with both CAP and repressor bound $P_{CR} = K_{C1}K_{O1}[C][R]\omega_{C1O1}/Q$. Taken individually, these probability expressions are of little use without detailed information about the values of K_{C1} and K_{O1} and the free concentrations of CAP and *lac* repressor in the equilibrium mixtures. However, a ratio of state probabilities can be used to determine the value of the cooperativity parameter, ω_{C1O1} (Hudson and Fried, 1990; Vossen *et al.*, 1996).

$$\frac{P_0 P_{CR}}{P_C P_R} = \omega_{C1O1}. \tag{14.11}$$

This ''partition function'' approach has been used to solve a wide range of binding distribution problems visualized by mobility-shift and/or footprinting methods (Fried and Crothers, 1981; Hudson and Fried, 1990; Senear and Brenowitz, 1991; Senear *et al.*, 1993; Vossen *et al.*, 1996; Gavigan *et al.*, 1999; Howard *et al.*, 2002; Chahla *et al.*, 2003).

14.8. A LOOK INTO THE FUTURE

Although this chapter has focused on the analysis of protein–DNA interactions at equilibrium, quantitative mobility-shift assays are not limited to such systems. Mobility-shift assays have been used for analyses of protein–RNA interactions (Rouault *et al.*, 1989; Long and Crothers, 1995; Kash and Menon, 1998; Wang *et al.*, 1998; Mucha *et al.*, 2002), DNA–DNA, RNA–RNA, and DNA–RNA interactions (Ferber and Maher, 1997; Mills *et al.*, 2002; Ueda and Roberts, 2004; Gaidamakov *et al.*, 2005), and protein–protein interactions (Stern and Frieden, 1993; Mackun and Downard, 2003; Park and Raines, 2004). The use of mobility-shift techniques for detection of protein–oligosaccharide interactions (Wu *et al.*, 2002; Seyfried *et al.*, 2005) suggest that it may be possible to extend the method to quantitative analyses of these systems as well.

Mobility-shift assays are also useful for the study of systems that are far from equilibrium (Fried and Crothers, 1981; Long and Crothers, 1995). The sequestration of reaction components in a gel matrix provides a convenient way to quench reactions (Gerstle and Fried, 1993). The enhanced stability of labile species in the gel environment (see earlier section) and the exceptional resolution offered by electrophoresis allow the detection and the characterization of reaction intermediates (Fried and Crothers, 1984b). To date, the use of manual sampling techniques has limited the use of mobility-shift assays to systems with long relaxation times, but it may be possible to adapt electrophoresis technology to quickly resolve samples obtained from a quenched-flow mixer (Fried, unpublished results). This may reduce the ''dead time'' of the technique by as much as two orders of magnitude, allowing access to a very wide range of kinetic problems.

The slab gel format that has been characteristic of the technique since its inception is by no means the only one that works. Capillary electrophoresis has been used successfully for mobility-shift assays of several experimental systems (Xian and Harrington, 1998; Mucha *et al.*, 2002). Although the mobility effects of electroendoosmosis and Joule heating remain as challenges to overcome (Plenert and Shear, 2003; Xuan and Li, 2005), the potential for automation, the small sample size, rapid resolution, and sensitive detection that are available with this technology suggest that it will attract more methods development and use in the future. Finally, microfluidic electrophoresis devices are under development (Hsiung *et al.*, 2005; Nagata *et al.*, 2005; Peeni *et al.*, 2005). Their ability to handle sample volumes in the nanoliter range facilitates the analysis of protein and nucleic acid components from very small samples of tissue or perhaps from single cells. Such spatial resolution should usher in a range of new applications for the electrophoresis mobility-shift assay.

REFERENCES

Amersham Biosciences (2002). Typhoon User's Guide Vol. 3.0, sections 6-1–6-22.

Bading, H. (1988). Determination of the molecular weight of DNA-bound protein(s) responsible for gel electrophoretic mobility shift of linear DNA fragments examplified with purified viral *myb* protein. *Nucl Acids Res* 16:5241–5248.

Bain, D. L. and Ackers, G. K. (1998). A quantitative cryogenic gel-shift technique for analysis of protein–DNA binding. *Anal Biochem* 258:240–245.

Barkley, M. (1981). Salt dependence of the kinetics of the *lac* repressor-operator interaction: role of nonoperator deoxyribonucleic acid in the association reaction. *Biochemistry* 20:3833–3842.

Berger, R., Duncan, M. R., and Berman, B. (1993). Nonradioactive gel mobility shift assay using chemiluminescent detection. *Biotechniques* 15:650–652.

Berman, J., Eisenberg, S., and Tye, B.-K. (1987). An agarose gel electrophoresis assay for the detection of DNA binding activities in yeast cell extracts. *Methods Enzymol* 155:528–537.

Bloomfield, V. A. and Lim, T. K. (1978). Quasi-elastic light scattering. *Methods Enzymol* 48:415–494.

Bolton, A. E. and Hunter, W. M. (1973). The labelling of proteins to high specific radioactivities by conjugation to a 125I-containing acylating agent. *Biochem J* 133:529–239.

Burchard, W. (1992). Static and Dynamic Light Scattering. In Harding, S. E., Sattelle, D. B. and Bloomfield, V. A. (eds), *Laser Light Scattering in Biochemistry*. The Royal Society of Chemistry, Cambridge, pp. 3–22.

Burnette, W. N. (1981). "Western blotting": electrophoretic transfer of proteins from sodium dodecyl sulfate–polyacrylamide gels to unmodified nitrocellulose and radiographic detection with antibody and radioiodinated protein A. *Anal Biochem* 112:195–203.

Calero, M. and Ghiso, J. (2005). Radiolabeling of amyloid-beta peptides. *Methods Mol Biol* 299: 325–348.

Carey, J. (1988). Gel retardation at low pH resolves *trp* repressor–DNA complexes for quantitative study. *Proc Natl Acad Sci U S A* 85:975–979.

Carey, J. (1991). Gel retardation. *Methods Enzymol* 208:103–117.

Chahla, M., Wooll, J., Laue, T. M., Nguyen, N., and Senear, D. F. (2003). Role of protein–protein bridging interactions on cooperative assembly of DNA-bound CRP–CytR–CRP complex and regulation of the *Escherichia coli* CytR regulon. *Biochemistry* 42:3812–3825.

Chen, H. and Chang, G. D. (2001). Simultaneous immunoblotting analysis with activity gel electrophoresis in a single polyacrylamide gel. *Electrophoresis* 22:1894–1899.

Chodosh, L. A. (1988). Mobility shift DNA-binding assay using gel electrophoresis. In Ausubel, F. M., Brent, R. Kingston, R. E., Moore, D. D, Seidman, J. G., Smith, J. A., and Struhl, K. (eds), *Current Protocols in Molecular Biology*. John Wiley and Sons, New York, pp. 12.12.11–12.12.10.

Clark, W. A., Izotova, L., Philipova, D., Wu, W., Lin, L., and Pestka, S. (2002). Site-specific ^{32}P-labeling of cytokines, monoclonal antibodies, and other protein substrates for quantitative assays and therapeutic application. *Biotechniques* 33 Suppl:S76–S87.

Crothers, D. M. and Drak, J. (1992). Global features of DNA structure by comparative gel electrophoresis. *Methods Enzymol* 212:46–71.

Crothers, D. M., Gartenberg, M. R., and Schrader, T. E. (1991). DNA bending in protein–DNA complexes. *Methods Enzymol* 208:118–146.

Daugherty, M. A. and Fried, M. G. (2003). Analysis of transcription factor interactions at sedimentation equilibrium. In: Adhya S. and Garges, S. (eds), *Methods in Enzymol, v. 370: RNA Polymerases and Associated Factors, Part C*. Elsevier, New York, pp. 349–369.

de Boer, A. R., Letzel, T., Lingeman, H., and Irth, H. (2005). Systematic development of an enzymatic phosphorylation assay compatible with mass spectrometric detection. *Anal Bioanal Chem* 381:647–655.

Delucchi, A. B., Jensen, K. A., and Chan, W. K. (2003). Synthesis of 32P-labelled protein probes using a modified thioredoxin fusion protein expression system in *Escherichia coli*. *Biomol Eng* 20:1–5.

Dhavan, G. M., Crothers, D. M., Chance, M. R., and Brenowitz, M. (2002). Concerted binding and bending of DNA by *Escherichia coli* integration host factor. *J Mol Biol* 315:1027–1037.

Ferber, M. J. and Maher, L. Jr. (1997). Quantitating oligonucleotide affinities for duplex DNA: footprinting vs electrophoretic mobility shift assays. *Anal Biochem* 244:312–320.

Fickert, T. and Muller-Hill, B. (1992). How lac repressor finds lac operator in vitro. *J Mol Biol* 226:59–68.

Fried, M. G. (1989). Measurement of protein–DNA interaction parameters by electrophoresis mobility shift assay. *Electrophoresis* 10:366–376.

Fried, M. G. and Bromberg, J. L. (1997). Factors that affect the stability of protein–DNA complexes during gel electrophoresis. *Electrophoresis* 18:6–11.

Fried, M. G. and Crothers, D. M. (1981). Equilibria and kinetics of lac repressor-operator interactions by polyacrylamide gel electrophoresis. *Nucl Acids Res* 9:6505–6525.

Fried, M. G. and Crothers, D. M. (1984a). Equilibrium studies of the cyclic AMP receptor protein–DNA interaction. *J Mol Biol* 172:241–262.

Fried, M. G. and Crothers, D. M. (1984b). Kinetics and mechanism in the reactions of gene regulatory proteins with DNA. *J Mol Biol* 172:263–282.

Fried, M. G. and Daugherty, M. A. (1998). Electrophoretic analysis of multiple protein–DNA interactions. *Electrophoresis* 19:1247–1253.

Fried, M. G. and Garner, M. M. (1998). The electrophoretic mobility shift assay (emsa) for detection and analysis of protein–DNA interactions. In: Tietz, D. (ed.) *Molecular Biology Methods and Applications*. Elsevier, New York, pp. 239–271.

Fried, M. G., Kanugula, S., Bromberg, J. L., and Pegg, A. E. (1996). DNA binding mechanisms of O^6-alkylguanine-DNA alkyltransferase: stoichiometry and effects of DNA base composition and secondary structures on complex stability. *Biochemistry* 35:15295–15301.

Fried, M. G. and Liu, G. (1994). Molecular sequestration stabilizes CAP–DNA complexes during polyacrylamide gel electrophoresis. *Nucl Acids Res* 22:5054–5059.

Fried, M. G., Stickle, D., Vossen, K., Adams, C., MacDonald, D., and Lu, P. (2002). Role of macromolecular hydration in lac repressor-DNA interactions. *J Biol Chem* 277:50676–50682.

Gaidamakov, S. A., Gorshkova, I. I., Schuck, P., Steinbach, P. J., Yamada, H., Crouch, R. J., and Cerritelli, S. M. (2005). Eukaryotic RNases H1 act processively by interactions through the duplex RNA-binding domain. *Nucl Acids Res* 33:2166–2175.

Garner, M. M. and Revzin, A. (1981). A gel electrophoresis method for quantifying the binding of proteins to specific DNA regions: application to components of the *Escherichia coli* lactose operon system. *Nucl Acids Res* 9:3047–3060.

Garner, M. M. and Revzin, A. (1982). Stoichiometry of catabolite activator protein/adenosine cyclic 3′,5′-monophosphate interactions at the *lac* promoter of *Escherichia coli*. *Biochemistry* 21:6032–6036.

Garner, M. M. and Revzin, A. (1986). The use of gel electrophoresis to detect and study nucleic acid-protein interactions. *Trends Biol Sci* 11:395–396.

Gavigan, S. A., Nguyen, T., Nguyen, N., and Senear, D. F. (1999). Role of multiple CytR binding sites on cooperativity, competition, and induction at the *Escherichia coli* udp promoter. *J Biol Chem* 274:16010–16019.

Gerstle, J. T. and Fried, M. G. (1993). Measurement of binding kinetics using the gel electrophoresis mobility shift assay. *Electrophoresis* 14:725–731.

Gotoh, O., Wada, A., and Yabuki, S. (1979). Salt-concentration dependence of melting profiles of lambda phage DNAs: evidence for long-range interactions and pronounced end effects. *Biopolymers* 18:805–824.

Hall, D. and Minton, A. P. (2003). Macromolecular crowding: qualitative and semiquantitative successes, quantitative challenges. *Biochim Biophys Acta* 1649:127–139.

Hendrickson, W. and Schleif, R. F. (1984). Regulation of the *Escherichia coli* L-arabinose operon studied by gel electrophoresis DNA binding assay. *J Mol Biol* 178:611–628.

Heyduk, T. and Lee, J. C. (1990). Application of fluorescence energy transfer and polarization to monitor *Escherichia coli* cAMP receptor protein and *lac* promoter interaction. *Proc Natl Acad Sci U S A* 87:1744–1748.

Hill, T. L. (1985). *Cooperativity Theory in Biochemistry*. Springer-Verlag, New York.

Howard, V. J., Belyaeva, T. A., Busby, S. J. W., and Hyde, E. I. (2002). DNA binding of the transcription activator protein MelR from *Escherichia coli* and its C-terminal domain. *Nucl Acids Res* 30:2692–2700.

Hsieh, M. and Brenowitz, M. (1997). Comparison of the DNA association kinetics of the Lac repressor tetramer, its dimeric mutant LacIadi, and the native dimeric Gal repressor. *J Biol Chem* 272: 22092–22096.

Hsiung, S. K., Lin, C. H., and Lee, G. B. (2005). A microfabricated capillary electrophoresis chip with multiple buried optical fibers and microfocusing lens for multiwavelength detection. *Electrophoresis* 26:1122–1129.

Huang, C. Y. (1982). Determination of binding stoichiometry by the continuous variation method: the Job plot. *Methods Enzymol* 87:509–525.

Hudson, J. M. and Fried, M. G. (1990). Co-operative interactions between the catabolite gene activator protein and the *lac* repressor at the lactose promoter. *J Mol Biol* 214:381–396.

Jing, D., Agnew, J., Patton, W. F., Hendrickson, J., and Beechem, J. M. (2003). A sensitive two-color electrophoretic mobility shift assay for detecting both nucleic acids and protein in gels. *Proteomics* 3:1172–1180.

Jing, D., Beechem, J. M., and Patton, W. F. (2004). The utility of a two-color fluorescence electrophoretic mobility shift assay procedure for the analysis of DNA replication complexes. *Electrophoresis* 25:2439–2446.

Kash, J. C. and Menon, K. M. (1998). Identification of a hormonally regulated luteinizing hormone/human chorionic gonadotropin receptor mRNA binding protein. Increased mRNA binding during receptor down-regulation. *J Biol Chem* 273:10658–10664.

Kozlov, A. G. and Lohman, T. M. (2002). Kinetic mechanism of direct transfer of *Escherichia coli* SSB tetramers between single-stranded DNA molecules. *Biochemistry* 41:11611–11627.

Krämer, H., Amouyal, M., Nordheim, A., and Müller-Hill, B. (1988). DNA supercoiling changes the spacing requirement of two *lac* operators for DNA loop formation with *lac* repressor. *EMBO J* 7:547–556.

Krämer, H., Niemöller, M., Amouyal, M., Revet, B., von Wilcken-Bergmann, B., and Müller-Hill, B. (1987). Lac repressor forms loops with linear DNA carrying two suitably spaced lac operators. *EMBO J* 6:1481–1491.

Lane, D., Prentki, P., and Chandler, M. (1992). Use of gel retardation to analyze protein–nucleic acid interactions. *Microbiol Rev* 56:509–528.

L'Annunziata, M. F. (2005). Application Note CIA 002: Cerenkov Counting of ^{32}P—Instrument Performance Data. Packard Instrument Co., Meriden, CT, p. 5.

Laue, T. M. (1995). Sedimentation equilibrium as a thermodynamic tool. *Methods Enzymol* 259: 427–452.

Lobell, R. B. and Schlief, R. F. (1990). DNA looping and unlooping by AraC protein. *Science* 250: 528–532.

Lobell, R. and Schleif, R. (1991). AraC–DNA looping: orientation and distance-dependent loop breaking by the cyclic AMP receptor protein. *J Mol Biol* 218:45–54.

Lomholt, B. and Frederiksen, S. (1987). Detection of a few picograms of DNA on polyacrylamide gels by silver staining. *Anal Biochem* 164:146–149.

Long, K. S. and Crothers, D. M. (1995). Interaction of human immunodeficiency virus type 1 tat-derived peptides with TAR RNA. *Biochemistry* 34:8885–8895.

Lorand, L., Parameswaran, K. N., Velasco, P. T., Hsu, L. K., and Siefring, G. E. Jr. (1983). New colored and fluorescent amine substrates for activated fibrin stabilizing factor (Factor XIIIa) and for transglutaminase. *Anal Biochem* 131:419–425.

Mackun, K. and Downard, K. M. (2003). Strategy for identifying protein–protein interactions of gel-separated proteins and complexes by mass spectrometry. *Anal Biochem* 318:60–70.

Maillet, I., Lagniel, G., Perrot, M., Boucherie, H., and Labarre, J. (1996). Rapid identification of yeast proteins on two-dimensional gels. *J Biol Chem* 271:10263–10270.

Maxam, A. and Gilbert, W. S. (1977). A new method for sequencing DNA. *Proc Natl Acad Sci U S A* 74:560–565.

McGhee, J. and von Hippel, P. H. (1974). Theoretical aspects of DNA-protein interactions: co-operative and non-co-operative binding of large ligands to a one-dimensional homogeneous lattice. *J Mol Biol* 86:469–489.

Means, G. E. (1977). Reductive alkylation of amino groups. *Methods Enzymol* 47:469–478.

Means, G. E. and Feeney, R. E. (1995). Reductive alkylation of proteins. *Anal Biochem* 224:1–16.

Melançon, P., Burgess, R. R., and Record, M. T. Jr. (1983). Direct evidence for the preferential binding of *Escherichia coli* RNA polymerase holoenzyme to the ends of deoxyribonucleic acid restriction fragments. *Biochemistry* 22:5169–5176.

Melcher, K. and Xu, H. E. (2001). Gal80–Gal80 interaction on adjacent Gal4p binding sites is required for complete GAL gene repression. *EMBO J* 20:841–851.

Milev, S., Bosshard, H. R., and Jelesarov, I. (2005). Enthalpic and entropic effects of salt and polyol osmolytes on site-specific protein–DNA association: The Integrase Tn916–DNA complex. *Biochemistry* 44:285–293.

Mills, M., Arimondo, P. B., Lacroix, L., Garestier, T., Klump, H., and Mergny, J. L. (2002). Chemical modification of the third strand: differential effects on purine and pyrimidine triple helix formation. *Biochemistry* 41:357–366.

Mucha, P., Szyk, A., Rekowski, P., and Barciszewski, J. (2002). Structural requirements for conserved Arg52 residue for interaction of the human immunodeficiency virus type 1 trans-activation responsive element with trans-activator of transcription protein (49–57). Capillary electrophoresis mobility shift assay. *J Chromatogr* A 968:211–220.

Mukhopadhyay, J., Mekler, V., Kortkhonjia, E., Kapanidis, A. N., Ebright, Y. W., and Ebright, R. H. (2003). Fluorescence resonance energy transfer (FRET) in analysis of transcription-complex structure and function. *Methods Enzymol* 371:144–159.

Muller, J., Oehler, S. and Muller-Hill, B. (1996). Repression of lac promoter as a function of distance, phase and quality of an auxiliary lac operator. *J Mol Biol* 257:21–29.

Nagata, H., Tabuchi, M., Hirano, K., and Baba, Y. (2005). Microchip electrophoretic protein separation using electroosmotic flow induced by dynamic sodium dodecyl sulfate-coating of uncoated plastic chips. *Electrophoresis* 26:2687–2691.

Naritsin, D. B. and Lyubchenko, Y. L. (1991). Melting of oligodeoxynucleotides with various structures. *J Biomol Struct Dyn* 8:813–825.

Nordheim, A. and Meese, K. (1988). Topoisomer gel retardation: detection of anti-Z-DNA antibodies bound to Z-DNA within supercoiled DNA minicircles. *Nucleic Acids Res* 16:21–37.

Oehler, S., Eismann, E. R., Krämer, H., and Müller-Hill, B. (1990). The three operators of the *lac* operon cooperate in repression. *EMBO J* 9:973–979.

Olmsted, M. C., Anderson, C. F., and Record, M. T. (1989). Monte Carlo description of oligoelectrolyte properties of DNA oligomers: range of the end effect and the approach of molecular and thermodynamic properties to the polyelectrolyte limits. *Proc Natl Acad Sci U S A* 86:7766–7770.

Park, S. H. and Raines, R. T. (1997). Green fluorescent protein as a signal for protein–protein interactions. *Protein Sci* 6:2344–2349.

Park, S. H. and Raines, R. T. (2004). Fluorescence gel retardation assay to detect protein–protein interactions. *Methods Mol Biol* 261:155–160.

Peeni, B. A., Conkey, D. B., Barber, J. P., Kelly, R. T., Lee, M. L., Woolley, A. T., and Hawkins, A. R. (2005). Planar thin film device for capillary electrophoresis. *Lab Chip* 5:501–505.

Perri, S. D., Helinski, D. R., and Toukdarian, A. (1991). Interactions of plasmid-encoded replication initiation proteins with the origin of DNA replication in the broad host range plasmid RK2. *J Biol Chem* 266:12536–12543.

Plenert, M. L. and Shear, J. B. (2003). Microsecond electrophoresis. *Proc Natl Acad Sci U S A* 100: 3853–3857.

Qu, L., Li, X., Wu, G., and Yang, N. (2005). Efficient and sensitive method of DNA silver staining in polyacrylamide gels. *Electrophoresis* 26:99–101.

Rasimas, J. J., Pegg, A. E., and Fried, M. G. (2003). DNA-binding mechanism of O6-alkylguanine-DNA alkyltransferase. Effects of protein and DNA alkylation on complex stability. *J Biol Chem* 278:7973–7980.

Record, M. T., Anderson, C. F., and Lohman, T. M. (1978). Thermodynamic analysis of ion effects on the binding and conformational equilibria of proteins and nucleic acids: the roles of ion association or release, screening, and ion effects on water activity. *Q Rev Biophys* 11:103–178.

Record, M. T., Anderson, C. F., Mossing, M., and Roe, J.-H. (1985). Ions as regulators of protein–nucleic acid interactions in vitro and in vivo. *Adv Biophys* 20:109–135.

Roder, K. and Schweizer, M. (2001). Running-buffer composition influences DNA–protein and protein–protein complexes detected by electrophoretic mobility-shift assay (EMSA). *Biotechnol Appl Biochem* 33:209–214.

Rodgers, J. T., Patel, P., Hennes, J. L., Bolognia, S. L., and Mascotti, D. P. (2000). Use of biotin-labeled nucleic acids for protein purification and agarose-based chemiluminescent electromobility shift assays. *Anal Biochem* 277:254–259.

Rouault, T. A., Hentze, M. W., Haile, D. J., Harford, J. B., and Klausner, R. D. (1989). The iron-responsive element binding protein: a method for the affinity purification of a regulatory RNA-binding protein. *Proc Natl Acad Sci U S A*. 86:5768–5772.

Rye, H. S., Drees, B. L., Nelson, H. C., and Glazer, A. N. (1993). Stable fluorescent dye–DNA complexes in high sensitivity detection of protein–DNA interactions. Application to heat shock transcription factor. *J Biol Chem* 268:25229–25238.

Scatchard, G. (1949). The attractions of proteins for small molecules and ions. *Ann NY Acad Sci* 51: 660–672.

Schleif, R. (1992). DNA looping. *Annu Rev Biochem* 61:199–223.

Senear, D. F. and Brenowitz, M. (1991). Determination of binding constants for cooperative site-specific protein–DNA interactions using the gel mobility-shift assay. *J Biol Chem* 266:13661–13671.

Senear, D. F., Dalma-Weiszhausz, D. D., and Brenowitz, M. (1993). Effects of anomalous migration and DNA to protein ratios on resolution of equilibrium constants from gel mobility-shift assays. *Electrophoresis* 14:704–712.

Seyfried, N. T., Blundell, C. D., Day, A. J., and Almond, A. (2005). Preparation and application of biologically active fluorescent hyaluronan oligosaccharides. *Glycobiology* 15:303–312.

Shimba, N., Yamada, N., Yokoyama, K., and Suzuki, E. (2002). Enzymatic labeling of arbitrary proteins. *Anal Biochem* 301:123–127.

Sidorova, N. Y. and Rau, D. C. (2004). Differences between EcoRI nonspecific and ''star'' sequence complexes revealed by osmotic stress. *Biophys J* 87:2564–2576.

Smith, C. L. and Peterson, C. L. (2003). Coupling tandem affinity purification and quantitative tyrosine iodination to determine subunit stoichiometry of protein complexes. *Methods* 31:104–109.

Stern, R. V. and Frieden, E. (1993). Partial purification of the rat erythrocyte ceruloplasmin receptor monitored by an electrophoresis mobility shift assay. *Anal Biochem* 212:221–228.

Straney, D. C. and Crothers, D. M. (1985). Intermediates in transcription initiation from the *E. coli* lac UV5 promoter. *Cell* 43:449–459.

Torizawa, T., Shimizu, M., Taoka, M., Miyano, H., and Kainosho, M. (2004). Efficient production of isotopically labeled proteins by cell-free synthesis: a practical protocol. *J Biomol NMR* 30:311–312.

Towbin, H., Staehelin, T., and Gordon, J. (1989). Immunoblotting in the clinical laboratory. *J Clin Chem Clin Biochem* 27:495–501.

Towbin, H., Staehelin, T., and Gordon, J. (1992). Electrophoretic transfer of proteins from polyacrylamide gels to nitrocellulose sheets: procedure and some applications. *Biotechnology* 24:145–149.

Tsodikov, O. V., Holbrook, J. A., Shkel, I. A., and Record, M. T. Jr. (2001). Analytic binding isotherms describing competitive interactions of a protein ligand with specific and nonspecific sites on the same DNA oligomer. *Biophys J* 81:1960–1969.

Ueda, C. T. and Roberts, R. W. (2004). Analysis of a long-range interaction between conserved domains of human telomerase RNA. *RNA* 10:139–147.

Vallone, P. M. and Benight, A. S. (2000). Thermodynamic, spectroscopic, and equilibrium binding studies of DNA sequence context effects in four 40 base pair deoxyoligonucleotides. *Biochemistry* 39:7835–7846.

Vossen, K. M. and Fried, M. G. (1997). Sequestration stabilizes *Lac* repressor–DNA complexes during gel electrophoresis. 245:85–92.

Vossen, K. M., Stickle, D. F., and Fried, M. G. (1996). The mechanism of CAP-lac repressor binding cooperativity at the *E. coli* lactose promoter. *J Mol Biol* 255:44–54.

Wang, S., Huber, P. W., Cui, M., Czarnik, A. W., and Mei, H.-Y. (1998). Binding of Neomycin to the TAR Element of HIV-1 RNA Induces Dissociation of Tat Protein by an Allosteric Mechanism. *Biochemistry* 37:5549–5557.

Wu, H. M. and Crothers, D. M. (1984). Identification of the locus of sequence-directed and protein-induced DNA bending. *Nature* 308:509–513.

Wu, Z. L., Zhang, L., Beeler, D. L., Kuberan, B., and Rosenberg, R. D. (2002). A new strategy for defining critical functional groups on heparan sulfate. *FASEB J* 16:539–545.

Wyman, J. and Gill, S. J. (1990). *Binding and Linkage*. University Science Books, Mill Valley, CA.

Xian, J. and Harrington, M. G. (1998). Mobility shift electrophoresis of protein–DNA complexes by capillary electrophoresis. In Tietz, D. (ed.), *Nucleic Acid Electrophoresis*. Springer, Berlin, pp. 272–291.

Xuan, X. and Li, D. (2005). Band-broadening in capillary zone electrophoresis with axial temperature gradients. *Electrophroesis* 26:166–175.

Zerby, D. and Lieberman, P. M. (1997). Functional analysis of TFIID-activator interaction by magnesium-agarose gel electrophoresis. *Methods* 12:217–223.

Zimmerman, S. B. and Minton, A. P. (1993). Macromolecular crowding: biochemical, biophysical and physiological consequences. *Annu Rev Biophys Biomol Struct* 22:27–65.

15

Electrospray Ionization Mass Spectrometry and the Study of Protein Complexes

Alan M. Sandercock and Carol V. Robinson*

15.1. INTRODUCTION

Since the discovery over a decade ago that noncovalent protein–ligand interactions could be maintained in the gas phase of a mass spectrometer, the technique has been applied with great success to a number of problems in structural biology. These developments arose directly from the development in the late 1980s of ''soft'' methods of ionizing samples without fragmentation, and in particular electrospray ionization (ESI) (Fenn *et al.*, 1989). Mass spectrometry has now been applied to the study of a growing range of complexes, reaching molecular masses of megadaltons (MDa) in the case of ribosomes (Rostom *et al.*, 2000; Hanson *et al.*, 2003, 2004; Hanson and Robinson, 2004; Videler *et al.*, 2005).

In this chapter, we demonstrate the potential of mass spectrometry as a tool for the structural biologist, by describing some of the cases where mass spectrometry has provided unique insight. These range from straightforward determinations of stoichiometry from a molecular mass, to more subtle studies of binding equilibria and kinetics in complexes; many of these examples would be difficult to solve without the high mass resolution of modern mass spectrometers. Two additional benefits of mass spectrometry as a tool are the very low sample consumption required—which can be as low as microliters of micromolar solutions—and the speed at which analysis is performed; ions are detected within milliseconds of their release from solution. The former is clearly a boon where limited sample is available, while the latter means that, not only are data acquired rapidly, but also complexes can be detected at high speeds relative to rates of dynamic processes in solution.

A. M. SANDERCOCK • University of Cambridge, Department of Chemistry, Lensfield Road, Cambridge, CB2 1EW, UK C. V. ROBINSON • University of Cambridge, Department of Chemistry, Lensfield Road, Cambridge, CB2 1EW, UK and *Corresponding author: Tel: +44 1223 763846; Fax: +44 1223 763843; E-mail: cvr24@cam.ac.uk

The most obvious use of mass spectrometry for studying noncovalent protein complexes is as a tool to determine the molecular mass, and hence the stoichiometry of a complex. However, mass spectrometry also allows more detailed probing of a protein complex. One particularly useful experimental tool is tandem mass spectrometry, commonly referred to as MS/MS or MS^2 analysis. Tandem mass spectrometry is made possible by spectrometers that are capable of isolating specific sample ions in the gas phase, then exciting them to release dissociation products. The dissociation products are useful for confirming stoichiometry, particularly in heterogeneous samples, but can also give hints to the architecture of the complex.

In addition to molecular mass determinations, mass spectrometry also offers other, more subtle information on protein complexes, taking advantage of the high speed at which samples can be analyzed. An interesting application of mass spectrometry is in the study of the kinetics of protein complex formation and dissociation, or subunit exchange, where the appearance of new molecular mass species can be monitored after mixing (Fändrich *et al.*, 2000). A particularly powerful application can be achieved where a homooligomeric protein is available in both isotope-labeled (i.e., mass-labeled) and unlabeled forms; the labeled and unlabeled forms are essentially indistinguishable other than by mass, and the formation of mixed-isotope complexes can be followed over time.

15.2. HISTORY AND DEVELOPMENT OF MASS SPECTROMETRY AS A TOOL FOR STUDYING PROTEIN COMPLEXES

The key development that opened up the study of noncovalent complexes by mass spectrometry was the use of ESI (Fenn *et al.*, 1989), for which John Fenn was awarded a share of the 2002 Nobel Prize for chemistry. Together with its later refinement as ‘‘nanospray’’ ionization, the importance of ESI as a means of generating gas-phase ions arises from its ability to ionize large biological molecules, directly from aqueous solutions, without fragmentation of the ions. At first, it was mainly employed as a technique for studying small molecules, but it was soon realized that an ionization method gentle enough to produce ions of macromolecules without fragmenting their covalent structure might also allow maintenance of noncovalent interactions within gas-phase ions.

The earliest reports of noncovalent complexes involving proteins were studies of monomeric proteins bound to ligands. One of the earliest, in 1991, was the observation of substrate and product binding by lysozyme (Ganem *et al.*, 1991). Adding an enzyme substrate, the hexasaccharide of *N*-acetylglucosamine (NAG_6), led to observation of a species with a mass consistent with the $lysozyme_1–(NAG_6)_1$ complex, which altered over time to a $lysozyme_1–(NAG_4)_1$ complex as the substrate was cleaved by the enzyme. Binding of cleaved disaccharide was not observed, consistent with its much lower affinity for lysozyme; this suggests the observed species are complexes with specific binding interactions, transferred intact from solution into the gas phase, rather than a result of nonspecific binding during the electrospray process. Further evidence for the specific nature of the observed interactions was shown by a study with active site mutants of the acyl coenzyme A binding protein (ACBP) (Robinson *et al.*, 1996). Mutating tyrosine residues involved in specific interactions with the cofactor, as revealed by the crystal structure, led to reduced, or eliminated, binding of CoA observed by mass spectrometry.

Other early examples include myoglobin bound to its haem cofactor (Katta and Chait, 1991) and, in 1994, the first mass spectrum of a multimeric noncovalent protein complex—the 60-kDa avidin tetramer (Schwartz *et al.*, 1994), plus specific binding of its biotin and biotin maleimide ligands. Further examples of protein multimers studied by mass spectrometry were the hemoglobin tetramer and concanavalin A (Light-Wahl *et al.*, 1994), where a pH-induced dimer–tetramer transition was observed directly, and streptavidin (Schwartz *et al.*, 1995).

In recent years, the most impressive development has been in the sizes of complexes, which have been usefully studied by mass spectrometry. Many of the best examples have been taken from the cellular machinery of protein metabolism, such as ribosomes (Rostom *et al.*, 2000; Hanson *et al.*, 2003, 2004), RNA polymerase (Ilag *et al.*, 2004), and the chaperone complex GroEL (Rostom and Robinson, 1999; Sobott and Robinson, 2004; van Duijn *et al.*, 2005). Ribosomes represent the largest complexes successfully studied to date; the largest molecular mass to have been directly measured by mass spectrometry is that of the ribosome of the thermophilic bacterium *Thermus thermophilus*, at 2,325,463 $\pm$ 2,003 Da (Ilag *et al.*, 2005).

While it is remarkable that protein complexes of several hundred kilodaltons or greater can be maintained as gas-phase ions, it is also important to be able to address questions that are not easily answered by other techniques. One example that shows how mass spectrometry can complement the information already available is given by the recent demonstration of altered ribosome stalk stoichiometries in a group of thermophilic bacteria (Ilag *et al.*, 2005). The stalk is more dynamic than other regions of the ribosome, and is not observed in crystal structures. In mesophilic bacterial ribosomes, the stalk structure is pentameric, comprising two copies of a L7/L12 dimer bound to the protein L10. (L7 is N-terminally acetylated L12.) This structure had been thought to be universal (Gonzalo and Reboud, 2003), but mass spectra of the stalk complexes from *T. thermophilus*, *Thermus aquatica*, and *Thermotoga maritima* showed a heptameric stoichiometry of $L12_6L10$.

15.3. PRINCIPLES OF ESI MASS SPECTROMETRY OF PROTEIN COMPLEXES

15.3.1. Instrument Requirements

Mass spectrometry at its simplest has two requirements: (i) a means of generating gas-phase ions from a sample in the ''source'' region of the mass spectrometer and (ii) a means of separating these ions as a function of mass, or more correctly, the mass-to-charge ratio (m/z), in the ''analyzer'' stage of the instrument. Ions are detected by the m/z, rather than directly by mass, for reasons most easily comprehended for the simplest mass-analyzer, the time-of-flight (ToF) analyzer. Because the kinetic energy acquired by an ion as it crosses an electric field gradient is proportional to its charge, the square of its velocity is proportional to charge divided by mass. The ToF analyzer measures the time taken for an ion to travel a fixed distance, which is inversely proportional to its velocity, and therefore proportional to the square root of m/z. Another mass analyzer commonly used for protein applications is the quadrupole, which employs a set of parallel electrodes carrying a combination of DC and AC potentials to generate an oscillating electric field where only ions of specific m/z can maintain a stable trajectory.

The ToF analyzer is the ideal mass analyzer for large complexes, since, in principle it has an unlimited m/z detection range. In contrast, quadrupole, ion trap, and Fourier-transform ion cyclotron resonance (FT-ICR) mass analyzers each have a maximum technical m/z limit. The early studies of protein complexes were performed on ''standard'' models of mass spectrometer, but it has since been found that certain modifications to the instruments are beneficial for optimizing the maintenance and the detection of intact complexes (Sobott *et al.*, 2002b). Chief among these are the use of nanospray ionization (Chung *et al.*, 1999), a modification of ESI that allows even gentler ionization conditions to be used, and control of the pressure inside the instrument, which has a critical bearing on the transmission efficiency of ions to the instrument's detector (Chernushevich and Thomson, 2004). Another useful modification was to extend the quadrupole m/z range of quadrupole-ToF (Q-ToF) instruments, since ions of large biological macromolecules give m/z values that frequently fall above the limits of commercial instruments. Modifying the range of quadrupoles has allowed the transmission and the isolation of higher m/z ions, bringing complexes of several megadaltons within the working range for typical charge states of ions.

15.3.2. Electrospray Ionization

The ionization method most suited to proteins in aqueous solution is ESI, a ''soft'' ionization technique giving spectra dominated by intact molecular ions. The application of ESI to mass spectrometry of peptides and denatured proteins is nowadays routinely coupled to LC methods and automated. By contrast, the use of ESI-MS for native state proteins, and for their noncovalent complexes, currently requires a case-by-case approach to optimizing conditions.

In ESI, ions are generated by the application of high electric field strengths to the tip of a capillary containing a sample in polar solvent (Figure 15.1). For protein samples, ''positive electrospray'' is typically used, where the capillary is held at a high positive electrical potential; this gives rise to cationic adducts of the sample, such as $[M + nH]^{n+}$ or $[M + nNa]^{n+}$. The high field causes the exposed liquid surface to form an extended structure called a Taylor cone, from which charged droplets are ejected, trapping molecules of the analyte. The droplets evaporate and

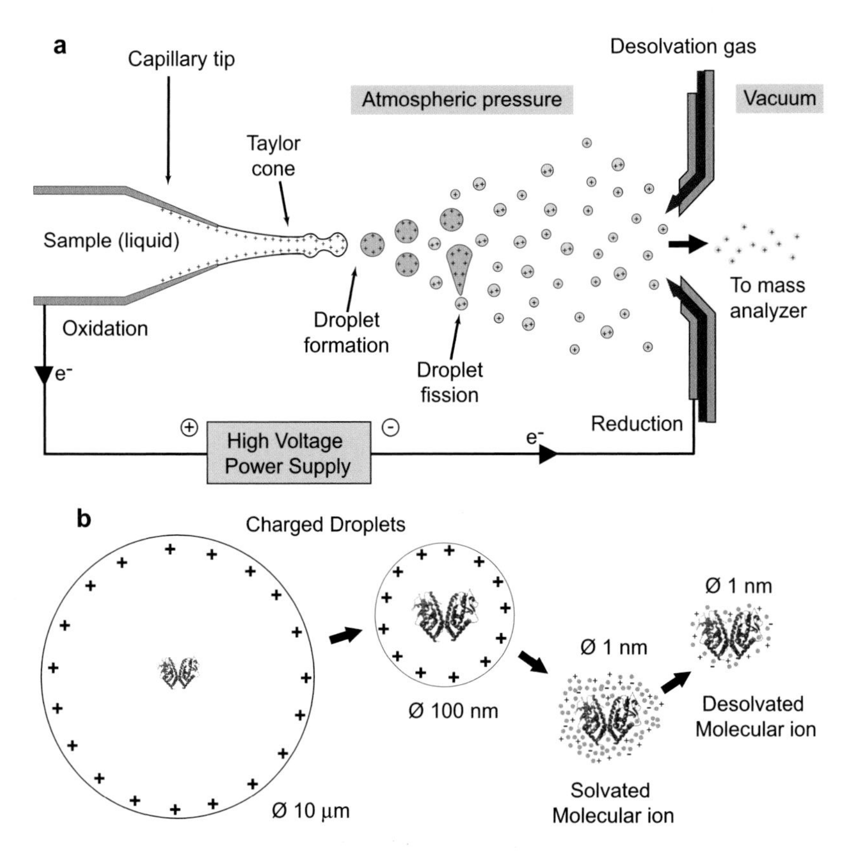

Figure 15.1. Electrospray ionization (ESI). (a) The sample is introduced via a capillary held at high voltage. The capillary diameter is ca. 0.5 mm in ''standard'' ESI and 1–10 μm in nanospray ionization. Due to the strong potential gradient, the exposed surface of the liquid carries a net charge and forms an extended structure called a Taylor cone, at the tip of which a fine spray of charged droplets is generated. Droplets lose solvent by evaporation, with the assistance of a drying gas, usually nitrogen or argon. This leads to an increase in surface charge density until the Rayleigh limit is reached, where Coulombic repulsion outweighs the surface tension and droplet fission occurs. (b) Sample molecules trapped in the droplets are released as solvated molecular ions as the solvent evaporates from the droplets. Under gentle conditions, the sample ions retain significant solvation shells (indicated by gray circles). Harsher ionization conditions lead to further desolvation, but may also lead to dissociation of native interactions within the sample ion.

break up because of electrostatic repulsion, often with the assistance of a heated drying gas, to yield gas phase, desolvated ions of the analyte. The ions enter the vacuum stages of the mass spectrometer, where they are focused and separated by manipulation of electrical potential gradients *en route* to a detector.

One unusual feature of ESI, compared with other ionization methods, is the production of ions of several different charge states for the same compound—that is, $[M + nH]^{n+}$ ions with a distribution of values of n. This phenomenon is particularly marked for proteins, and since ions are detected on the m/z scale, it leads to a "charge series" for each mass detected (Figure 15.2). It is simple to calculate the molecular mass—the m/z values of two consecutive peaks contain sufficient information to solve simultaneous equations in z and m, and this calculation is usually performed automatically by the data analysis software supplied with the instrument. It is tempting to convert the data from the m/z scale to a true mass scale for benefit of presentation, but while this has a benefit of simplifying the data, information contained in the charge state distribution is lost in this process. The distribution of charge states observed for a protein is (among other factors) a function of the number of solvent-exposed basic sites. This leads to a distinct difference in the charge state distributions for proteins sprayed from denaturing and nondenaturing solution conditions; the former tends to be more highly charged, and with a greater range of charge states than the latter (Figure 15.2). There is therefore some merit in examining m/z scale data.

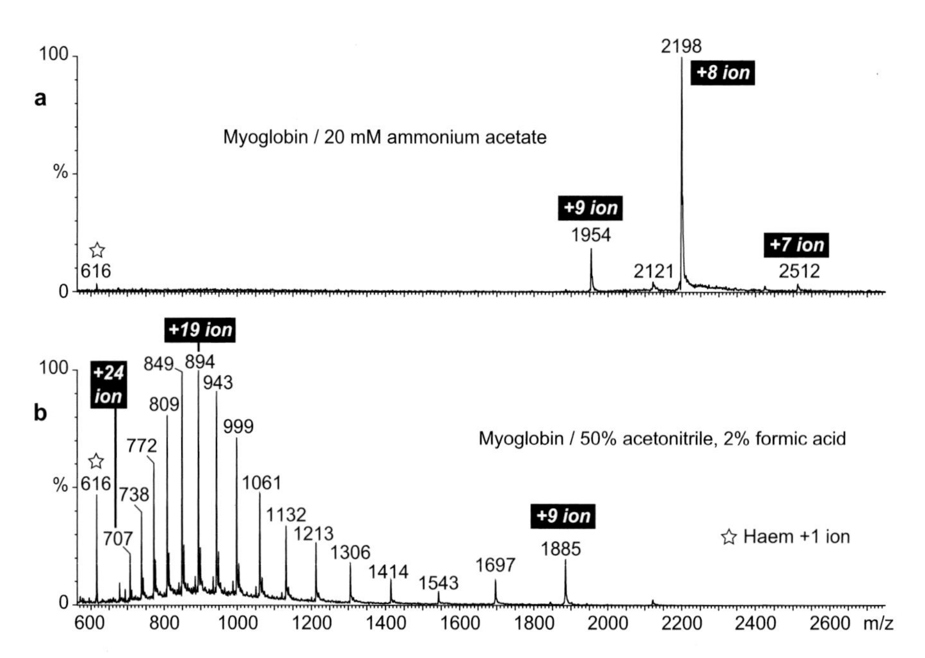

Figure 15.2. Mass spectra of horse-heart myoglobin under conditions where native interactions are retained (a) or disrupted (b). Proteins ionized by ESI show a charge series on the m/z scale, with denatured proteins exhibiting higher average charge state (hence lower m/z), and a wider distribution of charge states, compared with folded proteins. (a) Mass spectrum of myoglobin electrosprayed from ammonium acetate solution at neutral pH. A charge series is observed, as the protein populates three charge states significantly. The molecular mass calculated from these charge states shows retention of haem (observed mass = 17,577 Da; sequence mass = 16,951 Da + haem = 17,566 Da). (b) The observed mass for myoglobin electrosprayed from acidified 50:50 acetonitrile:water (16,959 Da) indicates dissociation of the haem cofactor, which is seen as a +1 ion at 616 Da. (As is conventional in mass spectrometry, the vertical axes show ion intensity as a percentage of the most intense ion.)

Many common buffer systems are incompatible with ESI, since the presence of high concentrations of salt ions suppresses the formation of sample ions. Since it is desirable to buffer native protein samples during the electrospray process to protect complexes from large pH changes during the ESI process, samples are usually buffer-exchanged against a volatile buffer solution, most commonly ammonium acetate, with a typical concentration range of 10 mM to 1 M. For protein complexes stable in ammonium acetate buffers, the exchange can be performed in bulk, e.g., by dialysis methods. Alternatively, exchange can be achieved rapidly and immediately before analysis, by washing on centrifugal filters or miniature gel filtration columns. Where other buffer components are required for maintaining a stable complex (e.g., metal ions, cofactors, or reducing agents), their concentrations must be kept to a minimum.

15.3.3. Nanospray

In order to maximize the generation of ions, and hence the signal intensity, protein samples for accurate mass determinations are often sprayed from solutions containing ca. 50% organic solvent to increase the volatility (usually methanol or acetonitrile) plus up to 1% formic acid. Acidified organic solutions, along with heated drying gases, are clearly undesirable when the interest is in the noncovalent interactions formed by the proteins, so a major challenge in native state protein mass spectrometry is to obtain satisfactory levels of desolvation from aqueous conditions only, without disrupting any interactions between the proteins. An important breakthrough in meeting this challenge was the adaptation to using nanoflow ESI, or ''nanospray,'' whereby the diameter of the electrospray capillary is reduced to a few microns (compared with ca. 0.5 mm typical of conventional ESI), typically using disposable glass or quartz capillaries coated on the outside with a thin layer of gold (Wilm and Mann, 1996). From the perspective of studying noncovalent complexes, the advantage of nanospray is that it produces smaller size charged droplets, which therefore desolvate with less energy input. Efficient desolvation of sample ions can therefore occur directly from aqueous solution, without heating or organic solvent, allowing retention of native interactions both in the solution phase before electrospray, and hence in the gas-phase ion. A second benefit of using nanospray is a reduction in the quantity of sample required. Typically a few microliters of protein solution are needed, at a concentration of some tens of micromolars—giving a total sample consumption on the order of picomoles of complex. (The sample concentration is best kept to a minimum, since high concentrations can lead to nonspecific association during ESI, when several molecules of sample become trapped within the same desolvating droplet.)

However, as a result of the gentle conditions, it is generally observed that proteins ionized from nondenaturing conditions retain solvent or buffer molecules, giving broadened peak widths and slightly heavier species than would be predicted from the sequence mass and isotope distribution of the protein(s) alone (Figure 15.3). The observed peaks represent masses of the order of $[M + n(\text{Buffer}) + zH]^{z+}$

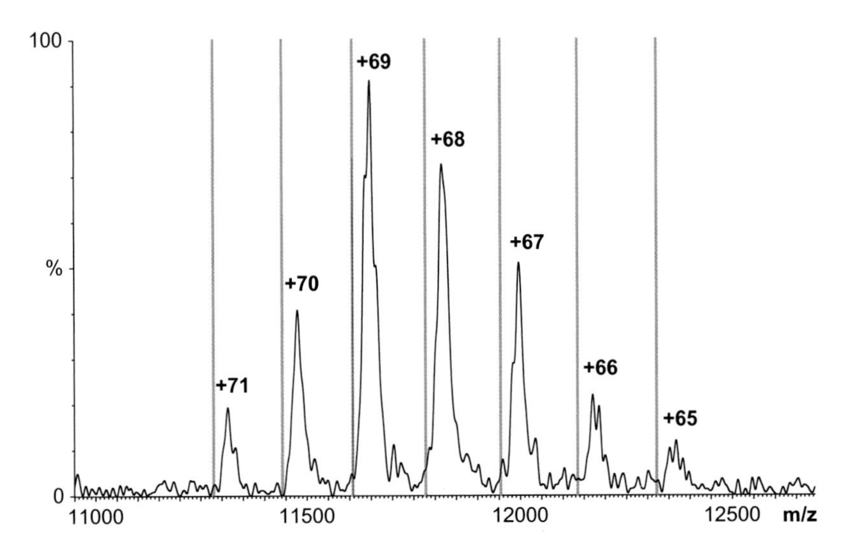

Figure 15.3. Increase in mass observed for noncovalent complexes due to buffer adducts, shown here for the GroEL 14-mer. Gray lines indicate the positions of ions expected from the atomic formula for the protein sequence ($C_{34664}H_{57792}N_{9660}O_{11228}S_{350}$). The mass calculated from the apex m/z values of charge states $+71$ to $+66$ is found to be 803,425 $\pm$ 202 Da, whereas the mass predicted from the protein sequence is 800,770 Da. The excess mass, 2,655 Da in this case, is highly dependent on the ionization conditions, since it is due to the amount of buffer retained by the complex.

instead of $[M + zH]^{z+}$. The amount and the distribution of these adducts is usually the major limit on the accuracy of the molecular mass determination, rather than the intrinsic resolution of the mass spectrometer (but note that the adducts can only lead to an increase in observed mass, giving the behavior of a systematic error rather than a random error). In spite of this, the accuracy is several orders of magnitude better than other techniques for protein mass determination.

15.3.4. The Role of Instrument Pressure in Mass Spectrometry of Native Protein Complexes

In seeking conditions for maximum signal intensity for large complexes, it was realized that the pressure gradient in the instrument's ion source or focusing stages was a key variable for detection of large ions (Rostom and Robinson, 1999; Tahallah *et al.*, 2001). Contrary to small molecule mass spectrometry, where it is normal to minimize the pressure inside the instrument in order to maximize the path lengths of ions between collisions, it was noted that the lowest pressures achieved by standard instruments were not ideal for large molecular mass ions. Instead, raising the pressure, e.g., by throttling the lines to rotary vacuum pumps, was found to improve the transmission of ions to the detector. Many recent studies of large complexes have been performed using instruments modified to allow control of pressures in the early vacuum stages by introducing an inert gas (Figure 15.4) (Sobott *et al.*, 2002b).

Increasing the pressure inside a mass spectrometer will increase the number of collisions experienced by ions on their way to the detector. Collisions have two effects on the protein ions. They can slow and deflect ion paths. They can also assist in "thermalizing" the ions; since the collisions are inelastic, they act to redistribute the translational kinetic energy gained by the ions from their acceleration along electrical potential gradients into both the thermal motions of the gas and the internal vibrations of the protein. These properties of gas-phase collisions are both useful for obtaining mass spectra of protein complexes. The observation of improved ion transmission and detection has been attributed to a phenomenon called "collisional cooling" (Chernushevich and Thomson, 2004). The early vacuum stages of mass spectrometers are fitted with ion guides, which manipulate electrical potentials in order to focus the ions entering from the source region into a narrow beam, able to pass through the small apertures that separate the differential pumping stages of the instrument (Krutchinsky *et al.*, 1998). This focusing requires the radial motions of ions around the focal axis to be minimized, a process which is less efficient for high mass species than it is for "small" ions. Modeling suggests the extra collisions with neutral gas caused by an increase in pressure assist by dampening these off-axis oscillations, thus assisting with ion beam focusing

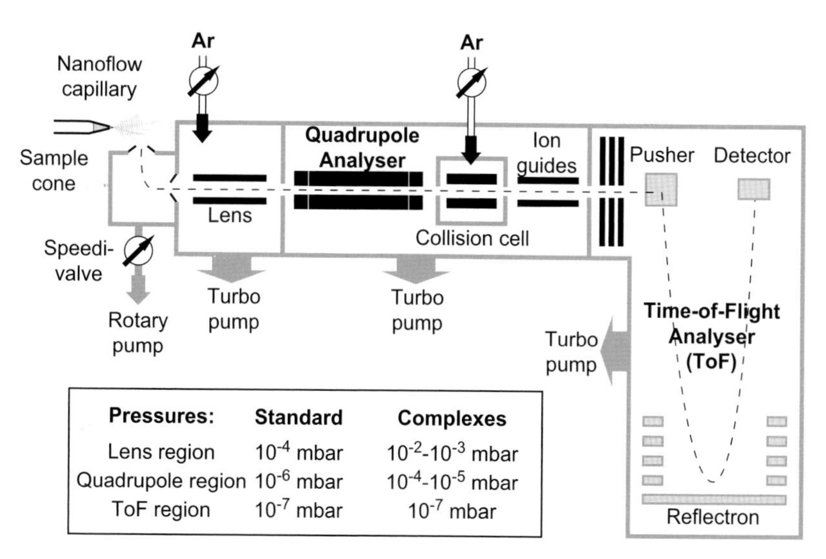

Figure 15.4. Schematic of a quadrupole-time-of-flight (Q-ToF) mass spectrometer modified for analyzing noncovalent complexes (Sobott *et al.*, 2002b). An altered pressure regime is achieved by an argon line into the ion lens region, and by throttling a speedivalve on the source region vacuum pump line. A modified quadrupole is used to allow transmission and isolation of higher m/z ions. The ion path through the instrument is indicated by the dashed line.

and hence with transmission of ions to subsequent stages of the instrument (Chernushevich and Thomson, 2004).

A second benefit of raised pressures is that it makes the desolvation process gentler. This can be seen by the increased retention of buffer molecules by complexes under increasing source region pressure and constant acceleration potentials (Figure 15.5) (Sobott and Robinson, 2004). This is a result of transfer of kinetic energy from the ions into the surrounding gas. The retention of buffer can be reduced by increasing the accelerating potentials to raise the energy of individual collisions, leading to a ''collisional cleaning'' effect, whereby the collisions provide sufficient energy to cause dissociation of bound water molecules. This process is a milder version of the processes involved in tandem mass spectrometry, where higher-energy collisions are used to dissociate full subunits from a complex (see later). Thus, the surplus buffer molecules can be removed, and the peak width reduced, by gentle acceleration of ions through a field of low-pressure collision gas.

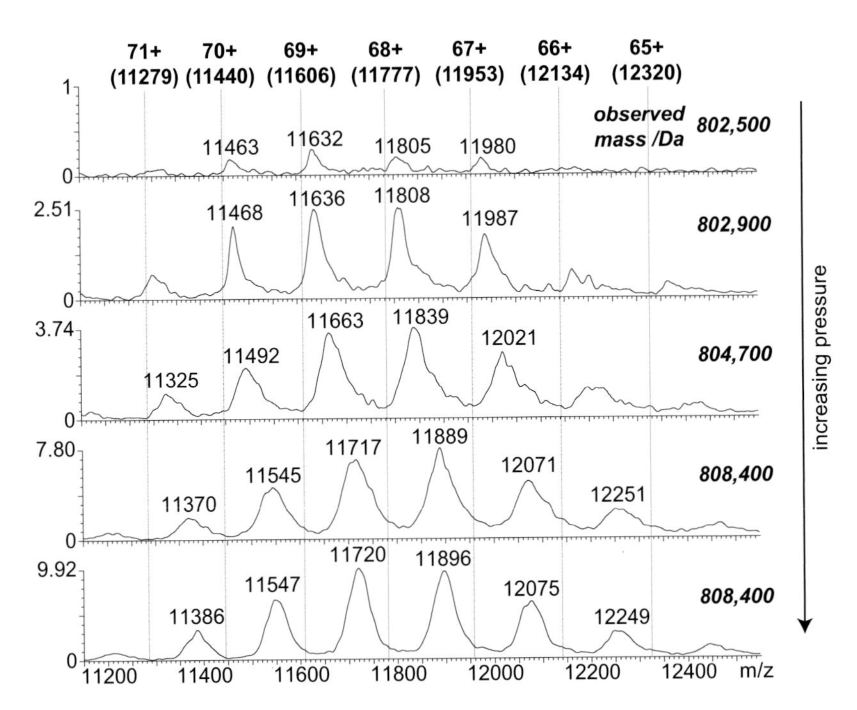

Figure 15.5. The effect of increasing pressure in the source region of the mass spectrometer, at constant acceleration potentials, on the ion intensity and retention of buffer molecules by the GroEL chaperonin complex. As the pressure is increased, transmission of the ions to the detector is improved, giving increased intensity (vertical axis). The collisions transfer kinetic energy from the ions to the surrounding gas, leading to reduced desolvation. This causes the ions to appear at higher m/z and to have broader peaks. (Reproduced with permission from Sobott and Robinson (2004). Copyright 2004 Elsevier.)

15.3.5. Tandem Mass Spectrometry

Tandem mass spectrometry (or MS/MS, MS^2) is a process made possible by mass spectrometers with two sequential mass analyzers separated by a collision cell; several types of instrument are capable of MS/MS analysis, but Q-ToF instruments are best suited for investigation of noncovalent protein complexes, although triple quadrupole instruments have also been used (Williams J. P. *et al.*, 2005). (MS/MS is also possible in instruments that temporarily store gas-phase ions, such as ion traps and Fourier-transform ion cyclotron resonance FT-ICR analyzers, but these are less amenable to large complexes.)

A Q-ToF MS/MS experiment uses the first quadrupole mass analyzer to isolate a ''parent'' ion of specific m/z from the mixture of ions present in the gas phase (Figure 15.6a); this ion is then thermally excited by acceleration into an inert gas, usually argon, held in a collision cell. As the energy of the collisions is increased, the parent ion is seen to dissociate, leading to the release of components of the activated complex (Figure 15.6b and c). This process is known as collision-induced

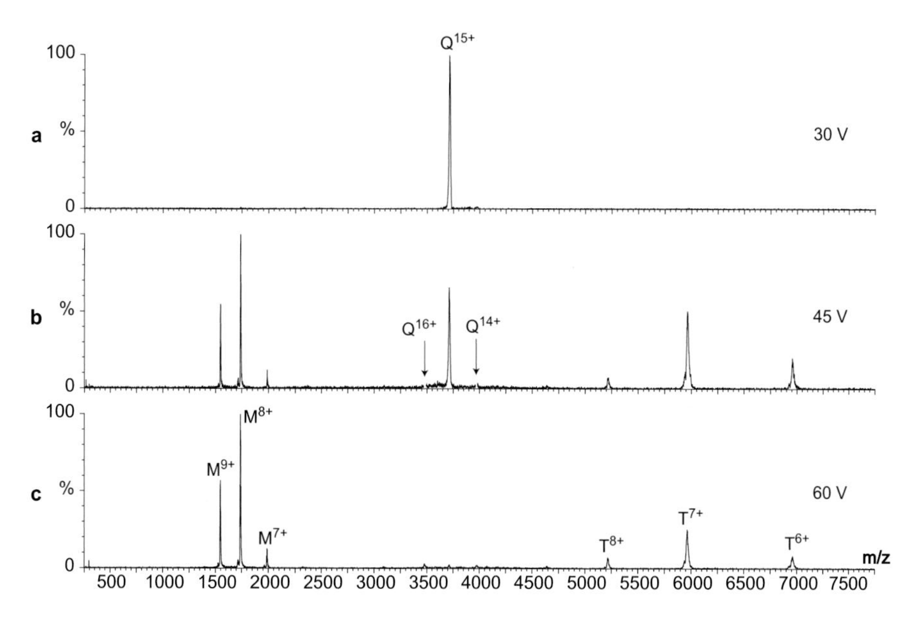

Figure 15.6. The MS/MS collision-induced dissociation (CID) pathway for a noncovalent complex, the 55-kDa transthyretin (TTR) tetramer, whereby a high-charge monomer is ejected from the complex. (a) The tetramer Q^{15+} ion is isolated by the quadrupole mass analyzer. At low collision energy, fragmentation does not occur, and the Q^{+15} ion reaches the detector intact. (b/c) Increased collision energies lead to asymmetric dissociation to monomers (M) and trimers (T). The total charge on the monomer and the trimer is equal the original charge on the tetramer. Also observed are weak ''charge-stripped'' tetramer ions (Q^{+16} and Q^{+14}) that result from dissociation of charged buffer molecules from the complex. (Reproduced with permission from Sobott *et al.* (2003). Copyright 2005 Elsevier.)

dissociation (CID). The ions released during this process are detected by the second, ToF, mass analyzer.

CID experiments performed on multiprotein complexes have revealed a common dissociation pathway for gas-phase noncovalent complexes (Sobott *et al.*, 2003), first observed for the break-up of avidin tetramers in the source region of the mass spectrometer (Light-Wahl *et al.*, 1994). Typically, the gas phase dissociation pathway of complex ions is different from dissociation pathways observed in solution, although it can reflect the spatial arrangement of subunits, and it is sensitive to the extent of interactions. Instead it is generally observed that CID leads to ejection of a single protein subunit from the oligomer, usually with disproportionately high charge (i.e., its m/z is much lower than that of the parent ion); this is termed ''asymmetric dissociation.'' A species with a mass corresponding to the remaining subunits of the complex can also be observed. The total charge states on the highly charged monomer and the ''stripped oligomers'' are conserved from the original charge of the parent ion.

Why should the dissociation pathway observed in CID differ from that in solution? The absence of bulk water is the key, since in solution, binding is in competition with interactions with water molecules. Without surrounding water, there is no hydrophobic effect to drive the burial of hydrophobic surfaces, which arises from the entropy benefit of releasing ordered water molecules at a surface. Since this is a major driving force in protein folding, the native structures of proteins are likely to be more thermodynamically unstable in the gas phase, compared with their solution counterparts. However, dissociation requires the breaking of electrostatic interactions within the complex, which are now stronger than in solution since the alternative stabilization provided by hydrogen bonding to water is lost. CID must provide the necessary energy for breaking these interactions, by thermal activation in the collision cell. It is intriguing to note that CID products from isolated ribosome ions lead to a bias toward the release of only proteins that were not in contact with the rRNA molecules; this accords with the higher proportion of charged electrostatic interactions expected in protein–RNA contacts, compared with protein–protein contacts (Hanson *et al.*, 2003). CID will also favor dissociation of subunits with the smallest interface areas with other subunits, making it a means of probing the geometry of large complexes.

Models for explaining the asymmetric charge distribution observed for CID products are based on the Coulombic repulsion between like charges on the surface of a dividing charged droplet (Schwartz *et al.*, 1995). As the thermal energy of the complex is increased by gas-phase collisions, the threshold will be reached where the energy is sufficient to break local electrostatic interactions within the complex, leading to loss of native inter- and intrasubunit interactions, and formation of some type of ''molten'' state. If one protein subunit of a complex was to substantially unfold in this state, the mobile charge carriers (e.g., H^+, NH_4^+, Na^+ etc.) would be able to redistribute over its extended surface area to minimize repulsion. This unfolding subunit will therefore accumulate a proportion of the overall charge in relation to its surface area. This charge will be disproportionately high in proportion to the mass of the subunit relative to the overall complex, since mass scales approximately with volume, leading to the observed high charges in the released subunit when it is eventually ejected.

The simplest use of MS/MS is as a means of confirming stoichiometry in a complex where this is ambiguous from just the molecular mass. This might be the case where one subunit of a heteromeric complex has a similar mass to another (within the mass error caused by retention of solvent molecules), or where one subunit has a mass close to a multiple of the mass of another. MS/MS is also useful where heterogeneity in a sample leads to overlapping ion series for different mass species, where it gives a means of checking the assignment of peaks in a spectrum.

An example where tandem mass spectrometry has revealed new information on a complex is the *Bacillus subtilis* protein *trp* RNA-binding attenuation protein (TRAP). TRAP is a regulatory protein that represses the expression of genes involved in the tryptophan biosynthesis pathway when free tryptophan levels are high. It crystallizes as a symmetrical, disc-shaped oligomer, with bound tryptophan ligands forming part of the subunit interfaces (Antson *et al.*, 1994). The outer surface of the disc forms an RNA-binding site that binds specific, repetitive nucleotide sequences found upstream of the transcribed *trp* operon, shown by crystal structures to be mediated by specific interactions with the bases and 2′-hydroxyl groups (Antson *et al.*, 1999; Hopcroft *et al.*, 2002). Crystal structures of TRAP show the protein crystallizes as an 11-mer (Antson *et al.*, 1994, 1999; Hopcroft *et al.*, 2002), and observations from ultracentrifugation were also consistent with this stoichiometry (Snyder *et al.*, 2004). However, mass spectrometry has shown that a minor population of 12-mer rings also exists, which were evidently not captured by the crystallization process (McCammon *et al.*, 2004). The 12-mer population was not easily resolved in MS-mode, since its ion series was obscured by that of the 11-mer. However, its presence was belied by MS/MS products, where the stripped oligomers produced by ejection of single subunits showed masses consistent with both 11-mers and 10-mers. The latter is the stripped product of the 11-mer, while the stripped 11-mers can only have come from 12-mer parent ions.

A more complicated example that demonstrates the scope of mass spectrometry for studying very heterogeneous complexes is given by the eye lens proteins αA-crystallin and αB-crystallin, which are members of the small heat shock family of molecular chaperone proteins. As well as acting as the major structural proteins of the eye lens, the crystallins also serve to protect the eye lens from clouding because of the accumulation of protein aggregates over time. Isolated crystallins are found to be highly polydisperse multimers, which has hampered attempts to characterize their structures. While size exclusion and multiangle laser light scattering have been used to calculate the approximate average size of the multimers, these techniques could not determine the size of minor species in the population or their distribution.

Applying tandem mass spectrometry to the study of the crystallins has allowed a more detailed characterization of the range of species present (Aquilina *et al.*, 2003, 2005). The characterization made use of the power of MS/MS to resolve an artifact arising from the m/z scale, where the presence of different size oligomers of a single protein can give rise to coincident ions when the charge is proportional to the oligomer size (i.e., oligomer ions of the formula $[P_n + inH]^{in+}$, where n is the number of subunits in the oligomer and i is the charge per subunit). For small oligomers, it is likely that other ions in the charge series will be resolved, allowing mass determination, but for larger species the adjacent ions in the charge series will be close to coincident and will not be resolved. Use of tandem mass spectrometry to select the coincident ions and strip a monomer by CID separates the product ions sufficiently to resolve the different oligomers present. This approach, when applied to αB-crystallin, revealed stoichiometries ranging from 10-mers (ca. 200 kDa) to 40-mers (ca. 800 kDa) within a single sample, with a 28-mer (ca. 560 kDa) being the most abundant oligomer (Figure 15.7). MS/MS analysis of αA-crystallin in its wild-type form and a C-terminally truncated form that accrues naturally over time, again revealed a distribution of stoichiometry sizes, with a mode of 26-mer for the full-length protein and 22-mer for the truncated form. Interestingly, αA-crystallin

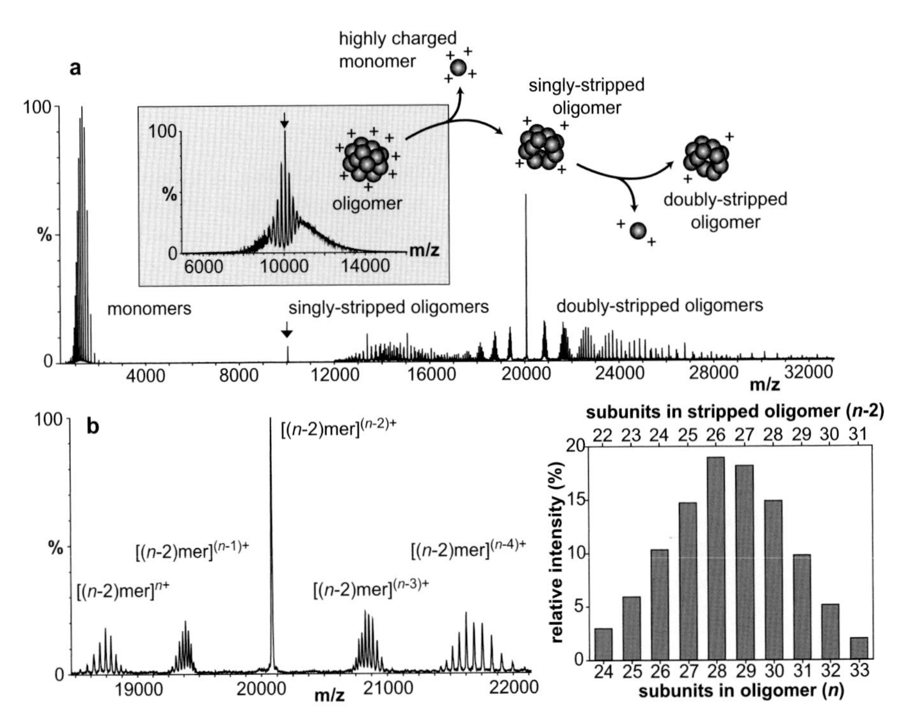

Figure 15.7. MS/MS analysis of polydisperse oligomers of αB-crystallin. (a) MS/MS spectrum showing highly charged monomers at low m/z, and singly and doubly stripped oligomers. (*Inset*) MS-mode spectrum, indicating the selected peak. (b) Expanded view of the doubly stripped oligomer region. Ions with two charges per subunit coincide exactly in m/z, while ions of other charge states are resolved. (c) Distribution of oligomer size within the parent ion, calculated from the intensities of ions for doubly stripped oligomers. (Reprinted with permission from Aquilina *et al.* (2003). Copyright 2003 National Academy of Sciences, USA.)

showed a bias toward even-numbered oligomer sizes, a bias that was enhanced by the truncation. This may indicate a dimeric substructure to the oligomers (Aquilina *et al.*, 2005). Further mass spectrometric evidence has shown that the dimeric substructure is lost on phosphorylation of a particular serine residue of αA-crystallin, despite the modification having no effect on the average size of the oligomeric assemblies (Aquilina *et al.*, 2004).

15.4. APPLICATIONS OF MASS SPECTROMETRY TO STRUCTURAL BIOLOGY

15.4.1. Defining Masses and Stoichiometries

The most obvious application of mass spectrometry in the study of large protein complexes is to use the unrivaled mass resolution to determine the stoichiometry of subunits within a complex. Unlike low-resolution techniques such as size exclusion or dynamic light scattering, the molecular masses of complexes of well over 100 kDa can be determined with sufficient accuracy to give absolute certainty of stoichiometries. For example, in spectra of thermophile ribosomes, careful analysis of the ion series assigned to the 30-S subunit was able to identify the absence of two protein subunits from the complex (Ilag *et al.*, 2005). One, named S1, is thought to be lost during purification, and the other, S6, appears as a dissociated protein at low m/z; the change in mass resulting from the dissociation of S6 is a decrease of 12 kDa from a total mass of 819 kDa—a change of less than 1.5%.

Even smaller mass differences can be observed by tandem mass spectrometry of ribosome ions. Isolation and CID of the 50-S ribosome subunit revealed dissociation of the protein L12, plus a second species 80 Da heavier, characteristic of a single phosphorylation, which was confirmed by phosphatase treatment. L12 is part of the labile stalk region, which is also seen as a separate complex in the mass spectrum. L12 released by CID from the free stalk did not show this phosphorylation, suggesting it is important in stabilizing stalk binding to the rest of the ribosome (Ilag *et al.*, 2005).

15.4.2. Beyond Mass Measurements—MS as a Tool for Studying Changes in Binding

Mass spectrometry is not only useful as a tool for mass determination. Its capability for rapid analysis and low sample requirements makes it an ideal tool for tracking changes in binding equilibria in response to changes in temperature or addition of new binding partners. The speed of analysis relative to equilibration also means mass spectrometry has the potential to resolve complexes in rapid equilibration between different stoichiometries. In these cases, size exclusion or ultracentrifugation may only detect an average structure, while crystallization may select for one structure from an equilibrating population.

A good example of an unexpected change in binding interactions detected by mass spectrometry was found in a study of the *Escherichia coli* RNA polymerase core enzyme binding to its σ^{70} cofactor, on addition of a regulator of σ^{70}, Rsd (Ilag *et al.*, 2004). Addition of recombinant σ^{70} to the core polymerase led to the expected binding of one equivalent of σ^{70} to give the *holo*-polymerase. The Rsd protein has previously been shown to interact with σ^{70} by mass spectrometry (Westblade *et al.*, 2004). On addition of Rsd to the *holo*-polymerase, a new species appeared, with a molecular mass that showed it to be σ_1^{70}: Rsd_1, hence identifying the stoichiometry of this complex. However, close examination of the peaks in the mass spectrum arising from the core polymerase showed formation of a second interaction, a 1:1 complex between Rsd and the core polymerase. The change in mass because of Rsd binding (adding 18 kDa to a 389-kDa complex) was easily resolved with the aid of collisional cleaning processes—an increase in mass of less than 5%. Rsd had originally been identified as a σ^{70}-binding protein, present in nongrowing cell populations. Originally it was proposed to sequester σ^{70}, so as to facilitate binding of other σ-factors and hence alter gene expression profiles; the mass spectrometry evidence shows this model to be simplistic, and suggests Rsd may have a more subtle role in controlling polymerase activity.

One interesting use of mass spectrometry to detect a change in binding equilibria in response to altered conditions is shown by a study of the RXR:RAR nuclear receptor heterodimer (Sanglier *et al.*, 2004). Nuclear receptors are ligand-activated, DNA-bound transcription factors that bind either repressor or activator proteins according to the occupancy of an agonist-binding site in their C-terminal ligand binding domains (Nagy and Schwabe, 2004). RXR and RAR are two members of the nuclear receptor class that form a heterodimer *in vivo* and are activated by binding of the agonist 9-*cis*-retinoic acid. Mass spectrometry of the RXR–RAR ligand-binding domains heterodimer, in the presence of a peptide comprising a receptor-binding motif from a repressor protein, showed binding of the peptide to receptor heterodimer. Addition of agonists led to disruption of the receptor–repressor protein interaction, leading to loss of the peptide mass from heterodimer observed in the mass spectrum (Sanglier *et al.*, 2004).

An impressive example of using mass spectrometry to follow the activity of a large protein complex is the recent study of the GroEL chaperonin complex binding and folding a chemically denatured protein substrate (van Duijn *et al.*, 2005). GroEL forms a 801-kDa 14-mer structure in solution, and its successful transfer to the gas phase for mass spectrometry was first reported in 1999 by Rostom and Robinson. The GroEL structure forms a double ring with two deep cavities, which are capped in its complex with its co-chaperonin GroES, where un-/misfolded proteins are given the opportunity to fold correctly. In these cavities, a structural transition induced by ATP hydrolysis is proposed to alter the inner surface from a predominantly hydrophobic, to a predominantly hydrophilic one, forcing the trapped protein chain to re-equilibrate and fold properly. In the recent study (van Duijn *et al.*, 2005), GroEL was studied in conjunction with gp31, a bacteriophage T4 GroES-analog that also binds to GroEL in a nucleotide-dependent manner, and with a urea-denatured sample of the bacteriophage gp23 capsid protein. In the absence of the cochaperonin, GroEL–gp23 complexes were seen with stoichiometries 14:0, 14:1, and 14:2, suggesting both of the cavities of GroEL could accommodate gp23 simultaneously. Addition of gp31 and MgADP led to formation of a single mass ternary complex, with a mass consistent with GroEL with one bound gp23 and one cavity capped by gp31 (although whether this was the same cavity as was occupied by gp23 could not be stated from the mass alone). Replacement of ADP for ATP, led to the appearance in the spectrum of hexamers of the previously unfolded gp23 monomers, along with peaks corresponding to *apo*-GroEL. Thus mass spectrometry has been used to provide a direct visualization of catalytic activity in a chaperone ternary complex greater than 800 kDa.

15.4.3. Temperature-Induced Changes in Binding

Most of the preceding examples of using mass spectrometry for structural biological applications have related to cases where changes in interactions of proteins have been related to the introduction of new binding partners. However, in order to characterize the thermodynamics of binding interactions more fully, it is useful to be able to examine samples over a range of temperatures. The effect of temperature on a noncovalent protein complex observed by mass spectrometry was noted in 1994, where the instrument set-up allowed heating of a metal capillary downstream of the electrospray source (Light-Wahl *et al.*, 1994). In this case, the ions were heated after transfer to the gas phase, and the observed dissociation pathway was identical to that achieved by CID.

For understanding the thermodynamics of a complex, it is more useful to study the effect of temperature while the sample is still in solution. This has been made possible by the development of a temperature-controlled source block for holding the sample within a nanospray needle (Figure 15.8a) (Benesch *et al.*, 2003). This source was used to follow thermal denaturation of hen-egg lysozyme by the increase in charge states observed as the protein unfolds, and found a midpoint that agreed closely with more established experimental methods. The dissociation of subunits from a multimeric protein, the 12-mer of *Ta*HSP16.9, was also demonstrated. Here, the heat-denaturation pathway was more complicated, with the appearance of both monomeric and dimeric species, and also of a 14-mer species (Figure 15.8b). The appearance of a dimeric species indicates that the dissociation is taking place in solution, not the gas phase, and the observed change in oligomeric behavior with temperature in a heat shock protein may reflect an aspect of its biological role. It is difficult to see how this information on oligomer size could have been obtained by methods other than mass spectrometry.

15.4.4. Subunit Exchange

The technique of subunit exchange is a very powerful extension of the capabilities of mass spectrometry, and is a means of measuring the kinetics of complex

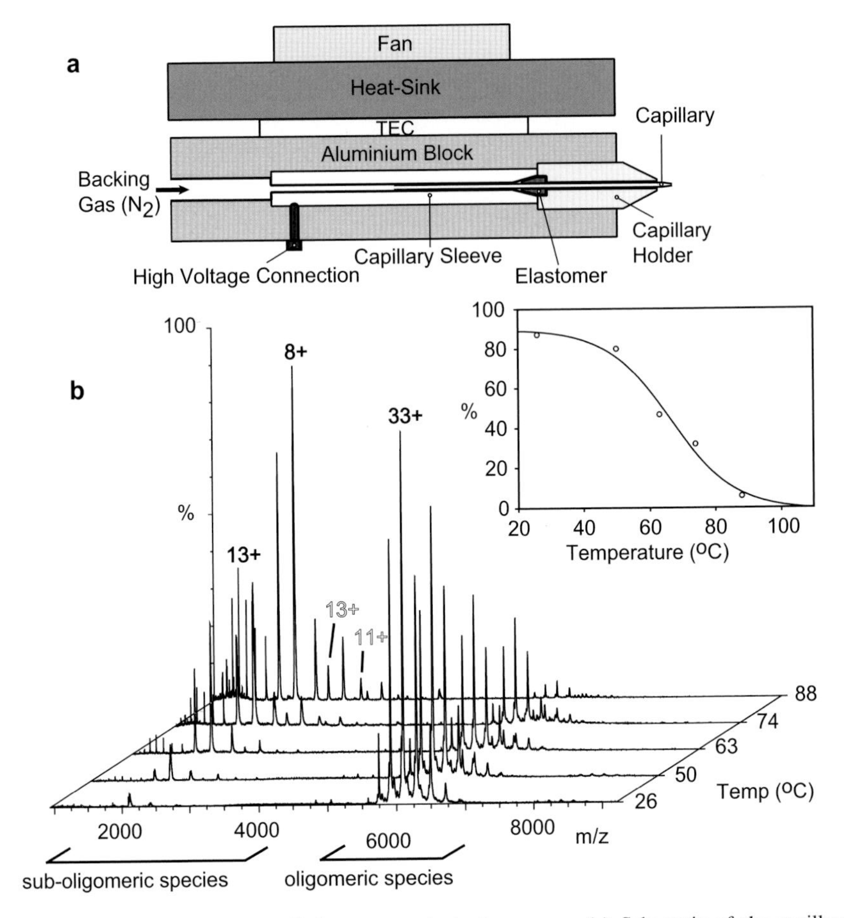

Figure 15.8. A temperature-controlled nanospray ionization source. (a) Schematic of the capillary holder (TEC = thermoelectric cooler). (b) Effect of increasing temperature on a small heat shock protein (sHSP) dodecamer (*Ta*HSP16.9) measured by mass spectrometry. As the temperature is increased, the uniform dodecamer population dissociates predominantly to monomers (solid numbering) and dimers (hollow numbering), but a small population of 14-mers is also observed. (*Inset*) Decrease in the dodecamer signal intensity as a function of temperature. (Reprinted with permission from Benesch *et al*. (2003). Copyright 2003 American Chemical Society.)

assembly and disassembly in systems where the dissociation constants are sufficiently small that only the complex is readily detected. Subunit exchange can be detected by mass spectrometry wherever exchange causes a change in mass; either heterooligomer formation, or in the mixing of unlabeled and, e.g., ^{15}N-labeled protein.

A good example of the use of mass spectrometry in measuring subunit exchange kinetics in heterooligomer formation is given by a study of oligomerization between two members of the small heat shock protein (sHSP) family (Sobott *et al.*, 2002a). The dodecameric sHSPs HSP18.1 from *Pisum sativum* (pea) and HSP16.9 from *Triticum aestivum* (wheat), which are both members of the same class of cytosolic sHSPs, were mixed, and the formation of heterooligomers was followed by mass spectrometry. Since the subunits of pea and wheat proteins differ in mass by just over 1 kDa, the mixed oligomers with different compositions are distinguishable by mass spectrometry. By tracking the formation of these mixed species as a function of time, it was possible to show that oligomers exchanged subunits on a rapid timescale, and that the pathway for exchange in this system proceeded via a dimeric substructure (Sobott *et al.*, 2002a).

Subunit exchange has also been measured by mass spectrometry in the α-crystallins, discussed earlier with regard to their subunit polydispersity. The two isoforms αA-crystallin and αB-crystallin form a heteromeric complex *in vivo*, and because of the difference in mass between the two types of protein subunits, the formation of mixed oligomers can be followed as a change in mass over time. The formation of mixed oligomers of αA-crystallin and αB-crystallin from the pure proteins was completed over a period of about 30 min; however, mixing of the truncated form of αA-crystallin with either the full-length αA-crystallin or with αB-crystallin over the same period was not detected, indicating reduced subunit dynamics in the truncated form (Figure 15.9) (Aquilina *et al.*, 2005).

The systems highlighted earlier describe the use of subunit exchange in cases where a heterooligomer forms between two protein subunits of different mass. It would be useful to extend this method to the study of homooligomeric proteins where the role of subunit dynamics is thought to be important, but for this to be possible there needs to be a "mass label" in order to distinguish molecules within a

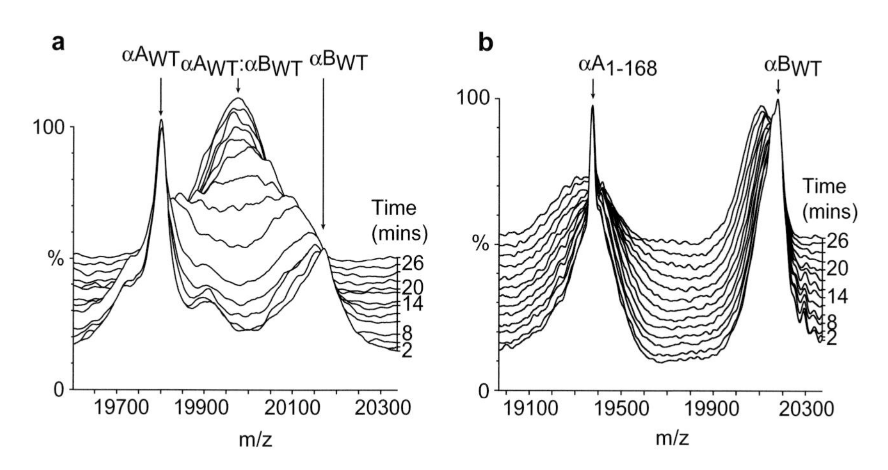

Figure 15.9. Subunit exchange within the α-crystallins measured by mass spectrometry. (a) Exchange between αA-crystallin and αB-crystallin occurs substantially within 30 min. (b) A C-terminally truncated form of αA crystallin shows reduced subunit dynamics, with no measurable heterooligomers formed with αB-crystallin on the same timescale. (Reprinted with permission from Aquilina *et al.* (2005). Copyright 2005 American Society for Biochemistry and Molecular Biology, Inc.)

mixed population. The most effective way of achieving this is through the use of isotopic labels, by bacterial expression in enriched media, a well-established process for producing proteins for NMR studies. The key benefit of isotopic labeling is that the mass label does not affect binding interactions.

The application of mass spectrometry to subunit exchange in labeled proteins is a very recent development. One example involves the human protein transthyretin (TTR), a thyroid hormone-binding protein, sequence variants of which are frequently involved in familial amyloid diseases. MS has been used to study the exchange of subunits within wild-type TTR tetramer, in the aggressively aggregation-prone mutant L55P, and in the formation of WT-L55P heterotetramers. These data have been used to compare models for the subunit exchange mechanism within TTR (Keetch *et al.*, 2005). Given that subunit exchange is thought to be important in the roles of a range of proteins, it seems likely that this technique is to be given increasing prominence in the future.

15.5. SUMMARY

In this chapter, we have highlighted some of the ways mass spectrometry is applied to answer questions on protein interactions, and the developments that have made this possible. Mass spectrometry has always benefited from high resolution and very low sample requirements compared with many other techniques. Recent progress in instrument design has increased the size limits to allow direct mass measurements to be performed on complexes in the megadalton range. When combined with a better understanding of how noncovalently bound protein complexes behave as gas-phase ions, the use of mass spectrometry permits a deeper insight into high-mass complexes, especially when coupled to the power of tandem mass spectrometry to probe within the structures of large complexes. This has proved particularly usefully in cases where heterogeneity in a system places limits on crystallographic methods, or where other solution techniques have insufficient resolution. The most exciting recent developments are applications of mass spectrometry to observe dynamic processes such as subunit exchange, where the combination of resolution and fast data acquisition gives mass spectrometry a unique advantage over other techniques. We look forward to seeing increased use of these methods to study a broad range of problems in protein chemistry in the near future.

REFERENCES

Antson, A. A., Brzozowski, A. M., Dodson, E. J., Dauter, Z., Wilson, K. S., Kurecki, T., Otridge, J., and Gollnick, P. (1994). 11-Fold symmetry of the *trp* RNA-binding attenuation protein (TRAP) from *Bacillus subtilis* determined by X-ray analysis. *J Mol Biol* 244:1–5.

Antson, A. A., Dodson, E. J., Dodson, G., Greaves, R. B., Chen, X., and Gollnick, P. (1999). Structure of the *trp* RNA-binding attenuation protein, TRAP, bound to RNA. *Nature* 401:235–242.

Aquilina, J. A., Benesch, J. L. P., Bateman, O. A., Slingsby, C., and Robinson, C. V. (2003). Polydispersity of a mammalian chaperone: mass spectrometry reveals the populations of oligomers in alpha B-crystallin. *Proc Natl Acad Sci USA* 100:10611–10616.

Aquilina, J. A., Benesch, J. L., Ding, L. L., Yaron, O., Horwitz, J., and Robinson, C. V. (2004). Phosphorylation of alphaB-crystallin alters chaperone function through loss of dimeric substructure. *J Biol Chem* 279:28675–28680.

Aquilina, J. A., Benesch, J. L. P., Ding, L. L., Yaron, O., Horwitz, J., and Robinson, C. V. (2005). Subunit exchange of polydisperse proteins: mass spectrometry reveals consequences of αA-crystallin truncation. *J Biol Chem* 280:14485–14491.

Benesch, J. L. P., Sobott, F., and Robinson, C. V. (2003). Thermal dissociation of multimeric protein complexes by using nanoelectrospray mass spectrometry. *Anal Chem* 75:2208–2214.

Chernushevich, I. V., and Thomson, B. A. (2004). Collisional cooling of large ions in electrospray mass spectrometry. *Anal Chem* 76:1754–1760.

Chung, E. W., Henriques, D. A., Renzoni, D., Morton, C. J., Mulhern, T. D., Pitkeathly, M. C., Ladbury, J. E., and Robinson, C. V. (1999). Probing the nature of interactions in SH2 binding interfaces—evidence from electrospray ionization mass spectrometry. *Protein Sci* 8:1962–1970.

Fändrich, M., Tito, M. A., Leroux, M. R., Rostom, A. A., Hartl, F. U., Dobson, C. M., and Robinson, C. V. (2000). Observation of the noncovalent assembly and disassembly pathways of the chaperone complex MtGimC by mass spectrometry. *Proc Natl Acad Sci USA* 97:14151–14155.

Fenn, J. B., Mann, M., Meng, C. K., Wong, S. F., and Whitehouse, C. M. (1989). Electrospray ionization for mass-spectrometry of large biomolecules. *Science* 246:64–71.

Ganem, B., Li, Y.-T., and Henion, J. D. (1991). Observation of noncovalent enzyme-substrates and enzyme-product complexes by ion-spray mass spectrometry. *J Am Chem Soc* 113:7818–7819.

Gonzalo, P. and Reboud, J. P. (2003). The puzzling lateral flexible stalk of the ribosome. *Biol Cell* 95:179–193.

Hanson, C. L., Fucini, P., Ilag, L. L., Nierhaus, K. H., and Robinson, C. V. (2003). Dissociation of intact *Escherichia coli* ribosomes in a mass spectrometer. Evidence for conformational change in a ribosome elongation factor G complex. *J Biol Chem* 278:1259–1267.

Hanson, C. L. and Robinson, C. V. (2004). Protein–nucleic acid interactions and the expanding role of mass spectrometry. *J Biol Chem* 279:24907–24910.

Hanson, C. L., Videler, H., Santos, C., Ballesta, J. P., and Robinson, C. V. (2004). Mass spectrometry of ribosomes from *Saccharomyces cerevisiae*: implications for assembly of the stalk complex. *J Biol Chem* 279:42750–42757.

Hopcroft, N. H., Wendt, A. L., Gollnick, P., and Antson, A. A. (2002). Specificity of TRAP–RNA interactions: crystal structures of two complexes with different RNA sequences. *Acta Crystallogr D Biol Crystallogr* 58:615–621.

Ilag, L. L., Videler, H., McKay, A. R., Sobott, F., Fucini, P., Nierhaus, K. H., and Robinson, C. V. (2005). Heptameric (L12)6/L10 rather than canonical pentameric complexes are found by tandem MS of intact ribosomes from thermophilic bacteria. *Proc Natl Acad Sci USA* 102: 8192–8197.

Ilag, L. L., Westblade, L. F., Deshayes, C., Kolb, A., Busby, S. J., and Robinson, C. V. (2004). Mass spectrometry of *Escherichia coli* RNA polymerase: interactions of the core enzyme with sigma70 and Rsd protein. *Structure (Camb)* 12:269–275.

Katta, V. and Chait, B. T. (1991). Observation of the heme–globin complex in native myoglobin by electrospray-ionization mass spectrometry. *J Am Chem Soc* 113:8534–8535.

Keetch, C. A. et al., (2005). *J Biol Chem* 50:41667–41674.

Krutchinsky, A. N., Chernushevich, I. V., Spicer, V. L., Ens, W., and Standing, K. G. (1998). Collisional damping interface for an electrospray ionization time-of-flight mass spectrometer. *J Am Soc Mass Spectrom* 9:569–579.

Light-Wahl, K. J., Schwartz, B. L., and Smith, R. D. (1994). Observation of the noncovalent quaternary associations of proteins by electrospray-ionization mass-spectrometry. *J Am Chem Soc* 116: 5271–5278.

McCammon, M. G., Hernández, H., Sobott, F., and Robinson, C. V. (2004). Tandem mass spectrometry defines the stoichiometry and quaternary structural arrangement of tryptophan molecules in the multiprotein complex TRAP. *J Am Chem Soc* 126:5950–5951.

Nagy, L. and Schwabe, J. W. R. (2004). Mechanism of the nuclear receptor molecular switch. *Trends Biochem Sci* 29:317–324.

Robinson, C. V., Chung, E. W., Kragelund, B. B., Knudsen, J., Aplin, R. T., Poulsen, F. M., and Dobson, C. M. (1996). Probing the nature of noncovalent interactions by mass spectrometry. A study of protein-CoA ligand binding and assembly. *J Am Chem Soc* 118:8646–8653.

Rostom, A. A., Fucini, P., Benjamin, D. R., Juenemann, R., Nierhaus, K. H., Hartl, F. U., Dobson, C. M., and Robinson, C. V. (2000). Detection and selective dissociation of intact ribosomes in a mass spectrometer. *Proc Natl Acad Sci USA* 97:5185–5190.

Rostom, A. A. and Robinson, C. V. (1999). Detection of the intact GroEL chaperonin assembly by mass spectrometry. *J Am Chem Soc* 121:4718–4719.

Sanglier, S., Bourget, W., Germain, P., Chavant, V., Moras, D., Gronemeyer, H., Potier, N., and van Dorsselaer, A. (2004). Monitoring ligand-mediated nuclear receptor-coregulator interactions by noncovalent mass-spectrometry. *Eur J Biochem* 271:4958–4967.

Schwartz, B. L., Bruce, J. E., Anderson, G. A., Hofstadler, S. A., Rockwood, A. L., Smith, R. D., Chilkoti, A., and Stayton, P. S. (1995). Dissociation of tetrameric ions of noncovalent streptavidin complexes formed by electrospray ionization. *J Am Soc Mass Spectrom* 6:459–465.

Schwartz, B. L., Lightwahl, K. J., and Smith, R. D. (1994). Observation of noncovalent complexes to the avidin tetramer by electrospray-ionization mass-spectrometry. *J Am Soc Mass Spectrom* 5:201–204.

Snyder, D., Lary, J., Chen, Y., Gollnick, P., and Cole, J. L. (2004). Interaction of the *trp*-RNA-binding attenuation protein (TRAP) with anti-TRAP. *J Mol Biol* 338:669–682.

Sobott, F., Benesch, J. L. P., Vierling, E., and Robinson, C. V. (2002a). Subunit exchange of multimeric protein complexes—real-time monitoring of subunit exchange between small heat shock proteins by using mass spectrometry. *J Biol Chem* 277:38921–38929.

Sobott, F., Hernández, H., McCammon, M. G., Tito, M. A., and Robinson, C. V. (2002b). A tandem mass spectrometer for improved transmission and analysis of large macromolecular assemblies. *Anal Chem* 74:1402–1407.

Sobott, F., McCammon, M. G., and Robinson, C. V. (2003). Gas-phase dissociation pathways of a tetrameric protein complex. *Int J Mass Spectrom* 230:193–200.

Sobott, F. and Robinson, C. V. (2004). Characterising electrosprayed biomolecules using tandem-MS—the noncovalent GroEL chaperonin assembly. *Int J Mass Spectrom* 236:25–32.

Tahallah, N., Pinkse, M., Maier, C. S., and Heck, A. J. R. (2001). The effect of the source pressure on the abundance of ions of noncovalent protein assemblies in an electrospray ionization orthogonal time-of-flight instrument. *Rapid Commun Mass Spectrom* 15:596–601.

van Duijn, E., Bakkes, P. J., Heeren, R. M. A., van den Heuvell, R. H. H., van Heerikhuizen, H., van der Vries, S. M., and Heck, A. J. R. (2005). Monitoring macromolecular complexes involved in the chaperonin-assisted protein folding cycle by mass spectrometry. *Nat Methods* 2:371–376.

Videler, H., Ilag, L. L., McKay, A. R., Hanson, C. L., and Robinson, C. V. (2005). Mass spectrometry of intact ribosomes. *FEBS Lett* 579:943–947.

Westblade, L. F., Ilag, L. L., Powell, A. K., Kolb, A., Robinson, C. V., and Busby, S. J. (2004). Studies of the *Escherichia coli* Rsd-sigma70 complex. *J Mol Biol* 335:685–692.

Williams, J. P. et al., (2005). *Biochemistry* 44:8282–8290.

Wilm, M. and Mann, M. (1996). Analytical properties of the nanoelectrospray ion source. *Anal Chem* 68:1–8.

16

Sedimentation Velocity in the Study of Reversible Multiprotein Complexes

Peter Schuck

16.1. INTRODUCTION

Analytical ultracentrifugation (AUC) is one of the classic methods in the study of biological macromolecules. Developed by Svedberg in the 1920s, it was a central tool to show, for the first time, the existence of macromolecules in solution (Svedberg and Fahraeus, 1926). Owing to the commercial instrument developed by Pickels (Elzen, 1988), it was in widespread use in the following decades to study the physicochemical properties of proteins and nucleic acids, including their size, shape, and reversible complex formation (Schachman, 1959). Although in the 1970s and 1980s AUC use in mainstream biochemistry laboratories went into a temporary decline, preparative centrifuges had been spawned by AUC and continued to represent a basic tool of most biochemical laboratories (Elzen, 1988). In the 1990s, with the increasing availability and sophistication of recombinant protein expression, with the general interest in protein interactions emerging, and with improved AUC instrumentation and modern computational capabilities making the study of the sedimentation of interacting systems more accessible, AUC experienced a renaissance, as it was recognized that the ultracentrifuge could provide important and unique information on protein interactions (Schachman, 1989, 1992). In this book, two chapters are devoted to describe the capabilities of AUC for the biophysical study of complex interacting systems; sedimentation velocity (SV) AUC, which examines the time course of sedimentation, and sedimentation equilibrium (SE) AUC (see Chapter 10), which is concerned with the thermodynamic analysis of the final macromolecular distribution.

P. SCHUCK • National Institute of Biomedical Imaging and Bioengineering, National Institutes of Health, Bethesda, MD, USA. Tel: +1 301 4351950; Fax: +1 301 4801242; E-mail: pschuck@helix.nih.gov

What makes AUC particularly attractive for the study of complex protein interactions is, first, the conceptual simplicity of the approach—only a gravitational force is applied to the macromolecular mixtures in free solution. The driving force for migration is directly linked to the molar mass. AUC permits the highly quantitative interpretation based on first principles, firmly rooted in the theory of multi-component solution thermodynamics. Second, the size-dependent migration reveals the number and the size of macromolecular components and their different complexes. The ability to discern easily the existence of multiple different reversible protein complexes that may be formed free in solution and to assess their molar masses and stoichiometry (and in some cases, energetics and kinetics of complex formation) is a unique feature of AUC. Third, for studying protein interactions, a key advantage compared with many other separation methods is that the larger complexes will not be separated from the free constituents. Since the faster-sedimenting reversibly formed complexes always remain in a bath of the slower-sedimenting unbound protein, they continue to participate in association–dissociation reactions, which makes reversible, transient complexes more accessible for study by AUC. A consequence of this is that in sedimentation equilibrium, simultaneous sedimentation and chemical equilibria are studied (see Chapter 10) whereas the time course of sedimentation in SV can be governed by coupled migration and chemical reactions, such as those described by Gilbert–Jenkins theory (Gilbert and Jenkins, 1956). This is a very rich source of information on the kinetic and the energetic determinants of protein interactions.

It seems that, so far, no fundamental limitations on the number of interacting protein components or the number of sedimenting species have been encountered in the field of SV AUC, even though for many different reasons, the level of detail accessible may currently be limited. Many studies have addressed ternary protein interactions, mixed heterogeneous interactions of self-associating proteins, and extended associations with multiple mixed species, which have been found to play important roles in understanding protein functions in many biological systems. Conformational states may be distinguished by virtue of altered hydrodynamic properties, opening SV to the study of ligand-linked conformational changes and protein allostery. In addition, filamenting proteins have been the traditional subjects of SV, owing to the large macromolecular size range of AUC.

Many variations of sedimentation velocity AUC have been developed over the last eight decades, and many generations of physical biochemists have continuously extended the theoretical basis, AUC technology, and the practical applications for the study of proteins, nucleic acids, and polymers, and their interactions with solvent or other macromolecules. Even a partial overview of the work of the many laboratories contributing to the present state-of-the-art or any attempt to systematically reflect the historic development is far outside the scope of this chapter. Nevertheless, it is impossible not to reflect to some extent on historical development in order to provide an introduction to the main current approaches and a description of the recent advances for studying protein interactions. In addition, necessarily, only very selective examples from the large AUC literature could be included to illustrate the approaches. For more comprehensive information and further reading, the interested reader is referred to the monographs (Svedberg and Pedersen, 1940; Schachman, 1959; Fujita, 1975; Elzen, 1988; Harding *et al.*, 1992; Schuster and Laue, 1994; Scott *et al.*, 2005) and other reviews (Correia, 2000; Lebowitz *et al.*, 2002; Schuck, 2003; Cole, 2004; Howlett et al., 2006). Likewise, practical information on conducting SV experiments are not covered, but detailed information can be found in the references (Balbo and Schuck, 2005; Schuck, 2006d).

This chapter reviews first the experimental idea of AUC and the basic theory of macromolecular sedimentation, as described by the Lamm equation (Lamm, 1929), the discrimination of diffusion from sedimentation, and the analysis of sedimentation data by computing continuous sedimentation coefficient distributions. The following section will address the study of protein conformation and conformational changes, and highlight relationships to other biophysical approaches that can be used synergistically, including dynamic light scattering (DLS), fluorescence correlation spectroscopy (FCS) (Chapter 1), circular dichroism (CD), fluorescence anisotropy (Chapter 13), and small angle scattering (Chapter 11). Finally, the presence of reversible reactions arising from protein conformational equilibria, self-association, or heteroassociation is considered. Strategies for how to determine the equilibrium constant, kinetic rate constants, and the hydrodynamic shape of the complexes are reviewed, and known limitations are described. The approaches discussed will include the techniques of direct boundary modeling, multisignal sedimentation coefficient distributions, and Gilbert–Jenkins theory-based isotherm analysis. Finally, some examples of applications of SV to the study of biological systems, containing binary and ternary protein interactions, are presented and referenced.

16.2. EXPERIMENTAL SET-UP

Conceptually, the analytical ultracentrifuge is a device that applies a centrifugal field and permits us to optically probe in real-time the distribution of the macromolecules in solution. This is achieved by placing the sample in an assembly with sector-shaped cross section and with quartz or sapphire windows on top and bottom, such that when the assembly is inserted in the rotor, the centrifugal field is directed radially and parallel to the walls of the sample container, and light can transverse the sample throughout the solution column in the direction perpendicular to the plane of the rotor (i.e., parallel to the axis of rotation) (Figure 16.1). The optical detection is synchronized with the revolution of the rotor. A second sector of identical shape is placed next to the sample sector, and is filled with buffer to serve as an optical reference.

Typical centrifugal fields in SV are a few 10^5 g (much lower centrifugal fields are used for sedimentation equilibrium, see Chapter 10). In these fields, most proteins will sediment with a rate much larger than their diffusional transport, such that a diffusionally broadened migrating boundary is formed (Figure 16.2). Due to the radial geometry providing larger cross sections of the solution column

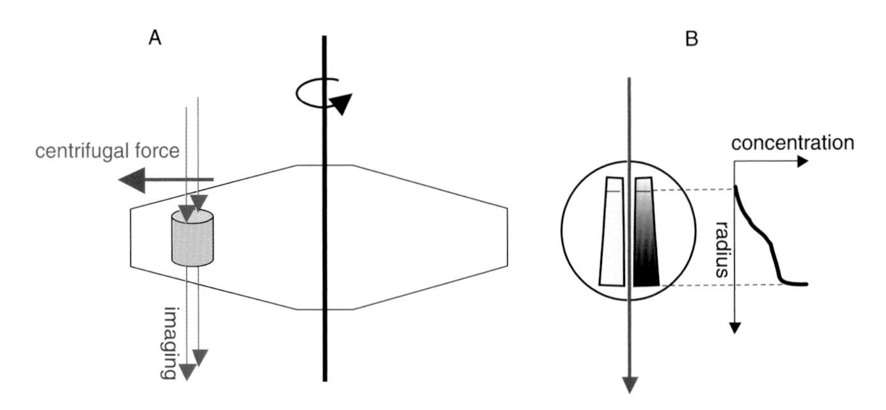

Figure 16.1. Principle of the analytical ultracentrifuge. (A) A sample holder (gray) consisting of a resin with sector-shaped cavities enclosed from bottom and top with optically transparent windows (not shown) is inserted in a rotor (thin black contour indicates a cross section) located at a radius 6–7 cm from the center of rotation (solid black line). The optical illumination of each sector is indicated as the two blue arrows, either from the flashlamp of the absorbance system or the laser beams of the interference optical system. The illumination and the optical detection are triggered to be synchronous with the rotor revolution, such that it takes place only at an angular position of the rotor during which the sample is aligned in the light path of the optical system. After traveling through the sample parallel to the axis of rotation, and perpendicular to the centrifugal force (indicated as red arrow), light enters a scanning or imaging system (not shown). (B) Cross section of the sample cell in a plane perpendicular to the axis of rotation. The two sectors filled with solvent (left) and sample (right) are sector-shaped, such that no trajectory of macromolecular sedimentation intersects the walls of the sample column, preventing convection. The centrifugal force causes redistribution of the macromolecules, and the optical detection system records traces of concentration versus radius (solid black line). The data acquisition is repeated and the concentration profiles are refreshed on a timescale that can be referred to as ''real-time'' relative to the sedimentation process.

with increasing distance from the center of rotation, the migrating macromolecules will experience some dilution with time. For small proteins or low rotor speeds, back-diffusion from the bottom of the solution column is observed. By adjusting the rotor speed, molecules in the size range from at least 10^3 to 10^8 Da can be conveniently studied. The entire sedimentation experiment takes a few hours, dependent on protein size.

It should be noted that, in contrast to some techniques in preparative density gradient sedimentation where a lamella of sample is placed on top of a protein-free buffer, no artificial supporting density gradient is applied in the standard SV experiment. Dilute buffer components will automatically establish a slight density gradient (typically on the order of 0.0001 $\mathrm{g\,mL^{-1}}$ or less), which is sufficient for stable macromolecular sedimentation.

Practical considerations, such as the choice of rotor speed, sample concentration, or the relative merits of the different detection systems (see later) for particular problems, are beyond the scope of the current review. Generally, both the resolution of species and the precision of the sedimentation coefficient increase with higher

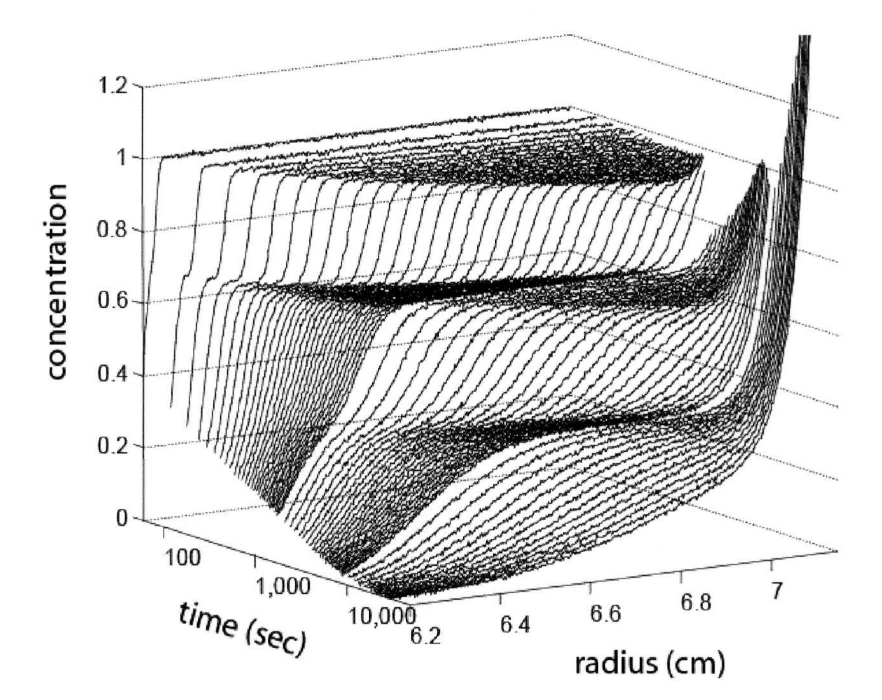

Figure 16.2. Calculated radial sedimentation profiles as a function of time for a mixture of globular species of 10 kDa (1 S), 100 kDa (5.9 S), and 1 MDa (28 S), superimposed with noise resembling realistic signal–noise ratio. The timescale is logarithmic, in order to allow visualizing the different timescales of the sedimentation processes for the different size species. In addition to the different sedimentation velocity, the different degree of boundary spreading from diffusion can be discerned, with the 1 MDa species exhibiting relatively little diffusion because of a low diffusion coefficient and the short timescale of its rapid sedimentation, whereas the 10-kDa species exhibits larger root-mean-square displacement from diffusion than sedimentation, hence the broad boundary features at the longest times. From inspecting the plateau signals (i.e., the signals at radii higher than the leading edge of the boundary), it can be seen that slight dilution takes place with time, even in regions not yet reached by the sedimentation boundary. This is due to the macromolecular sedimentation, throughout the whole solution column, taking place in radial direction, which lowers the lateral intramolecular distance and hence the drop in concentration. The end (or bottom) of the solution column at 7.2 cm is not shown; the molecules sedimenting to the bottom of the solution column form very steep concentration gradients there. Only for the medium-sized and small proteins, do we see the steep concentration increase at the bottom. This is due to the higher diffusion coefficient of the smaller species opposing the accumulation of material at the bottom. This part of the traces is referred to as "back-diffusion."

rotor speed, since at higher rotor speeds the experimental time and hence the diffusional spread are diminished. For more information on how to plan, conduct, and analyze an SV experiment, the reader is referred to the extensive protocol by Balbo and Schuck (2005) and the internet resources provided by Schuck (2006b), and the references cited therein. However, some experimental factors and preparative requirements of SV will be discussed, insofar as they shed light on the principle, and on the practical applicability of this technique, or relate to the theoretical approximations made in the data analysis.

Experimentally limiting factors for the accuracy and the resolution of SV include the upper limit of the rotor speed imposed by the currently commercially available rotors, the signal–noise ratio of the optical detection, the temperature control of the rotor, and the possible convection during the SV run. It should be noted that these factors clearly do not impact most studies of protein interactions, but they represent hurdles in some advanced boundary shape analyses, for example, for hydrodynamically resolving conformational ensembles of proteins. Convection is a macroscopic mixing of the solution that masks the microscopic macromolecular migration, and can be caused by temperature gradients and lateral density gradients arising from local accumulation of macromolecules at imperfectly radial contours of the sample column. Only a few very specialized analysis methods [those using an experimental scan as initial condition (Cox, 1966; Schuck *et al.*, 1998), and methods using differential mass balance to determine the weight-average sedimentation coefficients (Baldwin, 1953)] do not imply the complete history of sedimentation, and all interpretation methods that are common currently will be biased by a transient convection phase, even though data from this part of the experiment may be discarded. A partial compensation is possible by treating the starting point of sedimentation, the meniscus position, as an unknown parameter to be determined in the data analysis by back-extrapolation from the progression of the later scans. The temperature contribution to convection can be minimized after sufficient temperature equilibration of the rotor before the run, which usually takes $\sim$1 h, whereas the density gradients can be minimized by careful alignment of the sector-shaped column. The precision of sedimentation coefficients currently is routinely better than 1%.

The highest pressure at the bottom of the sample column is less than $\sim$30 MPa, which is sufficiently small for the compressibility of water and the proteins to be neglected in routine experiments. They should be considered, however, for the most accurate characterization of the sedimentation parameters, in particular, of small solutes (Schuck, 2004b). Likewise, pressure effects on protein–protein associations at the pressures accessible in AUC are rare [a well-known exception is the depolymerization of tubulin at high rotor speeds (Marcum and Borisy, 1978)], and observed only when significant changes of the partial-specific volume occur during complex formation (Schachman, 1959; Harrington and Kegeles, 1973) (see Chapter 9). These changes in volume will result in a rotor-speed-dependent association constant. Such pressure effects on the equilibrium constant of protein interactions are assumed to be absent in the following discussion.

Current commercial systems are a dual-beam absorption spectrophotometer that scans in radial direction an absorbance difference between sample and reference sector, and a Rayleigh interferometry system that images with high precision the refractive index differences between sample and reference. Both provide a spatial resolution of $\sim$10 μm, and a time resolution ranging from 10 s to a few minutes, dependent on the number of samples in the rotor (up to eight is possible) and the column lengths, which are usually $\sim$10 mm for a 300–400-μL sample. The optical pathlength is usually 12 mm, but can be easily shortened to 3 mm with a specialized sample cell. In the absorbance system, the wavelength can be adjusted between 200 and 800 nm, which allows for the selective detection of solutes, such as aromatic amino acids or other intrinsic and extrinsic chromophores of proteins. The quality of the absorbance measurement is generally slightly lower than typical benchtop dual-beam spectrophotometers, and the range of optical density that can be reliably measured is typically between 0.01 and 1.5 OD (assuming the absence of a steep extinction gradient at the detection wavelength). The interference optical system is not selective, but provides higher time resolution and highly sensitive detection ($<10^{-7}$ refractive index units), which allows studying proteins and other macromolecules at concentrations down to the low microgram per milliliter range. It has some disadvantages regarding characteristic systematic baseline offsets that require additional terms when fitting the data (see later). A fluorescence optical system was developed by Schmidt and Riesner (1992), and more recently by Laue *et al.* (1997). (Studies using the latter are still to occur in the literature).

Typically, 0.1–0.5 mg of protein is used in an AUC experiment, although this can vary greatly depending on the study. Frequently, the experiment is set up with a series of solutions at different concentrations or molar ratios, in order to shift the chemical equilibria. (For systems with very slow chemical rate constants, it can be nontrivial to establish chemical equilibrium before the start of the sedimentation; this is not a problem in sedimentation equilibrium AUC.) The purity should be >95%, but contaminants outside the size range of the proteins under study and their complexes are frequently tolerable. Due to the differential migration of the sedimenting species, the SV experiment itself can provide a stringent measure of purity from the size distribution of the sample under study. For example, trace amounts of oligomeric aggregates can be reliably detected and quantified (Shire *et al.*, 2004). No detailed knowledge of the protein concentration is necessary, since the measurement of protein concentrations is part of the AUC detection.

There are generally few restrictions on buffer composition, but SV requires the consideration of a few principles that may not be equally important in other techniques. (1) In order to suppress long-range electrostatic interactions, the buffer should contain at least 10 mM in ionic strength. (2) High concentrations of additives that lead to significant changes in density and viscosity of the buffer, such as a few percent glycerol, make AUC experiments much more difficult to interpret. For instance, at the high centrifugal fields, these buffer components themselves sediment and can form a dynamic density gradient, which can lead to a time- and position-dependent buoyancy of the macromolecules and can significantly influence the macromolecular sedimentation (Schuck, 2004a). At salt concentrations <0.2 M, this effect is usually negligible. (3) Preferential interaction with solvent components, as described in Chapter 9, also can contribute to the protein sedimentation behavior. Effects of preferential hydration on sedimentation properties of proteins can be minimized and are usually negligible in buffers with a density close to water (Lebowitz *et al.*, 2003). (4) Refractive index gradients from sedimenting buffer components will be clearly visible in the interference optical detection system, even, for example, in common phosphate-buffered saline, where the resulting density gradients are usually negligible. Therefore, it is very useful to make a careful buffer exchange against the buffer used as an optical reference. This can be accomplished, for example, by gel filtration or exhaustive equilibrium dialysis. The signal from optically unmatched buffer components can be accounted for in the data analysis, but this can slightly degrade the precision of the macromolecular sedimentation parameters.

Variations of the experimental configuration have been developed. These include dynamic overlaying techniques which allow, for example, the formation of a protein lamella and following its sedimenting through buffer (''analytical zone'' or ''band'' centrifugation) (Vinograd *et al.*, 1963; Lebowitz *et al.*, 1998). This approach appears only slightly less quantitatively precise due to possible initial mixing, and it is more parsimonious in sample consumption. Most importantly, it allows for buffer exchange of the protein, and for the protein to come in contact with the new buffer solution only at the start of the centrifugation. This is exploited, for example, in ''active enzyme centrifugation,'' where at the start of centrifugation a lamella of enzyme is formed over a solution containing substrate. Enzymatic activity is followed by a change in the product absorption spectrum (Cohen *et al.*, 1967; Cohen and Claverie, 1975; Rochet *et al.*, 2000).

The following discussion is restricted to conventional long-column boundary sedimentation in the configuration described earlier. All computational approaches described are implemented in the software SEDFIT and SEDPHAT (Schuck, 2006b,c).

16.3. BASIC PRINCIPLES OF SEDIMENTATION VELOCITY ANALYSIS

The sedimentation coefficient s describes the linear velocity u of a macromolecule resulting from a centrifugal field $\omega^2 r$ [Eq. (16.1)]. At the balance of sedimentation force, buoyancy force, and frictional force, making use of the Stokes–Einstein relationship $D = kT/f$, we obtain

$$s = \frac{u}{\omega^2 r} = \frac{M(1-\bar{v}\rho)}{N_A f}, \tag{16.1a}$$

$$\frac{s}{D} = \frac{M(1-\bar{v}\rho)}{RT}, \tag{16.1b}$$

where f denotes the macromolecular translational frictional coefficient, M and $\bar{v}$ are the molar mass and the partial-specific volume of the protein, R is the gas constant, k is the Boltzmann constant, N_A is Avogadro's number, and T is the absolute temperature (Svedberg and Pedersen, 1940). s is a molecular constant, commonly expressed in Svedberg units, abbreviated S, with 1 S = 10^{-13} s. Equation (16.1b) is termed the Svedberg equation and relates sedimentation and diffusion coefficients with the buoyant molar mass $M(1-\bar{v}\rho)$. Generally, the buoyancy term $(1-\bar{v}\rho)$ can be calculated based on predictions of $\bar{v}$ from amino acid composition. (This method and related methods to account for protein modifications, such as glycosylation, or bound detergent, are described in detail in Chapter 10.) It is important to note that—in contrast to equilibrium AUC—the molar mass is not measured directly in SV, but inferred from the ratio of sedimentation and diffusion coefficients, which are the quantities based on directly discernable features of the sedimentation boundary—translation and diffusional spread. Heterogeneity and chemical reactions between sedimenting species can contribute to the boundary spread, and the Svedberg equation does not hold for the observed average migration and average boundary spread.

Given the molar mass of a macromolecule and its density (or partial-specific volume, $\bar{v}$, respectively), Eq. (16.1) allows to calculate the Stokes radius and the hydrodynamic frictional ratio f/f_0, which can be interpreted in terms of hydrodynamic model shapes or compared with predictions from hydrodynamic theory for a given structure (see later). For this purpose, it is frequently useful to transform the measured s-value, s_{exp}, to an equivalent s-value that would be observed under standard conditions of water at 20°C, $s_{20,\mathrm{w}}$

$$s_{20,\mathrm{w}} = s_{\mathrm{exp}} \left(\frac{\eta_{\mathrm{exp}}}{\eta_{20,\mathrm{w}}} \right) \left(\frac{1-\bar{v}\rho_{20,\mathrm{w}}}{1-\bar{v}\rho_{\mathrm{exp}}} \right), \tag{16.2}$$

correcting for contributions arising from buffer viscosity and density. If a measurement of the diffusion coefficient or independent knowledge on the frictional ratio is not available, it is not possible to estimate the protein molar mass from its s-value alone. However, limits for the molar mass can be deduced from reasonable assumptions of possible f/f_0 values, which usually range between 1.3 for nearly globular hydrated proteins and 2.0 for very elongated and/or glycosylated proteins.

Due to the radial dependence of the centrifugal force, the linear velocity of a sedimenting particle is not constant. Integrating the left-hand side of Eq. (16.1a), the trajectory $r(t)$ of a single ''ideal'' nondiffusing particle that is at the meniscus r_{m} at time 0 can be described as

$$r(t) = r_{\mathrm{m}} \exp(s\omega^2 t). \tag{16.3}$$

A simple data analysis can proceed by identifying the evolution of an apparent boundary midpoint with this trajectory and determining the slope of a plot of $\log(r(t)/r_{\mathrm{m}})$ against time.

A more detailed description of sedimentation considers the ensemble of molecules loaded in the centrifugal cell. With the macromolecular flux of sedimentation proportional to the concentration, $j_{\mathrm{s}} = cs\omega^2 r$, and the diffusional flux proportional to the concentration gradient, $j_{\mathrm{d}} = -D\partial c/\partial r$, the continuity equation in radial coordinates leads to the Lamm equation

$$\frac{\partial \chi}{\partial t} + \frac{1}{r}\frac{\partial}{\partial r}\left[r\left(s\omega^2 r\chi - D\frac{\partial \chi}{\partial r} \right) \right] = 0, \tag{16.4}$$

for an ideally sedimenting species in the absence of chemical reactions (using the symbol $\chi(r,t)$ to describe the radial- and time-dependent macromolecular distribution) (Lamm, 1929). More detailed expressions for sedimentation in multi-component systems can be found in Fujita (1975) (the case of sedimentation with a cosolvent is discussed in Chapter 9). In the absence of closed analytical solutions, numerous important special case approximations have been derived by many researchers over the past eight decades; these are beyond the scope of the present review (Faxén, 1929; Archibald, 1947; Fujita, 1975). The availability of computers has stimulated the development of numerical or complex analytical approximations (Dishon *et al.*, 1966; Cox, 1969; Goad and Cann, 1969; Claverie *et al.*, 1975; Holladay, 1979; Cox and Dale, 1981; Philo, 1997; Schuck, 1998; Schuck *et al.*, 1998; Behlke and Ristau, 2002; Cao and Demeler, 2005; Dam *et al.*, 2005), with finite element solutions on a resting (Claverie *et al.*, 1975) or moving (Schuck, 1998) frame of reference at present being the most commonly applied. Initial conditions usually reflect uniform loading of the sample, but can be adapted to mimic synthetic boundaries generated in experiments with overlaying sample and solution volumes in different geometries, including analytical zone centrifugation, and in some cases can be based on experimental scans (Cox, 1966; Schuck *et al.*, 1998; Schuck, 2006b). The boundary conditions are for impermeable walls at the meniscus and the bottom of the solution column, but it can be computationally advantageous to use a boundary condition of a permeable bottom for modeling experiments where no back-diffusion is present (Dam *et al.*, 2005).

The Lamm equation can now be solved on desktop PCs efficiently enough to permit nonlinear regression of experimental data. This has enabled a substantial advance of SV AUC over the last decade. This includes extensions to model data with Lamm equations considering solvent compressibility (Schuck, 2004b), dynamic density gradients from the sedimentation of cosolutes (Schuck, 2004a), sedimentation of particle size distributions (Schuck, 2000), and coupled sedimentation of reacting systems (Goad and Cann, 1969; Urbanke *et al.*, 1980; Kindler, 1997; Stafford and Sherwood, 2004; Dam *et al.*, 2005). Lamm equation modeling will be described in detail later.

Integration of Eq. (16.4) shows that the change in the amount of material above a reference radius r_{p} (i.e., at $r < r_{\mathrm{p}}$) reveals a weight-average s-value, s_{w},

$$-\frac{1}{\omega^2 r_{\mathrm{p}}^2 c_{\mathrm{p}}} \frac{\mathrm{d}}{\mathrm{d}t} \int_m^{r_{\mathrm{p}}} \chi(r,t) r \, \mathrm{d}r = s_{\mathrm{w}} \tag{16.5}$$

where r_{p} is an arbitrarily chosen reference point in a plateau region where $\partial\chi(r_{\mathrm{p}})/\partial r = 0$ and c_{p} is the concentration at this point (Schachman, 1959; Fujita, 1975; Schuck, 2003). It is termed ''weight average'' or ''signal average,'' since it can be shown that even in the presence of heterogeneity and chemical reactions the mass balance on the left-hand side of Eq. (16.5) leads to a thermodynamically well-defined weight average of all detected macrosolutes (see later). This is of great practical importance, since the determination of s_{w} does not require the interpretation of the boundary shape. The integral can be determined either numerically on the experimental scans or based on fitted boundary models that faithfully represent the area of the experimental boundary. The former approach requires a reference radius to be chosen that is located in a plateau region for all scans, whereas the second approach can be generalized and only requires a subset of scans to exhibit a solution plateau (Schuck, 2003). The differential sedimentation coefficient distribution of Lamm equation solutions, $c(s)$, described later, provides the most precise determination of s_w (Schuck, 2003).

In order to extract more quantitative information from the SV experiment than just the weight-average s-value or the s-values from discernable boundary components, methods for interpreting the boundary spread were developed. To a first approximation, while ignoring the radial-dependent force and the sector-shaped solution geometry, the boundary shape is a convolution of a step function with a Gaussian, with the former resulting from migration and the latter from diffusion. In a more careful derivation, Faxén (1929) provided expressions that can be used to determine an ''apparent'' diffusion coefficient from the observed boundary spread (Fujita, 1975). While very important historically and conceptually, the use of this approach to determine diffusion coefficients (and buoyant molar mass values) is of limited accuracy for several reasons and has been superseded. Most importantly, it cannot be extended to interpret SV of heterogeneous mixtures or of interacting systems. In the 1950s, it was known from experimental data and from theoretical considerations by Gilbert (1959) and Gilbert and Jenkins (1956, 1959) that the boundary shapes can be profoundly affected by chemical reactions on the timescale of sedimentation as will be described later. Transformation of data subsets exploiting the Faxén approximation or Eq. (16.3) can be used to generate sedimentation coefficient distributions and/or derive diffusion coefficients, such as in the radial derivative method dc/dr (Signer and Gross, 1934; Baldwin and Williams, 1950), the time-derivative distribution dc/dt (Stafford, 1992), and the van Holde–Weischet extrapolation to determine an integral distribution (van Holde and Weischet, 1978). They can be of diagnostic value in detecting sample heterogeneity or interactions (Demeler *et al.*, 1997). A discussion of their relationships and resolution was given by Schuck *et al.* (2002).

The concept of directly fitting all sedimentation data with an explicit Lamm equation model of the macromolecular sedimentation process, sometimes referred to as ''direct boundary modeling,'' was envisioned four decades ago and is now firmly established (Cox, 1965; Claverie *et al.*, 1975; Frigon and Timasheff, 1975a; Holladay, 1979; Gilbert and Gilbert, 1980; Todd and Haschemeyer, 1981; Schuck *et al.*, 1998). The experimental data basis for direct boundary modeling is rich, typically 10^4–10^5 data points of signal-to-noise ratio 10^2–10^3, and can cover from early profiles showing initial partial depletion at the meniscus to late profiles exhibiting only the trailing edge of the diffusionally broadened boundary close to the bottom of the solution column.

A significant obstacle to the precise least-squares modeling of SV data from interference optical detection system, which is often best for many SV experiments due to the highest density of data points and highest frequency of scans, is the systematic time-dependent radially uniform offsets, $\beta(t)$, and the systematic time-independent radial baseline profile $b(r)$, that are superimposed to the refractive index signals from solute redistribution. These signal contributions can be substantial and can exceed the macromolecular signal. They arise from the high sensitivity of the interferometric imaging system to optical path length differences in the nanometer range, and consequently the susceptibility to vibrations and surface roughness of the optical elements. A combination of alignment of scans in the air-to-air region above the solution column and time differencing of scans has been introduced by Stafford (1992). We have shown more recently that it is possible, and statistically advantageous, to consider appropriate systematic baseline terms directly as unknowns to be determined in the least-squares fit (Schuck and Demeler, 1999), without further increasing the flexibility of the model (Schuck, 2003). For sets of scans encompassing equal radial range, they can be efficiently calculated using algebraic separation of linear parameters, and be removed from the signal to reveal the macromolecular sedimentation. Such a radial baseline profile is also now commonly used to compensate for imperfections in the absorbance optical detection. For clarity, the baseline terms are omitted in the following presentation.

The Lamm equation in the form of Eq. (16.4) can describe experimental sedimentation data remarkably very well (i.e., strictly within the statistical noise of the data acquisition), if the sample consists of a single species and is of sufficient purity. However, this condition is not trivial, since the spread sedimentation boundary is quite susceptible to macromolecular heterogeneity. Figure 16.2A and B illustrates the superposition of the signals of two species that migrate at slightly different rates, such that the midpoints of their boundaries are separated by less than the diffusional spread. In this case, the observed superposition exhibits a boundary midpoint migrating approximately at the average s-value, but a boundary spread that is much larger than any of the individual species. If the heterogeneity is unrecognized and the boundary is fitted with a single-species model, this results in an increased apparent diffusion coefficient, which when inserted in the Svedberg equation (16.1), would lead to an underestimate of the apparent molar mass. This effect is opposite to that encountered in DLS, where the greater sensitivity to higher molar mass species leads to a z-average diffusion coefficient measured from the mixture. In fact, it has been demonstrated that the apparent molar mass arrived at by combining the diffusion coefficient from DLS and the weight-average s-value from SV in the Svedberg equation, in comparison with the apparent molar mass from boundary spreading in SV, can provide a very stringent test for macromolecular homogeneity (Schönfeld *et al.*, 1998).

Figure 16.3 illustrates the effect of heterogeneity on the evolution of sedimentation profiles. Shown in Figure 16.3A are sedimentation profiles of two species of 90 and 115 kDa with s-values of 5.6 and 6.5 S, respectively (Figure 16.3A), corresponding to globular proteins with frictional ratios 1.3. The superposition can be modeled superficially with a single-species model (red line in Figure 16.3B and residuals in Figure 16.3C). However, the deviations are significantly above the noise of data acquisition. This is due to the migration from diffusion proceeding with $\sqrt{t}$, and sedimentation—as well as differential sedimentation—$\sim t$. In 1978, on this basis, van Holde and Weischet (1978) developed a data transformation $G(s)$ to remove $\sqrt{t}$ components from the evolution of a subset of absorbance scans. Unfortunately, even though it is designed to account for sample heterogeneity, it is fundamentally still based on a single-species Faxén approximation and fails to resolve mixtures with overlapping sedimentation boundaries (Figure 16.3E) (Schuck *et al.*, 2002). This makes the approach more suitable for the study of larger (noninteracting) species, where diffusion is negligible compared with sedimentation.

As a general direct boundary modeling approach to deconvolute diffusion from differential sedimentation of heterogeneous mixtures, in 2000, Schuck proposed a sedimentation coefficient distribution, $c(s)$, defined as

$$a(r,t) \cong \int c(s)\chi_1(s, D(s), r, t)\,\mathrm{d}s, \tag{16.6}$$

where $a(r,t)$ denotes the experimental signal, $\chi_1(s, D, r, t)$ denotes the normalized Lamm equation solution for a single species, and the integral $c(s)\mathrm{d}s$ represents the loading concentration of material with s-values ranging from s to $s + \mathrm{d}s$ (Schuck, 2000). $c(s)$ can be computed by least-squares methods, and to avoid instabilities associated with Fredholm integral equations, it is combined with maximum entropy regularization (Schuck, 2000; Schuck *et al.*, 2002; Dam and Schuck, 2004). A similar strategy was previously developed by Provencher (1982) in the program CONTIN for the analysis of autocorrelation functions from DLS. Following Bayesian concepts, in the absence of prior knowledge, it results in the simplest distribution that is statistically acceptable with the noise level of the experimental data, adhering to the principle of Occam's razor. This means that $c(s)$ will exhibit peaks only for protein species with statistically significant signal contributions and otherwise suppress the noise, which permits to fully exploit the exquisite sensitivity of SV concentration profiles for trace components. If prior knowledge is available, it can be incorporated in various ways into the numerical evaluation of Eq. (16.6) (Schuck, 2005). Likewise, Eq. (16.6) can be modified to account for solvent compressibility (Schuck, 2004b) and dynamic density gradients generated by high concentrations of sedimenting cosolutes (Schuck, 2004a).

The question that arises in Eq. (16.6) is how to relate the diffusion coefficient to the sedimentation coefficient, $D(s)$, which is required for maintaining a one-parametric distribution. In principle, an arbitrary user-defined realtionship can be used, optionally implemented in the software SEDFIT in the form of a look up table with (s,M) data pairs for interpolation. However, the scaling law

$$D(s) = \frac{\sqrt{2}}{18\pi} kT\, s^{-1/2} \left(\eta \left(\frac{f}{f_0} \right)_{\mathrm{w}} \right)^{-3/2} \left(\frac{1 - \bar{v}\rho}{\bar{v}} \right)^{1/2}, \tag{16.7}$$

is frequently a good choice for mixtures of proteins (Schuck *et al.*, 2002). It relates the diffusion to the sedimentation coefficient based on a single, weight-average

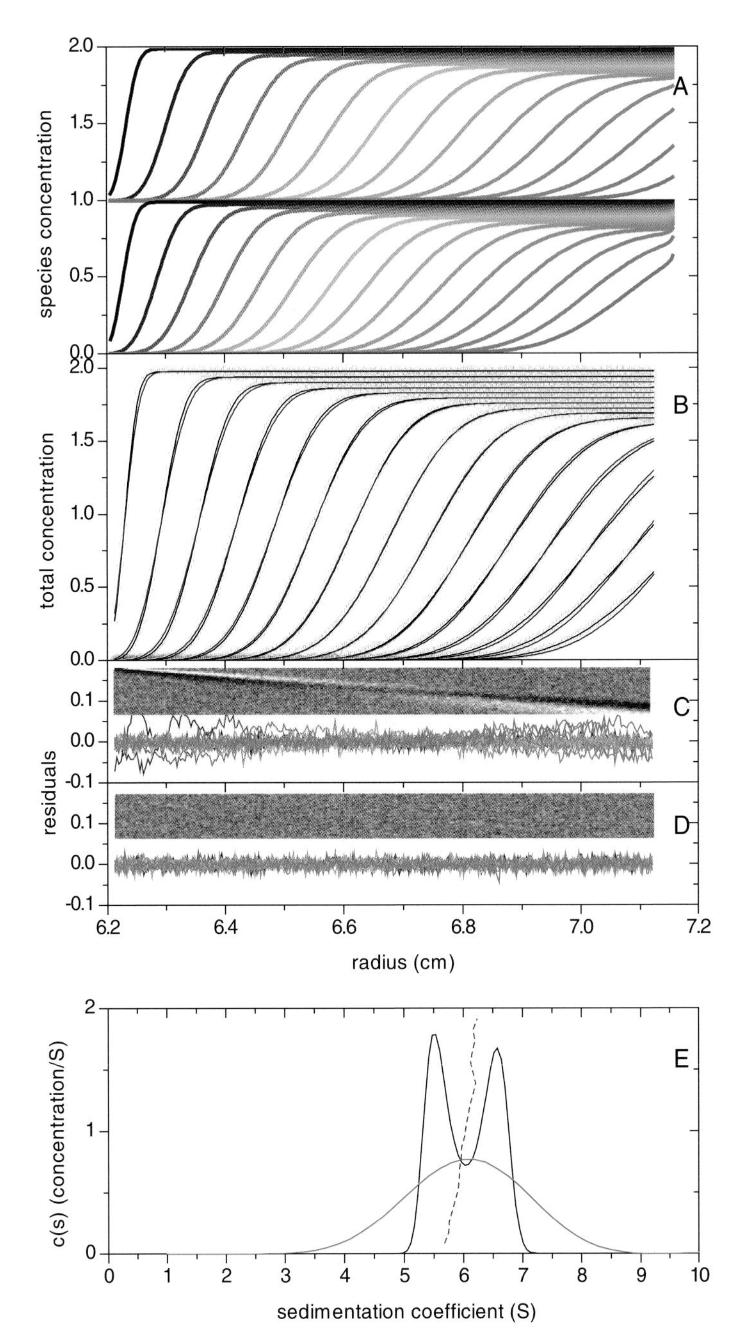

Figure 16.3. Illustration of the basic problem of distinguishing differential migration from diffusion, the effect of heterogeneity on a single-species interpretation, and the resolution of the diffusional deconvolution of $c(s)$ boundary model. (A) Sedimentation profiles for a 90-kDa, 5.6-S species and a

frictional coefficient $(f/f_0)_w$. The latter serves as scaling parameter that can be derived from the experimental data by nonlinear regression, and is usually well defined (Schuck *et al.*, 2002). The approximation of Eq. (16.7) is favored by several factors, including the narrow range of f/f_0 values for folded proteins and the low size dependence of D. Further, it was observed that the precise choice of $(f/f_0)_w$ has little influence on the peak position and area in the $c(s)$ distribution, but mostly governs the resolution. This is illustrated in Figure 16.3D and E. In this example, the $c(s)$ distribution resolves the two species, and arrives at the correct frictional ratio $(f/f_0)_w = 1.30$ (1.28–1.33). This can be compared with the value of $f/f_0 = 1.05$ that corresponds to the boundary spread in the impostor single-species model mentioned earlier (Figure 16.3B and C), which can be rejected simply on the grounds that such a low f/f_0 value is impossible for a hydrated protein. In an extreme case, if the $(f/f_0)_w$ value in $c(s)$ is constrained to infinity, which corresponds to the limiting approximation of nondiffusing species, an apparent sedimentation coefficient distribution is obtained [termed (ls-$g^*(s)$) (Schuck and Rossmanith, 2000)]. (ls-$g^*(s)$ is similar to the results of the ''time-derivative'' (Stafford, 1992) or radial derivative (Signer and Gross, 1934) data transforms, introduced earlier, that scale r to s at each time point via Eq. (16.3), but ls-$g^*(s)$ is arrived at here by direct least-squares modeling of the data.) In this case, the sedimentation coefficient distribution is convoluted approximately by a Gaussian that arises from the uncompensated diffusion, but the average peak position is maintained.

Figure 16.3. (*Continued*) 115-kDa, 6.5-S species (offset), both corresponding to globular proteins with hydrated frictional ratios of 1.30. The colors indicate the evolution in time, with early scans in red and late scans in blue. (B) Circles show the superposition of the two species, with 0.010 Gaussian noise. The red lines are the best fit of an impostor single-species model, which would result from a model in which the underlying heterogeneity is not recognized. It results in an s-value of 6.0 S and an apparent molar mass of 74 kDa, which would correspond to an apparent frictional ratio of 1.05. The blue lines are the best fit from $c(s)$ sedimentation coefficient distribution model, which permits heterogeneity and determines a weight-average frictional ratio from nonlinear regression of the data. It results in a frictional ratio of 1.30 ± 0.03. (C) Residuals of the impostor single-species fit, which has a root-mean-square deviation of 0.016. In order to highlight the systematic misfit of the sedimentation boundary, we have introduced a bitmap representation of the residuals, with radius and time corresponding to pixel columns and rows, respectively, and the magnitude of deviation proportional to the gray scale of the pixel (Dam and Schuck, 2004). With this tool, systematic misfit of the boundary is recognized as a diagonal structure across the bitmap. (D) Residuals of the $c(s)$ fit, with a root-mean-square deviation of 0.010. (E) The best-fit $c(s)$ distribution (black line) shows two peaks resolving the two species. For comparison, a $c(s)$ distribution is calculated with f/f_0 fixed at ∞ (green line), which corresponds to an apparent sedimentation coefficient distribution (ls-$g^*(s)$) (Schuck and Rossmanith, 2000), where diffusion is not deconvoluted. This results in apparent sedimentation coefficient distributions that reflect the true sedimentation coefficient distribution approximately convoluted by Gaussians. Additional asymmetric convolution by a hyperbola segment occurs with the apparent sedimentation coefficient distribution determined by the time-derivative transformation (Schuck, 2003). The blue dotted line represents the van Holde–Weischet distribution (van Holde and Weischet, 1978), an integral distribution, here scaled to reflect the total loading signal. The overall diffusion is approximately deconvoluted but the species cannot be resolved because of the mixed composition of all boundary levels (Schuck *et al.*, 2002), resulting in a sloped line that only qualitatively reflects the heterogeneity.

Once the relationship $D(s)$ has been established with the quantity $(f/f_0)_w$, the Svedberg equation (16.1) can be used to scale $c(s)$ into a molar mass distribution $c(M)$. This is more susceptible to the precise value of f/f_0 for each species, but if the distribution exhibits a single major peak, the precision is typically better than $\pm 10\%$. This is usually sufficient to identify the protein oligomeric state. $c(M)$ can be more appropriate than a discrete species model if the protein exhibits microheterogeneity, such as from statistical glycosylation or from conformational mixtures (Kornblatt and Schuck, 2005). This was illustrated, for example, in the study of the oligomeric state of a natural killer cell receptor (NKR) fragment by Dam and Schuck (2004). Clearly, the separation and precision of mass spectrometry is far superior (see Chapters 5 and 15), but sedimentation provides a quantitative representation of the relative amounts of populations of macromolecules in solution (which may or may not be advantageous, see Chapter 15), maintaining noncovalent interactions, in equilibrium with solvent, and, in some cases, in the presence of small ligands driving the association.

A generalization of $c(s)$ to a two-dimensional distribution of $c(s, f/f_0)$ has recently been proposed (Brown and Schuck, 2006). It is representative of a size-and-hydrodynamic shape distribution of the sedimenting macromolecules and can be expressed, for example, as $c(s, M)$ or $c(M, R_S)$. Even though the additional dimension describing f/f_0 (or M) does not offer high resolution, the hydrodynamic resolution in sedimentation coefficients is usually close to that of the $c(s)$ distribution. It has the advantage of an analysis completely free of scaling relationships and not requiring knowledge on the sedimenting material.

The $c(s)$ approach has been applied in many studies (Schuck, 2006a). First, Perugini *et al.* (2000) used SV with $c(s)$ analysis in the study of the monomer–tetramer–octamer self-association of apolipoproteins E3 and E4. The robustness of the $c(s)$ distribution with regard to different levels of regularization and different frictional ratio values was examined. It was found that the apoE4 isoform, which is implicated in Alzheimer's disease, has stronger propensity to self-associate, and that the oligomers of both forms undergo phospholipid-induced dissociation to monomers. Another example is the study of Solovyova *et al.* (2004) on the oligomeric state and assembly intermediates of the ring-forming bacterial toxins, pneumolysin and perfringolysin O. Pneumolysin was found to be initially predominantly monomeric, whereas perfingolysin O was observed to be in a slowly converting monomer–dimer equilibrium. After determining the s-values of monomer, dimer, and with the aid of small angle X-ray scattering data, low-resolution bead models for the solution structure of these molecules could be constructed. Further, these authors were able to detect trace levels of higher oligomers forming with time, which corresponded well with partial rings, as predicted from hydrodynamic modeling (Solovyova *et al.*, 2004).

A deviation from the hydrodynamic scaling law [Eq. (16.7)] is useful, over certain ranges of s-values, for example, for populations of proteins of known mass in different conformations [which has been applied to the study of the conformational changes of rotavirus nonstructural protein NSP2 (Schuck *et al.*, 2000)], or for distributions of proteins with identical Stokes radius [such as apoferritin and ferritin (Schuck, 2000)]. In addition, in the absence of intrinsic microheterogeneity, $c(s)$ can be structured as a piecewise continuous distribution, where $c(s)$ peaks are replaced by discrete species, for which the molar mass can be determined more precisely (Boukari *et al.*, 2004). For example, the last approach was taken also in a study of the formation of the antitermination complex by Greive *et al.* (2005). The binary and the ternary interactions of NusB, NusE, and boxA RNA were studied by SV, and after determining the sedimentation coefficient distributions $c(s)$ and $c(M)$, peaks were identified to correspond to complex species. Their molar masses were determined using a hybrid model that combines piecewise continuous distributions with discrete Lamm equation solutions. For the triple mixture, the molar mass of the ternary complex was determined and found to be consistent with a stoichiometry of 1:1:1 (Greive *et al.*, 2005). The special case of neglecting diffusion altogether can be a good approximation in the study of very large species, such as the size distribution of apoC-II amyloid fibrils (MacRaild *et al.*, 2003).

In essence, the $c(s)$ approach fully exploits the centrifugal separation and provides a distribution of sedimentation coefficients where the average, scaled extent of diffusion is calculated from the experimental data, and used for deconvolution. When studying protein mixtures, this usually allows one to resolve different species, small molecular weight fragments, and aggregates, and to permit calculation of their sedimentation coefficients with high precision (Figure 16.4). In particular for proteins >10 kDa, the resolution and the precision can far exceed that from diffusion-based methods, such as size-exclusion chromatography or DLS.

As mentioned earlier, the presence of chemical reactions that are rapid on the timescale of sedimentation can result in significant changes of the boundary profiles, which will translate to different characteristic peak structures of $c(s)$. The interpretation of $c(s)$ for protein interactions that take place on the timescale of sedimentation is discussed later.

For studies of heterogeneous protein mixtures, the $c(s)$ approach has been extended very recently to the global modeling of signals from different optical systems or different wavelengths (Balbo *et al.*, 2005).

$$a_\lambda(r,t) \cong \sum_{k=1}^{K} \varepsilon_{k\lambda} \int c_k(s)\chi_1\left(s, \left(\frac{f}{f_0}\right)_{k,\mathrm{w}}, r, t\right) \mathrm{d}s$$
$$\lambda = 1 \ldots \Lambda,\ K \leq \Lambda,\ \det(\varepsilon_{k\lambda}) \neq 0, \qquad (16.8)$$

using predetermined extinction coefficients $\varepsilon_{k\lambda}$ of K macromolecular components at Λ different signals. This produces multiple distributions $c_k(s)$ that reflect the sedimentation coefficient distributions of sedimenting particles containing the individual components. Integration of $c_k(s)$ leads to the molar loading concentrations

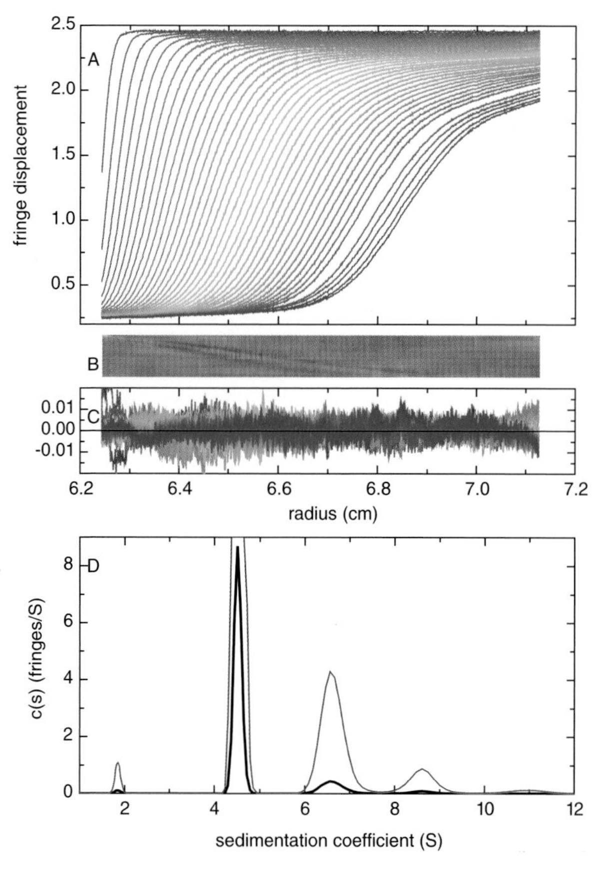

Figure 16.4. Experimental sedimentation velocity (SV) data and $c(s)$ analysis of bovine serum albumin and its oligomers. (A) Sequence of interference optical fringe displacement profiles (calculated systematic noise contribution removed) over a time range from 190 s (red) to 7,844 s (blue). (B) and (C) Residuals bitmap and overlay, respectively, of the $c(s)$ analysis, which resulted in a root-mean-square deviation of 0.0036 fringes. (D) $c(s)$ distribution (black) and tenfold magnified $c(s)$ (blue). The monomer (~4.5 S), dimer (~6.5 S), trimer (~8.5 S), and higher aggregate peak, along with a small proteolytic degradation product or contamination at 2 S, and sedimentation of unmatched buffer salts at 0.4 S (not shown) can be resolved, as a result of diffusional deconvolution of the evolution of boundary shapes.

of each species. Typically, data from the interference optical system can be easily acquired concurrently with absorbance optical scans at one or more wavelengths. Even in the absence of extrinsic labels or chromophores in the visible spectrum, different proteins may be distinguished in favorable cases on the basis of their different composition of aromatic amino acids alone. An example is shown in Figure 16.5. The spectroscopic and the hydrodynamic separation are orthogonal, which can lead to a significantly higher overall resolution (Figure 16.5).

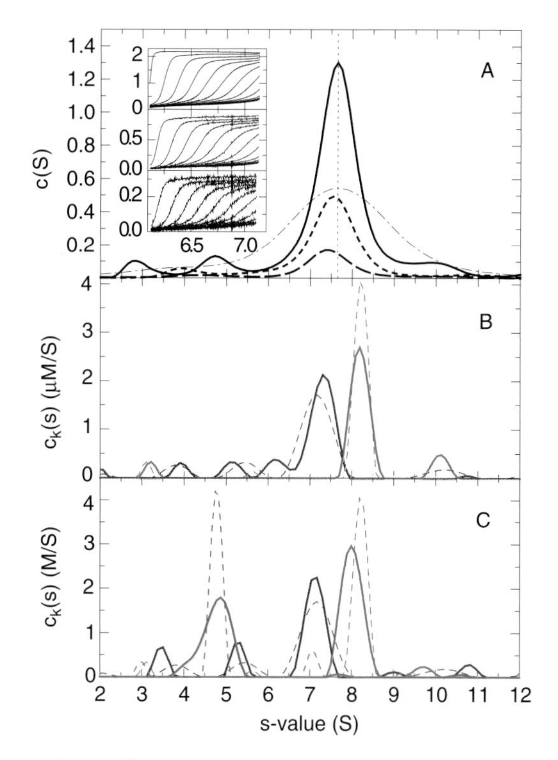

Figure 16.5. Sedimentation coefficient distributions derived from sedimentation velocity profiles of a mixture of IgG and aldolase. The inset shows the raw sedimentation signals acquired at different time points with interference, absorbance at 280 nm and absorbance at 250 nm (top to bottom), at a rotor speed of 48,000 rpm at 26°C. (The signal profiles are shown in units of fringes, OD_{280}, and OD_{250}, respectively, as a function of radius in centimeter.) (A) $c(s)$ distributions calculated separately for the refractive index data (solid line), the absorbance at 280 nm (dotted line), and the absorbance at 250 nm (dashed line). For comparison, the apparent sedimentation coefficient distribution ($ls\text{-}g^*(s)$) (Schuck and Rossmanith, 2000) without diffusional deconvolution applied to the interference data set is shown (thin dash-dotted line). To facilitate comparison of the peak positions, the vertical dotted line indicates the peak of the $c(s)$ distribution from the interference data. (B) Global multiwavelength analysis and decomposition into the component sedimentation coefficient distributions $c_k(s)$ for IgG sample (red) and aldolase (blue). The extinction coefficients for these two components at the three signals were predetermined from separate experiments with IgG and aldolase alone, which resulted in the sedimentation coefficient distributions indicated by the dotted lines. (C) Calculated component sedimentation coefficient distributions from a mixture (solid lines) of IgG (red), aldolase (blue), and bovine serum albumin (BSA) (green, right ordinate), and comparison with the distributions obtained from the individual proteins (dotted lines). Reproduced from Balbo *et al.* (2005).

16.4. ASSESSING CONFORMATION OF PROTEINS AND PROTEIN COMPLEXES

As mentioned earlier, if the molar mass of a protein and its partial-specific volume are known, the sedimentation coefficients can be used to calculate the translational frictional coefficient, or the frictional ratio. Since a single number is not sufficient to predict a three-dimensional structure, no unambiguous protein shape can be determined. However, the frictional ratio obtained is of high accuracy; for instance, a study comparing hydrodynamic information for a macromolecular complex obtained from FCS and SV showed excellent agreement (Boukari *et al.*, 2004). Theories and computational approaches are available to predict the hydrodynamic frictional ratio for any given structure, assuming boundary conditions for the solvent flow and levels of hydration (see, e.g., Yamakawa and Fujii, 1973; Garcia de la Torre and Bloomfield, 1981; Douglas *et al.*, 1994; Garcia De La Torre *et al.*, 2000; Aragon, 2004; Rai *et al.*, 2005). This topic is discussed and reviewed in Chapter 1. Beyond the necessity for knowing the protein partial-specific volume [which may be determined, for example, by SE AUC (Chapter 10)], some additional aspects can arise in the interpretation of SV experiments, which will be discussed later.

In the literature, the interpretation of sedimentation coefficients in terms of protein structure are many and of different types. Structural predictions are most commonly derived from X-ray crystallography. Rai *et al.* (2005) have compared predictions from X-ray crystallography and NMR, and highlighted the role of side-chain flexibility contributing to the translational and rotational hydrodynamic frictional coefficient. If an experimentally observed *s*-value is not consistent with the structural prediction, this has been taken as an indication that the overall average solution conformation is different from that in the crystal (Li *et al.*, 2005). For protein complexes or protein domains where alternative orientations can be hypothesized, their corresponding theoretical *s*-values may be compared with experimentally observed data (Che *et al.*, 2005). Similarly, a comparison of predicted and experimental *s*-values can be used as a constraint in the refinement of predictions of the solution structure by small angle scattering (see Chapter 11) (Scott *et al.*, 2002; Durchschlag and Zipper, 2003; Solovyova *et al.*, 2004; Hu *et al.*, 2005).

Conversely, for the study of the energetics of reversible protein interactions when the *s*-value of the complex cannot be well determined experimentally (see later), constraints based on theoretical predictions derived from structural models can be used, which may lead to a well-defined thermodynamic analysis, for example, of the weight-average *s*-values as a function of loading concentration (Rivas *et al.*, 2000; Dam *et al.*, 2003).

In the absence of high-resolution structural data, simple models for the protein shapes are routinely used. For globular proteins, they are typically the hydrodynamically equivalent prolate or oblate ellipsoids. For flexible macromolecules, such as protein fibrils or nucleic acids, models for wormlike chains have been applied (MacRaild *et al.*, 2003; Koenderink *et al.*, 2005). The utility of ellipsoidal models increases for self-associating proteins if *s*-values for oligomers in addition to those of the monomers are available, since it allows qualitatively assessing the merits of side-by-side versus end-to-end association, and the merit of oblate versus prolate ellipsoidal shapes for the proteins. Frequently, only one model shape is consistent with the *s*-values of all species, without postulating major conformational rearrangements. For example, in a recent study of the malarial surface protein

MSP3 (which is of yet unknown structure and function), it was shown that specific protein fragments are hydrodynamically equivalent with highly elongated prolate ellipsoids undergoing end-to-end association to form oligomers. Qualitatively, the same results were obtained independent of the assumed level of hydration (Burgess *et al.*, 2005). Similarly, in an SV study of the rotavirus protein NSP2, it was shown that NSP2 is an octamer that can undergo conformational changes and coexist with tetramers, both dependent on ligand binding. Only an oblate model with side-by-side association was consistent with the measured frictional ratios of both the tetramers and the octamers (Schuck *et al.*, 2000). (This view was later confirmed in the doughnut-shaped crystal structure of the octamer (Jayaram *et al.*, 2002), except for the central tunnel—a feature that cannot be anticipated from translational friction properties, even though it could be taken into account if such prior knowledge exists.) An example for another use of hydrodynamic models is the study of the two-stage self-association of the molecular chaperone gp57A of bacteriophage T4, where direct modeling of the sedimentation profiles with Lamm equation solutions provided estimates of s-values of the oligomeric species, and the predicted hydrodynamic shapes were used as a criterion to discriminate between different association models (see later) (Ali *et al.*, 2003).

An important area of SV application is the study of ligand-induced conformational changes in proteins. If the ligand is small relative to the protein, the structural changes in the protein can lead to an increased hydrodynamic friction, which may counteract the additional centrifugal force from the increased mass of the complex. A classical example is aspartate transcarbamylase, which is a hexameric enzyme that exhibits a 3% decrease of the s-value when liganded with N-phosphonacetyl)-L-aspartate (PALA) (Howlett and Schachman, 1977; Cohen *et al.*, 1985). This change is very significant and can be very reproducibly observed, which highlights that the high precision of SV can compensate for the fact that the translational friction coefficient is intrinsically not a highly shape-sensitive parameter. This ligand-induced change in the s-value of PALA provided a unique access to study fundamental questions of protein allostery (Howlett *et al.*, 1977). It is possible to resolve mixtures of proteins in different conformations if the conformational transition is slow on the timescale of sedimentation (Werner and Schachman, 1989). Distinguishing the timescale of conformational transitions, or the presence of conformational mixtures, without prior knowledge is difficult. It requires high protein concentrations, very precise detection, and is favored by large molar mass and, of course, large differences in hydrodynamic friction.

Studies of ligand-induced conformational changes in SV can be synergistically combined with CD. Even when the overall hydrodynamic friction may remain unchanged, conformational changes may be accompanied by changes in secondary structure observed in CD. Conversely, it can easily be imagined that conformational changes consisting of rearrangements of protein domains may not have a significant far-UV CD signature, but exhibit large changes in hydrodynamic friction. CD is also preparatively compatible, as it uses concentrations in a range accessible by SV and is nondestructive, such that it can be performed immediately before SV.

If the conformational changes from ligand binding do not lead to an increase in hydrodynamic friction that exceeds the increase in gravitational force, or even lead to a decrease in hydrodynamic friction, detection can be difficult from the observed s-value alone. Ligand binding alone, in the absence of a conformational change of the protein, would lead to an increased s-value, and the magnitude of this increase may be difficult to predict, as it may depend on the location of the binding site and the partial-specific volume of the ligand. In this case, the use of DLS or FCS can be extremely useful; in the absence of a conformational change, the protein Stokes radius can be expected to increase in the complex state. However, if ligand binding makes the protein hydrodynamically more compact, a decrease in the Stokes radius would be measured, which should coincide with an increase in the s-value in SV. This has been observed, for example, for the conformational change in the iron regulatory protein-2 (IRP2) induced by binding of its cognate regulatory RNA (Yikilmaz *et al.*, 2005), and in the conformational change of rotavirus NSP2 induced by nucleotide binding (Schuck *et al.*, 2000).

On a different scale of conformational changes, SV has frequently been applied to the study of protein folding, to monitor the unfolding transition, for example, in the presence of chaotropic reagents, via changes in the oligomeric state and/or differences in the hydrodynamic radius of folding intermediates or the unfolded state (Doster and Hess, 1981; Consler and Lee, 1988; Lee *et al.*, 2003; Garrido *et al.*, 2005).

16.5. SEDIMENTATION OF INTERACTING SYSTEMS

Protein interactions reveal themselves by altered sedimentation behavior, as compared with the noninteracting proteins. The changes can be very substantial. They can arise from forces other than centrifugal force causing correlations or anticorrelations of the particle distribution, from changes in the mass of the sedimenting particles, from chemical conversion of species during the sedimentation process, or from a concentration-dependent population of otherwise ideally sedimenting species. There are different theoretical and analytical frameworks for the characterization of protein interactions. The one most suitable for a given problem depends on the type of the interaction studied, the amount and type of prior knowledge, as well as the possibility to make assumptions of strict monodispersity of the samples. In the order of increasing detail and assumptions, we will describe the following: the use of virial and nonideality coefficients, the interpretation of isotherms of the concentration-dependent weight-average s-values, the use of $c(s)$ profiles and isotherms from Gilbert–Jenkins theory, the application of multisignal $c(s)$ analysis for heterogeneous interactions, and the fitting of sedimentation data with solutions of systems of Lamm equation coupled with reaction terms. In many cases, these approaches may be used concurrently. It should be noted that this is not a historical sequence or one of increasing importance or utility. Although very significant progress has been made throughout the last decade, all of these different strands have been envisioned or had their foundations laid in the 1960s or earlier, and they have developed, essentially simultaneously, with changing emphasis at different times.

16.5.1. Virial Coefficients, Nonideality Coefficients

The description of macromolecular interactions with virial coefficients does not imply particular structural features, such as the formation of well-defined complexes of certain stoichiometry, but is more generally related to correlated particle motions in solution, i.e., molecules are preferentially found in, or excluded from, the vicinity of each other (compare section 9.4.3). This description can be applied to attractive and repulsive interactions alike, even though in AUC this description has traditionally been used mostly to describe repulsive interactions. The consideration of repulsive interactions has a long tradition in AUC, in part, perhaps, because several decades ago experiments with a Schlieren optical system requiring higher protein concentrations were more common. At high protein concentrations, repulsive interactions between sedimenting proteins cause a reduction in the sedimentation coefficient, which can be expressed as

$$s = \frac{s_0}{1 + k_S c} \tag{16.9}$$

with the nonideality coefficient for sedimentation k_S (Schachman, 1959). Typical magnitudes for k_S are in the order of 10 mL g^{-1} or less, but larger values can be expected for asymmetric molecules (Schachman, 1959; Creeth and Knight, 1965; Rowe, 1992). Negative values are indicative of protein self-association. Similarly, the mutual diffusion coefficient shows a concentration dependence

$$D = D_0(1 + k_D c). \tag{16.10}$$

It can be shown that

$$k_D \approx 2A_2M - k_S \tag{16.11}$$

(with the second virial coefficient A_2) which indicates that the concentration dependence of diffusion coefficient arises from both the thermodynamic nonideality and the hydrodynamic nonideality (Solovyova *et al.*, 2001). More detail on this topic can be found in Section 9.8.4.

The shapes of the sedimentation profiles generally show steeper boundaries for repulsive nonideal sedimentation. Dishon *et al.* (1967) simulated Lamm equation solutions for concentration-dependent sedimentation with locally concentration-dependent s and D values. Creeth determined that at highly nonideal conditions, a time-independent boundary shape will be adopted, and derived analytical limiting expressions for the boundary shape in the case $k_D = 0$ (Creeth, 1964; Dishon *et al.*, 1967). Currently, finite element solutions of the Lamm equation with concentration-dependent s and D for the common centrifugal geometry are available for nonlinear regression of experimental data (Solovyova *et al.*, 2001). Sedimentation profiles from repulsive nonideality are distinctly different from and generally cannot be modeled with a linear combination of ideal Lamm equation solutions. More complex sedimentation patterns can be obtained in multicomponent mixtures at high concentrations. This includes the classic Johnston–Ogston effect (Johnston and Ogston, 1946), which is a boundary inversion in the slowly sedimenting component due to the differential concentration dependence in the presence and absence (below and above the boundary) of a faster sedimenting component. A rigorous extension of the sedimentation coefficient distribution analysis techniques to the consideration of nonideal mixtures has not been reported.

The discovery that protein crystallization is favored in solution conditions where proteins exhibit slightly negative second virial coefficients (George and Wilson, 1994), indicative of weakly attractive interactions, has stimulated recent interest in methods to determine the second virial coefficient. Solovyova *et al.* (2001) have shown that the analysis of the sedimentation boundary shapes for halophilic malate dehydrogenase can provide estimates of k_D and k_S, thus A_2 [Eq. (16.11)], consistent with those derived from small-angle neutron scattering.

In the following description, unless otherwise noted, the presence of moderate or strong interactions is assumed, such that protein concentrations are below the level where repulsive interactions become significant (i.e., typically $<$1–2 mg mL^{-1} for moderately globular proteins in buffers of sufficient ionic strengths), and the interactions can be described in the structural picture of the formation of discrete protein complexes with well-defined stoichiometries.

16.5.2. Isotherms of Weight-Average *s*-Values

As introduced earlier, the weight-average sedimentation coefficient s_w is based on a mass balance that considers the change in the total amount of material above (at smaller radii than) an arbitrarily chosen plane in the plateau of the sample. As such, it is an experimentally well-defined quantity that depends on the sedimentation coefficients of all components, but s_w is completely independent of the shape of the boundary, and it does not imply any details of the nature of their interactions.

These enter in the interpretation of the isotherm because of the concentration dependence of $s_w(c)$. A plot of $s_w(c)$ can be used, for example, to determine the nonideality coefficient k_S for single components [Eq. (16.9)]. For multicomponent systems, the integration of a set of Lamm equations coupled with reaction terms leads to

$$s_w(c_1, \ldots, c_N) = \frac{\sum_i s_i(c_1, \ldots, c_N)c_i}{\sum_i c_i}, \tag{16.12}$$

where the summation is over all macromolecular species in solution. Analogous expressions can be derived for a "signal-average" *s*-value when a selective detection method is applied (Rivas *et al.*, 1999). Clarification of the correct reference concentration is important to avoid proportional errors in the determination of the equilibrium constants. Conceptually, since the depletion of the accounted material takes place at the reference radius in the plateau, the relevant concentration is that in the plateau. Due to the radial dilution caused by the sector-shaped solution column (typically ~20–30%), interacting systems may exhibit a small time dependence of s_w if dissociation occurs during the dilution (Svedberg and Pedersen, 1940).

As mentioned earlier, precise estimates of s_w can be obtained from integration of any of the differential sedimentation coefficient distributions, such as $c(s)$, provided that their faithful representation of the sedimentation profiles is established. This permits a large number of experimental scans to be included in the analysis (thus providing high precision), and provides the opportunity to discriminate the sedimenting species not participating in the interaction of interest. Since the sedimentation coefficient distributions imply the entire time course of sedimentation, the relevant concentrations are the time-averaged plateau concentration during sedimentation, which are closer to the loading concentration. Correction factors for this situation have been derived by Schuck (2003). If the dissociation of species is slow compared with the sedimentation process, the relevant concentration clearly is the loading concentration.

The isotherm of $s_w(c)$ represents a binding isotherm, and can provide the basis for a thermodynamic analysis of an interaction (Oncley *et al.*, 1952; Steiner, 1954). For example, for a self-associating system, Eq. (16.12) can be written as

$$\begin{aligned} s_w(c_{tot}) &= \sum_i \frac{s_{0,i}}{1 + \sum_j k_{S,ij} K_j c_1^j} K_i c_1^i / c_{tot}, \\ &\cong \frac{1}{1 + k_S c_{tot}} \sum_i s_{0,i} K_i c_1^i / c_{tot}, \end{aligned} \tag{16.13}$$

where $s_{0,i}$ are the species sedimentation coefficients at infinite dilution, $k_{S,ij}$ their mutual hydrodynamic nonideality coefficients, and K_i the association constants (with $K_1 = 1$). The first equation is taking into account also possible cross terms between k_S of different species (Schachman, 1959). However, because the values of $k_{S,ii}$ cannot easily be determined separately for each species, the second equation makes the assumption that the hydrodynamic nonideality coefficients for all species can, in a first approximation, be described by an average value (Frigon and Timasheff, 1975b). This seems to be a reasonable approximation at not too high concentrations, or if the different species are not too dissimilar in shape, or for moderately weak associations where the largest species dominate the populations at higher concentration. At concentrations <1 mg mL^{-1}, frequently, nonideality does not significantly contribute to the sedimentation of globular proteins, and the coefficients k_S may usually be omitted ($k_S c_{tot} \ll 1$).

An example of the shapes of the isotherms is given in Figure 16.6. Well-conditioned fitting of the experimental data requires high precision in s_w values, and a concentration range covering well both the conditions where the free species are predominant and those where the complex species are close to saturation. From the change in populations, the energetics of the interaction may be characterized

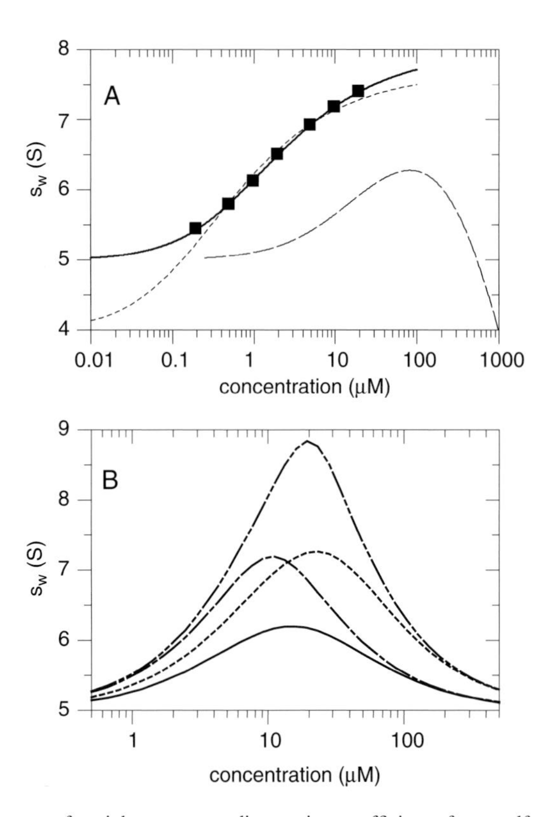

Figure 16.6. Isotherms of weight-average sedimentation coefficients for a self-association (A) and heteroassociation titration (B). (A) Reversible self-association of a 5-S monomer to form a 8-S dimer, with an equilibrium constant $K_D = 2\ \mu M$. Squares indicate the s_w values for loading concentrations of 0.2, 0.5, 1, 2, 5, 10, and 20 μM, which are experimentally accessible, for example, for a typical 100-kDa protein and data acquisition at multiple wavelengths in the far UV. The solid line is the expected isotherm. The short-dashed line is an isotherm for a monomer with 4 S, a dimer with 7.68 S, and an equilibrium constant $K_D = 5\ \mu M$. This highlights that for a precise determination of K_D, it is important either to acquire data over a large concentration range, or to use other hydrodynamic prior knowledge on the s-values of the different species. The long-dashed line indicates the s_w isotherm at higher concentrations (for weaker associations of the same molecule), with repulsive nonideality as expected for spherical particles (Schuck, 2003). (B) shows the isotherms for the titration of a 5-S protein at constant 10 μM with a second 5-S protein, forming a 1:1 complex of 8 S with a K_D of 10 μM (solid line) or 1 μM (dash-dotted line), and forming both 1:1, 8-S and 2:1, 10-S complexes with equivalent microscopic binding constants with $K_D = 10\ \mu M$ (dotted line) or 1 μM (dash-dot-dotted line). The maximum of the titration is characteristically shifted to higher concentrations for higher stoichiometries.

and K_D be determined. For heterogeneous interactions, if the concentration series is a titration series, we get a characteristic maximum for the stoichiometry of the interaction (Figure 16.6B). This is related to the continuous variation analysis described in Chapter 14. Further, if there are characteristic signals for each component, the concentration dependence of the signal-average s-values at all detection wavelengths or signals can be modeled globally, providing additional information on the stoichiometry of the interaction.

It should be noted that for self-associations, such spectral information is not available, and the $s_w(c)$ data do not themselves reveal the mode of interaction (e.g., the number and the size of the complex species). This needs to be introduced in the analysis when formulating the model. However, if saturation is nearly achieved, the limiting s-value can be identified, and the s-value of the largest species determined (Correia, 2000). Frequently, this is taken as a good indicator of the stoichiometry of this complex because of the limited range of possible frictional ratios of most folded proteins (see earlier). Likewise, it may be possible to deduce the nature of the smallest species from the observed limiting s-value at low concentrations. Alternatively, prior knowledge on the oligomeric state may be introduced, stemming, for example, from sedimentation equilibrium studies (see Chapter 10). In this case, the hydrodynamic shape of the complex may be determined, revealing gross structural features of the complex and the assembly (see earlier). In some cases, both the nature of the complex and its s-value may be available, for example, from sedimentation under conditions of strong complex formation or from hydrodynamic predictions.

In 1975, Frigon and Timasheff have used the analysis of weight-average s-values as a function of concentration in the study of the magnesium-induced self-association of calf brain tubulin (Frigon and Timasheff, 1975a). They found that modeling $s_w(c)$ data supported an isodesmic self-association model with a ring-closure step. Since the s-value of the intermediate oligomers cannot be separately determined, the approximation was made that the s-values scale with the two-thirds power of the monomer s-value, which implies the same hydrodynamic friction of all species. Similarly, the nonideality coefficient k_S was assumed to be identical to that of the monomer. The binding constants and the ring size could be determined, the latter being consistent with electron microscope pictures (Frigon and Timasheff, 1975a). The authors proceeded to determine the dependence of the self-association constant on magnesium concentration, temperature, and pressure, and arrived at a detailed characterization of enthalpy, entropy, heat capacity, and molar volume changes of the self-association (Frigon and Timasheff, 1975b). This study highlights the great utility of SV in the study of extended association reactions with multiple species. In particular, if complementary structural information is available, the sedimentation velocity can allow a detailed thermodynamic characterization of the interaction. More recently, Correia and colleagues have determined binding constants of the ligand-induced self-association of tubulin by analysis of the $s_w(c)$ isotherm for the vinca alkaloids vinorelbine and vinflunine (Lobert *et al.*, 1998).

Recently, Rivas *et al.* (2000) have characterized the thermodynamics of the ligand-induced indefinite self-association of the cell division protein FtsZ, a bacterial homolog of tubulin, by expressing the concentration dependence of the average molar mass from sedimentation equilibrium AUC as a decaying isodesmic (isoenthalpic) self-association (Chatelier, 1987), and exploited the known binding parameters and known dimensions of FtsZ monomer to extract from the $s_w(c)$

isotherms from SV information about the shape of the different oligomeric species. The hydrodynamic model consisted of a bead model for different possible geometries, with the best model being that of a single-stranded linear oligomer. Later, SV studies were conducted of FtsZ in the presence of GTP (using a GTP regeneration system), conditions under which cooperative assembly of FtsZ takes place (Gonzalez *et al.*, 2005). The sedimentation coefficient distribution $c(s)$ at different concentrations (see above) exhibited bimodal shapes, with the hydrodynamic parameters quantitatively consistent with cyclization of single-stranded oligomers, structures corroborated by AFM and EM images. It was concluded that cyclization could be the basis for the observed cooperativity of the FtsZ assembly, and may be related to the bacterial septation ring (Gonzalez *et al.*, 2005).

Other recent examples for the analysis of $s_w(c)$ isotherm include systems that do not have extended associations. This includes studies on the driving forces for the assembly of the tetrameric malate dehydrogenase (Irimia *et al.*, 2003), the monomer–dimer self-association equilibrium of the Epstein–Barr virus protease (Buisson *et al.*, 2001), the trimer–dodecamer self-association of the anti-TRAP protein (Snyder *et al.*, 2004), and the stoichiometry and affinity of the solution interaction between the NKR Ly49C and the major histocompatibility complex (MHC) class I molecules (Dam *et al.*, 2003). In the last study, a 1:2 (NKR dimer:MHC) stoichiometry was observed, consistent with the crystal structure, from which s-values for the 1:1 and 1:2 complex could be predicted to aid in the thermodynamic analysis of the $s_w(c)$ data (Dam *et al.*, 2003). In addition, a sedimentation coefficient distribution $c(s)$ with a bimodal reaction boundary was observed, reminiscent of the broad sedimentation coefficient distribution expected from the hydrodynamic separation of two complex species (see below).

16.5.3. Direct Modeling of the Sedimentation Profiles as $c(s)$ Distributions of Ideal Lamm Equation Solutions

In the field of SV, there has been a long-standing intense theoretical and practical interest in the interpretation of the boundary shapes of interacting systems. They provide characteristic information on the type of interaction, which is not used in the consideration of the weight-average sedimentation coefficients. Due to the gradient established in the boundary, the evolution of the boundary shape reflects the state of the protein mixture over a wide concentration range, much like gradients in sedimentation equilibrium AUC. For example, Figure 16.7A shows the sedimentation profiles for different self-associating systems, all of which have the same weight-average s-value. As can be expected, the sedimentation of systems interacting more slowly than the timescale of sedimentation resembles the sum of the boundaries of each species. This holds true also for the sedimentation profiles of heteroassociations (Figure 16.7B). However, it can be clearly discerned that the patterns for reactions faster than the timescale of sedimentation are intriguingly different: Even for systems containing more than one complex, only two boundary components exist (black solid line in Figure 16.7B). For fast two-component heterogeneous associations, one boundary is at a position expected for one of the unbound species (the ''undisturbed'' boundary) and one at a position intermediate to that of the complex species and the faster-sedimenting unbound species (the ''reaction boundary'') (Figure 16.7B). Further, the species that is mimicked by the undisturbed boundary is dependent on the relative concentration of components, but it does not follow simply the molar excess. Finally, it can be shown that both free species occur, in different ratios, in the fast reaction boundary sedimenting alongside the complex. Thus, uncritical interpretation of the sedimentation boundary shapes, neglecting the existence of a fast reaction and mistaking the s-values of reaction boundaries as those of species participating in the reaction, can be very misleading (Fujita, 1975). It is equally clear that the experimentally observed boundary shapes are very rich in information.

The apparently ''anomalous'' sedimentation behavior of rapidly reacting systems can be understood by considering the molecules in a complex that travels at a higher velocity, but due to its finite lifetime dissociates into free species that continue to travel slower (Figure 16.7C). Fundamentally different from a zonal configuration like in chromatography, in conventional SV, the faster species always migrate through a bath of the slower ones. Therefore, the free species can reassociate to form complexes. The average distance traveled by the molecules after a certain time will reflect an average sedimentation velocity between that of the unbound and the complex species. Quantitatively, the average velocity measured will depend on the time average of the molecules existing in free and complex forms, which is—following ergodic theorem—determined by the population average that is driven by mass action law. Complicating is the fact that, due to the different sedimentation velocities of all species, spatial gradients of each components are established in the migrating boundary. As a consequence, (infinitesimal) reaction boundary components with different molar ratios will form, which migrate with different s-values and reflect different cosedimenting mixtures in steady state at different total concentrations and molar ratios. Typically, a range of steady-state conditions producing boundary components at a range of s-values exist. Additional transport occurs from coupled diffusion.

From this perspective, it can be understood that sedimentation coefficient distributions are a natural framework for the quantitative and the qualitative descriptions of the sedimentation boundary shapes. Importantly, the sedimentation coefficient distributions are derived from and are descriptive of the data; they do not imply any reaction scheme, but can be a basis for further diagnostics and quantitative interpretation. Historically, apparent sedimentation coefficient distributions, i.e., sedimentation coefficient distributions approximately convoluted by Gaussians arising from diffusion (see earlier), were almost directly observed in experiments with the Schlieren optical system. (It gives a signal proportional to the gradient dc/dr, which can be transformed with Eq. (16.3) to the differential sedimentation coefficient distribution dc/ds^* (Signer and Gross, 1934), similar to the dc/dt transformation (Stafford, 1992).) The newer $c(s)$ distributions essentially are a tool to obtain a sedimentation coefficient distribution free from diffusion broadening.

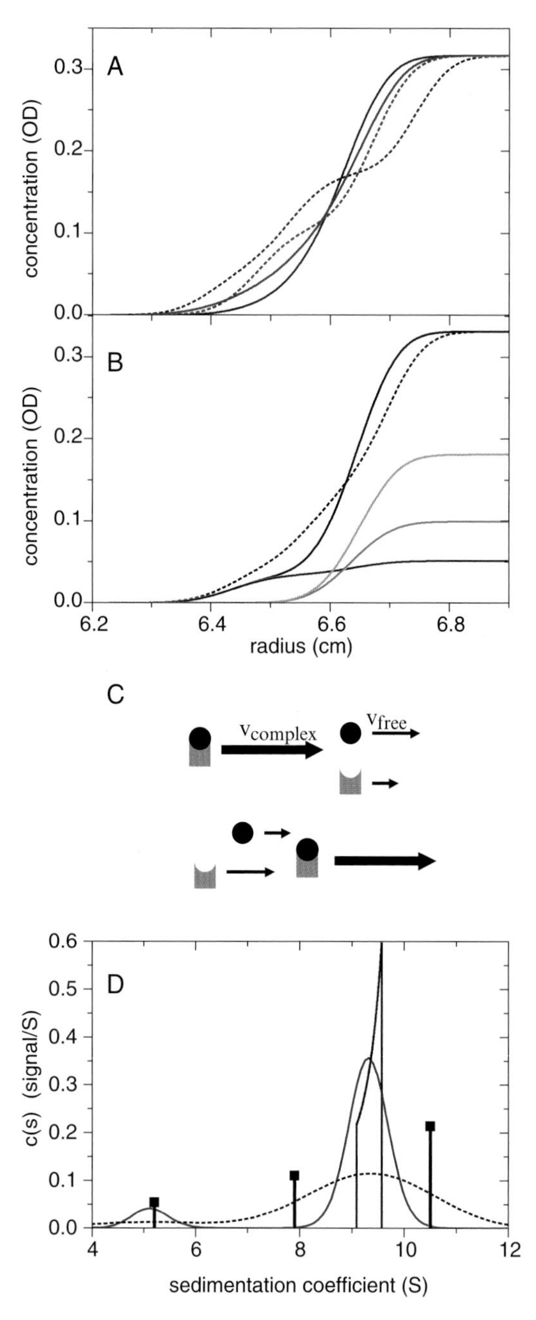

Figure 16.7. Boundary shapes and sedimentation coefficient distributions for mixtures of rapidly interacting proteins. (A) Self-association: Examples of boundary shapes after 45 min sedimentation at 50,000 rpm of a fast ($k_{off} = 0.1/s$) and slow ($k_{off} = 10^{-5}/s$) monomer–dimer system (red solid and

In the 1950s, Gilbert and Jenkins (Gilbert, 1959; Gilbert and Jenkins, 1959) derived a general theory for the coupled migration of chemically reacting systems. They have solved the Lamm equation in rectangular geometry and linear force for a rapidly reacting system for the limit of infinite time, and determined the corresponding stable sedimentation velocity distributions, termed ''asymptotic boundaries'' (red line in Figure 16.7D). This limit corresponds to the limit of diffusion-free migration, since migration from diffusion decreases relative to migration from sedimentation with $\sim 1/\sqrt{t}$. For example, for the case of molecules A and B forming a 1:1 complex C, the asymptotic velocity profile can be calculated by solving the system

$$(u - u_a)\frac{\mathrm{d}a}{\mathrm{d}u} = (u - u_b)\frac{\mathrm{d}b}{\mathrm{d}u} = (u - u_c)\frac{\mathrm{d}c}{\mathrm{d}u}, \quad \text{with} \quad ab = Kc,$$
$$a + c = a_0, \quad b + c = b_0, \tag{16.14}$$

where u is the linear velocity of an infinitesimal fraction of the reaction boundary (in steady state), u_a, u_b, and u_c are the linear sedimentation velocity of species A, B, and C, respectively, and a_0 and b_0 are the total loading concentrations of both components. For self-associating and heteroassociating systems, Gilbert and Jenkins have given analytical and numerical expressions for predicting the amplitudes, shapes, and composition of the undisturbed and reaction boundary as a function of loading concentrations of the components. The Gilbert–Jenkins theory was extremely influential, and it constitutes the foundation for understanding the migration of interacting protein systems in any transport method, including chromatography, electrophoresis, and ultracentrifugation.

Figure 16.7. (*Continued*) dotted lines, respectively) at a concentration of three times K_D. Analogous shapes for fast and slow interactions of a monomer–dimer–tetramer system (solid and dotted blue lines, respectively). All traces reflect the same weight-average sedimentation coefficient s_w, but exhibit very different boundary shapes. (B) Boundary shapes of a heterogeneous association after 45 min sedimentation at 50,000 rpm of a 80-kDa, 5.2-S protein and a 150-kDa, 7.9-S protein forming a 10.5-S complex at equimolar concentrations of threefold K_D. For a rapid interaction, a bimodal boundary is observed (black solid line), for a slow interaction a trimodal boundary is formed (dotted black line), both broadened by diffusion. (Generally, the effect of diffusion is that of reducing the characteristic features, an effect that increases at lower molar mass species). The location of both free and complex species is shown as red, blue, and magenta lines, respectively. It can be discerned that the boundary location of the complex is colocalized with gradients in the free species. (C) Schematics for association and dissociation of the molecules during the sedimentation. During the time the molecules are in complex form, their migration is faster than in the free forms. (D) Sedimentation coefficient distributions for the system shown in B. The s-values and relative population of the species in the loading mixture is shown as squares with dropped lines. The transformation of the boundary shown in B for the rapid interaction into an apparent sedimentation coefficient distribution (ls-$g^*(s)$) is indicated as dotted line. The diffusion-deconvoluted $c(s)$ distribution with features broadened by maximum entropy regularization is the solid blue line, and the asymptotic boundary shape predicted with Gilbert–Jenkins theory is shown in red (scaled by a factor 2). Integration of the blue fast peak and the red distribution gives identical amplitudes and signal-average s-values of the reaction boundary (Dam and Schuck, 2005).

For sedimentation of small molecules interacting with much faster sedimenting macromolecules, Krauss *et al.* (1975) solved the Lamm equations of rapidly reacting systems in a different approximation than Gilbert and Jenkins. With the approximation that the spatial gradient of the small molecule is negligible, it was shown that the interacting system sediments like the solution of a single Lamm equation of a noninteracting ''effective'' species with an average s-value and D-value, and this ''constant-bath'' approximation quantitatively predicts the concentration dependence of both s and D for the reaction boundary. Recently, we have shown that the constant bath theory holds in excellent approximation, even for the interaction of similar-sized proteins (Dam *et al.*, 2005). Based on this observation, and on the finding of the constant bath approximation that diffusion proceeds approximately normal in the reaction boundary, it is apparent that the sedimentation coefficient distribution $c(s)$ of ideal Lamm equation solutions is also a good description of the reaction boundary, and that the property of deconvoluting diffusion is maintained when applied to the sedimentation of rapidly interacting protein mixtures (Dam *et al.*, 2005). As a consequence, the $c(s)$ distribution derived from the experimental sedimentation profiles constitutes an approximation of the diffusion-free asymptotic boundaries from Gilbert–Jenkins theory (Dam and Schuck, 2005; Schuck, 2005), even though the precise shapes are not identical, in part due to the limited signal–noise ratio of the data and the maximum entropy regularization employed in calculating the $c(s)$ distribution.

The significance of this is that the integration of the $c(s)$ peaks determining the s-value and amplitude of both the undisturbed and the reaction boundary can be modeled with the predictions from Gilbert–Jenkins theory. This provides a robust quantitative basis for exploiting the bimodal boundary shape of rapidly interacting systems. For example, the analysis of the concentration dependence of both the amplitude and the s-value of both boundary components, in global analysis including the weight-average s-values, could be globally fitted to a 1:2 interaction model of the Ly49C dimer–MHC interaction given by Dam and Schuck (2005) (Figure 16.8). This approach provides highly useful additional information as compared with the $s_w(c)$ isotherm analysis alone. First, in a titration configuration, the transition point, where the species representing the undisturbed boundary switches, contains characteristic information about the stoichiometry of the interaction. (This is similar to the maximum of the $s_w(c)$ isotherm or the continuous variation analysis.) Second, the reaction boundary is always closer in sedimentation velocity to the complex species than the weight-average s-value, and as a consequence, this provides a theoretically well-founded quantitative approach to improve the determination of the s-value of the complex.

As with the $s_w(c)$ isotherms, the analysis of the s-value of the reaction boundary using Gilbert–Jenkins theory can naturally exploit differences of spectroscopic properties of interacting proteins in heterogeneous interactions. If the sedimentation process is monitored at different signals or wavelengths (e.g., using the interference optics and the absorbance optics at a characteristic wavelength), the resulting $c(s)$ distributions can be integrated to give isotherms at each of the signals,

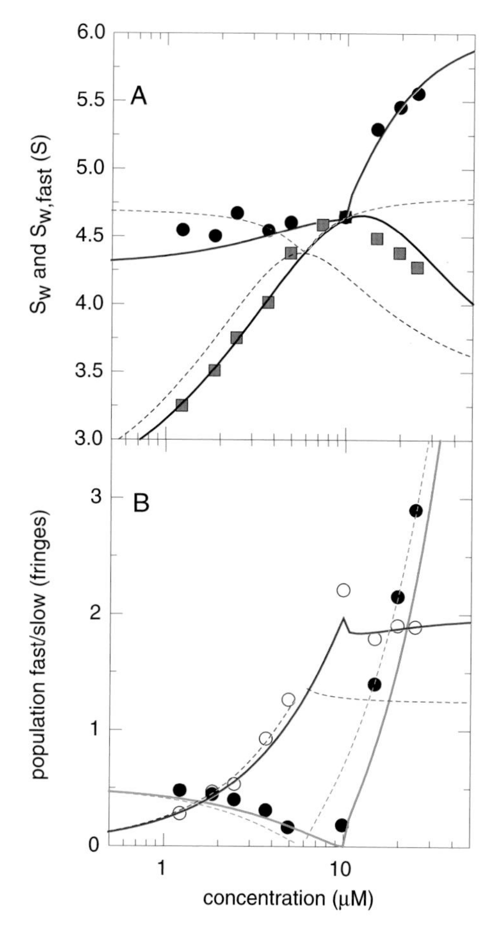

Figure 16.8. Analysis of the reaction boundaries of Ly49C receptor interacting with major histocompatibility complex (MHC) molecules, applying Gilbert–Jenkins theory to the integrated $c(s)$ sedimentation coefficient distributions from a titration of a 5-μM receptor with varying concentrations of MHC molecules (Dam and Schuck, 2005). (A) s-values of the reaction boundary (solid circles) and conventional weight-average sedimentation coefficients (gray squares) measured at different loading concentrations. Theoretical isotherms from a global fit assuming a maximal 1:2 receptor dimer–MHC stoichiometry are shown as solid red and black lines, respectively. Analogous fits assuming a 1:1 stoichiometry of the interaction are shown as dotted lines. (B) Amplitudes of the reaction boundary (open circles) and undisturbed boundary (solid circles) as a function of concentration, and best-fit isotherms assuming 1:2 (solid lines) and 1:1 (dotted lines) stoichiometry. For experimental details, see Dam and Schuck (2005).

which can be globally fitted taking into account the different extinction coefficients (or signal increments) of the different macromolecular components. This provides additional information about the stoichiometry of the interaction.

So far, we have discussed the application of $c(s)$ to systems with fast chemical kinetics relative to the sedimentation process. Figure 16.9 illustrates how the $c(s)$

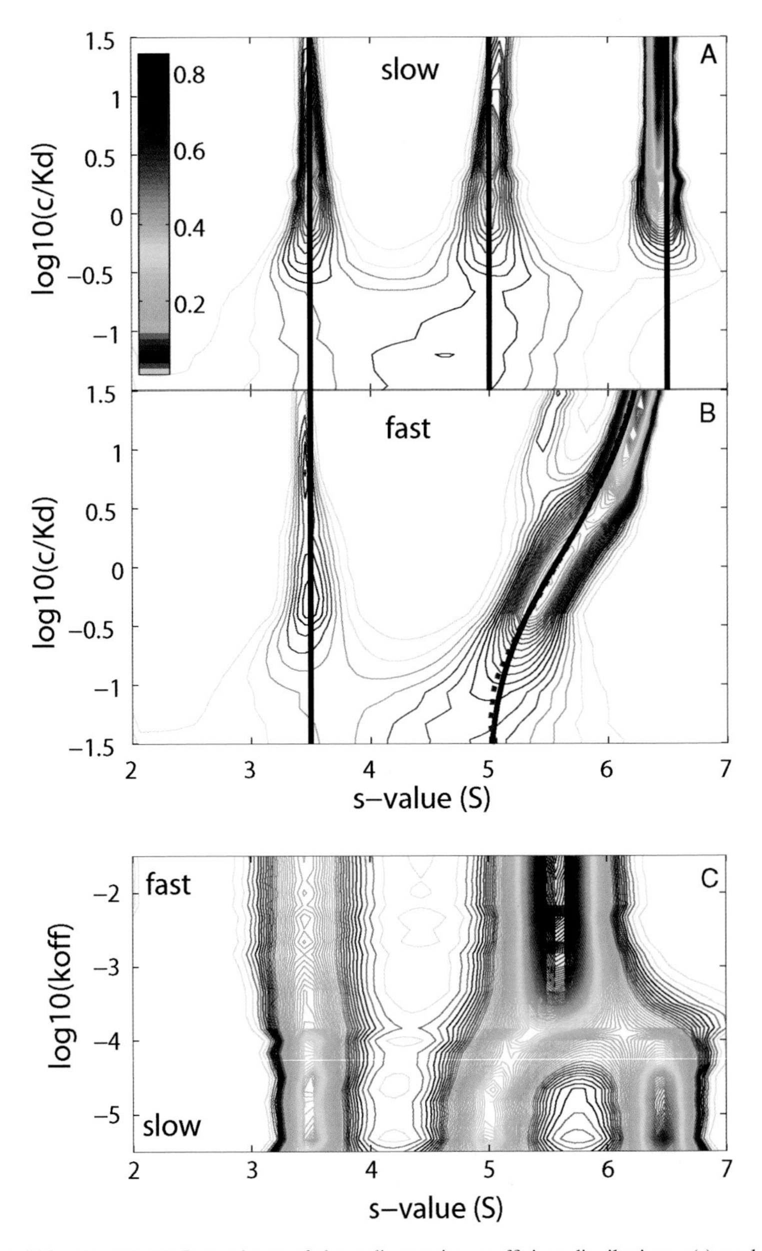

Figure 16.9. (A) and (B) Dependence of the sedimentation coefficient distributions $c(s)$ on loading concentration for a slow and rapid heterogeneous association. Lamm equation solutions were calculated for a 40-kDa, 3.5-S protein interacting with a 60-kDa, 5-S protein to form a 6.5-S complex, at equimolar concentrations and with a reaction rate of 10^{-5}/s (A) and 0.01/s (B) at 50,000 rpm. Sedimentation coefficient distributions $c(s)$ were calculated and normalized relative to the loading concentration, and plotted a contour maps dependent on $\log_{10}(c/K_D)$. In this view, relative heights of $c(s)$ peaks are indicated by color, showing, for example, the increasing fraction of complex at higher concentration. A K_D of 5 μM

distributions, obtained for a protein–protein interaction at different concentrations, depend on the timescale of the reactions. There is only a narrow range of rate constants where a transition between fast and slow characteristics takes place (Figure 16.9C). As discussed, for rapid reactions on the timescale of sedimentation ($k_{off} > 0.01 s^{-1}$), two boundaries appear, with (at least) one exhibiting a concentration-dependent peak location. (For some special cases, for which the presence of a fast reaction will also be obvious, see Dam and Schuck (2005).) Only in the two extremes where the concentrations establish either essentially stoichiometric complex formation or no complex formation at all, do the $c(s)$ peaks reflect the participating species. On the other hand, for slow reactions ($k_{off} < 0.0001 s^{-1}$), little chemical conversion takes place during the sedimentation process, and the deconvolution of diffusion in $c(s)$ provides for peak positions that always reflect the underlying sedimenting species, which remain at a concentration-independent position (provided sufficient signal–noise ratio is available for their hydrodynamic discrimination). It should be noted, however, that it requires concentrations very high above K_D or very slow reactions, for the observed peak s-values to be sufficiently close to the species s-values to permit hydrodynamic modeling of the species on the basis of the peak s-values.

For slow reactions, the $c(s)$ distribution also reveals the number of species. In many cases, this reveals extended multistage associations, such as in the temperature-dependent association of the nonstructural protein NSP2 from the tsE rotavirus mutant (Figure 16.10), the self-association of the heat shock protein Hsp 27 (Lelj--Garolla and Mauk, 2005), the extended heteroassociation of the trp RNA-binding attenuating protein (TRAP) with anti-TRAP (Snyder *et al.*, 2004), and others.

Even though weak associations have frequently limited stability and high k_{off}, many proteins have been studied recently that exhibit weak binding and slow kinetics, which is indicative of a large activation energy, frequently caused by conformational changes induced by binding or flexibility of the recognition site (Willcox *et al.*, 1999). This topic and the role of naturally unstructured proteins are discussed in more detail in Chapter 8. The superposition of $c(s)$ distributions from SV experiments at different loading concentrations can be a very robust diagnostics

---

Figure 16.9. (*Continued*) was assumed, with realistic values for protein extinction and signal–noise of detection, highlighting the problem that the species at very low concentrations cannot be well resolved. At higher concentrations with sufficient signal–noise ratio, for slow reactions on the timescale of sedimentation (A), peaks can be discerned that reflect the sedimenting species (black vertical lines indicate the species s-value), with the mass action law shifting their relative populations. For fast reactions on the timescale of sedimentation (B), an undisturbed and a reaction boundary is observed, with the s-value of the latter being concentration dependent. The theoretical s-values of both boundary components from Gilbert–Jenkins theory are indicated as bold black lines. The dotted red line is the signal average over the reaction boundary component determined from $c(s)$ analysis. Similarly, a distinction between slow and fast reactions can be made for self-associating systems from the concentration dependence of the $c(s)$ peaks (Balbo and Schuck, 2005). (C) Dependence on the chemical off-rate constant at equimolar concentrations $c_A = c_B = K_D$. The transition between the fast and the slow situation occurs over a narrow range of rate constants, which is comparable with the total time of the experiment.

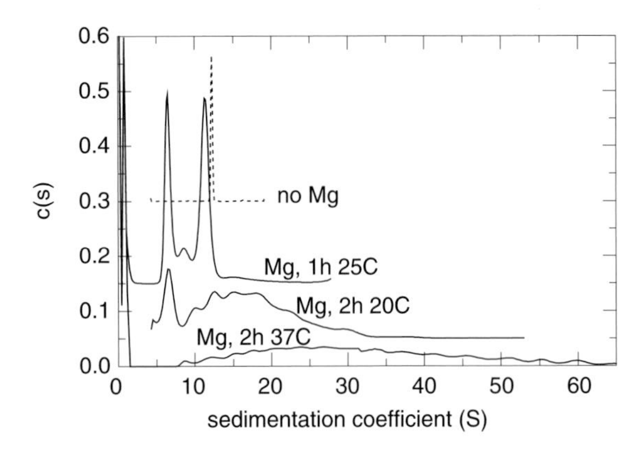

Figure 16.10. Temperture and Mg^{2+}-dependent self-association of the rotavirus nonstructural protein NSP2 from the temperature-sensitive variant tsE (Taraporewala *et al.*, 2002). NSP2 is an octameric single-stranded protein involved in the packaging of the viral RNA into the nascent capsids. It undergoes an Mg^{2+} and ligand-dependent conformational change, exhibiting a hydrodynamically more compact conformation in the presence of nucleotides, and octamer–tetramer dissociation in the presence of Mg^{2+} (Schuck *et al.*, 2000). In addition, NSP2 from tsE at nonpermissive temperatures shows a slow aggregation of the tetramers into a broad range of larger oligomers, which appear to be highly elongated structures, possibly fibers, as judged from the evolution with time of the static and dynamic light scattering (DLS) (Taraporewala *et al.*, 2002). Under these conditions, tsE NSP2 exhibits less RNA-binding activity, and the tsE mutant produces mostly empty particles. Reproduced with permission from Taraporewala *et al.* (2002).

not only for the presence, but also for the timescale of an interaction. Dependent on the timescale, the $c(s)$ analysis of the boundary shape provides either the relative contributions of the different species, which can be modeled directly with mass action law, or it provides the isotherms of the amplitude and s-values of the undisturbed and reaction boundaries, which can be modeled on the basis of Gilbert–Jenkins theory. Both can be analyzed in a global fit jointly with the weight-average s-values $s_w(c)$, to determine s-values of the complex and the affinity constant. As with most techniques, the determination of the affinities depends on the ability to detect both free and complex species, which can be achieved in SV for interactions typically ranging from 10 nM to 100 μM.

16.5.4. Multisignal $c_k(s)$

For moderately slow interactions, where the complexes are relatively stable during the timescale of sedimentation, peaks in $c(s)$ reflect the sedimentation of different species in solution, and the emergence of new peaks in the heterogeneous protein mixtures reflects the formation of complexes. It can be difficult to determine the stoichiometry of the complex, since the s-value is only an indirect measure of the mass because of the shape dependence of the hydrodynamic friction. This is even more difficult for systems exhibiting extended mixed association, where multiple complexes of different size and stoichiometry are formed. In order to overcome this limitation, a second, spectroscopic dimension of discrimination can be exploited in SV when using simultaneously multiple detection systems or wavelengths to observe the sedimentation process. Howard Schachman first exploited differences in the absorbance spectra of interacting molecules to characterize protein–small molecule interactions via constituent sedimentation coefficients (Steinberg and Schachman, 1966). The concept of integrating multiple signals was recently applied in a more detailed boundary shape analysis in the form of the diffusion-deconvoluted multisignal sedimentation coefficient distribution $c_k(s)$. This analysis was introduced earlier, and shown to allow following separately the sedimentation coefficient distributions of each component in the mixture. This provides knowledge of the composition of protein components of a certain sedimenting species, which enables us to determine the stoichiometry of the complex, and, in turn, also its molar mass and hydrodynamic shape.

Minor *et al.* (2005) have applied this new technique to study the mechanism for assembly of complexes of vitronectin and plasminogen activator inhibitor-1 (PAI-1), which play an important role in controlling cell adhesion and pericellular proteolysis, the response of tissue to injuries. It was shown that PAI-1 and vitronectin form initial complexes of 1:1 and 2:1 stoichiometries, with the latter being the building block for the assembly of a large number of higher-order complexes. In mixtures of vitronectin and PAI-1, several complexes can be hydrodynamically distinguished, which are formed in a concentration-dependent fashion. With extrinsically labeled protein and using the multiwavelength $c_k(s)$, it was possible to identify a 6.5-S species to be the 2:1 complex, which is oligomerizing to form the larger species, which are thought to be multivalent in the interaction with cell surface receptors and therefore have enhanced adhesive properties (Minor *et al.*, 2005).

A different application of multisignal SV can be found in the study of Arthos *et al.* (in preparation) on the interactions of HIV envelope protein gp120 with CD4 and DC-SIGN receptor, which are important in understanding the pathway of viral entry. Figure 16.11B shows the sedimentation coefficient distributions $c_k(s)$ from the triple mixture. They reveal coexisting binary and ternary complexes—a 1:1 complex of a (monomeric) envelope protein gp120 variant with CD4 at ~7 S, and a 1:1:1 complex of gp120, CD4, and DC-SIGN at ~8.5 S (Arthos *et al.*, in preparation). Due to the relatively small size of CD4, its participation in a triple complex would be difficult to detect on the basis of the observed s-values. The conformational changes and cooperativity of binding were further characterized by CD and SPR, as shown and discussed in detail by Arthos *et al.* (in preparation). A different application of multi-signal SV is the characterization of the oligomerization of ternary adaptor protein complexes, which are important in the formation of signaling particles after T-cell activation (Houtman *et al.*, 2006).

16.5.5. Modeling with Lamm Equation Solution with Reaction Terms

From a theoretical perspective, perhaps the most attractive approach to characterize protein interactions in SV is to globally model the sedimentation profiles

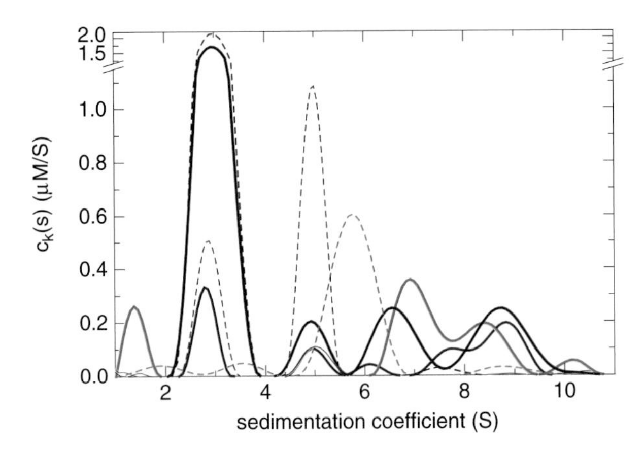

Figure 16.11. Examples of multisignal $c_k(s)$ analysis of triple protein mixtures. $c_k(s)$ analysis of the triple mixture of a viral glycoprotein (green), its cognate receptor (blue), and an antigen-recognition receptor fragment (red) are shown as solid lines. The analogous distributions of each protein alone are shown as dotted lines. For details, see Arthos *et al.* (in preparation).

directly with numerical solutions of the coupled Lamm equations of the interacting system

$$\frac{\partial \chi_i}{\partial t} + \frac{1}{r}\frac{\partial}{\partial r}\left[r\left(s_i\omega^2 r\chi_i - D_i\frac{\partial \chi_i}{\partial r}\right)\right] = q_i, \qquad (16.15)$$

where q_i denotes the time-dependent local reaction flux. The precise form of Eq. (16.15) and, in particular, its reaction fluxes require specification of the number and the stoichiometry of species. It can be implemented either for instantaneous equilibria following mass action law, or considering finite reaction kinetics. In the kinetic case, for example, for a simple bimolecular reaction to form a 1:1 complex, q_i becomes $q_1 = q_2 = -q_3 = -q$, with $q = k_{\mathrm{on}}\chi_1\chi_2 - k_{\mathrm{off}}\chi_3$. Although the numerical solution of Eq. (16.15) has been possible for several decades (Cox, 1969; Cohen and Claverie, 1975; Cox and Dale, 1981), and a qualitative comparison with experimental sedimentation profiles has been used as a data interpretation tool equally long (Frigon and Timasheff, 1975a; Gilbert and Gilbert, 1980), the global nonlinear modeling of Lamm equation solutions [Eq. (16.15)] to experimental data has become computationally practically feasible only recently (Urbanke *et al.*, 1980; Kindler, 1997; Schuck, 1998, 2003; Stafford and Sherwood, 2004; Dam *et al.*, 2005).

The formulation of the sedimentation model by the Lamm equation (16.15) is undoubtedly the most precise and detailed approach, since it does not involve approximations and can explicitly account for reaction kinetics. However, the practical application poses the following hurdles.

First, it requires *a priori* knowledge on the number and the stoichiometry of complexes formed. This is frequently a nontrivial problem, in particular for the extended or multistage associations for which SV is a particularly powerful tool. For simple reactions, it may be possible to develop the model by trial and error. In practice, a more subtle, but frequently very important problem is the consideration of contaminating species, which cannot be easily deduced from a trial-and-error approach. Fortunately, Lamm equation modeling can be applied after the sedimentation coefficient distribution analysis described earlier, which displays the presence of contaminating species outside the size range of interest, reveals the type of interaction, and provides parameter estimates for the binding and kinetic constants, as well as estimates for each species sedimentation coefficient. In this sense, the Lamm equation modeling constitutes a step of refinement of the data analysis. An example of this analysis can be found in Figure 16.12, which shows the sedimentation data of peptides derived from the adaptor proteins SLP-76 and PLCγ1. The multisignal $c_k(s)$ analysis reveals the presence of a 1:1 complex (Balbo *et al.*, 2005), which is confirmed with a direct fit with Lamm equation solutions (Dam *et al.*, 2005), resulting in very low estimates for the chemical off-rate constant, consistent with the expected slow reaction kinetics (Houtman *et al.*, 2004).

Second, the calculated boundary shapes vary significantly for different values of kinetic rate constants only in a narrow kinetic range, where the chemical reaction is taking place on the same timescale as the sedimentation experiment. For typical SV experiments, this range can be described by chemical off-rate constants from 10^{-3} to 10^{-4}/s. (This transition is displayed in Figure 16.9C, even though the information content on kinetics is somewhat higher, since the frictional ratio parameter in the $c(s)$ deconvolution will absorb some features of boundary shapes stemming from finite reaction kinetics.) Outside this range, a rapid reaction may be clearly distinguishable from a slow reaction, but the precise rate constants may not be discriminated within the available signal–noise ratio of the experimental data (Dam *et al.*, 2005).

Third, the detailed interpretation of the boundary shapes can be very susceptible to imperfections that cause artificial broadening of the sedimentation boundary. Similar to the determination of the apparent diffusion coefficients (or apparent molar mass) indicated earlier, a reliable estimate of the kinetic rate constants will depend on the absence of impurities and microheterogeneity. (This may arise from, e.g., conformational mixtures, heterogeneous glycosylation, fractions of only partially folded and/or binding incompetent proteins, or microheterogeneity in binding properties (Cann, 1986).) In particular, this can be expected to exacerbate the problem of ill-conditioned analysis involving chemical rate constants outside the most sensitive range by generating systematic bias, which would not be detected through a statistical error analysis (Dam *et al.*, 2005).

So far, the practical experience and critical error analysis are limited (Gelinas *et al.*, 2004; Sontag *et al.*, 2004; Dam *et al.*, 2005; Yikilmaz *et al.*, 2005). With regard to kinetic analysis, the strength of this approach appears to be the detailed study of relatively well-defined systems with reaction kinetics described by complex lifetimes of $\sim$1 h. For systems outside this kinetic range, the information obtained seems to be largely redundant to that from the interpretation of the

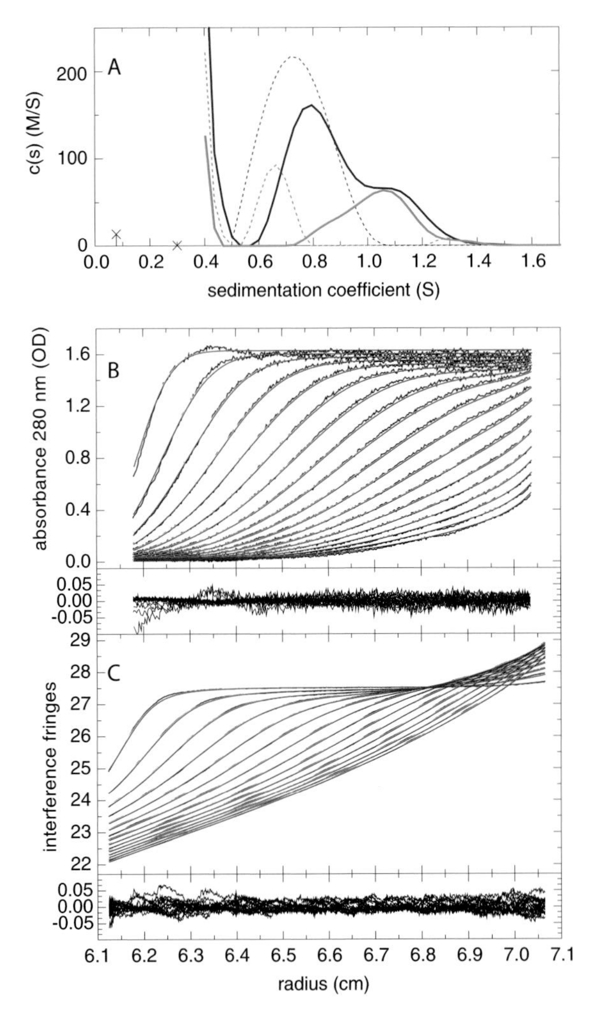

Figure 16.12. Multiwavelength sedimentation coefficient distribution (A) and Lamm equation modeling (B and C) of the interaction of peptides derived from the adaptor protein SLP-76 (11.7 kDa, 0.60 S) and PLC-γ (7.4 kDa, 0.75 S) which form complexes with 1:1 stoichiometry. SLP-76 contains only one tyrosine and no tryptophan residues, allowing its spectral discrimination from PLC-γ. (A) Calculated $c_k(s)$ traces of SLP-76 (green) and PLC-γ (blue) from individual experiments (dotted lines) and in the mixture (solid lines). The interference optical data are superimposed by the sedimentation of a small buffer component (likely predominantly optically unmatched NaCl) that can be modeled well as discrete species at 0.055 S. For details, see Balbo *et al.* (2005). The distribution from the mixture clearly shows a peak at ~1.1 S with 1:1 stoichiometry. (B) and (C) show raw sedimentation profiles, corrected for systematic noise contributions (black lines) obtained with the absorbance (B) and interference (C) optical system. The red lines are the calculated distributions from direct modeling a system of Lamm equations with a reaction term, with $s_{AB} = 1.05\,\text{S}, K_D = 10\,\mu\text{M}$, and $k_{off} = 10^{-5}/\text{s}$, with an root-mean-square deviation (rmsd) of 0.0137 signal units (OD or fringes). The reaction kinetics was not very well determined, with rmsd of 0.0155 signal units for an instantaneous reaction, 0.0150 for $k_{off} = 10^{-4}/\text{s}$, and 0.0137 for $k_{off} < 10^{-5}/\text{s}$, and the best-fit value of 0.0136 for $k_{off} = 1.6 \times 10^{-6}/\text{s}$. For details, see Dam *et al.* (2005).

isotherms based on diffusion-deconvoluted sedimentation coefficient distributions, which naturally account for all sedimenting species and provide an opportunity to exclude contaminants from biasing the results. Like the $c_k(s)$ distribution, direct Lamm equation modeling can naturally exploit differences in extinction spectra of the different components in heterogeneous interactions. Both approaches may be extended to systems involving proteins with multiple binding sites and/or conformational states. One can speculate that, likely, the specific power of direct kinetic Lamm equation modeling will emerge most clearly when future experimental configurations are included in the analysis that do not start in chemical equilibrium, as in active enzyme centrifugation (Cohen and Claverie, 1975).

A good approach to avoid the limitations of kinetic analysis for fast reactions is to regard the reaction as in instantaneous equilibrium, in which case the reaction fluxes of Eq. (16.15) will be determined by the coupled sedimentation obeying mass action law. This was applied to several systems (Lewis *et al.*, 2002; Gelinas *et al.*, 2004). For fundamental reasons, this approximation appears to be particularly advantageous for rapidly self-associating systems (Tarabykina *et al.*, 2001; Lewis *et al.*, 2002; Ali *et al.*, 2003). Here, the determination of the reaction scheme is more difficult because no difference in the spectral properties is available, and only a single concentration parameter can be varied in the titration. Further, making use of Gilbert theory for self-associations analogous to the application of Gilbert–Jenkins theory for heteroassociations appears not easily possible, because Gilbert theory predicts no undisturbed components and, instead, for sedimentation to display only reaction boundaries. To make use of Gilbert theory would require experimentally determined diffusion-deconvoluted sedimentation coefficient distributions that mimic the predicted asymptotic boundaries precisely with regard to detailed shape. As shown earlier, $c(s)$ distributions approximate the theoretical asymptotic boundaries for heteroassociations well in their bimodal structure, relative height, and average s-value of the undisturbed and reaction boundary, but not their precise shape. This is due to the limitations in the signal–noise ratio, which can be recognized from the fact that maximum entropy regularization results in smooth $c(s)$ peaks for both boundary components. Another approach to derive diffusion-deconvoluted asymptotic boundaries from experimental data of self-associating systems was proposed in 1977 by Winzor *et al.*, but did not find much application subsequently, most likely due to similar limitations. Clearly, mass balance analyses from the weight-average $s_w(c)$ isotherm are a valid and important tool for self-associations. However, in contrast to rapid heterogeneous associations, rapid self-associations do not provide a robust quantitative approach to boundary shape analysis, short of direct Lamm equation modeling. The use of $s_w(c)$ isotherms and Lamm equation modeling for rapidly reversible self-associating systems is discussed in detail by Schuck (2003).

An example for the application of this approach is the study of the rapid trimer–hexamer–dodecamer self-association of the molecular chaperone gp57A of the bacteriophage T4 (Ali *et al.*, 2003) (Figure 16.13). The gp57A is a small protein of 79 amino acids, which facilitates the long and short tail fiber formation, and is

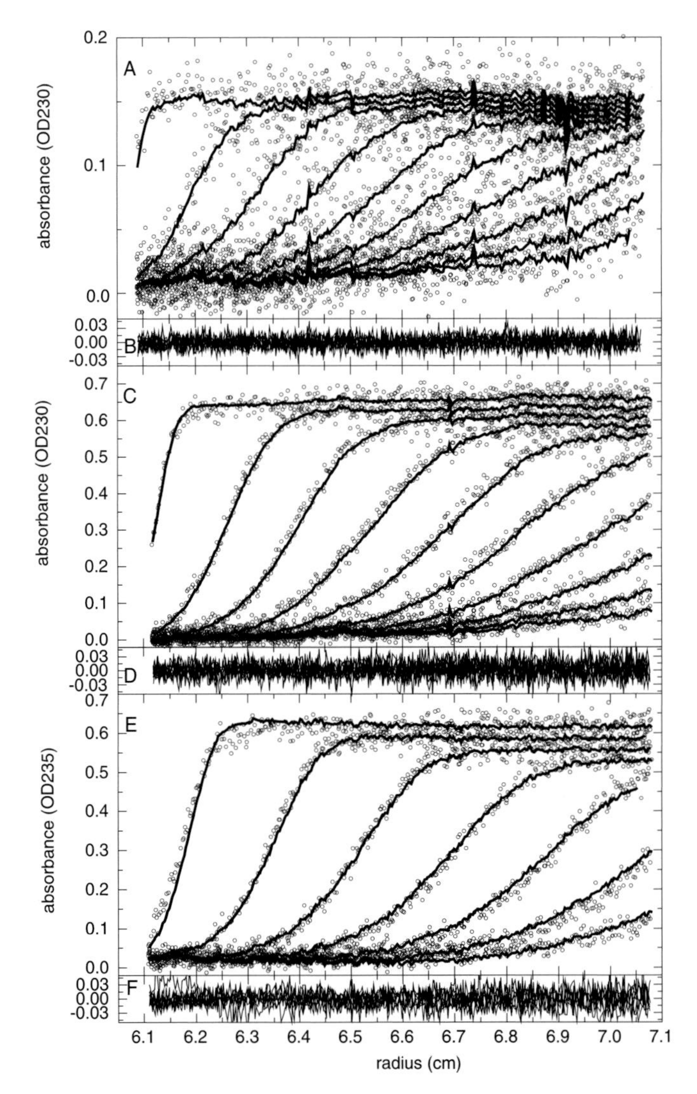

Figure 16.13. Self-association of the molecular chaperone gp57A of the bacteriophage T4. Shown are the raw sedimentation velocity data and global direct Lamm equation fit with a model for a trimer–hexamer–dodecamer self-association in instantaneous chemical equilibrium. Shown are only every tenth experimental absorbance profile (circles) obtained at a rotor speed of 50,000 rpm at 20°C at pH 8.0, at concentrations of 0.1 mg/mL (A) 0.52 mg/mL (C), and 2.1 mg/mL (E). The scans shown are taken from equivalent time points for each loading concentration. This highlights the faster sedimentation at higher concentrations. A global fit with a trimer–hexamer–dodecamer model was achieved with a larger set of sedimentation data at loading concentrations of 0.1, 0.31, 0.52, 1.05, and 2.1 mg/mL, which resulted in binding constants of $K_{3-6} = 3.24 \times 10^4 \mathrm{M}^{-1}$ and $K_{6-12} = 7.64 \times 10^3 \mathrm{M}^{-1}$, with sedimentation coefficients $s_3 = 2.16\,\mathrm{S}, s_6 = 3.25\,\mathrm{S}$, and $\mathrm{s}_{12} = 4.59\,\mathrm{S}$. The solid lines show the best-fit sedimentation distributions (including time invariant noise components). The residuals of the fits corresponding to the data in (A), (C), and (E) are shown in the (B), (D), and (F), respectively. Reproduced with permission from Ali *et al.* (2003).

unusual in its highly reversible dissociation and unfolding transitions. In this study, stopped-flow CD indicated the presence of fast association. In sedimentation equilibrium, the dependence of the weight-average molecular weight as a function of protein concentration indicated an extended self-association, but the association scheme could not be determined. A global fit of the SV profiles at different concentrations by direct Lamm equation modeling allowed us to determine that the association likely proceeds as trimer–hexamer–dodecamer association, as judged from both the quality of fit and the likelihood of the hydrodynamic shapes of the oligomeric species implied by the best-fit s-values. This conclusion was consistent with the results from differential scanning calorimetry, where a global fit to the asymmetric bimodal temperature unfolding curves observed at different concentrations was best described with a hexamer–trimer–monomer transition, and with the results from refolding studies by stopped-flow CD, which indicated the second association stage to be a bimolecular reaction (trimer–hexamer transition) (Ali *et al.*, 2003). This example highlights the opportunity provided by hydrodynamic separation and boundary spread analysis of rapidly self-associating systems with direct Lamm equation modeling in SV, and the advantage of its combination with orthogonal biophysical methods.

16.6. CONCLUSIONS

Sedimentation velocity is a very versatile technique. This review was limited to the currently most commonly applied approaches for studying protein interactions. The experiment is conceptually very simple, and a detailed first principle-based analysis is possible. Significant advances in the computational analysis have made increasingly complex interacting systems accessible for study, including many interactions with triple protein complexes and extended multistage associations. SV provides hydrodynamic discrimination of complexes by observing macroscopic migration and concentration gradients, but in a configuration that leaves the complexes in a bath of the free components at all times, such that reversibly formed complexes can be dynamically maintained.

SV provides several sources of information on self-association and heterogeneous associations of proteins in free solution. Very important in comparison with other techniques is the ability to discern multiple coexisting complexes from the diffusion-deconvoluted migration pattern. Generally, the resolution is based on the sedimentation coefficient s, which is dependent on frictional shape and roughly on the two-thirds-power of the particle radius, a stronger dependence than that found in purely diffusion-based methods. The boundary shapes report on the species diffusion coefficients, D, that together with s reveals the species molar mass. Orthogonal to this, spectral differences in the interacting components can allow to determine the composition of complexes. The observed s-values make the hydrodynamic shape of the complexes an accessible parameter. The interpretation of the sedimentation boundary shapes as a function of loading composition enables estimating the order of magnitude of the kinetic rate constants of complex formation. Even though highly pure protein preparations are typically required, the experiments display all sedimenting materials, and the analysis can be conducted such that it is relatively tolerant to impurities outside the sedimentation range of the proteins of interest. Data analysis techniques can be chosen that balance possible preparative imperfections, prior knowledge, and the detail of the derived parameters.

SV is well compatible with regard to sample consumption and buffer conditions for use in conjunction with several other biophysical techniques. This includes the use as stringent control experiments, for example, to determine oligomeric state and purity of analyte protein in SPR, but also to provide orthogonal information useful to shed light on different aspects of protein interactions. For example, conformational changes can be assessed synergistically if SV is combined with CD and DLS (or FCS). Hydrodynamic modeling establishes relationships with several hydrodynamic techniques for rotational and translational diffusion and small-angle scattering, which can help to assess low-resolution solution structure of protein complexes. Thermodynamic information from the isotherms of titrations in ITC, surface-binding equilibria in SPR, or specific CD signatures of protein conformational changes in the complex, as well as sedimentation equilibrium profiles, can be analyzed in parallel in global analysis with the isotherms of s-values from SV.

ACKNOWLEDGMENT

I am grateful to Drs John Burgner and Christine Ebel for critical reading of the manuscript.

REFERENCES

Ali, S. A., Iwabuchi, N., Matsui, T., Hirota, K., Kidokoro, S., Arai, M., Kuwajima, K., Schuck, P., and Arisaka, F. (2003). Rapid and dynamic association equilibrium of a molecular chaperone, gp57A, of bacteriophage T4. *Biophys J* 85:2606–2618.

Aragon, S. (2004). A precise boundary element method for macromolecular transport properties. *J Comput Chem* 25:1191–1205.

Archibald, W. J. (1947). A demonstration of some new methods of determining molecular weights from the data of the ultracentrifuge. *J Phys Colloid Chem* 51:1204–1214.

Balbo, A., Minor, K. H., Velikovsky, C. A., Mariuzza, R., Peterson, C. B., and Schuck, P. (2005). Studying multi-protein complexes by multi-signal sedimentation velocity analytical ultracentrifugation. *Proc Natl Acad Sci USA* 102:81–86.

Balbo, A. and Schuck, P. (2005). Analytical ultracentrifugation in the study of protein self-association and heterogeneous protein–protein interactions. In: Golemis, E. and Adams, P. D. (eds), *Protein–Protein Interactions*. Cold Spring Harbor Laboratory Press, Cold Spring Harbor, New York, pp. 253–277.

Baldwin, R. L. (1953). Sedimentation coefficients of small molecules: methods of measurement based on the refractive-index gradient curve. The sedimentation coefficient of polyglucose A. *Biochem J* 55:644–648.

Baldwin, R. L. and Williams, J. W. (1950). Boundary spreading in sedimentation velocity experiments. *Am Chem Soc* 72:4325.

Behlke, J. and Ristau, O. (2002). A new approximate whole boundary solution of the Lamm differential equation for the analysis of sedimentation velocity experiments. *Biophys Chem* 95:59–68.

Boukari, H., Nossal, R., Sackett, D. L., and Schuck, P. (2004). Hydrodynamics of nanoscopic tubulin rings in dilute solution. *Phys Rev Lett* 93:098106.

Brown, P. and Schuck, P. (2006). Macromolecular size-and-shape distributions by sedimentation velocity analytical ultracentrifugation. *Biophys J* 90:4651–4661.

Buisson, M., Valette, E., Hernandez, J. F., Baudin, F., Ebel, C., Morand, P., Seigneurin, J. M., Arlaud, G. J., and Ruigrok, R. W. (2001). Functional determinants of the Epstein-Barr virus protease. *J Mol Biol* 311:217–228.

Burgess, B. R., Schuck, P., and Garboczi, D. N. (2005). Dissection of merozoite surface protein 3, a representative of a family of *Plasmodium falciparum* surface proteins, reveals an oligomeric and highly elongated molecule. *J Biol Chem* 280:37236–37245.

Cann, J. R. (1986). Effects of microheterogeneity on sedimentation patterns of interacting proteins and the sedimentation behavior of systems involving two ligands. *Methods Enzymol* 130:19–35.

Cao, W. and Demeler, B. (2005). Modeling analytical ultracentrifugation experiments with an adaptive space-time finite element solution of the Lamm equation. *Biophys J* 89:1589–1602.

Chatelier, R. C. (1987). Indefinite isoenthalpic self-association of solute molecules. *Biophys Chem* 28:121–128.

Che, M. M., Boja, E. S., Yoon, H. Y., Gruschus, J., Jaffe, H., Stauffer, S., Schuck, P., Fales, H. M., and Randazzo, P. A. (2005). Regulation of ASAP1 by phospholipids is dependent on the interface between the PH and Arf GAP domains. *Cell Signal* 17:1276–1288.

Claverie, J.-M., Dreux, H., and Cohen, R. (1975). Sedimentation of generalized systems of interacting particles. I. Solution of systems of complete Lamm equations. *Biopolymers* 14:1685–1700.

Cohen, R. and Claverie, J. M. (1975). Sedimentation of generalized systems of interacting particles. II. Active enzyme centrifugation—theory and extensions of its validity range. *Biopolymers* 14: 1701–1716.

Cohen, R., Giraud, B., and Messiah, A. (1967). Theory and practice of the analytical centrifugation of an active substrate–enzyme complex. *Biopolymers* 5:203–225.

Cohen, R. E., Foote, J., and Schachman, H. K. (1985). On conformational changes in the regulatory enzyme aspartate transcarbamoylase. *Curr Top Cell Regul* 26:177–190.

Cole, J. L. (2004). Analysis of heterogeneous interactions. *Methods Enzymol* 384:212–232.

Consler, T. G. and Lee, J. C. (1988). Domain interaction in rabbit muscle pyruvate kinase. I. Effects of ligands on protein denaturation induced by guanidine hydrochloride. *J Biol Chem* 263:2787–2793.

Correia, J. J. (2000). Analysis of weight average sedimentation velocity data. *Methods Enzymol* 321: 81–100.

Cox, D. J. (1965). Computer simulation of sedimentation in the ultracentrifuge. I. Diffusion. *Arch Biochem Biophys* 112:249–258.

Cox, D. J. (1966). Sedimentation of an initially skewed boundary. *Science* 152:359–361.

Cox, D. J. (1969). Computer simulation of sedimentation in the ultracentrifuge. IV. Velocity sedimentation of self-associating solutes. *Arch Biochem Biophys* 129:106–123.

Cox, D. J. and Dale, R. S. (1981). Simulation of transport experiments for interacting systems. In: Frieden, C. and Nichol, L. W. (eds), *Protein–Protein Interactions*. Wiley, New York.

Creeth, J. M. (1964). Approximate steady state condition in ultracentrifuge. *Proc Royal Soc A* 282:403.

Creeth, J. M. and Knight, C. G. (1965). On the estimation of the shape of macromolecules from sedimentation and viscosity measurements. *Biochim Biophys Acta* 102:549–558.

Dam, J., Guan, R., Natarajan, K., Dimasi, N., Chlewicki, L. K., Kranz, D. M., Schuck, P., Margulies, D. H., and Mariuzza, R. A. (2003). Variable MHC class I engagement by Ly49 NK cell receptors revealed by the crystal structure of Ly49C bound to H-2Kb. *Nat Immunol* 4:1213–1222.

Dam, J. and Schuck, P. (2004). Calculating sedimentation coefficient distributions by direct modeling of sedimentation velocity profiles. *Methods Enzymol* 384:185–212.

Dam, J. and Schuck, P. (2005). Sedimentation velocity analysis of protein–protein interactions: sedimentation coefficient distributions $c(s)$ and asymptotic boundary profiles from Gilbert–Jenkins theory. *Biophys J* 89:651–666.

Dam, J., Velikovsky, C. A., Mariuzza, R., Urbanke, C., and Schuck, P. (2005). Sedimentation velocity analysis of protein–protein interactions: Lamm equation modeling and sedimentation coefficient distributions $c(s)$. *Biophys J* 89:619–634.

Demeler, B., Saber, H., and Hansen, J. C. (1997). Identification and interpretation of complexity in sedimentation velocity boundaries. *Biophys J* 72:397–407.

Dishon, M., Weiss, G. H., and Yphantis, D. A. (1966). Numerical solutions of the Lamm equation. I. Numerical procedure. *Biopolymers* 4:449–455.

Dishon, M., Weiss, G. H., and Yphantis, D. A. (1967). Numerical simulations of the Lamm equation: III. Velocity centrifugation. *Biopolymers* 5:697–713.

Doster, W. and Hess, B. (1981). Reversible solvent denaturation of rabbit muscle pyruvate kinase. *Biochemistry* 20:772–780.

Douglas, J. F., Zhou, H. X., and Hubbard, J. B. (1994). Hydrodynamic friction and the capacitance of arbitrarily shaped objects. *Phys Rev E Stat Phys, Plasmas, Fluids, Related Interdisciplinary Topics* 49:5319–5331.

Durchschlag, H. and Zipper, P. (2003). Modeling the hydration of proteins: prediction of structural and hydrodynamic parameters from X-ray diffraction and scattering data. *Eur Biophys J* 32: 487–502.

Elzen, B. (1988). *Scientists and Rotors. The Development of Biochemical Ultracentrifuges*, Dissertation. University Twente, Enschede.

Faxén, H. (1929). Über eine Differentialgleichung aus der physikalischen Chemie. *Ark Mat Astr Fys* 21B:1–6.

Frigon, R. P. and Timasheff, S. N. (1975a). Magnesium-induced self-association of calf brain tubulin. I. Stoichiometry. *Biochemistry* 14:4559–4566.

Frigon, R. P. and Timasheff, S. N. (1975b). Magnesium-induced self-association of calf brain tubulin. II. Thermodynamics. *Biochemistry* 14:4567–4573.

Fujita, H. (1975). *Foundations of Ultracentrifugal Analysis*. John Wiley & Sons, New York.

Garcia de la Torre, J. G. and Bloomfield, V. A. (1981). Hydrodynamic properties of complex, rigid, biological macromolecules: theory and applications. *Q Rev Biophys* 14:81–139.

Garcia de La Torre, J., Huertas, M. L., and Carrasco, B. (2000). Calculation of hydrodynamic properties of globular proteins from their atomic-level structure. *Biophys J* 78:719–730.

Garrido, F., Gasset, M., Sanz-Aparicio, J., Alfonso, C., and Pajares, M. A. (2005). Rat liver betaine-homocysteine *S*-methyltransferase equilibrium unfolding: insights into intermediate structure through tryptophan substitutions. *Biochem J* 391:589–599.

Gelinas, A. D., Toth, J., Bethoney, K. A., Stafford, W. F., and Harrison, C. J. (2004). Mutational analysis of the energetics of the GrpE.DnaK binding interface: equilibrium association constants by sedimentation velocity analytical ultracentrifugation. *J Mol Biol* 339:447–458.

George, A. and Wilson, W. W. (1994). Predicting protein crystallization from a dilute solution property. *Acta Crystallogr D Biol Crystallogr* 50:361–365.

Gilbert, G. A. (1959). Sedimentation and electrophoresis of interacting substances. I. Idealized boundary shape for a single substance aggregating reversibly. *Proc Royal Soc London A* 250:377–388.

Gilbert, G. A. and Gilbert, L. M. (1980). Ultracentrifuge studies of interactions and equilibria: impact of interactive computer modelling. *Biochem Soc Trans* 8:520–522.

Gilbert, G. A. and Jenkins, R. C. (1956). Boundary problems in the sedimentation and electrophoresis of complex systems in rapid reversible equilibrium. *Nature* 177:853–854.

Gilbert, G. A. and Jenkins, R. C. (1959). Sedimentation and electrophoresis of interacting substances. II. Asymptotic boundary shape for two substances interacting reversibly. *Proc Royal Soc A* 253:420–437.

Goad, W. B. and Cann, J. R. (1969). Theory of sedimentation of interacting systems. *Ann N Y Acad Sci* 164:172–182.

Gonzalez, J. M., Velez, M., Jimenez, M., Alfonso, C., Schuck, P., Mingorance, J., Vicente, M., Minton, A. P., and Rivas, G. (2005). Cooperative behavior of *Escherichia coli* cell-division protein FtsZ assembly involves the preferential cyclization of long single-stranded fibrils. *Proc Natl Acad Sci U S A* 102:1895–1900.

Greive, S. J., Lins, A. F., and von Hippel, P. H. (2005). Assembly of an RNA–protein complex. Binding of NusB and NusE (S10) proteins to boxA RNA nucleates the formation of the antitermination complex involved in controlling rRNA transcription in *Escherichia coli*. *J Biol Chem* 280:36397–36408.

Harding, S. E., Rowe, A. J., and Horton, J. C. (eds) (1992). *Analytical Ultracentrifugation in Biochemistry and Polymer Science*. Royal Society of Chemistry, Cambridge.

Harrington, W. F. and Kegeles, G. (1973). Pressure effects in ultracentrifugation of interacting systems. *Methods Enzymol* 27:106–345.

Holladay, L. A. (1979). An approximate solution to the Lamm equation. *Biophys Chem* 10:187–190.

Houtman, J. C., Higashimoto, Y., Dimasi, N., Cho, S., Yamaguchi, H., Bowden, B., Regan, C., Malchiodi, E. L., Mariuzza, R., Schuck, P., Appella, E., and Samelson, L. E. (2004). Binding specificity of multiprotein signaling complexes is determined by both cooperative interactions and affinity preferences. *Biochemistry* 43:4170–4178.

Houtman, J.C., Yamaguchi, H., Barda-Saad, M., Braiman, A., Bowden, B., Appella, E., Schuck, P., and Samelson, L.E. (2006). Oligomerization of signaling complexes by the multipoint binding of Grb2 to both LAT and Sos1. *Nature Struct Mol Biol* 13:798–805.

Howlett, G. J., Blackburn, M. N., Compton, J. G., and Schachman, H. K. (1977). Allosteric regulation of aspartate transcarbamoylase. Analysis of the structural and functional behavior in terms of a two-state model. *Biochemistry* 16:5091–5100.

Howlett, G. J. and Schachman, H. K. (1977). Allosteric regulation of aspartate transcarbamoylase. Changes in the sedimentation coefficient promoted by the bisubstrate analogue *N*-(phosphonacetyl)-L-aspartate. *Biochemistry* 16:5077–5083.

Howlett, G.J., Minton, A.P., and Rivas, G. (2006). Analytical ultracentrifugation for the study of protein association and assembly. *Curr Opin Chem Biol* 10:430–436.

Hu, Y., Sun, Z., Eaton, J. T., Bouloux, P. M., and Perkins, S. J. (2005). Extended and flexible domain solution structure of the extracellular matrix protein anosmin-1 by X-ray scattering, analytical ultracentrifugation and constrained modelling. *J Mol Biol* 350:553–570.

Irimia, A., Ebel, C., Madern, D., Richard, S. B., Cosenza, L. W., Zaccai, G., and Vellieux, F. M. (2003). The oligomeric states of *Haloarcula marismortui* malate dehydrogenase are modulated by solvent components as shown by crystallographic and biochemical studies. *J Mol Biol* 326:859–873.

Jayaram, H., Taraporewala, Z., Patton, J. T., and Prasad, B. V. (2002). Rotavirus protein involved in genome replication and packaging exhibits a HIT-like fold. *Nature* 417:311–315.

Johnston, J. P. and Ogston, A. G. (1946). A boundary anomaly found in the ultracentrifugal sedimentation of mixtures. *Trans Faraday Soc* 42:789–799.

Kindler, B. (1997). Akkuprog: Auswertung von Messungen chemischer Reaktionsgeschwindigkeit und Analyse von Biopolymeren in der Ultrazentrifuge. Phd. Thesis, University Hannover.

Koenderink, G. H., Planken, K. L., Roozendaal, R., and Philipse, A. P. (2005). Monodisperse DNA restriction fragments II. Sedimentation velocity and equilibrium experiments. *J Colloid Interface Sci* 291:126–134.

Kornblatt, J. A. and Schuck, P. (2005). Influence of temperature on the conformation of canine plasminogen: an analytical ultracentrifugation and dynamic light scattering study. *Biochemistry* 44:13122–13131.

Krauss, G., Pingoud, A., Boehme, D., Riesner, D., Peters, F., and Maass, G. (1975). Equivalent and non-equivalent binding sites for tRNA on aminoacyl-tRNA synthetases. *Eur J Biochem* 55: 517–529.

Lamm, O. (1929). Die Differentialgleichung der Ultrazentrifugierung. *Ark Mat Astr Fys* 21B(2):1–4.

Laue, T. M., Anderson, A. L., and Weber, B. W. (1997). Prototype fluorimeter for the XLA/XLI analytical ultracentrifuge. In: Cohn, G. E. and Soper, S. A. (eds), *Ultrasensitive Biochemical Diagnostics II. SPIE Proceedings*, Vol. 2985. SPIE, Bellingham, WA, pp. 196–204.

Lebowitz, J., Lewis, M. S., and Schuck, P. (2002). Modern analytical ultracentrifugation in protein science: a tutorial review. *Protein Sci* 11:2067–2079.

Lebowitz, J., Lewis, M. S., and Schuck, P. (2003). Back to the future: a rebuttal to Henryk Eisenberg. *Protein Sci* 12:2649–2650.

Lebowitz, J., Teale, M., and Schuck, P. (1998). Analytical band centrifugation of proteins and protein complexes. *Biochem Soc Trans* 26:745–749.

Lee, H. J., Lu, S. W., and Chang, G. G. (2003). Monomeric molten globule intermediate involved in the equilibrium unfolding of tetrameric duck delta2-crystallin. *Eur J Biochem* 270:3988–3995.

Lelj-Garolla, B. and Mauk, A. G. (2005). Self-association of a small heat shock protein. *J Mol Biol* 345:631–642.

Lewis, R. J., Scott, D. J., Brannigan, J. A., Ladds, J. C., Cervin, M. A., Spiegelman, G. B., Hoggett, J. G., Barak, I., and Wilkinson, A. J. (2002). Dimer formation and transcription activation in the sporulation response regulator Spo0A. *J Mol Biol* 316:235–245.

Li, J., Correia, J. J., Wang, L., Trent, J. O., and Chaires, J. B. (2005). Not so crystal clear: the structure of the human telomere G-quadruplex in solution differs from that present in a crystal. *Nucleic Acids Res* 33:4649–4659.

Lobert, S., Ingram, J. W., Hill, B. T., and Correia, J. J. (1998). A comparison of thermodynamic parameters for vinorelbine- and vinflunine-induced tubulin self-association by sedimentation velocity. *Mol Pharmacol* 53:908–915.

MacRaild, C. A., Hatters, D. M., Lawrence, L. J., and Howlett, G. J. (2003). Sedimentation velocity analysis of flexible macromolecules: self-association and tangling of amyloid fibrils. *Biophys J* 84:2562–2569.

Marcum, J. M. and Borisy, G. G. (1978). Sedimentation velocity analyses of the effect of hydrostatic pressure on the 30 S microtubule protein oligomer. *J Biol Chem* 253:2852–2857.

Minor, K. H., Schar, C. R., Blouse, G. E., Shore, J. D., Lawrence, D. A., Schuck, P., and Peterson, C. B. (2005). A mechanism for assembly of complexes of vitronectin and plasminogen activator inhibitor-1 from sedimentation velocity analysis. *J Biol Chem* 31:28711–28720.

Oncley, J. L., Ellenbogen, E., Gitlin, D., and Gurt, F. R. N. (1952). Protein–protein interactions. *J Phys Chem* 56:85–92.

Perugini, M. A., Schuck, P., and Howlett, G. J. (2000). Self-association of human apolipoprotein E3 and E4 in the presence and absence of phopholipid. *J Biol Chem* 275:36758–36765.

Philo, J. S. (1997). An improved function for fitting sedimentation velocity data for low molecular weight solutes. *Biophys J* 72:435–444.

Provencher, S. W. (1982). CONTIN: a general purpose constrained regularization program for inverting noisy linear algebraic and integral equations. *Comp Phys Comm* 27:229–242.

Rai, N., Nollmann, M., Spotorno, B., Tassara, G., Byron, O., and Rocco, M. (2005). SOMO (SOlution MOdeler) differences between X-ray- and NMR-derived bead models suggest a role for side chain flexibility in protein hydrodynamics. *Structure (Camb)* 13:723–734.

Rivas, G., Lopez, A., Mingorance, J., Ferrandiz, M. J., Zorrilla, S., Minton, A. P., Vicente, M., and Andreu, J. M. (2000). Magnesium-induced linear self-association of the FtsZ bacterial cell division protein monomer. The primary steps for FtsZ assembly. *J Biol Chem* 275: 11740–11749.

Rivas, G., Stafford, W., and Minton, A. P. (1999). Characterization of heterologous protein–protein interactions via analytical ultracentrifugation. *Methods: Comp Methods Enzymol* 19: 194–212.

Rochet, J. C., Brownie, E. R., Oikawa, K., Hicks, L. D., Fraser, M. E., James, M. N., Kay, C. M., Bridger, W. A., and Wolodko, W. T. (2000). Pig heart CoA transferase exists as two oligomeric forms separated by a large kinetic barrier. *Biochemistry* 39:11291–11302.

Rowe, A. J. (1992). The concentration dependence of sedimentation. In: Harding, S. E., Rowe, A. J., and Horton, J. C. (eds), *Analytical Ultracentrifugation in Biochemistry and Polymer Science*. Royal Society of Chemistry, Cambridge, pp. 394–406.

Schachman, H. K. (1959). *Ultracentrifugation in Biochemistry*. Academic Press, New York.

Schachman, H. K. (1989). Analytical ultracentrifugation reborn. *Nature* 341:259–260.

Schachman, H. K. (1992). Is there a future for the ultracentrifuge? In: Harding, S. E., Rowe, A. J., and Horton, J. C. (eds), *Analytical Ultracentrifugation in Biochemistry and Polymer Science*. Royal Society of Chemistry, Cambridge, pp. 3–15.

Schmidt, B. and Riesner, D. (1992). A fluorescence detection system for the analytical ultracentrifuge and its application to proteins, nucleic acids, viroids and viruses. In: Harding, S. E., Rowe, A. J., and Horton, J. C. (eds), *Analytical Ultracentrifugation in Biochemistry and Polymer Science*. Royal Society of Chemistry, Cambridge, pp. 176–207.

Schönfeld, H.-J., Pöschl, B., and Müller, F. (1998). Quasi-elastic light scattering and analytical ultracentrifugation are indispensable tools for the purification and characterization of recombinant proteins. *Biochem Soc Trans* 26:753–758.

Schuck, P. (1998). Sedimentation analysis of noninteracting and self-associating solutes using numerical solutions to the Lamm equation. *Biophys J* 75:1503–1512.

Schuck, P. (2000). Size distribution analysis of macromolecules by sedimentation velocity ultracentrifugation and Lamm equation modeling. *Biophys J* 78:1606–1619.

Schuck, P. (2003). On the analysis of protein self-association by sedimentation velocity analytical ultracentrifugation. *Anal Biochem* 320:104–124.

Schuck, P. (2004a). A model for sedimentation in inhomogeneous media. I. Dynamic density gradients from sedimenting co-solutes. *Biophys Chem* 108:187–200.

Schuck, P. (2004b). A model for sedimentation in inhomogeneous media. II. Compressibility of aqueous and organic solvents. *Biophys Chem* 187:201–214.

Schuck, P. (2005). Diffusion-deconvoluted sedimentation coefficient distributions for the analysis of interacting and non-interacting protein mixtures. In: Scott, D. J., Harding, S. E., and Rowe, A. J. (eds), *Modern Analytical Ultracentrifugation: Techniques and Methods*. Royal Society of Chemistry, Cambridge, pp. 26–50.

Schuck, P. (2006a). http://www.analyticalultracentrifugation.com/references.htm.

Schuck, P. (2006b). www.analyticalultracentrifugation.com.

Schuck, P. (2006c). www.analyticalultracentrifugation.com/sedphat/sedphat.htm.

Schuck, P. (2006d). www.nih.gov/od/ors/dbeps/PBR/AUC.htm.

Schuck, P. and Demeler, B. (1999). Direct sedimentation analysis of interference optical data in analytical ultracentrifugation. *Biophys J* 76:2288–2296.

Schuck, P., MacPhee, C. E., and Howlett, G. J. (1998). Determination of sedimentation coefficients for small peptides. *Biophys J* 74:466–474.

Schuck, P., Perugini, M. A., Gonzales, N. R., Howlett, G. J., and Schubert, D. (2002). Size-distribution analysis of proteins by analytical ultracentrifugation: strategies and application to model systems. *Biophys J* 82:1096–1111.

Schuck, P. and Rossmanith, P. (2000). Determination of the sedimentation coefficient distribution by least-squares boundary modeling. *Biopolymers* 54:328–341.

Schuck, P., Taraporewala, Z., McPhie, P., and Patton, J. T. (2000). Rotavirus nonstructural protein NSP2 self-assembles into octamers that undergo ligand-induced conformational changes. *J Biol Chem* 276:9679–9687.

Schuster, T. M. and Laue, T. M. (eds) (1994). *Modern Analytical Ultracentrifugation*. Birkhauser, Boston.

Scott, D. J., Grossmann, J. G., Tame, J. R., Byron, O., Wilson, K. S., and Otto, B. R. (2002). Low resolution solution structure of the apo form of *Escherichia coli* haemoglobin protease Hbp. *J Mol Biol* 315:1179–1187.

Scott, D. J., Harding, S. E., and Rowe, A. J. (eds) (2005). *Modern Analytical Ultracentrifugation: Techniques and Methods*. Royal Society of Chemistry, Cambridge.

Shire, S. J., Shahrokh, Z., and Liu, J. (2004). Challenges in the development of high protein concentration formulations. *J Pharm Sci* 93:1390–1402.

Signer, R. and Gross, H. (1934). Ultrazentrifugale Polydispersitätsbestimmungen an hochpolymeren Stoffen. *Helv Chim Acta* 17:726.

Snyder, D., Lary, J., Chen, Y., Gollnick, P., and Cole, J. L. (2004). Interaction of the trp RNA-binding attenuation protein (TRAP) with anti-TRAP. *J Mol Biol* 338:669–682.

Solovyova, A. S., Nollmann, M., Mitchell, T. J., and Byron, O. (2004). The solution structure and oligomerization behavior of two bacterial toxins: pneumolysin and perfringolysin O. *Biophys J* 87:540–552.

Solovyova, A., Schuck, P., Costenaro, L., and Ebel, C. (2001). Non-ideality by sedimentation velocity of halophilic malate dehydrogenase in complex solvents. *Biophys J* 81:1868–1880.

Sontag, C. A., Stafford, W. F., and Correia, J. J. (2004). A comparison of weight average and direct boundary fitting of sedimentation velocity data for indefinite polymerizing systems. *Biophys Chem* 108:215–230.

Stafford, W. F. (1992). Boundary analysis in sedimentation transport experiments: a procedure for obtaining sedimentation coefficient distributions using the time derivative of the concentration profile. *Anal Biochem* 203:295–301.

Stafford, W. F. and Sherwood, P. J. (2004). Analysis of heterologous interacting systems by sedimentation velocity: curve fitting algorithms for estimation of sedimentation coefficients, equilibrium and kinetic constants. *Biophys Chem* 108:231–243.

Steinberg, I. Z. and Schachman, H. K. (1966). Ultracentrifugation studies with absorption optics. V. Analysis of interacting systems involving macromolecules and small molecules. *Biochemistry* 5:3728–3747.

Steiner, R. F. (1954). Reversible association processes of globular proteins. V. The study of associating systems by the methods of macromolecular physics. *Arch Biochem Biophys* 49:400–416.

Svedberg, T. and Fahraeus, R. (1926). A new method for the determination of the molecular weight of the proteins. *J Am Chem Soc* 48:320–438.

Svedberg, T. and Pedersen, K. O. (1940). *The Ultracentrifuge*. Oxford University Press, London.

Tarabykina, S., Scott, D. J., Herzyk, P., Hill, T. J., Tame, J. R., Kriajevska, M., Lafitte, D., Derrick, P. J., Dodson, G. G., Maitland, N. J., Lukanidin, E. M., and Bronstein, I. B. (2001). The dimerization interface of the metastasis-associated protein S100A4 (Mts1): in vivo and in vitro studies. *J Biol Chem* 276:24212–24222.

Taraporewala, Z. F., Schuck, P., Ramig, R. F., Silvestri, L., and Patton, J. T. (2002). Analysis of a temperature-sensitive mutant rotavirus indicates that NSP2 octamers are the functional form of the protein. *J Virol* 76:7082–7093.

Todd, G. P. and Haschemeyer, R. H. (1981). General solution to the inverse problem of the differential equation of the ultracentrifuge. *Proc Natl Acad Sci USA* 78:6739–6743.

Urbanke, C., Ziegler, B., and Stieglitz, K. (1980). Complete evaluation of sedimentation velocity experiments in the analytical ultracentrifuge. *Fresenius Z Anal Chem* 301:139–140.

van Holde, K. E. and Weischet, W. O. (1978). Boundary analysis of sedimentation velocity experiments with monodisperse and paucidisperse solutes. *Biopolymers* 17:1387–1403.

Vinograd, J., Bruner, R., Kent, R., and Weigle, J. (1963). Band centrifugation of macromolecules and viruses in self-generating density gradients. *Proc Natl Acad Sci USA* 49.

Werner, W. E. and Schachman, H. K. (1989). Analysis of the ligand-promoted global conformational change in aspartate transcarbamoylase. Evidence for a two-state transition from boundary spreading in sedimentation velocity experiments. *J Mol Biol* 206:221–230.

Willcox, B. E., Gao, G. F., Wyer, J. R., Ladbury, J. E., Bell, J. I., Jakobsen, B. K., and van der Merwe, P. A. (1999). TCR binding to peptide-MHC stabilizes a flexible recognition interface. *Immunity* 10:357–365.

Winzor, D. J., Tellam, R., and Nichol, L. W. (1977). Determination of the asymptotic shapes of sedimentation velocity patterns for reversibly polymerizing solutes. *Arch Biochem Biophys* 178:327–332.

Yamakawa, H. and Fujii, M. (1973). Translational friction coefficient of wormlike chains. *Macromolecules* 6:407–415.

Yikilmaz, E., Rouault, T. A., and Schuck, P. (2005). Self-association and ligand-induced conformational changes of iron regulatory proteins 1 and 2. *Biochemistry* 44:8470–8478.

Index

Printed in Singapore